P8-AUW-983

ESSENTIALS OF RADIO—ELECTRONICS

ELECTRONIC BOOKS BY
Morris Slurzberg and William Osterheld

Essentials of Electricity for Radio and Television 2/e
Essentials of Radio—Electronics 2/e
Essentials of Television (with Elmo N. Voegtlin)

ESSENTIALS OF RADIO— ELECTRONICS

MORRIS SLURZBERG, B.S. in E.E., M.A.

WILLIAM OSTERHELD, B.S. in E.E., M.A.

SECOND EDITION

McGRAW-HILL BOOK COMPANY, INC.
New York   Toronto   London   1961

ESSENTIALS OF RADIO—ELECTRONICS

Copyright © 1961 by the McGraw-Hill Book Company, Inc.   Printed in the United States of America.   All rights reserved.   This book, or parts thereof, may not be reproduced in any form without permission of the publishers.

ESSENTIALS OF RADIO

Copyright 1948, by the McGraw-Hill Book Company, Inc.

Library of Congress Catalog Card Number: 60-15288

THE MAPLE PRESS COMPANY, YORK, PA.

PREFACE

The field of electronics has expanded at a phenomenal pace, and predictions are that it will continue to grow at a rapid rate for some time to come. Applications of electronic principles cover an expansive field including (1) communications, (2) scientific research, (3) therapeutics, (4) business, (5) industry, (6) automation, (7) safety control, (8) computers, (9) guided missiles, (10) telemetry, (11) aeronautics, (12) astronautics, (13) cybernetics.

Although many of the circuits used in the applications of electronics are quite diversified and complex they generally have two points in common: (1) They employ one or more of the basic circuit elements such as resistors, inductors, capacitors, vacuum tubes, gas tubes, and transistors. (2) They employ one or more of the basic circuit applications of these basic circuit elements. In order to understand the many complex circuits used in electronics, it is necessary to have a thorough knowledge of the basic circuit elements and their basic circuit applications. With this background it is then only necessary to study the new combination of circuit elements and their circuit applications in order to understand the operation of the complex circuits that are used in modern applications of electronics.

The purpose of this text is to present, at an intermediate level, a comprehensive study of the principles of operation of vacuum tubes and transistors, their basic circuits, and the application of these circuits to (1) audio-frequency sound systems, and (2) radio receiver applications. A chapter on test equipment and test procedures as applied to receiver circuits is included to provide an introductory knowledge to these subjects.

This book is intended for: (1) students studying radio and/or basic electronics in a technical institute, junior college, trade and vocational schools, and industrial training programs; (2) persons not attending any regular school but who wish to study the subject at home on an intermediate level. This book is also intended to provide the background necessary for further study of electronics in fields such as the high-frequency and ultrahigh-frequency circuit applications which presuppose a knowledge of the basic detector, amplifier, oscillator, and rectifier circuits for low-frequency applications as presented in this text.

The following features, not generally found in any one book, have been incorporated in this text.

1. A minimum knowledge of mathematics is required. Most of the mathematics involves only the use of addition, subtraction, multiplication, division, and square root.

2. Examples are used throughout the book to illustrate the applications of the equations and principles discussed in the text. All major equations are followed by an illustrative example. The values used in the examples represent actual practical values. Examples of complex as well as simple circuits are illustrated for d-c circuits, a-c circuits, vacuum-tube circuits, and transistor circuits.

3. The principle of operation of the various circuit elements and the analysis of the operation of electric, vacuum-tube, and transistor circuits are explained according to the electron theory.

4. In recognition of the value of visual instruction, drawings are used to illustrate each principle as it is presented. Where important features of the parts used in radio and electronic circuits cannot be readily shown by diagrams, illustrations of actual commercial products are used.

5. The operation of the circuits used to perform basic functions of vacuum tubes and transistors, namely, detection, amplification, oscillation, and rectification, is explained in great detail in order to illustrate the purpose and action of each circuit element in the composite circuit. In the explanation of the operation of these circuits a wide variety of currently used tubes and transistors has been employed.

6. Composite circuits of a-m receivers are analyzed in detail with each part and its approximate numerical value listed; in one case, the function of each part is also listed. Three receiver circuits are presented, namely, (1) five-tube trf (tuned-radio-frequency) receiver, (2) five-tube, a-c/d-c, superheterodyne receiver, (3) nine-tube, a-c operated, superheterodyne receiver including such features as an r-f amplifier stage, push-pull a-f output stage, and electron-ray tuning indicator.

7. The principles and operation of f-m receivers are explained in a separate chapter. Complete receiver circuits are analyzed for (1) f-m tuner, (2) f-m receiver, (3) a-m/f-m receiver.

8. Three chapters are devoted to transistors. They explain (1) theory of operation of transistors, (2) basic circuit applications, and (3) composite radio receiver circuits.

9. The principles of high-fidelity sound reproducing systems are presented in a separate chapter. This includes discussions of microphones, amplifiers, controls for tone, volume, and loudness, loudspeakers, baffles and enclosures, corrective networks, and stereophonic systems.

10. Ten appendixes provide general reference data and serve as useful tools when working with electronic circuits and problems. The appendix of drawing symbols, together with an illustration of the circuit element that it represents, contains approximately 100 items. These drawing symbols conform, to a reasonable degree, with the recognized standards

of (1) IRE, (2) EIA, (3) ASA, or (4) common usage by the electronic industry. The appendix of equations generally used in radio and electronics is compiled from (1) the body of the text, and (2) basic electric-circuit equations. To facilitate cross reference between this appendix and the text, wherever possible the equation numbers corresponding to their location in the body of the text have been included with the appendix listing.

11. As an aid to the instructor and a challenge to the more interested student, there are numerous questions and/or problems at the ends of the chapters. Tho values used in the problems have been carefully selected and represent actual practical values.

12. Answers for approximately one-half of the problems are provided in the appendix as a guide to the conscientious students and to those persons using the book without the benefit and aid of a classroom instructor.

Numerous industrial organizations have been of great assistance in providing illustrations and technical information regarding their products, and this service is greatfully acknowledged. These organizations are Aerovox Corporation; American Phenolic Corporation; American Telephone and Telegraph Company; Bliley Electric Company; Delco Radio Division, General Motors Corporation; Federal Telephone and Radio Corporation; General Electric Company; The Hammarlund Manufacturing Company, Inc.; The Hickok Electrical Instrument Company; International Resistance Company; Jensen Radio Manufacturing Company; Maguire Industries, Inc.; P. R. Mallory & Company, Inc.; Ohmite Manufacturing Company; Philco Corporation; Radio Corporation of America; Shure Brothers, Inc.; Simpson Electric Company; Solar Mfg. Corp.; Standard Transformer Corporation; Texas Instruments, Incorporated; Trimm Manufacturing Company, Inc.; Westinghouse Electric Corporation; Weston Instruments Division, Daystrom, Inc.

Morris Slurzberg
William Osterheld

CONTENTS

Chapter 16. Test Equipment 627

Systems of Testing. Combination Meters. Special-purpose Voltmeters. Signal Generators. Cathode-ray Tubes. Cathode-ray Oscilloscope. Tube Testers. Transistor Testers.

APPENDIX

CHAPTER 1

INTRODUCTION TO RADIO

Much of the progress in civilization may be attributed to man's ability to communicate with his fellow man and thereby transmit his thoughts to them. Radio is one of the modern means of communication and is a branch of the field of science called *electronics*. Electronics has enabled scientists to develop means of examining the germ structure of bacteria and to see through fog and the utter blackness of night. It is used to detect poisons, to guarantee food values, to control manufacturing processes, and to protect life and property. Electronics has opened up a new field to science and is constantly contributing to a better world.

1-1. Forms of Communication. Man's chief means of receiving communication are the senses of sight and hearing. These are commonly referred to as the *audible* and *visible* means of communication.

Visual Means of Communication. The sense of sight has long been useful to man, first as a means of warning him of approaching dangers and later to enable him to receive messages in written form. Examples illustrating the progress of visible methods of communication are the hieroglyphics of the ancient Egyptians, the smoke signals of the American Indians, printed matter such as newspapers, books, etc., photography, motion pictures, and television.

Audible Means of Communication. The sense of hearing has been used for communication through many stages of civilization. It served to warn man of approaching dangers long before the development of modern communication systems. As civilization progressed, the audible methods of communication went through numerous stages as illustrated by the development of languages to convey thoughts, the development of devices such as the telegraph, telephone, wireless, and radio to transmit messages over greater distances, and the development of devices such as the phonograph, tape recorders, and sound motion pictures to make it possible to keep a record of the message.

1-2. Other Applications of Electronics. The principles of electronics are used for applications other than those of radio communications, such as (1) television, (2) industrial processes, (3) instruments, and (4) therapeutics.

Television. This branch of communications deals with the transmission and reception of visual images over great distances. Television deals with the study of lenses, light, colorimetry, electronic scanning, the iconoscope or electric eye of the television camera, and the kinescope which reproduces the picture at the television receiver. In addition, all the basic principles of electronic circuits used in radio receivers and transmitters are also used in television apparatus.

Industrial Applications. The applications of electronics and vacuum tubes to industry are many and varied. The basic circuits of many of these applications are similar to those used in radio, the only difference being the use to which they are put. For example, the principles of resonance are used to control the thickness, quality, weight, and moisture content of a material. Amplifier circuits are used to increase the intensity of weak current impulses produced by phototubes and cause them to operate relays controlling circuits of door openers, lighting systems, power systems, safety devices, etc.

Principles other than those common to radio are also used, such as stroboscopic lighting, which can cause fast-moving objects to appear motionless or make their movements appear similar to the slow motion of motion pictures. This principle makes it possible to study the movements of various parts of a machine under their operating conditions. It is also used for high-speed photography applications in order to arrest the motion of fast-moving objects. Phototubes and photocells are used to control lighting, operate protective devices, open doors, etc.

These are but a few of the many applications of the principles of electronics to industry. Further developments indicate that many new applications of electronics will result in numerous additional uses.

Instruments. The principles of electronics have made it possible to measure quantities that up to now have been impossible to measure. The vacuum-tube voltmeter, cathode-ray oscillograph, resonant circuit checkers, and signal generators are but a few of the many new types of instruments that have become synonymous with radio and electronics. The use of these instruments has become as valuable for checking electronic circuits as the ammeter and voltmeter are for checking electric circuits. The electron microscope, electron telescope, etc., have become valuable aids to scientists.

Therapeutics. Medical doctors and scientists in the field of therapeutics (i.e., the treatment of diseases) are constantly finding new uses for the principles of electronics to aid in treating and preventing physical ailments. Their instruments include: (1) X rays, which are used for treatment of skin disorders and acute infections as well as for taking pictures of internal structures; (2) ultraviolet lamps for arresting harmful mold and bacteria; (3) short-wave diathermy units for healing sprains and fractures; (4) electrocardiographs for measuring heartbeats; (5) inductotherm units used to generate artificial fever; and (6) oscillographs for illustrating muscle actions. These are but a few of the many applications of electronics in the field of therapeutics.

1-3. Sound.   Radio is a means of sending information through space from one point to another without any wires connecting the two points. The information may be either a sound wave produced by the voice or some musical instrument, or a wave so interrupted that it is broken into a combination of long and short groups that correspond to the characters of the Morse code.

Sound is the sensation produced in the brain by sound waves.   It makes use of one of our five senses, namely, that of hearing.   If a sound is made by a person, by a musical instrument, or by any other means, the air about it is set into vibration.   These air vibrations are called *sound waves*.   When sound waves strike the eardrum of a person, the eardrum too will vibrate in a similar manner.   The auditory nerves will be stimulated and will communicate the sensation of sound to the brain.

The vibrations of the reeds in a harmonica, the skin on a drum, the strings on a violin, or the cone of a radio loudspeaker will all send out various sound waves.   These waves will produce different sounds, depending on the number of vibrations that the wave makes per second.   The number of complete waves or vibrations created per second is called the *frequency* of the sound and is expressed as the number of cycles per second.

If the sound is loud enough to be heard by the human ear, it is said to be *audible*.   Its pitch will vary with the frequency.   High-frequency waves produce sounds having a high pitch, and low-frequency waves produce sounds of low pitch.

1-4. Frequency Ranges of Sound Waves.   The range of frequencies that the human ear is capable of hearing varies with individuals, the lower limit being approximately 20 cps (cycles per second) and the upper limit 20,000 cps.   Table 1-1 lists a few common audible sounds and their approximate frequency range (including harmonics).

TABLE 1-1. APPROXIMATE FREQUENCY RANGE OF COMMON SOUNDS, CPS

Human voice	80–10,000
Piano	30– 8,000
Violin	200– 8,000
Trombone	80– 7,000
Clarinet	150–10,000
Flute	250–10,000
Piccolo	500–10,000

Although code signals may be sent at any audio frequency, a 1,000 cps signal has been found least fatiguing and permits each dot, dash, or space to be quickly distinguished.

Sound waves travel only comparatively short distances and at approximately 1,130 ft per sec.   In order to be carried through air over long distances, the sound waves are converted into electrical waves of corresponding frequencies and applied to a high-frequency carrier wave at the radio transmitter.

Sound waves are also referred to in terms of the length of a wave. Figure 1-1 shows a tuning fork producing sound waves of 256 cps.   At this frequency 1 cycle is completed in $\frac{1}{256}$ sec, and, since sound waves travel

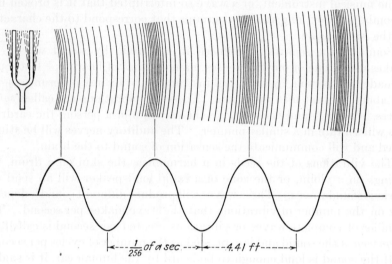

FIG. 1-1. Propagation of a sound wave of 256 cps in air.

at approximately 1,130 ft per sec, the length of one wave is $\frac{1130}{256}$ or 4.41 ft.

Example 1-1.   The frequency range of a piano is from 25 to 8,000 cycles.   (a) What is the range of wavelengths in feet?   (b) In meters?   (c) If the sound waves are converted to electrical waves by a microphone, what is the frequency range of the electric currents?

Given: Sound waves = 25–8,000 cycles   Find: (a) Wavelengths, ft
                                                    (b) Wavelengths, meters
                                                    (c) Frequency range of electric currents, cycles

Solution:

(a)   Wavelength, 25 cycles $= \dfrac{\text{ft per sec}}{\text{cps}} = \dfrac{1{,}130}{25} = 45.2$ ft

   Wavelength, 8,000 cycles $= \dfrac{\text{ft per sec}}{\text{cps}} = \dfrac{1{,}130}{8{,}000} = 0.14125$ ft

(b)   NOTE: 1 meter = 39.37 inches = 3.28 ft

   Wavelength, 25 cycles $= \dfrac{\text{wavelength, ft}}{3.28} = \dfrac{45.2}{3.28} = 13.7$ meters

   Wavelength, 8,000 cycles $= \dfrac{\text{wavelength, ft}}{3.28} = \dfrac{0.14125}{3.28} = 0.043$ meter

(c)   25 to 8,000 cycles (same frequencies as the sound waves)

1-5. Radio Waves.   Radio transmitting stations convert sound waves to electrical impulses.   The electrical impulses representing the original sound waves are sent out by the use of high-frequency alternating currents. The high-frequency currents produce magnetic and electric fields that radiate in all directions over long distances. The magnetic and electric fields produced by this means are called *radio waves*. The strength and frequency of the radio wave is dependent on the high-frequency alternating current producing it and will vary in the same manner as the alternating current.

An a-c wave rises from zero to its maximum value, then diminishes to zero and reverses its direction at fixed intervals. Figure 1-2 shows that an a-c wave completes 1 cycle after it has made two alternations, one in the positive direction and one in the negative direction. The maximum amount of current occurs at each 90° and 270° instant of the a-c cycle.

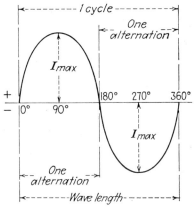

Fig. 1-2. An a-c wave.

1-6. Wavelength, Frequency. *Speed of Radio Waves.* Radio waves travel at the same speed as light waves, or 186,000 miles per sec. In some radio calculations the metric system is used, and the speed of the radio waves is then expressed in meters per second.

Example 1-2. Radio waves travel at the rate of 186,000 miles per sec. What is the rate in (*a*) feet per second, (*b*) meters per second? NOTE: One mile = 5,280 ft; also, 1 meter = 3.28 ft.

Given: Miles per sec = 186,000   Find: (*a*) Ft per sec
        Ft per mile = 5,280   (*b*) Meters per sec
        Ft per meter = 3.28

Solution:

(*a*) Ft per sec = 186,000 × 5,280 ≅ 982,000,000

(*b*) Meters per sec = $\dfrac{982,000,000}{3.28}$ ≅ 300,000,000

NOTE: ≅ means *is approximately equal to.*

Wavelength and Frequency Definitions. WAVELENGTH. The distance that the radio wave travels in the time of one cycle is called its *wavelength;* it is expressed in meters and is represented by the symbol λ, a letter of the Greek alphabet pronounced "lambda."

FREQUENCY. The number of cycles per second of a radio wave is called its *frequency* and is represented by the letter *f*. The frequency is often referred to as a number of *cycles* instead of cycles per second; this is merely an abbreviation and it should be remembered that it really means cycles per second.

Wavelength and Frequency Calculations. WAVELENGTH. If the frequency of a wave is known, the distance it will travel in 1 cycle can be calculated by

$$\lambda = \frac{300,000,000}{f \ (in \ cps)} \qquad (1\text{-}1)$$

where λ = wavelength, meters

f = frequency, cps

KILOCYCLES AND MEGACYCLES. The frequencies of the common radio waves are of high values, that is, in the hundreds of thousands or millions of cycles per second. For convenience these frequencies are generally expressed in kilocycles or megacycles and abbreviated as kc and mc respectively. *Kilo-* is a prefix meaning thousand; hence a kilocycle is equal to 1,000 cycles, which actually means 1,000 cycles per second. The prefix *mega-* means million; hence a megacycle means 1,000,000 cycles per second.

When radio frequencies are expressed in kilocycles or megacycles, Eq. (1-1) becomes

$$\lambda = \frac{300,000}{f \ (in \ kc)} \qquad (1\text{-}1a)$$

$$\lambda = \frac{300}{f \ (in \ mc)} \qquad (1\text{-}1b)$$

Example 1-3. What is the wavelength of radio station WMCA, which operates on an assigned frequency of 570 kc?

Given: f = 570 kc   Find: λ

Solution:

$$\lambda = \frac{300,000}{f} = \frac{300,000}{570} = 526.3 \ meters$$

FREQUENCY. Equations (1-1), (1-1a), and (1-1b) can be transposed to solve for frequency instead of wavelength, and become

$$f \ (in \ cps) = \frac{300,000,000}{\lambda} \qquad (1\text{-}2)$$

$$f \ (in \ kc) = \frac{300,000}{\lambda} \qquad (1\text{-}2a)$$

$$f \ (in \ mc) = \frac{300}{\lambda} \qquad (1\text{-}2b)$$

Example 1-4. If by definition a short radio wave is one whose wavelength does not exceed 200 meters, what is the lowest frequency at which a short-wave radio receiver may operate?

Given: λ = 200 meters   Find: f

Solution:

$$f = \frac{300,000}{\lambda} = \frac{300,000}{200} = 1,500 \ kc$$

If the wavelength in meters is changed to our more commonly used unit of feet, it should provide a better understanding of the length of the radio waves.

Example 1-5.   What is the length in feet of one radio wave of the broadcast station referred to in Example 1-3?

Given: $\lambda = 526.3$ meters   Find: Length, ft

Solution:

$$\text{Length} = \text{meters} \times 3.28 = 526.3 \times 3.28 = 1{,}726 \text{ ft}$$

The solution of Example 1-5 indicates that each wave transmitted by station WMCA is 1,726 feet long, or approximately one-third of a mile.

Knowing that radio waves travel 186,000 miles per sec, the time required for a radio wave to get from one place to another can be readily calculated.

Example 1-6.   How long does it take a radio wave to travel from New York to San Francisco, a distance of approximately 2,600 miles?

Given: Distance = 2,600 miles   Find: Time, t
         Rate = 186,000 miles per sec

Solution:

$$t = \frac{\text{miles}}{186{,}000} = \frac{2{,}600}{186{,}000} = 0.0139 \text{ sec}$$

Example 1-6 shows that it takes only about 0.014 sec for a radio program broadcast from New York to travel to San Francisco.

1-7. Modulated Radio-frequency Waves.   *Need for Modulating the R-F Wave.*   Sound waves in air can be heard only over relatively short distances, but radio waves can be used over very long distances.   Therefore, the transmission of intelligence (i.e., code, voice, music) is accomplished by combining electrical impulses corresponding to the intelligence signals with a constant high-frequency wave generated by the transmitting equipment.   The process of combining the audio- and radio-frequency (a-f and r-f) waves is called *modulation*.

Types of Waves.   The following definitions are from the Institute of Radio Engineers (IRE) standards.   The term *signal* is defined as the form or variation with time of a wave whereby the information, message, or effect is conveyed in communication.   A *signal wave* is a wave the form of which conveys a signal.   A *carrier wave* is the unmodulated component of a signal wave.   A *modulated wave* is a wave of which either the amplitude, frequency, or phase is varied in accordance with a signal.   A *sideband* is a band of frequencies on either side of the carrier frequency, produced by the process of modulation.

Modulation.   The carrier wave of a radio transmitting system is a wave of constant value of frequency and amplitude.   (The carrier wave by itself will not produce any sound at the loudspeaker of the ordinary broadcast receiver.)   The transmission of intelligence occurs when the carrier wave is modulated by a signal.   The IRE definition of modulation states that *modulation* is the process by which some characteristic of a periodic wave is varied with time in accordance with a signal.   Modulation is generally accomplished by combining a signal and a carrier wave. The signal is also referred to as a *modulating wave*, and the resultant wave is called the *modulated wave*.

Although many methods of modulation can be devised, the only ones of practical value at present are amplitude modulation, frequency modulation, and phase modulation. Only amplitude modulation (abbreviated a-m) and frequency modulation (abbreviated f-m) are used in radio broadcast transmission.

Mathematically, a carrier wave may be considered as

$$e = E_m \sin (2\pi ft + \theta) \tag{1-3}$$

From Eq. (1-3) it can be seen that the waveform can be varied by three factors, namely, E_m, f, and θ. *Amplitude modulation* occurs when modulation is obtained by varying only the voltage E_m. *Frequency modulation* occurs when modulation is obtained by varying only the frequency f. *Phase modulation*, which is very similar to frequency modulation, occurs when modulation is obtained by varying only the phase relationship represented by θ.

1-8. Amplitude Modulation. *A-M Wave.* An amplitude-modulated wave is defined by the IRE standards as one whose envelope contains a component similar to the waveform of the signal to be transmitted. In amplitude modulation the amplitude of the carrier wave is varied by the strength of the signal, which is the modulating quantity. The effect of amplitude modulation can be seen from a study of Figs. 1-3 and 1-4. Figure 1-3a represents a high-frequency carrier wave of constant amplitude and frequency. The wave of Fig. 1-3b represents an a-f (audio-frequency) signal of sine-wave form; the sine-wave form is used here to show the effect of modulation more clearly. Figure 1-3c shows the result obtained by modulating the carrier wave (*a*) with the modulating wave (*b*). Further examination of Fig. 1-3c will show that the outline of the modulated carrier wave is similar in form to the modulating wave; accordingly, this outline is commonly called the *modulation envelope*.

Per Cent of Modulation. In amplitude modulation it is common practice to refer to the per cent of modulation, usually designated as M. Actually this is a means of expressing the degree to which the signal modulates the carrier wave. The per cent of modulation is proportional to the ratio of the maximum values of the signal and carrier waves, or

$$M = \frac{\text{maximum value of signal}}{\text{maximum value of carrier}} \times 100 \tag{1-4}$$

This may be expressed in terms of Fig. 1-3 as

$$M = \frac{B}{A} \times 100 \tag{1-5}$$

where M = per cent of modulation
  B = maximum value of modulating wave, volts
  A = maximum value of carrier wave, volts
The effect of different amounts of modulation upon the carrier wave is
shown by the various illustrations in Fig. 1-3. As the maximum undistorted power output of a transmitter is obtained with 100 per cent modulation, it is generally desirable to operate with such a fully modulated

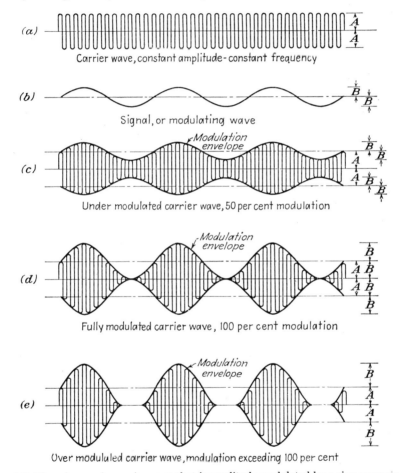

(a) Carrier wave, constant amplitude-constant frequency

(b) Signal, or modulating wave

(c) Under modulated carrier wave, 50 per cent modulation

(d) Fully modulated carrier wave, 100 per cent modulation

(e) Over modulated carrier wave, modulation exceeding 100 per cent

FIG. 1-3. Waveshapes of a carrier wave that is amplitude-modulated by a sine-wave signal.

carrier wave. If the modulation is less than 100 per cent, the power
output is reduced, even though the power of the carrier wave has not been
reduced. If the modulation exceeds 100 per cent, the output of the
transmitter will be a distorted version of the original modulating wave.
  Sidebands. During the process of modulation a heterodyne action
takes place, and as a result two additional frequencies appear. These

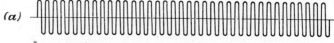

(a)

Carrier wave, constant amplitude-constant frequency

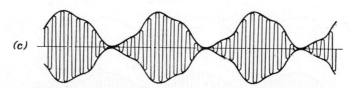

(b)

Voice signal, or modulating wave

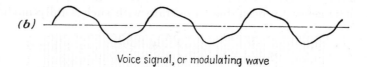

(c)

Modulated carrier wave, 100 per cent modulation at signal peaks

Fig. 1-4. Waveshapes of a carrier wave that is amplitude-modulated by a voice signal.

new frequencies are the result of the heterodyning action and are beat frequencies whose values are equal to the sum and difference of the

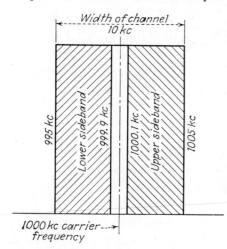

Fig. 1-5. Illustration of the sidebands and the channel width for a 1,000-kc carrier wave that is being amplitude-modulated by audio frequencies ranging from 100 to 5,000 cycles.

carrier frequency and the modulating frequency. The value of frequency equal to the sum of the carrier and modulating frequencies is called the *upper side frequency,* and the value equal to the difference of the carrier and modulating frequencies is called the *lower side frequency.* In radio broadcasting the modulating frequency varies continually and over a considerable range. Accordingly, the single value of upper side frequency referred to above is replaced by a band of frequencies, called the *upper sideband,* whose width is equal to the difference between the maximum and minimum values of the modulating frequencies. Likewise, the single value of lower side frequency will be replaced by a *lower sideband.* For example, if a 1,000-kc carrier wave is modulated by audio signals varying between 100 and 5,000 cycles, the maximum upper side frequency is 1,005 kc and the minimum lower side frequency is 995 kc. Under this condition, the transmitter requires a 10-kc channel extending from 995 to 1,005 kc (see Fig. 1-5).

1-9. Frequency Modulation. *F-M Wave.* In frequency modulation, the amplitude of the modulated wave is maintained at a constant strength, namely, the same value as the unmodulated carrier wave. The frequency of the modulated wave is varied in proportion to the amplitude of the modulating signal, and at a rate determined by the frequency of the modulating signal. This is illustrated by Figs. 1-6 and 1-7. From Fig. 1-6 it can be seen that the frequency of the modulated wave increases as the signal voltage increases and that it decreases as the signal voltage decreases. Comparison of Figs. 1-6b and 1-6c shows that the variation in frequency is determined only by the amplitude of the signal, and that the rate of variations in frequency is determined by the frequency of the

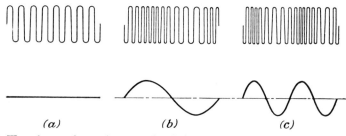

(a)   (b)   (c)

Fig. 1-6. Waveshapes of a carrier wave that is frequency-modulated by a sine-wave signal. (a) No modulating signal. (b) 500-cycle modulating signal. (c) 1,000-cycle modulating signal.

signal. Figure 1-7 is another method of illustrating the effect of the amplitude of the signal upon the frequency of the modulated wave.

Frequency Deviation. The frequency of an f-m transmitter without any signal input is referred to as the *resting frequency* or the *center frequency*. This value corresponds to the assigned frequency of the transmitter. When a signal is applied, the variation in frequency either above or below the resting frequency is called the *frequency deviation*, and the total variation in frequency is called the *carrier swing*. For example, in Fig. 1-7 the resting frequency is 5 cycles, the deviation for the weak signal is 2 cycles, and the carrier swing is 4 cycles. With the strong signal, the deviation is 4 cycles and the carrier swing is 8 cycles. The values of frequency shown in Fig. 1-7 are not practical values, but for convenience of illustration have been made of very low values. Actually, f-m broadcast transmitters operate at frequencies of 88 to 108 mc.

Modulation. With f-m systems, variations in the amplitude of the signal produce variations in the frequency of the modulated carrier; therefore, the degree of modulation is expressed in terms of the frequency deviation. A frequency deviation of 75 kc is equivalent to 100 per cent modulation in the a-m system.   Under this condition, the total carrier swing will be 150 kc.   To avoid interference between stations on adjacent channels, the assigned frequencies are kept at least 200 kc apart.

The frequency deviation is sometimes expressed in terms of the ratio of maximum frequency deviation to the maximum audio frequency being transmitted; this is called the *deviation ratio*.   For example, if an f-m transmitter operates with a maximum frequency deviation of 75 kc and reproduces audio signals up to 15 kc, its deviation ratio is 5.

Sidebands.   The sidebands present in f-m transmission are determined by the amplitude of the modulating signal.   While the sideband frequencies are apparently unlimited, present f-m transmission is based on a 75-kc

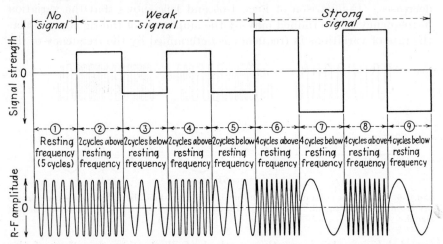

FIG. 1-7.   Waveshapes of a carrier wave that is frequency-modulated by square-wave signal voltages.   (*From N. M. Cooke and L. Markus, "Electronics Dictionary," McGraw-Hill Book Company, Inc., New York, 1945.*)

frequency deviation.   For example, with a given signal frequency the frequency deviation, and hence the number of sidebands, is dependent upon the amplitude of the signal.   If the value of the deviation ratio is approximately 5, the sidebands above the maximum frequency are so small that they may be ignored.   Under this condition, the channel width for f-m transmission should be approximately double the value of the maximum frequency deviation.   On this basis, a 150-kc carrier swing can reproduce a-f signals up to 15,000 cycles.

1-10. General Picture of Radio Transmission and Reception.   Figure 1-8 presents a simple picture of the operations required to send a signal wave out into space and to have it received miles away.   The top line shows the essential portions of a radio transmitter and receiver.   The first unit is the *microphone;* here the audible sound waves are picked up and changed into electrical impulses.   The signal waves from the microphone are very weak, and hence these signals are first amplified by send-

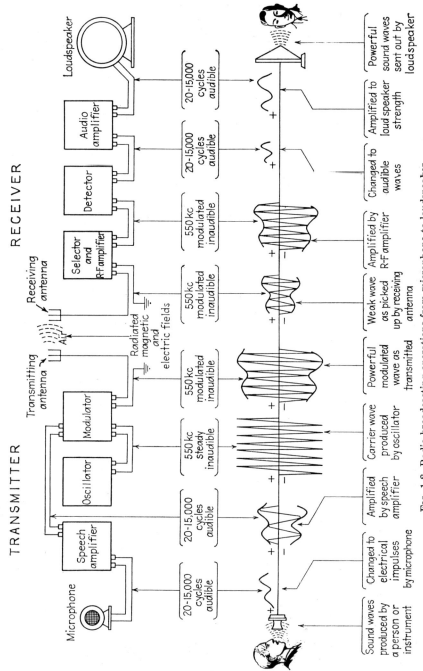

FIG. 1-8. Radio-broadcasting operations from microphone to loudspeaker.

ing them from the microphone to the *speech amplifier*. The next section, called the *oscillator*, sets up the carrier wave of the transmitter, in this example 550 kc. This is followed by the *modulator*, which receives energy from both the oscillator and the speech amplifier. At the modulator the signal waves from the speech amplifier are superimposed on the *carrier wave*, and this modulated carrier wave is then sent out into space by the transmitting antenna.

The *receiving antenna* is affected by the magnetic and electric fields set up in space by the transmitting antenna, and if the *selector* or tuning portion of the receiver is set for the proper frequency (in this example 550 kc), a workable amount of electrical energy enters the receiver. The amount of energy is small and must be increased in strength at this point by the *radio-frequency amplifier*. The selector and the r-f amplifier are shown in a single section because these two operations are generally combined. The next section, labeled *detector*, might also be called the *demodulator*, because at this point the audio waves are separated from the carrier wave. The audio waves coming from the detector are too weak to operate a loudspeaker and therefore must be sent through an *audio-frequency amplifier* before going on to the *loudspeaker*.

The second line indicates the frequency of the wave as it enters and leaves the various parts of the transmitter and receiver operated at a frequency assumed to be 550 kc. The third line is a diagrammatic representation of these frequencies. Examination of the figure will show that every step performed in the transmitter is also performed in the receiver but in reversed order, starting with the sound waves entering the microphone at the transmitter and ending with similar sound waves leaving the loudspeaker of the receiver. The fourth line summarizes the purpose of each part of the radio transmitter and receiver.

To prevent interference among modulated waves, the FCC (Federal Communications Commission) assigns definite frequencies to the transmitting or broadcasting stations. For example, the frequency of the carrier wave of station KFI, Los Angeles, is 640 kc; WLAC, Nashville, 1,510 kc; WLS, Chicago, 890 kc; and WOR, New York, 710 kc.

1-11. Functions of a Receiver. A radio receiver makes it possible to listen to the audio signals sent out from any one of the transmitters within the range of the receiver. The basic operations are the same for a simple crystal receiver as for a receiver using vacuum tubes or transistors. Through the study of the operations of the crystal receiver the basic functions can easily be understood; to include vacuum tubes and transistors at this point would only complicate the subject. Once the fundamental operations of a radio receiver are thoroughly understood, they can be applied to receivers using vacuum tubes or transistors.

The essential functions of a receiver are as follows: (1) to *receive* the various waves sent out by radio transmitters, (2) to *select* the desired radio wave and to exclude all others, (3) to *separate* the signal wave from the modulated carrier wave, and (4) to *convert* the signal waves to sound waves.

When a radio wave cuts through a conductor, it induces a voltage in that conductor and causes free electrons to be drawn toward the positive terminal of the antenna circuit, thus setting up an electric current in that conductor.   In a radio receiver this conductor is called the *antenna*.

At any instant, a number of different radio waves are cutting through the receiving antenna, and each induces a voltage in the antenna.   Since the signal of only one station should be heard at one time, some provision must be made for separating the desired signal from the others.   The part of the receiver performing this function is called the *selector* or the *tuner*. When a receiver having good selectivity is properly tuned, the modulated carrier wave of only the desired station is utilized.   The next function of the receiver is to separate the signal wave from the modulated carrier

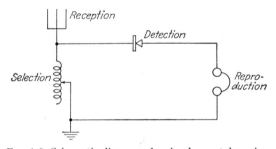

FIG. 1-9. Schematic diagram of a simple crystal receiver.

wave.   This function is called *detection*, and the part of the receiver doing this work is called the *detector*.

A sound may be heard when air is set into motion and vibrates in a manner corresponding to the frequency of that sound.   The final function of the receiver therefore is to convert the signal waves to sound waves. This is accomplished by causing a diaphragm to vibrate in accordance with the variations in strength and frequency of the signal waves.   This function is performed by either *earphones* or a *loudspeaker*.

These four functions are performed by the simple receiving circuit shown in Fig. 1-9 in the following manner: (1) the antenna receives the radio waves; (2) the desired signal is selected by adjusting the variable tuning coil; (3) the crystal detector allows only the audio component of the modulated wave to flow through it; (4) the earphones change the electrical impulses of the signal waves into sound waves.

1-12. Reception. *Types of Receiving Antennas.*   An antenna is a wire or system of wires installed at sufficient height above the ground so that it is free from any surrounding objects.   Three simple types of antenna are shown in Fig. 1-10.

Purpose of the Antenna.   The purpose of the antenna is to intercept as much of the r-f power radiated by the transmitting station as is possible.   The amount of voltage induced in an antenna will depend on the power of the transmitter, the distance from the transmitter to the receiver, and the length, location, and direction of the receiving and transmitting antennas.

Although the power of a transmitter may be 50,000 watts or more and the antenna voltage may be as high as 50,000 volts, only a few microwatts of power are picked up by the receiving antenna, and the value of the induced voltage is only in microvolts.   The voltage induced in a receiving

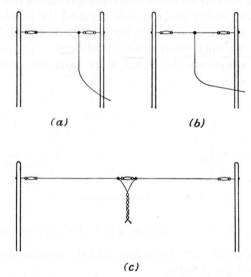

(a)   *(b)*

(c)

Fig. 1-10. Three types of simple antennas.   (*a*) Inverted *L* with end-connected lead-in. (*b*) *T* type with center-connected lead-in.   (*c*) Doublet type with center-connected lead-in.

antenna decreases as the distance from the transmitting antenna increases. Because of this the signal from a low-power station cannot be heard over great distances.

Factors Affecting the Manner of Construction of an Antenna.   In order to intercept the maximum amount of energy from the transmitted wave, the receiving antenna should be constructed in such a manner that its horizontal portion will be in the direction that will enable it to intercept the maximum amount of energy from the transmitted wave.   The direction of the receiving antenna is no longer as important a factor as in the past because of the increased power of the modern transmitters, and because of the increased over-all sensitivity of the modern receivers. Many portable receivers contain a loop antenna mounted in the receiver case.   The effect of the relative directions of the receiving and transmitting antennas on these receivers can sometimes be observed by changing the position of the receiver, thus changing the relative position of the receiving and transmitting antennas. Although the maximum signal input is received from one station for a certain position of the receiver, the signal from another station may be very weak for the same position.

The antennas shown in Fig. 1-10 should be mounted higher than the surrounding buildings; otherwise some radio waves may be prevented from cutting the antenna. The length of an antenna is generally determined by the building on which it is to be mounted. If the antenna is too long, it may be necessary to connect a capacitor in series with it.

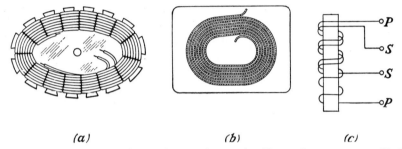

(a)   (b)   (c)

FIG. 1-11. Loop antennas used in modern receivers. (a) Air-core loop antenna. (b) Air-core loop antenna. (c) Ferrite (iron) loopstick antenna with bifilar-wound secondary.

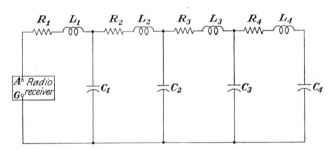

FIG. 1-12. Electrical characteristics of the antenna circuit.

The best operation of a radio receiver can only be obtained when it is used with a properly constructed antenna. In a simple receiver using either a crystal, tube, or transistor, successful operation depends upon the strength and quality of the input signal, since there is no provision made for amplification.

1-13. Electrical Characteristics of the Antenna. *Equivalent Circuit of the Antenna.* If a receiving antenna is divided into a number of equal lengths, each division can be represented by a resistor, an inductor, and a capacitor as shown in Fig. 1-12. The effect of the inductances L_1, L_2, L_3, L_4, . . . is produced by the high-frequency currents induced in the antenna. The capacitors C_1, C_2, C_3, C_4, . . . represent the capacitances existing between the antenna acting as one plate of a capacitor, the ground as the other plate, and the air as the dielectric. The resistance of the wire and the resistance of the ground are represented by the resistors R_1, R_2, R_3, R_4, etc. The total resistance, inductance, and capacitance of the antenna circuit may be calculated as follows:

$$R_A = R_1 + R_2 + R_3 + R_4 + \cdots \qquad (1\text{-}6)$$
$$L_A = L_1 + L_2 + L_3 + L_4 + \cdots \qquad (1\text{-}7)$$
$$C_A = C_1 + C_2 + C_3 + C_4 + \cdots \qquad (1\text{-}8)$$

The antenna may be represented by its equivalent series resonant circuit, Fig. 1-13. When the frequency of a modulated wave being received is the same as the resonant frequency of the antenna, the amount of signal current will vary inversely with the total resistance of the circuit. It is therefore essential that the resistance should be as low as practicable. The relationships among the frequency of resonance, inductance, and capacitance of a tuned circuit are expressed by the following equations.

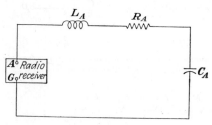

$$f_r = \frac{159}{\sqrt{LC}} \qquad (1\text{-}9)$$

$$C = \frac{25{,}300}{f_r^2 L} \qquad (1\text{-}10)$$

$$L = \frac{25{,}300}{f_r^2 C} \qquad (1\text{-}11)$$

Fig. 1-13. Equivalent circuit of the antenna.

where f_r = resonant frequency, kc
C = capacitance, μf
L = inductance, μh

Example 1-7. A typical receiving antenna circuit whose length is approximately 100 ft has an equivalent circuit capacitance of 200 $\mu\mu$f and an equivalent circuit inductance of 50 μh. What is the resonant frequency of this antenna circuit?

Given: $C_A = 200\ \mu\mu$f   Find: f_r
$L_A = 50\ \mu$h

Solution:

$$f_r = \frac{159}{\sqrt{L_A C_A}} = \frac{159}{\sqrt{50 \times 200 \times 10^{-6}}} = 1{,}590 \text{ kc}$$

Fundamental Frequency of the Antenna. If the antenna of Example 1-7 is connected to a crystal detector, a pair of earphones, and the ground, as shown in Fig. 1-14a, the signal waves sent out by a station transmitting on a carrier frequency of 1,590 kc will be heard in the earphones. The frequency of resonance of an antenna is called the *fundamental frequency of the antenna.* The value of the fundamental frequency will vary with the length and height of the antenna and should correspond to the frequency range of the radio waves to be received. For best results, the electrical characteristics of an antenna circuit should be such that its fundamental frequency will be slightly higher than the maximum frequency for which it is to be used.

1-14. Variable-inductance Tuning. The antenna circuit of Example 1-7 and Fig. 1-14 has a fundamental frequency of 1,590 kc and can receive the signal from a station transmitting on this frequency. In

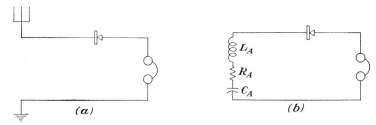

FIG. 1-14. A simple crystal receiver circuit. (*a*) Circuit diagram showing the antenna, crystal detector, earphones, and ground. (*b*) The equivalent circuit.

order to receive the signals of any other stations, the resonant frequency of the receiving circuit must be changed to correspond with the frequency of the desired station. This may be done by varying either the capacitance or the inductance of the antenna circuit. The process of selecting the signals from any one particular station is called *tuning*.

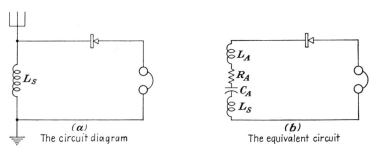

FIG. 1-15

Example 1-8. What frequency of resonance would the receiving circuit in Example 1-7 have, if a 450-μh inductance is connected in series with the antenna circuit as shown in Fig. 1-15?

Given: $C_A = 200\ \mu\mu\text{f}$ Find: f_r
$\qquad\quad L_A = 50\ \mu\text{h}$
$\qquad\quad L_S = 450\ \mu\text{h}$

Solution:

$$L_T = L_A + L_S = 50 + 450 = 500\ \mu\text{h}$$
$$f_r = \frac{159}{\sqrt{L_T C_A}} = \frac{159}{\sqrt{500 \times 200 \times 10^{-6}}} = 503\ \text{kc}$$

If the 450-μh coil is made adjustable so that any amount of inductance from zero to 450 μh can be connected in series with the antenna circuit, the receiver can then be tuned for any frequency between 503 and 1,590 kc.

1-15. Variable-capacitance Tuning. Tuning may also be accomplished by varying the capacitance and keeping the inductance constant as shown in Fig. 1-16. When the capacitor's rotor plates are completely out of mesh, the capacitance is theoretically zero. With the capacitor plates in this position, no capacitance is added to the circuit, and therefore it will tune to the same frequency as it would without the capacitor. When the rotor plates of the capacitor C are brought into mesh, the capacitance of the circuit is increased, and the resonant frequency is decreased.

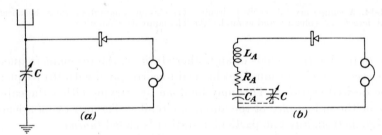

FIG. 1-16. A simple crystal receiver circuit using a variable capacitor for tuning. (*a*) The circuit diagram. (*b*) The equivalent circuit.

Example 1-9. What capacitance must the variable capacitor in Fig. 1-16 have if it is to be used with the antenna circuit of Example 1-7 and provide the same frequency range?

Given: $L_A = 50 \ \mu h$ Find: C
$C_A = 200 \ \mu\mu f$
Frequency range = 503 to 1,590 kc

Solution:

$$C_T = \frac{25,300}{f_r{}^2 L_A} = \frac{25,300}{503 \times 503 \times 50} = 0.002 \ \mu f$$
$$C = C_T - C_A = 0.002 - 0.0002 = 0.0018 \ \mu f$$

1-16. Two-circuit Tuner. *Method of Isolating the Resistance in the Antenna Circuit.* In the receiving circuits just described, the antenna circuit was a part of the tuning circuit. The effects of the inductance and the capacitance upon the tuning circuit were described, but the effect of the resistance was not considered. Resistance in a series tuned circuit decreases the current and broadens its tuning as is illustrated by the curves *A*, *B*, and *C* of Fig. 1-17. These curves represent a series tuned circuit with three different values of circuit resistance but with each circuit tuned to a resonant frequency of 1,500 kc. The circuit with the least resistance, curve A, provides a higher signal current and sharper tuning than the circuits represented by curves B and C.

By using the transformer principle, the resistance of the antenna

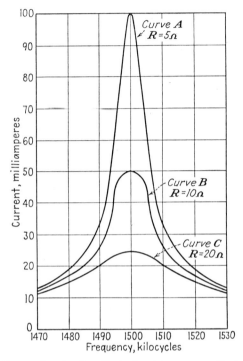

FIG. 1-17. Curves showing the effect of resistance upon the tuning circuit.

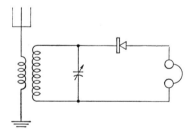

FIG. 1-18. Circuit diagram of a simple crystal receiver using a two-circuit tuner and a variable capacitor.

circuit can be isolated from the tuning circuit (see **Fig. 1-18**). The primary winding of the transformer is connected between the antenna and the ground and the secondary winding is connected to the detector and earphones. A variable capacitor is connected across the secondary winding in order to tune the circuit. This type of tuning circuit, called a *two-circuit tuner*, provides better selectivity and a higher signal current output than the methods previously described.

The magnetic field about the primary winding will vary in accordance with the high-frequency voltage variations induced in the antenna circuit. This magnetic field links the secondary winding and sets up an induced voltage that corresponds to the variations of voltage in the antenna circuit. The variable capacitor and the secondary winding of the transformer form a series resonant circuit, whose frequency of resonance is determined by the capacitance of the variable capacitor and the inductance of the secondary winding of the transformer.

Example 1-10. A variable capacitor having a maximum capacitance of 250 $\mu\mu$f is used in a radio receiver to tune the secondary of the r-f transformer to 500 kc. What value of inductance must the secondary winding have?

$$\text{Given:} \quad C = 250 \ \mu\mu\text{f} \qquad\qquad \text{Find:} \quad L_S$$
$$f_r = 500 \text{ kc}$$

Solution:

Using Eq. (1-11)

$$L_S = \frac{25{,}300}{f_r{}^2 C} = \frac{25{,}300}{500 \times 500 \times 250 \times 10^{-6}} = 404.6 \ \mu\text{h}$$

Ratio of Primary and Secondary Voltages. The ratio of the voltages across the primary and secondary windings of an r-f transformer used in a tuning circuit is dependent upon the ratio of the number of turns, the coefficient of coupling, and the Q of the tuned circuit. At resonance, the voltage across the inductor of a series tuned circuit is Q times greater than the applied voltage. Since the secondary is part of a series tuned circuit and its circuit Q is generally high, the secondary voltage will be much higher than that resulting from transformer action alone. This gain in voltage is another advantage of the two-circuit tuner.

1-17. Crystal Detectors. *Detector Action.* The purpose of a detector is to separate the signal wave from the modulated carrier wave. The manner in which this is accomplished may be seen from Fig. 1-19. Curve *a* represents the modulated wave as it is transmitted by the broadcasting station and also as it is received by the radio receiver. This is also a typical representation of the output of the tuning circuit. The form of the sound wave is represented by the envelope formed by joining the peaks of each individual cycle of the modulated wave. This envelope is the same for both halves of the cycle but is opposite in direction. Therefore, in order to obtain the sound wave either half of the a-c cycle may be used.

A crystal offers very little resistance to the flow of current in one direction and a very high resistance to the flow of current in the opposite direction. Crystals can therefore be used as rectifiers of alternating currents and, since detection is a process of rectification, they can be used as detectors.   Curves *b* and *c* of Fig. 1-19 represent the wave after it has been rectified.

Construction and Operation of a Crystal Detector.   The crystal detector consists of a crystal and a very fine wire called a *catwhisker*.   Various mineral substances such as galena, carborundum, silicon, iron pyrites, etc., may be used as the crystal—the most popular are galena and carborundum.

When a high-frequency modulated voltage is applied to the crystal detector, the catwhisker and crystal are alternately made positive and negative for each half-cycle.   When the crystal is made positive with respect to the catwhisker, the resistance of the crystal is very low and

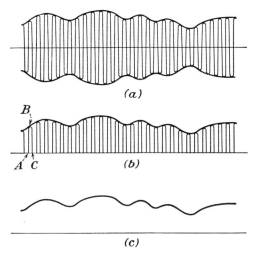

(a)

(b)

(c)

Fig. 1-19. Waveform of a radio signal at various stages in a receiver.   (*a*) As received by the antenna.   (*b*) Rectified current waveform at the detector.   (*c*) Average current output from the detector.

current will flow.   During the half-cycle that the crystal is negative with respect to the catwhisker, the resistance of the crystal is very high and very little current will flow.   This action is repeated for each cycle of the high-frequency alternating current applied to the detector.   The resultant wave is shown in Fig. 1-19*b*.   The high-frequency varying current will therefore cause a flow of current in only one direction.   A small amount of current may flow in the opposite direction, the amount varying with the quality of the crystal, but it is generally ignored since its ratio to the rectified current is small.   A good crystal detector will almost entirely eliminate the flow of current in one direction.

1-18. Audible Device.   Some form of audible device, such as an earphone or loudspeaker, must be used to convert the electrical current impulses of the signal wave to sound waves.

Construction of Earphones. Earphones operate on the electromagnetic principle. Figure 1-21*b* is a cross-sectional diagram showing the construction of a typical earphone. Two permanent magnets, N and S, are used as cores for the coils C_1 and C_2. A curved permanent magnet D connects these two pole pieces and forms a U-shaped magnet, which provides more uniform action than two single magnets. Each coil is wound with several thousand turns of very small insulated copper wire, generally No. 40 or smaller. The two coils are connected in series so that the signal current will pass through both windings. The magnetic field produced by this current will either aid or oppose the constant field of the permanent magnet, depending on the direction in which the

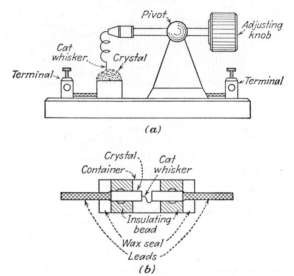

FIG. 1-20. Crystal detectors. (*a*) Adjustable type. (*b*) Sealed fixed-position type.

current flows through the coils. A thin, flexible, soft-iron diaphragm E, approximately 0.005 inch thick, is mounted above the pole pieces and very close to them. The edge of the diaphragm rests on the case A and is held fixed by the cap B.

Operation of Earphones with Alternating Current. When no current flows through the coils, the magnetism of the pole pieces attracts the diaphragm and bends it slightly. This initial deflection is shown as position 1 in Fig. 1-21*b*. If an alternating current is caused to flow through the coils, this force of attraction will be increased for one-half of the cycle and decreased during the other half. During the half-cycle when the current aids the magnetic field of the permanent magnet, the diaphragm will be attracted to position 2. When the current is reversed, its field opposes the permanent magnet's original field and the force of attraction is decreased to a value less than that required to produce

the initial deflection.   The diaphragm will therefore spring to position 3.
At the end of the cycle no current is flowing, so the diaphragm returns
to position 1.

The diaphragm therefore makes a complete vibration for each cycle
of an alternating current.   Its motion will depend on the variation of
current flowing through the coils.   The rapid vibration of the diaphragm
sets the surrounding air into motion, and sound waves will be produced.
If the frequency of the current flowing through the coils is within the
audible range, the sound waves produced by the vibrations of the diaphragm (which are of the same frequency) may be heard by the human
ear.

Operation of Earphones with Varying Unidirectional Current.   It is
not necessary to have alternating current flowing through the coils to
produce vibrations of the diaphragm.   A varying unidirectional current

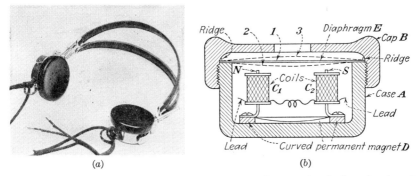

Fig. 1-21. Earphones. (*a*) A set of earphones. (*b*) Cross-sectional view showing the
construction of an earphone.

will cause the diaphragm to vibrate in the same manner that the unidirectional current varies.   By referring to Fig. 1-19*b*, it can be seen that at
the instant *A* no current is flowing from the detector, and the diaphragm
will remain at its initial position.   At the instant *B* the maximum
amount of current is flowing, and the strength of the pole pieces will be
increased; the diaphragm is then deflected to position 2.   At the instant
C the current is again zero, and the diaphragm should return to its
initial position.   The inertia of the diaphragm is too great to follow
these rapid variations in current changes; hence before the diaphragm
has a chance to return to its initial position, the current is again forcing
it in the opposite direction.   The diaphragm of the audible device,
therefore, is not actuated by the instantaneous variations in current of
each cycle, but by the average of these instantaneous currents.   The
curve shown at *c* in Fig. 1-19 is identical in every respect to the ones in
a and *b* except for intensity.   A small capacitor is generally connected
across the earphones to provide a low-impedance path for the r-f currents.

Figure 1-22 shows the complete circuit diagram for a crystal detector circuit and the type of signal current flowing through each of its parts.

Action of the R-F Bypass Capacitor. The small capacitor connected across the earphones provides an easy path for the r-f currents and thus allows only the varying direct current to flow through the earphones. It also increases the strength of the signal current flowing through the audible device in the following manner. The capacitor is charged during the half-cycle that current flows from the detector and discharges

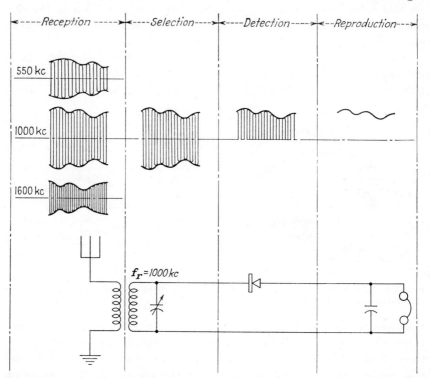

Fig. 1-22. Diagram of a crystal detector receiving circuit and the type of current flowing through each of its parts.

during the half-cycle when no current flows. Since it cannot discharge the current through the crystal, it must discharge it through the earphones. This action causes current to flow through the earphones during the interval when no current is flowing from the crystal. It thereby maintains a varying unidirectional current through the phones, the value of which is almost equivalent to the peak value of the r-f variations in current. This capacitor therefore eliminates the r-f pulsations and increases the a-f voltage across the phones. The value of capacitance used should not be too large; otherwise it will bypass the a-f currents as well as the r-f currents.

QUESTIONS

1. Name six examples of (*a*) the visual method of communication, (*b*) the audible method of communication.

2. Name six types of applications, other than radio, that use the principles of electronics. List specific applications in each.

3. Define (*a*) sound, (*b*) sound waves, (*c*) frequency of sound waves, (*d*) pitch.

4. Explain what occurs that produces the sensation of sound when sound waves strike the human ear.

5. How are radio waves produced?

6. What are the two components of radio waves?

7. How does the speed of radio waves compare with (*a*) light waves? (*b*) electricity? (*c*) sound?

8. Define (*a*) wavelength, (*b*) frequency, (*c*) cycle, (*d*) kilocycle, (*e*) megacycle.

9. Define (*a*) signal, (*b*) signal wave, (*c*) carrier wave, (*d*) modulation.

10. Define (*a*) modulating wave, (*b*) modulated wave, (*c*) sideband, (*d*) modulation envelope.

11. Name and define three methods of modulation.

12. (*a*) What is meant by per cent of modulation? (*b*) Why is it desirable to operate an a-m transmitter with 100 per cent modulation?

13. (*a*) Explain why a 10-kc channel is required by an a-m transmitter whose modulating frequency has an upper limit of 5,000 cycles. (*b*) If the upper limit of the modulating frequency is increased to 10,000 cycles, why must the operating channel of the transmitter be increased to 20 kc? (*c*) What would be the effect if the transmitter in (*b*) were operated with a channel of less than 20 kc?

14. What are the essential characteristics of an f-m wave?

15. Define the following terms: (*a*) center frequency, (*b*) resting frequency, (*c*) frequency deviation, (*d*) carrier swing, (*e*) deviation ratio.

16. (*a*) What is the range of a-f signals that can be transmitted with the f-m system? (*b*) What is the usual frequency deviation permitted with the f-m system? (*c*) What is the carrier-frequency swing for the conditions in (*a*) and (*b*)?

17. What are the essential portions of a transmitter?

18. What function does each portion of the transmitter perform?

19. What are the essential portions of a receiver?

20. What function does each portion of the receiver perform?

21. What is done to prevent interference among the signals from neighboring radio stations?

22. How do the basic operations of a simple crystal receiver compare with those of a complex modern receiver using five or more tubes and/or transistors?

23. Name the four essential functions of a receiver.

24. Explain briefly each of the four essential functions of a receiver.

25. (*a*) Define antenna. (*b*) Name and describe the construction of three simple types of antenna.

26. What factors determine the amount of power induced in a receiving antenna?

27. How do the power and voltage at a receiving antenna compare with the power and voltage at the transmitting antenna?

28. What is the advantage of a loop antenna?

29. (*a*) Describe the electrical characteristics of an antenna. (*b*) Describe its equivalent electrical circuit.

30. What is meant by the fundamental frequency of an antenna?

31. What factor determines the desired electrical characteristics of the antenna circuit?

32. Explain the principle of variable-inductance tuning.

33. Explain the principle of variable-capacitance tuning.

34. (*a*) Describe the two-circuit tuner. (*b*) State two advantages of the two-circuit tuner.

35. (*a*) What is the purpose of the detector? (*b*) Explain the action of the detector.

36. What property of a crystal makes it possible to use a crystal as a detector?

37. Explain the operation of a crystal detector when a high-frequency modulated voltage is applied to it.

38. Explain the basic principle of operation of the earphones.

39. Explain the operation of an earphone when the following types of current are flowing in its coils: (*a*) alternating current, (*b*) varying unidirectional current.

40. What is the purpose of connecting a capacitor in parallel with the earphones?

41. Explain the action of the r-f bypass capacitor.

42. (*a*) Should the capacitance of the r-f bypass capacitor be large or small? (*b*) Explain why.

PROBLEMS

1. The frequency of the sound waves produced by middle C on a piano is 256 cycles. (*a*) What is the wavelength of the sound in feet? (*b*) In meters?

2. A musical note of 256 cycles (Prob. 1) is picked up by a microphone and changes from sound waves to electrical impulses. (*a*) What is the wavelength of the electrical impulses in feet? (*b*) In meters? (NOTE: Electricity and radio waves travel at the rate of 186,000 miles per sec.)

3. Radio programs are often presented to studio audiences as well as to the radio audience. (*a*) How long does it take the sound waves to reach a listener in the studio audience seated 200 ft away? (*b*) How long does it take the program to reach a listener at the loudspeaker of a radio receiver 500 miles away? (*c*) Which listener hears the program first?

4. How far would a sound wave travel in the time that it takes a radio wave to travel 500 miles?

5. If the musical notes of a violin range from 200 to 8,000 cycles (vibrations) per sec, what is its range of wavelength?

6. If the musical notes of a piano range from 30 to 8,000 cycles (vibrations) per sec, what is its range of wavelength?

7. If the shrill sound of an insect has a frequency of 12,000 cycles, what is the wavelength in feet?

8. If the sound of a creaking door has a frequency of 15,000 cycles, what is its wavelength in feet?

9. What is the frequency of the carrier wave of a transmitter whose wavelength is 526 meters?

10. What is the frequency of the carrier wave of a transmitter whose wavelength is 206.8 meters?

11. What are the wavelengths of a television transmitter whose frequencies are 61.25 and 65.75 mc?

12. What are the wavelengths of a television transmitter whose frequencies are 211.25 and 215.75 mc?

13. A certain short-wave transmitter operates on a wavelength of 38.4 meters. What is its frequency?

14. A certain f-m station operates on an assigned frequency of 95.5 mc. (*a*) What is its wavelength in meters? (*b*) What is its wavelength in feet?

15. A certain f-m station operates on an assigned frequency of 99.5 mc. (*a*) What is its wavelength in meters? (*b*) What is its wavelength in feet?

16. A certain radio station operating on an assigned frequency of 1,050 kc is transmitting a violin solo. (*a*) If the notes of the violin range from 200 to 8,000 vibrations per sec, what is the range of the a-f electrical current impulses? (*b*) How many cycles does the carrier-wave current make for each cycle of the lowest frequency note? (*c*) How many cycles does the carrier-wave current make for each cycle of the highest-frequency note?

17. How many cycles does the current of a 95.5-mc carrier wave make in the time that it takes a 5,000-cycle a-f current to complete 1 cycle?

18. How many cycles does the current of a 550-kc carrier wave make in the time that it takes a 5,000-cycle a-f current to complete 1 cycle?

19. How many cycles does the current of the 550-kc carrier wave (Prob. 18) make in the time that it takes a 50-cycle a-f current to complete 1 cycle?

20. How long does it take the radio waves of a transmitter located at Chicago to reach San Francisco, a distance of approximately 1,600 miles?

21. How long does it take the radio waves of a transmitter located at San Francisco to reach Honolulu, Hawaii, approximately 2,500 miles away?

22. How long does it take the radio waves of a transmitter located at New York to reach Melbourne, Australia, approximately 10,000 miles away?

23. How far would a sound wave travel in the time that it takes the radio waves of Prob. 21 to travel 2,500 miles?

24. How far would a sound wave travel in the time that it takes the radio waves of Prob. 22 to travel 10,000 miles?

25. What is the per cent of modulation of an a-m wave if the maximum values of the signal and carrier waves are 4 and 5 volts respectively?

26. What is the per cent of modulation of an a-m wave if the maximum values of the signal and carrier waves are 25 and 30 volts respectively?

27. What is the carrier-frequency swing of an f-m signal when the frequency deviation is (*a*) 40 kc, (*b*) 60 kc?

28. What is the carrier-frequency swing of an f-m signal when the frequency deviation is (*a*) 35 kc, (*b*) 70 kc?

29. If the deviation ratio of an f-m signal is 5, what are the corresponding maximum a-f signals in Prob. 27?

30. If the deviation ratio of an f-m signal is 5, what are the corresponding maximum a-f signals in Prob. 28?

31. The inductance of an antenna circuit is equal to 60 μh, and its capacitance is equal to 188 $\mu\mu$f. What is the resonant frequency of this antenna circuit?

32. The antenna of Prob. 31 is reconstructed so that its inductance is reduced to 40 μh, and its capacitance is increased to 282 $\mu\mu$f. What is the resonant frequency of this antenna circuit?

33. It is desired that the fundamental frequency of an antenna circuit be equal to 15,000 kc. If the capacitance of the antenna circuit is equal to 22.5 $\mu\mu$f, what value of inductance is required?

34. It is desired that the fundamental frequency of an antenna circuit be equal to 6,000 kc. If the inductance of the antenna circuit is equal to 8 μh, what value of capacitance is required?

35. The resonant frequency of an antenna circuit having an inductance of 40 μh is equal to 1,600 kc. What is the capacitance of this circuit?

36. The resonant frequency of an antenna circuit having a capacitance of 140 $\mu\mu$f is equal to 1.6 mc. What is the inductance of this circuit?

37. The capacitance of an antenna circuit is equal to 50 $\mu\mu$f, and its inductance is 5.12 μh. What is its resonant frequency?

38. What resonant frequency will the antenna circuit of Prob. 31 have if a 390-μh inductance coil is connected in series with the antenna circuit?

39. What value of inductance must be connected in series with the antenna circuit of Prob. 32 in order that it will be resonant at 550 kc?

40. What is the resonant frequency of an antenna circuit having a capacitance of 200 $\mu\mu$f and an inductance of 10 μh when a 40-μh inductance coil is connected in series with the antenna circuit?

41. What value of inductance must be connected in series with an antenna circuit having an inductance of 5 μh and a capacitance of 100 $\mu\mu$f in order to make it resonant at 4.1 mc?

42. A variable capacitor is connected in the antenna circuit as shown in Fig. 1-16. The inductance of the antenna circuit is 124 μh, and its capacitance is 188 $\mu\mu$f. (*a*) What must the maximum value of the variable capacitor be in order to obtain resonance at 550 kc? (*b*) What is the frequency range of this circuit?

43. Repeat Prob. 42 for an antenna circuit whose inductance is 136 μh and whose capacitance is 281 $\mu\mu$f.

44. A variable capacitor having a maximum capacitance of 320 $\mu\mu$f is connected across the secondary of an r-f transformer to tune the circuit to 500 kc. What value of inductance is required in the secondary winding?

45. It is desired to obtain various frequency bands by using plug-in coils and a variable capacitor whose maximum capacitance is 260 $\mu\mu$f. What value of inductance is required for the secondary of each coil in order to obtain a minimum frequency of (*a*) 500 kc for range *A*, (*b*) 1,500 kc for range *B*, (*c*) 4 mc for range *C*?

46. A variable capacitor having a maximum capacitance of 320 $\mu\mu$f is used with a set of three plug-in coils in order to obtain three frequency bands. What value of inductance is required for the secondary of each coil in order to obtain a minimum frequency of (*a*) 550 kc for range *A*, (*b*) 1,600 kc for range *B*, (*c*) 4.2 mc for range *C*?

47. A tuned circuit having an inductance of 328 μh is resonant at 550 kc when its variable capacitor is at its maximum value of 260 $\mu\mu$f. What value of capacitance must be connected in series with the variable capacitor in order that the circuit will be resonant at 1,500 kc when its maximum capacitance is being used?

48. A variable capacitor having a maximum capacitance of 350 $\mu\mu$f is connected across the secondary of an r-f transformer in order to tune the circuit to a minimum frequency of 550 kc. (*a*) What value of inductance is required in the secondary circuit? (*b*) What is the maximum frequency of the tuning circuit if its minimum capacitance is 25 $\mu\mu$f?

CHAPTER 2 ✓

VACUUM TUBES

The vacuum tube may be called the heart of electronics because it plays a vital part in the operation of electronic equipment. The importance of the vacuum tube lies in its ability to operate efficiently over a wide range of frequencies and to control almost instantaneously the flow of millions of electrons. Vacuum tubes have many uses in communications as well as in industrial, therapeutic, and other types of electronic equipment. In radio circuits vacuum tubes are used as (1) voltage amplifiers, (2) power amplifiers, (3) detectors, (4) oscillators, (5) frequency converters, (6) rectifiers, (7) regulators, and (8) modulators.

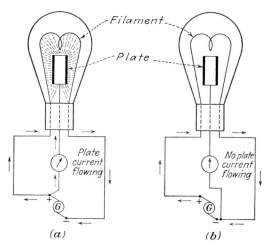

Fig. 2-1. Edison effect. (a) Plate connected to the positive terminal of the generator. (b) Plate connected to the negative terminal of the generator.

2-1. Edison Effect. While experimenting with the incandescent lamp, Thomas A. Edison constructed a lamp in which he placed an additional electrode (Fig. 2-1). When the filament was heated by the current from a generator and the electrode was connected to the positive terminal (Fig. 2-1a), the galvanometer indicated a current flow from the plate to the filament. When the electrode was connected to the negative terminal (Fig. 2-1b), no current flow was indicated on the galvanometer. This phenomenon is known as the *Edison effect*. It should be noted that the direction of "current flow" (used in early references) is opposite to the direction of "electron flow."

31

Professor J. A. Fleming of England later utilized the principles of the Edison effect and developed the *oscillation valve*. He also proved that a unidirectional current would flow in the cold electrode circuit even if an alternating current were used to heat the filament. This valve, which consisted of a filament and a plate, is the simplest type of vacuum tube and is still used, although it is now commonly referred to as a *diode*. Many present-day receivers use a diode as a detector, and the rectifier tubes in a-c-operated receivers are also diodes.

2-2. The Cathode. *Purpose of the Cathode.* When any substance is heated, the speed of the electrons revolving about their nucleus is increased, and some electrons acquire sufficient speed to break away from the surface of the material and go off into space. This action,

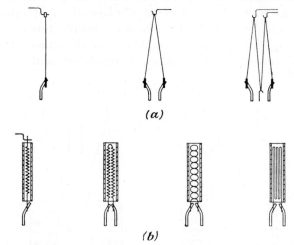

Fig. 2-2. Types of cathodes. (*a*) Directly heated cathodes. (*b*) Indirectly heated cathodes.

which is accelerated when the substance is heated in a vacuum, is utilized in vacuum tubes to produce the necessary electron supply. When used for this purpose, the substance is called the *cathode*. All vacuum tubes contain a cathode and one or more electrodes mounted in an evacuated envelope, which may be a glass bulb or a compact metal shell.

Purpose of the Heater. The cathode is an essential part of a vacuum tube because it supplies the electrons necessary for its operation. The electrons are generally released by heating the cathode. The purpose of a heater in a vacuum tube is to radiate heat when an electric current flows through it. The amount of heat that is radiated is dependent on the material of which the conductor is made and the amount of current flowing in the conductor. The source of power used to supply current for heating the cathode is called the *A power supply*.

Directly Heated Cathodes. A directly heated cathode, called the *filament type*, is one in which the heater is also the cathode. Materials that are good conductors are found to be poor electron emitters; therefore directly heated cathodes must be operated at high temperatures in order to emit a sufficient number of electrons. Various kinds of materials are used in the construction of directly heated cathodes.

TUNGSTEN. Pure tungsten filaments, used as cathodes, are poor emitters and require a relatively large amount of filament power. They are, however, very rugged and are used where high plate voltages are necessary.

THORIATED TUNGSTEN. This type of filament is made from tungsten bars that have been impregnated with thorium and has an electron emission many times greater than pure tungsten. Also, because less power is required to heat the filament, thoriated tungsten is more economical than pure tungsten. Thoriated tungsten filaments are generally used for heating power tubes.

OXIDE COATING. The oxide-coated filament is made by applying successive coatings of calcium, barium, or strontium oxides, used separately or mixed, on a nickel-alloy wire. It has long life, is very efficient, and requires relatively little filament power to emit a sufficient number of electrons.

Directly heated cathodes require a comparatively small amount of heating power and are used in almost all the tubes designed for battery operation. Alternating-current-operated tubes seldom use the filament-type cathode.

Indirectly Heated Cathodes. Some materials emit more electrons than others yet require less heat. Most of these materials are poor conductors and therefore cannot be used for directly heated cathodes. An *indirectly heated cathode* is one in which the emitting material is coated on a thin sleeve which is heated by radiation from a heater placed inside the sleeve and insulated from it. Its advantages may be enumerated as follows:

1. A material that will radiate heat with the least amount of current can be used for the heater; therefore a material that is a good electron emitter can be used for the cathode.

2. The insulation between the heater and cathode and the shielding effect of the sleeve minimize the a-c hum and other electrical interference.

3. The regulation of a rectifier tube is improved because the spacing between the cathode and plate is decreased.

4. The gain of a tube, when used as an amplifier, is increased because the spacing between the cathode and grid is decreased.

2-3. Heater Connections. Directly heated cathodes, or the heaters of indirectly heated cathodes, may be connected in series, parallel, or series-parallel. The choice of connection used is determined by the number of tubes, the voltage of the power supply, the voltage of the heaters, the current rating of the heaters, and the circuit design.

Dropping-resistor Connections. When the rated voltage of each heater is equal to the supply voltage, they are connected in parallel (Fig. 2-3*a*). If the voltage of the power supply is higher than the rated voltage of the tubes, it may be reduced by connecting a resistor in series with the power supply (Fig. 2-3*b*). If one or more tubes are rated at a lower voltage than the others, a dropping resistor is connected in series with its heater

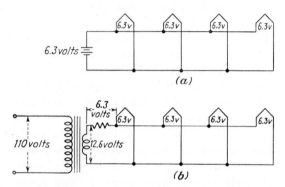

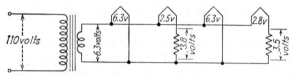

Fig. 2-3. Circuits with heaters (or directly heated cathodes) connected in parallel. (*a*) Voltage of the heaters equal to the voltage of the power supply. (*b*) Using a voltage-dropping resistor to compensate for the difference in power-supply and heater voltages.

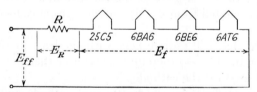

Fig. 2-4. A separate voltage-dropping resistor used to compensate for the difference between the voltage of the power supply and the voltage of individual heaters.

Fig. 2-5. Method of using a series voltage-dropping resistor in order to operate the heaters directly from a 110-volt line.

or filament to decrease the voltage at the heater to its rated value (Fig. 2-4).

In communication equipment using transformerless power supplies, the heaters of the tubes are usually connected in series. In many of these circuits the types of tubes used are selected so that the sum of the rated heater voltages is equal to the applied voltage. When the sum of the rated heater voltages is less than the line voltage, a resistor is connected in series with the heater circuit to cause a drop in voltage equal to the difference between the line voltage and the sum of the rated heater voltages (Fig. 2-5).

Example 2-1.   A small radio receiver employs a selenium rectifier and the following tubes: 6BE6, 6BA6, 6AT6, and 25C5.   If the heaters of the tubes are connected in series, what value resistor must be connected in series with the heater circuit in order to operate the receiver from a 120-volt line?

Given: 6BE6, 6BA6, 6AT6, 25C5   Find: R
E_{ff} = 120 volts

Solution:

See Table 2-1.

TABLE 2-1. VALUES FROM A TUBE MANUAL

Tube	Heater voltage, volts	Heater current, amp
6BE6	6.3	0.3
6BA6	6.3	0.3
6AT6	6.3	0.3
25C5	25	0.3
Total	43.9	

$$E_R = E_{ff} - E_f = 120 - 43.9 = 76.1 \text{ volts}$$
$$R = \frac{E_R}{I_f} = \frac{76.1}{0.3} = 253.6 \text{ ohms}$$
$$P = I^2R = 0.3^2 \times 253.6 = 22.8 \text{ watts}$$

As resistors are rated for their capacity in unrestricted air, a resistor rated at approximately twice the computed wattage should be used when the ventilation is restricted, as is usually the case in radio receivers.   Thus the resistor for Example 2-1 should be rated at 50 watts.

Shunt-resistor Connections.   In cases where all the tubes are not rated at the same current, the series resistance necessary is calculated by using the highest current rating.   Additional resistors must be connected across the heaters of those tubes of lower current rating in order that all the heaters will draw their rated current (Fig. 2-6).

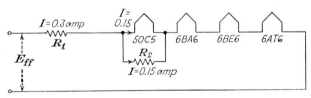

FIG. 2-6. Use of a shunt resistor to supply the heater of one tube with a current less than the value of the line current.

Example 2-2.   If in Example 2-1 a 50C5 tube is substituted for the 25C5, what value of resistor must be used in the line, and what value of resistor should be connected across the heater of the 50C5 in order that its rated current of 0.15 amp will flow through the heater?

Given: Circuit of Fig. 2-6   Find: R_1
          R_2

Solution:

See Table 2-2.

TABLE 2-2. VALUES FROM A TUBE MANUAL

Tube	Heater voltage, volts	Heater current, amp
6BE6	6.3	0.3
6BA6	6.3	0.3
6AT6	6.3	0.3
50C5	50	0.15
Total	68.9	

$$R_1 = \frac{E_{R1}}{I_{R1}} = \frac{120 - 68.9}{0.3} = 170.3 \text{ ohms}$$

$$P_{R1} = I_{R1}^2 R_1 = 0.3^2 \times 170.3 = 15.3 \text{ watts}$$
(Use a 25-watt resistor)

$$R_2 = \frac{E_{R2}}{I_{R2}} = \frac{50}{0.15} = 333 \text{ ohms}$$

$$P_{R2} = I_{R2}^2 R_2 = 0.15^2 \times 333 = 7.5 \text{ watts}$$
(Use a 15-watt resistor)

Parallel-series Connections. The filaments or heaters of tubes may also be arranged in parallel-series combinations to suit the requirements of the tubes and the source of power supply used. For example, if a number of tubes with 6-volt heaters are to be operated from a 12-volt battery, two tubes of similar current rating may be connected in series, and any number of series-connected pairs may be connected in parallel (Fig. 2-7).

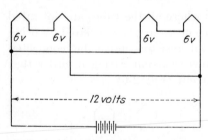

FIG. 2-7. Parallel-series operation of heaters.

Controlled-warm-up-time Tubes. Because of the low resistance value of heaters when cold, the current in the heater circuit is much greater than the normal rated value of the heater current of the tubes for the first minute of operation. This high current may cause one or more of the heaters to burn out. To prevent tube burnouts due to this effect, a series of tubes with a controlled warm-up time have been developed. These tubes are used in many series string heater circuits.

2-4. Diodes. *The Plate.* A vacuum tube having a cathode and one other electrode is called a *diode*. The second electrode is called the *plate* or *anode*. If a positive voltage is applied to the plate, the electrons emitted from the cathode, being negative, will be attracted to the plate. These electrons will flow through the external plate battery circuit as indicated in Fig. 2-8. This flow of electrons is called the *plate current.* If the polarity of the plate is made negative, the electrons will be forced back to the cathode, and no current will flow in the plate circuit.

When an alternating current is applied to the plate, its polarity will be positive during every other half-cycle. As electrons will flow to the plate only when it is positive, the current in the plate circuit will flow in only one direction, and the current is called a *rectified current* (Fig. 2-9).

Diodes are used in radio receivers as detectors, and in a-c-operated receivers and transmitters to convert alternating current to direct

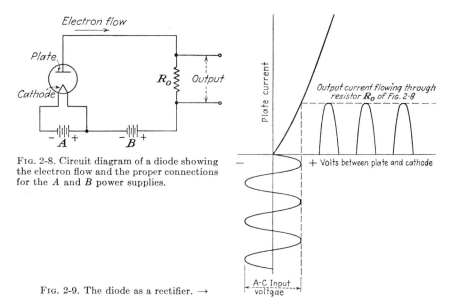

Fig. 2-8. Circuit diagram of a diode showing the electron flow and the proper connections for the *A* and *B* power supplies.

Fig. 2-9. The diode as a rectifier. →

current. Rectifier tubes having one plate and one cathode are called *half-wave rectifiers,* because the rectified current only flows during one-half of the cycle.

Duodiodes. Rectifier tubes having two plates and one or two cathodes have their plates connected into the external circuit so that each plate will be positive for opposite halves of the cycle. As one or the other of the plates will always be positive, current will flow during each half-cycle in the external plate circuit. These tubes are called *full-wave rectifiers,* because current flows during both half-cycles.

Space Charge. The number of electrons drawn to the plate depends on the number given off by the cathode and by the plate voltage. If the plate voltage is not high enough to draw off all the electrons emitted by the cathode, those not drawn off will remain in space. These electrons form a repelling force to the other electrons being given off by the cathode, thus impeding their flow to the plate. This repelling action is called the *space charge*. Space charge can be reduced by decreasing the space between the plate and cathode.

Saturation Current. Increasing the plate voltage will increase the plate current until all the electrons given off by the cathode are drawn to the plate. Further increase in plate voltage will not increase the plate current, as no more electrons can be drawn off to it (Fig. 2-10). The point on the curve at which the current has reached its highest value is called the *saturation point*, and the plate current for this condition is called the *saturation current* or *emission current*.

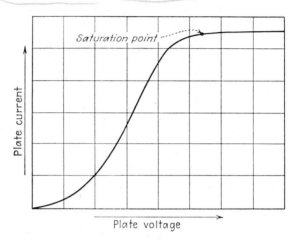

Fig. 2-10. Variation of plate current with changes in plate voltage.

2-5. The Triode. *Action of the Control Grid.* When a third electrode is used, the tube is called a *triode*. This third element is called the *control grid* and consists of a spiral winding or a fine-mesh screen extending the length of the cathode and placed between the cathode and plate. The circuit connections and the direction of electron flow for a triode are shown in Fig. 2-11. If the grid is made more negative than the cathode, some of the electrons going toward the plate will be repelled by the grid, thus reducing the plate current. By making the grid still more negative, it is possible to reduce the plate current to zero. This condition is called *cutoff*.

Grid Bias. The third electrode is called the *control grid*, because it controls the number of electrons allowed to flow from the cathode to the plate. The source of power used to keep the grid negative is called the *C power supply*, and the amount of *negative voltage* used is called the *grid bias*.

Relation between the Grid Bias and the Plate Current. If a varying signal voltage is applied to the grid, the number of electrons flowing in the plate circuit will vary in the same manner as the signal voltage. If the triode in the circuit of Fig. 2-11 operates with a grid bias of 3 volts (Fig. 2-12*b*), a steady plate current I_b will flow (Fig. 2-12*d*). If an alternating signal voltage of 1 volt (Fig. 2-12*a*) is impressed across the input terminals of the grid circuit, the voltage on the grid of the tube will vary from -2 to -4 volts as shown in Fig. 2-12*c*. As the number of electrons flowing to the plate is controlled by the negative potential on the grid, the plate current will vary in the same manner as the signal voltage. This is illustrated by the varying plate current shown in Fig. 2-12*d*, maximum plate current flowing when the grid is least negative and minimum plate current flowing when it is most negative.

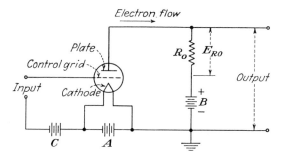

FIG. 2-11. Circuit diagram for a triode showing the electron flow and the proper connections for the A, B, and C power supplies.

Phase Relation between the Varying Grid Voltage, Varying Plate Current, and Varying Plate Voltage. Since the plate current of a tube varies in the same manner as the signal voltage that is applied to its grid, the plate-current variations are in phase with the grid-voltage variations. The output of the tube circuit is taken off at the plate, or top of resistor R_o, and the ground. The voltage available between these two points is equal to the B supply voltage minus the voltage drop at the resistor R_o. If the tube has a grid bias of 3 volts, a steady plate current I_b will flow, and the output voltage will have a value indicated as the steady voltage E_b (Fig. 2-12*e*). When a signal is applied to the grid, the positive portions of the signal voltage will cause an increase in the plate current, which will increase the voltage drop across the output resistor R_o and thus will decrease the output voltage e_b (Fig. 2-12*e*). The negative portions of the signal voltage will decrease the plate current, reduce the voltage drop across R_o, and increase the output voltage e_b. Thus an increase in signal voltage will produce a decrease in the output voltage, or a decrease in signal voltage will produce an increase in the output voltage. The output-voltage variations are therefore 180 degrees out of phase with the signal voltage variations.

✓*Action of the Grid with a Positive Voltage.*   If the value of the signal voltage is such that it will make the grid positive, the grid will act in the same manner as a plate and will draw electrons to it, thus causing a current to flow in the grid circuit.   This condition can be avoided by using a grid bias whose value is larger than the maximum amount of input signal voltage that is applied.

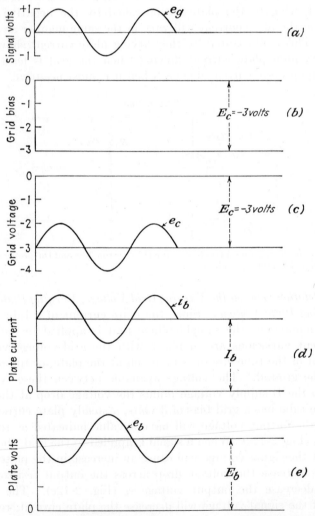

Fig. 2-12. Illustration of the effect of the signal voltage upon the grid voltage and plate current.   (a) Signal voltage (instantaneous values of the varying component of the grid voltage).   (b) Grid bias (average or quiescent value of the grid voltage).   (c) Grid voltage (instantaneous total grid voltage).   (d) Plate current I_b (average or quiescent value of plate current) and plate current i_b (instantaneous total plate current).   (e) Plate voltage E_b (average or quiescent value of plate voltage) and plate voltage e_b, also output voltage (instantaneous total plate voltage).

2-6. Characteristic Curves of a Triode. *Curve and Tabular Representation of Operating Characteristics.* The major operating characteristics of a vacuum tube are used to identify the electrical features and operating values of the tube. These values may be listed in tabular form or plotted on graph paper to form a curve. When given in curve form, they are called *characteristic curves.* These curves are used to determine the performance of a tube under any operating condition. The tube's constants can also be calculated from these curves. Tabular-form listings as found in most tube manuals are limited, as they list only the values of the characteristics for one or two of the operating conditions commonly used for that particular tube.

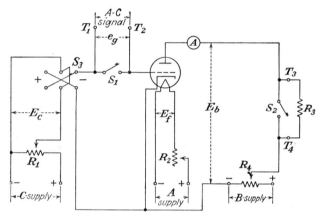

Fig. 2-13. Diagram of a test circuit for obtaining static or dynamic characteristics for triodes.

Methods Used to Obtain Operating Characteristics. Tube characteristics are obtained from electrical measurements of the tube with definite values of voltage applied to the various electrodes. Characteristic curves may be further classified as to the condition of the circuit when these values are obtained. *Static characteristics* are obtained by varying the direct voltages applied to the electrodes, and with no load applied to the plate. *Dynamic characteristics* approximate the performance of a tube under actual working conditions. They are obtained by applying an alternating voltage to the grid circuit and inserting a load resistance in the plate circuit, the direct voltages on all electrodes being adjusted to the desired values. By varying the load resistance and the direct voltages on the electrodes, the characteristics for any operating condition of the tube can be obtained. By using the circuit shown in Fig. 2-13, either the static or dynamic characteristics of a triode can be obtained. If static characteristics are desired, switches S_1 and S_2 are closed. For dynamic characteristics, switches S_1 and S_2 are opened, a signal voltage is applied at the input terminals T_1 and T_2, and a load resistance R_3 is connected across the output terminals T_3 and T_4.   Rated filament voltage can be obtained by adjusting the rheostat R_2 in the A power supply. The voltage applied to the plate of the tube may be varied by means of the potentiometer R_4 in the B power supply.   The voltage on the grid

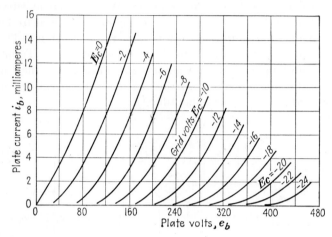

FIG. 2-14. Plate characteristic curves of a triode showing the variation in plate current for changes in plate voltage.

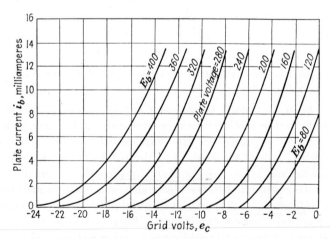

FIG. 2-15. Transfer characteristic curves of a triode showing the variation in plate current for changes in grid voltage.   These curves have been cross-plotted from Fig. 2-14.

of the tube may be made positive or negative by means of the reversing switch S_3, and its value may be adjusted by varying the potentiometer R_1 in the C power supply.

Plate-voltage–Plate-current and Grid-voltage–Plate-current Characteristic Curves.   Values for plotting characteristic curves may be obtained by adjusting the heater voltage to its rated value and recording the current

i_b flowing in the plate circuit when either the plate voltage e_b is varied and the grid voltage E_c is kept constant, or when the grid voltage is varied and the plate voltage is held constant. Curves showing the variation in plate current for changes in plate voltage with constant steps of grid voltage are called *plate characteristic curves*. A typical set, or family, of these curves, also called $e_b i_b$ or $E_p I_p$ curves, is shown in Fig. 2-14. Curves showing the variation in plate current for changes in grid voltage with constant steps of plate voltage are called *grid-plate transfer characteristic curves*. A typical set of these curves, also called $e_c i_b$ or $E_g I_p$ curves, is shown in Fig. 2-15.

 Tube Constants. Vacuum-tube characteristics are often referred to as the constants of the tube. The constants most commonly used are the *amplification factor* $\mu(mu)$, the *plate resistance* r_p, and the *control-grid-to-plate transconductance* g_m. Control-grid-to-plate transconductance is usually referred to as just *transconductance*. Transconductance is also known as *mutual conductance*.

2-7. Amplification Factor. *Definition.* The amplification factor of a tube may be defined as a measure of the relative ability of the grid and the plate to produce an equal change in the plate current. Mathematically it is defined as the ratio of the change in plate voltage to a change in control-grid voltage for a constant value of plate current, with the voltages applied to all other electrodes maintained constant; or

$$\mu = \frac{de_b}{de_c} \, (i_b - \text{constant}) \tag{2-1}$$

where μ = amplification factor
  d = change or variation in value
  de_b = change in instantaneous total plate voltage, volts
  de_c = change in instantaneous total control-grid voltage necessary to produce the same effect upon the plate current as would be produced by de_b, volts
  i_b = instantaneous total plate current, amp

Method of Obtaining the Amplification Factor. The amplification factor may be calculated from values obtained from the plate characteristic curves such as those shown in Fig. 2-14.

Example 2-3. What is the amplification factor of the tube whose characteristic curves are shown in Fig. 2-14 when operated at its normal heater voltage and with a plate voltage of 160 volts and a grid bias of 4 volts?

  Given: e_b = 160 volts Find: μ
    E_c = −4 volts
Solution:

  From Fig. 2-14

    i_b = 7.5 ma when e_b = 160 volts and E_c = −4 volts

Further examination of Fig. 2-14 shows that if the plate voltage is increased from 160 volts to 200 volts, the grid-bias voltage would have to be increased from -4 to -6 volts in order to maintain the plate current at 7.5 ma. Thus the amplification factor is

$$\mu = \frac{de_b}{de_c} = \frac{200 - 160}{6 - 4} = \frac{40}{2} = 20$$

2-8. Plate Resistance. *Definition.* The plate resistance of a tube may be defined as the resistance to the flow of alternating current offered by the path between the cathode and plate. This is sometimes called the *dynamic plate resistance* or the *a-c plate resistance*. The value of the plate resistance will depend upon the values of the grid and plate voltages being applied to the tube. Mathematically it is defined as the ratio

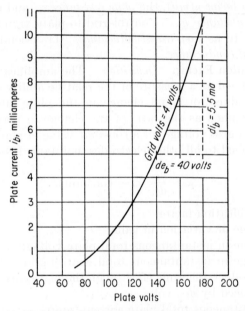

FIG. 2-16. Method of determining the plate resistance of a tube.

of a change in plate voltage to the corresponding change produced in the plate current with the grid voltage maintained constant, or

$$r_p = \frac{de_b}{di_b} \quad (e_c - \text{constant}) \tag{2-2}$$

where r_p = dynamic plate resistance, ohms

    di_b = change in instantaneous total plate current, amp

    e_c = instantaneous total grid voltage, volts

Method of Obtaining the Plate Resistance. The plate resistance is obtained by determining the variation in plate current for a specific change in plate voltage with the control-grid voltage maintained at a constant value. The values for calculating the plate resistance can be obtained from the plate characteristic curves by constructing a triangle whose base extends equal amounts above and below the operating value of plate voltage as shown in Fig. 2-16.

Example 2-4. What is the plate resistance of the tube whose characteristic curves are shown in Fig. 2-14, when being operated with 160 volts applied to the plate and with a grid bias of 4 volts?

$$\text{Given:}\quad e_b = 160 \text{ volts} \qquad\qquad \text{Find:}\quad r_p$$
$$E_c = -4 \text{ volts}$$

Solution:

Using the curve for −4-volt grid, construct a triangle whose base extends 20 volts above and below the operating plate voltage, or from 140 to 180 volts (see Fig. 2-16). The altitude of the triangle represents the variation in plate current produced by the change in plate voltage, or 5.5 ma as indicated on Fig. 2-16. The plate resistance therefore is

$$r_p = \frac{de_b}{di_b} = \frac{40}{0.0055} = 7{,}272 \text{ ohms}$$

The solution of Example 2-4 gives the dynamic or a-c plate resistance of the tube. It should be noted that this is not the same value as would be obtained by dividing the operating plate voltage by its corresponding plate current. Such a value is known as the *static resistance* or the *d-c resistance* of the tube, but it is seldom used in the study of tubes and their circuits. The plate resistance is sometimes referred to as the *impedance* of the tube, the *internal resistance* of the tube, the *dynamic plate resistance*, or the tube's *a-c resistance*.

2-9. Transconductance. *Definition.* The transconductance of a tube may be defined as the ratio of the change in plate current to a change in the control-grid voltage when all other tube-element voltages are kept constant. Expressed mathematically,

$$g_m = \frac{di_b}{de_c} \, (e_b - \text{constant}) \tag{2-3}$$

where g_m = transconductance, mhos

e_b = instantaneous total plate voltage, volts

Method of Obtaining the Transconductance. The transconductance is obtained by determining the variation in plate current for a specific change in grid voltage with the plate voltage maintained at a constant value. The values for calculating the transconductance can be obtained from the grid-plate transfer characteristic curves by constructing a triangle whose base extends equal amounts above and below the operating value of the grid voltage, as shown on Fig. 2-17.

Example 2-5. What is the transconductance of the tube whose characteristics are shown in Fig. 2-15, when being operated with 160 volts applied to the plate and with a grid bias of 4 volts?

$$\text{Given:} \quad e_b = 160 \text{ volts} \qquad\qquad \text{Find:} \quad g_m$$
$$E_c = -4 \text{ volts}$$

Solution:

Using the curve for 160 volts, construct a triangle whose base extends 0.5 volt above and below the operating grid voltage or from -3.5 to -4.5 volts (Fig. 2-17).

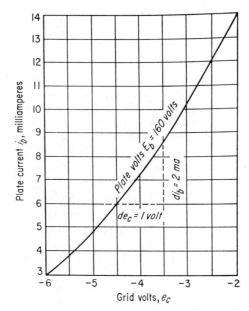

FIG. 2-17. Method of determining the transconductance of a tube.

The altitude of the triangle represents the variation in plate current produced by the change in grid voltage, or 2.5 ma as indicated on Fig. 2-17. The transconductance, therefore, is

$$g_m = \frac{di_b}{de_c} = \frac{0.0025}{1} = 0.0025 \text{ mho}$$

The unit of conductance is the mho, and its name is obtained by spelling ohm backwards. For convenience of numbers, the transconductance is often expressed in micromhos. Thus, the answer to Example 2-5 may be expressed as 2,500 micromhos (abbreviated as 2,500 μmhos).

Equation (2-3) shows that a tube that produces a relatively large change in plate current for a small change in grid voltage will have a relatively high value of transconductance, and, since such conditions are generally desired, tubes having a high value of transconductance are preferred.

2-10. Relation among Amplification Factor, Plate Resistance, and Transconductance. A definite relation exists among the three tube constants, and this relation is developed mathematically in the following steps:

Step 1: $$\mu = \frac{de_b}{de_c} \qquad (2\text{-}1)$$

Step 2: $$\mu = \frac{de_b}{de_c} \times \frac{di_b}{di_b} \qquad (2\text{-}4)$$

$$\mu = \frac{di_b}{de_c} \times \frac{de_b}{di_b} \qquad (2\text{-}4a)$$

Step 3: Substituting Eqs. (2-3) and (2-2) in (2-4a)

$$\mu = g_m r_p \qquad (2\text{-}5)$$

$$g_m = \frac{\mu}{r_p} \qquad (2\text{-}5a)$$

$$r_p = \frac{\mu}{g_m} \qquad (2\text{-}5b)$$

Example 2-6. What is the transconductance of the triode section of a 12SQ7 duodiode-triode tube when operated at such values of voltage that its amplification factor is 100 and its plate resistance is 91,000 ohms?

Given:　Tube = 12SQ7　　　　　　　　　Find: g_m
$\mu = 100$
$r_p = 91{,}000$ ohms

Solution:

$$g_m = \frac{\mu}{r_p} = \frac{100}{91{,}000} \cong 0.0011 \text{ mho, or } 1{,}100\mu \text{ mhos}$$

Example 2-7. What is the plate resistance of a 6AF4 high-frequency triode tube when operated at such values of voltage that its amplification factor is 15 and its transconductance is 6,600 μmhos?

Given:　Tube = 6AF4　　　　　　　　　Find: r_p
$\mu = 15$
$g_m = 6{,}600$ μmhos

Solution:

$$r_p = \frac{\mu}{g_m} = \frac{15}{6{,}600 \times 10^{-6}} = 2{,}272 \text{ ohms}$$

2-11. Voltage Amplification per Stage. The maximum theoretical value of voltage amplification per amplifier stage is indicated by the amplification factor of the tube. However, this theoretical maximum value cannot be obtained in practical amplifier circuits, as the voltage amplification is limited by the plate resistance of the tube and the impedance of the load.

If the signal output of the tube is taken as the voltage across the output or load resistor R_o (Fig. 2-18), the output voltage e_p will be proportional to the resistance R_o and the plate current i_p. The voltage amplification of the circuit will be

$$VA = \frac{e_p}{e_g} \qquad (2\text{-}6)$$

From Fig. 2-18*b*

$$e_p = i_p R_o \tag{2-7}$$

also

$$i_p = \frac{\mu e_g}{R_o + r_p} \tag{2-8}$$

Substituting Eq. (2-8) for i_p in Eq. (2-7)

$$e_p = \frac{\mu e_g R_o}{R_o + r_p} \tag{2-9}$$

Substituting Eq. (2-9) for e_p in Eq. (2-6)

$$\mathrm{VA} = \frac{\mu R_o}{R_o + r_p} \tag{2-10}$$

where VA = voltage amplification of stage

     e_p = instantaneous value of varying component of plate voltage, volts

     e_g = instantaneous value of varying component of grid voltage, volts

     R_o = resistance of output or load resistor, ohms

Equation (2-10) represents the voltage amplification per stage for a circuit using a vacuum tube and a pure resistance output load R_o (Fig.

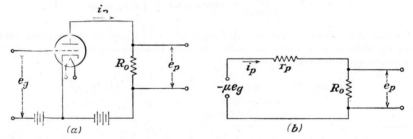

Fig. 2-18. Simple triode amplifier. (*a*) The circuit diagram. (*b*) The equivalent plate circuit diagram.

2-18). The load may not always be pure resistance but may sometimes be an inductance or capacitance load.

Relation among r_p, R_o, and the Voltage Amplification. In order to obtain the maximum voltage amplification per stage with tubes having approximately the same amplification factor, a tube having the lowest plate resistance should be used. This is shown by Eq. (2-10), which indicates that for a definite plate load the voltage amplification will increase as the plate resistance is decreased. (This should not be confused with the maximum power output, which occurs when R_o is equal to r_p as is explained in Chap. 7.)

Example 2-8. Two tubes, each with an amplification factor of 40 but with plate resistances of 4,000 and 8,000 ohms respectively, are alternately placed in a circuit whose load resistance has a value of 8,000 ohms. Which tube produces the greater voltage amplification for the circuit?

Given: R_o = 8,000 ohms   Find: VA for tube 1
         Tube 1: μ = 40; r_p = 4,000 ohms   VA for tube 2
         Tube 2: μ = 40; r_p = 8,000 ohms

Solution:

Tube 1 $VA = \dfrac{\mu R_o}{R_o + r_p} = \dfrac{40 \times 8,000}{8,000 + 4,000} = 26.66$

Tube 2 $VA = \dfrac{\mu R_o}{R_o + r_p} = \dfrac{40 \times 8,000}{8,000 + 8,000} = 20$

Thus the tube with the lower value of r_p produces the greater voltage amplification for the circuit.

Equation (2-10) also shows that for a given value of plate resistance and amplification factor of a tube, the voltage amplification of a circuit will increase if the value of the load resistor is increased.

Example 2-9.   A tube whose amplification factor is 40 and whose plate resistance is 4,000 ohms is alternately used in circuits containing load resistors of 4,000 and 8,000 ohms respectively. With which circuit does the tube produce the greater voltage amplification?

Given: μ = 40   Find: VA for circuit 1
         r_p = 4,000 ohms   VA for circuit 2
         Circuit 1: R_o = 4,000 ohms
         Circuit 2: R_o = 8,000 ohms

Solution:

Circuit 1 $VA = \dfrac{\mu R_o}{R_o + r_p} = \dfrac{40 \times 4,000}{4,000 + 4,000} = 20$

Circuit 2 $VA = \dfrac{\mu R_o}{R_o + r_p} = \dfrac{40 \times 8,000}{8,000 + 4,000} = 26.66$

Thus the circuit with the higher value of R_o produces the greater voltage amplification.

2-12. Relation between the Transconductance and the Operating Performance of a Tube.

Vacuum-tube amplifiers are ordinarily operated so that small variations in voltage of the grid or input circuit will produce large current variations in the plate or output circuit.   Since it is usually desired to keep the plate resistance low, the transconductance must be high if a high value of amplification factor is desired.   The transconductance is very useful when comparing the relative merits and performance capabilities of tubes designed for the same service.   A comparison of the transconductance of a power-output tube with that of a tube used as a converter would, however, have no practical value.   Generally, the value of transconductance is accepted as the best single figure of merit for vacuum-tube performance.

2-13. The Tetrode. *The Screen Grid.*

When a fourth electrode is used, the tube is called a *tetrode;* it is also referred to as a *screen-grid tube* or a *four-electrode tube.*   The fourth electrode is known as the *screen grid* and consists of a spiral-wound wire or a screen, slightly coarser than that used for the control grid, placed between the plate and the control grid of the tube. The circuit connections and the direction of electron flow for a tetrode are shown in Fig. 2-20.

Interelectrode Capacitance. Capacitance is present whenever two conductors are separated by an insulator. In any tube, the electrodes act as conductors and the space between the electrodes acts as an insulator; therefore a capacitance will exist between each pair of electrodes. These

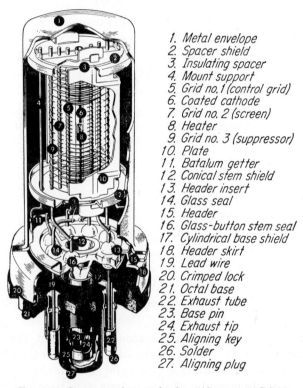

1. Metal envelope
2. Spacer shield
3. Insulating spacer
4. Mount support
5. Grid no. 1 (control grid)
6. Coated cathode
7. Grid no. 2 (screen)
8. Heater
9. Grid no. 3 (suppressor)
10. Plate
11. Batalum getter
12. Conical stem shield
13. Header insert
14. Glass seal
15. Header
16. Glass-button stem seal
17. Cylindrical base shield
18. Header skirt
19. Lead wire
20. Crimped lock
21. Octal base
22. Exhaust tube
23. Base pin
24. Exhaust tip
25. Aligning key
26. Solder
27. Aligning plug

FIG. 2-19. Structure of a metal tube.   (*Courtesy of RCA.*)

capacitances, known as interelectrode capacitances, may form undesired paths through which current can flow (Fig. 2-21).

Effect of the Interelectrode Capacitance. The capacitance between the grid and plate is generally the most troublesome. This capacitance causes some of the energy of the plate circuit to be applied to the grid circuit. The energy transferred from the output to the input circuit is called *feedback*. If the output circuit contains resistance load, the feedback voltage will be 180 degrees out of phase with the input voltage and therefore will reduce the effect of the input signal; this is called *degeneration* or *negative feedback*. When the output circuit load is inductive, the feedback voltage will be in phase with the input voltage and therefore will increase the effect of the input signal; this is called *regeneration* or *positive feedback.*

The amount of feedback caused by the interelectrode capacitance is usually very small at audio frequencies because the capacitive reactance at these frequencies is very high. At radio frequencies, however, the capacitive reactance becomes much lower, and the feedback may reach an amount sufficient to cause trouble in the operation of the circuit.

Positive feedback has the advantage of increasing the gain of a circuit but has the disadvantage of causing distortion. Negative feedback has the advantage of reducing distortion but has the disadvantage of causing a reduction in the voltage gain of the circuit. If either regeneration or degeneration gets out of control, the tube no longer operates successfully as an amplifier.

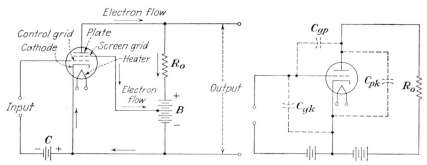

Fig. 2-20. Circuit diagram for a tetrode showing the electron flow and the proper connection for the screen grid.

Fig. 2-21. Interelectrode capacitances of a triode.

Elimination of Feedback by Use of a Screen Grid. The capacitance between the control grid and the plate of a triode can be reduced to a negligible amount by adding a fourth electrode or *screen grid.* The screen grid is mounted between the control grid and the plate and acts as an electrostatic shield between the two, thus reducing the control-grid-to-plate capacitance. Connecting a bypass capacitor between the screen grid and the cathode will increase the effectiveness of this shielding action.

Another advantage of the tetrode is that it can be operated over a wide range of plate voltage, but only for values of voltage greater than the screen-grid voltage, with comparatively little change in the plate current. This is made possible by constructing the screen grid with comparatively large spacing between its wires, thereby permitting most of the electrons drawn toward the screen grid to pass through it and on to the plate. As the screen grid is operated at a comparatively high potential and because of the spacing of its wires, it produces an electrostatic force pulling electrons from the cathode to the plate. At the same time, the screen grid shields the electrons between the cathode and the screen grid from the plate and therefore the plate exerts very little electrostatic force on electrons near the cathode. Hence, as long as the plate voltage is higher than the screen-grid voltage, the plate current depends largely on the screen-grid voltage and very little on the plate voltage. The fact that the plate current of the tetrode is largely independent of the plate voltage makes it possible to obtain much higher amplification with the tetrode than with a triode. The low control-grid-to-plate capacitance makes it possible to obtain this high amplification without plate-to-control-grid feedback and the resultant instability.

The screen grid also reduces the space charge. Being situated between the control grid and the plate and having a positive potential, the electrons coming from the cathode receive added acceleration on their way to the plate, and the tendency to form a space charge is reduced.

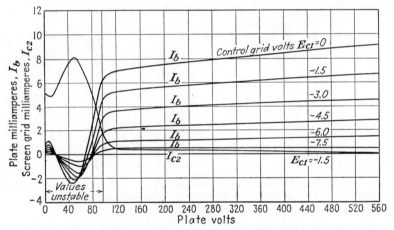

Fig. 2-22. Family of plate characteristic curves for a tetrode.

Characteristic Curves for a Tetrode. A family of plate characteristic curves for a tetrode is shown in Fig. 2-22. Each curve represents the variations in plate current with changes in plate voltage for a different value of grid bias; the screen-grid voltage was kept at 90 volts for all conditions. The dip that occurs in the curve when the plate voltage is less than the screen-grid voltage is caused by the secondary emission of electrons at the plate. Secondary emission occurs when electrons emitted from the cathode, called *primary emission*, strike other electrodes with sufficient force to dislodge other electrons. The electrons liberated in this manner are secondary to the original cathode or primary emission, and this effect is therefore called *secondary emission*. If the screen-grid voltage is higher than the plate voltage, the secondary electrons will be attracted by the screen grid and will flow in that circuit. The dip in the curve occurs because the plate current is decreased by the amount of the secondary emission that finds its way into the screen-grid circuit. When the plate voltage reaches the value of the screen-grid voltage, the secondary electrons find it more difficult to reach the screen grid, since the plate itself now attracts some of the secondary electrons.   When the plate voltage exceeds the screen-grid voltage, practically all the secondary electrons return to the plate, and the curve approaches a horizontal line.

Constants of the Tetrode.   The constants for the tetrode are found by constructing a triangle on the characteristic curve about the operating point desired in the same manner as for triodes.   It can be observed that a small change in grid voltage produces a very large change in plate current compared with the change in plate voltage required to produce an equal change in the plate current.   The amplification factor of the tube is therefore very high.   It can also be noted that a large change in plate voltage produces very little change in plate current; hence the plate resistance will also be very high.   The transconductance, however, is equal to or slightly lower than for a triode of similar construction.

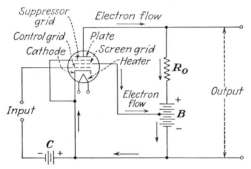

Fig. 2-23. Circuit diagram for a pentode showing the electron flow and the proper connections for the screen grid and the suppressor grid.

2-14. The Pentode.   *The Suppressor Grid.*   A tube with five electrodes—namely, a cathode, three grids, and a plate—is called a *pentode* or a *five-electrode tube.*   The fifth electrode is an extra grid called the *suppressor grid.*   The electrodes of the pentode are arranged with the cathode at the center and surrounded by the control grid, the screen grid, the suppressor grid, and the plate, in the order named.   The circuit connections and the direction of electron flow for the pentode are shown in Fig. 2-23.

The suppressor grid consists of a spiral-wound wire or a coarse-mesh screen placed between the screen grid and the plate.   When the various grids of the tube are in the form of a screen, the control grid is of a very fine mesh so that small changes in control-grid voltage will produce relatively large changes in plate current and consequently will produce a high value of transconductance for the tube.   The screen grid is of a somewhat coarser mesh so that it will not appreciably affect the flow of electrons to the plate, its purpose being largely to reduce the control-grid-to-plate capacitance.   The suppressor grid is of a coarse mesh so that it will not retard the flow of electrons to the plate while serving its function of preventing the secondary emission from reaching the screen grid.   In some pentodes the suppressor grid is internally connected to the cathode, while in others the suppressor grid is brought out as a separate terminal.

Action of the Suppressor Grid.   In pentodes, the suppressor grid is added to prevent the secondary electrons from traveling to the screen grid.   In order to accomplish this the suppressor grid is connected directly to the cathode.   Being at cathode potential, the suppressor grid is negative with respect to the plate, and because it is close to the plate it will repel the secondary electrons and drive them back into the plate.

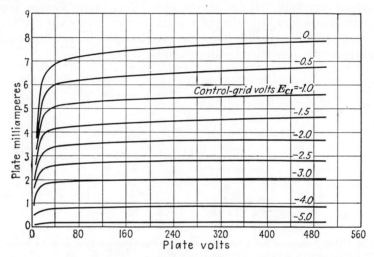

Fig. 2-24. Family of plate characteristic curves for a pentode.

Characteristic Curves for a Pentode.   A family of curves for a typical pentode is shown in Fig. 2-24.   Comparing this set of curves with the curves for a tetrode will show that the unstable portion of the tetrode curves is eliminated.   It is therefore possible to have a larger plate-voltage swing with a small variation in plate current.

Constants of the Pentode.   The tube constants for a pentode are determined in the same manner as for the triode and tetrode, that is, by constructing triangles about the operating point desired.   The amplification factor and the plate resistance are higher for the pentode than for a comparable tetrode, and the transconductance is about the same as for a similar triode or tetrode.

Pentode tubes can be used as either voltage or power amplifiers.   In power pentodes, a higher power output is obtained with lower grid voltages.   Pentodes used as r-f amplifiers give a high voltage amplification when used with moderate values of plate voltage.

2-15. Variable-mu or Supercontrol Tubes. A triode or pentode with its control grid constructed in such a manner that the amplification factor of the tube will vary with a change in grid bias is called a *variable-mu tube*, *supercontrol tube*, or a *remote-cutoff tube*.

Cutoff with Ordinary Grid Structure. If the grid bias on a tube is steadily increased, it will eventually reach a value that will reduce the plate current to zero, and it is then said that the tube has reached *cutoff*. In a tube with the ordinary grid construction, that is, one in which the turns of the spiral-wound control grid are equally spaced, increasing the grid bias causes the plate current to decrease very rapidly to cutoff (Fig. 2-25).   This type of grid construction produces a tube with a practically constant amplification factor for all values of grid bias and is called a *sharp-cutoff* or *constant-mu tube*.

Effects of Sharp Cutoff. A tube with a sharp cutoff is limited to use in circuits with relatively small changes in grid voltage.   In circuits that have a large signal voltage, sharp cutoff would produce distortion in the form of cross modulation and modulation distortion.

Cross modulation or *cross talk* is the effect produced when the signal from a second station is heard in addition to the signal of the desired station.   Cross modulation is generally caused in the first stage of r-f amplification.

Modulation distortion is the effect produced when the signal voltage drives the tube beyond cutoff and thereby distorts the desired audio signal.   Modulation distortion is generally caused by the last stage of i-f amplification.

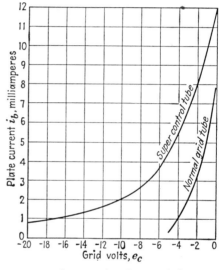

Fig. 2-25. Comparative characteristic curves of tubes with supercontrol grid and with uniformly spaced grid.

Action of the Supercontrol Tube. The characteristics of the supercontrol tube enable the tube to handle both large and small input signals with a minimum amount of distortion over a wide range of signal voltage. The control grid is spiral-wound with its turns close together at the ends but with considerable space between turns at the middle (Fig. 2-26).   For weak signals and a low grid bias, the tube operates practically the same as one with a uniformly spaced grid.   With larger signal inputs, more grid bias is required.   The increased grid bias will produce a cutoff effect at the section of the cathode enclosed by the ends of the grid because of the close spacing of the grid wires.   The plate current and other tube characteristics are now dependent on the electron flow through the center section of the grid, where the turns are spaced farther apart. The wide spacing changes the gain of the tube and enables it to handle large signals with a minimum amount of distortion.

Figure 2-25 shows the characteristic curves for a sharp cutoff and a supercontrol tube. The rate of plate-current change is approximately the same in both tubes for small values of grid-bias voltage, while for large values of grid bias the plate current decreases at a much lower rate for the supercontrol tube. This low rate of change enables the tube to handle large signals satisfactorily. The variable mu permits the tube to be used in automatic-volume-control (avc) circuits.

2-16. The Beam Power Tube. The beam power tube is constructed so that the electrons flowing to the plate are made to travel in concentrated beams. As it is capable of handling larger

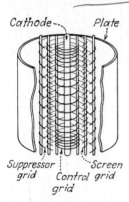

Fig. 2-26. Cross-sectional view showing the construction of the electrodes of a supercontrol amplifier tube.

amounts of power it is called a *beam power tube.* This tube contains a cathode, a control grid, a screen grid, and a suppressor.

Action of the Beam Power Tube. The control grid and the screen grid are of the spiral-wound wire construction and their respective turns are placed so that each turn of the screen grid is shaded from the cathode by a turn of the control grid. This arrangement of the grid wires causes the electrons to flow in directed paths between the turns of the screen grid. Beam-forming plates (Fig. 2-27) are added to aid in producing the desired beam effect and to prevent stray electrons from the plate from flowing into the screen-grid circuit. This results in the tube having a relatively low value of screen-grid current. The beam-forming plates are operated at cathode potential by connecting them directly to the cathode. By increasing the spacing between the screen grid and the plate, a space charge is set up in this area. This space charge repels the secondary electrons emitted from the plate and forces them back into the plate.

Characteristics of a Beam Power Tube. The characteristic curves of the beam power tube (Fig. 2-28) are similar to those of the pentode. The beam power tubes provide a straighter curve at the lower plate voltages than the pentode and hence will have less chance of producing distortion. The amplification factor is high when compared with triodes and low when compared with tetrodes and pentodes. The plate resistance is high but not so high as that of pentodes. The transconductance is generally higher than for any other type of tube.

The combined effects of the directed concentrated beam of electrons,

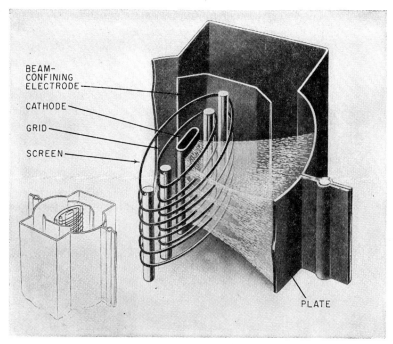

FIG. 2-27. Internal structure of a beam power tube. (*Courtesy of RCA.*)

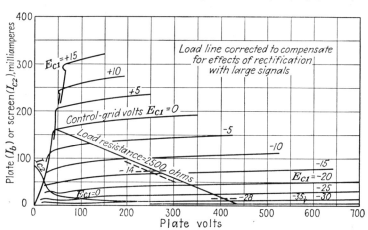

FIG. 2-28. Family of plate characteristic curves for a beam power tube.

the suppressor action of the space charge, and the low value of the screen-grid current result in a tube of high power output, high efficiency, and high power sensitivity.

2-17. Multiunit Tubes. When a tube contains within one envelope two or more groups of electrodes associated with independent electron streams, it is called a *multiunit tube*. The type 6X4 is a duodiode consisting of two half-wave rectifier units in a single envelope. The pentagrid converter is also a multiunit tube consisting of a triode oscillator and a pentode mixer. Advancements in circuit and tube designs have resulted in numerous combinations in multiunit tubes. In general, the combinations are easily identified by the names given to the tubes, such as duplex-diode, twin-triode, duplex-diode-triode, diode-triode-pentode, and rectifier-beam power amplifier (Fig. 2-29). In most cases a single cathode is used for all the units in a tube, although in some types separate cathodes are provided for each unit.

An example of the multiunit type tube is the 1D8-GT, which is a diode-triode-pentode. This tube performs three functions, namely, the diode is used as a detector and automatic volume control; the triode is used as an a-f amplifier; and the pentode is used as a power output tube.

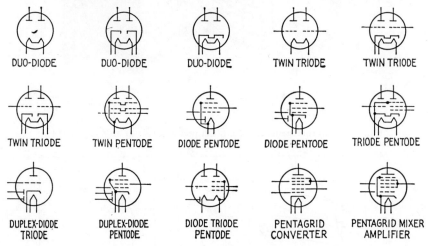

| DUO-DIODE | DUO-DIODE | DUO-DIODE | TWIN TRIODE | TWIN TRIODE |

| TWIN TRIODE | TWIN PENTODE | DIODE PENTODE | DIODE PENTODE | TRIODE PENTODE |

| DUPLEX-DIODE TRIODE | DUPLEX-DIODE PENTODE | DIODE TRIODE PENTODE | PENTAGRID CONVERTER | PENTAGRID MIXER AMPLIFIER |

Fig. 2-29. Base diagrams for various types of multiunit tubes.

2-18. New Types of Tubes. The large number of tube types that has been produced is the result of: (1) new applications, (2) a wide range of heater voltages, (3) improvements in tube design, (4) a demand for tubes requiring less space, and (5) changes made for uniformity or standardization. The ever-increasing demand for smaller electronic equipment has resulted in the development of smaller tubes. These tubes have a glass envelope and are made as a small tubular bulb. They are known under various names such as *dwarf, bantam, miniature, doorknob,* and *acorn* tubes; also the very small tube is called the *nuvistor.*

2-19. Tube Bases and Socket Connections. *Need for a Means of Identification.* Because of the large number of tube types that are manufactured and because of the variation in the number of electrodes and types of tube bases used, it became necessary to establish a system of identifying the socket connections and the tube electrodes with which thcy are to make contact. In diagrams of circuits that include tubes, it is common practice to show the socket connections, which in turn correspond to the connecting pins in the base of the tube to be used. References to socket connections and tube-pin numbering are made for bottom views of sockets and tubes. The arrangement of tube-base and socket designations has been standardized by the EIA (Electronic Industries Association), and the two systems described here represent the methods used for numbering tube pins and socket connections.

Methods of Identifying Socket Connections. The method of numbering the socket or tube-base connections for the early tube types is shown in

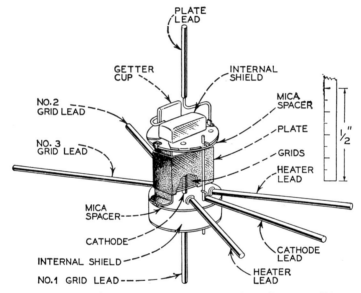

Fig. 2-30. Internal structure of an acorn pentode. (*Courtesy of RCA.*)

Fig. 2-32. In this system, the filament or heater pins of the tube and the corresponding holes in the socket are larger in diameter than the others and are generally shown at the bottom of the diagram. The lower left-hand pin is designated as number 1, and the remaining pins are numbered consecutively in a clockwise direction. The order in which the tube elements are arranged varies with the tube types and may be obtained from a tube manual.

Metal and other octal-base tubes all use the same eight-pin socket. The *octal socket,* as it is commonly called, has eight equally spaced holes arranged in a circle. All of the holes in the socket are of the same size and, in order to ensure correct placement of the tube in the socket, a large center hole with an extra notch is provided. The socket connection to the left of the centering notch is designated as number 1, and the remaining connections are numbered consecutively up to 8. The pins in the base of the tube are all of equal size and are arranged in a circle. A large insulated pin, provided with a centering key to fit the notch on the socket, is located in the center of the tube base. Some tubes using octal sockets have only six or seven pins, while others have eight. The six or seven pin bases merely omit one or two of the eight pins according to the number used. This, however, does not alter the numbering of the socket con-

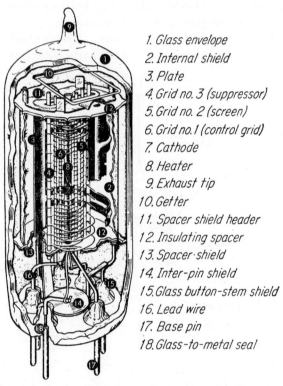

1. *Glass envelope*
2. *Internal shield*
3. *Plate*
4. *Grid no. 3 (suppressor)*
5. *Grid no. 2 (screen)*
6. *Grid no. 1 (control grid)*
7. *Cathode*
8. *Heater*
9. *Exhaust tip*
10. *Getter*
11. *Spacer shield header*
12. *Insulating spacer*
13. *Spacer·shield*
14. *Inter-pin shield*
15. *Glass button-stem shield*
16. *Lead wire*
17. *Base pin*
18. *Glass-to-metal seal*

FIG. 2-31. Miniature-type tube with a button base (enlarged to about twice actual size). (*Courtesy of RCA.*)

nections. The order in which the tube elements are arranged varies with tube types and may be obtained from a tube manual. The loktal socket has an 8-pin base similar to the octal socket with the following differences: (1) the holes are smaller, and (2) provision is made for locking the tube in the socket.

The development of acorn and miniature types of tubes also resulted in the introduction of a new means of connection for the acorn type and a new socket for the miniature type. The acorn tube has its leads brought out of the side of the tube, as shown in Fig. 2-30. The miniature tube (Fig. 2-31) has a thin glass base, called a *button base*, and its socket is called a *button socket*. Miniature tubes are made with either a seven- or nine-pin base (Fig. 2-32).

2-20. Tube Type Numbers. The type number of a tube is intended to give information concerning its construction and application. The early tubes were numbered consecutively and their numbers had no particular significance. The numbers assigned to later tubes provide a means of identifying some of the tube characteristics. The tube number consists of at least three units. The first unit is a number, the second consists of one or more letters, and the third is a number. In some cases a letter or group of letters is added as a fourth unit.

The first unit is used to represent the filament or heater voltage. It is always expressed as a whole number, and when the rated voltage contains a decimal value its decimal numbers are dropped. An exception to this

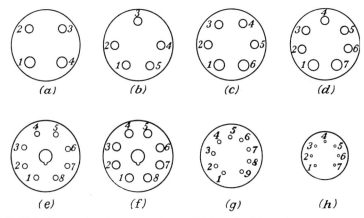

Fig. 2-32. Tube base and socket connections. (*a*) Standard four-pin base. (*b*) Standard five-pin base. (*c*) Standard six-pin base. (*d*) Standard seven-pin base. (*e*) Eight-pin loktal base. (*f*) Eight-pin octal base. (*g*) Nine-pin miniature base. (*h*) Seven-pin miniature base.

is 2-volt battery tubes, which are represented by the number 1 in order to avoid conflict with the 2.5-volt tubes, which bear the number 2. Another exception is the series of tubes whose first unit designation is 7. Although these tubes are nominally rated at 7 volts, they are generally operated at 6.3 volts.

The second unit is a letter separating the first and third units, which are numbers. One version of the significance of this unit is that the letters at the beginning of the alphabet represent amplifiers and detectors while letters at the end of the alphabet represent rectifiers. The number of tube types has increased so rapidly that double letters have to be used, and it has become rather difficult to attach much significance to this unit. When the letter *S* is added, as in the case of the 6SQ7, it indicates that the tube is single-ended, that is, one with all of its leads brought out through the base.

The third unit of the numbering system appears to have at least two versions of its significance. One is that it represents the number of elements in the tube, and the second is that it represents the number of useful leads.

A fourth unit has been added since the introduction of metal tubes and the small glass tubes.   A three-unit type number indicates a metal tube. The fourth unit is a letter or group of letters used to indicate the constructional features of the tube and has the following significance.   *G* indicates a glass tube and octal base; *GT* denotes a tubular glass tube and octal base; *A*, *B*, *C*, *D*, *E*, and *F* assigned in that order signifies a later and modified version which can be substituted for any previous version but not vice versa; *X* indicates that the base is made of special low-loss material; *Y*, that the base is made of special intermediate-loss material.

The following example will aid in understanding the tube numbering system.   In the designation of the 6SA7-GT tube, the first unit (6) indicates that the heater voltage is approximately 6 volts; actually it is rated at 6.3 volts.   In the second unit (SA), the letter *S* indicates a single-ended tube.   The tube is used as a pentagrid converter and in this case the code letter *A* does not bear any significance to its application.   The third unit (7) indicates that the tube has either seven elements or seven useful leads or possibly both.   From a tube manual it can be seen that both versions will apply in the case of the type 6SA7-GT tube.   The fourth unit is represented by the letters GT, which indicate that the tube is of the small tubular glass type.

QUESTIONS

1. What are the characteristics of a vacuum tube that make it one of the most important contributions to the fields of communication and industrial electronics?

2. What is meant by the Edison effect?

3. What was Professor Fleming's contribution to the advancement of vacuum-tube design?

4. Define the following terms: (*a*) cathode, (*b*) filament, (*c*) heater, (*d*) directly heated cathode, (*e*) indirectly heated cathode.

5. Explain the purpose and theory of operation of (*a*) the cathode, (*b*) the heater.

6. Compare the operating characteristics of tungsten, thoriated tungsten, and oxide-coated filaments when used as directly heated cathodes.

7. What are the advantages of directly heated cathodes and where are they generally used?

8. Explain four advantages of indirectly heated cathodes.

9. (*a*) What is meant by a dropping resistor?   (*b*) What is its purpose?   (*c*) How is it connected in the circuit?

10. What conditions in the heater circuit of a radio receiver make it necessary to connect (*a*) a resistor in series with the heater or filament of a tube, (*b*) a resistor in parallel with the heater or filament of a tube?

11. How should the power rating of a resistor used in the heating circuit of a receiver compare with its actual power loss?

12. (*a*) What is meant by a controlled-warm-up-time tube?   (*b*) What is its purpose?   (*c*) How does it serve this purpose?

13. Explain the following terms: (a) plate, (b) diode, (c) duodiode, (d) space charge, (e) emission current, (f) saturation current.

14. Explain the rectifier action of a diode.

15. Compare the half-wave and full-wave rectifier characteristics.

16. Explain the following terms: (a) control grid, (b) triode, (c) grid bias.

17. Draw the circuit diagram for a triode showing the connections for the A, B, and C power supplies. Indicate the direction of the electron flow on the diagram.

18. Explain the action of the triode.

19. Explain the action of a triode for the following conditions: (a) grid bias but no signal input, (b) grid bias and a varying signal input, (c) positive grid voltage.

20. In the circuit of Fig. 2-11, what is the phase relation between the following quantities: (a) e_g and i_b? (b) e_g and e_b? (c) e_L and i_b?

21. Why should the fixed grid bias of a tube always be greater than the maximum value of the input signal voltage?

22. What are the advantages of a family of characteristic curves over the tabular listing of tube characteristics?

23. Explain the difference between the static and dynamic characteristics of a vacuum tube.

24. How are the static characteristics of a tube obtained?

25. How are the dynamic characteristics of a tube obtained?

26. (a) What is meant by a family of plate characteristic curves? (b) What is meant by a family of grid-plate transfer characteristic curves?

27. (a) What is meant by the tube constants? (b) Name the three tube constants.

28. (a) What does the amplification factor of a tube represent? (b) How is it obtained?

29. What factors affect the amplification factor of a tube?

30. (a) What does the plate resistance of a tube represent? (b) How is it obtained?

31. What factors affect the plate resistance of a tube?

32. (a) What does the transconductance of a tube represent? (b) How is it obtained?

33. What factors affect the transconductance of a tube?

34. What is the mathematical relation between the amplification factor, transconductance, and the plate resistance of a tube?

35. What is meant by the voltage amplification per stage?

36. (a) How does the maximum possible voltage amplification of a triode amplifier stage compare with the amplification factor of the tube? (b) Explain why.

37. (a) How does an increase or decrease in the plate resistance affect the voltage amplification of the stage? (b) How does an increase or decrease in the load resistance affect the voltage amplification of the stage?

38. What is the relation between the transconductance of a tube and its operating characteristics?

39. (a) What is a tetrode? (b) By what other names is it also known?

40. Describe the construction and location of the screen grid.

41. (a) What is meant by interelectrode capacitance? (b) Why is it necessary to keep these capacitances at a minimum?

42. What is meant by (a) regeneration, (b) degeneration?

43. How does the addition of a screen grid eliminate feedback?

44. What is meant by secondary emission?

45. What effect does secondary emission have upon the operating characteristics of a tetrode?

46. How do the amplification factor and plate resistance for tetrodes compare with those for triodes?

47. What is a pentode?

48. Describe the construction and location of the suppressor grid.

49. How does the addition of a suppressor grid reduce the effect of secondary emission?

50. How do the characteristic curves for a pentode compare with those for a tetrode?

51. How do the amplification factor and plate resistance for pentodes compare with those for a triode and tetrode?

52. Where are pentodes generally used?

53. (*a*) What is a supercontrol tube?   (*b*) What are its constructional features?

54. Explain the meaning of the following terms: (*a*) sharp cutoff, (*b*) remote cutoff, (*c*) cross modulation, (*d*) modulation distortion.

55. Explain the action of the supercontrol tube.

56. What are the advantages of variable-mu tubes and where are they generally used?

57. Describe the beam power tube.

58. In a beam power tube, what factors are responsible for its high power output and high efficiency?

59. What is meant by multiunit tubes and what are their advantages?

60. What important developments have been made in the design of the newer type of tubes?

61. Describe two systems of tube-pin and socket-connection numbering.

62. Describe the system used to designate the various types of tube.

PROBLEMS

1. The tube complement of a certain radio receiver includes a 12BE6, 12BA6, 12AT6, and 50C5; all heaters are rated 150 ma.   (*a*) If the heaters of the tubes are connected in series, what value of resistance must be connected in series with the heaters in order to operate the receiver on a 115-volt line?   (*b*) What minimum wattage rating should the resistor have, assuming that its ventilation is restricted?

2. A receiver with a tube complement of a 12BE6, 12BA6, 12AT6, 12V6, and 35W4 has its heaters connected in series and uses an additional series-connected resistor in order that the receiver may be operated on a 115-volt line.   Because the heater of the 12V6 tube is rated at 225 ma and the heaters of all the other tubes are rated at 150 ma, it is necessary to include a shunt (parallel-connected) resistor.   (*a*) Draw a circuit diagram showing the series-connected resistor, the heaters connected in a sequence that will permit the use of a single parallel-connected resistor, and the parallel-connected resistor.   (*b*) What value of series resistor is required?   (*c*) What value of shunt resistor is required?   (*d*) What minimum wattage rating is recommended for the resistors in parts (*b*) and (*c*)?

3. A receiver with a tube complement of a 1T4, 1R5, 1T4, 1U5, and 3V4 has its heaters connected in parallel.   The 3V4 tube is connected for 1.4-volt operation and requires 100 ma; each of the other tubes requires 1.4 volts and 50 ma.   It is desired to operate the heaters from a 2.2-volt storage cell with a series-connected resistor.   (*a*) Draw a circuit diagram showing the connections necessary for the series resistor. (*b*) What value of resistor is required?   (*c*) What minimum wattage rating should the resistor have?

4. It is desired to operate the heaters of a four-tube receiver from a power supply consisting of three 1.5-volt cells connected in series.   The heaters are connected to form a parallel-series circuit consisting of a 1R5 and a 1U4 connected in series to form one of the parallel members and a 3V4 and a 1U5 connected in series to form the second parallel member. The 3V4 tube is connected for 2.8-volt operation and requires 50 ma; each of the other tubes requires 1.4 volts and 50 ma. (*a*) Draw a circuit diagram showing the resistors and connections required for this circuit. (*b*) How much resistance is required in the circuit with the 1R5 and 1U4? (*c*) How much resistance is required in the circuit with the 3V4 and 1U5? (*d*) What wattage rating is recommended for the resistors in (*b*) and (*c*) respectively?

5. A certain radio receiver has its heaters connected in parallel to a 12.6-volt power supply. It is desired to substitute a 6BE6 for a 12BE6 in this circuit. (*a*) Draw a circuit diagram showing the resistor that must be added in order to make this change. (*b*) What value of resistor is required? (*c*) What is the minimum wattage rating recommended for this resistor?

6. A certain radio receiver includes a number of 6.3-volt, 300-ma tubes in its series-connected heater circuit. It is desired to substitute a tube with a 12.6-volt 150-ma heater in place of one of the 6.3-volt tubes. (Assume that the current in the heater circuit will remain at 300 ma.) (*a*) Draw a circuit diagram showing the resistor that must be added in order to make this change. (*b*) What value of resistor is required? (*c*) What is the minimum wattage rating recommended for this resistor?

7. A certain radio receiver has all its heaters connected in series and is supplied by a 115-volt line. By adding two resistors, a 6BE6 tube could be substituted for the 12BE6 used in the circuit. (*a*) Draw a circuit diagram showing the two resistors required in order to make this change. (Assume that the tube being changed is the first one in the series circuit.) (*b*) What is the value of each of the resistors? (*c*) What is the minimum wattage rating recommended for each of the resistors?

8. A certain radio receiver has all of its heaters connected in series and is to be operated from a 115-volt a-c line. By adding a resistor, a 35C5 tube could be substituted for the 50C5 used in the circuit. (*a*) Draw a circuit diagram showing the resistor that must be added to the heater circuit in order to make this change. (*b*) What value of resistor is required? (*c*) What is the minimum wattage rating recommended for the resistor?

9. The data in Table 2-3 were obtained from a test of a type 6AT6 tube operated at its rated heater voltage. Plot the plate characteristic curves from the test data.

TABLE 2-3. TEST DATA—TYPE 6AT6

e_b, volts	i_b, ma		
	$E_c = 0$	$E_c = -1$	$E_c = -2$
0			
25	0.6		
50	1.3	0.15	
75	2.1	0.45	
100	2.9	0.9	0.1
125	3.7	1.4	0.3
150	4.6	2.05	0.65
175	5.4	2.75	1.0
200	. . .	3.5	1.5
225	. . .	4.3	2.0
250	. . .	5.1	2.65
275	. . .	. . . .	3.4
300	. . .	. . . .	4.15
325	. . .	. . . .	5.0

10. What is the amplification factor of the tube represented by the curves of Fig. 2-14 if it is operated with 200 volts on its plate, and with a grid bias of 6 volts? (Note: Use the curves of Fig. 2-14 for the solution of this and the following problems.)

11. What is the amplification factor of the tube used in Prob. 10 when its plate voltage is increased to 220 volts, the grid bias remaining at 6 volts?

12. What is the amplification factor of the tube used in Prob. 10 when its plate voltage is decreased to 160 volts, the grid bias remaining at 6 volts?

13. What is the amplification factor of the tube used in Prob. 10 when the grid bias is increased to 8 volts and the plate voltage is kept at 200 volts?

14. What is the amplification factor of the tube used in Prob. 10 when the grid bias is decreased to 4 volts and the plate voltage is kept at 200 volts?

15. What is the plate resistance of the tube represented by the curves of Fig. 2-14 if it is operated with 200 volts on its plate, and with a grid bias of 6 volts?

16. What is the plate resistance of the tube used in Prob. 15 when its plate voltage is increased to 220 volts, the grid bias remaining at 6 volts?

17. What is the plate resistance of the tube used in Prob. 15 when its plate voltage is decreased to 160 volts, the grid bias remaining at 6 volts?

18. What is the plate resistance of the tube used in Prob. 15 when the grid bias is increased to 8 volts and the plate voltage is kept at 200 volts?

19. What is the plate resistance of the tube used in Prob. 15 when the grid bias is decreased to 4 volts and the plate voltage is kept at 200 volts?

20. (*a*) What is the transconductance of the tube represented by the curves of Figs. 2-14 and 2-15 if it is operated with 200 volts on its plate, and with a grid bias of 6 volts? (Solve by obtaining data directly from the curves of Fig. 2-15.) (*b*) Check the answer to part (*a*) by use of Eq. (2-5*a*) and the answers of Probs. 10 and 15.

21. (*a*) What is the transconductance of the tube used in Prob. 20 when its plate voltage is increased to 220 volts, the grid bias remaining at 6 volts? (Solve by obtaining data directly from the curves of Fig. 2-15.) (*b*) Find the value by use of Eq. (2-5*a*) and the answers of Probs. 11 and 16.

22. (*a*) What is the transconductance of the tube used in Prob. 20 when its plate voltage is decreased to 160 volts, the grid bias remaining at 6 volts? (Solve by obtaining data directly from the curves of Fig. 2-15.) (*b*) Find the value by use of Eq. (2-5*a*) and the answers of Probs. 12 and 17.

23. (*a*) What is the transconductance of the tube used in Prob. 20 when its grid bias is increased to 8 volts and the plate voltage is kept at 200 volts? (Solve by obtaining data directly from the curves of Fig. 2-15.) (*b*) Find the value by use of Eq. (2-5*a*) and the answers of Probs. 13 and 18.

24. (*a*) What is the transconductance of the tube used in Prob. 20 when its grid bias is decreased to 4 volts and the plate voltage is kept at 200 volts? (Solve by obtaining data directly from the curves of Fig. 2-15.) (*b*) Find the value by use of Eq. (2-5*a*) and the answers of Probs. 14 and 19.

25. The type 6AT6 tube, whose curves were plotted in Prob. 9, when operated with 200 volts at its plate and with a grid bias of 2 volts has a plate resistance of 50,000 ohms and an amplification factor of 70. What is the transconductance of the tube under these conditions?

26. When the tube of Prob. 25 is operated with 100 volts at its plate and with a grid bias of 1 volt, its plate resistance is 53,000 ohms and its transconductance is 1,400 μmhos. What is the amplification factor of the tube under these conditions?

27. When the tube of Prob. 25 is operated with 200 volts at its plate and with a grid bias of 1 volt, its transconductance is 2,500 μmhos and its amplification factor is 80. What is the plate resistance of the tube under these conditions?

28. The tube represented by Fig. 2-14 when operated with 200 volts at its plate and with a grid bias of 6 volts has an amplification factor of 20 (Prob. 10) and a plate resistance of 7,500 ohms (Prob. 15). If the tube is used in an amplifier stage, what is the voltage amplification of the stage when the value of the load resistance is (a) 1,500 ohms? (b) 7,500 ohms? (c) 15,000 ohms? (d) If a 1.5-volt signal is applied to the grid of the tube, what is the output signal voltage for (a), (b), and (c)?

29. The tube represented by Fig. 2-14 when operated with 220 volts at its plate and with a grid bias of 6 volts has an amplification factor of 20 (Prob. 11) and a plate resistance of 7,270 ohms (Prob. 16). If the tube is used in an amplifier stage, what is the voltage amplification of the stage when the value of the load resistor is (a) 1,500 ohms? (b) 7,500 ohms? (c) 15,000 ohms? (d) If a 1.5-volt signal is applied to the grid of the tube, what is the output signal voltage for (a), (b), and (c)?

30. The tube represented by Fig. 2-14 when operated with 160 volts at its plate and with a grid bias of 6 volts has an amplification factor of 20 (Prob. 12) and a plate resistance of 11,700 ohms (Prob. 17). If the tube is used in an amplifier stage, what is the voltage amplification of the stage when the value of the load resistor is (a) 1,500 ohms? (b) 7,500 ohms? (c) 15,000 ohms? (d) If a 1.5-volt signal is applied to the grid of the tube, what is the output signal voltage for (a), (b), and (c)?

31. The tube represented by Fig. 2-14 when operated with 200 volts at its plate and with a grid bias of 8 volts has an amplification factor of 19 (Prob. 13) and a plate resistance of 12,500 ohms (Prob. 18). If the tube is used in an amplifier stage, what is the voltage amplification of the stage when the value of the load resistor is (a) 1,500 ohms? (b) 7,500 ohms? (c) 15,000 ohms? (d) If a 1.5-volt signal is applied to the grid of the tube, what is the output signal voltage for (a), (b), and (c)?

32. The tube represented by Fig. 2-14 when operated with 200 volts at its plate and with a grid bias of 4 volts has an amplification factor of 19 (Prob. 14) and a plate resistance of 7,700 ohms (Prob. 19). If the tube is used in an amplifier stage, what is the voltage amplification of the stage when the value of the load resistor is (a) 1,500 ohms? (b) 7,500 ohms? (c) 15,000 ohms? (d) If a 1.5-volt signal is applied to the grid of the tube, what is the output signal voltage for (a), (b), and (c)?

33. A certain high-mu triode (6AB4) is operated with such values of plate and grid voltages that produce a plate resistance of 15,000 ohms and an amplification factor of 60. What value of load resistance is required to produce a voltage amplification of (a) 30? (b) 45?

34. A certain high-mu triode (6AV6) is operated at such values of plate and grid voltages that produce a plate resistance of 80,000 ohms and an amplification factor of 100. What value of load resistance is required to produce a voltage amplification of (a) 50? (b) 75?

35. If a 20-mv signal is applied to the input of the amplifier tube of Prob. 33, what value of load resistance is required to produce a 700-mv output signal?

36. If a 12-mv signal is applied to the input of the amplifier tube of Prob. 34, what value of load resistance is required to produce an output signal of 0.8 volt?

CHAPTER 3

AMPLITUDE-MODULATED DETECTOR CIRCUITS

One of the functions of a radio receiver is the demodulation of a modulated radio wave picked up by the receiving antenna. This function is called *detection*. In Chap. 1, it was shown that a-m detection involves two operations: (1) rectification of the modulated wave; (2) elimination of the r-f component of the modulated wave.

3-1. Detection. *Detector Action.* The average value of a modulated radio wave for 1 cycle of the a-f (audio-frequency) wave is zero, and therefore the average change of current during the same period is zero (Fig. 3-1a). The r-f (radio-frequency) waves produced during one a-f cycle are referred to as a *wave train*. A modulated r-f wave consists of a number of consecutive wave trains, and therefore the average change of current of a modulated r-f wave will always be zero. If the modulated r-f wave is rectified, one-half of the wave is eliminated, and the average change in current for each cycle of each wave train will no longer be zero (Fig. 3-1b). The changes in current will be similar to the a-f signal that modulates the r-f carrier wave at the transmitter.

Because the electromechanical devices used to produce audible sound waves cannot respond to the rapid variations in current of an r-f wave, it is necessary to remove the r-f component of the demodulated a-f wave. The modulation envelope, formed by joining the peaks of each of the r-f cycles, varies in the same manner as the signal impressed upon the r-f carrier wave. The a-f component of the modulated wave is represented by the a-f variations of the modulation envelope. The current flowing through the detector circuit will be equal to the average value of the current (Fig. 3-1b). Its variations are identical in all respects, except intensity, to the variations in current represented by the modulation envelope.

The Vacuum Tube as a Detector. Because current can flow in the plate circuit of a vacuum tube only when the plate is positive with respect to the cathode, vacuum tubes can be used as rectifiers of alternating currents. As detection involves the function of rectification, vacuum tubes may be used as detectors.

Factors That Determine the Type of Detector Circuit to Be Used. There are numerous tubes and a variety of circuits that can be used with such tubes to perform the function of detection. Factors to be taken into consideration when determining which tube and circuit to use are: (1) sensitivity, (2) signal-handling ability, (3) fidelity of reproduction.   Fidelity is a very important factor in a radio receiver, and the detector circuit should be capable of reproducing the sound waves with a minimum amount of distortion.   Radio receivers designed for broadcast reception employ high-gain amplifier circuits, and hence sensitivity is not an important consideration.   In these receivers the r-f signal is amplified before it reaches the detector, which should therefore be capable of handling a large input signal.

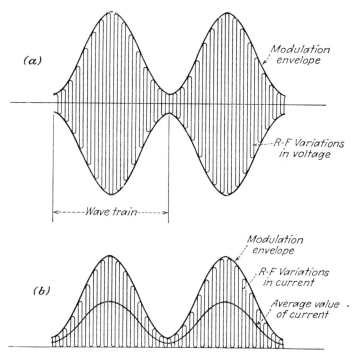

Fig. 3-1. Detector action.   (a) Input signal voltage.   (b) Current through the detector.

In portable receivers, the over-all physical size is an important factor. In order to obtain small units, r-f amplifier circuits are sometimes omitted and the detector circuits should then have a high degree of sensitivity.

If a receiver is to be used for code reception only, the fidelity of the detector circuit can generally be ignored.   For portable receivers of code signals the important factor is sensitivity.   For a fixed unit, where sufficient amplifier circuits can be used, the main consideration is the power-handling ability of the detector circuit.

3-2. Diode Detection.   *Action of a Diode Detector.*   The simplest vacuum-tube detector circuit is obtained by using a two-element tube as a half-wave rectifier (Fig. 3-2).   When the tuning circuit L_2C_2 is in resonance with a desired input signal, an r-f voltage is developed across the tuned circuit. This voltage is applied to the plate-cathode circuit of the tube through the diode bypass capacitor C_1 and the diode load resistor R_1.

When the plate of the diode is positive, it attracts electrons emitted from the cathode, and these electrons will return to the cathode through the external circuit consisting of the secondary winding L_2 and the output resistor R_1. The path taken by the electrons is indicated by the arrows on Fig. 3-2. No current flows through the external plate circuit during the time that the signal voltage makes the plate negative with respect to the cathode, and thus the current in R_1 will flow in only one direction.

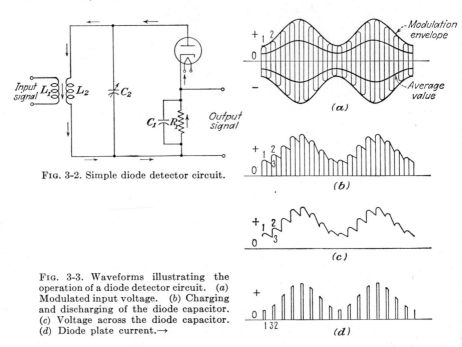

Fig. 3-2. Simple diode detector circuit.

Fig. 3-3. Waveforms illustrating the operation of a diode detector circuit. (*a*) Modulated input voltage. (*b*) Charging and discharging of the diode capacitor. (*c*) Voltage across the diode capacitor. (*d*) Diode plate current.→

Capacitor C_1 and resistor R_1 eliminate the r-f pulsations and increase the a-f voltage developed across R_1 in the following manner. During the initial half of the first positive half-cycle of the applied r-f voltage, shown at 0–1 on Fig. 3-3, C_1 charges to the peak value indicated by point 1. The applied r-f voltage, continuing its cycle, then rapidly diminishes to zero. As the r-f voltage starts decreasing from its first positive peak value, C_1 starts to discharge through R_1 but at a very slow rate as indicated by points 1 to 3 on Fig. 3-3. The time constant of this RC circuit is very long compared to the short interval required for the r-f voltage to change from the positive peak value at 1 to the next positive peak at 2. The voltage on the capacitor, therefore, will decrease only slightly during this interval. Because of this capacitor action, the voltage on the cathode will be kept more positive than the voltage applied to the plate. When the signal voltage is lower than the voltage charge on C_1, electrons will cease to flow in the plate circuit; hence no current will flow in the plate circuit during the interval 1 to 3.   During the positive half of the second r-f cycle, current will again flow in the plate circuit when the signal voltage exceeds the voltage at which the charge on the capacitor holds the cathode.   The capacitor will then be charged to the peak value of the second positive half-cycle.   This action will be repeated for each succeeding r-f cycle, thus causing the voltage across the capacitor to follow the peak values of the applied r-f voltage.   The a-f modulation is therefore reproduced at the capacitor as indicated in Fig. 3-3c.

The voltage across R_1 and C_1 will be a pulsating voltage representing the positive half of the modulated r-f voltage, whose average value has been increased by C_1.   The combination of the diode, C_1, and R_1 changes the r-f signal input voltage to a pulsating voltage.

The capacitor C_1 is called the *r-f bypass* or *r-f filter capacitor*, because it smooths out the r-f pulsations at the diode load resistor.   The value of the capacitor depends on the frequency—the higher the frequency the smaller the amount of capacitance required.   For broadcast reception, 250 $\mu\mu$f is generally used.

Example 3-1.   A detector circuit similar to Fig. 3-2 uses a 0.25-megohm resistor for R_1 and a 250-$\mu\mu$f capacitor for C_1.   (*a*) What is the time constant of R_1C_1?   (*b*) If the resonant frequency of L_2C_2 is 710 kc, what time is required for the r-f wave to complete 1 cycle?   (*c*) How many times greater is the time constant of R_1C_1 than the time of 1 cycle of the r-f wave?

$$\text{Given:}\quad R = 0.25 \text{ megohm} \qquad\qquad \text{Find:}\quad (a)\ t$$
$$C = 250\ \mu\mu\text{f} \qquad\qquad\qquad\qquad (b)\ t_1$$
$$f = 710 \text{ kc} \qquad\qquad\qquad\qquad (c)\ \frac{t}{t_1}$$

Solution:

(*a*) $t = RC = 0.25 \times 10^6 \times 250 \times 10^{-12} = 62.5 \times 10^{-6}$ sec

(*b*) $t_1 = \dfrac{1}{f} = \dfrac{1}{0.710 \times 10^6} = \dfrac{10^{-6}}{0.710} = 1.408 \times 10^{-6}$ sec

(*c*) $\dfrac{t}{t_1} = \dfrac{62.5 \times 10^{-6}}{1.408 \times 10^{-6}} = 44.3$ times greater

Example 3-2.   The circuit of Example 3-1 also acts as a filter circuit.   (*a*) What is the reactance of C_1 to a 710-kc r-f current?   (*b*) How does the reactance to the 710-kc r-f current compare with the value of R_1?   (*c*) Which path will the r-f currents take? (*d* What is the reactance of C_1 to a 400-cycle a-f current?   (*e*) How does the reactance to the 400-cycle a-f current compare with the value of R_1?   (*f*) Which path will the a-f currents take?

$$\text{Given:}\quad R = 0.25 \text{ megohm} \qquad\quad \text{Find:}\quad (a)\ X_C$$
$$C = 250\ \mu\mu\text{f} \qquad\qquad\qquad\qquad\ (b)\ \frac{R}{X_C}$$
$$\text{r-f} = 710 \text{ kc} \qquad\qquad\qquad\quad (c)\ \text{Path of r-f current}$$
$$\text{a-f} = 400 \text{ cycles} \qquad\qquad\qquad (d)\ X_C$$
$$(e)\ \frac{X_C}{R}$$
$$(f)\ \text{Path of a-f current}$$

Solution:

(a) $X_C = \dfrac{159,000}{fC} = \dfrac{159,000}{710 \times 10^3 \times 250 \times 10^{-6}} = 895$ ohms

(b) $\dfrac{R}{X_C} = \dfrac{250,000}{895} = 279$ (R is 279 times greater than X_C)

(c)   The r-f currents will take the capacitor path.

(d) $X_C = \dfrac{159,000}{fC} = \dfrac{159,000}{400 \times 250 \times 10^{-6}} = 1,590,000$ ohms

(e) $\dfrac{X_C}{R} = \dfrac{1,590,000}{250,000} = 6.36$ (X_C is 6.36 times greater than R)

(f)   The a-f currents will take the resistor path.

Full-wave Diode Detectors. Diode detectors are essentially half-wave rectifiers and do not contribute to the amplification of the signal as do some other methods of detection. Two diodes, or a duplex diode, can be connected in a detector circuit so that current will flow in either plate during opposite halves of each r-f signal input cycle (Fig. 3-4). Current

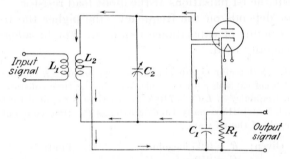

FIG. 3-4. Duplex-diode full-wave detector circuit.

will flow continually in one direction through the output resistor, and full-wave detection is obtained. The advantage gained with full-wave detection is that the circuit may be balanced so that no carrier-frequency signal flows to the grid of the tube in the following amplifier stage. Figure 3-4 shows a duplex-diode full-wave detector circuit and the paths taken by the electrons flowing from the cathode to each plate during alternate half-cycles.

Advantages and Disadvantages of Diode Detectors. The advantages of the diode detector are (1) its ability to handle large signal voltages, and (2) its low distortion factor. A disadvantage of the diode detector is that during part of the positive half of each r-f cycle current flows through the coil in the tuned circuit. This current causes a loading effect on the tuned circuit and produces the same results as when a resistance is connected in series with a tuned circuit. Adding resistance to a tuned circuit decreases the gain and the selectivity of the circuit; thus the diode detector is characterized by its low sensitivity.

Modern receivers generally employ some r-f amplification and the desired selectivity and gain are obtained before the signal reaches the detector circuit.   The low sensitivity of the diode detector is therefore not of great importance in these receivers.   Because diode detection pro-duces very little distortion and permits the use of simple avc circuits without an additional voltage supply, diode detection is used widely in broadcast receivers.

Practical Diode-detector Circuits.   Practical detector circuits use filter circuits to prevent the r-f voltages from reaching the output.   The filter circuits in Fig. 3-5 consist of R_1 and C_3.   In Fig. 3-5a, the diode load

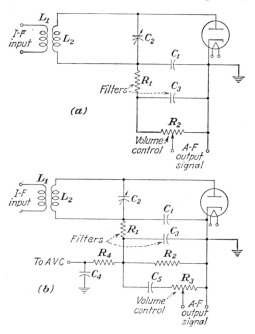

Fig. 3-5. Diode-detector circuits.   (a) Diode circuit with filter.   (b) Diode circuit with provision for obtaining voltage for automatic volume control.

resistance consists of the sum of R_1 and R_2; the output, however, is taken off R_2 only.   While C_1 will bypass most of the r-f current, the output filter capacitor C_3 is added to bypass any r-f current from R_2 that may have entered into the R_1 path.   In Fig. 3-5b, the diode load resistance consists of R_1 and R_2.   Capacitors C_1 and C_3 serve the same purpose as in the circuit of Fig. 3-5a.   Coupling capacitor C_5 also blocks the d-c component from the output circuit.   In these circuits, the capacitance of C_1 and C_3 are generally equal.   In the circuit of Fig. 3-5a, the value of R_2 is generally several times the value of R_1.   In the circuit of Fig. 3-5b, the value of R_3 is generally several times the value of R_1.   The r-f filter, however, reduces the useful rectified output by the amount of the voltage drop at R_1.   The adjustable capacitor C_2 is used in superheterodyne receivers to tune the L_2C_2 circuit to the intermediate frequency of the receiver, usually at about 456 kc.

Example 3-3.   The circuit elements shown in Fig. 3-5a have the following values: $R_1 = 50,000$ ohms, $R_2 = 250,000$ ohms, $C_1 = 100$ $\mu\mu$f, $C_3 = 100$ $\mu\mu$f.   (*a*) What impedance does C_1 offer to a 456-kc i-f current?   (*b*) Which path will the i-f current take?   (*c*) Will any of the i-f current flow into the R_1 path?   (*d*) What impedance does C_3 offer to any 456-kc i-f current?   (*e*) What purpose does C_3 serve?   (*f*) Neglecting the effect of C_1 and C_3, what per cent of the a-f voltage developed across R_1 and R_2 is available at the output terminals?

Given: $R_1 = 50,000$ ohms   Find: (*a*) X_{C1}
              $R_2 = 250,000$ ohms   (*b*)   Path of i-f current
              $C_1 = 100$ $\mu\mu$f   (*d*) X_{C3}
              $C_3 = 100$ $\mu\mu$f   (*f*)   Per cent
              i-f $= 456$ kc

Solution:

(*a*) $X_{C1} = \dfrac{159,000}{fC} = \dfrac{159,000}{456 \times 10^3 \times 100 \times 10^{-6}} = 3,486$ ohms

(*b*)   The i-f current will take the capacitor path.

(*c*)   Yes.   A small amount.

(*d*)   Same as (*a*). $X_{C3} = 3,486$ ohms

(*e*)   Provides a low-impedance path for any i-f current that was not bypassed by the capacitor C_1.

(*f*)   % a-f voltage at $R_2 = \dfrac{i_{af} R_2}{i_{af}(R_1 + R_2)} \times 100$

Because the current is the same in both the numerator and the denominator, then

$$\% \text{ of } E_{a\text{-}f} = \frac{R_2}{R_1 + R_2} \times 100 = \frac{250,000}{50,000 + 250,000} \times 100 = 83.3$$

Use of Multiunit Tubes as Diode Detectors.   A multiunit tube, such as a duplex-diode-triode, is frequently used as a detector and amplifier. The two diodes may be connected together to form a simple diode half-wave detector.   The triode is used to amplify the rectified signal.   Two circuits using a duplex-diode-triode are shown in Fig. 3-6.   In these circuits, R_1 is the diode load resistor.   The bias voltage for the triode section is obtained by the cathode-bias resistor R_3.   The bypass capacitor C_3 keeps the a-f current out of the cathode-bias resistor.   Capacitor C_4 blocks the d-c bias of the cathode from the grid.   Capacitor C_5 bypasses any r-f current from the grid circuit to the cathode and thus prevents any r-f voltage from reaching the grid of the tube.

3-3. Plate Detection.   *Theory of Plate Detection.*   Plate detection uses the cutoff characteristics of a tube to accomplish rectification. Because rectification takes place in the plate circuit of the tube, this type of circuit is called a *plate detector*.   The grid is biased almost to cutoff, so that practically no current flows in the plate circuit when the signal applied to the grid is zero.   C batteries were formerly used to obtain the necessary grid bias, and because of this the plate detector is also called a *C-bias detector*, a *grid-bias detector*, or simply a *bias detector*.

Theoretical Operating Point of a Bias Detector.   With the correct value of negative grid voltage, a tube can be made to operate at the negative bend of its curve (point A in Fig. 3-7). If a tube is operated at or near this point, it will act as a rectifier. Each negative half of the input signal is practically eliminated, and the unidirectional current flowing in the plate circuit varies in the same manner as the positive halves of the input signal. The sharper the negative bend of the curve and the nearer the grid bias is adjusted to the value corresponding to this bend, the more perfect will be the elimination of the negative half-cycle of the input signal voltage.

Practical Operating Point of a Grid-bias Detector. Theoretically, a tube used as a grid-bias detector should be operated at cutoff. Under this

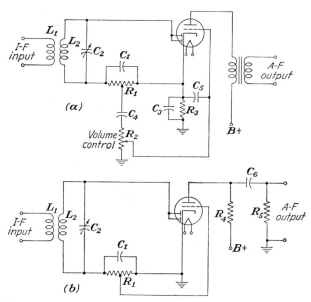

FIG. 3-6. Diode detector circuits using duplex-diode triodes. (a) Using cathode bias (resistor R_3) for the triode section. (b) Using diode bias (portion of the diode load resistor R_1) for the triode section of the tube.

condition the current in the plate circuit with no signal voltage applied will be zero (Fig. 3-9a). In actual practice, however, the voltage on the grid is set at slightly less than cutoff value, so that a small amount of current (100 to 600 μa) will flow in the plate circuit when the signal applied to the grid of the tube is zero (Fig. 3-9c). If the value of the negative grid voltage is too large, the tube will be operated beyond cutoff, and part of the positive half-cycle of each a-c wave impressed on the input circuit will be eliminated in addition to the entire negative half-cycle (Fig. 3-9b). As the rectified signal is not a true representation of the modulated input signal voltage, the resulting output is distorted. If the value of the negative grid voltage is too small, the tube will be operated before cutoff, and a portion of each negative half-cycle will appear in the output (Fig. 3-9c). The rectification is not complete and the resulting output will be distorted.

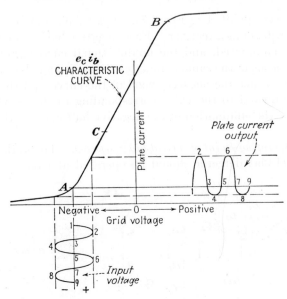

FIG. 3-7. Operating conditions of a plate circuit detector.

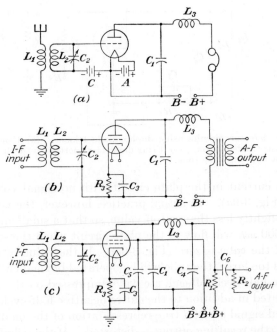

FIG. 3-8. Plate detector circuits. (*a*) Grid bias obtained by use of a C battery. (*b*) Grid bias obtained by use of a cathode resistor. (*c*) Grid bias obtained by use of a cathode resistor.

R-F Filter Circuits.   The output of the detector tube is a unidirectional current varying in accordance with the modulated r-f input (Fig. 3-1b). A filter circuit is needed to separate the a-f and r-f components of the rectified wave.   An r-f bypass capacitor is connected between the plate and cathode to smooth out the high-frequency variations in the output circuit.   As inductance tends to oppose changes in the amount of current, an r-f choke coil may be connected in the plate circuit to aid the filter action of the r-f bypass capacitor.   Some receivers use two capacitors connected to the r-f plate choke coil to form the more effective *low-pass π-type filter* circuit shown in Fig. 3-8c.

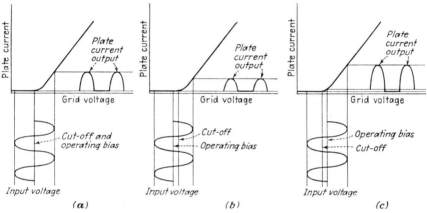

FIG. 3-9. Operating points of a grid-bias detector.   (a) At cutoff.   (b) Beyond cutoff. (c) Before cutoff.

Example 3-4.   The r-f filter circuit of Fig. 3-8b has a 500-$\mu\mu$f capacitor at C_1 and an 80-mh choke at L_3.   (a) What impedance does the capacitor offer to a 456-kc i-f current?   (b) What impedance does the choke offer to a 456-kc i-f current?   (c) What impedance does the capacitor offer to a 500-cycle a-f current?   (d) What impedance does the choke offer to a 500-cycle a-f current?   (e) Which path will the i-f currents take?   (f) Which path will the a-f currents take?

$$\text{Given:}\quad C = 500\ \mu\mu\text{f}\qquad \text{Find:}\quad (a)\ Z$$
$$L = 80\ \text{mh}\qquad\qquad (b)\ Z$$
$$f = 456\ \text{kc}\qquad\qquad (c)\ Z$$
$$(d)\ Z$$
$$(e)\ \text{Path of i-f current}$$
$$(f)\ \text{Path of a-f current}$$

Solution:

(a) $Z = \dfrac{159{,}000}{fC} = \dfrac{159{,}000}{456 \times 10^3 \times 500 \times 10^{-6}} = 697$ ohms

(b) $Z = 2\pi fL = 2 \times 3.14 \times 456 \times 10^3 \times 80 \times 10^{-3} = 229{,}094$ ohms

(c) $Z = \dfrac{159{,}000}{fC} = \dfrac{159{,}000}{500 \times 500 \times 10^{-6}} = 636{,}000$ ohms

(d) $Z = 2\pi fL = 2 \times 3.14 \times 500 \times 80 \times 10^{-3} = 251.2$ ohms

(e)   The i-f currents will take the capacitor path.

(f)   The a-f currents will take the inductor path.

Plate Load.   The plate circuit contains a loading device through which the plate current must flow.   The voltage developed across this device will equal the product of the plate current and the impedance of the output load.   The plate load may be a resistor, primary winding of a transformer, choke coil, earphones, loudspeaker, or some other device.

Amount of Grid Bias Required.   Plate detectors are comparatively insensitive to weak signals, have a high distortion factor, and are capable of handling large signal voltages.   To avoid driving the grid positive, the largest signal voltage applied to the grid of a tube should not exceed the value of the grid bias.

When a signal is applied to the grid, the voltage at the grid will depend upon the value of the grid bias and the input signal voltage.   The voltage at the grid will increase during the negative halves of the input signal and decrease during the positive halves.   The upper and lower values of the grid voltage indicate its range and are called the *grid-voltage swing.* These values may be expressed mathematically as

$$e_{c \cdot \text{max}} = E_c - E_{g \cdot \text{max}} \qquad (3\text{-}1)$$
$$e_{c \cdot \text{min}} = E_c + E_{g \cdot \text{max}} \qquad (3\text{-}2)$$

where e_c = instantaneous total grid voltage, volts

E_c = average or quiescent value of grid voltage, volts

$E_{g \cdot \text{max}}$ = maximum value of varying component of grid voltage, volts

Example 3-5.   A tube is operated with a grid bias of 9 volts.   What is the grid-voltage swing when an a-c signal is applied to the input circuit if the maximum value of the signal voltage is (*a*) 6 volts? (*b*) 9 volts? (*c*) 10 volts?

Given: $E_c = -9$ volts   Find: (*a*) $\left\{ \begin{array}{l} e_{c \cdot \text{max}} \\ e_{c \cdot \text{min}} \end{array} \right.$

(*a*) $E_{g \cdot \text{max}} =$ 6 volts   (*b*)

(*b*) $E_{g \cdot \text{max}} =$ 9 volts   (*c*)

(*c*) $E_{g \cdot \text{max}} =$ 10 volts

Solution:

(*a*) $e_{c \cdot \text{max}} = E_c - E_{g \cdot \text{max}} = -9 - 6 = -15$ volts
$e_{c \cdot \text{min}} = E_c + E_{g \cdot \text{max}} = -9 + 6 = -3$ volts

(*b*) $e_{c \cdot \text{max}} = E_c - E_{g \cdot \text{max}} = -9 - 9 = -18$ volts
$e_{c \cdot \text{min}} = E_c + E_{g \cdot \text{max}} = -9 + 9 = 0$ volts

(*c*) $e_{c \cdot \text{max}} = E_c - E_{g \cdot \text{max}} = -9 - 10 = -19$ volts
$e_{c \cdot \text{min}} = E_c + E_{g \cdot \text{max}} = -9 + 10 = +1$ volt

Factors that determine the efficiency of rectification of a plate detector are the grid bias, plate voltage, and signal voltage.   The value of grid bias to be used will depend on the plate voltage and the tube used.   These values can be obtained from standard tube manuals.   The value of the grid bias used will determine the amount of signal voltage that can be applied to the tube without causing distortion in the output circuit.

Example 3-6.   A certain triode tube (6C5) is used as a grid-bias detector and is operated with 100 volts on its plate.   It is recommended that the grid bias be approximately 7 volts and that the plate current be adjusted to 0 2 ma for zero signal condition.   What is the greatest amount of signal voltage that can be applied to the input circuit without causing distortion?

Given: Triode grid-bias detector   Find: Maximum input signal
E_b = 100 volts
E_c = −7 volts

Solution:

$$e_{s\cdot\text{max}} = E_c = 7 \text{ volts}$$

NOTE: In order to avoid distortion, the grid should never be driven positive; therefore the maximum input signal should not exceed 7 volts.

3-4. Automatic Grid Bias.   *Methods of Obtaining the Required Amount of Grid Bias.*   In the early days of radio, the grid bias was obtained by use of a C battery (Fig. 3-8a) or from the voltage divider in the power supply.   The grid bias for the various tubes in a receiver can also be obtained by causing the plate current to flow through a resistor, called a *cathode-bias resistor*, which is placed in the cathode circuit.

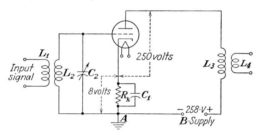

FIG. 3-10. Method of obtaining grid bias by use of a cathode-bias resistor.

Obtaining Grid Bias by Use of a Cathode-bias Resistor.   In the circuit shown in Fig. 3-10, the grid bias for the tube is obtained by placing a resistor R_k between the cathode and the negative side of the B supply, or ground.   Automatic grid bias is obtained by making the d-c component of the cathode current flow through R_k.   (In a triode the cathode current is equal to the plate current, while in a pentode the cathode current is equal to the sum of the plate and screen grid currents.)   The voltage drop across R_k is determined by the value of its resistance and the amount of current flowing through it.   The value of resistance required for R_k can be obtained by

$$R_k = \frac{E_c}{I_k} \tag{3-3}$$

where R_k = value of cathode-bias resistance, ohms
E_c = grid bias, volts
I_k = average or quiescent value of cathode current, amp

From Fig. 3-10, it can be seen that point A will be negative with respect to the cathode.   The grid, being connected to point A through the secondary of the input transformer, will be negative with respect to the cathode by the amount of the voltage drop across R_k.

The grid-bias voltage provided by the cathode-bias resistor is dependent upon the amount of cathode current, which in turn is dependent on the value of the plate voltage used.   If the plate voltage is increased, the plate and cathode currents will increase and the amount of grid bias furnished by the cathode-bias resistor will increase.   If the plate voltage is decreased, the cathode current will decrease and the amount of grid bias provided by the cathode-bias resistor will decrease.   Therefore, the correct value of grid-bias voltage is automatically maintained regardless of changes in plate voltage.   This method of obtaining the grid-bias voltage is called *automatic self-biasing.*

A-F Bypass Capacitor.   The cathode-bias resistor must be shunted by a large capacitor (usually 1 μf or larger) in order to bypass the a-c component of the cathode current, which would otherwise cause the voltage across the grid circuit to vary continually.   This capacitor is indicated as C_3 in Fig. 3-8 and as C_1 in Fig. 3-10.

Example 3-7.   A practical method of obtaining the grid bias for the detector tube of Example 3-6 is by connecting a resistor into the circuit between the cathode and ground (Fig. 3-10).   (*a*) What value of cathode-bias resistor is necessary for the conditions of Example 3-6?   (*b*) How much power is consumed by this resistor?   (*c*) What power rating should the resistor have?

Given: E_c = −7 volts   Find: (*a*) R_k
    I_k = 0.2 ma   (*b*) P
                                (*c*) Rating

Solution:

(*a*) $R_k = \dfrac{E_c}{I_k} = \dfrac{7}{0.2 \times 10^{-3}} = 35,000$ ohms

(*b*) $P = I_k{}^2 R_k = (0.0002)^2 \times 35,000 = 0.0014$ watt

(*c*)   0.25 watt

Example 3-8.   (*a*) What value of bypass capacitance should be used with the cathode-bias resistor of Example 3-7 if the resistor is to offer at least 100 times more impedance to a 500-cycle a-f current than the capacitor?   (*b*) What standard rating and type of capacitor is recommended for this application?

Given: R_k = 35,000 ohms   Find: (*a*) C
    f = 500 cps   (*b*) Rating, type

Solution:

(*a*) $X_C = \dfrac{35,000}{100} = 350$ ohms (max)

$C = \dfrac{159,000}{fX_c} = \dfrac{159,000}{500 \times 350} = 0.91\ \mu\text{f}$

(*b*)   A 1-μf, 10-volt, electrolytic capacitor.

Plate, Grid, and B Supply Voltages.   In Fig. 3-10, it should be observed that the voltage of the B supply must be equal to the sum of the plate voltage and the voltage drop at the cathode-bias resistor.   In this case it is equal to 250 volts plus 8 volts, or 258 volts.   It is assumed that there is no drop in voltage at the load impedance, which may be the primary winding of a transformer, a loudspeaker, or other device having a negligible d-c resistance.   Plate voltages are measured between the cathode and plate terminals, and grid voltages are measured between grid and cathode terminals.   The d-c resistance of the input tuning coil L_2 is negligible, and therefore its voltage drop can be disregarded.

3-5. Grid Detection.   *Grid Resistor-capacitor Detector Circuits.*   The most sensitive type of detector circuit is the *grid detector*, sometimes called the *grid-rectification detector* or the *grid-leak detector*.   Because of its high sensitivity, this circuit was frequently used in early radio designs.   With

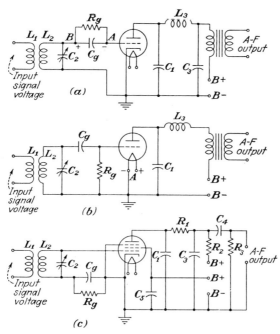

Fig. 3-11. Grid-leak detector circuits.   (*a*) Grid resistor connected in the grid side of the grid circuit.   (*b*) Grid resistor connected from the grid of the tube to the cathode.   (*c*) Grid resistor connected in the cathode side of the grid circuit.

the development of improved r-f amplifiers, sensitivity no longer remained an important factor, and this type of detector circuit is seldom used in receivers designed for broadcast reception.   Three grid-detection circuits are shown in Fig. 3-11.

Principle of Grid Detection.   The principle of detector operation is the same for all three circuits of Fig. 3-11.   The grid resistor R_g and the capacitor C_g form an impedance that corresponds to the load impedance C_1 and R_1 of the diode detector in Fig. 3-2.   The grid circuit, consisting of the source of input signal voltage, the grid and cathode of the tube, and the resistor-capacitor combination R_gC_g, operates as a half-wave rectifier in the same manner as a diode detector.

Circuit Action.   In grid-circuit detection, a triode is used simultaneously as a diode detector and a triode amplifier.   The grid and cathode of the triode operate as a diode detector and at the same time serve as the input circuit of the triode amplifier.   To understand the operation of this type of detection, it is necessary to study the instantaneous action of the grid both as a plate for the diode detector and as a grid for the triode amplifier.

When the signal input is zero, the voltages on the grid and the cathode are the same, that is, zero.   The voltage on the plate will cause a steady stream of electrons to flow from the cathode to the plate.   As the grid is in the path of this electron flow, some electrons will strike it and make the grid negative with respect to the cathode.   The grid capacitor C_g (Fig. 3-11a) becomes charged, plate A negatively and plate B positively.   This charge remains on the capacitor, and the resultant current flowing through R_g causes a constant bias on the grid.   The amount of current flowing through R_g is very small; therefore the voltage drop across it is only a fraction of a volt.   Consequently, when the tube is considered as a triode amplifier, it will have a small amount of grid bias.

During the positive half-cycles of the modulated r-f input signal, additional current will flow through the grid resistor.   This increases the voltage drop across R_g, thus increasing the grid bias.   The current flowing in the plate circuit will therefore decrease.   As the modulated r-f signal voltage varies with succeeding r-f cycles, the voltage drop across R_g will vary accordingly.   The current in the plate circuit will decrease in the same manner as the voltage drop across R_g increases.   The shift in grid bias and the resulting plate current is shown in Fig. 3-12.

The grid capacitor has a low value of capacitance and will pass only the high-frequency signal input.   The current is rectified by means of the diode-detector action of the grid and cathode.   The rectified signal current will then flow through R_g, thus causing a voltage to be developed across it.   This voltage is applied to the input (grid-cathode) circuit of the tube which may then be considered as operating as an amplifier.   The various stages of grid-circuit detection are shown in Fig. 3-12.

Filter Circuit.   In grid-detector circuits, the grid operates with a negative voltage that varies at radio frequencies, and consequently the plate current will vary in the same manner (see Fig. 3-12a).   The filter circuit, consisting of C_1, C_3, and L_3 (Fig. 3-11a), will smooth out the high-frequency variations, and the output of the filter will be a unidirectional current varying at an a-f rate similar to the average plate current indicated on Fig. 3-12a.

Operating Voltage of Grid Detectors.   In analyzing the operation of the grid-circuit detector its two functions can be considered separately.   The detector or grid-cathode circuit may be analyzed from inspection of the grid characteristic curves in Fig. 3-13, which shows that the tube is being operated near the lower bend of the curve, and hence rectification or detection is accomplished.   The amplifying action involves the grid, plate, and cathode, and may be analyzed from examination of the grid-plate transfer characteristic curve.   As grid-circuit detectors operate with practically zero grid bias, the operating point on the grid-plate transfer characteristic curve is determined by the plate voltage.   In order to obtain maximum amplification with minimum distortion, the operating point as

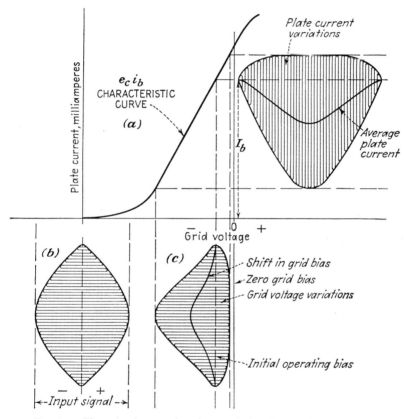

Fig. 3-12. Plate circuit operating characteristics of a grid-leak detector.

an amplifier should be as near as possible to the middle point of the straight portion of the curve, point C in Fig. 3-7.

Characteristics of Grid Detection.   The current in the grid circuit of a grid detector does not vary in the same proportion as the variations of the input signal voltage.   Figure 3-13 shows a typical grid characteristic curve.   Because of the small grid bias, the tube will operate on the curved portion of the curve.   Under this condition, the variation in grid current is practically proportional to the square of the grid-voltage variations, and grid-circuit detectors are therefore often referred to as *square-law detectors.* Because of the square-law variation, the strength of the input signal will be increased, but it will also be distorted. Grid detection is therefore characterized by its high sensitivity and its high distortion factor.

3-6. Power Detection. *Power Detectors.* Modern receivers use high-gain amplifier circuits and a number of tuned circuits before the detector in order to obtain the desired gain and selectivity. The detector circuit

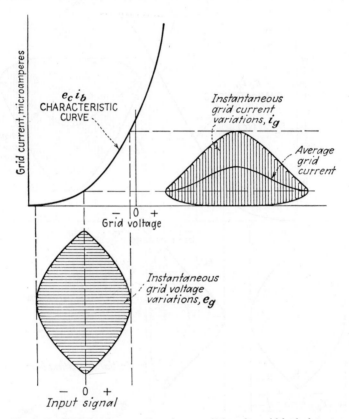

Fig. 3-13. Grid circuit operating characteristics of a grid-leak detector.

used in these receivers must be capable of handling large signal voltages without producing distortion. This type of detector circuit is called a *power detector* and may be a diode, grid-circuit, or plate-circuit detector.

Because the input signal voltage should never drive the grid of plate detectors positive, the power-handling ability of a plate detector is determined by the amount of grid bias used. In grid-circuit detection, the amount of signal voltage that can be applied without causing distortion in its amplifying action is determined by the amount of voltage required to operate the tube as an amplifier on the straight portion of its grid-plate transfer characteristic curve.   The input signal voltage of power detectors is generally greater than 1 volt.

Grid- and Plate-circuit Power Detectors.   The operating voltages and external circuits for power detectors are different from those used for detectors having a weak signal input.   For grid-circuit detection, the plate voltage is generally very high, and the values of the grid capacitor and grid resistor are lower than those used in detectors having a weak signal input.   A grid-circuit power detector is shown in Fig. 3-14a. With zero signal input, the plate current would be very high because the tube is operating with zero grid bias.   In order to keep the plate current at a safe value, the plate voltage is reduced by means of the resistor R_1.

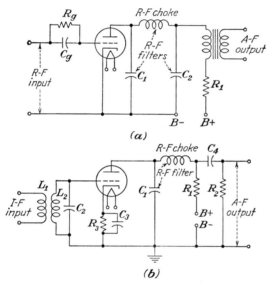

FIG. 3-14. Power-detector circuits.   (a) Grid-circuit power detection.   (b) Plate-circuit power detection.

When an input signal is applied, the grid becomes negative, thus lowering the plate current.   This decrease in plate current lowers the voltage drop across R_1 and restores the plate voltage to normal.

A plate-circuit power detector is shown in Fig. 3-14b.   The value of grid bias to be used is determined by the plate voltage and the resistance of the plate circuit.   In general, it is desirable to use as high a plate voltage as is practical in order to produce a high degree of amplification.   In plate-circuit rectification, the tube is operated with a grid bias of approximately one-tenth of the voltage applied to the plate.

3-7. Heterodyne Detection.   *Detection of Continuous-wave Code Signals.*   Code signals are transmitted by interrupting the carrier wave at definite intervals corresponding to the dots and dashes of the Morse code.   When a dot or dash is being transmitted, an r-f signal of constant amplitude is being sent out.   This method of transmission is called *continuous-wave* or *c-w transmission*.

If a c-w signal is applied to any of the detector circuits previously described, rectification of the r-f wave will take place, and a series of interrupted currents whose average values will be of constant amplitude will appear at the output side of the detector circuit.   If this output current is applied to earphones or a loudspeaker, the only sound it can produce is a click, which may occur at the start and finish of each dot or dash. In order to reproduce c-w signals, *heterodyne detection* may be used.

Principle of Heterodyne Action.   Heterodyne action is the result of combining two alternating voltages of different frequencies in a common detector in order to obtain a signal voltage of a new frequency value.   Actually, heterodyne action produces signals of two new frequency values: (1) the sum of the two original frequencies, and (2) the difference of the two original frequencies.   For heterodyne detection, only the difference value is used.   When the signal waves of two frequencies are combined by heterodyne action, the envelope of the resultant wave will vary in amplitude at the new frequency rate.   The range of the amplitude swing of the new voltage is determined by the sum and difference of the two voltages being combined.   Figure 3-15 shows two signal waves, *A* and *B*, and the resultant voltage wave *C*.   The frequency of signal *A* is 8 cycles, and that of signal *B* is 12 cycles; the frequency of the resultant wave is $12 - 8$, or 4 cycles.   The maximum voltage of signal *A* is 2 volts, and that of *B* is 4 volts.   The envelope of the resultant wave will therefore vary in voltage from $E_B - E_A$ to $E_B + E_A$.   In this case, it will vary from $4 - 2$ to $4 + 2$, or from 2 to 6 volts.

The values of frequency and voltage used here to explain heterodyne action are not practical values but were chosen to provide an easier means of illustration.   This same action, however, holds for signals of higher frequencies and different voltages, such as are found in the r-f circuits of receivers.

Beat Frequency.   The frequency at which the amplitude of the resultant wave varies is called *the beat frequency*, or *difference frequency*.   The process of producing these beats by combining two waves of different frequencies in a nonlinear circuit is called *heterodyning*.   The average value of the rectified heterodyne signal will vary in amplitude at the beat frequency.   The frequency of an a-c wave can therefore be changed by heterodyne action.   This action has a number of applications in the field of radio communications.   An inaudible c-w signal can be made audible by combining it with another wave and producing a beat frequency that lies within the audible sound range.   This is known as *heterodyne code reception*.   In this type of reception, the incoming signal is usually combined with a frequency of such a value that the beat frequency produced will be approximately 1,000 cycles.

I-F Amplifiers.   Another use of heterodyne action is to convert the frequency of an input signal to a definite intermediate frequency. Amplification of the signal can thus be made more efficient as the amplifiers can then be designed for a definite frequency. In this method, the i-f signal is adjusted to the frequency for which the amplifier is designed instead of the amplifier being made to amplify signals over a wide range of frequencies. Superheterodyne receivers use this principle in the i-f amplifier stages.

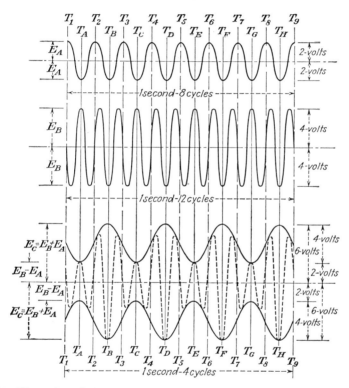

Fig. 3-15. Illustration of the heterodyne action resulting when two waves of unequal amplitude and frequency are combined.

Heterodyne Detector Circuits.   The simple heterodyne-detector circuit shown in Fig. 3-16 has two fundamental parts, a plate detector and a tuned-grid oscillator. The oscillator circuit is adjusted to a frequency slightly above or below the frequency of the input signal. The current in the plate circuit of the oscillator VT_2 is made to flow through coil L_3, which is inductively coupled to coil L_2 in the grid circuit of the plate detector VT_1. In this manner, both the input signal from coil L_4 and the oscillator output from coil L_3 are applied to the grid of the plate detector. As a result of the heterodyne action that takes place, the output of the detector circuit will contain an alternating voltage whose frequency is equal to the difference of the input frequencies, or the beat frequency.

Example 3-9. A c-w signal is transmitted at a frequency of 600 kc. To what frequency must the oscillator circuit of a receiver be adjusted if the receiver is to produce a 1,000-cycle audio signal from the received signal?

$$\text{Given:} \quad f_{\text{c-w}} = 600 \text{ kc} \qquad\qquad \text{Find:} \quad f_{\text{osc}}$$
$$f_{\text{a-s}} = 1 \text{ kc}$$

Solution:

$$f_{\text{osc}} = f_{\text{c-w}} \pm f_{\text{a-s}} = 600 \pm 1 = 601 \text{ or } 599 \text{ kc}$$

Whether the oscillator frequency is higher or lower than the frequency of the c-w signal is not important, as it is the difference between the two

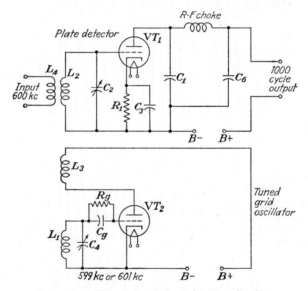

Fig. 3-16. Simple heterodyne detector circuit.

frequencies that determines the frequency of modulation or beat frequency. A more efficient method, generally used in small receivers, uses a single tube for both the detector and oscillator and is called an *autodyne detector*. The energy of the oscillator circuit can be transferred to the detector by coupling the oscillator to the plate circuit of the detector instead of to the grid circuit. These two circuits can be coupled either capacitively or inductively.

3-8. Regenerative Detectors. *Principle of Regenerative Detection.* The output of any triode detector circuit will vary with the signal voltage impressed on the grid of the tube. If the resistance of the resonant grid circuit is lowered, the circuit Q will increase, thereby increasing the input signal applied to the grid of the tube. The ratio between the input signal applied to the grid and the input signal received is increased, thus increasing the sensitivity of the detector circuit.

The same effect can be obtained by returning part of the energy of the plate circuit to the grid circuit. If this feedback is of the proper phase relation, the input voltage on the grid will be increased. This will increase the ratio between the initial input voltage applied to the grid and the output voltage across the load, thus increasing the sensitivity of the circuit. This is the principle of the regenerative detector. Figure 3-17 shows two regenerative-detector circuits. These two diagrams indicate that the basic circuit of the regenerative detector is somewhat similar to the tuned-grid oscillator section of Fig. 3-16.

Action of a Regenerative-detector Circuit. When an input signal is applied to coil L_3 of Fig. 3-17, a voltage is induced in coil L_2. This voltage will be at its maximum value when the grid circuit is resonant for the frequency of the signal being received. When this voltage is impressed on the grid of the tube, it will cause the plate current flowing through coil L_1 to vary. This change in current in L_1 causes the magnetic field about coil L_1 to vary in the same manner and causes a voltage to be induced in coil L_2. If the induced voltage is in phase with the voltage due to the input signal, the feedback voltage will be added to the signal voltage, thus causing a still greater change in the amount of plate current. This effect is repeated until some limiting action takes place. If the original feedback voltage is slightly less than the original signal voltage, the amount of increase in voltage and current will decrease with each energy transfer from plate to grid, until the action becomes stabilized. If the coupling between the plate and grid coils is such that the original amount of feedback voltage is greater than the original input signal voltage, the resistance of the circuit is made negative. After a number of successive transfers of energy from the plate to the grid circuit, the circuit will break into oscillation. The amplification of the signal voltage at the grid will increase with increases in the amount of feedback, reaching a maximum at the point where the detector starts to oscillate.

Methods of Controlling the Amount of Regeneration. The amount of regeneration may be controlled by: (1) adjusting the amount of coupling between the grid and plate coils, (2) varying the amount of capacitance C_1 between the plate coil and the cathode (Fig. 3-17a), and (3) varying the amount of voltage applied to the plate by means of the resistor R_1 (Fig. 3-17b). By varying C_1 or R_1 to a point just below oscillation, the feedback to the grid circuit may be adjusted so that the amount of regeneration will remain constant.

Disadvantages of Regenerative Circuits. Regeneration is an economical means of obtaining r-f amplification. Because regenerative circuits have a number of disadvantages, this type of circuit is not generally used in commercial receivers. Among its disadvantages are: (1) critical adjustments are required to obtain the proper amount of regeneration, (2) squeals will result from accidentally allowing the tube to break into oscillation, and (3) the detector acts as a transmitter when it is oscillating, and these oscillations may be heard by other nearby receivers tuned to the same frequency.

3-9. Autodyne Detection. *Principle of Autodyne Detection.* Because regeneration can be increased to a point where a circuit will break into oscillation, it is possible to obtain heterodyne reception of c-w signals with the use of a single vacuum tube. An oscillating detector used for

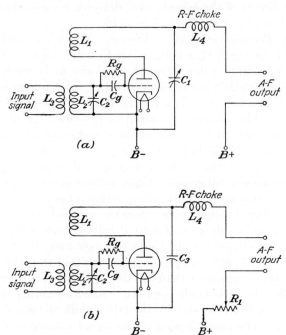

Fig. 3-17. Regenerative detector. (*a*) Capacitor controlled. (*b*) Resistor controlled.

this purpose is called an *autodyne detector.* Its circuit is the same as for the regenerative-detector circuits (Fig. 3-17).

The tuned-grid circuit produces local oscillations, and the beat frequency will be equal to the difference between the frequency of the c-w signal and the local oscillations. A beat frequency of 1,000 cycles is commonly used. By tuning the grid circuit to a frequency of 1,000 cycles above or below the frequency of the incoming c-w signal, a 1,000-cycle beat frequency is obtained.

Advantages of Autodyne Detection. The advantages of autodyne detection may be listed as follows: (1) A single tuning control is used instead of two, as required for heterodyne detection. (2) The circuit is simple. (3) Its sensitivity and selectivity are very high compared with other

methods of detection.   (4) A single-tube autodyne detector can produce
as much power as can be obtained from two stages of r-f amplification
followed by a heterodyne detector.

3-10. Superregeneration. *Critical Operating Point of Regenerative
Detectors.*   In a regenerative detector, the amplification due to regeneration will build itself up to a point where the circuit breaks into oscillation.
At this point, the output is the greatest because the effective resistance
of the circuit is zero and the losses are at a minimum.   This operating
point is very critical, and therefore it is very difficult to maintain the circuit at the desired operating point.   A slight change in any part of the
circuit will cause the circuit to break into oscillation.

Principle of Superregeneration.   Superregeneration is a means whereby
the effective resistance of the circuit is kept at zero.   This is accomplished
by varying the regeneration from an oscillatory to a nonoscillatory condition at a low r-f rate.   During the oscillatory interval, the oscillations

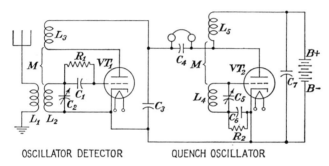

OSCILLATOR DETECTOR   QUENCH OSCILLATOR

FIG. 3-18. Separately quenched superregenerative detector circuit.

will build up, only to be suppressed by the low frequency that is applied
to the plate.   This causes the circuit to go in and out of oscillation at a
periodic rate that is equal to the frequency being applied to the plate.
This frequency, called the *quench frequency,* is slightly higher than the
audio frequencies, generally 20 to 25 kc.

Separately Quenched Superregenerative Circuit.   Superregeneration may
be obtained by: (1) using a separate oscillator to supply the quench frequency, or (2) intermittently blocking the grid at the desired frequency.
The first method, employing a separately quenched circuit, is shown
in Fig. 3-18.   This circuit is similar to the regenerative circuit, except
that the plate of the oscillating detector VT_1 is supplied with a low
r-f voltage of constant frequency in place of the steady direct voltage.
During the interval that the plate of VT_1 is positive, electrons will flow
in its plate circuit.   During the opposite half-cycle, the plate of VT_1 is
negative, and no electrons will flow, and the oscillations will die out during this interval.   The average voltage of the oscillations across the tuned
input circuit L_2C_2 will depend on the value of the signal voltage in this

circuit. If this signal is rectified, the plate current in VT_1 will vary in accordance with the envelope of its input grid signal. If a pentode is used, the quench frequency is applied to the screen grid instead of to the plate.

Self-quenched Superregenerative Circuits. A self-quenching superregenerative detector is similar to the basic regenerative circuits shown in Fig. 3-17 except for the values of the grid capacitor and grid resistor. Both of these values are increased considerably, thus increasing the discharge time for the grid capacitor. The high value of grid resistance blocks the electrons from the grid and causes (1) a decrease in the plate current, and (2) a decrease in the signal input to the grid. The charge on the grid capacitor gradually leaks off through the grid resistor, reduces the grid bias, and permits the tube to amplify and build up oscillations as before. The frequency of this self-quenching action is determined by the values of the grid resistor and grid capacitor.

Uses of Superregenerative Circuits. Superregenerative detector circuits are used in portable code receivers. As the ratio of quench frequency to signal frequency should be approximately 1 to 1,000, this type of circuit can only be used for frequencies over 25 mc. At the average operating point, the effective resistance of the circuit is practically zero; thus high amplification and sensitivity are obtained.

When the input signal is zero, the electron flow in the tube is irregular and a hissing noise is heard in the output circuit. This noise becomes insignificant when strong input signals are applied. The selectivity of a receiver using superregenerative detection is poor, but this is not important at the high frequencies for which this type of detection is used. The superregenerative detector circuit also radiates a signal that interferes with the reception of nearby receivers. Superregenerative detectors are used primarily for the reception of signals whose frequencies are too high for other methods of detection.

3-11. Automatic Volume Control. *Purpose of Automatic Volume Control.* In operating a receiver, it is sometimes found that the volume of the output varies without making any change in the setting of the volume control. This is called *fading*, and is most likely to be encountered when receiving the signals of distant stations. Many receivers use a special circuit, called an *automatic-volume-control* or *avc circuit*, to counteract this effect. These special circuits, usually associated with the detector circuits, are further classified as *simple* avc and *delayed* avc circuits. In a simple avc circuit the avc action is present at all times regardless of whether the incoming signal is weak or strong. In a delayed avc circuit, the avc action occurs only when strong signals are being received.

Principle of AVC Circuits. The principle of automatic volume control is based on varying the bias applied to the control grids of one or more of the tubes preceding the detector. When the avc bias is applied to a supercontrol or variable-mu tube, an increase in signal strength will cause an increase in the grid bias and thereby reduce the voltage amplification of the circuit.

Undesired changes in volume that make automatic volume control desirable appear as fluctuations in signal strength at both the input and output side of the detector circuit.   Automatic volume control is achieved by applying a direct voltage to the control grids of the supercontrol tubes. This voltage, obtained from the detector circuit, is negative with respect to ground and acts as a bias on the grids of the tubes to which it is applied. Increases in signal strength will increase the value of this avc bias and cause a reduction in the strength of the signal applied to the detector. Conversely, a decrease in signal strength will decrease the avc bias voltage

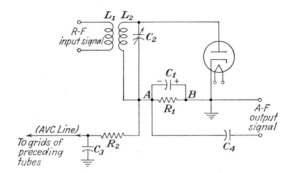

FIG. 3-19. Diode detector with simple automatic volume control.

and cause an increase in the signal applied to the detector.   In this manner, the avc bias tends to maintain automatically the output of the detector circuit at a constant level.

Simple AVC Circuits.   The input signal at a diode detector causes current to flow in the detector circuit during part of the positive half of each cycle.   This current produces a voltage at the resistor R_1 and also charges the capacitor C_1 (Fig. 3-19).   During the remainder of each cycle, C_1 supplies current to the output circuit and produces output signal variations that are a replica of the modulation envelope of the r-f input signal.

The charge on C_1 will make point A negative with respect to ground. This is so because the tube permits electrons to flow only in the direction of A toward B; furthermore, as electrons always flow from negative toward positive, point A must be negative with respect to point B or ground.   The voltage at A will be a pulsating voltage varying in strength according to the a-f component of the input signal (Figs. 3-1b and 3-3c).

The voltage at point A, being negative with respect to ground and proportional to the signal strength, is a good source for the avc bias.   This voltage, however, cannot be used directly as the avc bias, because it varies in the same manner as the a-f component of the signal.   Applying this voltage to the grids of the preceding tubes would neutralize the variations in the strength of the modulation envelope, with the result that the output of the detector circuit would consist of variations of audio frequencies but all of a constant strength.   In order to compensate for this undesirable effect, the filter circuit R_2C_3 (Fig. 3-19) is added.   The voltage from point A is now required to charge capacitor C_3 through resistor R_2.   The avc bias is then dependent on the amount of charge on C_3.   By properly choosing the values of R_2 and C_3, it is possible to control the charging rate of the capacitor so that the variations due to the a-f component of the signal will have no appreciable effect but that variations due to fading and other undesirable causes will be effective.

The time constant of the R_2C_3 circuit should be of sufficient length so that the lowest value of a-f sound to be reproduced will not cause any appreciable amount of charge on C_3.   On the other hand, the time constant should not be too long, for then the avc would not compensate for sudden changes due to fading, etc.   A time constant of approximately 0.1 to 0.2 sec is generally used.

Example 3-10.   The avc filter circuit of a certain receiver consists of a 2-megohm resistor and a 0.1-μf capacitor.   (*a*) What is the time constant of this circuit?   (*b*) What per cent of the maximum voltage charge possible will be attained in the time that the current from 1 cycle of a 50-cycle a-f signal will flow in the detector circuit (assuming that a constant voltage was being applied)?   (*c*) How many cycles of the 50-cycle note would be completed in one time constant?   (*d*) If a sine-wave voltage, whose maximum value is equal to the constant voltage assumed in part (*b*), were applied to the detector circuit, would it require the same amount of time, more time, or less time for the capacitor to reach 63.2 per cent of its final charge?   Why?

Given: R = 2 megohms   Find: (*a*) t
   C = 0.1 μf   (*b*) Per cent E
   f = 50 cycles   (*c*) Cycles
                 (*d*) t

Solution:

(*a*) $t = RC = 2 \times 10^6 \times 0.1 \times 10^{-6} = 0.2$ sec

(*b*)   As current can flow only during the positive half of the signal, the maximum time of current flow in one cycle of the 50-cycle a-f signal is 0.01 sec.   This corresponds to one-twentieth of a time constant, and from Appendix IX the charge produced will be approximately 5 per cent of the maximum voltage charge possible when $k = \frac{1}{20}$ or 0.05.

(*c*)   Time for 1 cycle of the 50-cycle a-f current,

$$t_1 = \frac{1}{f} = \frac{1}{50} = 0.02 \text{ sec}$$

Cycles completed in one time constant,

$$\text{No. of cycles} = \frac{t}{t_1} = \frac{0.2}{0.02} = 10$$

(*d*)   A much greater amount of time will be required because the value of the sine-wave voltage is equal to the voltage of part (*b*) at its maximum point only and then only for a very small period of time.

The avc bias may be applied to the grids of the tubes preceding the detector, such as the i-f amplifiers, r-f amplifiers, and mixers or converters. Successful avc operation is generally attained when the avc bias is applied to two or more tubes.

Delayed AVC Circuits.   The simple avc circuit applies some amount of avc bias at all times.   This is undesirable when the receiver is tuned to a weak station, as it will reduce the amount of volume available at the loudspeaker.   In order to overcome this effect, the circuit may be arranged so that the avc bias must first overcome some fixed amount of voltage before any avc bias can be applied to the grids of the preceding tubes.   This type of circuit is called the *delayed avc circuit.*   The delay referred to is not related to time but represents a value of voltage at which the automatic volume control first becomes effective.

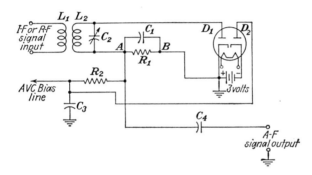

Fig. 3-20. Diode detector with delayed automatic volume control.

The circuit shown in Fig. 3-20 illustrates the principle of delayed avc operation.   The tube used is a duplex diode, and the left half of the tube D_1, with its associated circuit, acts in the same manner as the diode detector of Fig. 3-19.   The right half of the tube D_2 provides the delayed avc action of the circuit.   The plate of diode D_2 is positive with respect to its cathode and hence a current will flow in the circuit composed of D_2, R_2, R_1, and the 3-volt battery.   Since the voltage drop at D_2 will be very small, C_3 will charge to practically 3 volts and will produce an avc voltage of approximately -3 volts.   Furthermore, this current will produce a negative voltage at A, which will apply a negative voltage to the plate of D_1. Because R_2 is usually five or more times greater than R_1, this voltage is not enough to prevent the flow of current in D_1 during the positive halves of the input signal.

The rectified signal current flowing in the circuit of diode D_1 will still produce a voltage across R_1 with terminal A negative with respect to ground.   With weak signals, this voltage may well be less than 3 volts, and the avc line will be supplied with a constant 3 volts provided by the battery in the circuit of the diode D_2.

When strong signals are being received, the voltage at R_1 may exceed 3 volts, and C_3 will become charged to the voltage developed across R_1; the voltage of the avc line will correspond to the voltage at C_3. When the voltage at C_3 exceeds 3 volts, the plate of D_2 becomes negative with respect to the cathode, and the electron flow in this circuit will cease.

The above discussion shows that for weak input signals the avc bias remains constant at 3 volts, and the receiver can be operated with its maximum sensitivity and gain. With strong input signals, the avc bias will exceed 3 volts, and the volume will be maintained at a constant level by the avc circuit.

Practical Delayed AVC Circuit. In the circuit of Fig. 3-21, the detector and avc functions are performed by part of a duplex-diode-triode type tube. Plate D_1 of the duplex-diode section of the tube acts as the plate

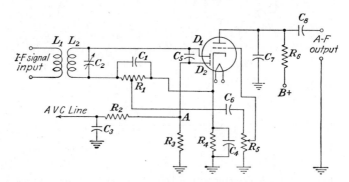

FIG. 3-21. Diode detector using a multiunit tube. Delayed automatic volume control obtained from cathode bias.

for the diode detector. Plate D_2, which is fed by capacitor C_5, supplies the avc bias. The cathode-biasing resistor R_4 provides the bias for the triode section of the tube. It also makes the avc diode plate D_2 negative with respect to the cathode and thereby provides the means of obtaining the delayed automatic volume control without the use of a battery. The diode load resistor R_1 is connected directly to the cathode, and hence the detector plate D_1 is at cathode potential. Because of this, rectification will take place in the detector circuit during part of the positive half of each cycle of the input signal received, whether from a weak station or from a strong one. The rectified current flowing through R_1 produces a unidirectional voltage at R_1, which varies in the same manner as the input signal. This output voltage is applied to the triode or amplifier portion of the tube through the coupling capacitor C_6 and the manual volume control R_5.

The avc plate D_2 is connected to ground through its load resistor R_3. With zero input signal voltage, D_2 will be negative with respect to the cathode by the amount of cathode bias produced by R_4, usually 2 or 3 volts. Under this condition, no current will flow in the D_2 plate circuit and hence no avc voltage is developed.   When a signal is being received, a voltage will be applied to the avc plate D_2 by means of the coupling capacitor C_5.   When this voltage exceeds the value of the cathode bias, D_2 will be positive with respect to the cathode, and a rectified current will flow in the avc circuit consisting of D_2, R_3, and R_4.   A voltage will then be developed across the avc load resistor R_3, point A being negative with respect to ground.   The avc filter network R_2C_3 eliminates the a-f voltage variations and passes on to the avc line any variations due to fading and changes in signal strength when tuning in a new station. Thus, the delayed avc circuit results in having the maximum signal received from weak stations and automatic volume control applied only to the strong stations.

QUESTIONS

1. Explain the fundamental process of each of the two operations necessary in any detector circuit.

2. Define each of the following terms: (*a*) wave train, (*b*) r-f carrier, (*c*) audio component, (*d*) modulated wave, (*e*) modulation envelope.

3. What operating features of vacuum tubes made possible their universal use as detectors in modern radio equipment?

4. Explain what is meant by each of the following terms: (*a*) sensitivity, (*b*) signal handling ability, (*c*) fidelity of reproduction.

5. Which factors are most important in selecting a detector circuit for a receiver that is to be used for: (*a*) signals from a distant broadcasting station, (*b*) signals from a powerful local broadcasting station, (*c*) code?

6. Explain the basic principle of a diode detector.

7. (*a*) Explain the purpose of the r-f bypass capacitor.   (*b*) How is it connected in the circuit?   (*c*) What is its approximate value?

8. (*a*) What is meant by a full-wave diode detector?   (*b*) What methods are used to obtain this type of detection?   (*c*) What are its advantages and disadvantages?

9. (*a*) What are the advantages and disadvantages of diode detectors?   (*b*) Where are diode detectors generally used?

10. Where are each of the following parts used and what is their purpose: (*a*) r-f filter resistor, (*b*) a-f bypass capacitor, (*c*) diode load resistor?

11. Why do r-f filter circuits reduce the useful rectified output of a detector circuit?

12. (*a*) How are multiunit tubes used as diode detectors?   (*b*) What is the main advantage of using multiunit tubes?

13. (*a*) Explain the basic principle of plate detection.   (*b*) What other names are also applied to this method of detection?

14. Explain the following terms: (*a*) cutoff, (*b*) grid bias, (*c*) positive bend, (*d*) negative bend, (*e*) bias detectors.

15. Explain the operation of a grid-bias detector when operated at its (*a*) negative bend, (*b*) positive bend.

16. How is the output of a plate detector affected when the grid bias is adjusted (*a*) too far beyond its cutoff point, (*b*) too far before its cutoff point?

17. What are the advantages and disadvantages of plate detectors?

18. (*a*) Explain the purpose of the r-f choke coil.   (*b*) How is it connected in the circuit?

19. What is the relation between the grid-voltage swing and the operating grid voltage?

20. (*a*) Describe three methods used to obtain the correct amount of grid-bias voltage. (*b*) Which of these methods is generally used?

21. Explain why the use of a cathode resistor to obtain the required grid bias is referred to as the *automatic* or *self-biasing method*.

22. How is the value of the cathode-bias resistor determined?

23. (*a*) Explain the purpose of the a-f bypass capacitor. (*b*) How is it connected in the circuit? (*c*) What is its approximate value?

24. Explain the principle of operation of grid detection.

25. How does the action and circuit of a grid detector compare with a half-wave diode detector?

26. Explain the circuit action of the grid capacitor.

27. Explain the circuit action of the grid resistor.

28. (*a*) What are the advantages and disadvantages of grid detectors? (*b*) Where are grid detectors generally used?

29. (*a*) What is meant by power detection? (*b*) What are its main advantages?

30. How do grid-circuit and plate-circuit power detectors differ from the regular grid-circuit and plate-circuit detectors?

31. Explain why c-w signals cannot be detected by using diode-, grid-, or plate-detector circuits.

32. Explain the principle of heterodyne action.

33. What is meant by (*a*) beat frequency, (*b*) heterodyning?

34. Explain two applications of heterodyne action.

35. (*a*) What are the essential parts of a heterodyne circuit? (*b*) How does this circuit operate?

36. Explain the action of a regenerative detector circuit.

37. Explain three methods of controlling the amount of regeneration.

38. What are the advantages and disadvantages of regenerative detectors?

39. (*a*) Explain the principle of operation of an autodyne detector. (*b*) How does it compare with regenerative and heterodyne detectors?

40. (*a*) What are the advantages of autodyne detectors? (*b*) Where are they generally used?

41. Explain the principle of superregeneration.

42. Explain the methods used to obtain superregeneration.

43. What is meant by the quench frequency?

44. How do the circuits for a regenerative detector and a self-quenching superregenerative detector compare?

45. What are the advantages and disadvantages of superregeneration?

46. Where are superregenerative circuits used?

47. What is the purpose of automatic volume control?

48. What is the principle of operation of avc circuits?

49. (*a*) What is meant by simple automatic volume control? (*b*) What is meant by delayed automatic volume control? (*c*) Explain the meaning of the word *delayed* in the term *delayed automatic volume control*.

50. (*a*) Can automatic volume control be applied to any type of tube? (*b*) Explain.

51. Describe the action of the simple avc circuit.

52. (*a*) What does the fundamental avc filter network consist of? (*b*) Why is it necessary? (*c*) What value of time constant is recommended for this circuit?

53. (*a*) To how many tubes should the avc bias be applied? (*b*) Where are these circuits located with respect to the detector? (*c*) To which circuits may this avc bias be applied?

54. (*a*) What is the disadvantage of simple avc? (*b*) How does delayed avc overcome this disadvantage?

55. Explain the action of an avc circuit using a duplex-diode tube and a battery to provide the delayed voltage.

56. (a) Why is the use of a battery undesirable in delayed avc circuits? (b) What method is used to eliminate the need of a battery?

57. Describe the action of a delayed avc circuit using a duplex-diode–triode tube.

PROBLEMS

1. A certain diode-detector circuit, similar to Fig. 3-2, uses a 0.5-megohm resistor for the diode load resistance R_1 and a 100-$\mu\mu$f capacitor for the diode bypass capacitor C_1. (a) What is the time constant of this RC circuit? (b) If the resonant frequency of the tuned circuit L_2C_2 is 1,500 kc, what time is required for the r-f wave to complete 1 cycle? (c) How many times greater is the time constant of the RC circuit than the time of 1 cycle of the r-f wave?

2. The circuit of Prob. 1 also acts as a filter circuit. (a) What is the reactance of the capacitor to a 1,500-kc r-f current? (b) How does the reactance to the 1500-kc r-f current compare with the value of the diode load resistance? (c) What path will the r-f currents take? (d) What is the reactance of the capacitor to a 500-cycle a-f current? (e) How does the reactance to the 500-cycle a-f current compare with the value of the diode load resistance? (f) Which path will the a-f currents take?

3. A certain diode-detector circuit, similar to Fig. 3-2, uses a 300,000-ohm resistor for the diode load resistance R_1 and a 250-$\mu\mu$f capacitor for the diode by-pass capacitor C_1. (a) What is the time constant of this RC circuit? (b) If the resonant frequency of the tuned circuit L_2C_2 is 500 kc, what time is required for the r-f wave to complete 1 cycle? (c) How many times greater is the time constant of the RC circuit than the time of 1 cycle of the r-f wave?

4. The circuit of Prob. 3 also acts as a filter circuit. (a) What is the reactance of the capacitor to a 500-kc r-f current? (b) How does the reactance to the 500-kc r-f current compare with the value of the diode load resistance? (c) What path will the r-f currents take? (d) What is the reactance of the capacitor to a 500-cycle a-f current? (e) How does the reactance to the 500-cycle a-f current compare with the value of the diode load resistance? (f) Which path will the a-f currents take?

5. The circuit elements shown in Fig. 3-5a have the following values: $R_1 = 100,000$ ohms, $R_2 = 400,000$ ohms, $C_1 = 100$ $\mu\mu$f, $C_3 = 100$ $\mu\mu$f. (a) What impedance does the capacitor C_1 offer to a 465-kc i-f current? (b) Which path will the i-f current take? (c) Will any of the i-f current flow into the R_1 path? (d) What impedance does the capacitor C_3 offer to any 465-kc i-f current? (e) What purpose does C_3 serve? (f) Neglecting the effect of the capacitors C_1 and C_3, what per cent of the a-f voltage developed across R_1 and R_2 is available at the output terminals?

6. The circuit elements shown in Fig. 3-5a have the following values: $R_1 = 50,000$ ohms, $R_2 = 200,000$ ohms, $C_1 = 250$ $\mu\mu$f, $C_3 = 250$ $\mu\mu$f. (a) What impedance does the capacitor C_1 offer to a 455-kc i-f current? (b) Which path will the i-f current take? (c) Will any of the i-f current flow into the R_1 path? (d) What impedance does the capacitor C_3 offer to any 455-kc i-f current? (e) What purpose does C_3 serve? (f) Neglecting the effect of the capacitors C_1 and C_3, what per cent of the a-f voltage developed across R_1 and R_2 is available at the output terminals?

7. The r-f filter circuit of Fig. 3-8a has a 100-$\mu\mu$f capacitor at C_1 and a 30-mh choke at L_3. (a) What impedance does the capacitor offer to a 1,500-kc r-f current? (b) What impedance does the choke offer to a 1,500-kc r-f current? (c) What impedance does the capacitor offer to a 500-cycle a-f current? (d) What impedance does the choke offer to a 500-cycle a-f current? (e) Which path will the r-f currents take? (f) Which path will the a-f currents take?

8. The r-f filter circuit of Fig. 3-8a has a 500-$\mu\mu$f capacitor at C_1 and a 125-mh

choke at L_3.   (a) What impedance does the capacitor offer to a 455-kc i-f current?
(b) What impedance does the choke offer to a 455-kc i-f current?   (c) What impedance does the capacitor offer to a 500-cycle a-f current?   (d) What impedance does
the choke offer to a 500-cycle a-f current?   (e) Which path will the i-f currents take?
(f) Which path will the a-f currents take?

9. A tube is operated with a grid bias of 3 volts.   What is the grid-voltage swing
when an a-c signal applied to the input circuit has a voltage of (a) 1 volt?   (b) 3 volts?
(c) 4 volts?

10. A tube is operated with a grid bias of 6 volts.   What is the grid-voltage swing
when an a-c signal applied to the input circuit has a voltage of (a) 2 volts?   (b) 4 volts?
(c) 6 volts?

11. A certain triode (6C5) is used as a bias detector and is operated with 250 volts
on its plate.   It is recommended that the grid bias be approximately 17 volts.   What
is the greatest amount of signal voltage that can be applied to the input circuit without
causing distortion?

12. A certain pentode (6J7) is used as a bias detector and is operated with 250 volts
on its plate and 100 volts on the screen grid.   It is recommended that the grid bias be
4.3 volts.   What is the greatest amount of signal voltage that can be applied to the
input circuit without causing distortion?

13. A certain triode when used as a bias detector is to be operated with a grid bias
of 12 volts.   What value of cathode resistance is required if with zero input signal
the plate current is (a) 0.1 ma? (b) 0.25 ma? (c) 0.4 ma?

14. A certain triode when used as a bias detector is to be operated with a grid bias
of 14 volts.   What value of cathode resistance is required if with zero input signal
the plate current is (a) 0.1 ma? (b) 0.2 ma? (c) 0.35 ma?

15. It is recommended that the plate current of the bias detector of Prob. 11 be
adjusted to 0.2 ma when the input signal is zero.   (a) What value of cathode resistor
is necessary to provide the recommended grid-bias voltage?   (b) How much power is
consumed by this resistor?   (c) What power rating should the resistor have?

16. It is recommended that the cathode current of the bias detector of Prob. 12 be
adjusted to 0.43 ma when the input signal is zero.   (a) What value of cathode resistor
is necessary to provide the recommended grid-bias voltage?   (b) How much power is
consumed by this resistor?   (c) What power rating should the resistor have?

17. The self-biased detector of Prob. 15 uses a cathode-bias resistor of 85,000 ohms.
(a) What value of bypass capacitor should be used with this resistor if it is desired
that the resistor offer at least 100 times more impedance to a 500-cycle a-f current than
the capacitor?   (b) What standard rating and type of capacitor is recommended for
this application?

18. The self-biased detector of Prob. 16 uses a cathode-bias resistor of 10,000 ohms.
(a) What value of bypass capacitor should be used with this resistor if it is desired
that the resistor offer at least 100 times more impedance to a 500-cycle a-f current
than the capacitor?   (b) What standard rating and type of capacitor is recommended
for this application?

19. The operating characteristics of a certain self-biased detector are provided in
Probs. 11, 15, and 17.   If a 250,000-ohm load resistor is connected in its plate circuit,
what value of voltage must the B power supply provide in order to maintain 250 volts
between the cathode and plate when the plate current is 0.2 ma?

20. The operating characteristics of a certain self-biased detector are provided in
Probs. 12, 16, and 18.   If a 220,000-ohm load resistor is connected in its plate circuit,
what value of voltage must the B power supply provide in order to maintain 250 volts
between the cathode and plate when the cathode current is 0.43 ma and the plate
current is 0.35 ma?

21. Using the plate characteristic curves for the triode represented by Fig. 2-14, calculate the value of cathode-bias resistance necessary to operate the tube with the following plate and grid voltages: (a) $E_b = 240$, $E_c = -8$; (b) $E_b = 200$, $E_c = -6$; (c) $E_b = 160$, $E_c = -4$; (d) $E_b = 100$, $E_c = -2$.

22. Using the grid-plate transfer characteristic curves for the triode represented by Fig. 2-15, calculate the value of cathode-bias resistance necessary to operate the tube with the following plate voltages and plate currents (milliamperes): (a) $E_b = 200$, $I_b = 7.5$; (b) $E_b = 120$, $I_b = 7.5$; (c) $E_b = 160$, $I_b = 7.5$.

23. (a) How much power is consumed by each cathode resistor in Prob. 21? (b) What power ratings should the resistors have?

24. (a) How much power is consumed by each cathode resistor in Prob. 22? (b) What power ratings should the resistors have?

25. Using the plate characteristic curves for the pentode represented by Fig. 2-24, calculate the value of cathode-bias resistance necessary to operate the tube with the following plate and grid voltages (the screen-grid voltage is kept at 100 volts): (a) $E_b = 120$, $E_c = -1$; (b) $E_b = 240$, $E_c = -2$; (c) $E_b = 320$, $E_c = -2.5$; (d) $E_b = 400$, $E_c = -3$. (NOTE: Screen-grid current is 0.5 ma.)

26. (a) How much power is consumed by each cathode resistor in Prob. 25? (b) What power rating should each resistor have?

27. (a) Plot the curves required to show the resultant wave when a 6-cycle sine-wave voltage whose maximum value is 4 volts is combined by heterodyne action with a 9-cycle sine-wave voltage whose maximum value is 8 volts. (b) What is the beat frequency of these two waves? (c) What is the voltage swing of the beat frequency?

28. A c-w signal transmitted on a frequency of 2,000 kc is being received by an autodyne detector. At what frequency must the oscillator circuit of the receiver be set in order that the frequency in the output circuit will be (a) 500? (b) 1,000? (c) 1,500 cycles?

29. The avc filter circuit shown in Fig. 3-19 has the following constants: $R_2 = 1$ megohm, $C_3 = 0.1$ μf. (a) What is the time constant of this circuit? (b) What per cent of the maximum voltage charge possible will be attained in the time that the current from 1 cycle of a 100-cycle a-f signal will flow in the detector circuit (assuming that a constant voltage was being applied)? (c) How many cycles of the 100-cycle note would be completed in one time constant?

30. The avc filter circuit shown in Fig. 3-19 has the following constants: $R_2 = 2$ megohms, $C_3 = 0.05$ μf. (a) What is the time constant of this circuit? (b) What per cent of the maximum voltage charge possible will be attained in the time that the current from 1 cycle of a 50-cycle a-f signal will flow in the detector circuit (assuming that a constant voltage was being applied)? (c) How many cycles of the 50-cycle note would be completed in one time constant?

31. The delayed avc circuit shown in Fig. 3-21 has the following constants: $R_1 = 0.5$ megohm, $R_2 = 1$ megohm, $R_3 = 2$ megohms, $R_4 = 2,500$ ohms, $C_1 = 250$ μμf, $C_3 = 0.1$ μf, $C_4 = 10$ μf. When the signal input is zero, the current in the cathode circuit is 1.2 ma. (a) What is the magnitude and the polarity of the voltage between D_2 and the cathode when the input signal is zero? (b) How much current flows in R_3 under the condition in part (a)? (c) What is the voltage of the avc line under the condition of part (a)? (d) To what voltage must the charge on the capacitor C_5 raise the plate D_2 in order to produce a current flow in R_3? (e) What are the magnitude and polarity of the voltage developed at point A when a current of 1μa flows through R_3? (f) What voltage will the avc line have under the condition in part (e)? (g) What is the time constant of the avc filter? (h) What is the time constant of the diode load resistor R_1 and the diode bypass capacitor C_1? (i) What impedance does the capacitor C_4 offer to the lowest a-f current, assuming it to be 50 cycles? (j) What purpose does the capacitor C_4 serve?

CHAPTER 4

TUNING CIRCUITS

Tuning is the process of adjusting the capacitance or inductance of a tuned circuit in order to select the signals of a desired station. Tuning circuits, therefore, form an essential part of all radio receivers. Selecting a desired signal is only one of three important functions performed by the tuning circuit. In addition, it must reject all undesired signals, and in many instances it also increases the voltage of the desired signal before passing it on to the following circuit.

4-1. Tuning. *Operating Characteristics.* The ability of a radio receiver to accomplish each of its three functions is referred to as its sensitivity, selectivity, and fidelity.

Sensitivity is a measure of the ability of a receiver to reproduce, with satisfactory volume, weak signals received by the antenna. It may further be defined as the minimum strength of signal input required to produce a specified a-f power output at the loudspeaker; it is generally expressed either in microvolts or in decibels below 1 volt.

Selectivity is a measure of the ability of a receiver to reproduce the signal of one desired station and to exclude the signals from all others. The selectivity of a receiver (or a tuning circuit) is generally expressed in the form of a graph, also called a *response curve*, showing the signal strength at its resonant frequency and the variation in signal current when the frequency is varied a specified amount above and below the frequency of resonance.

Fidelity is a measure of the ability of a receiver to reproduce faithfully all the frequencies present in the original signal. The fidelity of a receiver is generally expressed in the form of a graph showing the ratio of the actual output to the output at a standard audio frequency of 400 cycles. For good fidelity of reproduction, the bandwidth as shown by the selectivity graphs should be great enough to accommodate all the frequencies of the signal to be reproduced.

The Response Curve. The process of selecting the carrier wave of a desired station is called *tuning*. This may be accomplished by adjusting one or more components of a series tuned circuit so that its resonant frequency will be equal to that of the desired carrier wave. The impedance of the tuned circuit at resonance will be at its minimum value; therefore, the current in the tuned circuit produced by the desired station will be at its maximum value. As the resonant frequency of the tuning circuit is varied, either above or below the frequency of the desired station, the impedance of the circuit will increase and the signal current of the desired station will therefore decrease.　A graph showing the current in a tuned circuit at resonance and the decrease in current at frequencies off resonance is called a *response curve* (Fig. 4-1).

Ideal Response Curve.　A transmitted wave is made up of the carrier wave modulated by two sidebands whose frequencies are equal to the frequency of the carrier wave plus or minus the frequency of the audio signal. (For general commercial a-m broadcast transmission, the maximum frequency of the audio signal is 5 kc.) The sidebands of a modulated carrier wave will therefore vary up to 5 kc above and below its

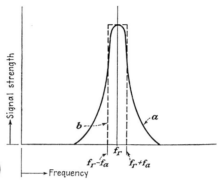

FIG. 4-1. Relation between actual and ideal response curves. (*a*) Actual response curve. (*b*) Ideal response curve.

carrier frequency.　For example, a broadcasting station operating on a carrier frequency of 1,000 kc will have sideband frequencies ranging from 995 to 1,005 kc.

In order to reproduce the signals as transmitted, the ideal response curve should have a flat top and straight sides, so that it may pass a 10-kc band for commercial a-m broadcast signals (Fig. 4-1).　This ideal can be closely approximated by proper use of resonant circuits.　Three methods of increasing the fidelity, sensitivity, and selectivity of a receiver are: (1) increasing the circuit Q, (2) using two or more tuned circuits, and (3) using a bandpass amplifier.

Radio Channels.　Each transmitting station requires a band 10 kc wide for commercial a-m broadcasting stations; this band is called a *radio channel.*　In order to prevent interference between two stations operating on adjacent channels there should be a difference of at least 10 kc in their carrier frequencies (Fig. 4-2).　For the same reason, the tuning circuit of a receiver should be capable of selecting signals from stations approximately 10 kc apart without any interference.　The broadcasting range extending from 550 to 1,600 kc could accommodate 106 channels. There are more than 1,000 stations assigned to these channels.　Under this crowded condition more than one station may be operating on the same frequency; also the sidebands from adjacent channels may overlap the signals of more than one station.　This interference between stations transmitting on the same frequency or operating partly within the same channel may produce a hum, whistle, or cross talk in the receiver.

In order to prevent interference between stations, the FCC has set up a zoning system and assigns the carrier frequency to broadcasting stations so that the interference is reduced to a minimum. Furthermore, stations are licensed as to the amount of power they may use, and in some cases they are also limited as to the hours during which they may broadcast. For example, 13 stations in the United States are assigned to the 710-kc channel. Four of these, namely, WOR at New York, KIRO at Seattle, KMPC at Los Angeles, and WGBS at Miami employ high-power transmitters; but because they are located great distances from one another, they do not interfere with each other and may therefore operate on a continuous time schedule. On the other hand, over 150 stations in the United States are assigned to the 1,240-kc channel, with as many as ten stations in the same state assigned to the same channel. The licensed

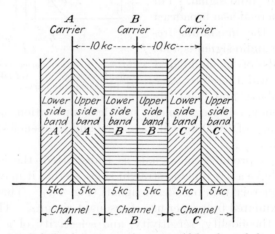

FIG. 4-2. Relation between carrier wave, sidebands, and channels.

power of these stations is kept low and some of the stations are restricted to operating at only certain hours of the day.

The ability of a receiver to minimize interference depends on the selectivity of its tuned circuits. In order to eliminate interference from adjacent stations, the selectivity will have to be increased. This reduces the width of the response curve and decreases the fidelity of the receiver, as the high notes are not reproduced. For example, if the width of the response curve is reduced to 8 kc, only 4 kc of the 5-kc sidebands as transmitted will be reproduced. Under this condition all audio signals between 4,000 and 5,000 cycles will not be reproduced by the receiver.

4-2. Circuit Elements. *Types of Variable Capacitors.* Although tuning can be accomplished by varying either the value of the inductance or capacitance of the tuning circuit, the method most commonly used is by varying the capacitance. Three types of variable capacitors are used in tuning. They are called *straight-line-capacity, straight-line-wavelength,* and *straight-line-frequency* capacitors; they are abbreviated as slc, slw, and slf respectively.

With an slc-type capacitor, the capacitance increases directly with the amount of rotation of its movable plates (Fig. 4-3a). For example, if the rotor plates are one-quarter in mesh with the stator, its capacitance will be one-quarter of its total value; if the rotor plates are one-half in mesh with the stator, its capacitance will be one-half of its total value, etc. If an slc capacitor is used in the tuning circuit of a broadcast receiver, most of the stations will appear on one-half of the dial (see Fig. 4-3a), because the resonant frequency of a tuned circuit does not vary in a direct ratio

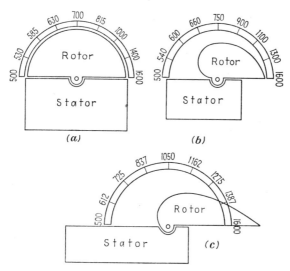

Fig. 4-3. Frequency distribution obtained with three types of variable capacitors. (a) Straight-line capacity. (b) Straight-line wavelength. (c) Straight-line frequency.

with changes in its capacitance. The upper half of the frequency band (1,075 to 1,600 kc) will appear on approximately only one-eighth of the dial; hence when this type of capacitor is used it is difficult to separate the signals from adjacent stations in the upper half of the frequency band.

In the slw capacitor the area of the rotor plates is reduced on the side that first enters into mesh with the stator plates, so that the wavelength of the tuned circuit increases in a direct ratio with the amount of rotation of the movable plates. With this type of rotor plate the capacitance increases very slowly at first but increases at a faster rate as the plates go further into mesh. If an slw capacitor is used in the tuning circuit of a broadcast receiver, most of the stations will appear on approximately three-fourths of the dial (Fig. 4-3b). Furthermore, the stations in the upper half of the frequency band will appear on approximately one-third of the dial.

With the slw capacitor, it was still difficult to separate signals from stations of carrier frequencies greater than 1,100 kc. This difficulty was overcome by using rotor plates shaped so that the resonant frequency of the tuned circuit will vary directly with the amount of rotation of the capacitor's rotor plates (Fig. 4-3c). Such capacitors are called *straight-line-frequency* capacitors. With this type of capacitor the resonant frequency of the tuned circuit varies in direct proportion to the amount of rotation of the variable capacitor.

The preceding discussion of variable capacitors is based on 180 degrees of rotation of the rotor plates. By increasing the amount of rotation to

(a)

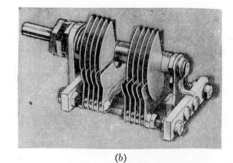

(b)

Fig. 4-4. Commercial types of capacitors. (*a*) Straight-line capacity. (*b*) Straight-line wavelength. (*c*) Straight-line frequency. (*Courtesy of National Company, Inc.*)

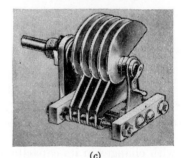

(c)

270 degrees the capacitor can be made more compact. The selectivity obtainable is also improved as the 1,050-kc range (550 to 1,600 kc) is spread over 270 mechanical degrees instead of 180 degrees. A commercial form of this type of capacitor is shown in Fig. 4-4c. Three additional methods used to obtain compactness for slf capacitors are: (1) using a greater number of rotor plates of a smaller surface area, (2) using thinner plates and a smaller air gap, and (3) using semicircular rotor plates and varying the design of the stator plates.

R-F Tuning Coils. The secondary winding of an r-f transformer provides the fixed inductance that is used with the variable capacitor to form the tuned circuit of a radio receiver. In this type of circuit the input signal flows through the primary winding of the r-f transformer and by means of the mutual inductance between the two windings the signal is transferred to the secondary. The variable capacitor and the secondary winding form a series resonant circuit, which must be tuned to the frequency of the desired station. The value of the inductance required in the secondary winding will depend on the frequency range desired and the value of the capacitance of the variable capacitor that is used. [Increasing the inductance of the secondary coil increases the amount of voltage developed across this winding, thus increasing the gain of the tuned circuit.] The selectivity of the tuned circuit is also increased, as increasing the inductance increases both the L/C ratio and the circuit Q

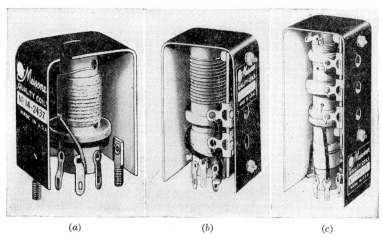

(a)   (b)   (c)

FIG. 4-5. Commercial types of r-f coils. (a) Single-band coil. (b) Dual-band coil. (c) Triple-band coil. (*Courtesy of Meissner Manufacturing Division, Maguire Industries, Inc.*)

(if the resistance remains constant). Several types of commercial r-f coils are shown in Fig. 4-5.

The amount of inductance that should be used is such a value that when combined with the distributed capacitance of the secondary winding and the minimum capacitance of the tuning capacitor it will form a resonant circuit whose frequency of resonance is equal to or greater than the highest frequency desired. For broadcast reception this frequency is approximately 1,600 kc. The maximum value of the variable capacitor must be equal to the amount of capacitance required to adjust the frequency of resonance of the tuned circuit to the minimum desired frequency, generally 550 kc.

Example 4-1. It is desired to determine the inductance required in the secondary winding of an r-f transformer, and it is assumed that the distributed capacitance of the winding and the circuit wiring is 23 $\mu\mu f$ (considered acting in parallel with the tuning capacitor). (a) What value of inductance is required in order to obtain resonance at 1,600 kc if the minimum value of the variable capacitor is 17 $\mu\mu f$? (b)

What value of capacitance must the capacitor have if it is desired to tune in stations as low as 550 kc with the coil used in part (a)?

$$\text{Given:} \quad C_D = 23 \ \mu\mu\text{f} \qquad\qquad \text{Find:} \quad (a)\ L$$
$$C_V = 17 \ \mu\mu\text{f (min.)} \qquad\qquad\qquad\quad (b)\ C_V$$
$$f = 1,600 \text{ kc } (a)$$
$$f = 550 \text{ kc } (b)$$

Solution:

(a) $\quad L = \dfrac{25,300}{f_r^2 C} = \dfrac{25,300}{1,600^2 \times (17 + 23)10^{-6}} = 246 \ \mu\text{h}$

(b) $\quad C_T = \dfrac{25,300}{f_r^2 L} = \dfrac{25,300}{550^2 \times 246} = 0.0003399 \ \mu\text{f or } 340 \ \mu\mu\text{f}$

$\qquad C_V = C_T - C_D = 340 - 23 = 317 \ \mu\mu\text{f}$

4-3. Short Waves. *Frequency Ranges.* Short-wave radio communication makes use of waves whose frequencies are higher than those used for commercial a-m broadcast transmission and reception. Thus stations operating on frequencies above 1,600 kc are often called *short-wave stations.* Their carrier waves are generally expressed in either megacycles or meters; for example, a station operating at 30,000 kc may be referred to as operating on either the 30-mc band or on the 10-meter band. For frequencies above 1,600 kc, various bands are allocated to the different types of communication service, such as amateur, aircraft, ship to shore, police, point-to-point, and long-distance commercial telephony.

Characteristics of Short-wave Communication. In general, the operating characteristics of short-wave and broadcast communication are similar. However, there are features of short-wave communication that are not common to those used in broadcast. These characteristics may be listed as follows: (1) Less power is required to transmit high-frequency signals over great distances than is needed for low-frequency signals. (2) The distances reached by short-wave signals during the daylight hours will become greater as the frequency of transmission is increased. (3) Nighttime reception may be poor, especially for signals of frequencies greater than 13 mc. (4) The reception in terms of signal strength is very irregular, that is, it may be strong one day and weak the next or it may even vary during the same day. (5) Short-wave signals may be received perfectly at distant points from the transmitter and not be received at all at certain localities that may be near the transmitter, this phenomenon being known as *skipping.*

Short-wave Bands. If a portion of the short-wave range extending from 1,600 to 60,000 kc is considered, it will be seen that this range covers a span of 58,400 kc. If this span were to provide channels of 10-kc width, there would be 5,841 channels for this span alone, as compared to 106 channels for the broadcast range. This is one of the reasons for using the short-wave channels for the transmission of frequency modulation and television signals.

For broadcast reception, it is possible to use a single coil and capacitor for each stage of tuning. Because of the width of the frequency range to be covered and the large number of stations that can be accommodated, a system employing a single coil and capacitor is not practical for short-wave reception. If a short-wave receiver were to cover a range of from 1.6 to 60 mc with a single range and had a dial marked with 100 divisions to cover 180 degrees of rotation of the variable capacitor, it would mean that for an slf capacitor each division would represent 584 kc. If such a system were used to tune stations operating on 10-kc channels, 58 channels may be tuned in or out by moving the dial just one division. The

TABLE 4-1

Band	Set A	Set B
Broadcast	530–1,550 kc	550–1,600 kc
1	1.5–4.2 mc	1.5–5.5 mc
2	4.0–11.5 mc	5.4–15.5 mc
3	11–23 mc	15–42 mc

minimum capacitance of the capacitor, plus the distributed capacitance of the coil and the stray capacitances of the wiring, makes it impossible to design a single combination of coil and capacitor that will tune this entire range satisfactorily. In order to correct this condition, the short-wave frequency span is divided into a number of bands. The frequency limits of each of these bands will vary with the type of receiver and the manufacturer, as is illustrated in Table 4-1. A separate coil designed to cover the frequency range for each band desired is generally used with a common variable capacitor.

FIG. 4-6. Typical plug-in coil.

4-4. All-wave Receivers. *Variable Inductance.* In radio receivers made to tune the broadcast band and one or more short-wave bands, the capacitor and inductor used in the tuning circuit are constructed so that either one or both can be adjusted to the value required for the band desired. One method of varying the inductance is by use of a series of plug-in coils, each coil being designed for a definite frequency range. The value of inductance required for these coils is determined by the maximum capacitance of the tuning capacitor and the lowest frequency of the desired band, as illustrated in the following examples.

Example 4-2. A 320-μμf variable capacitor having a minimum capacitance of 13.5 μμf is used in the tuning circuit of the all-wave receiver listed as set A in Table 4-1. (*a*) If the distributed capacitance of the coil is neglected, find the inductance of the secondary for each of the coils required. (*b*) What is the highest resonant frequency obtainable for each band?

Given: $C_{max} = 320 \ \mu\mu f$ Find: (a) L_S for each coil
              $C_{min} = 13.5 \ \mu\mu f$ (b) Maximum f_r for each band

Solution:

(a) $L_S = \dfrac{25,300}{f_{min}{}^2 C_{max}}$

$L_{S \cdot B} = \dfrac{25,300}{530^2 \times 320 \times 10^{-6}} = 281 \ \mu h$

$L_{S \cdot 1} = \dfrac{25,300}{1.5^2 \times 10^6 \times 320 \times 10^{-6}} = 35.1 \ \mu h$

$L_{S \cdot 2} = \dfrac{25,300}{4.0^2 \times 10^6 \times 320 \times 10^{-6}} = 4.93 \ \mu h$

$L_{S \cdot 3} = \dfrac{25,300}{11^2 \times 10^6 \times 320 \times 10^{-6}} = 0.653 \ \mu h$

(b) $f_{max} = \dfrac{159}{\sqrt{LC_{min}}}$

$f_{B \cdot max} = \dfrac{159}{\sqrt{281 \times 13.5 \times 10^{-6}}} = 2,581 \ kc$

$f_{1 \cdot max} = \dfrac{159}{\sqrt{35.1 \times 13.5 \times 10^{-6}}} = 7.30 \ mc$

$f_{2 \cdot max} = \dfrac{159}{\sqrt{4.93 \times 13.5 \times 10^{-6}}} = 19.5 \ mc$

$f_{3 \cdot max} = \dfrac{159}{\sqrt{0.653 \times 13.5 \times 10^{-6}}} = 53.5 \ mc$

In Example 4-2 the maximum frequency for each band is much greater than that listed for set A in Table 4-1, because the distributed capacitances of the coil and the circuit have been neglected. How much the stray capacitance affects the frequency range of the tuned circuit may be seen from the results of the following example.

Example 4-3. If the tuning circuit used in Example 4-2 has a distributed circuit capacitance of 20 $\mu\mu f$, what will the minimum and the maximum frequencies be for each band of the receiver?

Given: $C_{max} = 320 \ \mu\mu f$ Find: f_{min} and f_{max} for each band
              $C_{min} = 13.5 \ \mu\mu f$
              $C_D = 20 \ \mu\mu f$

Solution:

$$f_{min} = \frac{159}{\sqrt{L(C_{max} + C_D)}}$$

$$f_{max} = \frac{159}{\sqrt{L(C_{min} + C_D)}}$$

Broadcast band:

$$f_{min} = \frac{159}{\sqrt{281 \times (320 + 20) \times 10^{-6}}} = 514 \ kc$$

$$f_{max} = \frac{159}{\sqrt{281 \times (13.5 + 20) \times 10^{-6}}} = 1,639 \ kc$$

Band 1:

$$f_{min} = \frac{159}{\sqrt{35.1 \times (320 + 20) \times 10^{-6}}} = 1.45 \ mc$$

$$f_{max} = \frac{159}{\sqrt{35.1 \times (13.5 + 20) \times 10^{-6}}} = 4.63 \ mc$$

Band 2:

$$f_{min} = \frac{159}{\sqrt{4.93 \times (320 + 20) \times 10^{-6}}} = 3.88 \text{ mc}$$

$$f_{max} = \frac{159}{\sqrt{4.93 \times (13.5 + 20) \times 10^{-6}}} = 12.4 \text{ mc}$$

Band 3:

$$f_{min} = \frac{159}{\sqrt{0.653 \times (320 + 20) \times 10^{-6}}} = 10.7 \text{ mc}$$

$$f_{max} = \frac{159}{\sqrt{0.653 \times (13.5 + 20) \times 10^{-6}}} = 34 \text{ mc}$$

In Example 4-3, the frequency limits of each band are approximately the same as those listed for set A in Table 4-1. The values of capacitance and inductance used in these problems do not necessarily represent actual values but have been selected to illustrate their application in obtaining the various frequency bands. The maximum value of the tuning capacitor is too high for the receiver to be able to tune high frequencies if a coil with a reasonable number of turns is to be used. Tuning capacitors having a lower value of maximum capacitance are made with fewer plates and therefore usually also have a lower value of minimum capacitance. For this reason a smaller capacitor, generally about 140 $\mu\mu$f or less, is used in the tuning circuits of short-wave receivers.

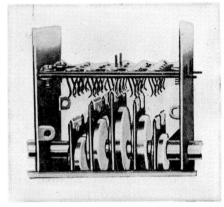

FIG. 4-7. Commercial band switch. (*Courtesy of Hammarlund Manufacturing Company, Inc.*)

Band Switching with Inductors. Changing from one band of frequencies to another may be accomplished by interchanging plug-in type coils. A simpler process for changing from one frequency band to another is to have the primary and secondary windings for each band wound on a single coil form (Fig. 4-5). By means of a band-selector switch (Fig. 4-7), any set of primary and secondary windings may be connected into or out of the tuning circuit. These switches are designed so that the amount of coupling introduced between the tuned circuits of the different stages is negligible. The switch contacts are usually arranged so that the unused coils are short-circuited. If the coils not actually in use are not short-circuited, the distributed capacitance of these coils may cause them to become resonant at some frequency, each within its own band, thereby coupling impedances into the coils that are being used. The switch should provide: (1) a low-resistance contact, and (2) some means whereby the connections between the coils and the switch are made as short as possible.

Band Switching with Capacitors.   If a single coil is to cover the entire frequency range, the capacitance of the tuned circuit for each band desired can be adjusted by connecting a fixed capacitor in series with the

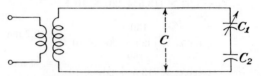

Fig. 4-8. Increasing the frequency range of a tuned circuit by use of a fixed capacitor.

tuning capacitor (Fig. 4-8).   The fixed capacitance reduces the capacitance of the circuit, thereby increasing its resonant frequency.   The inductance required may be found by

$$L = \frac{25,300}{f_1^2 C_1} \tag{4-1}$$

$$L = \frac{25,300}{f_2^2 C} \tag{4-2}$$

where L = inductance of tuning circuit, μh

f_1 = frequency to which circuit will tune without fixed series capacitor, kc

f_2 = frequency to which circuit will tune with fixed series capacitor, kc

C_1 = capacitance of tuning capacitor, μf

C = capacitance of series circuit formed by tuning capacitor and fixed capacitor, μf

As the inductance is the same in Eqs. (4-1) and (4-2), then

$$\frac{25,300}{f_1^2 C_1} = \frac{25,300}{f_2^2 C} \tag{4-3}$$

dividing both sides by 25,300

$$\frac{1}{f_1^2 C_1} = \frac{1}{f_2^2 C} \tag{4-4}$$

and

$$C = \frac{f_1^2 C_1}{f_2^2} \tag{4-5}$$

Equation (4-5) is very useful in solving for the amount of capacitance necessary for any desired frequency range when only one value of inductance is to be used.   The value of capacitance C_2 that must be connected in series with the original tuning capacitor C_1 in order to obtain the required capacitance C may be determined by

$$C_2 = \frac{C C_1}{C_1 - C} \tag{4-6}$$

Example 4-4.　A certain tuned circuit using a 200-$\mu\mu$f tuning capacitor tunes from 500 to 1,500 kc.　What value of series capacitance C_2 (see Fig. 4-8) is required in order to provide a second band whose minimum frequency will be 1,500 kc?

Given:　$f_1 = 500$ kc　　　　　　　Find:　C_2
　　　　　$f_2 = 1,500$ kc
　　　　　$C_1 = 200\ \mu\mu$f

Solution:

$$C = \frac{f_1{}^2 C_1}{f_2{}^2} = \frac{500 \times 500 \times 200}{1,500 \times 1,500} = 22.22\ \mu\mu\text{f}$$

$$C_2 = \frac{CC_1}{C_1 - C} = \frac{22.22 \times 200}{200 - 22.22} = 25\ \mu\mu\text{f}$$

Example 4-4 shows that at 1,500 kc the capacitance of this tuned circuit must be 22.22 $\mu\mu$f.　In order for the circuit to be resonant at 1,500 kc without the series capacitor, the minimum capacitance of the tuning capacitor must be 22.22 $\mu\mu$f.　The maximum resonant frequency obtainable when the series capacitor is used will depend on the value of the inductance, which remains constant, and the capacitance of the series circuit formed by the minimum capacitance of the tuning capacitor C_1 and the fixed capacitor C_2 (Fig. 4-8).

Example 4-5.　The tuning circuit of Example 4-4 is capable of tuning from 500 to 1,500 kc with its tuning capacitor alone, the capacitor having a maximum capacitance of 200 $\mu\mu$f and a minimum capacitance of 22.22 $\mu\mu$f.　The frequency range of the circuit may be extended by connecting the capacitor C_2 of Fig. 4-8 into the circuit as indicated in Example 4-4.　(*a*) What value of inductance is required for this circuit? (*b*) What is the minimum resonant frequency of the circuit when a 25-$\mu\mu$f capacitor is connected in series with the tuning capacitor (see Fig. 4-8)?　(*c*) What is the maximum resonant frequency of the circuit when a 25-$\mu\mu$f capacitor is connected in series with the tuning capacitor?　(*d*) What is the frequency range of the tuning circuit for both bands?

Given:　　$C_{1\text{-max}} = 200\ \mu\mu$f　　　　Find:　(*a*) L
　　　　　　$C_{1\text{-min}} = 22.22\ \mu\mu$f　　　　　　(*b*) $f_{r\text{-min}}$ band 2
　　　$f_r\text{-band 1} = 500$ to 1,500 kc　　　　(*c*) $f_{r\text{-max}}$ band 2
　　　　　　$C_2 = 25\ \mu\mu$f　　　　　　　　　(*d*) Frequency range of both bands

Solution:

(*a*)　$L = \dfrac{25,300}{f_{min}{}^2 C_{1\text{-max}}} = \dfrac{25,300}{500 \times 500 \times 200 \times 10^{-6}} = 506\ \mu\text{h}$

(*b*)　$C_{max} = \dfrac{C_{1\text{-max}} C_2}{C_{1\text{-max}} + C_2} = \dfrac{200 \times 25}{200 + 25} = 22.22\ \mu\mu\text{f}$

　　　$f_{min} = \dfrac{159}{\sqrt{LC_{max}}} = \dfrac{159}{\sqrt{506 \times 22.22 \times 10^{-6}}} = 1,500\ \text{kc}$

(*c*)　$C_{min} = \dfrac{C_{1\text{-min}} C_2}{C_{1\text{-min}} + C_2} = \dfrac{22.22 \times 25}{22.22 + 25} = 11.76\ \mu\mu\text{f}$

　　　$f_{max} = \dfrac{159}{\sqrt{LC_{min}}} = \dfrac{159}{\sqrt{506 \times 11.76 \times 10^{-6}}} = 2,061\ \text{kc}$

(*d*)　Frequency range = 500 to 2,061 kc

Effect of Distributed Capacitances.　In the preceding examples and discussion the distributed capacitance of the coil and the tuning circuit was neglected in order to simplify the explanations and the examples. In actual practice these capacitances must be taken into consideration because they increase the minimum and maximum capacitance of the tuning circuit and hence change its frequency range.

Example 4-6.   What is the frequency range of each band of the tuning circuit of Example 4-5 if the distributed capacitance of the coil and circuit is to be considered and its value is 20 $\mu\mu$f?   (NOTE: Use the value of inductance found in part (*a*) of Example 4-5.)

Given: $C_{1 \cdot \max} = 200 \ \mu\mu$f   Find: Frequency range, band 1
 $C_{1 \cdot \min} = 22.22 \ \mu\mu$f Frequency range, band 2
 $C_2 = 25 \ \mu\mu$f
 $C_D = 20 \ \mu\mu$f
 $L = 506 \ \mu$h

Solution:

Band 1:

$$C_{\max} = C_{1 \cdot \max} + C_D = 200 + 20 = 220 \ \mu\mu\text{f}$$

$$f_{\min} = \frac{159}{\sqrt{LC_{\max}}} = \frac{159}{\sqrt{506 \times 220 \times 10^{-6}}} = 476 \text{ kc}$$

$$C_{\min} = C_{1 \cdot \min} + C_D = 22.22 + 20 = 42.22 \ \mu\mu\text{f}$$

$$f_{\max} = \frac{159}{\sqrt{LC_{\min}}} = \frac{159}{\sqrt{506 \times 42.22 \times 10^{-6}}} = 1{,}088 \text{ kc}$$

Band 2:

$$C_{\max} = C_{\max \cdot 2} + C_D = 22.22 + 20 = 42.22 \ \mu\mu\text{f}$$

$$f_{\min} = \frac{159}{\sqrt{LC_{\max}}} = \frac{159}{\sqrt{506 \times 42.22 \times 10^{-6}}} = 1{,}088 \text{ kc}$$

$$C_{\min} = C_{\min \cdot 2} + C_D = 11.76 + 20 = 31.76 \ \mu\mu\text{f}$$

$$f_{\max} = \frac{159}{\sqrt{LC_{\min}}} = \frac{159}{\sqrt{506 \times 31.76 \times 10^{-6}}} = 1{,}254 \text{ kc}$$

Example 4-6 shows that a small amount of distributed capacitance in the tuned circuit will decrease its frequency range considerably.   It is

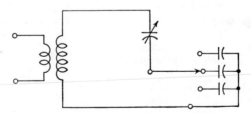

Fig. 4-9. Use of a rotary switch and several fixed capacitors to obtain several frequency bands.

therefore important that the stray capacitances in a tuned circuit be kept at a minimum.

By using several fixed capacitors of different values and a rotary switch, a number of different frequency bands can be obtained.   A circuit using this system is shown in Fig. 4-9.

Band Switching Using Both Inductors and Capacitors. Although the frequency range of a tuned circuit may be increased by changing the amount of either its capacitance or its inductance, a number of receivers change both. This method is much more expensive, since a greater number of parts is required and the wiring of the selector switch becomes quite complex. With this type of receiver it is possible to obtain slf reception on all bands.

4-5. Bandspread. *Frequency Range of the Individual Bands.* The use of separate coils or fixed capacitors provides a means of covering the broadcast band and a number of short-wave bands with one receiver. To eliminate the possibility of a gap between two adjacent bands, the coils are usually designed to overlap the extreme frequencies of the adjacent bands. Generally, the maximum resonant frequency of each band is chosen as some multiple of its minimum frequency. Broadcast receivers use a ratio of 3 to 1, while for high-frequency bands ratios of less than 1.5 to 1 are sometimes used.

To facilitate tuning, the tuning range for each band should occupy practically the entire scale of the dial. Because of the varying widths of these bands, special tuning circuits are used to obtain the correct maximum-minimum capacitance ratio for each band. This process, called *bandspreading,* is accomplished by connecting a small adjustable capacitor in series, parallel, or combination with the main tuning capacitor (Fig. 4-10).

Trimmers. The operation of these auxiliary capacitors is based on the principle that the capacitance of a circuit is increased by connecting capacitors in parallel and decreased by connecting them in series.

Example 4-7. A 20-$\mu\mu$f auxiliary capacitor, C_2 of Fig. 4-10a, is connected in parallel with a tuning capacitor C_1 having a range of 10 to 100 $\mu\mu$f. What is the range of capacitance of the combined circuit?

Given: $C_2 = 20 \ \mu\mu$f   Find: C_{min} to C_{max}
$C_{1 \cdot min} = 10 \ \mu\mu$f
$C_{1 \cdot max} = 100 \ \mu\mu$f

Solution:

$C_{min} = C_{1 \cdot min} + C_2 = 10 + 20 = 30 \ \mu\mu$f
$C_{max} = C_{1 \cdot max} + C_2 = 100 + 20 = 120 \ \mu\mu$f
Range = 30 to 120 $\mu\mu$f

Example 4-8. What is the range of the combined circuit of Example 4-7 if a 100-$\mu\mu$f auxiliary capacitor is used in place of the 20-$\mu\mu$f capacitor?

Given: $C_2 = 100 \ \mu\mu$f   Find: C_{min} to C_{max}
$C_{1 \cdot min} = 10 \ \mu\mu$f
$C_{1 \cdot max} = 100 \ \mu\mu$f

Solution:

$C_{min} = C_{1 \cdot min} + C_2 = 10 + 100 = 110 \ \mu\mu$f
$C_{max} = C_{1 \cdot max} + C_2 = 100 + 100 = 200 \ \mu\mu$f
Range = 110 to 200 $\mu\mu$f

In Example 4-7, adding a 20-$\mu\mu$f auxiliary capacitor in parallel with the 10- to 100-$\mu\mu$f tuning capacitor increases its range to 30 to 120 $\mu\mu$f. Example 4-8 shows that adding a 100-$\mu\mu$f auxiliary capacitor in parallel with the tuning capacitor increases the range to 110 to 200 $\mu\mu$f.   In each case the greatest per cent of increase occurs at the minimum value of capacitance.   If the auxiliary capacitor is made adjustable between the values of 20 and 100 $\mu\mu$f, then a large number of minimum and maximum values of capacitance can be obtained to produce a corresponding bandspread.   Such an adjustable capacitor is called a *trimmer*.   Trimmers have the same effect on a tuned circuit as does distributed capacitance.

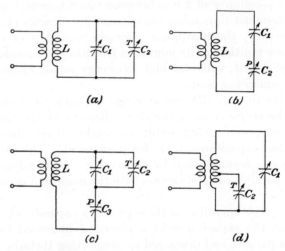

Fig. 4-10. Trimmer and padder capacitors connected to produce bandspread and band compression.

The effectiveness of the trimmer capacitor is shown by Examples 4-2 and 4-3, where the effect of including a distributed capacitance of 20 $\mu\mu$f showed a change in the broadcast frequency limits from 530 to 2,581 kc to 514 to 1,639 kc.

Padders.   In order that a desired frequency range may be obtained, it is sometimes desirable to restrict the maximum capacitance of the tuning circuit without greatly changing its minimum value.   To accomplish this, an auxiliary capacitor, C_2 of Fig. 4-10b, is connected in series with the secondary coil L and the tuning capacitor C_1.   This capacitor, called a *padder*, is also made adjustable so that the minimum capacitance of the tuning circuit can be kept fairly constant.

Example 4-9.   A 20- to 200-$\mu\mu$f adjustable capacitor (C_2 of Fig. 4-10b) is connected in series with the tuning capacitor C_1, having a range of 10 to 100 $\mu\mu$f.   What is the range of capacitance of the combined circuit if the adjustable capacitor is set at 20 $\mu\mu$f?

Given: $C_2 = 20$ $\mu\mu$f   Find: C_{min} to C_{max}
$C_{1 \cdot min} = 10$ $\mu\mu$f
$C_{1 \cdot max} = 100$ $\mu\mu$f

Solution:

$$C_{min} = \frac{C_{1 \cdot min} C_2}{C_{1 \cdot min} + C_2} = \frac{10 \times 20}{10 + 20} = 6.66 \ \mu\mu f$$

$$C_{max} = \frac{C_{1 \cdot max} C_2}{C_{1 \cdot max} + C_2} = \frac{100 \times 20}{100 + 20} = 16.66 \ \mu\mu f$$

Range = 6.66 to 16.66 $\mu\mu f$

Example 4-10. What is the range of the combined circuit of Example 4-9 if the padder capacitor is set at 200 $\mu\mu f$?

Given: $C_2 = 200 \ \mu\mu f$ Find: C_{min} to C_{max}

                  $C_{1 \cdot min} = 10 \ \mu\mu f$

                  $C_{1 \cdot max} = 100 \ \mu\mu f$

Solution:

$$C_{min} = \frac{C_{1 \cdot min} C_2}{C_{1 \cdot min} + C_2} = \frac{10 \times 200}{10 + 200} = 9.52 \ \mu\mu f$$

$$C_{max} = \frac{C_{1 \cdot max} C_2}{C_{1 \cdot max} + C_2} = \frac{100 \times 200}{100 + 200} = 66.66 \ \mu\mu f$$

Range = 9.52 to 66.66 $\mu\mu f$

Example 4-9 shows that adding a 20-$\mu\mu f$ capacitance in series with the 10- to 100-$\mu\mu f$ tuning capacitor decreases its range to 6.66 to 16.66 $\mu\mu f$. Example 4-10 shows that adding a 200-$\mu\mu f$ capacitance in series with the tuning capacitor decreases the range to 9.52 to 66.66 $\mu\mu f$. In each case the greatest per cent of decrease occurs at the maximum value of capacitance; hence the padding capacitor has comparatively little effect on the minimum capacitance of the tuning circuit but definitely controls its maximum capacitance.

Radio receivers may use both trimmers and padders (Fig. 4-10c) in order to obtain the desired bandspread. The desired frequency limits are obtained by adjusting these two capacitors to the correct values. Sometimes the trimmer is connected across only part of the secondary coil *L*, as shown in Fig. 4-10d. The amount of bandspread increases as the tap is made closer to the bottom of the coil.

Example 4-11. The tuning circuit of the oscillator section of a superheterodyne receiver is to be adjusted so that it will always tune 465 kc higher than the tuning circuit of the receiver, whose frequency limits are to be 530 kc and 1,550 kc. A circuit similar to that shown in Fig. 4-10c is to be used, and the circuit elements are to have the following values: $L = 200 \ \mu h$, $C_{1 \cdot min} = 12.5 \ \mu\mu f$, $C_{1 \cdot max} = 250 \ \mu\mu f$, and $C_2 = 6.5 \ \mu\mu f$. The distributed capacitance of the circuit is 15 $\mu\mu f$. (a) What value of padder capacitor C_3 is required when the tuning circuit is adjusted to its minimum frequency? (b) Using the padder capacitance as calculated in part (a), what is the resonant frequency of the oscillator section when the tuning circuit is adjusted to its maximum frequency?

Given: $f = 530$ kc to 1,550 kc   Find: (a) C_3

                $C_1 = 12.5$ to 250 $\mu\mu f$ (b) $f_{osc \cdot max}$

                $C_2 = 6.5 \ \mu\mu f$

                $C_D = 15 \ \mu\mu f$

                $L = 200 \ \mu h$

Solution:

(a) $f_{osc\cdot min} = f_{min} + 465 = 530 + 465 = 995$ kc

$$C_{f\cdot min} = \frac{25,300}{(f_{osc\cdot min})^2 L} = \frac{25,300}{995^2 \times 200} = 127.7 \ \mu\mu f$$

$$C_3 = \frac{(C_{1\cdot max} + C_2)(C_{f\cdot min} - C_D)}{(C_{1\cdot max} + C_2) - (C_{f\cdot min} - C_D)} = \frac{256.5 \times 112.7}{256.5 - 112.7} = 201 \ \mu\mu f$$

(b) $$C_{min} = \frac{(C_{1\cdot min} + C_2)C_3}{(C_{1\cdot min} + C_2) + C_3} + C_D = \frac{19 \times 201}{19 + 201} + 15 = 32.35 \ \mu\mu f$$

$$f_{osc\cdot max} = \frac{159}{\sqrt{LC_{min}}} = \frac{159}{\sqrt{200 \times 32.35 \times 10^{-6}}} = 1,976 \text{ kc}$$

The maximum frequency of the oscillator as calculated in Example 4-11 is equal to 1,976 kc instead of 2,015 kc (1,550 + 465). By adjusting the trimmer capacitor C_2, the correct value of maximum frequency can be obtained. As any change in the value of trimmer capacitance will have only a slight effect on the maximum value of capacitance of the entire circuit, the minimum frequency will remain practically the same.

FIG. 4-11. Bandspread tuning capacitor. (*Courtesy of Meissner Manufacturing Division, Maguire Industries, Inc.*)

Bandspread Tuning. Dials used for tuning generally have a rotation of 180 mechanical degrees. With such a dial, the amount of change in kc for each degree of rotation will vary with the amount of bandspread. On the broadcast band, 550 to 1,600 kc, each degree would represent a change of 1,050 divided by 180, or approximately 6 kc. As there should be at least a 10-kc difference between adjacent stations, there is no problem in tuning a desired station on the broadcast band.

On the short-wave bands, tuning a desired station increases in difficulty as the frequency of the band is increased. This can be seen by observation of the frequency bands in set A of Table 4-1. For this receiver the approximate kilocycle change for each degree of rotation will be 15 kc for band 1, 41.6 kc for band 2, and 66.6 kc for band 3. Two methods are used to spread the amount of kilocycle change over a greater portion of the dial, one mechanical and the other electrical.

The *mechanical* method employs a geared dial that causes the variable capacitor to move at a slower rate than the dial. The greater the ratio between the two gears, the slower will be the movement of the variable capacitor. However, there is a limit to the amount the gear ratio can be increased without producing a backlash.

The *electrical* method utilizes the principle of the trimmer capacitor to obtain bandspread. A midget or micro variable capacitor, depending on the frequency band, is substituted for C_2 (Fig. 4-10). This capacitor is called the *bandspread capacitor*. The main tuning capacitor is used to adjust the tuning circuit to approximately the desired frequency. The bandspread capacitor is used to obtain the exact frequency, as a considerable movement of this capacitor will change the resonant frequency by

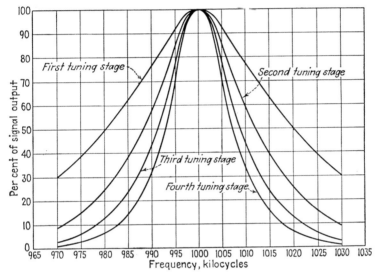

Fig. 4-12. Graph illustrating the manner in which the selectivity of a receiver increases with additional stages of tuning.

only a fraction of a kilocycle. To receive signals from a station transmitting at 16.28 mc, the main tuning capacitor C_1 would be adjusted to approximately 16 mc, and the bandspread capacitor would be used to tune the circuit to 16.28 mc. Bandspread tuning is usually found only in receivers specially designed for amateurs' use.

4-6. Multiple Stages. *Selectivity of Multiple Stages.* A single tuning stage does not always provide sufficient selectivity for satisfactory reception; hence in order to obtain the desired selectivity a number of similar tuning stages are sometimes used. The selectivity of the receiver will then be dependent on the selectivity of each stage and the number of stages used. The extent to which the selectivity is improved by increasing the number of tuned circuits is shown in Fig. 4-12. The frequency of resonance for each tuned circuit is 1,000 kc. The signal output at the resonant frequency is taken as 100 per cent. The signal output at all other frequencies will be a definite percentage of this output. If the characteristics of each tuning circuit are the same, the percentage of the input signal, for all frequencies above and below resonance, can be represented by the response curve for the first stage of tuning. For example, at 5 kc off resonance, that is, 995 or 1,005 kc, the input signal is reduced to 92 per cent, as represented by the response curve for the first stage of tuning. The output for the second stage for this frequency would be equal to 92 per cent of 92 per cent, or 84.6 per cent. The output from the third stage for the same frequency would be 92 per cent of 84.6 per cent, or 77.8 per cent; for the fourth stage the output would be 92 per cent of 77.8 per cent, or 71.6 per cent. This procedure can be followed for any value of frequency.

Number of Tuning Stages to Be Used. Theoretically, it would seem that any number of tuning stages could be added, but practically this is not so, since other factors must be taken into consideration. In order to obtain good fidelity from commercial a-m broadcast stations, the width of the band passed (measured at 0.707 of the maximum value) should not be less than 10 kc. Adding too many stages may decrease the width of the band passed below this value, thus decreasing the fidelity of reception.

In modern receivers, the tuning circuit forms a definite part of each stage of r-f amplification and the complete unit is generally called a *stage of r-f amplification.* A number of methods is used to couple one stage of r-f amplification to another; the choice of method used will depend on the type of r-f amplifier tube and circuit used. The total amplification desired and its relation to the amplification produced by each tube, coupling device, and tuning circuit must also be considered. The circuit used is another factor to be considered. The simple trf (tuned-radio-frequency) receiver generally has two stages of r-f amplification before the detector. Superheterodyne receivers do not depend on the r-f stages to obtain the necessary amplification before detection, but use i-f (intermediate frequency) amplifiers for this purpose. Although the number of r-f tuning circuits in superheterodyne receivers will vary, it is general practice to use one or more tuning circuits in addition to the oscillator section.

4-7. Single-control Tuning. *Ganged Tuning.* Since the coil, capacitor, and wiring for each of the tuning circuits in a trf receiver are usually similar to one another, the dial setting for a given station is approximately the same for each of the controls. It is therefore possible to mechanically couple the rotors of all the variable capacitors with one another so that they can be rotated simultaneously with a single control; this is called *ganged tuning.* In superheterodyne receivers, the variable capacitors are also ganged so that they may be operated by a single control.

Multiple Capacitors. A multiple or ganged capacitor consists of two or more separate capacitors built into a single frame so that all the rotors can be operated from a common shaft (Fig. 4-13).   In order to reduce the capacitance between adjacent stator sections, a flat metal plate is mounted between each section.   This plate is connected to ground and thus acts as an electrostatic shield.

For some positions of the rotor all its weight is concentrated on one side.   If the rotor contains a large number of plates this unequal distribution of weight will tend to move the rotor out of the position to which

(a)

(b)

Fig. 4-13. Commercial types of multiple, or gang capacitors. (a) Two-gang capacitor. (b) Three-gang capacitor. (c) Capacitor with opposed rotor and stator for perfect counterbalancing. [(a) and (b) *Courtesy of Meissner Manufacturing Division, Maguire Industries, Inc.; (c) courtesy of Hammarlund Manufacturing Company, Inc.*]

(c)

it had been adjusted.   To overcome this difficulty the rotor is balanced mechanically by mounting half of the stator and rotor plates 180 mechanical degrees from each other (Fig. 4-13).

4-8. Equalizing the Tuning Circuits.   *Tracking.*   Although the single-control tuning system rotates each tuning capacitor by the same amount, the electrical adjustment, while approximately equal, is not exactly the same for each tuning circuit.   This electrical difference is due to the slight mechanical differences in the capacitors, coils, and wiring of the tuning circuits.   In order to obtain the maximum fidelity, selectivity, and sensitivity for a receiver, all the tuning circuits must track together over the entire range of the receiver.   This means that the resonant frequency for each of the tuning circuits must be exactly the same for all positions of the tuning control.   To obtain perfectly matched coils and capacitors and to have the wiring for each tuning circuit exactly the same requires precision manufacture.   This procedure is too expensive and also is impractical for mass production.   It is therefore necessary to employ some means of adjusting each circuit.

Methods Used to Align the Tuning Circuits.   The inductance of the coils and wiring for all tuning circuits in a receiver is constant for all positions of the tuning control.   Each tuning circuit can be assembled and wired so that the values of inductance, for all practical purposes, are close enough to each other to be considered as being the same.   It is only necessary to make adjustments for differences in capacitance between

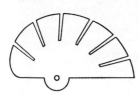

FIG. 4-14.   A slotted rotor plate.

each circuit.   One method used to compensate for differences in tuning circuit capacitance is to connect a small adjustable trimmer capacitor in parallel with the main tuning capacitor (Fig. 4-13).   The trimmer capacitors can be adjusted for only general capacitance differences between each tuning circuit but cannot be used to align each section for all positions of the tuning control.   It is general practice to adjust the capacitor sections for three positions, namely, the approximate center and each end of the frequency band.   Adjustments are made at these three positions and are then rechecked in order that all circuits track as closely as possible for all positions.

A more accurate method is to cut slots in the end rotor plates, as shown in Figs. 4-13 and 4-14.   These plates generally have four or more sections, and the capacitance for each section can be changed by bending the plate at that section.   If the plate is bent toward the stator, the capacitance is increased, and if it is bent away from the stator, the capacitance will be decreased.   The tuning control is adjusted for some frequency within a section of the rotor plate, and that section is adjusted to obtain maximum signal response for that frequency.   As any adjustments made for one value of frequency will not affect other frequency adjustments, maximum signal response can be obtained for as many positions as there are rotor-plate sections.   With this system, the tuning circuits can be made to track for all positions of the tuning control.

Some receivers use a combination of both of these methods.   In these receivers the trimmers are used to compensate for general differences in capacitance and the slotted rotor plates are used to align the capacitor sections for various positions of the tuning control.

4-9. Automatic Tuning.   In the methods of tuning described in the preceding discussion, the selecting of stations on a receiver is accomplished by rotating the tuning dial to the desired position by hand.   This is sometimes called *manual tuning*. Another method of tuning a receiver is by merely pushing a button or a lever. With this system, the receiver may automatically be tuned to the stations predetermined for each tuning button. This is called *automatic tuning*. The number of tuning buttons on a receiver varies with the manufacturer and the model. Automatic tuning is used extensively in radio receivers designed for use in automobiles, and these receivers usually have five or six buttons.

Methods of Obtaining Automatic Tuning Control. Although numerous systems have been devised for obtaining automatic tuning, they may be classified into three general types: (1) mechanically operated manual types, (2) tuned-circuit substitution types, and (3) motor-operated types.

Mechanically Operated Manual Types. In these types of automatic tuning, the shaft of the tuning capacitor is turned to the preset desired position by pressing a button, key, or some type of lever. This type of tuning may be further subdivided into five common methods of operation, namely, linear, rocker bar, rotary, indent, and flash. The *linear* method employs a series of cams and levers to obtain the station selection. Pushing a button produces motion in about the same manner as pushing on a typewriter key and thereby causes the shaft of the tuning capacitor to be rotated to its preset position. The *rocker-bar* method employs a series of pushbuttons, which translate the motion to preset positions of the tuning capacitor. The *rotary* type uses a dial similar in appearance to the dial of the modern telephone. The *indent* method, also known as *spot tuning*, uses a steel ball that is pressed in a groove of a soft metal cylinder at preset positions and thereby provides indents to aid the manual tuning. The *flash* method, also known as *light-indicator tuning*, does not use any button arrangement but is tuned by the ordinary manual tuning. When the tuning dial is in a position corresponding to one of the preset flash-tuned stations, a dial light is caused to light up in back of a transparent marker indicating the station to which the receiver is tuned.

Tuned-circuit Substitution Types. In these types of automatic tuning, a number of precalibrated tuned circuits are connected to a pushbutton type of selector switch. The precalibrated tuned circuits are generally tuned by means of mica trimmer capacitors, permeability tuned coils, or a combination of both. In most cases where the tuned-circuit substitution method of automatic tuning is used, the ordinary manual tuning is also provided. However, in a few instances, the manual type of tuning is omitted, and such receivers can only be tuned to those stations for which specific tuned circuits are provided.

Motor-operated Types. In these types of automatic tuning, the shaft of the tuning capacitor is turned to the position required for a desired station by means of a small electric motor. The tuning of desired stations is obtained by a station-selector switch or a number of pushbuttons and generally requires the use of a selecting commutator or some other device for stopping the motor at the desired point. (See also Art. 15-9.)

Auxiliary Controls Required with Automatic Tuning. Adding automatic tuning to a receiver generally requires including a number of extra auxiliary circuits or controls, the most important of which are (1) transfer circuit or mechanism to change from continuous or manual tuning to automatic tuning, (2) an audio silencing or muting provision that will silence the receiver when the automatic tuning mechanism is changing the receiver from one station to another, (3) a station-selecting commutator mechanism for stopping the motor at the correct position for station reception with the motor-operated types.

QUESTIONS

1. (*a*) What is meant by tuning? (*b*) How is it accomplished?

2. Name and explain the three functions generally performed by tuning circuits.

3. What are sidebands?

4. (*a*) What is meant by an ideal response curve? (*b*) How does the fidelity of transmission affect the width of the ideal curve?

5. What methods are used to approximate the ideal response curve?

6. What is meant by a radio channel?

7. Why does increasing the selectivity of a receiver beyond a certain point decrease its fidelity?

8. (*a*) What factors in radio transmission may produce cross talk in a receiver? (*b*) How can this interference be minimized?

9. How does the frequency vary with the amount of rotation for each of the following types of variable capacitors: (*a*) slc? (*b*) slw? (*c*) slf?

10. Describe the methods used to obtain compactness in variable-capacitor construction.

11. What factors determine the amount of inductance required of the secondary winding of a tuned r-f transformer?

12. What types of radio service use the short-wave bands?

13. What five characteristics of short-wave communication are not found in ordinary broadcast communication?

14. Why is it not practical to use a single coil and capacitor to cover the entire short-wave band?

15. How are the frequency limits for each band determined?

16. (*a*) What is meant by an all-wave receiver? (*b*) What methods may be used to obtain the required tuning range?

17. If a single variable capacitor is used for tuning, what factors determine the amount of inductance required for each coil?

18. (*a*) Is it practical in an all-wave receiver to use a variable capacitor whose maximum value is similar to that used for ordinary broadcast reception? (*b*) Explain.

19. What are the advantages of using a band-selector switch in place of plug-in coils for changing the frequency band to be tuned?

20. What factors must be taken into consideration in designing a band switch?

21. Describe the method of obtaining the frequency range for each of the bands desired by using a single inductance coil and varying the capacitance.

22. What effect does the distributed capacitance of a tuning circuit have on the frequency limits for each band?

23. What is the advantage of changing both the capacitance and inductance for each frequency band?

24. (a) What is meant by bandspreading? (b) Why is it used? (c) How may it be accomplished?

25. (a) How is a trimmer capacitor connected in a circuit? (b) What is the basic principle in the application of trimmer capacitors?

26. If trimming means to cut down or decrease, does the addition of a trimmer capacitor trim: (a) the capacitance of the circuit, or (b) the resonant frequency of the circuit?

27. (a) How is a padder capacitor connected in a circuit? (b) What is the basic principle in the application of padder capacitors?

28. If padding means to build up or increase, does the addition of a padder capacitor pad (a) the capacitance of the circuit, or (b) the resonant frequency of the circuit?

29. When used in conjunction with bandspread coils, what is the purpose of (a) trimmers? (b) padders?

30. (a) What is meant by bandspread tuning? (b) Where is it used? (c) Why is it desirable?

31. Describe the mechanical method of obtaining bandspread tuning.

32. Describe the electrical method of obtaining bandspread tuning.

33. Explain how the selectivity of a receiver is increased by using multiple-stage tuning.

34. What determines the number of stages of tuning to be used?

35. What are the advantages of single-control tuning?

36. (a) What is a multiple variable capacitor? (b) What are its advantages?

37. What is meant by tracking?

38. (a) What are equalizing capacitors? (b) Where are they used? (c) Why are they used?

39. (a) How are slotted rotor plates used to align the tuning circuit? (b) What are their advantages?

40. What is meant by automatic tuning?

41. (a) What are three basic methods of obtaining automatic tuning? (b) Describe each.

PROBLEMS

1. A tuning circuit whose minimum frequency is to be 550 kc uses a 300-$\mu\mu$f variable capacitor whose minimum capacitance is 16 $\mu\mu$f. (a) What is the value of inductance that the secondary winding must possess? (b) What is the highest frequency obtainable with the inductance as calculated in part (a)?

2. How is the frequency range of the tuning circuit in Prob. 1 affected by a distributed circuit capacitance of (a) 10 $\mu\mu$f? (b) 20 $\mu\mu$f?

3. A tuning circuit whose minimum frequency is to be 530 kc uses a 150-$\mu\mu$f variable capacitor whose minimum capacitance is 13 $\mu\mu$f. (a) What is the value of inductance that the secondary winding must possess? (b) What is the highest frequency obtainable with the inductance as calculated in part (a)?

4. How is the frequency range of the tuning circuit in Prob. 3 affected by a distributed circuit capacitance of (a) 12 $\mu\mu$f? (b) 18 $\mu\mu$f?

5. The inductance of the secondary winding of a certain tuning circuit is 41.5 μh. The distributed capacitance of the circuit is 12 $\mu\mu$f. (a) What is the maximum value required of the capacitor if the minimum frequency is to be 1,500 kc? (b) What is the highest frequency obtainable if the minimum capacitance of the variable capacitor is 16 $\mu\mu$f?

6. A variable capacitor and an inductor (secondary winding of a transformer) having an inductance of 14.4 μh are to be used to tune a frequency band whose limits are to be 4 mc and 9 mc. The distributed capacitance of the circuit is 10 $\mu\mu$f. What are the maximum and minimum values required of the variable capacitor?

7. A 300-$\mu\mu$f variable capacitor having a minimum capacitance of 16 $\mu\mu$f is used in the tuning circuit of the all-wave receiver listed as set *A* in Table 4-1. If the distributed capacitance of the tuning circuit is neglected, what value of inductance is required of the secondary winding of each of the four coils in order to tune the lower frequency of their respective bands?

8. Using the values of inductance found in Prob. 7, what is the highest resonant frequency obtainable for each band?

9. If the tuning circuit of Prob. 7 has a distributed capacitance of 15 $\mu\mu$f and the same coils are used, what are the frequency limits of each band if the distributed capacitance is considered?

10. A 200-$\mu\mu$f variable capacitor having a minimum capacitance of 11 $\mu\mu$f is used in the tuning circuit of the all-wave receiver listed as set *B* in Table 4-1. If the distributed capacitance of the tuning circuit is neglected, what value of inductance is required of the secondary winding of each of the four coils in order to tune the lower frequency of their respective bands?

11. Using the values of inductance found in Prob. 10, what is the highest resonant frequency obtainable for each band?

12. If the tuning circuit of Prob. 10 has a distributed capacitance of 12 $\mu\mu$f and the same coils are used, what are the frequency limits of each band if the distributed capacitance is considered?

13. A certain tuned circuit using a 300-$\mu\mu$f variable capacitor tunes from 530 to 1,550 kc. What value of series capacitance is required in order to increase the minimum frequency from 530 to 1,500 kc?

14. What is the frequency range of the tuning circuit of Prob. 13 when the series capacitor as calculated in Prob. 13 is connected in the circuit? (Assume the minimum capacitance of the tuning capacitor to be 35.1 $\mu\mu$f.)

15. What is the frequency range of each band of the tuning circuit of Prob. 13, if the distributed capacitance of the coil and circuit is equal to 15 $\mu\mu$f?

16. A two-band receiver uses a 335-$\mu\mu$f variable capacitor to tune the broadcast band to a minimum frequency of 530 kc. A fixed capacitor is connected in series with the variable capacitor to adjust the tuning circuit for the short-wave band whose minimum frequency is to be 1.5 mc. The variable capacitor has a minimum capacitance of 12 $\mu\mu$f. (*a*) What value of series capacitance is required for the short-wave band? (*b*) What are the frequency limits of each band?

17. What are the frequency limits of each band of the receiver of Prob. 16, if the distributed capacitance of the tuned circuit is equal to 12 $\mu\mu$f and the values of all the circuit elements are the same as those of Prob. 16?

18. A two-band receiver uses a variable tuning capacitor whose maximum capacitance is 500 $\mu\mu$f and whose minimum capacitance is 16 $\mu\mu$f. The distributed capacitance is 10 $\mu\mu$f. The minimum frequencies of the two bands are 550 kc and 1.6 mc respectively. (*a*) What value of inductance is required of the secondary in order to obtain the minimum frequency for the broadcast band? (*b*) What value of series capacitance is required in order to obtain the minimum frequency for the short-wave band using the same inductor as in part (*a*)? (*c*) What is the maximum resonant frequency of each band?

19. A tuning circuit using a variable capacitor that has a maximum capacitance of 250 $\mu\mu$f and a minimum capacitance of 10 $\mu\mu$f has a frequency range of from 500 to 1,800 kc. The distributed capacitance of the circuit is 10 $\mu\mu$f. (*a*) What is the value of the trimmer capacitor required to decrease the higher limit of the frequency range to 1,600 kc? (*b*) What effect does the use of this trimmer capacitor have on the lower limit of the frequency band?

20. The upper and lower limits of a frequency band are 11.8 and 3.88 mc respectively. The maximum and minimum capacitances of the tuning capacitor are 140 $\mu\mu$f and 6 $\mu\mu$f. The distributed capacitance of the circuit is 10 $\mu\mu$f. (*a*) What is the value of the trimmer capacitor required to decrease the higher limit of the frequency range to 10 mc? (*b*) What effect does the use of this trimmer capacitor have on the lower limit of the frequency band?

21. The tuning circuit of the oscillator section of a superheterodyne receiver is to be adjusted so that it will always tune 455 kc higher than the tuning circuit of the receiver whose frequency limits are 550 and 1,630 kc. A circuit similar to that shown in Fig. 4-10c is used, and the circuit elements have the following values: $L = 100$ μh, $C_{1\text{-max}} = 320$ $\mu\mu$f, $C_{1\text{-min}} = 20$ $\mu\mu$f, and $C_2 = 20$ $\mu\mu$f. The distributed capacitance of the circuit is 20 $\mu\mu$f. (*a*) What value is required of the padder capacitor when the tuning circuit is adjusted to its minimum frequency? (*b*) Using the padder capacitor as calculated in part (*a*), what is the resonant frequency of the oscillator section when the tuning circuit is adjusted to its maximum frequency?

22. Neglecting the distributed capacitance of the circuit of Prob. 21, (*a*) what value of padder capacitance would appear to be required when the tuning circuit is adjusted to its minimum frequency? (*b*) Using the padder capacitance as calculated in part (*a*), what would the resonant frequency of the oscillator section appear to be when the tuning circuit is adjusted to its maximum frequency?

23. The tuning circuit of the oscillator section of a superheterodyne receiver is to be adjusted so that it will always tune 265 kc higher than the tuning circuit of the receiver, whose frequency limits are 550 and 1,625 kc. A circuit similar to that shown in Fig. 4-10c is used and the circuit elements have the following values: $L = 135$ μh, $C_{1\text{-max}} = 320$ $\mu\mu$f, $C_{1\text{-min}} = 13.5$ $\mu\mu$f, and $C_2 = 20$ $\mu\mu$f. The distributed capacitance of the circuit is 20 $\mu\mu$f. (*a*) What value is required of the padder capacitor when the tuning circuit is adjusted to its minimum frequency? (*b*) Using the padder capacitor as calculated in part (*a*), what is the resonant frequency of the oscillator section when the tuning circuit is adjusted to its maximum frequency?

24. The tuning circuit of the oscillator section of a superheterodyne receiver is to be adjusted so that it will always tune 455 kc higher than the tuning circuit of the receiver, whose frequency limits are 550 and 1,650 kc. A circuit similar to that shown in Fig. 4-10c is used, and the circuit elements have the following values: $L = 130$ μh, $C_{1\text{-max}} = 260$ $\mu\mu$f, $C_{1\text{-min}} = 10$ $\mu\mu$f, and $C_2 = 20$ $\mu\mu$f. The distributed capacitance of the circuit is 15 $\mu\mu$f. (*a*) What value is required of the padder capacitor when the tuning circuit is adjusted to its minimum frequency? (*b*) Using the padder capacitor as calculated in part (*a*), what is the resonant frequency of the oscillator section when the tuning circuit is adjusted to its maximum frequency?

CHAPTER 5

RADIO-FREQUENCY AMPLIFIER CIRCUITS

Amplification is the process of increasing the amplitude of a signal. An amplifier is a device that increases the voltage, current, and/or power of an input signal with the aid of vacuum tubes or transistors by furnishing the additional power from a separate power source. The signal input, which is used to control the output power of the amplifier, may come from the antenna, a preceding stage of amplification, a detector circuit, a microphone, a phonograph pickup, a tape-recorder pickup, a photocell, or a transmission line. An *r-f amplifier* is one designed to increase the amplitude of signals at radio frequencies. In this chapter, vacuum-tube r-f amplifiers are described in detail. However, many of the principles and characteristics of these amplifier circuits will also apply to the transistor amplifier circuits described in later chapters.

5-1. The Vacuum Tube as an Amplifier. *Voltage and Power Amplifiers.* The amplifying action of a vacuum tube may be utilized in radio circuits in a number of ways, the method used depending upon the results desired. Fundamentally there are two types of vacuum-tube amplifier circuits, namely, the voltage amplifier and the power amplifier. Likewise, there are two types of amplifier tubes, voltage amplifiers and power amplifiers. The primary objective of a voltage amplifier is to increase the voltage of the input signal without regard to the output power. The primary objective of a power amplifier is to increase the energy of the input signal without regard to the output voltage.

R-F Voltage and Power Amplifiers. The sensitivity of a radio receiver can be increased by amplifying the incoming signal before applying it to the detector circuit. The primary objective of these amplifiers is to increase the voltage in the r-f circuits; thus they are called *r-f voltage amplifiers.* R-f amplifiers are also used in transmitters to increase the amplitude of the high-frequency modulated signal before it is applied to the antenna. The primary objective of these amplifiers is to increase the energy of the modulated signal, and they are called *r-f power amplifiers.*

R-f amplifiers are used in superheterodyne receivers to amplify the voltage of a relatively narrow band of intermediate values of frequencies and are called *i-f amplifiers.* These amplifiers are very efficient, because they are designed to operate on a definite narrow frequency band. The frequency at which an i-f amplifier operates varies with the application,

128

for example, (1) 455 kc for an a-m broadcast receiver, (2) 10.7 mc for an f-m broadcast receiver, (3) approximately 45 mc for a television receiver. The width of the band of frequencies that an i-f amplifier must amplify and pass also will vary with the application, for example, (1) a 10-kc band for standard a-m broadcast signals, (2) a 20-kc band for high-fidelity a-m broadcast signals, (3) a 200-kc band for f-m broadcast signals, (4) a 6-mc band (41 to 47 mc) for television broadcast signals. An i-f amplifier for a standard a-m broadcast receiver designed to operate at 455 kc should amplify all signals between 455 + 5, or 460 kc, and 455 − 5, or 450 kc. The signals of all frequencies outside this 10-kc band should be reduced to a negligible value.

5-2. Classification of Amplifiers. *Class A Amplification.* Amplifiers have been described as being either voltage or power amplifiers. They may also be described in terms of their frequency range of operation, as: (1) direct-current (d-c) amplifiers, (2) audio-frequency (a-f) amplifiers, (3) intermediate-frequency (i-f) amplifiers, (4) radio-frequency (r-f) amplifiers, and (5) video-frequency (v-f) amplifiers.

Another common classification of amplifiers is based upon their operating characteristics. Thus they are also known as Class A, Class B, Class AB, or Class C amplifiers (Fig. 5-1). A *Class A amplifier* is one in which the grid bias and the alternating input signal voltages are of such values that plate current flows in the output circuit at all times. A *Class B amplifier* is one in which the grid bias is made approximately equal to the cutoff value. In this case the plate current will be approximately zero for zero-signal input, and plate current will flow for approximately one-half of the input cycle. A *Class AB amplifier* is one in which the grid bias and the signal input voltages are of such values that plate current will flow for appreciably more than half but less than the complete time of the input cycle. A *Class C amplifier* is one in which the grid bias is considerably greater than cutoff value so that the plate current is zero for zero-signal input, and plate current will flow for appreciably less than one-half the time of the input cycle.

A further designation is made by adding the subscript 1 or 2, in which 1 indicates that grid current does not flow during *any* part of the input cycle, while 2 indicates that grid current does flow during *some* part of the input cycle. Thus an AB_1 amplifier does not draw grid current at any time while an AB_2 amplifier does take grid current during some part of the input cycle.

Most r-f, i-f, and a-f amplifiers used in radio receivers are voltage amplifiers operated as Class A; hence only this type will be considered in this chapter. The remaining classes of amplifier, generally used as power amplifiers, are described in a later chapter.

Class A Amplifiers. A vacuum-tube amplifier may be operated as Class A in order to reproduce the variations in the input signal, the reproduced signals appearing across the impedance in the plate or output circuit.   In the elementary voltage-amplifier circuit (Fig. 5-3) the variations in voltage across the output impedance Z_o will be a reproduction of the variations in the input signal voltage e_g impressed across the grid but of increased amplitude.

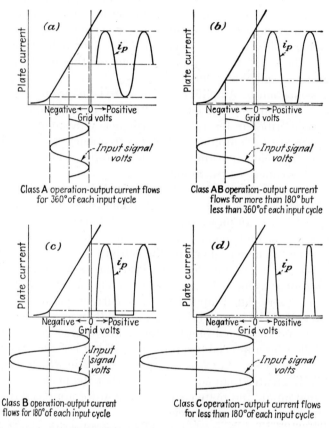

Fig. 5-1. Curves showing the relation between the current in the output circuit and the input signal for various classifications of amplifier operation.

Operating Grid Bias for a Class A Amplifier.   The grid bias required for Class A operation will depend upon the operating characteristics of the tube used and the voltage of the input signal.   Observation of the operating characteristics of a Class A amplifier (Fig. 5-2) indicates that the value of grid bias used should cause the tube to operate on only the straight portion AB of the grid-plate transfer characteristic curve.   The current in the plate circuit will then be an exact enlarged reproduction of the input signal variations that were applied to the grid.   Because the r-f component of the input signal remains practically constant, the impedance of the output circuit will also remain constant, and the variations in voltage developed across the plate load will therefore vary in the same manner as the variations in the plate current.

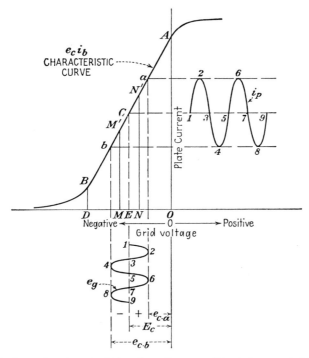

FIG. 5-2. Operating characteristics of a voltage-amplifier tube operated Class A.

Since driving the grid positive will cause distortion, the maximum permissible input signal voltage will depend on the extreme limits of negative grid voltage that will cause the tube to operate on the straight portion of its characteristic curve. In order that the maximum input signal voltage may be applied to a tube, the value of grid bias at which it is operated should be at the mid-point between the limits of the grid bias.

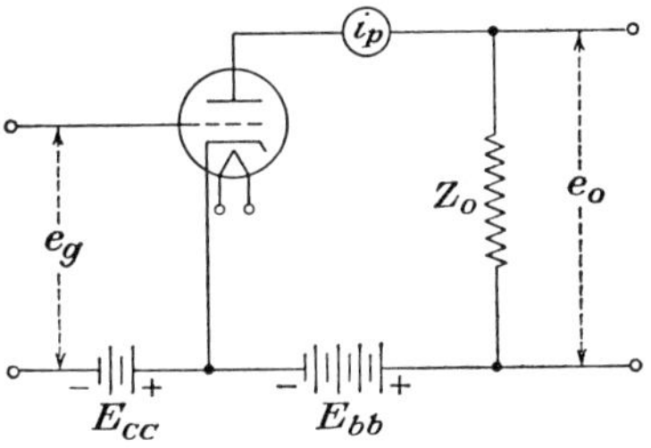

FIG. 5-3. Elementary voltage-amplifier circuit.

The limits of grid voltage for the tube whose operating characteristics are shown in Fig. 5-2 are represented by the points D and O. For maximum signal input the tube should be operated at point E. Increasing the grid bias to M or decreasing it to N will shift the operating point to M' or N'. In either case the limits of grid-voltage swing will be reduced, thus decreasing the amount of input signal voltage that can be applied to the grid of the tube without causing distortion.

The maximum input signal voltage that should be applied to a tube will be equal to either the amount of grid bias used, distance OE on Fig. 5-2, or the difference between the maximum amount of negative grid voltage that will cause the tube to operate on the straight portion of its curve and the grid-bias voltage at which the tube is operated, distance DE. The smaller of these two values will be the maximum input signal voltage that can be applied without causing distortion.

Example 5-1. A triode whose characteristic curves are shown in Fig. 2-15 is being operated as a Class A amplifier with 240 volts applied to its plate. What is the maximum input signal voltage that can be applied without producing distortion when the tube is being operated with a grid bias of (*a*) 8 volts? (*b*) 6 volts? (*c*) 4 volts?

NOTE: Assume that the straight portion of the curve extends beyond a grid bias of zero volts. This assumption will hold true for the characteristic curves of most Class A triode amplifier tubes.

<table>
<tr><td>Given:</td><td>E_b = 240 volts</td><td>Find:</td><td>(*a*) $E_{g \cdot m}$</td></tr>
<tr><td></td><td>(*a*) E_c = −8 volts</td><td></td><td>(*b*) $E_{g \cdot m}$</td></tr>
<tr><td></td><td>(*b*) E_c = −6 volts</td><td></td><td>(*c*) $E_{g \cdot m}$</td></tr>
<tr><td></td><td>(*c*) E_c = −4 volts</td><td></td><td></td></tr>
</table>

Solution:

The curve for E_b = 240 volts (Fig. 2-15) shows that the straight portion of the curve ends when the grid bias is approximately 10 volts. Thus

(*a*) $E_{g \cdot m}$ = 10 − 8 = 2 volts
(*b*) $E_{g \cdot m}$ = 10 − 6 = 4 volts
(*c*) $E_{g \cdot m}$ = 4 − 0 = 4 volts

Method of Obtaining Grid Bias. It was shown in Art. 3-4 that the grid bias required for the desired operating conditions of a detector tube can be obtained by means of a cathode-bias resistor. In a similar manner, it is also possible to obtain the proper grid bias for the desired operating conditions of a tube in an r-f amplifier circuit.

When a tube is used as an amplifier, the amount of direct current flowing in the plate circuit with zero-signal input can be obtained for any value of plate voltage and grid bias from either the plate characteristic or grid-plate transfer characteristic curves. The zero signal plate current for one or two values of plate voltage and grid bias may also be obtained from a standard tube manual. A bypass capacitor, generally in the order of 0.1μf, should be used in conjunction with the cathode-bias resistor in r-f circuits.

5-3. Distortion in Class A Amplifiers. *Distortion Caused by Operating the Tube on the Curved Portion of Its Characteristic Curve.* When the tube whose operating characteristics are shown in Fig. 5-2 is operated with a grid bias whose value is E_c, the variation in the input signal voltage

e_g will produce a grid-voltage swing whose limits are $e_{c.a}$ and $e_{c.b}$. With this variation in grid voltage, the tube will operate on the straight portion of the curve between a and b. The varying plate current i_p will therefore change in the same manner as the input voltage e_g.

When the tube is operated with a grid bias whose value is too near the negative bend of the curve, the output signal will become distorted. This is shown by the operating characteristics for this condition in Fig. 5-4. The grid bias $E_{c.1}$ causes the tube to operate about point c_1 on the

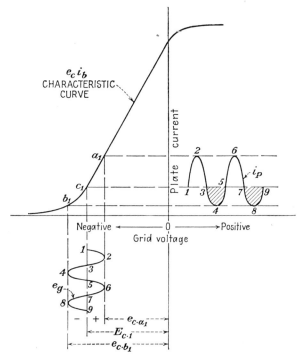

Fig. 5-4. Distortion in a Class-A-operated voltage-amplifier tube caused by using too large a bias.

curve. The input signal voltage e_g will produce a grid-voltage swing whose limits are $e_{c.a1}$ and $e_{c.b1}$. With this variation in grid voltage, the tube will operate on the portion of the curve between a_1 and b_1. The negative halves of each cycle of the input signal voltage will operate on the curved portion of the curve between b_1 and c_1. The plate current resulting from each of the negative half-cycles will therefore be distorted as illustrated by the shaded area (Fig. 5-4).

When the tube is operated with a grid bias whose value is too near the positive bend of the curve it will also cause distortion as illustrated in Fig. 5-5. The grid bias $E_{c.2}$ causes the tube to operate about point c_2 on the curve. The input signal e_g will produce a grid-voltage swing whose

limits are $e_{c.a2}$ and $e_{c.b2}$.   This variation in grid voltage causes the tube
to operate on the portion of the curve between a_2 and b_2.   The positive
halves of each cycle of the input signal voltage will operate on the curved
portion of the curve between c_2 and a_2.   The plate current resulting from
each of the positive half-cycles will therefore be distorted as illustrated
by the shaded area (Fig. 5-5).

Distortion Caused by Driving the Grid Positive.   When the grid is made
positive with respect to the cathode it will act in the same manner as the

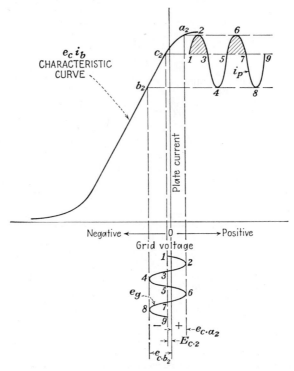

FIG. 5-5. Distortion in a Class-A-operated voltage-amplifier tube caused by using too small
a bias.

plate.   Some of the electrons emitted by the cathode will be attracted
to the grid, causing a current i_g to flow in the external grid circuit (Fig.
5-6).   This current must flow through the resistance R_g or any other
circuit element connected in this path, such as the secondary of a coupling
transformer.   The current i_g flowing through R_g produces a voltage drop
V_g in this circuit each instant that the grid is positive.   The effective
grid voltage at these instants will be equal to the applied voltage e_g
minus the voltage drop V_g.   The two operating conditions shown in Figs.
5-7 and 5-8 demonstrate how the input signal is distorted when the voltage
on the grid is made positive.   This distortion of the input signal voltage

will therefore produce distortion in the plate current as is illustrated by these diagrams.

Distortion of this type is produced by operating the tube with an incorrect amount of grid bias (Fig. 5-7) or by applying too large a signal to the input circuit (Fig. 5-8). For purposes of comparison the two characteristic curves used in Figs. 5-7 and 5-8 are the same as the one used in Fig. 5-2.

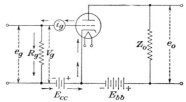

Distortion Caused by Operating the Tube with an Incorrect Grid Bias. When the tube is operated with a grid bias whose value is less than the maximum

FIG. 5-6. Flow of current in the grid circuit when the grid is positive with respect to its cathode.

value of the input signal voltage, the grid voltage will be driven positive during the portion of each cycle in which the positive value of the input signal is greater than the grid bias. The manner in which the output signal is distorted for this type of operation is shown in Fig. 5-7.

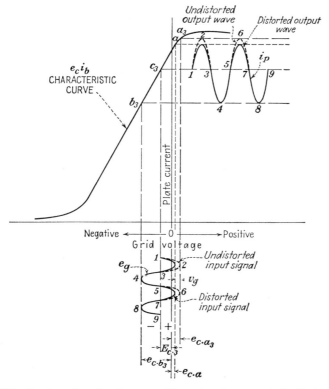

FIG. 5-7. Distortion in a class A voltage-amplifier tube when operated with incorrect grid bias so that the grid is driven positive during part of each cycle.

With a grid-bias voltage as indicated at $E_{c.3}$, the input signal e_g causes the tube to operate on the straight portion of the curve between a_3 and b_3. However, under this condition the grid is made positive at some instants. The voltage drop due to the current in the grid circuit during these intervals reduces the effective grid voltage during these instants, thus reducing the maximum positive grid voltage from $e_{c.a3}$ to $e_{c.a}$. This distortion of the input signal voltage causes the tube to operate between

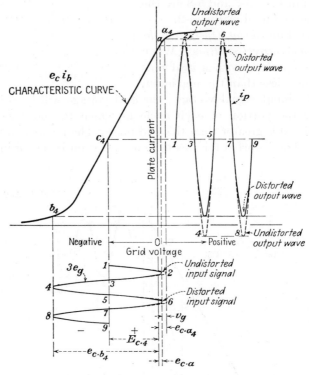

FIG. 5-8. Distortion in a Class A voltage-amplifier tube caused by applying too large an input signal.

b_3 and a on the curve. The output current will therefore be distorted in a similar manner, as is illustrated by the distorted output wave shown in Fig. 5-7.

Distortion Caused by Applying Too Large an Input Signal. When the tube is operated with its correct value of grid bias but the applied signal is too large, either half or both halves of the output signal may be distorted. The manner in which this type of operation causes distortion is shown in Fig. 5-8. The variation in input signal voltage $3e_g$ produces a grid-voltage swing whose limits are $e_{c.a4}$ and $e_{c.b4}$. With this variation in grid voltage, the tube will operate between points a_4 and b_4. During a part of each negative half-cycle the tube will operate on the negative bend of the curve, thus causing the plate current to be distorted during these intervals.   During a part of each positive half-cycle of the input signal, the grid is driven positive.   The voltage drop due to the current flowing in the grid circuit reduces the effective grid voltage, thus reducing the maximum positive grid-voltage swing from $e_{c.a4}$ to $e_{c.a}$.   This distortion of the input signal causes the tube to operate, for each positive half-cycle, between c_4 and a on the curve.   The current flowing in the plate circuit during each positive half-cycle will therefore be distorted.   Thus, for this operating condition distortion occurs during both the negative and positive half-cycles.

5-4. Voltage Amplification Produced by a Class A Amplifier. *Equivalent Amplifier Circuits Using Triodes.*   In Chap. 2 it was shown that the grid of a tube is μ times as effective in controlling the plate current as is the plate.   When an alternating signal voltage e_g is applied to the grid of a tube, the plate circuit may be considered as containing a generator of $-\mu e_g$ volts in series with the plate resistance r_p and the output impedance Z_o.   This principle is a basis for calculating the operating characteristics of an amplifier circuit.   The output voltage $-\mu e_g$ is negative because of the phase reversal (Art. 2-5).

The vacuum tube used in an amplifier circuit can therefore be considered as a generator whose output voltage is equal to $-\mu e_g$.   An equivalent electrical circuit can be drawn for an amplifier circuit by substituting a generator for the vacuum tube.   The equivalent electrical circuit for the elementary amplifier of Fig. 5-3 is shown in Fig. 5-9a.   This type of equivalent circuit is referred to as the *constant-voltage-generator* form and is very convenient for studying the operating characteristics of amplifier circuits using triodes.   The a-c component of the plate current flowing in this circuit is

$$i_p = \frac{\mu e_g}{Z_o + r_p} \tag{5-1}$$

The voltage developed across the load impedance by this current will then be

$$e_p = i_p Z_o \tag{5-2}$$

Substituting Eq. (5-1) in (5-2)

$$e_p = \frac{\mu e_g Z_o}{Z_o + r_p} \tag{5-3}$$

The voltage amplification of the circuit then becomes

$$\text{VA} = \frac{e_p}{e_g} = \frac{\dfrac{\mu e_g Z_o}{Z_o + r_p}}{e_g} \tag{5-4}$$

and

$$\text{VA} = \frac{\mu Z_o}{Z_o + r_p} \tag{5-5}$$

When the output impedance consists only of resistance, resistance-capacitance (RC)-coupled amplifier circuits, or a tuned amplifier circuit in which the reactive effects cancel one another so that the resultant impedance is only resistance, the output impedance Z_o will be equal to R_o. Under these conditions, Eqs. (5-1), (5-3), and (5-5) become

$$i_p = \frac{\mu e_g}{R_o + r_p} \qquad (2\text{-}8)$$

$$e_p = \frac{\mu e_g R_o}{R_o + r_p} \qquad (2\text{-}9)$$

$$\text{VA} = \frac{\mu R_o}{R_o + r_p} \qquad (2\text{-}10)$$

Equivalent Amplifier Circuit Using Pentodes. The maximum voltage amplification of a circuit using a low-mu triode can be made almost

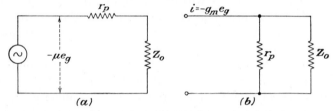

Fig. 5-9. Equivalent circuit for the elementary amplifier circuit of Fig. 5-3. (*a*) Constant-voltage-generator form. (*b*) Constant-current-generator form.

equal to its amplification factor because of the comparatively low values of plate resistance obtainable in low-mu triode-amplifier tubes. In amplifier circuits using high-mu triodes, tetrodes, and pentodes the values of plate resistance are so high that the amount of voltage amplification obtainable is but a fraction of the amplification factor of the tube used. The large difference between the voltage amplification of the circuit and the amplification factor of these tubes is due to the high values of the plate resistance and the comparatively low values of plate-load impedance that must be used for practical amplifier circuits. When the ratio between the plate resistance and the load impedance becomes great, the transconductance of the output circuit approaches the value obtained when the load impedance is practically zero. The voltage amplification of the circuit will then be dependent on the value of the tube's transconductance rather than on its amplification factor. This can be seen if Eqs. (5-1), (5-3), and (5-5) are expressed in terms of transconductance instead of amplification factor. Substituting Eq. (2-5) in Eq. (5-1) and regrouping the terms

$$i_p = g_m e_g \frac{r_p}{Z_o + r_p} \qquad (5\text{-}6)$$

Substituting Eq. (2-5) in Eq. (5-3) and regrouping, the voltage across the load impedance then becomes

$$e_p = g_m e_g \frac{r_p Z_o}{Z_o + r_p} \tag{5-7}$$

Substituting Eq. (2-5) in Eq. (5-5) and regrouping, the voltage amplification of the circuit then becomes

$$\text{VA} = g_m \frac{r_p Z_o}{Z_o + r_p} \tag{5-8}$$

By rearranging the terms in Eq. (5-8), the output impedance required to produce a definite value of voltage amplification may be found by

$$Z_o = \frac{\text{VA} r_p}{g_m r_p - \text{VA}} \tag{5-9}$$

Equation (5-7) shows that applying a signal voltage e_g to the input circuit of a tube may be considered the same as though the tube generated a current equal to $-g_m e_g$ that is made to flow through a parallel circuit formed by the plate resistance r_p and the load impedance Z_o. Thus an amplifier circuit may also be considered in the form shown in Fig. 5-9b, generally called the *constant-current-generator* form. This type of circuit is very convenient for studying the operating characteristics of amplifier circuits using high-mu tubes.

The impedance of a tuned circuit at its resonant frequency has the effect of only resistance. When the load in the plate circuit of an r-f amplifier tube is a tuned circuit, the effective output impedance Z_o at resonance will have the effect of only resistance and can therefore be added arithmetically to the plate resistance r_p in Eqs. (5-6), (5-7), and (5-8). All the equations developed in this article are interrelated, and the following examples will illustrate their applications.

Example 5-2.   A certain triode, used as a voltage amplifier, has a plate resistance of 85,000 ohms and an amplification factor of 100 when operated with 100 volts on its plate and a grid bias of 1 volt. What is the voltage amplification of an amplifier stage using such a tube under these operating conditions when the load resistance is 65,000 ohms?

Given: $r_p = 85,000$ ohms   Find: VA
$\quad\quad\quad R_o = 65,000$ ohms
$\quad\quad\quad \mu = 100$

Solution:

$$\text{VA} = \frac{\mu R_o}{R_o + r_p} = \frac{100 \times 65,000}{65,000 + 85,000} = 43.3$$

Example 5-3.   A 1-volt a-c signal (maximum value) is applied to the input side of the tube and amplifier circuit of Example 5-2.   (a) What amount of variation will this produce in the plate current? (*b*) What amount of voltage will be developed across the output load resistor?

$$\text{Given:} \quad e_g = 1 \text{ volt (max)} \qquad \text{Find:} \quad (a) \ i_p$$
$$\mu = 100 \qquad\qquad\qquad\qquad (b) \ e_p$$
$$r_p = 85,000 \text{ ohms}$$
$$R_o = 65,000 \text{ ohms}$$

Solution:

(*a*) $\quad i_p = \dfrac{\mu e_g}{R_o + r_p} = \dfrac{100 \times 1}{65,000 + 85,000} = 0.666 \text{ ma (max)}$

(*b*) $\quad e_p = \text{VA} \times e_g = 43.3 \times 1 = 43.3 \text{ volts (max)}$

or $\quad e_p = i_p R_o = 0.666 \times 10^{-3} \times 65,000 = 43.3 \text{ volts (max)}$

Example 5-4. A certain pentode is operated as a Class A voltage amplifier. The transconductance of the tube is 1,575 μmhos and the plate resistance is 700,000 ohms when the tube is operated with 100 volts on its plate, 100 volts on the screen grid, and with a grid bias of 3 volts. (*a*) If the plate load is a tuned circuit, what must its effective impedance be in order to obtain a voltage amplification per stage of 90? (*b*) How much will the plate current vary when a 20-mv a-c signal is applied to the grid circuit of the tube? (*c*) What amount of voltage will be developed across the output load resistor?

$$\text{Given:} \quad g_m = 1,575 \ \mu\text{mhos} \qquad \text{Find:} \quad (a) \ Z_0$$
$$r_p = 700,000 \text{ ohms} \qquad\qquad (b) \ i_p$$
$$\text{VA} = 90 \qquad\qquad\qquad\qquad (c) \ e_p$$
$$e_g = 20 \text{ mv}$$

Solution:

(*a*) $\quad Z_o = \dfrac{\text{VA} \ r_p}{g_m r_p - \text{VA}} = \dfrac{90 \times 700,000}{(1,575 \times 10^{-6} \times 700,000) - 90} = 62,222 \text{ ohms}$

(*b*) $\quad i_p = g_m e_g \dfrac{r_p}{Z_o + r_p} = \dfrac{1,575 \times 10^{-6} \times 20 \times 10^{-3} \times 7 \times 10^5}{62,222 + 7 \times 10^5} = 28.9 \ \mu\text{a}$

(*c*) $\quad e_p = i_p Z_o = 28.9 \times 10^{-6} \times 62,222 = 1.798 \text{ volts}$

or $\quad e_p = \text{VA} \ e_g = 90 \times 20 \times 10^{-3} = 1.8 \text{ volts}$

5-5. Radio-frequency Amplifier Circuits. *Classification of R-F Amplifier Circuits.*

Amplifier circuits are generally classified according to the method used to couple the output circuit of the amplifier tube to the input circuit of the tube in the following stage. Thus an amplifier may be either a *resistance-capacitance-coupled, impedance-coupled,* or *transformer-coupled* circuit. R-f amplifiers are generally either transformer-coupled or impedance-coupled. Resistance-capacitance coupling is used occasionally for r-f amplifiers when an untuned amplifier circuit is desired.

Voltage Amplification of Tuned R-F Amplifier Circuits. Equation (5-5) indicates that theoretically the voltage amplification, or gain per stage, can be increased by (1) using a tube with a higher amplification factor but with approximately the same value of plate resistance, (2) using a tube with a lower value of plate resistance but with the same amplification factor, (3) increasing the value of the load impedance.

Practically, these considerations will hold only for triodes.   High-gain r-f amplifiers seldom use triodes because of the undesired coupling produced by the interelectrode capacitances; r-f amplifiers generally use pentodes.

The plate resistance of r-f pentode voltage-amplifier tubes is very high, some of these tubes having a plate resistance in the order of a million ohms.   The amplification factor is likewise very high, and may be in the order of 1,000.   The structural design of these tubes produces practically complete electrostatic shielding between the plate and the control grid. The net result of all these factors is that greater amplification can be

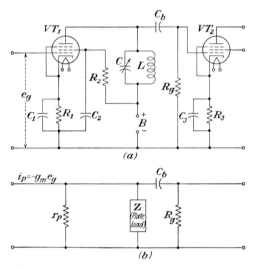

Fig. 5-10. R-f amplifier circuit using tuned-impedance coupling. (a) Actual circuit. (b) Equivalent circuit.

obtained with a pentode r-f voltage amplifier than with any other type of tube.

The plate load of a transformer-coupled r-f amplifier consists of the primary winding of the transformer and the impedance coupled from its tuned secondary circuit (Fig. 5-14a).   For the impedance-coupled r-f voltage amplifier in Fig. 5-10a, the plate load consists of only the tuned circuit.   In actual practice the load impedance of these circuits will be only a small fraction of the plate resistance; therefore when r_p and Z_o are considered as being in parallel with one another (Fig. 5-9b), the resultant impedance will be approximately equal to the load impedance. Equations (5-7) and (5-8) may then be simplified as

$$e_p = g_m e_g Z_o \qquad (5\text{-}10)$$
$$\text{VA} = g_m Z_o \qquad (5\text{-}11)$$

These two equations show that the output voltage and the voltage amplification will increase with increases in the output load impedance. As parallel tuned circuits offer a high impedance at their resonant frequency they are generally used as the plate load in r-f circuits.

Tuned Impedance-coupled R-F Amplifiers. An r-f amplifier circuit using impedance coupling is shown in Fig. 5-10a. The coupling element consists of the tuned LC circuit, which is also the plate load. When the tuned circuit is adjusted so that it is in resonance with a desired signal, maximum voltage will be developed across the tuned circuit. This voltage will vary in accordance with the r-f modulated input signal voltage. The varying voltage is coupled to the grid circuit of the next stage through capacitor C_b, which allows the varying signal current to flow through it and at the same time keeps the B voltage out of the grid circuit.

Because of the low resistance of inductor L, the d-c voltage drop across the inductor will be very small. The voltage at the plate of the tube will therefore be practically equal to the B power-supply voltage. At resonance the impedance of the parallel tuned circuit will be at its maximum value and will be equal to the product of the circuit Q and the inductive reactance. The reactance at this frequency $(X_L - X_C)$ will be zero, and the circuit will offer a pure high-resistance load to the plate of the tube.

Voltage Amplification of a Tuned Impedance-coupled R-F Amplifier. The constant-current-generator form of equivalent circuit for the tuned impedance-coupled amplifier circuit of Fig. 5-10a is shown in Fig. 5-10b. In actual circuits the value of the grid-leak resistance is made much higher than the parallel impedance of the resonant circuit. The impedance of the output circuit, with a fair degree of approximation, can therefore be considered as being equal to the parallel impedance of the tuned circuit, thus

$$Z_o \cong Z_{T \cdot r} = 2\pi fLQ \tag{5-12}$$

An approximate value of the voltage amplification at resonance may then be obtained by substituting Eq. (5-12) in Eq. (5-11)

$$\text{VA} \cong g_m 2\pi fLQ \tag{5-13}$$

Substituting $\dfrac{X_L}{R_L}$ for Q and X_L for $2\pi fL$ in Eq. (5-13)

$$\text{VA} \cong \frac{g_m X_L{}^2}{R_L} \tag{5-13a}$$

Substituting $2\pi fL$ for X_L in Eq. (5-13a)

$$\text{VA} \cong \frac{g_m (2\pi fL)^2}{R_L} \tag{5-14}$$

In Eqs. (5-13) and (5-14) R_L represents the high-frequency resistance of inductor L.   The exact resistance of this coil is difficult to calculate, as it depends in a complex manner upon the physical characteristics of the coil and the frequency of the current.   It is desirable that the voltage amplification of an amplifier be fairly uniform for the entire range of frequencies in any one band.   In order to obtain this condition, the coil must be of the correct size and shape and be wound with the correct size wire so that the high-frequency resistance of the circuit will vary in such

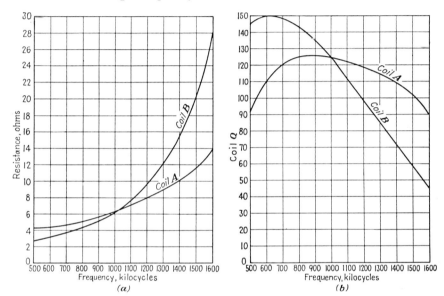

FIG. 5-11. Effect of frequency on the characteristics of a coil.   (*a*) Variation of coil resistance with frequency.   (*b*) Variation of coil Q with frequency.

a manner as to produce fairly uniform amplification for the band of frequencies being considered.   The curves in Fig. 5-11*a* illustrate the variation of resistance with frequency for two coils having the same value of inductance but with different physical characteristics.   The curves in Fig. 5-11*b* illustrate the variation of coil Q with frequency for these two coils.

The ratio of the effective Q of the amplifier circuit to the Q of the tuned circuit will depend upon the plate resistance for the tube used.   Because of the high plate resistance of high-mu triodes and pentodes, the Q of amplifier circuits using these tubes will be practically equal to the Q of the tuned circuit.   Because of the low plate resistance of low-mu triodes, the effective Q of a low-mu triode amplifier circuit will normally be much lower than the actual Q of the tuned circuit.

Example 5-5.   A certain pentode tube being used in a tuned impedance-coupled r-f amplifier similar to Fig. 5-10*a* has operating voltages that produce a transconductance of 750 μmhos. The coil has an inductance of 280 μh, and its resistance at 1,000 kc is 11.2 ohms. What is the voltage amplification of the circuit when the frequency of the input signal is 1,000 kc?

Given: $g_m = 750 \times 10^{-6}$ mho   Find: VA
$\quad\quad\quad L = 280 \times 10^{-6}$ henry
$\quad\quad\quad R = 11.2$ ohms
$\quad\quad\quad f = 10^6$ cycles

Solution:

$$VA = \frac{g_m(2\pi fL)^2}{R_L} = \frac{750 \times 10^{-6}(6.28 \times 10^6 \times 280 \times 10^{-6})^2}{11.2} = 207$$

Example 5-6. The inductance of an impedance-coupled r-f amplifier is 125.6 μh. The high-frequency characteristics of the coil used are illustrated by the curves drawn for coil B (Fig. 5-11). The transconductance of the tube used is 1,575 μmhos. What is the voltage amplification of the circuit at (*a*) 500 kc? (*b*) 1,000 kc? (*c*) 1,500 kc?

Given: $L = 125.6$ μh   Find: (*a*) VA at 500 kc
$\quad\quad\quad g_m = 1,575$ μmhos   (*b*) VA at 1,000 kc
$\quad\quad\quad$ Curves, Fig. 5-11   (*c*) VA at 1,500 kc

Solution:

$$VA = g_m 2\pi fLQ$$
$$Q_{500} = 145, \quad Q_{1,000} = 125, \quad Q_{1,500} = 58, \text{ from Fig. 5-11}b$$

(*a*) VA = $1,575 \times 10^{-6} \times 6.28 \times 500 \times 10^3 \times 125.6 \times 10^{-6} \times 145 = 90$
(*b*) VA = $1,575 \times 10^{-6} \times 6.28 \times 1,000 \times 10^3 \times 125.6 \times 10^{-6} \times 125 = 155$
(*c*) VA = $1,575 \times 10^{-6} \times 6.28 \times 1,500 \times 10^3 \times 125.6 \times 10^{-6} \times 58 = 108$

Untuned R-F Amplifier Circuits. Untuned r-f amplifier circuits do not have the high selectivity or the high amplifying qualities that are characteristic of the tuned circuit. Untuned circuits may be used to prevent the selectivity of an r-f amplifier from becoming too great. Because of the high-frequency discriminating qualities of a tuned amplifier, the selectivity may become too critical when several stages of r-f amplification are used. An untuned amplifier may be used to reduce the degree of selectivity; untuned impedance-coupled amplifiers (Fig. 5-12*a*) are generally used for this purpose. The circuit of such an amplifier is similar to the tuned impedance-coupled amplifier of Fig. 5-10*a* except for the r-f choke that is used as the coupling element in place of the tuned circuit. The constant-current-generator form of equivalent circuit for the untuned impedance-coupled amplifier circuit is shown in Fig. 5-12*b*.

RC-coupled R-F Amplifier Circuits. An *RC*-coupled r-f amplifier, shown in Fig. 5-13*a*, is another form of untuned r-f amplifier. The varying plate current flowing through R_c will produce a varying voltage drop across this resistor, thus varying the voltage applied to the grid of the following tube. The high d-c plate voltage is isolated from the grid circuit by means of C_b and R_g. The resistor R_g provides a path for discharging capacitor C_b, which might otherwise block the action of the tube when the capacitor becomes charged.

The equivalent circuits for this type of amplifier are shown in Figs. 5-13b and 5-13c.   The interelectrode capacitance between the plate and cathode and the stray capacitance of the wiring between the plate circuit and capacitor C_b is represented by C_p.   The interelectrode capacitance between the grid and cathode of the following tube and the stray capacitance of the wiring between capacitor C_b and this grid circuit is represented by C_g.

The capacitance C_p shunts R_c, thus causing some of the alternating output current from the plate of the tube to be bypassed through this capacitance.   Since the voltage applied to the grid of the following tube

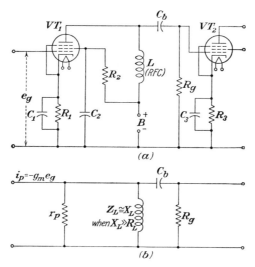

FIG. 5-12. Radio-frequency amplifier circuit using untuned-impedance coupling.   (a) Actual circuit.   (b) Equivalent circuit.

is dependent on the voltage developed across R_c and as this voltage is dependent on the amount of current flowing through it, the efficiency of this arrangement is decreased because of the capacitance C_p.   The capacitance C_g shunts R_g, thus further decreasing the efficiency of this type of coupling.   The impedances of C_p and C_g will be small, and the amount of current bypassed will be large.   Because the efficiency of this type of amplifier will be low when used in r-f circuits, it is not generally used in such circuits.

Modern receivers generally employ more than one stage of tuning in order to obtain the desired selectivity.   The transformers required to obtain this selectivity are also used to couple one circuit to another. *RC* coupling does not contribute anything to the gain and selectivity of the amplifier, whereas the transformer does.   *RC*-coupled amplifiers will amplify signals over a wide range of frequencies without making

adjustments to any of its component parts.   This feature is especially
adapted to a-f amplifier circuits, and to f-m and television r-f circuits.

R-F Amplifier Circuits Having a Tuned Secondary.   R-f amplifiers
with a tuned secondary are used extensively in radio receivers.   A great
deal of selectivity, fidelity, and amplification of the incoming signal will
depend on the design of the r-f amplifier.   A typical tuned-secondary
r-f amplifier circuit is shown in Fig. 5-14*a*.   This circuit differs from the
impedance-coupled amplifier only in the manner in which the tuned circuit

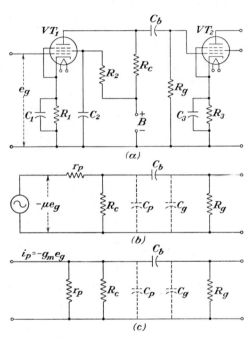

FIG. 5-13. Radio-frequency amplifier circuit using resistance-capacitance coupling.   (*a*)
Actual circuit   (*b*) Constant-voltage-generator form of equivalent circuit.   (*c*) Constant-
current-generator form of equivalent circuit.

is connected to the plate of the tube.   With the transformer-coupled
amplifier the tuned circuit is coupled inductively to the plate, and with
impedance coupling the tuned circuit is connected directly to the plate.
For this reason the circuit shown in Fig. 5-10*a* is sometimes referred to
as being *directly coupled*.   Although differing in details of analysis, the
actions of both of these will be the same.

For all practical purposes the equivalent circuit may be considered in
the constant-voltage-generator form as shown in Fig. 5-14*b* or in the constant-current-generator form of Fig. 5-14*c*.   In either case the equivalent
circuit is a simple coupled circuit whose output load impedance may be
found by applying the rules for analyzing coupled circuits.   For the

constant-voltage-generator form, the load impedance will be

$$Z_o = \frac{2\pi f M Q}{1 + \dfrac{(2\pi f M)^2}{R_S r_p}} \tag{5-15}$$

where $M = K\sqrt{L_P L_S}$.
Substituting this value of Z_o in Eq. (5-11)

$$VA = g_m \frac{2\pi f M Q}{1 + \dfrac{(2\pi f M)^2}{R_S r_p}} \tag{5-16}$$

When the amplifier tube is of the pentode type, its plate resistance is

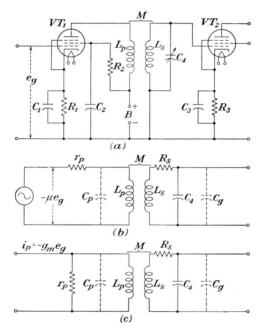

Fig. 5-14. Radio-frequency amplifier circuit having a tuned secondary. (a) Actual circuit. (b) Constant-voltage-generator form of equivalent circuit. (c) Constant-current-generator form of equivalent circuit.

so high that the voltage-amplification equation, to a fair degree of approximation, can be simplified as follows

$$VA \cong g_m 2\pi f M Q \tag{5-17}$$

The maximum amount of amplification will be obtained when the coupled impedance $\dfrac{(2\pi f M)^2}{R_S}$ (at resonance) and the plate resistance r_p are equal to each other. This seems to contradict the thought expressed earlier in this chapter that the voltage amplification increases continuously as the load impedance is made larger than the plate resistance. However, in circuits employing a transformer the output voltage is obtained at its secondary terminals and will be at its maximum value when the maximum amount of energy is transferred from the primary to the secondary. Actually, maximum energy transfer, and hence maximum voltage amplification, occurs when $\dfrac{(2\pi f M)^2}{R_S}$ is equal to r_p. It is only possible to obtain this condition for amplifier circuits using low-mu triodes. When triodes are used, transformer coupling provides a means of matching the output impedance Z_o with the plate resistance r_p in order that maximum voltage amplification may be obtained. In the case of pentodes, the plate resistance is so high that it is impossible to obtain a load impedance whose reactance is anywhere near this value. For amplifier circuits using pentodes, the amplification will be dependent upon the amount of coupling. This can be seen by observation of Eq. (5-17) and the fact that $M = K \sqrt{L_P L_S}$.

Example 5-7. A tuned-secondary r-f amplifier, similar to Fig. 5-14a, uses a pentode tube, and the voltages applied to its elements produce a transconductance of 650 μmhos. The values of the circuit elements are as follows: $L_P = 30$ μh, $L_S = 300$ μh, $R_S = 5$ ohms, $K = 0.3$, $f = 1,000$ kc. What is the voltage amplification of the circuit?

Given: $g_m = 65 \times 10^{-5}$ mho   Find: VA
$L_P = 3 \times 10^{-5}$ henry
$L_S = 30 \times 10^{-5}$ henry
$R_S = 5$ ohms
$f = 10^6$ cycles
$K = 0.3$

Solution:

$$VA = g_m 2\pi f M Q$$

$$= g_m 2\pi f K \sqrt{L_P L_S} \, \frac{2\pi f L_S}{R_S}$$

$$= 65 \times 10^{-5} \times 6.28 \times 10^6 \times 0.3 \sqrt{3 \times 10^{-5} \times 30 \times 10^{-5}}$$

$$\frac{6.28 \times 10^6 \times 30 \times 10^{-5}}{5} = 43.7$$

5-6. Intermediate-frequency Amplifier Circuits. *Advantages of I-F Amplifiers.* An i-f amplifier is an r-f amplifier circuit designed to amplify signals of a definite narrow band of frequencies instead of a wide band of frequencies. This type of amplifier is more efficient than the usual r-f amplifier, as it can be designed to produce optimum Q for the frequency at which it is to operate, thus producing maximum amplification.

In most radio receivers the i-f amplifier has a tuned primary and a tuned secondary. This type of circuit improves both the selectivity and fidelity of reception. The advantage of having both the primary and

secondary circuits tuned is shown in Fig. 5-15.   Curve A shows the
response of an amplifier having only one tuned circuit, and curve B shows
the response of an amplifier with two tuned circuits; curve B more nearly
approaches the ideal response shown by curve C.   Because of these
features the i-f amplifier is used extensively in radio receivers.

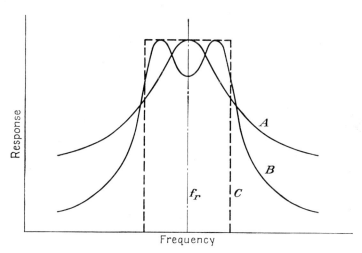

FIG. 5-15. Response curves of typical amplifier circuits.   Curve A, amplifier with a single
tuned circuit.   Curve B, amplifier with two tuned circuits that are tightly coupled.   Curve
C, the ideal response curve.

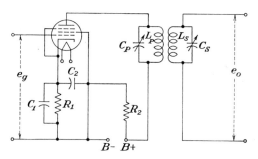

FIG. 5-16. Typical bandpass amplifier circuit as used for the i-f amplifier stage in superheterodyne receivers.

Bandpass Amplifiers.   In order to obtain high fidelity and high selectivity, the ideal response curve should have a flat top and straight sides.
This ideal can be closely approximated by using two resonant circuits
tuned to the same frequency and coupled to each other (Fig. 5-16).   This
circuit, called a *bandpass amplifier*, is used as an i-f amplifier in superheterodyne receivers.

Variable-coupling I-F Transformers.   The width of the frequency band
that will be passed will vary with the amount of coupling existing between

the primary and secondary windings. The effect of the coefficient of coupling upon the width of the band passed is expressed by

$$\text{Width of bandpass} = f_2 - f_1 = Kf_r \qquad (5\text{-}18)$$

While most receivers use coils having a fixed amount of coupling, some use i-f transformers that are provided with a means of varying the amount of coupling between the primary and secondary windings. One type of variable-coupling i-f transformer (Fig. 5-17a) has provisions for varying the mutual inductance between the primary and secondary windings

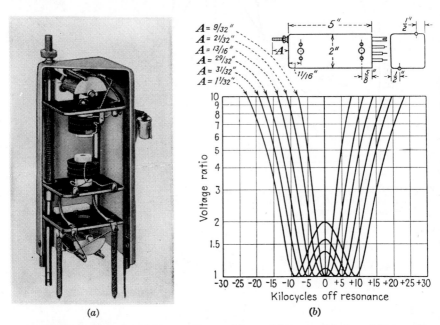

(a)   (b)

Fig. 5-17. Illustration of variable coupling used for an i-f stage. (a) A variable coupling i-f transformer. (b) Effect of various coupling values on the transmission characteristics of a single transformer as shown in (a).

throughout a wide range of values without otherwise affecting the circuit constants. Its approximate range is from one-third of critical coupling to more than three times critical coupling. Continuous variation between these limits may be controlled from the receiver panel by providing some form of mechanical arrangement for adjusting the relative positions of the coils. Where continuous variation is not required, the coupling may be adjusted to the desired value and locked at that point. The width of the bandpass should be equal to twice the value of the highest audio frequency being transmitted. The maximum values of audio frequencies used are approximately 5 kc for general a-m broadcast transmission and 10 kc for high-fidelity a-m transmission.

The slope of the sides of the response curve will vary with the Q of the circuit.   Increasing the value of Q will make the curve steeper and provide greater attenuation of the signal for frequencies outside of the desired band.   The slope of the response curve will also increase with the number of i-f transformers used.   In a single-stage i-f amplifier (Fig. 5-19) two transformers are used and in a two-stage i-f amplifier three transformers

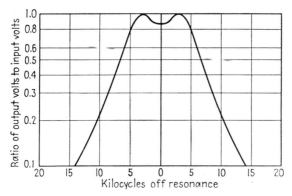

FIG. 5-18. Curves showing variation of the ratio of output volts to input volts with changes in frequency.

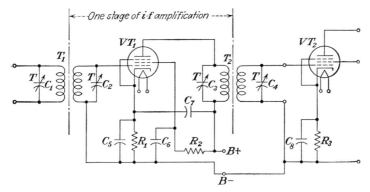

FIG. 5-19. A single-stage i-f amplifier circuit.

would be used.   The curves of Fig. 5-17b show the variation in the ratio of input to output voltage with changes in frequency and also show the effect of varying the coupling by changing the spacing between the primary and secondary windings.   Figure 5-18 is an enlargement of the curve of Fig. 5-17b for the case when A is $1\frac{3}{16}$ inch.   As a single stage of i-f amplification requires the use of a tube and two i-f transformers, the over-all selectivity of the stage must include the effect of both transformers.   The attenuation of the output voltage at frequencies off resonance, which is a measure of the selectivity, may be found for a single transformer at various conditions of coupling from the curves of Fig. 5-17*b*. If the two transformers used in a single-stage i-f amplifier are similar, the over-all attenuation in voltage can be found by squaring the value of the ordinate obtained from the response curve. In the case of a two-stage i-f amplifier using three identical transformers, the attenuation can be found by raising the value of the ordinate to the third power.

Example 5-8. A single-stage i-f amplifier uses two identical transformers with such a value of coupling that produces a response curve similar to Fig. 5-18. (*a*) What is the per cent of voltage output of the stage at 5 kc off resonance when the effect of coupling is considered? (*b*) What is the per cent of voltage output of the stage at 7.5 kc off resonance when the effect of coupling is considered?

Given: Curve, Fig. 5-18   Find: (*a*) Per cent of voltage
           (*a*) *f* off resonance = 5 kc   (*b*) Per cent of voltage
           (*b*) *f* off resonance = 7.5 kc
Solution:

(*a*)   Per cent of voltage at 5 kc off resonance for one transformer = 80 (from curve, Fig. 5-18)
        Per cent of voltage for the stage = $(0.8 \times 0.8) \times 100 = 64$
(*b*)   Per cent of voltage at 7.5 kc off resonance for one transformer = 40 (from curve, Fig. 5-18)
        Per cent of voltage for the stage = $(0.4 \times 0.4) \times 100 = 16$

Choice of the Intermediate-frequency Value. Various values of frequency are used for the i-f amplifiers of radio receivers. Most modern receivers, both single-band and multiband, use an i-f value of between 450 and 465 kc. In order to prevent image reception the choice of i-f value will depend on the number of tuned circuits in the receiver. I-f amplifiers are an important part of all superheterodyne receivers; therefore the factors affecting the choice of an i-f value will be considered in greater detail in Chap. 11.

Voltage Amplification of a Bandpass Amplifier. The extensive use of bandpass amplifier circuits in the i-f stages of superheterodyne receivers makes it desirable to recognize the factors that affect the voltage amplification or gain of this type of circuit. The voltage amplification of a bandpass amplifier circuit at a common resonant frequency for the two tuned circuits is expressed by

$$\text{VA} = g_m K \frac{2\pi f_r \sqrt{L_P L_S}}{K^2 + \dfrac{1}{Q_P Q_S}} \tag{5-19}$$

Example 5-9. An i-f amplifier stage is to operate at 460 kc, and its coils are adjusted so that the coefficient of coupling is 0.02. The circuit employs a pentode whose transconductance is 1,500 μmhos. The inductance of the primary and secondary windings are each 800 μh. The Q of the primary circuit is 100, and that of the secondary is 120. What is the voltage amplification of the circuit?

Given: $f_r = 460$ kc   Find: VA
$\quad\quad\quad K = 0.02$
$\quad\quad\quad g_m = 1,500$ μmhos
$\quad\quad\quad L_P = 800$ μh
$\quad\quad\quad L_S = 800$ μh
$\quad\quad\quad Q_P = 100$
$\quad\quad\quad Q_S = 120$

Solution:

$$VA = g_m K \frac{2\pi f_r \sqrt{L_P L_S}}{K^2 + \dfrac{1}{Q_P Q_S}}$$

$$= \frac{1,500 \times 10^{-6} \times 0.02 \times 6.28 \times 460 \times 10^3 \sqrt{800 \times 10^{-6} \times 800 \times 10^{-6}}}{0.02^2 + \dfrac{1}{100 \times 120}} = 143$$

Equation (5-19) shows that the voltage amplification is dependent upon the value of K. The maximum voltage amplification will occur when the coefficient of coupling is of such a value that will produce the maximum transfer of energy. This amount of coupling is called the *critical coupling* and occurs when the resistance coupled into the primary by the secondary is equal to the resistance of the primary, or

$$R_P = R_{P-S'} = \frac{(2\pi f M)^2 R_S}{Z_S^2} \tag{5-20}$$

But, at resonance $Z_S^2 = R_S^2$; hence

$$R_P = \frac{(2\pi f_r M)^2}{R_S} \tag{5-20a}$$

and

$$R_P R_S = K_c^2 (2\pi f_r)^2 L_P L_S \tag{5-20b}$$

$$R_P R_S = K_c^2 X_{L \cdot P} X_{L \cdot S} \tag{5-20c}$$

$$K_c^2 = \frac{R_P R_S}{X_{L \cdot P} X_{L \cdot S}} \tag{5-20d}$$

$$K_c^2 = \frac{1}{Q_P Q_S} \tag{5-20e}$$

$$K_c = \frac{1}{\sqrt{Q_P Q_S}} \tag{5-21}$$

The maximum possible amplification of the bandpass amplifier at its resonant frequency can then be found by substituting Eq. (5-21) in Eq. (5-19), which then becomes

$$VA_{max} = g_m \frac{1}{\sqrt{Q_P Q_S}} \frac{2\pi f_r \sqrt{L_P L_S}}{\left(\dfrac{1}{Q_P Q_S} + \dfrac{1}{Q_P Q_S}\right)} \tag{5-22}$$

By regrouping and simplifying

$$VA_{max} = g_m \pi f_r \sqrt{L_P L_S Q_P Q_S} \tag{5-23}$$

Example 5-10. What is the maximum possible voltage amplification of the amplifier circuit of Example 5-9?

$$\text{Given:} \quad f_r = 460 \text{ kc} \qquad\qquad \text{Find:} \quad \text{VA}_{max}$$
$$g_m = 1,500 \ \mu\text{mhos}$$
$$L_P = 800 \ \mu\text{h}$$
$$L_S = 800 \ \mu\text{h}$$
$$Q_P = 100$$
$$Q_S = 120$$

Solution:

$$\text{VA}_{max} = g_m \pi f_r \sqrt{L_P L_S Q_P Q_S}$$
$$= 1,500 \times 10^{-6} \times 3.14 \times 460 \times 10^3$$
$$\sqrt{800 \times 10^{-6} \times 800 \times 10^{-6} \times 100 \times 120} = 190$$

When the primary and secondary circuit Q's are equal, in addition to their values of inductance being equal, Eq. (5-23) may be further simplified, as

$$\text{VA}_{max} = g_m \pi f_r L_P Q_P \qquad\qquad (5\text{-}23a)$$

Comparing Eqs. (5-23a) and (5-13) will show that the voltage amplification of a bandpass amplifier circuit is only one-half that produced by an amplifier stage containing but a single tuned circuit. However, the advantage gained in the selectivity with the bandpass amplifier generally outweighs the disadvantage of reduced voltage amplification.

5-7. Multistage Radio-frequency Amplifier Circuits. *Need for Multistage R-F Amplifier Circuits.* In order to obtain the selectivity and amplification required in modern radio receivers, it is desirable to employ more than one stage of r-f amplification. Because a single stage of tuning does not provide sufficient selectivity for satisfactory reception, multistage tuning is used. Each tuning circuit generally also provides some r-f amplification; therefore if two or three stages of tuning are used, the receiver will consequently have two or three stages of r-f amplification.

Amplification of the desired signal can be accomplished before the detector, as in the case of r-f amplification, or after the detector, as in the case of a-f amplification. Early radio receivers used some r-f amplification and a considerable amount of a-f amplification. This practice was necessary because only triodes were available for r-f amplifier circuits, and it was impossible to produce any great amount of r-f amplification without producing feedback. Furthermore, a strong signal could not be applied to the detector circuit without causing distortion. The development of high-gain pentode r-f amplifier tubes, power detectors, and pentode power tubes made it possible to employ a greater amount of r-f amplification and less a-f amplification. Because a-f amplifiers amplify static and other undesired noises in the same proportion that they amplify the audio signal, the final output of the receiver will be noisy if a great amount of a-f amplification is used. The trend in modern radio design is to use a greater amount of r-f amplification and less a-f amplification.

Multistage R-F Amplifier Circuits.   A multistage amplifier circuit consists of two or more single-stage amplifier circuits coupled to each other. When more than two stages of r-f amplification are used, it is not necessary that they employ the same method of coupling throughout. A simple two-stage tuned transformer-coupled circuit is shown in Fig. 5-20. An

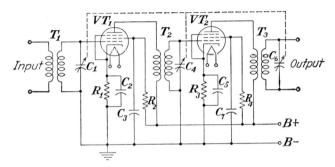

Fig. 5-20. A two-stage tuned transformer-coupled r-f amplifier circuit.

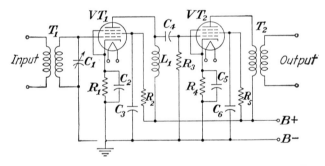

Fig. 5-21. An r-f amplifier circuit using both impedance coupling and transformer coupling.

r-f amplifier circuit using both impedance and transformer coupling is shown in Fig. 5-21.

Feedback.   Feedback occurs when a portion of the current present in one circuit is fed back to a preceding circuit. Feedback may take place through any of the various types of coupling such as capacitive, inductive, or resistive coupling. In amplifier circuits containing two or more stages, coupling may exist in each of these three types. Capacitive coupling may take place through capacitors, stray capacitance, or through the interelectrode capacitance of tubes. Inductive coupling may take place through the windings of coils, transformers, or chokes. Resistive coupling may take place through the amplifier resistors, the resistance of the wiring, or the resistance of the power supply.

Regeneration and Degeneration. When two circuits operate at the same frequency and have a common impedance, feedback may result. The energy that is fed back may have a regenerative or degenerative effect, depending upon the phase relation between the current in the input circuit and the current being returned.

If the energy being returned is in phase with the input signal energy of the amplifier stage, regeneration will result, and the output energy will be increased because of the additional amplification. A certain amount of regeneration can be used as an aid in obtaining an increase in amplification. However, undesired feedback may so strengthen the input signal at the grid circuit of the tube that it will cause the tube to oscillate and

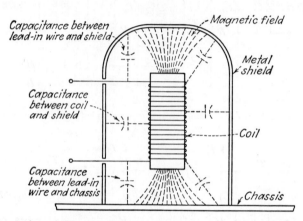

Fig. 5-22. Illustration of the capacitance between the coil and its lead-in wires and the shield and chassis. The manner in which the magnetic lines from the coil cut the shield and chassis is also shown.

result in unstable operation of the amplifier stage. ⟮Feedback that is controlled can be very useful to amplifiers, while uncontrolled feedback is harmful.⟯

If the energy being returned is out of phase with the input signal energy of the amplifier stage, degeneration will result and the output energy will be decreased. Ordinarily the effects of degeneration are not harmful unless they become excessive. In this case an additional stage of amplification may have to be added.

Methods Used to Control Undesired Regeneration. At radio frequencies, inductive and capacitive coupling between the tuned circuits of different stages is a common cause of regeneration. This cause of feedback is ordinarily controlled by shielding the various sections of the tuning capacitor from each other and by enclosing each coil in a copper or aluminum can (Fig. 5-22). Unless shielding is properly employed the output of the amplifier circuit will be reduced considerably. In Fig. 5-22, the winding and the shield form the plates of a capacitor and the space between them acts as a dielectric.   This capacitor acts in the same manner as the distributed capacitance of the coil, thus increasing the minimum capacitance of the tuning circuit and thereby decreasing its effective tuning range. Another effect of shielding is that produced by the magnetic field when a current flows in coils adjacent to shields.   The magnetic lines set up by the current flowing in the coil cut the metal shield and set up eddy currents in the shield.   These eddy currents produce a loss of energy and act in the same manner as increasing the resistance of the coil.

QUESTIONS

1. What is meant by amplification?

2. What is an r-f amplifier?

3. (a) Name and define two fundamental types of amplifiers.   (b) Where is each type used?

4. (a) What is an i-f amplifier?   (b) What are the frequencies commonly used for i-f amplifiers?

5. Name and define four classifications of amplifiers based upon their operating characteristics.

6. What is the fundamental difference between a Class AB_1 and a Class AB_2 amplifier?

7. Which class of amplifier operation is most commonly used for r-f and i-f amplifiers in radio receivers?

8. At what value of grid bias should a tube be operated when being used in a Class A amplifier?

9. What is the maximum value of input signal voltage that should be applied to a tube in terms of the grid bias and the shape of the characteristic curve?

10. What is meant by distortion in an amplifier circuit?

11. What are the causes of distortion of the positive portion of the input signal?

12. What are the causes of distortion of the negative portion of the input signal?

13. What is the effect on the fidelity of the output signal of an amplifier if the input signal drives the grid voltage to a positive value?

14. Why should the input signal voltage never exceed the value of the grid-bias voltage with zero signal?

15. (a) Describe the constant-voltage-generator form of equivalent vacuum-tube circuit.   (b) When is this type of equivalent circuit most useful?

16. (a) Describe the constant-current-generator form of equivalent vacuum-tube circuit.   (b) When is this type of equivalent circuit most useful?

17. Express the voltage amplification of a Class A voltage-amplifier circuit employing a triode in terms of the amplification factor of the tube, the plate-load impedance, and the plate resistance.

18. Express the voltage amplification of a Class A voltage-amplifier circuit employing a pentode in terms of the transconductance of the tube, the plate-load impedance, and the plate resistance.

19. (a) Name three methods of coupling amplifier circuits.   (b) Give an application of each.

20. Why are pentodes more commonly used in voltage-amplifier circuits than triodes?

21. Why are parallel tuned circuits generally used as the plate load of an r-f amplifier circuit employing a pentode?

22. What is the relation among the factors [Eq. (5-14)] affecting the voltage amplification of a tuned impedance-coupled r-f amplifier?

23. Why is it important to have the coils used for impedance-coupled amplifiers designed so that the coil Q remains nearly constant over the frequency range of the amplifier?

24. Where are untuned r-f amplifiers used?

25. (a) Are RC-coupled amplifiers commonly used in r-f amplifier circuits of radio receivers? (b) If so, or if not, why?

26. (a) What are the advantages obtained with an r-f amplifier circuit having a tuned secondary? (b) How does this circuit differ from the impedance-coupled amplifier?

27. What is meant by a direct-coupled amplifier?

28. What is the relation among the factors affecting the voltage amplification of an amplifier stage with a tuned secondary, employing a pentode, and as expressed by Eq. (5-17)?

29. (a) Under what condition may the maximum voltage amplification be obtained from an amplifier with a tuned secondary? (b) Is this attainable with triodes, pentodes, or both?

30. What are the advantages of using an i-f amplifier circuit over the usual r-f amplifiers?

31. What is meant by a bandpass amplifier?

32. How does the response curve of a bandpass amplifier compare with the ideal response curve?

33. (a) What is the purpose of variable-coupling i-f transformers? (b) How is variable coupling accomplished?

34. What are the factors that affect the voltage amplification of a bandpass amplifier?

35. Under what condition does the maximum voltage amplification of a bandpass amplifier occur?

36. (a) How does the voltage amplification of a bandpass amplifier compare with that of an amplifier containing only one tuned circuit? (b) What is the advantage of using a bandpass amplifier?

37. Why are multistage r-f amplifier circuits used?

38. Under what condition does feedback occur?

39. (a) How is regenerative feedback produced? (b) What is its effect?

40. (a) How is degenerative feedback produced? (b) What is its effect?

41. How can undesired regeneration be controlled?

42. Why is it necessary to exercise care in the use of shields?

PROBLEMS

1. A triode whose characteristic curves are shown in Fig. 2-15 is being operated as a Class A amplifier with 200 volts applied to its plate. What is the maximum amount of signal voltage that can be applied without producing distortion when the tube is operated with a grid bias of (a) 7 volts? (b) 6 volts? (c) 5 volts? (d) 4 volts? (NOTE: Assume that the straight portion of the curve ends at $E_c = -9$ volts.)

2. What is the maximum signal that can be applied to the tube of Prob. 1 without causing distortion if the tube is operated with 120 volts applied to its plate, and the grid bias is (a) 4 volts? (b) 3 volts? (c) 2 volts? (d) 1 volt? (NOTE: Assume that the straight portion of the curve ends at $E_c = -5$ volts.)

3. It is desired to obtain grid bias for the tube of Prob. 1 by means of a cathode-bias resistor. What is the value of resistance required and the amount of power consumed by the resistor for each value of grid bias?

4. It is desired to obtain grid bias for the tube of Prob. 2 by means of a cathode-bias resistor. What is the value of resistance required and the amount of power consumed by the resistor for each value of grid bias?

5. A type 6BJ6 pentode is being operated as an r-f amplifier with 250 volts at its plate and 100 volts at the screen grid. (*a*) What value of grid bias is recommended (from a standard tube manual)? (*b*) What are the plate and screen-grid currents for these operating voltages? (*c*) What value of cathode resistor is needed to provide the recommended grid bias? (*d*) How much power is consumed by this resistor?

6. A type 6SF5 tube is to be operated as a voltage amplifier with 250 volts supplied to its plate and operating through a load resistance of 66,000 ohms. (*a*) What is the voltage amplification of the circuit? (*b*) What value of grid bias is recommended for these operating conditions? (See a standard tube manual.) (*c*) If a signal with a maximum value of 2 volts is applied to the grid circuit, what value of output voltage is produced? (*d*) What is the maximum value of the varying component of the plate current?

7. A certain triode that is to be operated as a voltage amplifier has an amplification factor of 70 and a plate resistance of 58,000 ohms when its plate and grid voltages are 250 and −3 volts respectively. (*a*) What is the voltage amplification of this circuit if the plate load is a 58,000-ohm resistor? (*b*) What is the output voltage when a 1-volt input signal is applied? (*c*) What is the maximum value of the varying component of the plate current?

8. If it is desired to have the tube and circuit of Prob. 6 produce an output of 80 volts (maximum value) with the same values of plate voltage, grid bias, and input signal voltage, what value of resistance is required at the load?

9. If it is desired to have the tube and circuit of Prob. 7 produce an output of 45 volts (maximum value) with the same values of plate voltage, grid bias, and input signal voltage, what value of resistance is required at the load?

10. A type 12SK7 tube is to be operated as an r-f voltage amplifier with 250 volts supplied to its plate, 100 volts to the screen grid, and a grid bias of 3 volts. The load impedance is a tuned circuit that is resonant at the frequency being considered, thus producing a resistive effect of 40,000 ohms. What voltage amplification is produced by this circuit at its resonant frequency?

11. A type 12SK7 tube is to be operated as an r-f voltage amplifier with 100 volts supplied to its plate, 100 volts to the screen grid, and a grid bias of 1 volt. The load impedance is a tuned circuit that is resonant at the frequency being considered, thus producing a resistive effect of 40,000 ohms. (*a*) What voltage amplification is produced by this circuit at its resonant frequency? (*b*) If a signal with a maximum value of 0.8 volt is applied to the grid circuit, what value of output voltage is produced? (*c*) What is the maximum value of the varying component of the plate current?

12. A type 1T4 tube is to be operated as an r-f voltage amplifier with 90 volts supplied to its plate, 67.5 volts to the screen grid, and with zero grid bias. What impedance should the load have if it is desired to obtain a voltage amplification of 40 from this circuit?

13. A tuned impedance-coupled r-f amplifier uses a type 6BA6 tube operated with 250 volts on its plate, 100 volts on the screen grid, and 3 volts grid bias. The coil has an inductance of 100 μh and its high-frequency resistance is 10 ohms at 1,500 kc. What is the voltage amplification of the circuit at 1,500 kc?

14. A tuned impedance-coupled r-f amplifier uses a type 12BA6 tube operated with 100 volts on its plate, 100 volts on the screen grid, and a 1-volt grid bias. The coil has an inductance of 100 μh, and its high-frequency resistance is 10 ohms at 1,500 kc. What is the voltage amplification of the circuit at 1,500 kc?

15. The inductance of coil B (Fig. 5-11) is 125.6 μh. If the coil is used in an r-f amplifier circuit with a pentode whose transconductance is 2,000 μmhos, what is the voltage amplification of the circuit at (a) 500 kc? (b) 800 kc? (c) 1,200 kc? (d) 1,600 kc? [NOTE: Use Eq. (5-13).]

16. The inductance of coil A (Fig. 5-11) is 125.6 μh. If the coil is used in an r-f amplifier circuit with a pentode whose transconductance is 2,000 μmhos, what is the voltage amplification of the circuit at (a) 500 kc? (b) 800 kc? (c) 1,200 kc? (d) 1,600 kc? [NOTE: Use Eq. (5-13).]

17. The transformer used in the r-f amplifier circuit (Fig. 5-14a) of a trf receiver has the following values: $L_P = 50$ μh, $L_S = 250$ μh, $K = 0.2$. The tube used with the amplifier stage is a pentode with a transconductance of 2,000 μmhos. The high-frequency resistance of the coil varies in such a manner that the coil Q is 92 at 500 kc, 125 at 1,000 kc, and 100 at 1,500 kc. If it is assumed that the plate resistance of the tube is much greater than the coupled impedance, what is the approximate voltage amplification at (a) 500 kc? (b) 1,000 kc? (c) 1,500 kc? [NOTE: Use Eq. (5-17).]

18. The transformer used in the r-f amplifier circuit (Fig. 5-14a) of a trf receiver has the following values: $L_P = 100$ μh, $L_S = 250$ μh, $K = 0.3$, $R_S = 4$ ohms at 550 kc, $R_S = 8.5$ ohms at 1,000 kc, $R_S = 20$ ohms at 1,600 kc. The tube used is a type 1U4 with a transconductance of 900 μmhos. If it is assumed that the plate resistance of the tube is much greater than the coupled impedance, what is the voltage amplification of the stage at (a) 550 kc? (b) 1,000 kc? (c) 1,600 kc? [NOTE: Use Eq. (5-17).]

19. An r-f amplifier stage, using transformer coupling with untuned primary and tuned secondary, uses a type 6SK7 tube whose transconductance is 2,350 μmhos and whose plate resistance is 120,000 ohms. The transformer constants are $L_P = 200$ μh, $L_S = 340$ μh, $R_S = 14.5$ ohms at 1,000 kc, $K = 0.4$. What is the voltage amplification of the stage at 1,000 kc? [NOTE: Use Eq. (5-16).]

20. At what coefficient of coupling will the circuit of Prob. 19 produce the maximum amount of voltage amplification? (NOTE: This occurs when the coupled impedance at resonance is equal to the plate resistance of the tube.)

21. What is the voltage amplification of the circuit in Prob. 19 when the coefficient of coupling is (a) 0.7? (b) 0.8? (c) 0.9?

22. The curve of Fig. 5-18 shows that for a certain i-f transformer the attenuation at 5 kc off resonance reduces the output to 80 per cent of its maximum value. At 8 kc off resonance the output is only 35 per cent of its maximum value. What is the per cent output voltage of a stage using two identical transformers at (a) 5 kc off resonance? (b) 8 kc off resonance?

23. The attenuation of a certain i-f transformer reduces the output voltage to 90 per cent at 5 kc off resonance and to 60 per cent at 7.5 kc off resonance. If three such i-f transformers are used in a two-stage i-f amplifier, what is the per cent of reduction due to the transformers at (a) 5 kc off resonance? (b) 7.5 kc off resonance?

24. A certain 456-kc i-f amplifier stage uses a type 6BA6 tube whose transconductance is 4,400 μmhos. The constants of the i-f transformer are $L_P = L_S = 500$ μh, $Q_P = Q_S = 80$, $K = 0.025$. What is the voltage amplification of the circuit? [NOTE: See Eq. (5-19).]

25. (a) At what value of coupling will the voltage amplification of the circuit of Prob. 24 be maximum? [Eq. (5-21).] (b) What is the maximum voltage amplification? [Eq. (5-23).]

26. A certain 465-kc i-f amplifier stage uses a type 12BA6 tube whose transconductance is 4,300 μmhos. The constants of the i-f transformer are $L_P = L_S = 300$ μh, $Q_P = 64$, $Q_S = 100$, $K = 0.02$. What is the voltage amplification of the circuit? [NOTE: See Eq. (5-19).]

27. (*a*) At what value of coupling will the voltage amplification of the circuit of Prob. 26 be maximum? [Eq. (5-21).] (*b*) What is the maximum voltage amplification? [Eq. (5-23).]

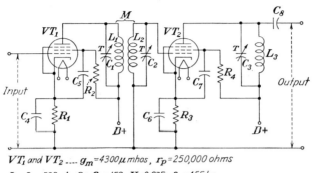

VT_1 and VT_2 ---- $g_m = 4300\,\mu mhos$, $r_p = 250,000$ ohms

$L_1 = L_2 = 500\,\mu h$, $Q_1 = Q_2 = 150$, $K = 0.035$, $f_r = 455\,kc$

z_3 ($L_3 C_3$ resonant at 455 kc) = 40,000 ohms

Fig. 5-23

28. The two-stage amplifier shown in Fig. 5-23 uses a double-tuned circuit in the first stage and a single-tuned circuit in the second stage to provide more uniform response at its output. The amplifier is to operate at 455 kc and the values of its circuit elements are indicated on the diagram (Fig. 5-23). (*a*) What voltage amplification is produced by the first stage? (*b*) What voltage amplification is produced by the second stage? (*c*) What is the over-all voltage amplification of the amplifier?

CHAPTER 6

AUDIO-FREQUENCY VOLTAGE-AMPLIFIER CIRCUITS

The signal delivered to a radio receiver by its antenna is generally only a few microvolts and its energy only a few micromicrowatts. The signal delivered to the loudspeaker may, however, be several volts and the energy may be several watts. The ratio of the strength of the signal at the loudspeaker to the strength of the signal at the antenna may be of the order of one billion. In order to obtain this large increase in voltage and/or power without producing distortion, or picking up and amplifying extraneous signals, several stages of amplification must be used. Radio-frequency amplifier circuits are most commonly used to obtain voltage amplification. Audio-frequency amplifier circuits are used to increase either the voltage, the power, or both; as voltage amplifiers they are operated as Class A. Power amplifiers are operated either as Class AB or Class B. Audio amplifiers are commonly classified according to the method of coupling that is used and hence are known as *resistance-capacitance-coupled, transformer-coupled,* or *impedance-coupled a-f amplifiers.*

6-1. Requirements of the Audio Amplifier. *Frequency Band.* Audio frequencies cover a band between 20 and 20,000 cycles. An *audio-frequency amplifier* is one that will amplify signals whose frequencies lie within this band. The frequency range of various audible sound waves is illustrated by the chart in Fig. 6-1.

Frequency Requirements of the Audio-frequency Amplifier. The sounds produced by a symphonic orchestra contain practically all the frequencies that are likely to be produced by any type of radio program. In order to obtain perfect fidelity of reproduction of the music produced by such an orchestra, sounds from 20 to 20,000 cycles should be reproduced. For the average radio receiver such fidelity of reproduction is neither obtainable nor necessary.

The sounds reproduced by a radio receiver should be essentially the same as those produced by the artists in the studio of the transmitting station. The more nearly the frequency range and the relative amplitudes of the signals reproduced by a receiver approach the frequency range and relative amplitudes of the sound waves as transmitted, the higher will be its *fidelity of reception.* A radio receiver that reproduces all frequencies from 60 to 5,000 cycles with the correct relative amplitudes is considered to be providing good-quality reproduction. Because of certain characteristics of the human ear, some changes in the quality

162

of the sound may take place without being detected by the average listener. For the average listener the correct reproduction of sounds having frequencies between 50 and 5,000 cycles is quite satisfactory.

The a-f amplifiers used in receivers designed for commercial a-m broadcast reception are usually designed to reproduce only those frequencies between 50 and 5,000 cycles. In receivers designed for high-fidelity a-m broadcast reception, the a-f amplifiers should be capable of reproducing audio signals of 30 to 10,000 cycles. In f-m broadcast receivers the a-f amplifiers should be capable of reproducing audio signals of 30 to 15,000 cycles.

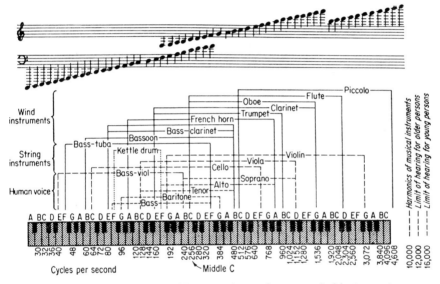

Fig. 6-1. Frequency ranges of human voices and some musical instruments.

Frequency Requirements of High-fidelity Amplifiers. The introduction of *high-fidelity* sound systems for reproducing musical selections recorded on discs and on tape has created a need for better amplifiers. The amplifier for a good high-fidelity system should amplify signals of at least 20 to 20,000 cycles with variations in response not exceeding ± 1 db. The frequency-response ratings for some high-fidelity amplifiers may be as great as ± 1 db from 5 to 150,000 cycles at 1-watt output. Typical amplifier frequency-versus-response curves are shown in Figs. 6-7, 6-10, 6-15, 6-21 and 6-26.

Intensity Required to Produce Audible Sounds. The intensity required to produce an audible sound varies with frequency. In order to produce all audible sounds with equal loudness, the intensity required will vary as shown in Fig. 6-2. The ordinates of these curves indicate the intensity required to produce a sound of any frequency whose loudness will be equal to that produced by a 1,000-cycle reference intensity level. The lowest curve (marked zero db) indicates the intensity required to produce a sound that is barely audible, sometimes called the *threshold of audibility*. The uppermost curve (120 db) indicates the intensity at which sound is not only heard but also felt, sometimes called the *threshold of feeling*. The numbers on each curve indicate the intensity of a signal in decibels above the minimum audible sound over the range of frequency indicated by the abscissa. These curves show that when the strength of the sound is relatively high, the intensity required to produce sounds of equal loudness will be comparatively uniform. When the strength of the sound is

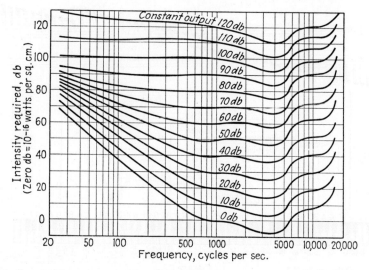

Fig. 6-2. Curves illustrating how the relative amount of power required to produce a sound of equal intensity varies with the frequency. (*Courtesy of Jensen Manufacturing Co.*)

relatively low, the intensity required will be fairly uniform between 1,000 and 5,000 cycles and increases rapidly as the frequency drops below 1,000 cycles or rises above 5,000 cycles. This relationship between intensity and frequency is very important in determining the amount of power that an amplifier must supply to the loudspeaker. An amplifier designed to amplify only sounds under 500 cycles should have a higher power rating than one designed to amplify only sounds between 500 and 5,000 cycles.

6-2. Use of Logarithms in Sound Measurements. *Relation of Sound Energy to Ear Response.* The operating characteristics of an audio amplifier are generally expressed in terms of its gain or loss in volume. The unit used to express this change is based on the ability of the human ear to respond to these changes.

In the rendition of a musical program, a symphonic orchestra will produce varying amounts of sound energy.   The amount of energy used in producing the loudest note may be many thousand times as great as that used to produce the lowest note.   However, the ear does not respond to these sounds in proportion to the energy used; the loudest note is not heard many thousands of times as loud as the lowest note.   Research has shown that the response will vary logarithmically.   In radio terminology, the ratio of any two levels of power is expressed in a unit called the *decibel*, commonly abbreviated db.   The number of decibels is equal to 10 times the logarithm of the ratio of the two levels or values of power. A knowledge of logarithms is essential in order to understand problems involving sound as related to amplifiers, speakers, microphones, etc.

Logarithms.   Logarithms are commonly used in engineering mathematics to facilitate mathematical computations.   Although numerous systems (or bases) of logarithms can be used, the common logarithm, that is, the logarithm to the base 10, is used most frequently and is the system used with the decibel.   By definition, *a common logarithm of a number is the exponent* (or power) *to which* 10 (called the base) *must be raised to produce the number.*   The logarithm of 100 will therefore be equal to 2, as 10^2 equals 100.   This may be expressed as

$$\log_{10} 100 = 2$$

As only the common system of logarithms is used in sound-level calculations, the reference to the base 10 can be omitted.   The expression may then be written as

$$\log 100 = 2$$

Since the logarithm of 10 is equal to 1 and the logarithm of 100 is equal to 2, it is evident that the logarithm of any number between 10 and 100 must be greater than 1 and less than 2.   Consequently the logarithm of a number will consist of two parts: (1) a whole number called the *characteristic*, and (2) a decimal called the *mantissa*.   For example, the logarithm of 50, which is equal to 1.699, is made up of the characteristic whose value is 1 and the mantissa whose value is 0.699.

The characteristic of any number greater than 1 is always positive; numerically it is equal to 1 less than the number of figures to the left of the decimal point.   The characteristic of any number less than 1 is always negative; numerically it is equal to 1 more than the number of zeros between the decimal point and the first significant figure.

Example 6-1.   What is the characteristic of the following numbers; (*a*) 18.3, (*b*) 183, (*c*) 18,300, (*d*) 1.83, (*e*) 0.183, (*f*) 0.00183?

Given: 18.3; 183; 18,300   Find: Characteristic
         1.83; 0.183; 0.00183

Solution:

	Number	Characteristic
(a)	18.3...............	1
(b)	183................	2
(c)	18,300................	4
(d)	1.83.............	0
(e)	0.183.............	−1
(f)	0.00183...........	−3

In solutions (e) and (f) of Example 6-1 the characteristic is shown as a negative quantity. The negative sign must be associated only with the characteristic, as the mantissas are always positive. One method of indicating that the negative sign applies only to the characteristic is to place the minus sign over the number instead of in front of it, as $\bar{1}$, $\bar{3}$, etc. Another more commonly used method is to add 10 to the characteristic and then write -10 after the logarithm (see Example 6-2).

The mantissa or decimal part of a logarithm is found by reference to a table of logarithms, which is a tabulation of mantissas. The mantissa is always a positive number and is determined from the significant digits in the number. A table of common logarithms of numbers is provided in Appendix VIII. It is to be understood that a decimal point is assumed in front of each of the values of mantissas in the table of Appendix VIII.

Example 6-2. What is the logarithm of: (a) 18,300? (b) 18.3? (c) 0.00183? (d) 650,000? (e) 1.25?

Given: 18,300; 18.3; 0.00183   Find: Logarithm
              650,000; 1.25

Solution:

See Table 6-1.

TABLE 6-1

	Number	Characteristic	Mantissa from Appendix VIII	Logarithm
(a)	18,300	4	.2625	4.2625
(b)	18.3	1	.2625	1.2625
(c)	0.00183	−3	.2625	$\bar{3}$.2625
				(or 7.2625 − 10)
(d)	650,000	5	.8129	5.8129
(e)	1.25	0	.0969	0.0969

It is sometimes necessary to find the number corresponding to a logarithm. The number corresponding to a logarithm is called an *antilogarithm* or *antilog* and may be found by working in the reverse order of finding the logarithm of a number. For example, the common logarithm of 100 is 2 and hence it is said that the antilog of 2 is 100.

Example 6-3.   What is the antilog of the common logarithm: (a) 3.8751?   (b) 0.0645? (c) 7.3711 − 10?

Given: (a) Log = 3.8751   Find: Antilog
         (b) Log ≈ 0.0645
         (c) Log = 7.3711 − 10

Solution:

(a)   Antilog of 3.8751

The logarithm should be divided into its characteristic (3) and its mantissa (0.8751). From Appendix VIII, tho mantissa corresponds to the number 750.   The characteristic 3 indicates that the number will have 3 |·1 or 4 figures to the left of the decimal point.   Thus, the antilog of 3.8751 is found to be 7,500.

(b)   Antilog of 0.0645

Characteristic = 0; mantissa = 0.0645.   Number corresponding to the mantissa = 116 (from Appendix VIII).   Number of places to the left of the decimal point = 0 + 1 = 1.   Thus, the antilog of 0.0645 is 1.16.

(c)   Antilog of 7.3711 − 10

Characteristic = 7 − 10 = −3; mantissa = 0.3711.   Number corresponding to the mantissa = 235 (from Appendix VIII).   With a characteristic of $\bar{3}$, there must be two zeros between the decimal point and the first significant number.   Thus, the antilog of 7.3711 − 10 is 0.00235.

6-3. Sound Measurements.   *The Decibel.*   The operating characteristics of an audio amplifier are generally expressed in terms of its gain or loss in volume.   The unit used to express this change in volume is based on the ability of the human ear to respond to these changes.   The unit most frequently used is the *bel*, named in honor of Alexander Graham Bell.   The bel is defined as the common logarithm of the ratio between two quantities and is a relative unit of measurement; it does not specify any definite amount of sound, power, voltage, or current.   The bel is too large a unit for general use and the *decibel*, which is one-tenth of a bel, is commonly used.   The decibel may be used to express the ratio between two values of either sound, power, voltage, or current.   The change in volume in any circuit, expressed in decibels, abbreviated db, can be found by the equation

$$db = 10 \log \frac{P_1}{P_2} \tag{6-1}$$

where P_1 and P_2 express the ratio of two values of power.

The decibel is a logarithmic unit and therefore represents a logarithmic change.   As the response of the human ear to sounds of varying intensity is logarithmic, regardless of the power level, the decibel provides a good means of expressing variations in sound measurements.   The smallest change in sound intensity that can be detected by the human ear is approximately 1 db, although the average person does not ordinarily detect changes under 2 or 3 db.

Example 6-4.   A certain triode being used in a radio receiver delivers a maximum undistorted power output of 3.2 watts to its loudspeaker.   What decibel gain in undistorted power will be obtained by substituting a tube with: (*a*) a maximum undistorted power output of 4.8 watts?   (*b*) a maximum undistorted power output of 6.4 watts?

Given: P = 3.2 watts   Find: (*a*) Decibel gain
P_a = 4.8 watts   (*b*) Decibel gain
P_b = 6.4 watts

Solution:

(*a*)   Gain = $10 \log \frac{P_a}{P}$ = $10 \log \frac{4.8}{3.2}$ = $10 \log 1.5$ = 10×0.1761 = 1.761 db

log 1.5 = 0.1761 (from Appendix VIII)

(*b*)   Gain = $10 \log \frac{P_b}{P}$ = $10 \log \frac{6.4}{3.2}$ = $10 \log 2$ = 10×0.301 = 3.01 db

log 2 = 0.301 (from Appendix VIII)

Example 6-4 shows that doubling the power increases the volume by only 3 db.   As this change is barely perceptible to the average listener, it would not be practical to substitute either of the tubes suggested in Example 6-4 in order to obtain a gain in volume.

Example 6-5.   What power output is required in order to produce a gain of 10 db over the 3.2 watts of output obtained with the tube used in Example 6-4?

Given: P_1 = 3.2 watts   Find: P_2
Gain = 10 db

Solution:

$$db = 10 \log \frac{P_2}{P_1}$$

$$\log \frac{P_2}{P_1} = \frac{db}{10} = \frac{10}{10} = 1$$

Antilog of 1 = 10 (from Appendix VIII).

$$\frac{P_2}{P_1} = 10$$
$$P_2 = P_1 \times 10 = 3.2 \times 10 = 32 \text{ watts}$$

A power output as large as 32 watts is generally not obtained with a single power output tube.   Power outputs as large as this are generally obtained in amplifier circuits by using a push-pull amplifier.

Voltage and Current Ratios.   The decibel is fundamentally a measure of power ratio; however, as voltage and current are functions of power, Eq. (6-1) can be transformed to express the change in volume in decibels for two different values of voltage or current output.

Substituting $\dfrac{E^2}{R}$ for P in Eq. (6-1), then

$$db = 10 \log \frac{\dfrac{E_1{}^2}{R_1}}{\dfrac{E_2{}^2}{R_2}} = 10 \log \left(\frac{E_1}{E_2}\right)^2 \frac{R_2}{R_1} \qquad (6\text{-}2)$$

or

$$db = 10 \log \left(\frac{E_1}{E_2}\right)^2 + 10 \log \frac{R_2}{R_1} \qquad (6\text{-}2a)$$

or

$$db = 20 \log \frac{E_1}{E_2} + 10 \log \frac{R_2}{R_1} \qquad (6\text{-}2b)$$

or

$$db = 20 \log \frac{E_1 \sqrt{R_2}}{E_2 \sqrt{R_1}} \qquad (6\text{-}2c)$$

Substituting I^2R for P in Eq. (6-1), then

$$db = 10 \log \frac{I_1{}^2 R_1}{I_2{}^2 R_2} \qquad (6\text{-}3)$$

or

$$db = 10 \log \left(\frac{I_1}{I_2}\right)^2 + 10 \log \frac{R_1}{R_2} \qquad (6\text{-}3a)$$

or

$$db = 20 \log \frac{I_1}{I_2} + 10 \log \frac{R_1}{R_2} \qquad (6\text{-}3b)$$

or

$$db = 20 \log \frac{I_1 \sqrt{R_1}}{I_2 \sqrt{R_2}} \qquad (6\text{-}3c)$$

For conditions where the impedances are equal, Eqs. (6-2c) and (6-3c) can be simplified as

$$db = 20 \log \frac{E_1}{E_2} \qquad (6\text{-}4)$$

$$db = 20 \log \frac{I_1}{I_2} \qquad (6\text{-}5)$$

Example 6-6.   The characteristics of a certain audio amplifier are such that a voltage amplification of 5 is obtained at 50 cycles, 15 at 1,500 cycles, and 30 at 5,000 cycles. Assuming the voltage amplification at 1,500 cycles as the reference level, what is the loss or gain in decibels at the other frequencies?

Given: $f_1 = 50$ cycles   Find: Change @ 50 cycles, db
              VA $= 5$ Change @ 5,000 cycles, db
              $f_2 = 1,500$ cycles
              VA $= 15$
              $f_3 = 5,000$ cycles
              VA $= 30$

Solution:

$$db \text{ @ 50 cycles} = 20 \log \frac{VA @ f_1}{VA @ f_2} = 20 \log \frac{5}{15} = 20 \log 0.333$$

$$= 20(9.5224 - 10) = -9.552$$

$$db \text{ @ 5,000 cycles} = 20 \log \frac{VA @ f_3}{VA @ f_2} = 20 \log \frac{30}{15} = 20 \log 2$$

$$= 20 \times 0.301 = 6.02$$

Example 6-6 indicates that this amplifier has poor fidelity. There is a loss of almost 10 db at the low frequencies and a gain of 6 db at the high frequencies. Such differences in volume are easily detected by the average listener.

Zero Reference Level. The decibel indicates the ratio between two quantities, namely two values of power, voltage, or current. Used in this manner, the decibel does not express any definite amount of power, voltage, or current but merely indicates the ratio between two magnitudes of any one of these quantities of measurement. In order to associate a decibel rating with a specific amount of power, voltage, or current it is necessary to establish a reference level for zero db. Although values of 1, 6, 10, 12.5, and 50 mw have been used at various times, only the 1- and 6-mw reference levels are now in common use.

Many applications of decibels to power ratings are based on the 6-mw zero reference level. This decibel rating may be referred to as (1) db 6m, (2) db (above 6 mw), (3) db (0 db = 6 mw), (4) db (reference level −6 mw), or (5) merely db. With the 6-mw reference level, power values greater than 6 mw are indicated as +db and power values less than 6 mw are indicated as −db. Meters designed to indicate decibels are calibrated for zero db when 1.73 volts is being fed into a 500-ohm load. Applying the power equation $P = E^2/R$ shows that this zero reference level corresponds to 6 mw. In some instances, the zero reference level for the decibel meter is stated to be 6 mw.

A zero reference level of 1 mw is also frequently used; the letter m is generally associated with the symbol, as db·m, to distinguish the 1-mw from the 6-mw reference level. Variations similar to those listed above for the 6-mw level may also be found in describing the 1-mw reference level.

When it is desired to express the voltage gain of an amplifier in decibels, as in Example 6-6, the symbol db·vg is sometimes used. In the application of decibels to specific amounts of voltage, a 1-volt reference level is used unless otherwise specified. The commonly used symbol for the 1-volt reference level is db·v.

Volume Units. The decibel ratings described above are to be used only with sine-wave signals. For complex waveforms, such as audio signals, the volume unit (abbreviated vu) is used. The *volume unit* expresses the decibel level of a complex wave above or below a certain reference level, namely that amount of power which produces a zero vu reading on a volume-unit meter. Meters designed to indicate volume units are calibrated for zero vu when a steady 1,000-cycle, 775-mv signal is applied to a 600-ohm load. This corresponds to 1 mw, and hence the vu meter is said to have a zero reading (or reference level) for a steady-state signal of 1 mw. The level of a sine-wave signal should be expressed only in db·m, never in vu.

Example 6-7.   A certain amplifier circuit is rated at 30 watts output.   (a) What is its decibel rating for a 6-mw reference level?   (b) What is its decibel rating for a 1-mw reference level?

Given: $P = 30$ watts   Find: (a) db·6m
                                              (b) db·m

Solution:

(a)   db·6m $= 10 \log \dfrac{P_1}{P_2} = 10 \log \dfrac{30}{0.006} = 10 \log 5{,}000 = 10 \times 3.6990 \cong 37$

(b)   db·m $= 10 \log \dfrac{P_1}{P_2} = 10 \log \dfrac{30}{0.001} = 10 \log 30{,}000 = 10 \times 4.4771 \cong 44.7$

Example 6-8.   What is the power output in watts of a microphone that has an output rating of (a) -43 db·m?   (b) -40 db·m?   (c) -50 db·m?   (d) Check the answers by reversing the procedure.

Given: (a) -43 db·m   Find: (a) P_o
           (b) -40 db·m   (b) P_o
           (c) -50 db·m   (c) P_o
                                               (d) Check answers

Solution:

$$\text{db·m} = 10 \log \frac{P_o}{P_R} \therefore \log \frac{P_o}{P_R} = \frac{\text{db·m}}{10} \text{ and } \frac{P_o}{P_R} = \text{antilog} \frac{\text{db·m}}{10}$$

thus, $P_o = P_R \text{ antilog} \dfrac{\text{db·m}}{10}$

(a) $P_o = P_R \text{ antilog} \dfrac{\text{db·m}}{10} = 0.001 \text{ antilog} \dfrac{-43}{10} = 0.001 \text{ antilog} -4.3000$

However, the value -4.3000 produced by dividing -43 db·m by 10 produces a negative mantissa as well as a negative characteristic.   As the mantissa values in Appendix VIII are for only positive values, a positive mantissa can be obtained by adding 10 and then subtracting 10, as

$$\begin{array}{r} 10.0000 - 10 \\ - \ 4.3000 \ \ \ \ \ \\ \hline 5.7000 - 10 \end{array}$$

thus, $P_o = 0.001 \text{ antilog } 5.7000 - 10 = 0.001 \times 0.0000501$
                   $= 0.0000000501$ watt or 0.05μw

Alternate Method of Solution:
    The use of negative characteristics can be avoided by placing the larger number of the power ratio in the numerator and then specifying that the answer represents a loss, or negative db·m.   For this condition,

$$\text{db·m} = 10 \log \frac{P_R}{P_o} \text{ and } P_o = \frac{P_R}{\text{antilog} \dfrac{(+)\text{db·m}}{10}}$$

Resolving part (a),

$$P_o = \frac{0.001}{\text{antilog } 4.3000} = \frac{0.001}{20{,}000} = 0.00000005 \text{ watt or } 0.05 \ \mu\text{w}$$

NOTE: Antilog of 4.3 was rounded off to 20,000 in place of 19,950.

(b) $P_o = 0.001 \text{ antilog } \dfrac{-40}{10} = 0.001 \text{ antilog } (6.0000 - 10)$

$= 0.001 \times 0.0001 = 0.0000001 \text{ watt or } 0.1 \ \mu\text{w}$

(c) $P_o = 0.001 \text{ antilog } \dfrac{-50}{10} = 0.001 \text{ antilog } (5.0000 - 10)$

$= 0.001 \times 0.00001 = 0.00000001 \text{ or } 0.01 \ \mu\text{w}$

(d)   Check on answers.

(a) $\text{db·m} = 10 \log \dfrac{P_o}{P_R} = 10 \log \dfrac{0.0000000501}{0.001} = 10 \log 0.0000501$

$= 10 \times \overline{5}.6998 = 10(5.6998 - 10) \cong 57 - 100 \cong -43$

By Alternate method,

$\text{Loss in db·m} = 10 \log \dfrac{P_R}{P_o} = 10 \log \dfrac{0.001}{0.05 \times 10^{-6}}$

$= 10 \log 20,000 = 10 \times 4.301 \cong 43 \text{ db·m}$

(b) $\text{db·m} = 10 \log \dfrac{P_o}{P_R} = 10 \log \dfrac{0.0000001}{0.001} = 10 \log 0.0001$

$= 10 \times \overline{4}.0000 = 10(6.0000 - 10) = -40$

(c) $\text{db·m} = 10 \log \dfrac{P_o}{P_R} = 10 \log \dfrac{0.00000001}{0.001} = 10 \log 0.00001$

$= 10 \times \overline{5}.0000 = 10(5.0000 - 10) = -50$

Example 6-9.   A certain audio amplifier has a hum and noise rating of 60 db below 3 volts.   What is the magnitude of the hum and noise voltage?

Given: db·3v = −60   Find: *E*

Solution:

$$\text{db·3v} = 20 \log \frac{E}{3}$$

Therefore $\log \dfrac{E}{3} = \dfrac{\text{db·3v}}{20} = \dfrac{-60}{20} = -3$

and $E = 3 \times \text{antilog} - 3 = 3 \times \text{antilog } (7.0000 - 10) = 3 \times 0.001$

$= 0.003 \text{ volt or } 3 \text{ mv}$

6-4. Methods of Coupling.   *Audio Amplifiers.*   Any system used to couple the output of one tube to the input of another tube must provide some means of preventing the high plate voltage of the one tube from affecting the grid bias of the next tube.   In order to obtain good fidelity and sensitivity, the coupling system should pass the audio signals with a minimum amount of change in frequency, amplitude, or phase.

The methods used to couple audio-amplifier stages include iron-core transformers; combinations of resistors and a capacitor; resistor, capacitor, and an inductor; and inductors and a capacitor.   Each of these methods of coupling will cause the audio signal to be distorted to some degree.   Each application must be considered individually to determine which coupling method is best to use.

6-5. Resistance-capacitance-coupled Amplifier.   *Basic Circuit Action.* *RC* (resistance-capacitance) coupling is obtained by connecting the plate circuit and grid circuit of two successive stages by means of two resistors and a capacitor, Figs. 6-3 and 6-5.   Any variations in the plate current in VT_1 and the *coupling resistor* R_c will produce corresponding variations in voltage across R_c.   These voltage variations will be applied to the grid of VT_2.

The *blocking capacitor* C_b prevents the high voltage that is applied to the plate of VT_1 from reaching the grid of VT_2.   A high positive voltage at the grid of a tube would cause a very high current to flow in the grid and plate circuits of the tube; this would overload and quickly damage the tube.   If this capacitor were used without resistor R_g, VT_2 would become blocked because the negative charges on the grid side of the capacitor would increase the bias sufficiently to cause the tube to operate beyond cutoff.   To prevent this blocking action a high resistance R_g, called a *grid leak*, is connected between the grid and cathode to provide a path for the accumulated electrons to leak off.

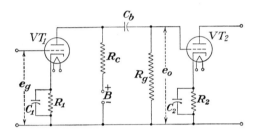

Fig. 6-3. Resistance-capacitance-coupled amplifier circuit using triodes.

Frequency Characteristics.   An important characteristic of the basic RC-coupled amplifier is the manner in which the amplification varies with the frequency.   Although the gain is not uniform for the entire a-f range, it is practically constant over a fairly wide range of frequencies, decreasing very rapidly at both the very low and very high frequencies (Fig. 6-7).   This variation in gain is due to the changes in impedance with frequency of capacitor C_b and the interelectrode capacitances C_{pk} of VT_1 and C_{gk} of VT_2.   The characteristics of the RC-coupled amplifier are best understood by studying the circuit actions at the low, intermediate, and high a-f ranges.

Equivalent Electrical Circuit of the Amplifier. RC coupling consists of two resistors and a capacitor connected as shown in Figs. 6-3 and 6-5   The impedance of the resistors remains constant over the entire a-f range, while the impedance of the capacitors varies inversely with changes in frequency.

The effects of a capacitor in a circuit are dependent on its impedance and the manner in which it is connected.   At low frequencies, the impedance is high, and its effect in a series circuit is important while its effect in a parallel circuit is negligible.   At high frequencies, the impedance is low, and its effect in a series circuit is negligible while its effect in a parallel circuit becomes important. Thus, at low frequencies only the series-connected capacitances need be considered, while at high frequencies only the parallel-connected capacitances need be considered. The characteristics of audio amplifiers are generally considered for the low, intermediate, and high a-f ranges. For calculating purposes, a single representative frequency of each range is used. For radio receiver

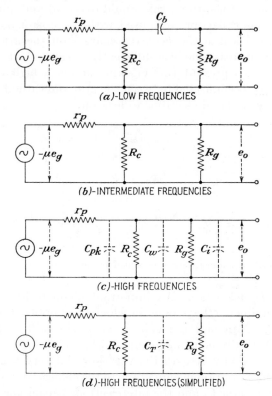

Fig. 6-4. Equivalent electrical circuits at various audio frequencies for an *RC*-coupled amplifier circuit using triodes.

circuits the three representative frequencies are 100 cycles for the low-frequency range, 1,000 cycles for the intermediate-frequency range, and 10,000 cycles for the high-frequency range. When determining the frequency-response characteristics of amplifiers for high-fidelity sound systems, the lower limit is carried down to at least 20 cycles, and the upper limit is extended to 20,000 cycles or higher.

Equivalent Amplifier Circuits for the Three A-F Ranges. The interelectrode capacitance C_{pk} of the amplifier tube acts as an impedance connected across the coupling resistor R_c (Figs. 6-4c and 6-6c). The interelectrode capacitance C_{gk} of the following tube acts as an impedance connected across the grid-leak resistor R_g; C_{gk} is a part of the input capacitance C_i shown in Figs. 6-4c and 6-6c.   The values of these capacitances are very small and their impedances at the low and intermediate frequencies are high.   The portion of the plate current that is bypassed by C_{pk} and C_{gk} at the low and intermediate frequencies is therefore very small.   Because this loss is small, the effect of these capacitances need only be considered at the high audio frequencies.

The blocking action of C_b is determined by its capacitance and the frequency of the signal being applied.   The signal voltage across R_g (Figs. 6-3 and 6-5) will be equal to the voltage across R_c minus the voltage drop across C_b.   At the low audio frequencies the impedance of C_b may become high enough to produce an appreciable voltage drop across this capacitor.   The increase in voltage across C_b causes a decrease in voltage across R_g, resulting in a decrease in the voltage applied to the grid of VT_2. The effect of the blocking capacitor must therefore be considered at the low audio frequencies.

The capacitance of C_b is generally of such a value that its impedance at 1,000 cycles is negligible in comparison to the resistance of R_g.   Practically all the signal voltage developed across R_c will then be applied to the grid of the following tube.   The effects of the blocking capacitor can therefore be ignored for the intermediate and high frequencies.

With these facts it is now possible to draw an equivalent electrical circuit of a stage of amplification for each frequency range.   Figures 6-4 and 6-6 show the equivalent electrical circuits for the triode and pentode amplifiers of Figs. 6-3 and 6-5 respectively.   At the high frequencies C_{pk} and C_i, together with the stray capacitance C_w due to the wiring (Figs. 6-4c and 6-6c), may be considered as a single capacitance C_T shunting the resistors as shown in Figs. 6-4d and 6-6d.

Factors Determining the Values of R_c, R_g, and C_b.   In order to obtain the maximum voltage gain, the value of the plate-load resistor R_c should be high.   The voltage at the plate of VT_1 will be equal to the B supply voltage minus the voltage drop $I_b R_c$.   Increasing the value of R_c increases this voltage drop, and decreases the voltage at the plate of VT_1.   In order to maintain the desired plate voltage, the B supply voltage would have to be increased.   If the value of R_c is made too high, the required B supply voltage will become prohibitive; hence the value of R_c is limited to some extent by the plate supply voltage.

Another factor limiting the extent to which R_c may be increased is the reduction in the gain of the amplifier at the high audio frequencies.   At these frequencies the effect of the series capacitance C_b may be ignored, but the effect of the shunting capacitance C_T (Art. 6-8) becomes important.   This shunting capacitance has a low impedance at the high audio frequencies and causes a reduction in the impedance of the parallel circuit formed by R_c, R_g, and C_T.   A decrease in this impedance will cause a decrease in the voltage at this parallel circuit, thereby causing a loss in amplification at the high frequencies. The division of current in the parallel combination depends on the impedance of its various members. Increasing R_c as a means of increasing the amplification does not produce the gain anticipated but instead results in a decrease in current through R_c, thereby decreasing the over-all gain of the circuit.

The low-frequency characteristics of an RC-coupled amplifier depend largely upon the ratio between the impedance of the blocking capacitor C_b and the resistance of the grid-leak resistor R_g. Increasing the value of C_b decreases its impedance, thus lowering the value of resistance required for R_g. The higher the value of R_g, the higher will be the over-all amplification. At the high frequencies part of the signal current will be bypassed by the interelectrode capacitance C_{gk} of VT_2; the portion of the signal current bypassed will depend upon the ratio between the resistance of R_g and the impedance of C_{gk}.

Practical Values of R_c, R_g, and C_b. The values of the coupling resistor, grid-leak resistor, and blocking capacitor will depend upon the tube used, the operating plate and grid voltages, and the frequency characteristics desired. The resistance of R_c will range from 50,000 ohms to 1 megohm. The plate resistance r_p of triodes is comparatively low and the value of resistance used for R_c is generally from two to five times the value of r_p. Pentodes have comparatively high plate resistance, and the value of R_c used with pentodes is generally in the order of 0.5 to 1 megohm.

The resistance of R_g will range from 100,000 ohms to 1 megohm. The value of R_g is dependent on the values of R_c and C_b and is generally from one to five times the value of R_c.

The capacitance of C_b will depend upon the frequency characteristics desired and ranges from 0.001 to 0.05 μf. This capacitor should have a high dielectric strength in order to prevent any of the plate voltage at VT_1 from leaking to the grid of VT_2. The coupling capacitor may be of mica, paper, or ceramic dielectric.

The ranges of values for R_c, R_g, and C_b listed above are very general. Specific values recommended for the various circuit elements of RC-coupled amplifiers using specific tube types may be obtained from most tube manuals.

6-6. Voltage Amplification for the Intermediate-frequency Range. RC-coupled a-f amplifiers are generally designed to give the maximum voltage gain for the intermediate frequencies. For amplifiers using triodes, the equivalent circuit for the intermediate frequencies is shown in Fig. 6-4b. For this circuit the output load impedance is equal to the impedance of the parallel circuit formed by R_c and R_g, or

$$Z_o = \frac{R_c R_g}{R_c + R_g} \tag{6-6}$$

The voltage amplification at intermediate or *middle* audio frequencies VA_M for a triode amplifier is obtained by substituting Eq. (6-6) for Z_o in the basic equation for voltage amplification with a triode as expressed by Eq. (5-5).

$$VA = \frac{\mu Z_o}{Z_o + r_p} \qquad (5\text{-}5)$$

Equation (5-5) may be rearranged to read

$$VA = \frac{\mu}{\dfrac{r_p}{Z_o} + 1} \qquad (6\text{-}7)$$

Substituting Eq. (6-6) for Z_o in Eq. (6-7)

$$VA_M = \frac{\mu}{\dfrac{r_p(R_c + R_g)}{R_c R_g} + 1} \qquad (6\text{-}8)$$

where VA_M is the voltage amplification at the middle- or mid-frequency value (also called the intermediate-frequency value).

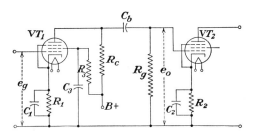

Fig. 6-5. Resistance-capacitance-coupled amplifier circuit using pentodes.

For amplifiers using pentodes, the equivalent circuit for intermediate frequencies is shown in Fig. 6-6b. The impedance of the load, that is, exclusive of the plate resistance r_p, is equal to the impedance of the parallel circuit formed by the coupling resistor R_c and the grid-leak resistor R_g and is expressed by Eq. (6-6). This may also be expressed as

$$\frac{1}{Z_o} = \frac{1}{R_c} + \frac{1}{R_g} \qquad (6\text{-}6a)$$

The voltage amplification at intermediate audio frequencies for a pentode amplifier can be found by substituting Eq. (6-6) or (6-6a) in the basic equation for the voltage amplification of a pentode, Eq. (5-8).

$$VA = g_m \frac{r_p Z_o}{Z_o + r_p} \qquad (5\text{-}8)$$

Equation (5-8) may be rearranged to read

$$VA = \frac{g_m}{\dfrac{1}{Z_o} + \dfrac{1}{r_p}} \tag{6-9}$$

If Eq. (6-6a) is substituted for $\dfrac{1}{Z_o}$ in Eq. (6-9)

$$VA_M = \frac{g_m}{\dfrac{1}{R_c} + \dfrac{1}{R_g} + \dfrac{1}{r_p}} \tag{6-10}$$

or
$$VA_M = g_m R_{eq} \tag{6-10a}$$

where $R_{eq} = \dfrac{1}{\dfrac{1}{R_c} + \dfrac{1}{R_g} + \dfrac{1}{r_p}}$

Example 6-10. A certain *RC*-coupled amplifier uses the triode section of a type 12AV6 tube, a 0.47-megohm coupling resistor, and a 0.47-megohm grid resistor. The tube constants are: $r_p = 80,000$ ohms, and $\mu = 100$. What is the voltage amplification of this circuit for the intermediate-frequency range?

Given: $R_c = 0.47$ megohm   Find: VA_M
$R_g = 0.47$ megohm
$r_p = 80,000$ ohms
$\mu = 100$

Solution:

$$VA_M = \frac{\mu}{\dfrac{r_p(R_c + R_g)}{R_c R_g} + 1} = \frac{100}{\dfrac{80,000(470,000 + 470,000)}{470,000 \times 470,000} + 1} = 74.5$$

Example 6-11. A certain *RC*-coupled amplifier uses the pentode section of a type 1U5 tube, a 0.47-megohm coupling resistor, and a 1-megohm grid resistor. The tube constants are: $r_p = 0.6$ megohm and $g_m = 625$ μmhos. What is the voltage amplification of this circuit for the intermediate-frequency range?

Given: $R_c = 0.47$ megohm   Find: VA_M
$R_g = 1$ megohm
$r_p = 0.6$ megohm
$g_m = 625$ μmhos

Solution:

$$VA_M = g_m R_{eq} = 625 \times 10^{-6} \times \frac{10^6}{4.8} = 130$$

where $R_{eq} = \dfrac{1}{\dfrac{1}{R_c} + \dfrac{1}{R_g} + \dfrac{1}{r_p}} = \dfrac{1}{\dfrac{1}{0.47 \times 10^6} + \dfrac{1}{10^6} + \dfrac{1}{0.6 \times 10^6}} = \dfrac{10^6}{4.8}$

6-7. Voltage Amplification for the Low-frequency Range. The voltage gain for the low audio frequencies will be less than that obtained for the intermediate-frequency range because of the voltage drop at the blocking capacitor C_b. The factor by which the gain is reduced is dependent largely upon the ratio of the resistance of R_g to the impedance of the series circuit consisting of C_b and R_g.   This factor is expressed mathematically as

$$K_L = \frac{1}{\sqrt{1 + \left(\dfrac{X_c}{R}\right)^2}} \tag{6-11}$$

where K_L = low-frequency factor of voltage amplification
  X_C = reactance of coupling capacitor C_b, ohms

$$R = R_g + \frac{r_p R_c}{r_p + R_c} = R_g + \frac{R_c}{1 + \dfrac{R_c}{r_p}}$$

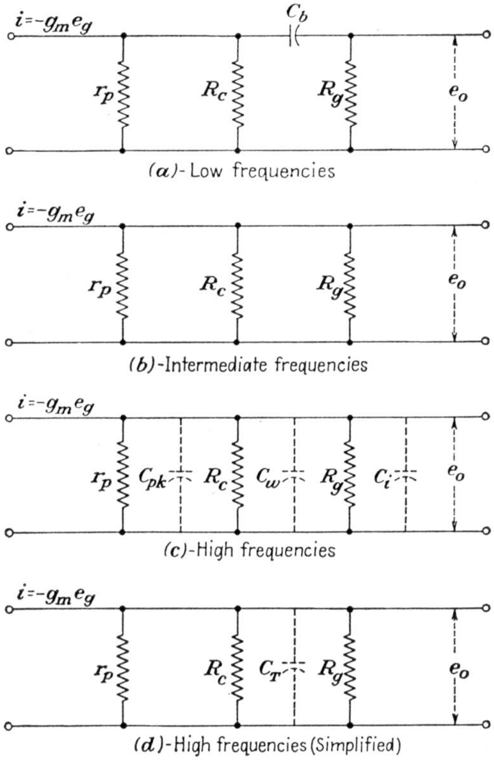

(a)- Low frequencies

(b)-Intermediate frequencies

(c)-High frequencies

(d)-High frequencies (Simplified)

Fig. 6-6. Equivalent electrical circuits at various audio frequencies for an RC-coupled amplifier circuit using pentodes.

The voltage amplification at low audio frequencies, VA_L, for triode amplifiers can be found by combining Eqs. (6-8) and (6-11), as

$$VA_L = K_L VA_M \tag{6-12}$$

or
$$VA_L = \frac{\mu}{\sqrt{1 + \left(\dfrac{X_C}{R}\right)^2 \left[\dfrac{r_p(R_c + R_g)}{R_c R_g} + 1\right]}} \tag{6-12a}$$

The voltage amplification at low frequencies for pentode amplifiers can be found by combining Eqs. (6-10a) and (6-11), as

$$VA_L = K_L VA_M \tag{6-12}$$

$$VA_L = \frac{1}{\sqrt{1 + \left(\dfrac{X_c}{R}\right)^2}} g_m R_{eq} \tag{6-13}$$

Examination of Eqs. (6-11) and (6-13) will show that when the frequency of the signal is such that X_c is equal to R, the voltage gain will be 70.7 per cent of the voltage amplification obtained for the intermediate-frequency range. When the frequency of the audio signal decreases so that X_c is twice the value of R, the voltage gain will be less than 50 per cent of the voltage amplification obtained for the intermediate frequency. When expressed in decibels, this would indicate -3 db when X_c is equal to R, and -7 db when X_c is twice the value of R. As changes in volume greater than 3 db are readily detected, the values of the resistors and the blocking capacitor should be carefully selected in order to obtain good sound reproduction.

Example 6-12.   If the amplifier of Example 6-10 uses a 0.005-μf coupling capacitor, what is the voltage amplification and the corresponding decibel loss of the circuit at: (*a*) 100 cycles? (*b*) 50 cycles? (*c*) 20 cycles?

Given: $VA_M = 74.5$ Find: (*a*) VA_L; decibel loss
           $R_c = 0.47$ megohm   (*b*) VA_L; decibel loss
           $r_p = 80,000$ ohms   (*c*) VA_L; decibel loss
           $R_g = 0.47$ megohm
           $C_b = 0.005$ μf
        (*a*) $f = 100$ cycles
        (*b*) $f = 50$ cycles
        (*c*) $f = 20$ cycles

Solution:

(*a*) $VA_L = K_L VA_M = 0.861 \times 74.5 = 64.2$

where $K_L = \dfrac{1}{\sqrt{1 + \left(\dfrac{X_c}{R}\right)^2}} = \dfrac{1}{\sqrt{1 + \left(\dfrac{318,000}{538,000}\right)^2}} = 0.861$

$$R = R_g + \frac{r_p R_c}{r_p + R_c} = 470,000 + \frac{80,000 \times 470,000}{80,000 + 470,000} \cong 538,000 \text{ ohms}$$

$$X_c = \frac{159,000}{fC} = \frac{159,000}{100 \times 0.005} = 318,000 \text{ ohms}$$

Loss $= 20 \log 0.861 = 20(9.9350 - 10) = -1.3$ db

(*b*) $VA_L = K_L VA_M = 0.645 \times 74.5 = 48$

where $K_L = \dfrac{1}{\sqrt{1 + \left(\dfrac{X_c}{R}\right)^2}} = \dfrac{1}{\sqrt{1 + \left(\dfrac{636,000}{538,000}\right)^2}} = 0.645$

$$X_c = \frac{159,000}{fC} = \frac{159,000}{50 \times 0.005} = 636,000 \text{ ohms}$$

$R = 538,000$ ohms [same as in part (*a*)]

Loss $= 20 \log 0.645 = 20(9.8096 - 10) = 3.8$ db

(c) $VA_L = K_L VA_M = 0.32 \times 74.5 = 23.8$

where $K_L = \dfrac{1}{\sqrt{1 + \left(\dfrac{X_C}{R}\right)^2}} = \dfrac{1}{\sqrt{1 + \left(\dfrac{1,590,000}{538,000}\right)^2}} = 0.32$

$X_C = \dfrac{159,000}{fC} = \dfrac{159,000}{20 \times 0.005} = 1,590,000$ ohms

$R = 538,000$ ohms [same as in part (a)]

Loss $= 20 \log 0.32 = 20(9.5051 - 10) = 9.9$ db

Example 6-13.   If the amplifier of Example 6-11 uses a 0.003-μf coupling capacitor, what is the voltage amplification and the corresponding decibel loss of the circuit at: (a) 100 cycles? (b) 50 cycles? (c) 20 cycles?

Given: $VA_M = 130$
$R_c = 0.47$ megohm
$r_p = 0.6$ megohm
$R_g = 1$ megohm
$g_m = 625$ μmhos
$c_b = 0.003$ μf
(a) $f = 100$ cycles
(b) $f = 50$ cycles
(c) $f = 20$ cycles

Find: (a) VA_L; decibel loss
(b) VA_L; decibel loss
(c) VA_L; decibel loss

Solution:

(a) $VA_L = K_L VA_M = 0.925 \times 130 = 120$

where $K_L = \dfrac{1}{\sqrt{1 + \left(\dfrac{X_C}{R}\right)^2}} = \dfrac{1}{\sqrt{1 + \left(\dfrac{530,000}{1,263,000}\right)^2}} = 0.925$

$R = R_g + \dfrac{r_p R_c}{r_p + R_c} = 10^6 + \dfrac{600,000 \times 470,000}{600,000 + 470,000} = 1,263,000$ ohms

$X_C = \dfrac{159,000}{fC} = \dfrac{159,000}{100 \times 0.003} = 530,000$ ohms

Loss $= 20 \log 0.925 = 20(9.9661 - 10) = 0.68$ db

(b) $VA_L = K_L VA_M = 0.766 \times 130 = 100$

where $K_L = \dfrac{1}{\sqrt{1 + \left(\dfrac{X_C}{R}\right)^2}} = \dfrac{1}{\sqrt{1 + \left(\dfrac{1,060,000}{1,263,000}\right)^2}} = 0.766$

$X_C = \dfrac{159,000}{fC} = \dfrac{159,000}{50 \times 0.003} = 1,060,000$ ohms

$R = 1,263,000$ ohms [same as in part (a)]

Loss $= 20 \log 0.766 = 20(9.8842 - 10) = 2.32$ db

(c) $VA_L = K_L VA_M = 0.43 \times 130 = 56$

where $K_L = \dfrac{1}{\sqrt{1 + \left(\dfrac{X_C}{R}\right)^2}} = \dfrac{1}{\sqrt{1 + \left(\dfrac{2,650,000}{1,263,000}\right)^2}} = 0.43$

$X_C = \dfrac{159,000}{fC} = \dfrac{159,000}{20 \times 0.003} = 2,650,000$ ohms

$R = 1,263,000$ ohms [same as in part (a)]

Loss $= 20 \log 0.43 = 20(9.6335 - 10) = 7.33$ db

6-8. Voltage Amplification for the High-frequency Range. The
voltage gain for the high audio frequencies will be less than that obtained

for the intermediate-frequency range because of the bypass path provided by the capacitance C_T (see Figs. 6-4d and 6-6d). For most practical purposes, the factor by which the voltage gain is reduced is dependent upon the ratio of the equivalent resistance of the parallel circuit formed by the plate resistance, coupling resistor, and grid-leak resistor to the reactance of the capacitance C_T that shunts these resistors. This factor is expressed mathematically as

$$K_H = \frac{1}{\sqrt{1 + \left(\frac{R_{eq}}{X_T}\right)^2}} \qquad (6\text{-}14)$$

or

$$K_H = \frac{1}{\sqrt{1 + (R_{eq}2\pi fC_T)^2}} \qquad (6\text{-}14a)$$

The voltage amplification at high audio frequencies VA_H for amplifiers using triodes can be found by combining Eqs. (6-14a) and (6-8)

$$VA_H = K_H VA_M \qquad (6\text{-}15)$$

or

$$VA_H = \frac{\mu}{\sqrt{1 + (R_{eq}2\pi fC_T)^2}\left(\frac{r_p(R_c + R_g)}{R_cR_g} + 1\right)} \qquad (6\text{-}15a)$$

The voltage amplification at high audio frequencies for amplifiers using pentodes can be found by combining Eqs. (6-14a) and (6-10a)

$$VA_H = K_H VA_M \qquad (6\text{-}15)$$

or

$$VA_H = \frac{g_m R_{eq}}{\sqrt{1 + (R_{eq}2\pi fC_T)^2}} \qquad (6\text{-}16)$$

Examination of the high-frequency factor [Eq. (6-14)] will show that when the frequency of the signal is such that X_T is equal to R_{eq}, the voltage gain will be 70.7 per cent of the voltage amplification obtained for the intermediate-frequency range. When the frequency of the signal increases so that X_T is reduced to one-half the value of R_{eq}, the voltage gain will be less than 50 per cent of the voltage amplification obtained at the intermediate audio frequencies. Expressed in decibels, these reductions in gain represent -3 and -7 db respectively. Thus, it is apparent that the interelectrode capacitances of the tube and the stray capacitance of the wiring have an important bearing on the fidelity of sound reproduction.

Factors Affecting the Value of the Shunting Capacitance. The total shunting capacitance C_T (Figs. 6-4d and 6-6d) is equal to the sum of the three separate shunting capacitances C_w, C_{pk}, and C_i; thus

$$C_T = C_w + C_{pk} + C_i \qquad (6\text{-}17)$$

where C_T = total shunting capacitance

C_w = stray capacitance of wiring

C_{pk} = interelectrode capacitance between plate and cathode of first tube

C_i = effective input capacitance of load—in this case the second tube

The stray capacitance due to the wiring is low in value, generally under 10 $\mu\mu f$. The capacitance C_{pk} of the tube may be obtained from a tube manual. The effective input capacitance of the load when the output is fed into a second tube is the combined effect of the capacitances C_{gk} and C_{gp} of the second tube. These two capacitances are in effect the same as two capacitors connected in parallel. However, their combined effect is not equal to their arithmetic sum because the voltage is not the same at each capacitance owing to the amplifying action, called the *Miller effect*, that takes place between the grid and plate circuits. The capacitance added by the Miller effect is due to the increase in the effective grid-cathode capacitance of a vacuum tube which is caused by the electrostatic charge induced on the grid by the plate through the grid-plate capacitance.

The effective input capacitance is explained by studying the charges at these two capacitances. The charge accumulated because of the capacitance C_{gk} of VT_2 is

$$Q_{gk} = C_{gk}e_o \qquad (6\text{-}18)$$

The charge accumulated as a result of the capacitance C_{gp} of VT_2 is

$$Q_{gp} = C_{gp}e_{gp} \qquad (6\text{-}19)$$

However, the difference of potential between the grid and plate is

$$e_{gp} = e_o - \text{VA}e_o \qquad (6\text{-}20)$$

But, as the plate voltage change of a tube is 180 degrees out of phase with its grid-voltage change, Eq. (6-20) may be expressed as

$$e_{gp} = e_o + \text{VA}e_o \qquad (6\text{-}20a)$$

or

$$e_{gp} = e_o(1 + \text{VA}) \qquad (6\text{-}20b)$$

where e_{gp} = difference of potential between grid and plate

e_o = voltage at grid of second tube

VA = voltage amplification of circuit

$\text{VA}e_o$ = difference of potential between plate and cathode

Substituting Eq. (6-20b) in (6-19)

$$Q_{gp} = C_{gp}e_o(1 + \text{VA}) \qquad (6\text{-}19a)$$

The total effective charge at the input of the second tube is equal to the sum of the grid-cathode and grid-plate charges, or

$$Q_i = Q_{gk} + Q_{gp} \tag{6-21}$$

also

$$C_i e_o = C_{gk} e_o + C_{gp} e_o (1 + \text{VA}) \tag{6-22}$$

and

$$C_i = C_{gk} + C_{gp}(1 + \text{VA}) \tag{6-23}$$

In Eq. (6-23), the term VA represents the voltage amplification of the circuit and normally is approximately one-half the amplification factor of the tube. In the term $C_{gp}(1 + \text{VA})$, the product of C_{gp} and VA represents a dynamic variation produced by the Miller effect.

Example 6-14. In the amplifier circuit of Example 6-10, the stray capacitance of the wiring is 5 $\mu\mu\text{f}$, and the capacitance C_{pk} of the first tube is 0.8 $\mu\mu\text{f}$. For the second tube (type 50C5), $C_{gk} = 13$ $\mu\mu\text{f}$, $C_{gp} = 0.64$ $\mu\mu\text{f}$, and the voltage amplification of the second stage is 30. (a) What is the total shunting capacitance of this amplifier circuit? (b) What is the value of the Miller-effect capacitance? What percentage is the Miller-effect capacitance of: (c) the input capacitance C_i? (d) the total shunting capacitance C_T?

Given:

$$C_w = 5 \ \mu\mu\text{f}$$
$$C_{pk} \text{ of } VT_1 = 0.8 \ \mu\mu\text{f}$$
$$C_{gk} \text{ of } VT_2 = 13 \ \mu\mu\text{f}$$
$$C_{gp} \text{ of } VT_2 = 0.64 \ \mu\mu\text{f}$$
$$\text{VA of } VT_2 = 30$$

Find: (a) C_T
       (b) $C_{\text{M·e}}$
       (c) Per cent of C_i
       (d) Per cent of C_T

Solution:

(a) $C_T = C_w + C_{pk} + C_i = 5 + 0.8 + 32.8 = 38.6 \ \mu\mu\text{f}$
     where $C_i = C_{gk} + C_{gp}(1 + \text{VA}) = 13 + 0.64(1 + 30) = 32.8 \ \mu\mu\text{f}$

(b) $C_{\text{M·e}} = C_{gp}\text{VA} = 0.64 \times 30 = 19.2 \ \mu\mu\text{f}$

(c) $\%C_i = \dfrac{C_{\text{M·e}}}{C_i} \times 100 = \dfrac{19.2}{32.8} \times 100 = 58.5$

(d) $\%C_T = \dfrac{C_{\text{M·e}}}{C_T} \times 100 = \dfrac{19.2}{38.6} \times 100 \cong 50$

Example 6-15. In the amplifier circuit of Example 6-11, the stray capacitance of the wiring is 5 $\mu\mu\text{f}$, and the capacitance C_{pk} of the tube is 2.4 $\mu\mu\text{f}$. For the second tube (type 3V4), $C_{gk} = 5.5$ $\mu\mu\text{f}$, $C_{gp} = 0.2$ $\mu\mu\text{f}$, and the voltage amplification of the second stage is 50. (a) What is the total shunting capacitance of this amplifier circuit? (b) What is the value of the Miller-effect capacitance? What percentage is the Miller-effect capacitance of: (c) the input capacitance C_i? (d) the total shunting capacitance C_T?

Given:

$$C_w = 5 \ \mu\mu\text{f}$$
$$C_{pk} \text{ of } VT_1 = 2.4 \ \mu\mu\text{f}$$
$$C_{gk} \text{ of } VT_2 = 5.5 \ \mu\mu\text{f}$$
$$C_{gp} \text{ of } VT_2 = 0.2 \ \mu\mu\text{f}$$
$$\text{VA of } VT_2 = 50$$

Find: (a) C_T
       (b) $C_{\text{M·e}}$
       (c) Per cent of C_i
       (d) Per cent of C_T

Solution:

(a) $C_T = C_w + C_{pk} + C_i = 5 + 2.4 + 15.7 = 23.1 \ \mu\mu\text{f}$
     where $C_i = C_{gk} + C_{gp}(1 + \text{VA}) = 5.5 + 0.2(1 + 50) = 15.7 \ \mu\mu\text{f}$

(b) $C_{\text{M·e}} = C_{gp}\text{VA} = 0.2 \times 50 = 10 \ \mu\mu\text{f}$

(c) $\%C_i = \dfrac{C_{\text{M·e}}}{C_i} \times 100 = \dfrac{10}{15.7} \times 100 = 63.5$

(d) $\%C_T = \dfrac{C_{\text{M·e}}}{C_T} \times 100 = \dfrac{10}{23.1} \times 100 = 43$

The results of Examples 6-14 and 6-15 show that the capacitance introduced by the Miller effect is a large percentage of the total shunting capacitance. Generally, the increase in capacitance due to the Miller effect is less for pentodes than for triodes because of the lower values of grid-plate capacitances in pentodes. Because the Miller effect results in higher values of C_T, the gain of an amplifier decreases at the higher frequencies.

Example 6-16.   What is the voltage amplification and the corresponding decibel loss of the amplifier circuit of Examples 6-10 and 6-14 at frequencies of: (*a*) 10,000 cycles? (*b*) 20,000 cycles? (*c*) 50,000 cycles? (*d*) 100,000 cycles?

Given: $VA_M = 74.5$ Find: (*a*) VA_H; decibel loss
$\quad\quad r_p = 80,000$ ohms   (*b*) VA_H; decibel loss
$\quad\quad R_g = 0.47$ megohm   (*c*) VA_H; decibel loss
$\quad\quad C_T = 38.6\ \mu\mu f$ (*d*) VA_H; decibel loss
$\quad\quad R_c = 0.47$ megohm
$\quad\quad$ (*a*) $f = 10,000$ cycles
$\quad\quad$ (*b*) $f = 20,000$ cycles
$\quad\quad$ (*c*) $f = 50,000$ cycles
$\quad\quad$ (*d*) $f = 100,000$ cycles

Solution:

(*a*) $VA_H = K_H VA_M = 0.990 \times 74.5 = 73.7$

where $K_H = \dfrac{1}{\sqrt{1 + \left(\dfrac{R_{eq}}{X_T}\right)^2}} = \dfrac{1}{\sqrt{1 + \left(\dfrac{59,700}{412,000}\right)^2}} = 0.990$

$R_{eq} = \dfrac{1}{\dfrac{1}{r_p} + \dfrac{1}{R_c} + \dfrac{1}{R_g}} = \dfrac{1}{\dfrac{1}{80,000} + \dfrac{1}{470,000} + \dfrac{1}{470,000}} \cong 59,700$ ohms

$X_T = \dfrac{159,000}{f C_T} = \dfrac{159,000}{10,000 \times 38.6 \times 10^{-6}} \cong 412,000$ ohms

Loss $= 20 \log 0.990 = 20(9.9956 - 10) = -0.088$ db

(*b*) $VA_H = K_H VA_M = 0.960 \times 74.5 = 71.5$

where $K_H = \dfrac{1}{\sqrt{1 + \left(\dfrac{R_{eq}}{X_T}\right)^2}} = \dfrac{1}{\sqrt{1 + \left(\dfrac{59,700}{206,000}\right)^2}} = 0.960$

$R_{eq} = 59,700$ [same as in part (*a*)]

$X_T = \dfrac{159,000}{f C_T} = \dfrac{159,000}{20,000 \times 38.6 \times 10^{-6}} = 206,000$ ohms

Loss $= 20 \log 0.960 = 20(9.9823 - 10) = -0.354$ db

(*c*) $VA_H = K_H VA_M = 0.813 \times 74.5 = 60.5$

where $K_H = \dfrac{1}{\sqrt{1 + \left(\dfrac{R_{eq}}{X_T}\right)^2}} = \dfrac{1}{\sqrt{1 + \left(\dfrac{59,700}{82,400}\right)^2}} = 0.813$

$R_{eq} = 59,700$ [same as in part (*a*)]

$X_T = \dfrac{159,000}{f C_T} = \dfrac{159,000}{50,000 \times 38.6 \times 10^{-6}} = 82,400$ ohms

Loss $= 20 \log 0.813 = 20(9.9101 - 10) = -1.80$ db

(d) $VA_H = K_H VA_M = 0.568 \times 74.5 = 42.3$

where $K_H = \dfrac{1}{\sqrt{1 + \left(\dfrac{R_{eq}}{X_T}\right)^2}} = \dfrac{1}{\sqrt{1 + \left(\dfrac{59,700}{41,200}\right)^2}} = 0.568$

$R_{eq} = 59,700$ [same as in part (a)]

$X_T = \dfrac{159,000}{fC_T} = \dfrac{159,000}{100,000 \times 38.6 \times 10^{-6}} = 41,200$ ohms

Loss $= 20 \log 0.568 = 20(9.7543 - 10) = -4.92$ db

Example 6-17.   What is the voltage amplification and the corresponding decibel loss of the amplifier circuit of Examples 6-11 and 6-15 at frequencies of: (a) 10,000 cycles?   (b) 20,000 cycles?   (c) 50,000 cycles?   (d) 100,000 cycles?

Given: $VA_M = 130$ Find: (a) VA_H; decibel loss

$R_{eq} = \dfrac{10^6}{4.8} \cong 208,000$ ohms   (b) VA_H; decibel loss

$C_T = 23.1 \ \mu\mu\text{f}$ (c) VA_H; decibel loss

(a) $f = 10,000$ cycles   (d) VA_H; decibel loss

(b) $f = 20,000$ cycles

(c) $f = 50,000$ cycles

(d) $f = 100,000$ cycles

Solution:

(a) $VA_H = K_H VA_M = 0.957 \times 130 = 124.4$

where $K_H = \dfrac{1}{\sqrt{1 + (R_{eq}2\pi fC_T)^2}}$

$= \dfrac{1}{\sqrt{1 + (208,000 \times 6.28 \times 10,000 \times 23.1 \times 10^{-12})^2}} = 0.957$

Loss $= 20 \log 0.957 = 20(9.9809 - 10) = -0.38$ db

(b) $VA_H = K_H VA_M = 0.855 \times 130 = 111$

where $K_H = \dfrac{1}{\sqrt{1 + (R_{eq}2\pi fC_T)^2}}$

$= \dfrac{1}{\sqrt{1 + (208,000 \times 6.28 \times 20,000 \times 23.1 \times 10^{-12})^2}} = 0.855$

Loss $= 20 \log 0.855 = 20(9.9320 - 10) = -1.36$ db

(c) $VA_H = K_H VA_M = 0.552 \times 130 = 71.7$

where $K_H = \dfrac{1}{\sqrt{1 + (R_{eq}2\pi fC_T)^2}}$

$= \dfrac{1}{\sqrt{1 + (208,000 \times 6.28 \times 50,000 \times 23.1 \times 10^{-12})^2}} = 0.552$

Loss $= 20 \log 0.552 = 20(9.7419 - 10) = -5.16$ db

(d) $VA_H = K_H VA_M = 0.315 \times 130 = 41$

where $K_H = \dfrac{1}{\sqrt{1 + (R_{eq}2\pi fC_T)^2}}$

$= \dfrac{1}{\sqrt{1 + (208,000 \times 6.28 \times 100,000 \times 23.1 \times 10^{-12})^2}} = 0.315$

Loss $= 20 \log 0.315 = 20(9.4983 - 10) = -10.03$ db

6-9. Circuit Characteristics of *RC*-coupled Amplifiers.   The results of Examples 6-10 to 6-17 have been plotted to produce the curves of Fig. 6-7, commonly called the *frequency-response curves*.   These curves show that both the triode and pentode amplifier circuits used in these examples produce uniform response from approximately 300 to 7,500 cycles. For a maximum deviation of 1.5 db, the response for the triode amplifier is 100 to 45,000 cycles and for the pentode amplifier 70 to 22,000 cycles. An outstanding characteristic of the *RC*-coupled amplifier is its good fidelity over a wide frequency range. The values used in Examples 6-10 to 6-17 are for typical *RC*-coupled amplifiers.

All the gain in the *RC*-coupled amplifier circuit is provided by the tube. The associated circuit elements do not add to the gain but rather reduce

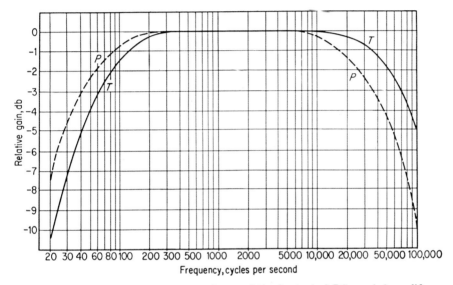

Fig. 6-7. Relative gain versus frequency characteristics for typical *RC*-coupled amplifiers. Curve *P* plotted from the values obtained for the pentode amplifier of Examples 6-11, 6-13, and 6-17.   Curve *T* plotted from the values obtained for the triode amplifier of Examples 6-10, 6-12, and 6-16.

the effective gain of the circuit so that the resultant amplification is less than the amplification factor of the tube.   However, the over-all amplification of an *RC*-coupled amplifier circuit is generally higher than can be obtained with the other coupling methods.   This is possible because of the high values of resistance used to match the high values of plate resistance of high-mu triode and pentode tubes.   Other advantages of this type of amplifier circuit are: (1) the parts are low in cost and require very little space; (2) as there are no coils or transformers in the circuit, there is very little pickup of undesirable currents from any a-c leads; thus the amount of nonlinear distortion is minimized.   A disadvantage of the *RC*-coupled amplifier is that a higher B supply voltage must be used in order to compensate for the voltage drop across the coupling resistor.

Because of the many advantages of *RC* coupling it is commonly used in audio-amplifier circuits.   The theory of this type of amplifier circuit has been presented in great detail because of this fact and also because its principle of operation is basic and is easily understood.   It therefore can serve as a comparison and can be used in the explanation of the operation of other types of amplifier circuits.

6-10. Impedance-coupled Amplifier. *Basic Circuit Action.* One method of eliminating the high voltage drop between the B power supply and the plate of the tube is to replace the coupling resistor of the *RC*-coupled amplifier with an iron-core choke having a high inductance and low resistance.   Amplifier circuits using this method of coupling are called *impedance-coupled amplifiers* and are similar to the *RC*-coupled amplifier except that an inductance coil L_c (Fig. 6-8) is substituted for the

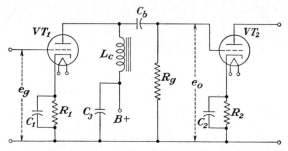

Fig. 6-8. Impedance-coupled amplifier circuit using triodes.

coupling resistor.   The voltage drop across the coupling impedance will be small as it is dependent only on the resistance of the coil and the plate current.   The voltage of the B power supply need then be only slightly higher than the required plate voltage of the tube used.

The impedance that the coupling coil offers to the signal current is dependent on its inductance and the frequency of the signal.   In order to obtain high impedance at low audio frequencies, the inductance of the coupling coil is made as high as is practicable.   A high impedance is desired in order to obtain a high value of voltage amplification.   The inductance of choke coils used as coupling impedances for a-f amplifiers ranges from 10 to 800 henrys.

Frequency Characteristics. The frequency response of impedance-coupled amplifiers is not so good as that obtained with *RC*-coupled amplifiers (see Fig. 6-10).   The decrease in gain is greater at both the low and high audio frequencies than for *RC*-coupled amplifiers.   This is due to the fact that the impedance of a resistor is fairly uniform at all frequencies and its distributed capacitance is negligible, while the impedance of a coil varies directly with frequency and the coil has an appreciable amount of distributed capacitance.   At low audio frequencies the impedance of the coupling coil will be relatively low, thus causing a decrease in the voltage gain at these frequencies.   The distributed capacitance of the coil increases the total shunt capacitance of the circuit, and at the high audio frequencies this increase in C_T causes more of the output current to be bypassed, thus further decreasing the voltage gain at these frequencies.

Voltage Amplification of the Impedance-coupled Amplifier. The method of determining the voltage amplification produced by the impedance-coupled amplifier is similar to that used for the RC-coupled amplifier circuit. The circuit characteristics are again observed at the low, intermediate, and high frequencies of the audio range. The equivalent

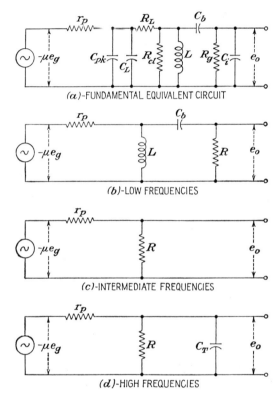

(a)-FUNDAMENTAL EQUIVALENT CIRCUIT

(b)-LOW FREQUENCIES

(c)-INTERMEDIATE FREQUENCIES

(d)-HIGH FREQUENCIES

Fig. 6-9. Equivalent electrical circuits at various audio frequencies for an impedance-coupled amplifier circuit using triodes.

electrical circuits for the three frequency ranges are given in Fig. 6-9. In these circuits R_L represents the resistance of the coupling impedance, C_L represents the distributed capacitance of the coupling unit, and R_{cl} represents the core loss of the coupling impedance. All other designations are the same as before.

At the low and intermediate frequencies, the reactance of the shunting capacitances is so high as compared with R that it may be disregarded at these frequencies. At the intermediate and high frequencies, the reactance of the inductor is much greater than R, and hence it may be disregarded at these frequencies. Thus the voltage amplification of the circuit at the intermediate frequencies [Eq. (6-24)] is dependent largely upon the resistances in the circuit because the effects of all the reactances are negligible at these frequencies. The voltage amplification at the low frequencies [Eq. (6-25)] decreases because of the reduction in the value of X_L shunting the load and the increase in the value of $X_{C.b}$ connected in series with R_g. The voltage amplification at the high frequencies [Eq. (6-26)] decreases because $X_{C.T}$ becomes relatively low at these frequencies, thereby causing a decrease in the voltage amplification. The equations for the approximate voltage amplification at the various frequencies are

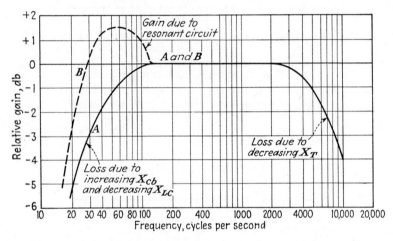

FIG. 6-10. Relative gain versus frequency characteristics for an impedance-coupled a-f amplifier. Curve A for a simple impedance-coupled circuit (Fig. 6-8), Curve B for a circuit with a resonant section [Fig. 6-11(a) and (b)].

$$VA_M \cong \frac{\mu R}{r_p + R} \tag{6-24}$$

also
$$VA_M \cong g_m R_{eq} \tag{6-10a}$$

$$VA_L \cong \frac{1}{\sqrt{1 + \left(\dfrac{R_{eq}}{X_L}\right)^2 + \left(\dfrac{R_{eq} X_C}{r_p R_g}\right)^2}} \, g_m R_{eq} \tag{6-25}$$

$$VA_H \cong \frac{1}{\sqrt{1 + \left(\dfrac{R_{eq}}{X_T}\right)^2}} \, g_m R_{eq} \tag{6-26}$$

or
$$VA_H \cong \frac{g_m R_{eq}}{\sqrt{1 + (R_{eq} 2\pi f C_T)^2}} \tag{6-26a}$$

where $R = \dfrac{R_{cl} R_g}{R_{cl} + R_g} \cong R_g$ when R_{cl} becomes so high that its effect is negligible

$R_{eq} = \dfrac{r_p R_{cl} R_g}{r_p R_{cl} + r_p R_g + R_{cl} R_g} \cong \dfrac{r_p R_g}{r_p + R_g}$ when R_{cl} is very high

R_{cl} = resistance which would produce an effect equivalent to core loss, ohms

X_C = reactance of blocking capacitor C_b, ohms

X_T = reactance of total shunting capacitance C_T, ohms

$C_T = C_{pk} + C_L + C_i + C_W$

Comparing Eqs. (6-24), (6-25), and (6-26a) with Eqs. (6-7), (6-13), and (6-16) will show that the equations for the voltage amplification of impedance-coupled amplifiers are very much similar to those of *RC*-coupled amplifiers.

Example 6-18.   The coupling coil of an impedance-coupled amplifier circuit similar to Fig. 6-8 has an inductance of 150 henrys and a resistance of 3,500 ohms.   The plate and grid-bias voltages applied to the low-mu tube used in this amplifier produce values of: I_b = 8 ma, r_p = 10,000 ohms, and g_m = 2,000 μmhos.   Other circuit values are: R_g = 500,000 ohms, C_b = 0.01 μf, and C_T = 200 $\mu\mu$f.   The core loss of the coupling unit is so low that R_{cl} is ignored.   (a) What is the impedance of the coupling coil at 50, 1,000, and 10,000 cycles (neglecting the effect of its resistance and distributed capacitance)?   (b) What voltage is required of the B power supply?   (c) What is the voltage amplification of the circuit at 50, 1,000, and 10,000 cycles?   (d) What is the gain in decibels at 50, 1,000, and 10,000 cycles?   (e) What is the decibel variation over a range of 50 to 10,000 cycles?

Given: $L = 150\ h$ Find: (a) X_L
$R_L = 3,500$ ohms   (b) E_{bb}
$I_b = 8$ ma   (c) VA
$r_p = 10,000$ ohms   (d) Decibel gain
$g_m = 2,000\ \mu$mhos   (e) Decibel variation
$R_g = 500,000$ ohms
$C_b = 0.01\ \mu$f
$C_T = 200\ \mu\mu$f

Solution:

(a) $X_L = 2\pi fL = 6.28 \times 50 \times 150 = 47,100$ ohms (at 50 cycles)
   $X_L = 2\pi fL = 6.28 \times 1,000 \times 150 = 942,000$ ohms (at 1,000 cycles)
   $X_L = 2\pi fL = 6.28 \times 10,000 \times 150 = 9,420,000$ ohms (at 10,000 cycles)

(b) $E_{bb} = E_b + I_b R_L = 250 + 0.008 \times 3,500 = 278$ volts

(c) $VA_{50} = \dfrac{1}{\sqrt{1 + \left(\dfrac{R_{eq}}{X_L}\right)^2 + \left(\dfrac{R_{eq}X_c}{r_p R_g}\right)^2}}\ g_m R_{eq}$

$= \dfrac{2,000 \times 10^{-6} \times 9,800}{\sqrt{1 + \left(\dfrac{9,800}{47,100}\right)^2 + \left(\dfrac{9,800 \times 318,000}{10,000 \times 500,000}\right)^2}} = 16.4$

where $R_{eq} = \dfrac{r_p R_g}{r_p + R_g} = \dfrac{10,000 \times 500,000}{10,000 + 500,000} = 9,800$ ohms

$X_L = 2\pi fL = 6.28 \times 50 \times 150 = 47,100$ ohms

$X_c = \dfrac{159,000}{fC} = \dfrac{159,000}{50 \times 0.01} = 318,000$ ohms

$VA_{1,000} = g_m R_{eq} = 2,000 \times 10^{-6} \times 9,800 = 19.6$

$VA_{10,000} = \dfrac{1}{\sqrt{1 + (R_{eq} 2\pi fC_T)^2}}\ g_m R_{eq}$

$= \dfrac{2,000 \times 10^{-6} \times 9,800}{\sqrt{1 + (9,800 \times 6.28 \times 10^4 \times 200 \times 10^{-12})^2}} = 19.4$

(d)   Gain = 20 × log 16.4 = 20 × 1.2148 = 24.296 db (at 50 cycles)
     Gain = 20 × log 19.6 = 20 × 1.2923 = 25.846 db (at 1,000 cycles)
     Gain = 20 × log 19.4 = 20 × 1.2878 = 25.756 db (at 10,000 cycles)
(e)   Variation = 25.8 − 24.3 = 1.5 db

Double-impedance Coupling. The gain of the impedance-coupled amplifier at the low frequencies may be increased by replacing both the coupling resistor and the grid resistor of the RC-coupled amplifier circuit with low-frequency choke coils (Fig. 6-11a); this circuit is called a *double-impedance-coupled amplifier.* The inductances of both coils are generally of the same value and combined with the blocking capacitor form a series tuned circuit that is resonant at a low a-f value. An increase in the voltage amplification will be obtained at this frequency.

A disadvantage of this circuit is that the use of two coils in parallel doubles the amount of distributed capacitance, thus further increasing the shunt capacitance of the circuit. The gain at the high audio frequencies will therefore be further decreased.

Parallel Plate Feed. In the impedance-coupled amplifier circuits of Figs. 6-8 and 6-11a the direct current from the B power supply must flow through the coupling coil. The flow of direct current through a coil having an iron core will decrease the permeability of the magnetic circuit, thus causing a decrease in the inductance of the coil. Although this decrease in inductance causes a decrease in the voltage gain over the entire a-f range, it is only at the low frequencies that any appreciable difference is obtained.

In order to prevent decreases in voltage gain due to changes in the inductance of the coil, the direct current is isolated from the coupling impedance by providing two parallel paths for the plate current, Fig. 6-11b. The resistor R_B provides the path for the d-c component or steady value of the plate current, and the coupling coil L_c and the blocking capacitor C_3 will carry only the signal or varying component of the plate current. This circuit, called an *impedance-coupled parallel-plate-feed amplifier,* is used whenever it is desirable to keep the d-c plate current out of the coupling impedance. At the resonant frequency of the series tuned circuit formed by L_c and C_3, the voltage gain will be increased since the voltage across the coupling coil will be comparatively high. Thus the voltage gain at any audio frequency can be increased by using values of L_c and C_3 that will make the series tuned circuit resonant for that frequency.

The response at frequencies above and below the resonant frequency will attenuate either slowly or rapidly, depending on the value of R_B. Increasing the value of R_B decreases the rate of attenuation, thus broadening the peak of the response curve. Decreasing this resistance will increase the rate of attenuation, thus making the peak of the response curve sharp. Figure 6-10 shows the variation in voltage gain over the a-f range. The curve shown in the dashed line represents the effect of the series tuned portion of the circuit. By choosing suitable values for

R_B, L_c, and C_3 (Fig. 6-11b), it is possible to improve the response at low audio frequencies and to broaden the entire response curve of the circuit.

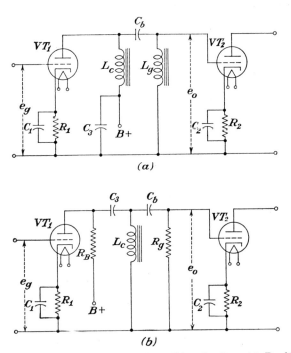

(a)

(b)

FIG. 6-11. Applications of impedance-coupled amplifier circuits. (*a*) Double-impedance-coupled amplifier circuit. (*b*) Impedance-coupled amplifier circuit with parallel plate feed.

Example 6-19.   The coupling coil used in an impedance-coupled audio-amplifier circuit similar to Fig. 6-11 has an inductance of 300 henrys.   What value of blocking capacitor is required to cause the tuned circuit to be resonant at (*a*) 82 cycles?   (*b*) 820 cycles?   (*c*) 8,200 cycles?

Given: $L_c = 300$ henrys   Find: (*a*) C at 82 cycles
(*b*) C at 820 cycles
(*c*) C at 8,200 cycles

Solution:

(*a*) $C = \dfrac{25,300}{f_r{}^2 L} = \dfrac{25,300}{(82 \times 10^{-3})^2 \times 300 \times 10^6} = 0.0125\ \mu f$

(*b*) $C = \dfrac{25,300}{f_r{}^2 L} = \dfrac{25,300}{(820 \times 10^{-3})^2 \times 300 \times 10^6} = 125\ \mu\mu f$

(*c*) $C = \dfrac{25,300}{f_r{}^2 L} = \dfrac{25,300}{(8,200 \times 10^{-3})^2 \times 300 \times 10^6} = 1.25\ \mu\mu f$

Circuit Characteristics.   In comparison with RC-coupled amplifiers, the impedance-coupled audio amplifier requires less B supply voltage, usually has a slightly greater gain per stage, but its response over the entire a-f range is not so uniform.   However, its disadvantage is sometimes offset by the fact that it is possible to peak the response of an impedance-coupled amplifier at any desired audio frequency.

6-11. Transformer-coupled Amplifier. *Basic   Circuit   Connections.*
Audio-amplifier circuits using a low-frequency (iron-core) transformer as
the coupling unit between two successive stages are called *transformer-
coupled audio amplifiers.*   As with the impedance-coupled amplifier
circuit, either series or parallel feed can be used.   Connecting one end of
the primary winding to the plate of the first tube and the other end to the
B power supply will produce series feed (Fig. 6-12a).   Parallel feed
(Fig. 6-14) is obtained by connecting resistor R_B between the plate and B
power supply; the primary winding, in series with capacitor C_b, is connected in parallel with R_B.

Frequency   Characteristics.   As   transformer-coupled   audio-amplifier
circuits generally employ triodes, the equivalent circuit can be considered
in the constant-voltage-generator form (see Fig. 6-12b).   A complete
analysis of this circuit is quite complex because of the many factors that
vary with the frequency such as the primary and secondary impedances,
the shunting effect due to the distributed capacitance of the windings, the
shunting effect of the core loss, and the coupled impedance effect.   However, an approximate solution can be obtained by omitting the shunting
effects of the distributed capacitance and the core loss and also omitting
the effects of coupled impedance.   The equivalent circuit may then be
simplified as shown in Fig. 6-12c.   In addition to the factors shown on
Fig. 6-12c, the voltage amplification is also dependent upon the coefficient
of coupling between the primary and secondary windings of the transformer.   In well-designed a-f transformers, the coefficient of coupling is
very close to unity, and hence this factor may be disregarded.   The
voltage amplification for any a-f value may be expressed as

$$\text{VA} = \mu n \frac{Z_P}{Z_P + r_p} \tag{6-27}$$

where n = transformer secondary-to-primary turns ratio
      Z_P = total equivalent impedance of primary winding

Voltage Amplification for the Intermediate-frequency Range.   At the low-
and intermediate-frequency values, the reactances of all of the shunting
capacitances shown in Fig. 6-12b are so high that their effects may be
disregarded.   The core loss in well-designed transformers is generally so
low that its effective resistance R_{cl} may also be disregarded; the equivalent circuit may then be further simplified as shown in Figs. 6-13a and
6-13b.   The output voltage will then be

$$e_o = \mu e_g \frac{X_{LP}}{\sqrt{R_1{}^2 + X_{LP}{}^2}} n \tag{6-28}$$

and the voltage amplification becomes

$$\text{VA}_M = \mu n \frac{X_{LP}}{\sqrt{R_1{}^2 + X_{LP}{}^2}} \tag{6-29}$$

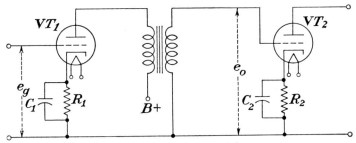

(a)-Simple transformer-coupled amplifier circuit

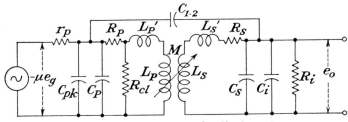

(b)-Complete equivalent plate circuit of Fig. *(a)*

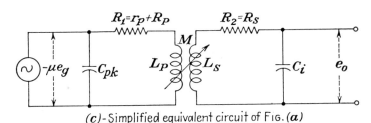

(c)-Simplified equivalent circuit of Fig. *(a)*

where r_p = Plate resistance of **VT₁**

C_{pk} = Plate-cathode capacitance of **VT₁**

C_P = Distributed capacitance of the primary winding

R_P = Resistance of the primary winding

R_{cl} = Equivalent resistance of the transformer's core loss

L_P' = Leakage inductance of the primary

L_P = Inductance of the primary

C_{1-2} = Distributed capacitance between the primary and secondary windings

L_S = Inductance of the secondary

L_S' = Leakage inductance of the secondary

R_S = Resistance of the secondary winding

C_S = Distributed capacitance of the secondary winding

C_i = Input capacitance of **VT₂**, see Eq.(6-23)

R_i = Input resistance of **VT₂**

Fig. 6-12. Simple transformer-coupled amplifier circuit using triodes.

where X_{LP} = reactance of primary winding

R_1 = equivalent primary circuit resistance = $\dfrac{r_p R_{cl}}{r_p + R_{cl}} + R_P$

NOTE: $R_1 \cong r_p + R_P$ when R_{cl} is large compared to r_p

R_{cl} = resistance producing an effect equal to the core loss

In well-designed transformers, the value of X_{LP} is much greater than R_1 and hence

$$VA_M \cong \mu n \qquad (6\text{-}30)$$

Voltage Amplification for the Low-frequency Range. At low frequencies, the equivalent circuit conditions as shown in Fig. 6-13a are the same as for

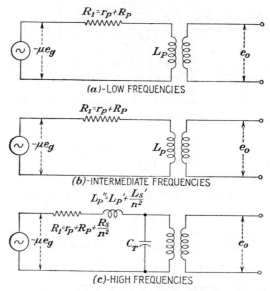

FIG. 6-13. Simplified equivalent electrical circuits at various audio frequencies for a transformer-coupled amplifier using triodes.

the intermediate frequencies. The voltage amplification will therefore be the same as expressed by Eq. (6-29). This may also be expressed as

$$VA_L = K_L VA_M \qquad (6\text{-}12)$$

where

$$K_L = \frac{X_{LP}}{\sqrt{R_1{}^2 + X_{LP}{}^2}} \qquad (6\text{-}31)$$

Well-designed transformers provide uniform response (± 2 db) for frequencies as low as 100 cycles with series-plate-feed circuits and as low as 30 cycles with parallel-plate-feed circuits. In order to obtain such uniform response, X_{LP} must be much greater than R_1, even at low frequencies. Consequently, the transformer must have a high value of inductance.

Voltage Amplification for the High-frequency Range. At the high frequencies, the effects of the shunting capacitances and the leakage inductances can no longer be disregarded.   Under these conditions, the equivalent circuit will be as shown in Fig. 6-13c.   The voltage amplification of the circuit will then be

$$VA_H = K_H VA_M \qquad (6\text{-}15)$$

and
$$K_H = \frac{X_T}{\sqrt{R_{1-2}{}^2 + (X_P'' - X_T)^2}} \qquad (6\text{-}32)$$

where X_T = total equivalent shunting reactance referred to primary
R_{1-2} = total equivalent resistance referred to primary
X_P'' – total equivalent leakage reactance referred to primary

Examination of Eq. (6-32) will show that K_H will be maximum when X_P'' is equal to X_T.   From Eq. (6-32) and the basic equation for frequency of resonance $f_r = \dfrac{1}{2\pi \sqrt{LC}}$, it can be shown that

$$K_{H \cdot max} = \frac{10^6 \sqrt{\dfrac{L_P''}{C_T}}}{R_{1-2}} \qquad (6\text{-}32a)$$

where L_P'' = inductance, henrys
C_T = capacitance, $\mu\mu$f

Comparison of the low- and high-frequency amplification factors of the transformer-coupled amplifier circuits with those of the RC-coupled circuits will show that they are similar in form.

Example 6-20.   In a certain transformer-coupled a-f amplifier circuit (Fig. 6-12a) the tube constants of VT_1 are: $C_{pk} = 3.6$ $\mu\mu$f, $\mu = 20$, and $r_p = 7{,}700$ ohms.   The input capacitance of VT_2 is 10 $\mu\mu$f.   The transformer constants are: $n = 3$; $L_P = 100$ henrys; $L_S = 900$ henrys; $R_P = 100$ ohms; $R_S = 900$ ohms; $L_P' = 0.15$ henry; $L_S' = 1.35$ henrys; total distributed capacitance referred to the primary $= 60$ $\mu\mu$f; assume the core-loss effect to be negligible.   The total stray capacitance of the circuit wiring referred to the primary is 10 $\mu\mu$f.   What voltage amplification is produced by the stage (consisting of VT_1 and the transformer) at (a) 1,000 cycles?   (b) 50 cycles? (c) 10,000 cycles?   (d) At what frequency will the maximum amplification occur? (e) What is the voltage amplification at this frequency?

Given: VT_1—$C_{pk} = 3.6$ $\mu\mu$f   Find: (a) VA @ 1,000 cycles
$\mu = 20$ (b) VA @ 50 cycles
$r_p = 7{,}700$ ohms   (c) VA @ 10,000 cycles
VT_2—$C_i = 10$ $\mu\mu$f   (d) f for VA$_{max}$
$n = 3$ (e) VA$_{max}$
$L_P = 100$ henrys
$L_S = 900$ henrys
$R_P = 100$ ohms
$R_S = 900$ ohms
$L_P' = 0.15$ henry
$L_S' = 1.35$ henrys
$C_D = 60$ $\mu\mu$f
$C_W = 10$ $\mu\mu$f

Solution:

(a) $\text{VA}_{1,000} = \mu n = 20 \times 3 = 60$

(b) $\text{VA}_{50} = \dfrac{X_{LP}\text{VA}_M}{\sqrt{R_1{}^2 + X_{LP}{}^2}} = \dfrac{31{,}400 \times 60}{\sqrt{7{,}800^2 + 31{,}400^2}} = 58.2$

where $R_1 \cong r_p + R_P = 7{,}700 + 100 = 7{,}800$ ohms
$X_{LP} = 2\pi f L_P = 6.28 \times 50 \times 100 = 31{,}400$ ohms

(c) $\text{VA}_{10,000} = \dfrac{X_T\text{VA}_M}{\sqrt{R_{1-2}{}^2 + (X_P'' - X_T)^2}} = \dfrac{97{,}333 \times 60}{\sqrt{7{,}900^2 + (18{,}840 - 97{,}333)^2}} = 74$

where $X_T = \dfrac{1}{2\pi f C_T} = \dfrac{1}{6.28 \times 10^4 \times 163.6 \times 10^{-12}} = 97{,}333$ ohms

$C_T = C_{pk\cdot 1} + C_D + C_w + n^2 C_i = 3.6 + 60 + 10 + 9 \times 10 = 163.6\ \mu\mu\text{f}$

$R_{1-2} = r_p + R_P + \dfrac{R_S}{n^2} = 7{,}700 + 100 + \dfrac{900}{9} = 7{,}900$ ohms

$X_P'' = 2\pi f L_P'' = 6.28 \times 10^4 \times 0.3 = 18{,}840$ ohms

$L_P'' = L_P' + \dfrac{L_S'}{n^2} = 0.15 + \dfrac{1.35}{9} = 0.3$ henry

(d) $f_r = \dfrac{1}{2\pi \sqrt{L_P'' C_T}} = \dfrac{1}{6.28 \sqrt{0.3 \times 163.6 \times 10^{-12}}} = 22{,}747$ cycles

(e) $\text{VA}_{\max} = \dfrac{10^6 \sqrt{\dfrac{L_P''}{C_T}}\,\text{VA}_M}{R_{1-2}} = \dfrac{10^6 \sqrt{\dfrac{0.3}{163.6}} \times 60}{7{,}900} = 325$

The low and high audio-frequency response of transformer-coupled amplifiers will decrease in the same manner and for similar reasons as the

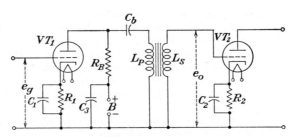

Fig. 6-14. Transformer-coupled amplifier circuit with parallel plate feed.

double-impedance-coupled amplifier. By using parallel feed instead of series, the frequency response curve can be peaked at either the low or high frequencies to obtain a more uniform response. The shape of the frequency response curve will, therefore, like the impedance-coupled amplifier, have a number of variations (see Fig. 6-15).

6-12. Audio-frequency Transformers. Audio transformers increase the voltage of the a-f signals in direct proportion to the turns ratio of the secondary and primary windings. This ratio may vary from 1 to 1 to as high as 6 to 1. In addition to the turns ratio, the voltage gain of the amplifier circuit will depend on the impedance of the primary winding.

In order to obtain maximum voltage amplification, the ratio of primary impedance to plate resistance should be high.

The impedance of the primary winding is dependent on its inductance and the frequency of the audio signal. In order to obtain a high inductance, a large number of turns is wound on a soft-iron core having a high permeability. Increasing the number of turns on the primary increases the number of turns required on the secondary. This in turn increases the size and the cost of the transformer. The size of the transformer may be kept at a minimum by winding both the primary and secondary with a small-size wire. Because of the large number of turns on the primary and secondary windings their distributed capacitance will be quite high. The shunting effect of this capacitance will decrease the gain, particularly at the high audio frequencies.

Decreasing the number of turns on both the primary and secondary windings and keeping their turns ratio high is another method of decreasing the size and cost of the transformer. The primary winding of this type of transformer has a low value of inductance. The impedance of the primary winding for low a-f signals will therefore be very low compared to the plate resistance of the tube. The voltage gain for low frequencies is thus reduced considerably.

The above discussion indicates that the frequency response of the transformer must be considered as well as its turns ratio. Maximum voltage amplification of a transformer-coupled amplifier cannot be obtained unless a well-designed transformer is used. To obtain high fidelity, design and construction considerations limit the turns ratio to approximately 3 to 1.

Example 6-21. An amplifier circuit similar to Fig. 6-12a uses a triode tube whose plate resistance is 10,000 ohms and whose amplification factor is 17. The transformer has a turns ratio of 2 to 1 and the impedance of the primary winding is 15,000 ohms at 50 cycles and 30,000 ohms at 100 cycles. (a) What is the voltage amplification at the intermediate frequency? (b) What is the voltage amplification at 50 cycles? (c) What is the voltage amplification at 100 cycles?

Given: $n = 2$ Find: (a) VA_M
           $\mu = 17$ (b) VA @ 50 cycles
           $r_p = 10{,}000$ ohms   (c) VA @ 100 cycles
           $Z_P = 15{,}000$ ohms @ 50 cycles
           $Z_P = 30{,}000$ ohms @ 100 cycles

Solution:

(a) $VA_M = \mu n = 17 \times 2 = 34$

(b) $VA_{50} = \mu n \dfrac{Z_P}{Z_P + r_p} = 34 \dfrac{15{,}000}{\sqrt{15{,}000^2 + 10{,}000^2}} = 28.3$

(c) $VA_{100} = \mu n \dfrac{Z_P}{Z_P + r_p} = 34 \dfrac{30{,}000}{\sqrt{30{,}000^2 + 10{,}000^2}} = 32.2$

Example 6-22. A transformer having a 3 to 1 turns ratio is substituted for the transformer used in Example 6-21. The impedance of this transformer is 7,500 ohms at 50 cycles and 15,000 ohms at 100 cycles. (*a*) What is the voltage amplification at the intermediate frequency? (*b*) What is the voltage amplification at 50 cycles? (*c*) What is the voltage amplification at 100 cycles?

Given: $n = 3$ Find: (*a*) VA_M
          $\mu = 17$ (*b*) VA_{50}
          $r_p = 10,000$ ohms   (*c*) VA_{100}
          $Z_P = 7,500$ ohms at 50 cycles
          $Z_P = 15,000$ ohms at 100 cycles

Solution:

(*a*) $VA_M = \mu n = 17 \times 3 = 51$

(*b*) $VA_{50} = \mu n \dfrac{Z_P}{Z_P + r_p} = 51 \dfrac{7500}{\sqrt{7,500^2 + 10,000^2}} = 30.6$

(*c*) $VA_{100} = \mu n \dfrac{Z_P}{Z_P + r_p} = 51 \dfrac{15,000}{\sqrt{15,000^2 + 10,000^2}} = 42.5$

Example 6-23.   What is the decibel loss at 50 cycles and 100 cycles for (*a*) the amplifier circuit used in Example 6-21? (*b*) The amplifier circuit used in Example 6-22?

Given: (*a*) $VA_M = 34$ Find: (*a*) Decibel loss @ 50 cycles
             $VA_{50} = 28.3$ Decibel loss @ 100 cycles
             $VA_{100} = 32.2$ (*b*) Decibel loss @ 50 cycles
        (*b*) $VA_M = 51$ Decibel loss @ 100 cycles
             $VA_{50} = 30.6$
             $VA_{100} = 42.5$

Solution:

(*a*)   Loss at 50 cycles $= 20 \log \dfrac{VA_{50}}{VA_M} = 20 \log \dfrac{28.3}{34} = 20 \log 0.832$

$= 20(9.9201 - 10) = -1.598$ db

Loss at 100 cycles $= 20 \log \dfrac{VA_{100}}{VA_M} = 20 \log \dfrac{32.2}{34} = 20 \log 0.947$

$= 20(9.9763 - 10) = -0.474$ db

(*b*)   Loss at 50 cycles $= 20 \log \dfrac{VA_{50}}{VA_M} = 20 \log \dfrac{30.6}{51} = 20 \log 0.600$

$= 20(9.7782 - 10) = -4.436$ db

Loss at 100 cycles $= 20 \log \dfrac{VA_{100}}{VA_M} = 20 \log \dfrac{42.5}{51} = 20 \log 0.833$

$= 20(9.9206 - 10) = -1.588$ db

The results obtained in Examples 6-21 and 6-22 show that increasing the turns ratio of the transformer increases the voltage amplification at the intermediate audio frequencies. However, if the increase in turns ratio is obtained by decreasing the number of turns on the primary, there will be very little change in amplification at the low audio frequencies. The results obtained in Example 6-23 for the amplifier circuit using the transformer having a 2-to-1 ratio indicate that the frequency response is fairly uniform as the changes in volume for the low audio frequencies are barely perceptible. On the other hand the frequency response for the amplifier circuit using the transformer having a 3-to-1 ratio is not as good because the decrease in inductance causes appreciable changes in volume at the low audio frequencies.

Figure 6-15 illustrates the variation of voltage gain with frequency in a transformer-coupled audio-amplifier circuit. Curve *A* represents the voltage gain of the circuit when using a high-grade transformer, and curve *B* shows the voltage gain of the same circuit when using a low-grade transformer. Using a high-grade transformer produces a fairly uniform response over a large portion of the a-f range; but using a low-grade

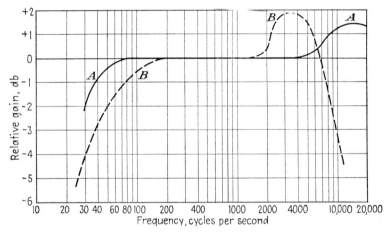

Fig. 6-15. Relative gain versus frequency characteristics for a transformer-coupled a-f amplifier. Curve *A* is for a circuit using a high-grade transformer; curve *B* is for a circuit using a low-grade transformer.

transformer causes the response to fall off at both the high and low audio frequencies.

Advantages of Using a Transformer as a Coupling Element. Audio transformers are used as a coupling element whenever a greater voltage amplification is required than can be obtained from the tube alone. By using a well-designed transformer, the voltage amplification for the intermediate audio frequencies will be approximately equal to the product of the amplification factor of the tube and the turns ratio. A fairly uniform frequency response can be obtained by using a transformer having a comparatively high value of inductance and limiting its distributed capacitance so that when combined with the transformer's inductance it will produce resonance at a high audio frequency (see Fig. 6-15).

Another advantage of this type of coupling is that the voltage drop across the transformer winding is negligible, thus making it possible to use a low-voltage B power supply. Disadvantages are the large size and the high cost of the transformers required to produce high fidelity.

6-13. Cathode-coupled Amplifier. *Basic Circuit Action.*   In the basic cathode-coupled amplifier circuit (Fig. 6-16) the signal is fed to the grid-cathode circuit, and the output is taken from across the unbypassed resistor R_k in the cathode circuit.   The plate may be connected directly to the B power supply or through resistor R_1, which must be bypassed by a capacitor C_2 that is large enough to effectively ground the plate to a-c signals.   This circuit is most commonly called a *cathode follower*, but is also known as a *cathode-coupled amplifier* and a *grounded-plate amplifier.*

When the input signal is positive, the plate (and cathode) current will increase and cause an increase in the output voltage developed across the cathode resistor R_k, hence the name *cathode follower*.   A decrease in the

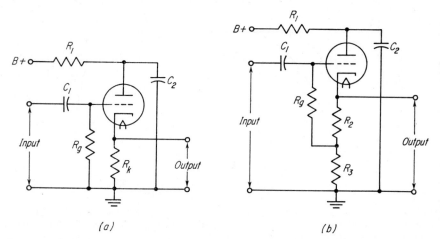

Fig. 6-16. Cathode-coupled amplifier circuits.   (*a*) Cathode resistor supplying all of the d-c bias.   (*b*) One method of obtaining reduced bias.

input signal will cause a corresponding decrease in the output signal. Thus, there is no reversal of phase between input and output signals as is the case in the RC-coupled amplifier where the output is taken from the external plate circuit.

Applying a positive-going signal to the input causes an increase in the voltage across R_k and makes the cathode more positive with respect to ground.   This is equivalent to making the grid more negative with respect to the cathode which is opposite to the effect produced by the original positive-going applied signal.   Thus, this circuit is a *degenerative-feedback amplifier*, also called an *inverse-feedback* or *negative-feedback amplifier.*

Voltage Amplification.   In order that the signal voltage applied to the grid may produce changes in the plate current, the voltage across R_k can never equal or exceed the signal voltage.   The voltage amplification of a cathode-follower stage is therefore always less than unity.   However, the cathode follower does provide a power gain and is capable of supplying relatively large values of signal current to low-impedance circuits. The voltage amplification of the cathode-follower circuit is expressed by

for triodes $$\text{VA} = \frac{\mu R_k}{r_p + R_k(\mu + 1)}$$ (6-33)

for pentodes $$\text{VA} = \frac{g_m R_k}{1 + g_m R_k}$$ (6-34)

In Eqs. (6-33) and (6-34) the denominator will always be greater than the numerator; hence the voltage amplification will always be less than 1. The higher the values are for μ and R_k with triodes, and g_m and R_k with pentodes, the closer the voltage amplification will approach unity.

Example 6-24.   A certain triode having an amplification factor of 20 and a plate resistance of 6,500 ohms is used in a cathode-follower circuit.   What is the voltage amplification of the circuit when the value of the cathode resistor is: (*a*) 22,000 ohms? (*b*) 2,200 ohms?   (*c*) 220 ohms?

Given: $\mu = 20$ Find: (*a*) VA
$r_p = 6,500$ ohms   (*b*) VA
(*a*) $R_k = 22,000$ ohms   (*c*) VA
(*b*) $R_k = 2,200$ ohms
(*c*) $R_k = 220$ ohms

Solution:

(*a*) $\text{VA} = \dfrac{\mu R_k}{r_p + R_k(\mu + 1)} = \dfrac{20 \times 22,000}{6,500 + 22,000 \times (20 + 1)} \cong 0.94$

(*b*) $\text{VA} = \dfrac{\mu R_k}{r_p + R_k(\mu + 1)} = \dfrac{20 \times 2,200}{6,500 + 2,200 \times (20 + 1)} \cong 0.834$

(*c*) $\text{VA} = \dfrac{\mu R_k}{r_p + R_k(\mu + 1)} = \dfrac{20 \times 220}{6,500 + 220 \times (20 + 1)} \cong 0.396$

Example 6-25.   A certain pentode having a transconductance of 6,000 μmhos is used in a cathode-follower circuit.   What is the voltage amplification of the circuit when the value of the cathode resistor is: (*a*) 10,000 ohms?   (*b*) 1,000 ohms?   (*c*) 100 ohms?

Given: $g_m = 6,000$ μmhos   Find: (*a*) VA
(*a*) $R_k = 10,000$ ohms   (*b*) VA
(*b*) $R_k = 1,000$ ohms   (*c*) VA
(*c*) $R_k = 100$ ohms

Solution:

(*a*) $\text{VA} = \dfrac{g_m R_k}{1 + g_m R_k} = \dfrac{6,000 \times 10^{-6} \times 10,000}{1 + 6,000 \times 10^{-6} \times 10,000} \cong 0.983$

(*b*) $\text{VA} = \dfrac{g_m R_k}{1 + g_m R_k} = \dfrac{6,000 \times 10^{-6} \times 1,000}{1 + 6,000 \times 10^{-6} \times 1,000} \cong 0.857$

(*c*) $\text{VA} = \dfrac{g_m R_k}{1 + g_m R_k} = \dfrac{6,000 \times 10^{-6} \times 100}{1 + 6,000 \times 10^{-6} \times 100} \cong 0.375$

Output Resistance.   The equivalent output circuit of the cathode follower (Fig. 6-17) shows that the output impedance consists basically of

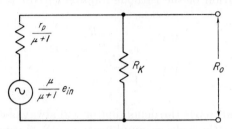

Fig. 6-17. Equivalent output circuit of the cathode-coupled amplifier of Fig. 6-16a.

two resistances connected in parallel, thus

$$R_o = \frac{\dfrac{r_p}{\mu + 1} R_k}{\dfrac{r_p}{\mu + 1} + R_k} \tag{6-35}$$

Transposing terms

$$R_o = \frac{r_p R_k}{r_p + R_k (\mu + 1)} \tag{6-35a}$$

when μ is much larger than unity,

$$R_o \cong \frac{r_p R_k}{r_p + \mu R_k} \tag{6-35b}$$

Substituting $g_m r_p$ for μ, then

$$R_o \cong \frac{R_k}{1 + g_m R_k} \tag{6-35c}$$

Example 6-26.   What is the output resistance of the cathode-follower circuit of Example 6-24?

Given: $\mu = 20$ Find: (a) R_o
                $r_p = 6{,}500$ ohms   (b) R_o
      (a) $R_k = 22{,}000$ ohms   (c) R_o
      (b) $R_k = 2{,}200$ ohms
      (c) $R_k = 220$ ohms

Solution:

(a) $R_o = \dfrac{r_p R_k}{r_p + R_k(\mu + 1)} = \dfrac{6{,}500 \times 22{,}000}{6{,}500 + 22{,}000 \times (20 + 1)} = 305$ ohms

(b) $R_o = \dfrac{r_p R_k}{r_p + R_k(\mu + 1)} = \dfrac{6{,}500 \times 2{,}200}{6{,}500 + 2{,}200 \times (20 + 1)} = 271$ ohms

(c) $R_o = \dfrac{r_p R_k}{r_p + R_k(\mu + 1)} = \dfrac{6{,}500 \times 220}{6{,}500 + 220 \times (20 + 1)} = 128$ ohms

Example 6-27.   What is the output resistance of the cathode-follower circuit of Example 6-25?

Given: $g_m = 6,000 \ \mu\text{mhos}$ Find: (a) R_o

            (a) $R_k = 10,000$ ohms   (b) R_o

            (b) $R_k = 1,000$ ohms   (c) R_o

            (c) $R_k = 100$ ohms

Solution:

(a) $R_o \cong \dfrac{R_k}{1 + g_m R_k} \cong \dfrac{10,000}{1 + 6,000 \times 10^{-6} \times 10,000} \cong 164$ ohms

(b) $R_o \cong \dfrac{R_k}{1 + g_m R_k} \cong \dfrac{1,000}{1 + 6,000 \times 10^{-6} \times 1,000} \cong 143$ ohms

(c) $R_o \cong \dfrac{R_k}{1 + g_m R_k} \cong \dfrac{100}{1 + 6,000 \times 10^{-6} \times 100} \cong 62.5$ ohms

The solutions of Examples 6-26 and 6-27 show that the output impedance of the cathode-follower circuit is always quite low even though the cathode resistor has a fairly high value.

Input Capacitance.   In the cathode-follower type of circuit, the large increase in the input capacitance due to the Miller effect (Art. 6-8) is eliminated by grounding the plate to a-c signals.   The equation for determining the input capacitance for the cathode follower is

$$C_i = C_{gp} + C_{gk}(1 - \text{VA}) \tag{6-36}$$

Because of the degenerative action of this type of circuit, the Miller-effect factor takes on a negative sign as compared with the positive effect in the grounded-cathode type of amplifiers; see Eq. (6-23).   Consequently, both the input capacitance C_i and the total shunting capacitance C_T are much lower for the cathode follower circuit.

Input Impedance.   The cathode-follower circuit has a high input impedance.   When the grid resistor R_g is returned to ground, as in Fig. 6-16a, the input impedance is equal to R_g.   However, if R_g is returned directly to the cathode, the input impedance will be increased by a factor $\dfrac{1}{1 - \text{VA}}$, and

$$R_i = \frac{R_g}{1 - \text{VA}} \tag{6-37}$$

Thus, with a voltage amplification of 0.9, the effective input resistance will be ten times R_g.   For the circuit of Fig. 6-16b,

$$R_i = \frac{R_g}{1 - \dfrac{\text{VA} \, R_3}{R_2 + R_3}} \tag{6-38}$$

Grid Bias of the Amplifier Tube.   The operating value of the grid bias is established by the average value of the plate-cathode current flowing through R_k.   Frequently the voltage developed across R_k is greater than the amount of grid-bias voltage desired.   In such instances the cathode resistance is made up of two resistors, R_2 and R_3 in Fig. 6-16b, and the grid resistor is returned to the junction of R_2 and R_3 in order to provide the desired value of grid bias.

Circuit Characteristics.   The characteristics of the cathode-follower circuit may be summarized as follows: (1) a voltage amplification of less than 1, (2) very low input capacitance, (3) very high input impedance, (4) low output impedance, (5) low amplitude distortion, (6) low-frequency distortion, (7) uniform frequency response over a very wide frequency band, (8) no phase inversion, (9) ability to supply relatively large signal currents to low impedance loads, (10) ability to handle high input voltages without overloading, (11) good circuit stability.

Applications.   A very important use of the cathode-follower circuit is as an impedance-changing or impedance-matching device.   Its high-input–low-output impedance characteristics make it an excellent means of coupling a signal from a high-impedance output to a low-impedance load, such as coupling the high-impedance output of an amplifier to a low-impedance interconnecting cable or transmission line.   It is frequently used in place of a transformer as it provides a more uniform response over a wider frequency range than a transformer-coupled amplifier.

Cathode-follower stages are frequently used in high-fidelity sound-amplifying systems because of their ability to (1) provide a more uniform response over a wider frequency range, (2) replace costly coupling transformers and at the same time improve the high-frequency response, (3) reduce the input capacitance of an amplifier stage and thereby improve its high-frequency response, and (4) reduce the nonlinear distortion.

The circuit of Fig. 6-18 shows a four-stage amplifier in which VT_1 and VT_3 are cathode-follower stages, and VT_2 and VT_4 are high-gain RC-coupled voltage-amplifying stages.   Using a cathode follower in the first stage presents a high impedance to the signal source which minimizes any undesirable loading effects on the source and permits the use of higher input signal voltages without introducing distortion.   It also reduces the effective input capacitance of VT_2 which, together with the inductor L_1, increases the high-frequency limit to well over 100,000 cycles with uniform response.   The second cathode-follower stage VT_3 presents a low-input capacitance to the preceding stage VT_2, thereby aiding in producing a wide-band uniform response and a minimum of phase shift. The low output impedance of the cathode follower prevents the input capacitance of the final stage VT_4 from causing attenuation of the signal at the high frequencies.   Some or all of these principles are used in many high-fidelity amplifying systems.

Phase-splitter Circuit.   A modification of the cathode-follower circuit is frequently used in high-fidelity amplifying systems as a phase splitter

(Fig. 6-19). In this circuit the signal is fed to the grid-cathode circuit of VT_1 through coupling capacitor C_1. The grid resistor is returned to the junction of R_3 and R_4 instead of to ground in order to obtain the

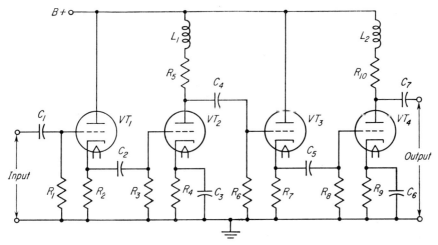

Fig. 6-18. Four-stage amplifier circuit using two cathode-follower stages VT_1 and VT_3 to couple the signals to two voltage-amplifier stages VT_2 and VT_4 respectively.

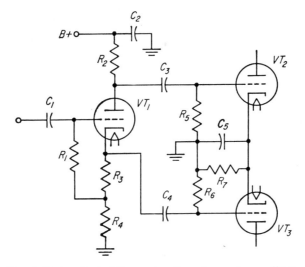

Fig. 6-19. A cathode-follower circuit used as a phase splitter.

proper amount of bias for VT_1. The plate-cathode current flowing through R_3 and R_4 produces a signal voltage that is in phase with the input signal; this signal is coupled through C_4 to the grid of VT_3. The same plate-cathode current of VT_1 flowing through R_2 produces a signal voltage of opposite polarity (or phase) at the plate of VT_1. This signal is coupled through C_3 to the grid of VT_2. Because two voltages 180 degrees out of phase are made available by this circuit, it is called a *phase-splitter* or a *paraphase amplifier*. In order to obtain a balanced input to VT_2 and VT_3, the value of R_2 is made approximately equal to the sum of R_3 and R_4. The a-c signal is returned to ground through a large capacitance ($10\mu f$) at C_2. Most of the advantages of the cathode follower will also apply to this circuit. The voltage amplification for each output of the phase-splitter circuit of Fig. 6-19, when $R_3 + R_4$ is equal to R_2, is

$$\text{VA} = \frac{\mu R_2}{(\mu + 2)R_2 + r_p} \tag{6-39}$$

6-14. Multistage Amplifier Circuits. *Need for Multistage Amplifier Circuits.* The final stage of a radio receiver, that is, the one that feeds the loudspeaker, is an a-f power amplifier stage. Power amplifiers (Chap. 7) require a higher signal input voltage than is available directly from the detector stage, and hence an a-f voltage amplifier stage is needed. This amplifier stage is also called the *driving stage* or the *driver*, as its purpose is to raise the signal voltage to an amount sufficient to drive the final output stage.

If the input signal voltage is very low, as may be the case in public-address systems, one or more additional stages of voltage amplification may be required. When several stages of amplification are connected in series they are said to be connected in *cascade*.

Voltage Amplification of Multistage Amplifier Circuits. The over-all voltage amplification of a multistage amplifier circuit is the ratio of the final output signal voltage to the input signal voltage at the first stage of the amplifier. In terms of voltage amplification of each stage individually, the over-all voltage amplification is

$$\text{VA}_T = \text{VA}_1 \times \text{VA}_2 \times \text{VA}_3, \ldots \tag{6-40}$$

When expressed in decibels, the over-all gain of a multistage amplifier circuit is

$$\text{db}_T = \text{db}_1 + \text{db}_2 + \text{db}_3, \ldots \tag{6-41}$$

Need for Decoupling with Multistage Amplifiers. When a common power supply is used for the various stages of a multistage amplifier circuit, the power supply acts as a common impedance to the various stages. Under this condition, coupling will exist between the various stages and may cause an appreciable amount of feedback. This feedback will be either regenerative or degenerative depending upon the phase relation of the feedback voltage, which to a large extent is governed by the number of stages employed. When an appreciable amount of feedback is present, especially at low frequencies, it may produce oscillations at the low frequencies, causing a noise in the loudspeaker called *motorboating*.

In order to obtain satisfactory operation of multistage amplifiers it may become necessary to provide special circuits to decouple one stage from another (Art. 6-19).

6-15. Feedback Amplifiers. *Positive and Negative Feedback.* Regeneration or degeneration occurs when part of the energy of the output circuit is returned to the input circuit. This principle is applied to an amplifier circuit by returning to its input circuit a voltage that has been obtained from its output circuit; this type of circuit is called a *feedback amplifier*. When the voltage returned to the input circuit is in phase with the signal voltage, it will increase the strength of the signal input of the amplifier, and the feedback is referred to as *positive feedback* or *regeneration*. When the voltage returned to the input circuit is 180 degrees out of phase with the signal voltage, it will decrease the strength of the signal input of the amplifier, and the feedback is referred to as *negative feedback*, *inverse feedback*, or *degeneration*.

Feedback amplifiers, as used in a-f amplifying systems, employ negative feedback. Although it may seem undesirable to operate an amplifier in a manner that does not produce the maximum possible gain for the circuit, the advantages of the negative-feedback amplifier outweigh the disadvantages of reduced gain. Among the advantages are (1) higher fidelity, (2) improved stability, (3) less amplitude distortion, (4) less harmonic distortion, (5) less frequency distortion, (6) less phase distortion, (7) lower ratio of noise level. Furthermore, by use of high-mu tubes or by using an additional stage of amplification it is possible to obtain the desired over-all gain for the amplifier circuit.

Principles of Feedback Amplifiers. The voltage amplification of a single-stage amplifier circuit is expressed by Eq. (2-6) as the ratio of the output signal voltage to the input signal voltage. From this equation, the output voltage may be expressed as

$$e_o = VAe_s \qquad (6\text{-}42)$$

In the study of feedback amplifiers, it is necessary to consider the fact that distortion due to noise, hum, etc., causes an additional signal to be set up in the output circuit, this additional signal being some definite percentage of the output voltage (Fig. 6-20a). The output of the amplifier circuit without feedback is then really equal to the sum of the output signal voltage and the distortion voltage, or

$$e_o + D = Ae_s + de_o \qquad (6\text{-}43)$$

where e_o = output signal without feedback, volts
D = amount of distortion without feedback, volts
A = voltage amplification of circuit
e_s = input signal, volts
d = distortion without feedback, expressed as decimal

Example 6-28. The input signal of an amplifier circuit having a voltage amplification of 80 is 1 volt, and the distortion due to noise and hum is 5 per cent. What are the values of the output signal and the noise and hum distortion voltages?

$$\text{Given:} \quad VA = 80 \qquad\qquad \text{Find:} \quad e_o$$
$$e_s = 1 \text{ volt} \qquad\qquad\qquad de_o$$
$$d = 0.05$$

Solution:

$$e_o = VAe_s = 80 \times 1 = 80 \text{ volts}$$
$$de_o = 0.05 \times 80 = 4 \text{ volts}$$

The principle of feedback amplifiers is illustrated in Fig. 6-20*b*. In this type of amplifier, a portion of the output voltage is returned to the input

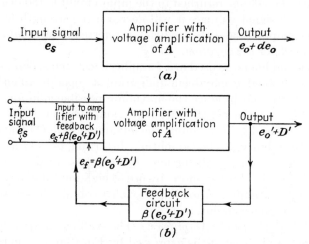

FIG. 6-20. Block diagrams illustrating the principles of feedback. (*a*) Amplifier circuit without feedback. (*b*) Amplifier circuit with feedback.

circuit either in phase with the input voltage for positive feedback, or 180 degrees out of phase with the input voltage for negative feedback. The amount of voltage that is fed back is generally expressed as a percentage of the output voltage. The decimal equivalent of this percentage is denoted by the Greek letter β. For positive feedback, β is positive, and for negative feedback it is a negative value. From the relation of the output and input signals as expressed in Eq. (6-43), the output of the feedback amplifier illustrated in Fig. 6-20*b* can be expressed as

$$e_o' + D' = A[e_s + \beta(e_o' + D')] + de_o' \qquad (6\text{-}44)$$

where e_o' = output signal with feedback, volts

$\quad D'$ = amount of distortion with feedback, volts

$\quad \beta$ = per cent of output voltage being fed back, decimal equivalent

NOTE: Terms marked $'$ indicate values when feedback has been considered.

Expanding and regrouping the terms of Eq. (6-44), then

$$Ae_s + de'_o = e'_o + D' - A\beta e'_o - A\beta D' \tag{6-44a}$$

and
$$Ae_s + de'_o = e'_o(1 - A\beta) + D'(1 - A\beta) \tag{6-44b}$$

From Eq. (6-44b) it can be shown that

$$Ae_s = e'_o(1 - A\beta) \tag{6-45}$$

and
$$de'_o = D'(1 - A\beta) \tag{6-46}$$

By rearranging the terms of Eq. (6-45) the output voltage can be expressed as

$$e'_o = e_s \frac{A}{1 - A\beta} \tag{6-47}$$

Also, by rearranging the terms of Eq. (6-46), the distortion voltage can be expressed as

$$D' = e'_o \frac{d}{1 - A\beta} \tag{6-48}$$

Principles of Negative-feedback Amplifiers.   Equation (6-47) shows that when negative feedback is used, that is, when β is negative, the output voltage of the amplifier will be reduced by a factor of $\dfrac{1}{1 + A\beta}$. Equation (6-48) shows that the distortion component of the output voltage is decreased by the same factor. By increasing the input signal voltage it is possible to restore the output voltage to its original value. Increasing the input signal will, however, also increase the distortion, but the distortion voltage will still be less than it was without negative feedback, since it is always a definite percentage of the output voltage.

The quantity $A\beta$, called the *feedback factor*, represents the ratio of the feedback voltage e_f to the input voltage at the amplifier, which is $e_s - e_f$ with negative feedback. This may be expressed mathematically as

$$A\beta = \frac{e_f}{e_s - e_f} \tag{6-49}$$

By rearranging the terms of this equation, it may be seen that

$$\frac{e_s}{e_f} = \frac{1 + A\beta}{A\beta} \tag{6-49a}$$

This indicates, for example, that if the feedback factor $A\beta$ has a value of 10, the feedback voltage e_f will then be 10 volts for each 11 volts of input signal e_s. This may be further interpreted as indicating that, for a circuit with a feedback factor of 10, it will require an input signal of 11 volts in order to produce the same output voltage as would be obtained from a 1-volt signal if the amplifier were used without feedback.

Example 6-29. A negative-feedback circuit is added to the amplifier circuit of Example 6-28. The proportion of the output voltage that is returned by feedback is 0.01. What are the values of the output signal voltage and the distortion voltage?

$$\text{Given:} \quad A = 80 \qquad\qquad \text{Find:} \quad e_o'$$
$$e_s = 1 \text{ volt} \qquad\qquad\qquad D'$$
$$d = 0.05$$
$$\beta = -0.01$$

Solution:

$$e_o' = e_s \frac{A}{1 - A\beta} = 1 \frac{80}{1 - 80(-0.01)} = 44.44 \text{ volts}$$

$$D' = e_o' \frac{d}{1 - A\beta} = 44.44 \frac{0.05}{1 - 80(-0.01)} = 1.23 \text{ volts}$$

Example 6-30. The results of Example 6-29 show that the addition of a feedback circuit to the amplifier of Example 6-28 has caused a reduction in the value of the output signal. (*a*) To what value must the input signal of the amplifier of Example 6-29 be raised in order to obtain the same amount of output signal voltage as was obtained without feedback as in Example 6-28? (*b*) Using the value of input signal obtained in part (*a*), what is the value of the distortion voltage?

$$\text{Given:} \quad e_o = 80 \text{ volts} \qquad\qquad \text{Find:} \quad e_s'$$
$$e_o' = 44.44 \text{ volts} \qquad\qquad\qquad D'$$
$$e_s = 1 \text{ volt}$$
$$A = 80$$
$$d = 0.05$$
$$\beta = -0.01$$

Solution:

(*a*) $\quad e_s' = e_s \dfrac{e_o}{e_o'} = 1 \times \dfrac{80}{44.44} = 1.80 \text{ volts}$

(*b*) $\quad D' = e_o'' \dfrac{d}{1 - A\beta} = 80 \times \dfrac{0.05}{1 - 80(-0.01)} = 2.22 \text{ volts}$

The values of output voltage and distortion voltage obtained in Examples 6-28 and 6-30 indicate that, although the same output voltage is obtained with negative feedback as without it, the distortion voltage with feedback is reduced to approximately 50 per cent of its former value. In a similar manner, the noise and hum voltages will also be reduced to about 50 per cent of their former values.

By increasing the value of the feedback factor, the proportion of the output voltage that is fed back is increased, thus further decreasing the magnitude of the distortion voltage. It is therefore possible to reduce the magnitude of the distortion voltage to a negligible value by controlling the amount of feedback voltage. As the over-all voltage amplification of the circuit is also reduced by an increase in the amount of feedback voltage, it is necessary to compensate for the loss in over-all voltage amplification if it falls below the value required of the amplifier.

6-16. Advantages and Limitations of Feedback Amplifiers. *Nonlinear Distortion and Phase Distortion.* Nonlinear distortion is caused by operating an amplifier tube over a nonlinear portion of its characteristic curve.   Phase distortion occurs when the input and output voltages of an amplifier are not exactly 180 degrees out of phase with each other.   The output of an amplifier having nonlinear distortion, phase distortion, or both can be considered as consisting of the amplified input signal voltage plus an added new signal.   These two types of distortion also will be reduced by negative feedback in the same manner as noise and hum.

Frequency Distortion.   Frequency distortion occurs when the gain of an amplifier varies with changes of frequency of the signal applied to the

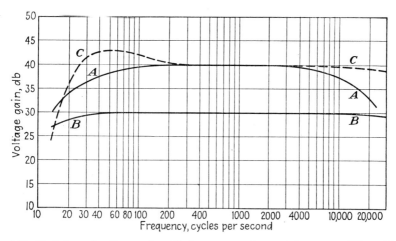

FIG. 6-21. Variation of voltage gain with frequency.   Curve *A* for an amplifier without feedback, Curve *B* for a similar amplifier with negative feedback, Curve *C* for an amplifier with balanced feedback.

grid circuit.   Because β is generally independent of the frequency, the variation in the gain of an amplifier due to changes in frequency will be reduced when negative feedback is used.   The relative effect of the frequency upon the gain of an amplifier with and without feedback is shown in Fig. 6-21.   By increasing the value of β, the gain of an amplifier can be made to be fairly uniform over a wide frequency range.   Because of this feature, negative-feedback amplifiers are used in high-fidelity audio amplifiers and in video-amplifier circuits where a wide-band amplifier is required.

Example 6-31.   The voltage amplification of a certain amplifier is 100 for a 400-cycle signal and drops to 10 for a 50-cycle signal.   (*a*) What is the change in output volume, expressed in decibels?   (*b*) If a negative-feedback circuit designed to feed back 4 per cent of the output signal is added, what is the change in volume, expressed in decibels, under this condition?

Given: $A_{400} = 100$ Find: (*a*) Change @ 50 cycles, db
$\qquad\quad\ A_{50} = 10$ $\qquad\qquad\ \ $ (*b*) Change @ 50 cycles, db
$\qquad\quad\ \ \ \beta = -0.04$

Solution:

(a) Change @ 50 cycles $= 20 \log \dfrac{A_{50}}{A_{400}} = 20 \log \dfrac{10}{100} = 20 \log 0.1$

$\qquad\qquad\qquad\qquad = 20(9.0000 - 10) = -20 \text{ db}$

(b) e_o' @ 400 cycles $= e_s \dfrac{A}{1 - A\beta} = e_s \dfrac{100}{1 - 100(-0.04)} = 20e_s$

$\qquad e_o'$ @ 50 cycles $= e_s \dfrac{A}{1 - A\beta} = e_s \dfrac{10}{1 - 10(-0.04)} = 7.14e_s$

$\qquad$ Change @ 50 cycles $= 20 \log \dfrac{A_{50}}{A_{400}} = 20 \log \dfrac{7.14}{20} = 20 \log 0.357$

$\qquad\qquad\qquad\qquad = 20(9.5527 - 10) = -8.94 \text{ db}$

Stability of Negative-feedback Amplifiers. The gain of an amplifier circuit may vary with (1) the tube used, (2) the operating voltages, (3) the load impedance. A change in one or more of these variables may cause a change in the gain of the amplifier, thus affecting its stability. The stability of an amplifier can be improved by using negative feedback. The manner in which negative feedback affects the stability of an amplifier can be shown mathematically by rearranging the terms of Eq. (6-47) to express the voltage amplification of the circuit with feedback.

$$\text{VA}' = \frac{e_o'}{e_s} = \frac{A}{1 - A\beta} \tag{6-50}$$

or

$$\text{VA}' = -\frac{1}{\beta}\left(\frac{1}{1 - \dfrac{1}{A\beta}}\right) \tag{6-50a}$$

When the feedback factor $A\beta$ is much greater than 1, the gain of a negative-feedback amplifier will then be approximately equal to

$$\text{VA}' \cong -\frac{1}{\beta} \tag{6-51}$$

Equations (6-50) and (6-50a) show that when the feedback factor $A\beta$ is much greater than 1, the voltage amplification of the circuit is dependent on the percentage of the output voltage returned, β, rather than on the voltage amplification of the amplifier. The value of β is generally dependent upon a resistance network whose component values are independent of the frequency, the voltages applied to the amplifier tube, and the tube's operating characteristics. The stability of negative-feedback amplifiers can be made to be comparatively high by adjusting the percentage of feedback to produce a high feedback factor.

Example 6-32. An amplifier circuit having a voltage amplification of 100 employs a negative-feedback circuit whose feedback factor is -19. (a) What is the over-all voltage amplification of the circuit as determined by use of Eq. (6-50)? (b) What is the approximate over-all voltage amplification of the circuit as determined by use of Eq. (6-51)?

$\qquad\qquad$ Given: $A\beta = -19$ $\qquad\qquad$ Find: (a) VA

$\qquad\qquad\qquad\qquad A = 100$ $\qquad\qquad\qquad\qquad$ (b) VA

Solution:

(a) $VA = \dfrac{A}{1 - A\beta} = \dfrac{100}{1 - (-19)} = \dfrac{100}{20} = 5$

(b) $VA \cong -\dfrac{1}{\beta} \cong \dfrac{-1}{-0.19} \cong 5.26$

   where $\beta = \dfrac{A\beta}{A} = \dfrac{-19}{100} = -0.19$

Limitations of Feedback Amplifiers. Any increase in the feedback factor will cause an increase in the proportion of the output voltage that is returned. The over-all voltage amplification of the circuit will then be lower than the voltage amplification of the same amplifier without feedback. In order to produce the same amplification as was obtained without feedback, it becomes necessary to increase the input voltage e_s by an amount $A\beta$ times the voltage of the input signal e_s. When the output with feedback, e'_o, is to be equal to the output without feedback, e_o, then the input signal e_s will have to be increased to a value e'_s. The voltage amplification of the circuit with feedback may then be expressed as

$$VA' = \frac{e'_o}{e'_s} = \frac{e_o}{e'_s} \tag{6-52}$$

and

$$e'_s = \frac{e_o}{VA'} \tag{6-53}$$

Substituting Eq. (6-50) for VA' in Eq. (6-53)

$$e'_s = \frac{e_o}{\dfrac{A}{1 - A\beta}} = \frac{e_o(1 - A\beta)}{A} \tag{6-53a}$$

Substituting $\dfrac{e_o}{e_s}$ for A in the denominator of Eq. (6-53a)

$$e'_s = \frac{e_o(1 - A\beta)}{\dfrac{e_o}{e_s}} = e_s(1 - A\beta) \tag{6-53b}$$

Thus, when β is negative, the input signal voltage must be increased by the factor $(1 + A\beta)$. This increased value of input signal voltage can be obtained by designing the preceding amplifier circuits to produce a higher voltage amplification or by using an additional amplifier stage. When the requirements of an amplifier are such that high fidelity is the paramount factor, any additional expense is warranted if the desired results are obtained.

In the preceding explanations of negative feedback it has been assumed that the feedback voltage was exactly 180 degrees out of phase with the input signal. In practical amplifier circuits, however, this condition is not obtained. The amount of lead or lag, with respect to the desired 180 degrees phase relation, is dependent upon the reactances of the coupling units, the interelectrode capacitances of the tubes, and the frequency of the input signal.   If the angle of lead or lag attains a value of 180 degrees, the feedback becomes positive and the circuit becomes unstable.

The phase shift caused by the reactances of the coupling units and the interelectrode capacitances cannot exceed 90 degrees for one amplifier stage.   Even though the phase shift of two stages is cumulative, it is unlikely that a 180-degree phase shift will be obtained in a two-stage amplifier.   However, it is possible to obtain such a condition in a three-stage amplifier.   Instability does not occur with positive feedback if the feedback factor is kept less than 1.   Limiting the feedback factor to such a low value will, however, limit the amount of feedback voltage that it is possible to return.   The maximum reduction of distortion, hum, and noise will therefore be decreased.   Thus negative feedback may readily be employed with one or two amplifier stages, but becomes rather difficult for three or more amplifier stages.

6-17. Negative-feedback Amplifier Circuits.   Negative feedback can be applied to a single-stage or multistage amplifier in a number of ways. Basically, feedback circuits may be divided into three general classes: (1) voltage-feedback circuits, that is, circuits that derive the feedback voltage directly from the output voltage of the amplifier; (2) current-feedback circuits, that is, circuits that derive the feedback voltage from a voltage drop produced by the output current flowing through a resistor; (3) a combination of both voltage and current feedback.

Principle of Voltage-controlled Feedback.   When the feedback circuit is connected so that the voltage returned to the input circuit is proportional to the voltage across the output load, the feedback circuit is said to be *voltage controlled.*   A simple amplifier circuit using voltage-controlled feedback is illustrated in Fig. 6-22.   The feedback voltage of this circuit is applied to the input of the amplifier between the cathode and the input voltage e_s'.   The portion e_f of the output voltage e_o' that is added to the input circuit is controlled by the voltage-dividing resistors, R_f and R_2, and the capacitor C_2.   The portion or fraction of the output voltage being fed back may be expressed mathematically as

$$\beta = \frac{R_f}{\sqrt{(R_f + R_2)^2 + X_C{}^2}} \qquad (6\text{-}54)$$

As the reactance of the capacitor C_2 is generally very low compared to $(R_f + R_2)$, the effect of the capacitor may be ignored; then

$$\beta \cong \frac{R_f}{R_f + R_2} \qquad (6\text{-}54a)$$

The capacitor C_2, connected in series with R_f and R_2 as shown in Fig. 6-22, is used as a blocking capacitor to prevent the high plate voltage from being applied to the input circuit. The polarity (and phase relation) of the feedback voltage e_f with respect to the input signal voltage e'_s can be determined by either of two methods of analysis: (1) by studying the a-c signal effect, (2) by observing the direction of electron flow in the circuit. In studying the a-c effect, it is most convenient to assume the conditions for one half-cycle of the input voltage, for example, the positive half-cycle as indicated on Fig. 6-22. Applying a positive signal to the grid of the tube causes an increase in the plate current. The increase in plate current causes an increase in the voltage drop at R_3 and as E_{bb} is constant, the voltage at the plate of the tube must decrease, as is indicated by the negative signal on the diagram. This negative signal will be transmitted through C_2 and R_2 to point A of the resistor R_f. The feedback voltage e_f,

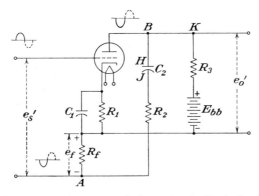

Fig. 6-22. A voltage-controlled negative-feedback circuit.

being equal to the voltage drop across R_f, will therefore be equivalent to adding a voltage source in series with the grid circuit. Furthermore, as the negative terminal A (that is, negative for the condition assumed above) is connected to the grid side of the grid-cathode circuit, the voltage e_f will be opposite in polarity to the input signal e'_s, and hence the feedback is negative. If the reactance of C_2 is small compared to the combined resistance of R_2 and R_f, the voltage e_f will be practically 180 degrees out of phase with the input signal voltage e'_s. In determining the polarity of the feedback voltage by observing the electron flow, the method of analysis is similar to the a-c signal analysis in most of the details. For example, consider the conditions for a half-cycle of the input signal e'_s. Using the positive half-cycle, it can be seen that a positive signal at the grid of the tube will produce a negative signal at the plate. Point B and consequently point H (Fig. 6-22) will however always be positive as they are connected to the positive terminal of E_{bb} via R_3. The negative signal at the plate then merely means that plate H of capacitor C_2 will become less positive, and hence C_2 must discharge some of its voltage. The path of electrons under this condition will be from J through R_2, R_f, C_1 (because the reactance of C_1 is small compared to R_1), cathode to plate, B to H. As electrons travel from negative to positive, point A of the resistor R_f will be negative. By the same reasoning as before, it can be seen that the feedback is negative. In the case of a negative signal at e'_s, the signal at B will be positive, and the capacitor C_2 will take on a higher voltage charge. The electron path will be from H B K, through R_3, E_{bb}, R_f, R_2 to J. The polarity at point A will now be positive, thus maintaining the condition of negative feedback.

Principle of Current-controlled Feedback. When the feedback circuit is connected so that the voltage returned to the input circuit is proportional to the current flowing through the output load, the feedback circuit is said to be *current controlled.* A simple amplifier circuit using

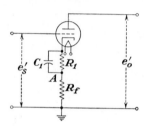

Fig. 6-23. A current-controlled negative-feedback circuit.

current-controlled feedback is illustrated in Fig. 6-23. In this circuit, the feedback voltage is obtained by connecting the cathode bypass capacitor C_1 so that it shunts only part of the cathode-bias resistor or by eliminating the by-pass capacitor C_1 entirely. The a-f output current will then flow through the feedback resistor R_f, causing the grid bias to vary with the changes in plate current. When the input signal e'_s is positive, the plate current, and consequently the current in R_f, will increase. The voltage drop across R_f will increase and make point A more positive. This makes the cathode more positive with respect to ground, which is equivalent to making the grid more negative. Thus a positive signal on the grid produces the condition necessary for negative feedback. When the input signal is negative, the plate current will decrease and consequently the voltage drop at R_f will decrease. The cathode therefore becomes less positive, which is equivalent to making the grid less negative (or more positive) and hence fulfills the requirement for negative feedback.

Principle of Feedback for Multistage Amplifiers. Feedback circuits for more than one amplifier stage become quite complex. An example of a two-stage negative-feedback amplifier is illustrated in Fig. 6-24. In this circuit both voltage-controlled and current-controlled feedback are present. However, by the choice of values of R_1, R_2, and R_f the voltage-controlled feedback is made the predominant source of feedback voltage, and hence the circuit is referred to as *voltage-controlled feedback.* The polarity of the feedback voltage at R_f can be checked by tracing the a-c signal for one half-cycle. Thus, a positive signal at the grid of VT_1 will cause a negative signal at the plate of VT_1. This negative signal is applied to the grid of VT_2 through C_b. The negative signal at the grid of VT_2 causes a positive signal at the plate of VT_2 which is transmitted to point A through C_2 and R_2. This positive signal at point A raises the cathode potential, which is equivalent to applying a negative signal to the grid, hence satisfying the requirement of negative feedback.

Another method of obtaining negative feedback for an amplifier circuit is shown in Fig. 6-25, in which the feedback voltage is taken from the secondary of the output transformer. Resistors R_f and R_2 form a voltage

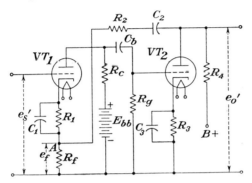

Fig. 6-24. A two-stage amplifier circuit illustrating the application of a voltage-controlled feedback circuit.

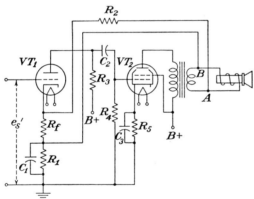

Fig. 6-25. Negative-feedback amplifier circuit with feedback voltage taken from the secondary of the output transformer.

divider that controls the amount of feedback. The correct connection for leads A and B is generally determined experimentally by first connecting them in one manner and then reversing them. One connection will produce degeneration and the other regeneration; the resulting output at the loudspeaker should indicate the correct connection.

Example 6-33. The resistors of the voltage divider of a voltage-controlled feedback amplifier circuit similar to that of Fig. 6-22 are $R_f = 5,000$ ohms, $R_2 = 50,000$ ohms. What is the approximate value of β?

Given: $R_f = 5,000$ ohms   Find: β
$R_2 = 50,000$ ohms

Solution:

$$\beta \cong \frac{R_f}{R_f + R_2} = \frac{5,000}{5,000 + 50,000} = 0.0909$$

6-18. Balanced-feedback Amplifier. In the RC-coupled amplifier, the voltage amplification tends to decrease at the very low and very high audio frequencies. The loss in amplification at the low frequencies is due to the high reactance of the coupling capacitor C_b at the low frequencies, which reduces the output voltage of the amplifier. The loss in amplification at the higher frequencies is due to the low reactance of the shunting

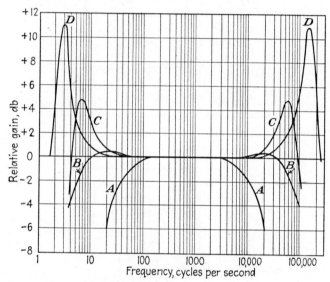

Fig. 6-26. Relative gain versus frequency characteristics of a two-stage RC coupled a-f amplifier with various amounts of feedback. Curve A, amplifier without feedback. Curve B, $A\beta$ equal to -2 at mid-frequency. Curve C, $A\beta$ equal to -10 at mid-frequency. Curve D, $A\beta$ equal to -50 at mid-frequency.

capacitance at these frequencies which thereby provides undesired paths for a large portion of the high-frequency signal currents. In addition to the reduction in the voltage amplification at the extreme frequencies, a large phase shift is introduced at the low and high frequencies.

If the feedback factor $A\beta$ is large and the phase shift approaches 180 degrees, positive feedback will occur, and the frequency-response characteristics will be affected as is illustrated in Fig. 6-26. These representative curves show that while an increase in the amount of feedback makes the voltage gain practically constant over a wider range of frequencies, it also produces peaks of voltage gain at the lower and upper ends of the frequency range. When the effects of these peaks of amplification are great enough to warrant special consideration, two methods of correction are available.

One method of reducing the effect of the peaks is by means of filter circuits as shown in Fig. 6-27. The *RC* filter network at the input side will neutralize the peaks that occur at the low frequency, and the *LR* network at the output side will neutralize the high-frequency peaks.

FIG. 0-27. A feedback amplifier with equalizing filter networks for neutralizing the amplification peaks.

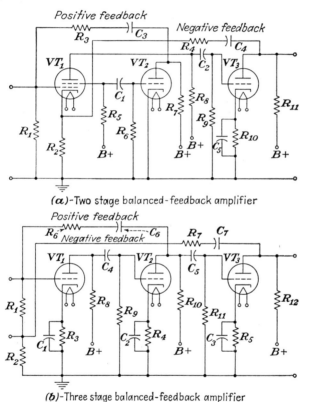

FIG. 6-28. Several methods of obtaining balanced feedback.

This method, called an *equalized circuit,* is useful only when the peaks are not too large.

A second method of reducing the effects of the peaks is by providing the amplifier with both positive and negative feedback. This type of amplifier, called a *balanced-feedback amplifier,* is shown in Fig. 6-28. With this type of circuit, a more nearly constant response is obtained for a multistage amplifier by designing the first portion of the amplifier to provide practically uniform response to all frequencies and then applying both positive- and negative-feedback voltages. The feedback voltages are generally of such proportions that the amount of positive feedback at the mid-frequency will cancel the negative feedback. At frequencies above and below this point, the amounts of positive and negative feedback will vary in a manner tending to neutralize any variations from the optimum response of the first portion of the amplifier and thus providing practically uniform response over a wider range of frequency.

An advantage of the balanced-feedback amplifier is that it provides practically uniform amplification over a wider range of frequency than an amplifier employing only negative feedback. A higher over-all gain is also obtained with the balanced-feedback amplifier as is illustrated in Fig. 6-21.

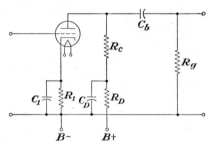

6-19. Decoupling Circuits. *Need for Decoupling Circuits.* The preceding article concerned circuits employing controlled feedback. It is possible that uncontrolled feedback, and therefore undesired feedback, may also be present in multistage amplifier circuits.

Fig. 6-29. A simple plate decoupling circuit.

Coupling may exist between circuits operating at the same frequency and having a common impedance. In many amplifier circuits, it is common practice to supply the plate and screen-grid voltages for a number of tubes from a single source of d-c power. This power supply then acts as a common impedance for all these circuits. The coupling existing among these circuits may be either regenerative or degenerative, depending upon the phase relationship. In a two-stage amplifier circuit the plate currents are 180 degrees out of phase with each other and the coupling between these circuits is degenerative, and therefore negative feedback will occur. The plate currents of the first and third stages of a three-stage amplifier are in phase with each other, and the coupling between these circuits is regenerative, and therefore positive feedback will occur.   If this undesired feedback is sufficient to produce any appreciable effects upon the operation of the amplifier it becomes necessary to decouple each stage of the amplifier from the other stages.

Decoupling Circuits. Decoupling may be accomplished by a simple decoupling network formed by capacitor C_D and resistor R_D in Fig. 6-29. The value of the decoupling capacitor must be high enough so that its reactance at the lowest audio frequency is considerably less than the total resistance of the decoupling resistor R_D and the internal resistance of the power supply R_B.   The undesired coupling through the common power supply is then reduced by the factor

$$K_D \cong \frac{X_D}{R_D + R_B} \tag{6-55}$$

where K_D = decoupling factor (approximate value when X_D is much less than $R_D + R_B$)

X_D = reactance of decoupling capacitor, ohms

R_D = resistance of decoupling resistor, ohms

R_B = internal resistance of power supply, ohms

The internal resistance of the power supply is generally very low in comparison to the value of the decoupling resistor and for all practical purposes can be ignored. Equation (6-55) may then be expressed as

$$K_D \cong \frac{X_D}{R_D} \tag{6-55a}$$

or

$$K \cong \frac{10^6}{2\pi f C_D R_D} \tag{6-55b}$$

Equation (6-55b) shows that the decoupling factor can be made smaller by increasing the value of either R_D or C_D. Increasing the value of R_D will increase the voltage drop at this resistor and thereby decrease the voltage at the plate of the tube. In order to maintain the plate voltage at the required value, it becomes necessary to increase the voltage of the power supply to compensate for the drop in voltage at the decoupling resistor.

In an RC-coupled audio-amplifier circuit (Fig. 6-29), the value of R_D is generally about one-fifth the value of R_c. The value of C_D varies from about 0.25 μf to 8μf. For r-f and i-f circuits the value of C_D varies from 0.01 μf to 0.1 μf.

In some amplifier circuits the amount of resistance required to obtain sufficient decoupling causes the voltage drop across the decoupling resistor to become quite high. In such circuits, a choke coil may be substituted for the decoupling resistor. Unless the voltage drop at the resistor becomes too high, resistors are used, as they are less expensive and require less space than a choke coil.

A three-stage audio amplifier using decoupling networks is shown in Fig. 6-30. The over-all decoupling of a multistage amplifier is equal to the product of the decoupling factors of each of the networks used. Hence, the over-all decoupling factor of a three-stage amplifier using two decoupling networks, similar to Fig. 6-30, becomes

$$K_{DT} = K_{D1}K_{D2} \tag{6-56}$$

or

$$K_{DT} \cong \frac{X_{D1}X_{D2}}{R_{D1}R_{D2}} \tag{6-56a}$$

$$K_{DT} \cong \frac{10^{12}}{(2\pi f)^2 C_{D1}C_{D2}R_{D1}R_{D2}} \tag{6-56b}$$

Example 6-34. A 1-μf capacitor and a 50,000-ohm resistor are connected to form a decoupling network. What is the approximate value of the decoupling factor for a 100-cycle a-f signal?

Given: $C_D = 1$ μf   Find: K_D
        $R_D = 50,000$ ohms
        $f = 100$ cycles

Solution:

$$K_D \cong \frac{10^6}{2\pi f C_D R_D} \cong \frac{10^6}{6.28 \times 100 \times 1 \times 50,000} \cong 0.0318$$

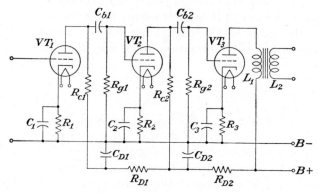

Fig. 6-30. Application of plate decoupling circuits to a three-stage amplifier.

Example 6-35. Two identical decoupling networks are connected in a three-stage amplifier similar to the circuit shown in Fig. 6-30. The values of the resistors and capacitors are the same as those used in Example 6-34. What is the value of the over-all decoupling factor for a 100-cycle a-f signal?

Given: $C_{D1} = C_{D2} = 1$ μf   Find: K_{DT}
      $R_{D1} = R_{D2} = 50,000$ ohms
      $f = 100$ cycles
      $K_{D1} = K_{D2} \cong 0.0318$ (from Example 6-34)

Solution:

$$K_{DT} = K_{D1}K_{D2} = 0.0318 \times 0.0318 = 0.001$$

The results obtained in Examples 6-34 and 6-35 show that a greater amount of over-all decoupling is obtained with a two-stage decoupling circuit than with a single decoupling section.

QUESTIONS

1. What is the purpose of an audio amplifier?

2. (a) What a-f range is generally considered acceptable in low-cost radio receivers? (b) What a-f range is desirable for high-fidelity reception?

3. (a) Is the same amount of energy required to produce sound waves at all audible frequencies? (b) Explain the reason for your answer to part (a). (c) How does this affect the design of an audio amplifier?

4. (a) In what manner does the human ear respond to sounds of various levels of energy? (b) What unit is used to express such variations?

5. (a) Define a common logarithm. (b) How is the characteristic of a logarithm determined? (c) How is its mantissa determined? (d) What is an antilog?

6. (a) Define the unit called the *bel*. (b) Define the decibel. (c) Explain the advantages of using the decibel.

7. Explain how the decibel may be used to express ratios of voltage and current as well as power.

8. (a) Why is the decibel only a relative unit of power measurement? (b) What amounts of power are used as the zero reference level in connection with radio receiver measurements?

9. (a) Define the volume unit. (b) Under what condition is the volume unit used?

10. (a) Name three methods of coupling used with audio amplifiers. (b) What precaution must be taken with regard to the plate and grid voltages?

11. (a) Explain the basic action of the resistance-capacitance method of coupling amplifiers. (b) In Fig. 6-3, what are the functions of R_c, C_b, and R_g? (c) What are the functions R_1 and C_1?

12. (a) Why is it necessary to consider the characteristics of the RC-coupled amplifier at several frequencies? (b) What representative values of frequency are commonly used to show the characteristics of an amplifier circuit?

13. (a) What is the advantage of drawing an equivalent electrical circuit for an RC-coupled amplifier stage? (b) With the aid of circuit diagrams, explain the variation in the equivalent circuits at the intermediate, low, and high frequencies.

14. (a) What factors affect the choice of values for R_c, R_g, and C_b? (b) Give the approximate range of values for these circuit elements in practical RC-coupled amplifier circuits.

15. Derive the equation for the voltage amplification of an RC-coupled triode amplifier for the intermediate-frequency range.

16. Derive the equation for the voltage amplification of an RC-coupled pentode amplifier for the intermediate-frequency range

17. Derive the equation for the approximate voltage amplification of an RC-coupled triode amplifier for the low-frequency range.

18. Derive the equation for the approximate voltage amplification of an RC-coupled pentode amplifier for the low-frequency range.

19. Derive the equation for the voltage amplification of an RC-coupled triode amplifier for the high-frequency range.

20. Derive the equation for the voltage amplification of an RC-coupled pentode amplifier for the high-frequency range.

21. Why is the maximum voltage amplification obtained at the intermediate frequency?

22. What is the cause of the attenuation or loss in amplification at the low frequencies?

23. What is the cause of the attenuation or loss in amplification at the high frequencies?

24. (a) Is the attenuation the same for pentode amplifiers as for triode amplifiers at the low and high frequencies? (b) Explain.

25. Upon what factors does the effective value of the shunting capacitance of an RC-coupled amplifier circuit depend?

26. What are the advantages of the RC-coupled amplifier that justify its extensive use?

27. What is the advantage of the impedance-coupled amplifier over the RC-coupled amplifier?

28. How does the frequency response of the impedance-coupled amplifier compare with that of the RC-coupled amplifier?

29. (a) How does the double-impedance-coupled amplifier compare with the RC-coupled amplifier? (b) What are its advantages and disadvantages?

30. (a) What is meant by parallel plate feed? (b) What are its advantages and disadvantages?

31. What factors affect the frequency response of impedance-coupled amplifiers?

32. Explain the difference in the connections for series feed and parallel feed as applied to a transformer-coupled amplifier circuit.

33. (a) Do transformer-coupled amplifier circuits generally use triodes or pentodes? (b) Why?

34. (a) What factors affect the frequency response of a transformer-coupled audio amplifier? (b) What factors are generally omitted for obtaining the approximate characteristics?

35. (a) What is a simple expression for the approximate voltage gain at the mid-frequency range of a transformer-coupled audio amplifier? (b) Why is this simple expression possible?

36. (a) What purpose, in addition to coupling, is served by the audio transformer? (b) What ratios of primary to secondary turns are used?

37. (a) Why is it important that the impedance of the primary winding matches the plate resistance of the tube with which it is associated? (b) What are the constructional features of the transformer that make it possible to obtain sufficiently high impedance?

38. (a) Is it desirable to obtain a high turns ratio for a transformer by decreasing the number of turns on the primary winding? (b) Why?

39. What are the advantages of transformer coupling?

40. (a) Describe the circuit characteristics of the basic cathode-coupled amplifier. (b) By what other names is this circuit also known?

41. Describe briefly the principle of operation of the basic cathode-coupled amplifier circuit.

42. (a) What is the approximate maximum voltage amplification obtainable with the cathode-follower stage? (b) What advantage of the cathode-follower stage outweighs its low voltage amplification?

43. (a) What are the input and output impedance characteristics of the cathode-coupled amplifier? (b) What are the input-capacitance characteristics of the cathode-coupled amplifier?

44. Give several applications of the cathode-follower circuit.

45. Describe the circuit actions of a cathode-follower circuit used as a phase splitter.

46. Why is it necessary to use a multistage a-f amplifier in a radio receiver?

47. What is meant by a driving stage?

48. (a) What is the relation of the over-all voltage amplification of a multistage amplifier circuit to the voltage amplification of its individual stages? (b) What is the relation of the over-all decibel gain of a multistage amplifier circuit to the gain of its individual stages?

49. (a) Why are decoupling circuits required with some multistage amplifier circuits? (b) What is meant by motorboating? (c) What causes motorboating in multistage amplifiers?

50. What is meant by feedback? (b) What is positive feedback? (c) What is negative feedback?

51. (a) What is meant by a feedback amplifier? (b) What are its advantages?

52. (a) What are some of the causes of distortion in the output of an amplifier? (b) How is distortion reduced by negative feedback?

53. (a) What is meant by the feedback factor? (b) How may the value of the feedback factor be interpreted in terms of input signal required?

54. (a) What is meant by nonlinear distortion? (b) What is meant by phase distortion? (c) What are their causes? (d) How may they be reduced?

55. (*a*) What is frequency distortion? (*b*) How can it be reduced?

56. Describe the effect of negative feedback upon the stability of an amplifier.

57. What must be done with the input signal of a negative-feedback amplifier if it is desired to obtain the same gain for the amplifier as it would have without feedback?

58. (*a*) What effect do the coupling units of a negative-feedback amplifier have upon the phase relation of the feedback voltage? (*b*) Is this effect more important in a single-stage, two-stage, or a three-stage amplifier? (*c*) Why?

59. Describe three methods of applying negative feedback to an amplifier.

60. Describe the function of the resistors R_2 and R_f and capacitor C_2 of Fig. 6-22.

61. (*a*) What is the polarity of point A of Fig. 6-22 during the time that the input signal is going through its negative half-cycle? (*b*) Give proof for your answer.

62. Explain how negative feedback is obtained in the circuit of Fig. 6-23.

63. (*a*) Which circuit elements of Fig. 6-24 form the feedback circuit? (*b*) Determine the polarity of the feedback signal being fed to point A during the negative half-cycle of the input signal. (*c*) Give proof for your answer.

64. Describe how the correct connection for negative feedback can be determined for the circuit of Fig. 6-25.

65. Describe two methods of correcting the response of a circuit that has pronounced peaks of amplification at the frequency extremes.

66. (*a*) What is meant by a balanced-feedback amplifier? (*b*) What are its advantages?

67. (*a*) How many feedback paths are there in the circuit of Fig. 6-28? (*b*) Identify the circuit elements of each feedback path. (*c*) Determine the polarity of each feedback signal during the positive half-cycle of the input signal. (*d*) Give proof for your answer.

68. (*a*) What is meant by a decoupling circuit? (*b*) Why are decoupling circuits necessary?

69. (*a*) What determines the rating of the decoupling capacitor? (*b*) What ranges of values are commonly used?

70. (*a*) What general proportion is sometimes used to determine the value of the decoupling resistor? (*b*) Under what condition would it be advisable to substitute a choke coil for the decoupling resistor?

PROBLEMS

1. Find the logarithms of the following numbers: (*a*) 180, (*b*) 2,750, (*c*) 8.75, (*d*) 12.5, (*e*) 5, (*f*) 98.5, (*g*) 35,000, (*h*) 0.00307, (*i*) 18.3, (*j*) 0.000986.

2. Find the logarithms of the following numbers: (*a*) 4.82, (*b*) 675, (*c*) 0.0000548, (*d*) 750,000, (*e*) 0.0377.

3. Find the antilog of the following common logarithms: (*a*) 2.4771, (*b*) 5.8779, (*c*) 0.7782, (*d*) 8.0294 − 10, (*e*) 6.3385 − 10.

4. Find the antilog of the following common logarithms: (*a*) 3.7007, (*b*) 0.9832, (*c*) 9.6085 − 10, (*d*) 7.3181 − 10, (*e*) 1.3444.

5. What decibel gain is obtained if a tube that delivers 4.3 watts of power is substituted in an amplifier circuit for a tube that delivers 2.2 watts of power?

6. A type 6K6 power amplifier tube, when operated with 250 volts on its plate and screen grid, and with a grid bias of 18 volts, can deliver 3.4 watts of power. If the plate and screen-grid voltages are reduced to 100 volts and the grid bias is reduced to 7 volts, the same tube can deliver only 0.35 watt. What is the corresponding decibel change in power?

7. A certain pentode is rated at 0.035 watt when its grid bias is 4.5 volts and with 45 volts applied to its plate and screen grid. The same tube is rated at 0.2 watt when its grid bias is 9 volts and 90 volts are applied to its plate and screen grid. What is the corresponding change in decibels?

8. What is the decibel rating of a public-address system whose power output is (*a*) 8 watts? (*b*) 15 watts? (*c*) 30 watts? (*d*) 70 watts? (Use 6 mw as the reference level.)

9. What is the decibel rating of a high-fidelity audio amplifier whose power output is (*a*) 8 watts? (*b*) 15 watts? (*c*) 30 watts? (*d*) 70 watts? (Use 1 mw as the reference level.)

10. A booster amplifier capable of delivering 100 watts of undistorted power has an over-all gain of 17 db. What amount of driving power is required to obtain this output?

11. What is the power output in watts of a microphone that has an output rating of −75 db·m?

12. What is the db·m rating of a microphone whose power output is 0.3 μw?

13. A certain audio amplifier capable of delivering 70 watts of power is said to have a hum and noise rating of 90 db·m below rated output. (*a*) What is the hum and noise power output? (*b*) How much hum and noise voltage would be developed if the output is fed into a 500-ohm resistor?

14. A certain audio amplifier has a hum and noise rating of 70 db·m below 15 watts. (*a*) What is the hum and noise power output? (*b*) How much hum and noise voltage would be developed if the output is fed into a 500-ohm resistor?

15. An amplifier has a signal current of 1 ma flowing through its 1,000-ohm input resistance. The amplifier increases the signal strength so that a 100-volt signal appears across its 10,000-ohm output resistance. Find (*a*) the voltage gain in decibels, (*b*) the current gain in decibels, (*c*) the power gain in decibels, (*d*) the decibel output above 6 mw, (*e*) the decibel output above 1 mw.

16. The following are the characteristics of a certain public-address system: gain (microphone input) 114 db, frequency response 65 to 9,000 cps ±2 db. (*a*) What is the over-all voltage amplification of the amplifier at mid-frequency if it is assumed that the gain at mid-frequency is 114 db? (*b*) What is the over-all voltage amplification at the extreme frequencies if the variation is ±2 db?

17. A certain transformer-coupled a-f amplifier produces a voltage amplification of 50 at 1,000 cps, 45 at 100 cps, and 25 at 10,000 cps. Using the mid-frequency as the reference level, what is the decibel gain or loss at (*a*) the low frequency? (*b*) the high frequency?

18. A certain *RC*-coupled a-f amplifier produces a voltage amplification of 50 at 1,000 cps, 48 at 100 cps, and 32 at 10,000 cps. Using the mid-frequency as the reference level, what is the decibel gain or loss at (*a*) the low frequency? (*b*) the high frequency?

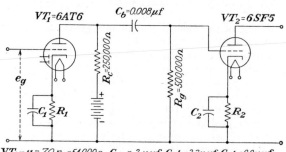

$VT_1 - \mu = 70\ r_p = 54,000\Omega,\ C_{gp} = 2\ \mu\mu f,\ C_{gk} = 2.2\mu\mu f,\ C_{pk} = 0.8\mu\mu f$
$VT_2 - \mu = 100,\ C_{gp} = 2.4\mu\mu f,\ C_{gk} = 4\mu\mu f,\ C_{pk} = 3.6\mu\mu f$
$C_w = Stray\ capacitance\ of\ wiring,\ etc. = 10\mu\mu f$

Fig. 6-31

19. A typical stage of *RC*-coupled a-f amplification is shown in Fig. 6-31. The effects of the cathode-bias resistor and its bypass capacitor are to be disregarded. What is the voltage amplification of the stage at: (*a*) 1,000 cps? (*b*) 100 cps? (*c*) 10,000 cps?

20. What is the decibel variation at the low and high audio frequencies over the intermediate frequency of the amplifier of Prob. 19?

21. A portion of the a-f amplifier of a battery-operated receiver is shown in Fig. 6-32. What is the voltage amplification of the stage at (*a*) 1,000 cps? (*b*) 100 cps? (*c*) 10,000 cps?

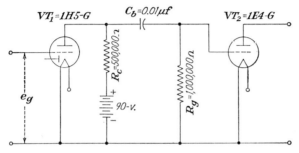

VT_1-μ =65, r_p=240,000Ω, C_{gp}= 1μμf, C_{gk}=1.1μμf, C_{pk}=5.8μμf
VT_2-μ=14.5,r_p=19,000Ω, C_{gp}=1.7μμf, C_{gk}=1.7μμf,C_{pk}=3 μμf
C_w=Stray capacitance of wiring, etc. =5μμf

Fig. 6-32

22. What is the decibel gain of the amplifier stage of Prob. 21 at the intermediate frequency?

23. A portion of the a-f amplifier of an automobile receiver is shown in Fig. 6-33. What is the voltage amplification of the first stage at (*a*) 1,000 cps? (*b*) 100 cps? (*c*) 10,000 cps?

24. What is the decibel variation at the low and high frequencies over the intermediate frequency of the amplifier stage in Prob. 23?

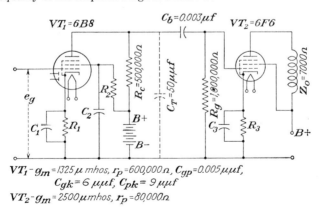

VT_1- g_m =1325μ mhos, r_p =600,000Ω, C_{gp}=0.005μμf,
C_{gk}= 6 μμf, C_{pk}= 9 μμf
VT_2-g_m= 2500μmhos, r_p =80,000Ω

Fig. 6-33

25. The plate load of a type 6F6 tube in the amplifier of Fig. 6-33 has an impedance of 7,000 ohms. (*a*) If the load impedance has unity power factor, what voltage amplification is produced by the second stage of the amplifier? (*b*) What is the approximate value of voltage amplification when calculated by means of Eq. (5-11)?

26. (*a*) What is the over-all voltage amplification of the amplifier of Fig. 6-33 at 1,000 cps? (*b*) What is the over-all decibel gain of this amplifier at 1,000 cps?

27. A high-gain two-stage amplifier circuit is shown in Fig. 6-34. What is the voltage amplification of the first stage at (*a*) 1,000 cps? (*b*) 100 cps? (*c*) 10,000 cps? (*d*) What is the voltage amplification of the second stage at the same three values of frequency?

28. What is the decibel variation at the low and high frequencies over the intermediate frequency of one stage of amplification in the amplifier of Prob. 27?

29. What is the over-all voltage amplification of the two-stage amplifier of Prob. 27 at (*a*) 1,000 cps? (*b*) 100 cps? (*c*) 10,000 cps?

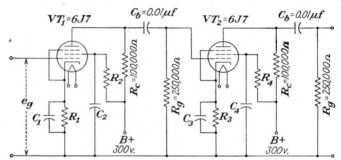

VT_1 and VT_2: $g_m = 1225\,\mu$ mhos, $r_p = 1{,}000{,}000\,\Omega$, $C_{gp} = 0.005\,\mu\mu f$, $C_{gk} = 7\,\mu\mu f$, $C_{pk} = 12\,\mu\mu f$
C_w = Stray capacitance of wiring, etc. = $10\,\mu\mu f$ (at each tube)

Fig. 6-34

30. What is the over-all decibel gain of the two-stage amplifier of Prob. 27 at (*a*) 1,000 cps? (*b*) 100 cps? (*c*) 10,000 cps?

31. The resistance and capacitance values given in Prob. 27 have been taken from a tube manual. The values recommended for the other resistors and capacitors shown on Fig. 6-34 can also be obtained from a tube manual. What values are recommended for (*a*) R_1? (*b*) R_2? (*c*) R_3? (*d*) R_4? (*e*) C_1? (*f*) C_2? (*g*) C_3? (*h*) C_4?

32. The voltage amplification of the amplifier in Prob. 27 can be increased by increasing the value of the coupling resistor R_c. (*a*) If R_c is increased to 250,000 ohms and R_g is increased to 500,000 ohms, what are the recommended values for R_1, R_2, R_3, R_4, C_b, C_1, C_2, C_3, and C_4? (*b*) What is the voltage amplification per stage at 1,000, 100, and 10,000 cps? (*c*) What is the over-all decibel gain of the amplifier at 1,000, 100, and 10,000 cps?

33. The amplifier of Prob. 32 has a loss of approximately 1.7 db at 100 cps. At what frequency will its loss be approximately 2 db? HINT 1: Rearranging the terms of Eq. (6-11) will produce a satisfactory equation, namely

$$f_{\min} = \frac{K_L}{2\pi R C_b \sqrt{1 - K_L{}^2}}$$

HINT 2: For -2 db, $K_L \cong 0.8$ for a single-stage amplifier; for a two-stage amplifier $K_L \cong 0.895$ per stage.

34. The amplifier of Prob. 32 has a loss of approximately 0.6 db at 10,000 cps. At what frequency will its loss be approximately 2 db? HINT 1: Rearranging the terms of

Eq. (6-14a) will produce a satisfactory equation, namely

$$f_{max} = \frac{\sqrt{1 - K_H^2}}{2\pi K_H R_{eq} C_T}$$

HINT 2: Same as Hint 2 for Prob. 33.

35. A single stage of a high-gain amplifier is shown in Fig. 6-35.   (a) From a tube manual determine the recommended values for C_b, R_1, C_1, R_2, and C_2.   (b) What is the voltage amplification produced by the stage at 1,000, 100, and 10,000 cps?   (c) What is the decibel gain at each of these frequencies?

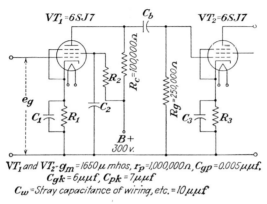

VT_1 and VT_2-g_m = 1650μ mhos, r_p = 1,000,000Ω, C_{gp} = 0.005μμf,
C_{gk} = 6μμf, C_{pk} = 7μμf
C_w = Stray capacitance of wiring, etc. = 10μμf

FIG. 6-35

36. What is the frequency range of the amplifier stage of Prob. 35 if ±2-db variation is permissible?

37. A single stage of an impedance-coupled amplifier is shown in Fig. 6-36.   The input tube is to be operated with 250 volts at its plate and with a grid bias of 8 volts. The total shunting capacitance of the circuit is 350 μμf.   Assume the core-loss effect to be negligible.   (a) What voltage is required of the B power supply if the plate current is 9 ma?   (b) What is the voltage amplification of the stage at 50, 1,000, and 10,000 cps?   (c) What is the decibel variation over a frequency range of 50 to 10,000 cps?

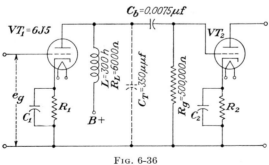

FIG. 6-36

38. A type 6J7 tube is substituted for the type 6J5 tube in the amplifier shown in Fig. 6-36.   Assume that all the necessary socket wiring changes are made; also that the screen grid is supplied with 100 volts and that the grid bias is reduced to 3 volts. Use 1 megohm as the value of the plate resistance.   (a) What voltage is required of the B power supply?   (b) What is the voltage amplification of the stage at 50, 1,000, and 10,000 cps? (*c*) What is the decibel variation over a frequency range of 50 to 10,000 cps? (*d*) Would this change be recommended and why? (*e*) Discuss the possibility of using the type 6J7 tube in this circuit as a triode by connecting the screen and suppressor grids to the plate.

39. The audio reactor used as the coupling element in an impedance-coupled a-f amplifier circuit similar to Fig. 6-8 has an inductance of 300 henrys. Its resistance is 6,250 ohms and its maximum current capacity is 10 ma. (*a*) What is the voltage drop at the reactor when it is used with a type 6C5 tube with 8 ma of plate current flowing? (*b*) What voltage is required of the B power supply if it is desired to maintain 250 volts at the plate of the tube? (*c*) What is the reactance of the choke to 100, 1,000, and 5,000 cps if its distributed capacitance is disregarded? (*d*) What is the reactance of the choke to 100, 1,000, and 5,000 cps if its distributed capacitance is 50$\mu\mu$f?

40. The audio choke used as a coupling unit in an impedance-coupled a-f amplifier circuit similar to Fig. 6-8 has an inductance of 150 henrys, a resistance of 3,500 ohms, and a current rating of 10 ma. (*a*) What is the voltage drop at the choke when 5 ma of plate current is flowing? (*b*) What voltage is required of the B power supply if it is desired to maintain 250 volts at the plate of the tube? (*c*) What is the reactance of the choke to 100, 1,000, and 5,000 cps if its distributed capacitance is disregarded? (*d*) What is the reactance of the choke to 100, 1,000, and 5,000 cps if its distributed capacitance is 20 $\mu\mu$f?

41. The audio reactor used as the coupling coil of an impedance-coupled a-f amplifier circuit similar to Fig. 6-11 has an inductance of 320 henrys. What value of blocking capacitor is required to make the tuned circuit resonant at (*a*) 100 cps? (*b*) 1,000 cps? (*c*) 5,000 cps?

42. The audio reactor used as the coupling coil of an impedance-coupled a-f amplifier circuit similar to Fig. 6-11 has an inductance of 150 henrys. What value of blocking capacitor is required to make the tuned circuit resonant at (*a*) 100 cps? (*b*) 1,000 cps? (*c*) 5,000 cps?

43. A single stage of a transformer-coupled amplifier with its circuit constants is given in Fig. 6-37. (*a*) What voltage is required at the B power supply in order to provide 250 volts at the plate of the first tube? (*b*) What is the voltage amplification at 50, 1,000, and 10,000 cps? (*c*) At what frequency will the maximum voltage amplification occur? (*d*) What is the maximum voltage amplification? (*e*) What is the decibel variation at 50 cps compared to 1,000 cps? (*f*) What is the decibel variation at 10,000 cps compared to 1,000 cps?

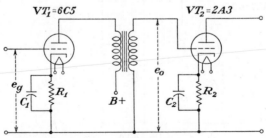

VT_1-E_b=250 volts, E_c=−8volts, C_{pk}=11$\mu\mu$f
VT_2-C_i=16$\mu\mu$f
Transformer-n=2.5, L_P=40h., L_S=250h., R_P=80Ω, R_S=500Ω,
$L_P{}'$=0.1 h., $L_S{}'$=0.625h., C_D=100$\mu\mu$f (at the primary),
Core-loss effect to be considered as negligible.
C_W= Stray capacitance of wiring, etc. = 10$\mu\mu$f (at the primary)

FIG. 6-37

44. In a certain transformer-coupled a-f amplifier circuit (Fig. 6-12a) the tube constants of VT_1 (6J5) are: $C_{pk} = 3.6$ $\mu\mu$f, $\mu = 20$, and $r_p = 7,700$ ohms. The input capacitance of VT_2 is 10 $\mu\mu$f. The transformer constants are $n = 3.16$, $L_P = 50$ henrys, $L_S = 500$ henrys, $R_P = 80$ ohms, $R_S = 800$ ohms, $L'_P = 0.1$ henry, $L'_S = 1$ henry. Total distributed capacitance referred to the primary $= 50$ $\mu\mu$f. Core-loss effect is assumed to be negligible. The total stray capacitance referred to the primary is 10 $\mu\mu$f. What voltage amplification is produced by the stage at (a) 1,000 cps? (b) 50 cps? (c) 10,000 cps? (d) At what frequency will the maximum voltage amplification occur? (e) What is the maximum voltage amplification?

45. One section of a twin triode (12AU7) is being used in a cathode-follower circuit to provide a low-impedance output. What is the voltage amplification of the circuit if the circuit constants are $\mu = 17$, $r_p = 7,700$ ohms, and $R_k = 27,000$ ohms?

46. In order to obtain a low-impedance output for a certain a-f preamplifier unit, one section of a type 12AU7 tube is used as a cathode-follower output stage. What is the voltage amplification of the stage if the tube is operated so that its circuit values are $\mu = 20$ and $r_p = 6,500$ ohms; the value of R_k is 47,000 ohms?

47. A certain pentode tube (6J7) is being used as a cathode follower to obtain a low-impedance output. If the transconductance of the tube is 1,225 μmhos, what is the voltage amplification of the stage when the cathode resistor has a value of (a) 5,000 ohms? (b) 50,000 ohms? (c) 500,000 ohms?

48. What is the equivalent output impedance of the cathode-follower stage of Prob. 45?

49. What is the equivalent output impedance of the cathode-follower stage of Prob. 46?

50. What is the equivalent output impedance of the cathode-follower stage of Prob. 47?

51. (a) What is the input capacitance of a triode used in a cathode-follower circuit if $C_{gk} = 3.4$ $\mu\mu$f, $C_{gp} = 3.4$ $\mu\mu$f, and VA $= 0.9$? (b) What is the input capacitance of this same tube used in a grounded-cathode amplifier circuit having a voltage amplification of 10? [Use Eq. (6-23).]

52. (a) What is the input capacitance of a triode (one section of a type 6CG7) used in a cathode-follower circuit if $C_{gk} = 2.3$ $\mu\mu$f, $C_{gp} = 4$ $\mu\mu$f, and VA $= 0.9$? (b) What is the input capacitance of the same tube used in a grounded-cathode amplifier circuit having a voltage amplification of 10? [Use Eq. (6-23).]

53. A certain cathode-follower circuit has a 0.5-megohm input resistor R_g returned directly to the cathode. What is the effective input impedance if the voltage amplification of the circuit is (a) 0.8? (b) 0.9? (c) 0.95.

54. A certain cathode-follower circuit has a 220,000-ohm input resistor R_g returned directly to the cathode. What is the effective input impedance if the voltage amplification of the circuit is (a) 0.8? (b) 0.85? (c) 0.92?

55. What is the effective input impedance of a cathode-follower circuit similar to Fig. 6-16b if $R_g = 0.5$ megohm, $R_2 = 5,000$ ohms, $R_3 = 5,000$ ohms, and VA $= 0.9$?

56. What is the effective input impedance of a cathode-follower circuit similar to Fig. 6-16b if $R_g = 0.5$ megohm, $R_2 = 3,000$ ohms, $R_3 = 7,000$ ohms, and VA $= 0.9$?

57. A cathode-follower phase-splitter circuit similar to Fig. 6-19 uses one-half of a twin triode (6CG7) having an amplification factor of 20 and a plate resistance of 7,700 ohms. The other circuit components have the following values: $R_2 = 27,000$ ohms, $R_3 = 3,000$ ohms, $R_4 = 24,000$ ohms. What is the voltage amplification for each output?

58. A cathode-follower phase-splitter circuit similar to Fig. 6-19 uses one-half of a twin triode (12AU7) having an amplification factor of 17 and a plate resistance of 7,700 ohms. The other circuit components have the following values: $R_2 = 22,000$ ohms, R_3 is omitted, $R_4 = 22,000$ ohms. What is the voltage amplification of each output?

59. A certain signal is being fed through three stages of amplification having gains of 20, 80, and 15 times respectively. (*a*) What is the total voltage amplification of the circuit? (*b*) What is the decibel gain of each stage? (*c*) What is the over-all decibel gain?

60. A certain signal is being fed through three stages of amplification having gains of 60, 30, and 12 times respectively. (*a*) What is the total voltage amplification of the circuit? (*b*) What is the decibel gain of each stage? (*c*) What is the over-all decibel gain?

61. A certain a-f amplifier is designed to produce an over-all voltage amplification of 4,500 with an input signal of 5 mv. The distortion in the amplifier is 5 per cent. (*a*) What is the output voltage? (*b*) What is the distortion in volts? (*c*) If a feedback circuit with $\beta = -0.01$ is added, what is the value of the feedback factor? (*d*) What is the output voltage with feedback added? (*e*) What is the distortion in volts with feedback added? (*f*) With this feedback, what value of input signal will be required in order to restore the output to its original value? (*g*) What is the distortion in volts with feedback and the increased input signal as found in part (*f*)?

62. A certain a-f amplifier is designed to produce an over-all voltage amplification of 3,600 with an input signal of 4 mv. The amplifier has 3 per cent distortion. (*a*) What is the output voltage? (*b*) What is the distortion in volts? (*c*) If a feedback circuit with $\beta = -0.005$ is added, what is the value of the feedback factor? (*d*) What is the output voltage when feedback is added? (*e*) What is the distortion in volts when feedback is added? (*f*) With this feedback, what value of input signal will be required in order to restore the output to its original value? (*g*) What is the distortion in volts with feedback and the increased input signal as found in part (*f*)?

63. A certain a-f amplifier has a voltage amplification of 16 and an input signal of 2.5 volts. (*a*) What is the output voltage? (*b*) If a feedback circuit with $\beta = -0.05$ is added, what is the value of the feedback factor? (*c*) What is the output voltage if no change is made in the input voltage? (*d*) What value of input signal will be required in order to restore the output voltage to its original value?

64. The voltage amplification of a certain amplifier circuit is 160 at 1,000 cps, but drops to 20 at 30 cps. (*a*) What is the change in gain expressed in decibels? (*b*) If a feedback circuit with $\beta = -0.10$ is added, what is the decibel change in gain between 1,000 and 30 cps?

65. The resistors of the voltage-divider network of a feedback circuit similar to Fig. 6-22 are: $R_f = 10,000$ ohms, and $R_2 = 90,000$ ohms. (*a*) What is the approximate value of β? (*b*) If the voltage amplification of the circuit is 160 without feedback, what is its amplification with feedback? (*c*) What is the approximate value of voltage amplification with feedback, using Eq. (6-51)?

66. The resistors of the voltage-divider network of a feedback circuit similar to Fig. 6-22 are: $R_f = 4,700$ ohms, and $R_2 = 150,000$ ohms. (*a*) What is the approximate value of β? (*b*) If the voltage amplification of the circuit is 150 without feedback, what is its amplification with feedback?

67. It is desired to add a feedback circuit, similar to Fig. 6-22, to an output tube being worked through a 7,000-ohm load. (*a*) What minimum resistance should the feedback network have if the ratio of feedback circuit impedance to load impedance is to be approximately 20 to 1? (*b*) If a feedback factor of 3 is desired, what standard resistance values should the two resistors have? (Assume $A = 20$.) (*c*) What standard size of capacitor is recommended if ($R_f + R_2$) should be approximately 20 times the reactance of the capacitor at a minimum frequency of 50 cps?

68. A 0.5-μf capacitor and a 100,000-ohm resistor are connected to form a decoupling network. What is the approximate value of the decoupling factor for a 50-cycle a-f signal?

69. An a-f amplifier circuit using a 250,000-ohm resistor as its plate load also has a decoupling resistor of 30,000 ohms and a decoupling capacitor of 0.25 μf. (*a*) What is the decoupling factor to a 50-cps signal? (*b*) What is the over-all decoupling factor for two such stages?

70. A certain radio receiver uses decoupling networks in the a-f and i-f circuits. The a-f decoupling network consists of a 15,000-ohm resistor and a 20-μf capacitor, and the i-f decoupling network consists of a 12,000-ohm resistor and a 0.05-μf capacitor. What is the decoupling factor for (*a*) a 100-cps a-f signal? (*b*) a 455-kc i-f signal?

CHAPTER 7 ✓

POWER-AMPLIFIER CIRCUITS

The prime purpose of voltage-amplifier circuits (Chaps. 5 and 6) is to produce a substantial increase in the output voltage for a given input voltage. The amplified voltage is then applied to the input circuit of a power-amplifier stage. Audio-frequency power amplifiers are used to provide the power required to operate a loudspeaker. Radio-frequency power amplifiers are used to produce the power required at the antenna of a transmitter.

7-1. Power-amplifier Circuits. *Classifications.* Power-amplifier circuits are used in both a-f and r-f circuits and are classified according to the frequency at which they are operated, namely, *a-f power amplifiers* or *r-f power amplifiers.* Whereas the tubes used in voltage-amplifier circuits are generally operated as Class A, the tubes used in power-amplifier circuits may be operated as Class A, Class B, Class AB, or Class C. Each of the two classifications of power amplifiers can therefore be subdivided according to the manner in which the tube is operated. All classes of tube operation may be used in r-f and a-f power amplifier circuits; however, Class C is used primarily in r-f power amplifiers.

Triodes, pentodes, and beam power tubes are used as power amplifiers and may be operated singly, in parallel, or in push-pull depending upon the amount of power output required. Although the principle of operation and the general construction of the vacuum tubes used as power amplifiers are similar to those used as voltage amplifiers, their operating characteristics are quite different. Voltage amplifier tubes are operated with a comparatively high value of impedance in the plate circuit in order to obtain a high output voltage because voltage gain is the paramount factor, and the power output is comparatively small. Power-amplifier tubes are operated with a lower value of plate-load impedance than voltage amplifier tubes because: (1) the plate resistance of these tubes is lower, (2) a larger power output is required, and (3) power output is the paramount factor, and the value of the output voltage is not important. Tubes used as power amplifiers must therefore be capable of carrying more current than voltage-amplifier tubes.

The Output Circuit. In a radio receiver the output of the power tube (or tubes) is fed to the voice coil of a loudspeaker. The impedance of the voice coil is generally in the order of 3 to 16 ohms and for practical purposes may be considered as being only resistance. Because of the

236

large difference between the resistance of the voice coil and the plate resistance of the power tube, the tube cannot be efficiently operated directly into the loudspeaker.   A transformer is used to couple the output of the power tube to the voice coil of the loudspeaker.

7-2. Output Transformer Circuit.   *Impedance Matching.*   The transformer used to couple the plate circuit of the power-output tube to the voice coil of the loudspeaker, generally called the *output transformer*, may be considered as an impedance-matching device.   The impedance reflected to the plate circuit by the voice coil can be controlled by the ratio of the primary to the secondary turns in the transformer.   Expressed mathematically,

$$R'_o = R_o \left(\frac{N_P}{N_S}\right)^2 \tag{7-1}$$

where R'_o = plate load resistance, ohms
R_o = resistance of voice coil, ohms
N_P = number of turns on primary winding
N_S = number of turns on secondary winding

Example 7-1.   A transformer having a primary-to-secondary turns ratio of 18 to 1 is used to couple a loudspeaker having a voice-coil resistance of 8 ohms to the output circuit of a power-amplifier tube.   What value of load resistance is presented to the plate circuit of the tube?

Given: $\dfrac{N_P}{N_S} = 18$ Find: R'_o

R_o = 8 ohms

Solution:

$$R'_o = R_o \left(\frac{N_P}{N_S}\right)^2 = 8 \times 18^2 = 2,592 \text{ ohms}$$

A good-quality transformer will accomplish the change in the magnitude of the impedance without producing any appreciable change in the phase relation.   If the impedance of the voice coil is considered as being pure resistance, the impedance reflected to the primary winding will also be pure resistance.

The Output Transformer.   The average radio receiver in the home is usually operated with an audio-power output of less than 1 watt.   Three watts of audio power generally exceeds the amount required to provide sufficient volume for the average person in a normal living room.   As output transformers are usually rated from 8 to 20 watts, this transformer is seldom operated at or near its rated output.

The output transformer used with a push-pull circuit must have a center-tapped primary and its core may be smaller than for a transformer used with single-tube operation.   This is due to the fact that the direct currents in the two halves of the primary winding flow in opposite directions and the resultant magnetization of the core is very low.

7-3. Vacuum-tube Characteristics.   *Use of Curves.*   The explanation
of the operation of circuits employing vacuum tubes is generally presented
with the aid of graphs or curves representing the characteristics of the
particular tube used (see Chaps. 2, 5, and 6).   The types of curves used

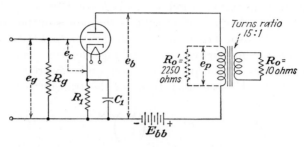

Fig. 7-1. Triode power amplifier and the reflected load R_o'.

for analyzing power-amplifier operation differ somewhat from those used
with voltage-amplifier analysis.

Effect of a Varying Input Signal on the Plate Voltage.   In the elementary
amplifier circuit (Fig. 7-2a) when the input signal is zero, the voltage on

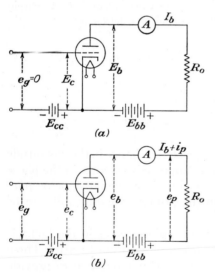

the grid of the tube will be equal to
the steady grid bias E_c.   The plate
current I_b, being dependent on the
grid bias, will also be steady.   The
voltage E_b at the plate of the tube
will then be equal to the plate-supply
voltage E_{bb} minus the voltage drop
across the output resistor R_o, or

$$E_b = E_{bb} - I_b R_o \qquad (7\text{-}2)$$

When an alternating signal voltage e_g is applied to the input circuit
of the amplifier, the voltage on the
grid of the tube will become more
negative during the negative half-
cycles of the input signal and less
negative during the positive half-
cycles.   This change in grid bias
causes a variation in the plate current, represented by i_p in Fig. 7-2b,

Fig. 7-2. Elementary amplifier circuit.
(a) With zero signal input. (b) With
an a-c signal input.

flowing through R_o.   Any change in the plate current will change the voltage drop at R_o, thus causing the voltage at the plate of the tube to vary.

Use of Static and Dynamic Curves.   The variation in plate current of a
voltage-amplifier tube is usually very small; hence the change in plate
voltage will also be small.   Thus, for all practical purposes, the operating

characteristics of a voltage-amplifier tube can be obtained from its static characteristic curves. The variation in plate current of a power-amplifier tube is comparatively high, thus causing the plate-voltage variations to be high. A new set of curves that will approximate the performance of a tube under actual working conditions must therefore be used. These curves, called the *dynamic characteristics*, show the performance of a tube when operating with an input signal and a plate-circuit load.

7-4. Dynamic Characteristics of a Triode Vacuum-tube Amplifier Circuit. *Load Line.* The dynamic characteristics of a tube can be obtained graphically from its static plate characteristic curves with the aid of a load line. The load line is drawn on the static characteristic curves and represents the variation of plate current with grid voltage for the type of load impedance used. If the load is resistive, the variation of plate current with grid voltage will be linear, and the load line will be straight. With an inductive or capacitive load, the load line will be an ellipse. Resistive loads may be an actual noninductive resistor (Fig. 7-3) or the reflected load from the secondary of a transformer (Fig. 7-1).

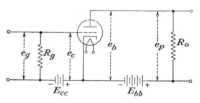

FIG. 7-3. Amplifier circuit using a triode.

Method of Plotting the Load Line for a Single-triode Class A Power Amplifier with a Resistive Load. The alternating plate-current output of a Class A power amplifier should have approximately the same waveform as that of the alternating voltage impressed on the input circuit of the tube. In order to obtain this type of output, the dynamic transfer characteristic curve (Fig. 7-5) must be practically a straight line for the complete range of the varying grid voltage. The location of the load line will be governed by such factors as: (1) the value of the plate-load resistance, (2) the voltage of the B power supply, (3) the maximum plate-dissipation rating of the tube, (4) the desired power output, (5) the allowable per cent of distortion. The method of locating the load line on the static plate characteristic curves is shown in the following group of examples based on the curves of a typical power triode shown in Fig. 7-4.

Zero-signal Bias. An approximate value of the grid bias for a triode power-amplifier tube is

$$E_c \cong \frac{0.68E_b}{\mu} \tag{7-3}$$

It is important to check whether the tube will be operated within its maximum plate power-dissipation rating with this value of grid bias.

Example **7-2.** What value of grid bias is recommended for the power-amplifier tube whose characteristics are given in Fig. 7-4?

Given: $E_b = 250$ volts   Find: E_c

$\mu = 4.2$

Solution:

$$E_c \cong \frac{0.68E_b}{\mu} \cong \frac{0.68 \times 250}{4.2} \cong 40.5 \text{ volts}$$

From the curves, when $E_c = -40.5$ volts and $E_b = 250$ volts, then $I_b = 80$ ma.   For this condition, the plate power dissipation (which is equal to the product of E_b and I_b) is equal to 250×0.08 or 20 watts.   As this exceeds the maximum plate-dissipation rating of 15 watts for this tube, a new value of grid bias must be selected so that the rated value of plate dissipation is not exceeded.   By simple mathematical deduction, if the plate voltage is maintained at 250 volts a plate current of 60 ma will just meet the maximum plate-dissipation rating of 15 watts.   The curves show that this will occur with a grid bias of approximately 43.5 volts.   Thus, the *recommended value* for E_c with zero-signal input is -43.5 volts.

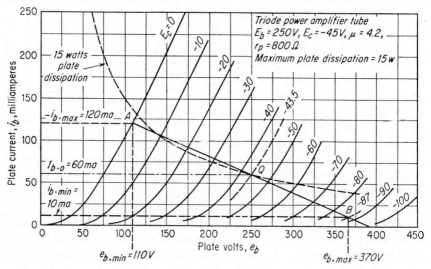

Fig. 7-4. Method of plotting the load line and of obtaining the dynamic operating characteristics for a triode vacuum tube.

Maximum-current Point.   The maximum operating plate current is often taken as twice the value of I_b for zero-signal input.   For the tube represented by Fig. 7-4 $I_{b\text{-max}}$ will be 2×60 or 120 ma.

Plotting the Load Line.   The load line, which is a straight line, can now be plotted from the available data which will locate two points on the line. The *quiescent* or zero-signal operating point Q is located along the curve for $E_c = -43.5$ volts where this curve crosses the plate-voltage value of $E_b = 250$ volts.   Point A is located where the curve for $E_c = 0$ crosses the plate-current value of 120 ma.   Points A and Q may now be connected by a straight line which is then extended through point B.

Operating Values Indicated by the Load Line.   Operating the tube with a grid bias of 43.5 volts limits the maximum signal input voltage to 43.5

volts in order to avoid driving the grid positive during any part of the input signal cycle. A 43.5-volt input signal will cause the grid voltage to swing between zero and -87 volts. With the maximum positive input signal the tube will be operating at point A, and with the maximum negative input signal the tube will be operating at point B.

Example 7-3. A triode power-amplifier tube whose characteristics are shown in Fig. 7-4 is operated with a 43.5-volt input signal. What is the amount of: (*a*) plate-current variation, (*b*) plate-voltage variation?

> Given: Curves, Fig. 7-4   Find: (*a*) Plate-current variation
> (*b*) Plate-voltage variation

Solution:

From curves: at point A—$i_b = 120$ ma $e_b = 110$ volts
at point B—$i_b = 10$ ma $e_b = 370$ volts

(*a*)   Plate-current variation $= i_{b\cdot A} - i_{b\cdot B} = 120 - 10 = 110$ ma
(*b*)   Plate-voltage variation $= e_{b\cdot B} - e_{b\cdot A} = 370 - 110 = 260$ volts

The phase relations among the input signal voltage e_g, the grid voltage e_c, the plate current i_b, and the plate voltage e_b are the same as shown in Fig. 2-12. If the plate load of the amplifier circuit consists of only resistance, the grid-voltage variations will be in phase with the plate-current variations, and the plate-voltage variations will be 180 degrees out of phase with the grid-voltage variations. As all amplifier circuits contain at least a small amount of undesired capacitance and inductance, this in-phase and 180-degree out-of-phase relationship is seldom obtained. The amount of phase difference from these theoretical conditions will depend upon the amount of capacitance and inductance present in the circuit and in some cases may be so small that it can be ignored.

Value of the Load Resistance. The value of the plate-load resistance R_o required to produce the characteristics shown in Fig. 7-4 is dependent upon the slope of the load line and is expressed by

$$R_o = \frac{e_{b\cdot max} - e_{b\cdot min}}{i_{b\cdot max} - i_{b\cdot min}} \tag{7-4}$$

Example 7-4. What value of plate-load resistance is required to produce the circuit characteristics corresponding to Fig. 7-4?

> Given: Curves, Fig. 7-4   Find: R_o

Solution:

$$R_o = \frac{e_{b\cdot max} - e_{b\cdot min}}{i_{b\ max} - i_{b\cdot min}} = \frac{370 - 110}{0.12 - 0.01} = 2{,}363 \text{ ohms}$$

Example 7-5. What value of B supply voltage will be required for the circuit whose characteristics are shown in Fig. 7-4 with a plate-load resistance as found in Example 7-4?

> Given: $E_{b\cdot o} = 250$ volts   Find: E_{bb}
> $I_{b\cdot o} = 60$ ma
> $R_o = 2{,}363$ ohms

Solution:

$$E_{bb} = E_b + I_{b \cdot o}R_o = 250 + 0.06 \times 2{,}363 = 392 \text{ volts}$$

Power Output.   The a-c power output of the amplifier circuit can be found by use of the peak-to-peak values of plate voltage and plate current, and the basic power equation $P = EI$.   In this equation the voltage and current must be effective values; hence, it is necessary to convert the peak-to-peak values of voltage and current to their effective values.   This is done by dividing the peak-to-peak values by the factor $2\sqrt{2}$.   The a-c power output can then be expressed as

$$P_o = \frac{(e_{b \cdot \max} - e_{b \cdot \min})}{2\sqrt{2}} \times \frac{(i_{b \cdot \max} - i_{b \cdot \min})}{2\sqrt{2}} \qquad (7\text{-}5)$$

or

$$P_o = \frac{(e_{b \cdot \max} - e_{b \cdot \min})(i_{b \cdot \max} - i_{b \cdot \min})}{8} \qquad (7\text{-}5a)$$

Example 7-6.   What is the value of the power output for the amplifier circuit whose characteristics are shown in Fig. 7-4?

Given: $e_{b \cdot \max} = 370$ volts   Find: P_o
         $e_{b \cdot \min} = 110$ volts
         $i_{b \cdot \max} = 120$ ma
         $i_{b \cdot \min} = 10$ ma

Solution:

$$P_o = \frac{(e_{b \cdot \max} - e_{b \cdot \min})(i_{b \cdot \max} - i_{b \cdot \min})}{8} = \frac{(370 - 110)(0.12 - 0.01)}{8} = 3.57 \text{ watts}$$

Nonlinear Distortion.   One cause of distortion of the output signal of an amplifier circuit is due to operating the amplifier tube on portions of the characteristic curves that are not perfectly straight.   Examination of the static characteristic curves of Fig. 7-4 and the dynamic characteristic curves of Fig. 7-5 will show that all of the lines have some amount of curvature.   A Class A power amplifier will therefore produce some distortion, since even the best operating portion of its characteristic curve is not actually a straight line.   This distortion, caused mainly by the second harmonic, can be calculated by

$$\% \text{ 2nd harmonic distortion} = \frac{(i_{b \cdot \max} + i_{b \cdot \min}) - 2I_{b \cdot o}}{2(i_{b \cdot \max} - i_{b \cdot \min})} \times 100 \qquad (7\text{-}6)$$

Example 7-7.   What is the per cent of second harmonic distortion for the amplifier circuit whose characteristics are shown in Fig. 7-4?

Given: Curves, Fig. 7-4   Find: % 2nd harmonic distortion

Solution:

$$\begin{aligned}
\text{2nd harmonic distortion} &= \frac{(i_{b \cdot \max} + i_{b \cdot \min}) - 2I_{b \cdot o}}{2(i_{b \cdot \max} - i_{b \cdot \min})} \times 100 \\
&= \frac{(120 + 10) - 2 \times 60}{2(120 - 10)} \times 100 = 4.54\%
\end{aligned}$$

7-5. Dynamic Characteristic Curves for a Triode Class A Power-amplifier Tube. Although the operating characteristics of a tube can be obtained from the static plate characteristic curves with the aid of a load line, a more complete interpretation can be acquired from a family of dynamic characteristic curves. A single dynamic characteristic curve represents the actual operating characteristics of a tube for a definite type and value of load impedance. The dynamic characteristic curve of a tube will therefore vary with the type and value of load impedance. Thus, a family of dynamic characteristic curves (Fig. 7-5) can be plotted from values obtained from the static plate characteristic curves and the load

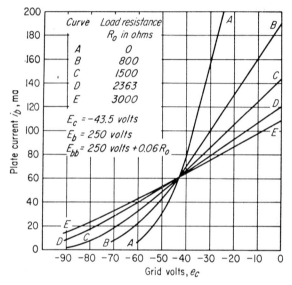

FIG. 7-5. Dynamic characteristics for a triode power amplifier tube (same tube as used in Fig. 7-4).

lines. The values used to plot curve D of Fig. 7-5 were obtained from Fig. 7-4 at the points where the load line intersects each of the static curves. Curves B, C, and E were obtained by first plotting curves similar to Fig. 7-4 with their corresponding values of plate-load resistance and then obtaining the values for the new curve in the same manner as for curve D. For comparison, the transfer static curve A ($R_o = 0$) has been drawn on the same graph; five sets of values for grid volts and plate current were obtained from Fig. 7-4 at the points where the line representing $e_b = 250$ volts intersects the static curves.

The family of curves in Fig. 7-5 shows that the slope of the transfer characteristic curve becomes steeper with a decrease in the plate load and flatter with an increase in the plate load. Thus, increasing the plate load will decrease the plate-current output. However, increasing the plate load also lengthens the straight portion of the transfer characteristic curve, thus making it possible to apply a larger signal voltage to the input circuit of the tube without producing excessive distortion.

7-6. Power Output of a Triode Class A Power-amplifier Tube. *Instantaneous Value.* The current that flows through the plate circuit of a tube consists of two parts: (1) a steady or d-c component I_b; (2) a varying or a-c component i_p (see Fig. 7-2). Only the varying component of the plate current produces sound variations at the loudspeaker. The steady plate current I_b does not contribute directly to the useful output of an amplifier circuit but does contribute toward the power loss of the circuit as it produces heat at both the load impedance and the plate of the tube.

The varying component i_p of the plate current flowing through the load resistor R_o produces a varying component of voltage e_p across R_o. The values of i_p and e_p are expressed by Eqs. (2-8) and (2-9).

$$i_p = \frac{\mu e_g}{R_o + r_p} \tag{2-8}$$

$$e_p = \frac{\mu e_g R_o}{R_o + r_p} \tag{2-9}$$

The instantaneous value of the a-c power output is

$$p_o = e_p i_p \tag{7-7}$$

Substituting Eqs. (2-8) and (2-9) for i_p and e_p in Eq. (7-7)

$$p_o = \frac{\mu e_g R_o}{R_o + r_p} \times \frac{\mu e_g}{R_o + r_p} = \frac{(\mu e_g)^2 R_o}{(R_o + r_p)^2} \tag{7-8}$$

A-C Power. The a-c power output can best be determined from values taken from the characteristic curves as in Example 7-6. However, if the nonlinear distortion is not too great, a reasonably approximate value of the a-c power output can be obtained by substituting the effective value of the signal voltage for the instantaneous value in Eq. (7-8). Then

$$P_o = \frac{(\mu E_g)^2 R_o}{(R_o + r_p)^2} \tag{7-9}$$

Principle of Maximum Power Output. Equations (2-8) and (2-9) show that for constant values of amplification factor and input signal voltage, the output voltage e_p will vary according to the factor $\dfrac{R_o}{R_o + r_p}$, and the plate current i_p will vary according to the factor $\dfrac{1}{R_o + r_p}$. Furthermore, with a fixed value of r_p, an increase in R_o will increase the output voltage e_p and decrease the output current i_p; conversely, a decrease in R_o will decrease the output voltage e_p and increase the output current i_p.

Because e_p and i_p do not change at the same rate, the power output may either increase or decrease with a change in R_o depending upon whether R_o is smaller or larger than r_p. The maximum power output will be obtained when R_o is equal to r_p as is shown in the following example.

Example 7-8.   Prove that the maximum amount of power output is obtained at the load of an amplifier circuit when the load resistance R_o is equal to the plate resistance r_p. Assume that the tube has a plate resistance of 8,000 ohms, a constant amplification factor of 20, and an input signal of 1 volt. Establish the proof by determining the power output with load resistance values of 2,000, 4,000, 6,000, 8,000, 10,000, 12,000, 18,000, and 24,000 ohms.

Given: $r_p = 8,000$ ohms   Find: $P_{o\text{-max}}$
$\mu = 20$
$e_g = 1$ volt
$R_o = 2,000, 4,000, 6,000, 8,000, 10,000,$
$12,000, 18,000, 24,000$ ohms

Solution:

TABLE 7-1.   SOLUTION OF EXAMPLE 7-8

R_o, ohms	$R_o + r_p$, ohms	e_p, volts [Eq. (2-9)]	i_p, ma [Eq. (2-8)]	p_o	
				mw [Eq. (7-7)]	watts [Eq. (7-8)]
2,000	10,000	4	2	8	0.008
4,000	12,000	6.666	1.666	11.11	0.01111
6,000	14,000	8.571	1.428	12.23	0.01224
8,000	16,000	10	1.25	12.5	0.0125
10,000	18,000	11.11	1.111	12.34	0.01234
12,000	20,000	12	1	12	0.012
18,000	26,000	13.84	0.769	10.64	0.01065
24,000	32,000	15	0.667	9.375	0.009375

The results set forth in Table 7-1 show that maximum power output is obtained when $R_o = 8,000$ ohms, which is also the value of r_p.

The principle illustrated in Example 7-8 can be applied to both voltage-amplifier circuits (as in Example 7-8) and power-amplifier circuits. This is called the principle of *maximum power transfer*. Amplifier circuits used in radio receivers are ordinarily not designed for maximum power transfer for two reasons: (1) In a voltage amplifier it is more important to obtain a high voltage output than the maximum power output. (2) In a power amplifier designed for maximum power output both the maximum plate-dissipation rating and the maximum allowable nonlinear distortion would be exceeded.

Graphical Representation of Maximum Power Output.   The values of power output for various values of R_o found in Example 7-8 have been plotted to produce the curve of Fig. 7-6. This curve shows that the output power increases rapidly as the load resistance is increased from zero and reaches its maximum value when the load and plate resistances are equal. Further increases in load resistance cause the output power to decrease, although at a much slower rate.

Mathematical Analysis of Maximum Power Output. When R_o is equal to r_p, the instantaneous value of the power output [Eq. (7-8)] then becomes

$$p_{o \cdot \text{max}} = \frac{(\mu e_g)^2}{4r_p} \tag{7-10}$$

If the effective value of the input signal is substituted for the instantaneous

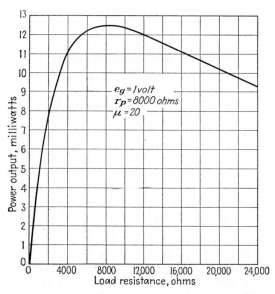

$e_g = 1 \, volt$
$r_p = 8000 \, ohms$
$\mu = 20$

FIG. 7-6. Power-output characteristics for a triode amplifier tube.

value in Eq. (7-10), the a-c power output of the circuit may then be found by use of the equation

$$P_{o \cdot \text{max}} = \frac{(\mu E_g)^2}{4r_p} \tag{7-11}$$

If the maximum value of the input signal is substituted for the effective value in Eq. (7-11), then

$$P_{o \cdot \text{max}} = \frac{(\mu E_{g \cdot \text{max}})^2}{8r_p} \tag{7-11a}$$

where $p_{o \cdot \text{max}}$ = instantaneous value of power output with maximum power transfer, watts

$P_{o \cdot \text{max}}$ = maximum a-c power output of amplifier circuit, watts

e_g = instantaneous value of input signal, volts

E_g = effective value of input signal, volts

$E_{g \cdot \text{max}}$ = maximum or peak value of the input signal, volts

Example 7-9.   A certain amplifier uses a triode that has an amplification factor of 4.2 and a plate resistance of 800 ohms.   It is operated as Class A and with a grid bias of 45 volts.   What is the maximum power output obtainable from this amplifier with an input signal having a peak value of 20 volts?

$$\text{Given:} \quad \mu = 4.2 \qquad\qquad \text{Find:} \quad P_{o \cdot max}$$
$$r_p = 800 \text{ ohms}$$
$$E_{g \cdot max} = 20 \text{ volts}$$

Solution:

$$P_{o \cdot max} = \frac{(\mu E_{g \cdot max})^2}{8r_p} = \frac{(4.2 \times 20)^2}{8 \times 800} = 1.102 \text{ watts}$$

Maximum Power Output vs. Nonlinear Distortion.   When operating a tube as a power amplifier, both the power output and the per cent of distortion should be considered.   Maximum power output is obtained when the load and plate resistances are equal.   Increasing the load resistance will decrease the curvature of the characteristic curve and thus decrease the amount of distortion.   It is common practice to sacrifice some of the power output in order to use a load resistance that will keep the amount of distortion at a satisfactory minimum.   The recommended maximum allowable distortion of audio amplifiers in radio receivers has been taken as 5 per cent.   In order to achieve this condition the value of R_o will be in the order of 2 to 3 times r_p; in Example 7-7 the nonlinear distortion was 4.54 per cent with R_o approximately three times r_p.

7-7. Ratings of Triode Power Amplifiers. *Power Amplification.* Amplifier tubes are generally operated with sufficient grid bias to prevent any current from flowing in its grid circuit.   Under this condition any current flowing through the grid resistor R_g must come from the driving source.   When the input signal is fed into an amplifier tube through a grid resistor R_g, the power used by this resistor represents the input power of the amplifier and is called the *driving power*.   The ratio of the a-c power output to the a-c power consumed in the grid circuit is called the *power amplification* of the amplifier circuit.

In contrast to the high power amplification obtained from a power-amplifier circuit, the voltage amplification obtained is very low.   Equations (7-8) to (7-11) show that the output power varies as the square of the input signal voltage; thus if the input signal voltage is tripled, the output power will be increased nine times.   For this reason, the input voltage to power-amplifier tubes should be comparatively high.   Power-amplifier tubes are therefore designed to operate with large input signal voltages without producing distortion.

Example 7-10.   A triode amplifier tube (Figs. 7-3 and 7-4) is being operated as a Class A_1 power amplifier with a 470,000-ohm grid input resistor R_g and a 2,400-ohm plate-load resistor R_o.   If the input signal E_g causes a current of 65 μa (effective value) to flow in R_g, find (*a*) the voltage amplification, (*b*) the power amplification.

Given: $\mu = 4.2$ Find: (a) VA
         $r_p = 800$ ohms   (b) PA
         $R_o = 2,400$ ohms
         $I_g = 65$ μa
         $R_g = 470,000$ ohms
Solution:

(a) $VA = \dfrac{\mu R_o}{R_o + r_p} = \dfrac{4.2 \times 2,400}{800 + 2,400} = 3.15$

(b) $P_i = I_g{}^2 R_g = (65 \times 10^{-6})^2 \times 470,000 = 0.001985$ watt
      $E_g = I_g R_g = 65 \times 10^{-6} \times 0.47 \times 10^6 = 30.55$ volts

      $P_o = \dfrac{(\mu E_g)^2 R_o}{(R_o + r_p)^2} = \dfrac{(4.2 \times 30.55)^2 \times 2,400}{(2,400 + 800)^2} = 3.85$ watts

      $PA = \dfrac{P_o}{P_i} = \dfrac{3.85}{0.001985} = 1,940$

Plate-circuit Efficiency. The efficiency of any power-supplying device is usually expressed as a percentage and represents the ratio of its power output to its power input. The power supplied to the plate circuit of a tube can be represented by the power that the tube releases from the B power supply and is equal to the product of the plate supply voltage and the average plate current at maximum signal input. The average plate current at maximum signal input may be slightly higher than that at zero signal input. This increase in current is produced by the nonlinear action of the tube. For most practical purposes, this difference in plate current can be ignored. The average plate current can then be taken as equal to the quiescent or operating plate current I_b. The per cent efficiency of the plate circuit can then be expressed as

$$\text{Plate circuit efficiency} = \frac{P_o}{E_{bb}I_b} \times 100 \qquad (7\text{-}12)$$

where P_o = a-c power output, watts
      E_{bb} = voltage of B power supply, volts
      I_b = average value of plate current, amp

Example 7-11. What is the plate-circuit efficiency of the power-output tube of Example 7-10 if the average plate current is 60 ma and the B supply is 392 volts?

Given: $P_o = 3.85$ watts   Find: Plate-circuit efficiency
         $I_b = 60$ ma
         $E_{bb} = 392$ volts
Solution:

$$\text{Plate-circuit efficiency} = \frac{P_o}{E_{bb}I_b} \times 100 = \frac{3.85 \times 100}{392 \times 0.06} = 16.3\%$$

Plate Efficiency. The plate-circuit efficiency is seldom used in rating power amplifiers; instead, the plate efficiency of the power-amplifier tube is generally used. The ratio of the a-c power output to the product of the average values of plate voltage and plate current at maximum signal input is called the *plate efficiency.*

$$\text{Plate efficiency} = \frac{P_o}{E_b I_b} \times 100 \qquad (7\text{-}13)$$

In Class A amplification maximum efficiency will be obtained when the maximum negative values of the varying components of the plate voltage e_p and the plate current i_p produce the lowest possible values of plate voltage e_b and plate current i_b, namely, zero. Under this condition, the maximum value of e_b will be twice the value of the operating plate voltage E_b and the maximum value of i_b will be twice the value of the operating plate current I_b. The peak values of E_p and I_p will thus be equal to E_b and I_b respectively. As the a-c power output (for resistive loads) is equal to the product of the effective values of E_p and I_p, then the a-c power output for the condition of maximum plate efficiency may be expressed as

$$P_o = \frac{E_b}{\sqrt{2}} \frac{I_b}{\sqrt{2}} = \frac{E_b I_b}{2} \qquad (7\text{-}14)$$

Substituting the value of P_o of Eq. (7-14) in Eq. (7-13) shows that the plate efficiency of a Class A amplifier cannot exceed 50 per cent. This is the theoretical maximum efficiency; in practical amplifier circuits the actual efficiency is much lower.

The plate efficiency is dependent upon the ratios of the maximum and minimum plate voltages and the maximum and minimum plate currents. These in turn are dependent upon the operating conditions of the tube, such as the plate voltage, the input signal voltage, and the load impedance. Using a load resistance whose value is approximately equal to the tube's plate resistance and increasing the plate voltage and the input signal voltage will increase the a-c power output. The plate efficiency of the amplifier circuit is therefore also increased.

The plate efficiency is generally low in amplifiers designed primarily for minimum distortion. The plate efficiency of triodes is lower than that of pentodes or beam power tubes. Increasing the amount of permissible distortion also increases the plate efficiency. Thus the efficiency of Class A_1 operation is the lowest, and the efficiency increases with Class A_2, Class AB_1, Class AB_2, and Class B operation.

Example 7-12. What is the plate efficiency of the tube and circuit of Example 7-11 if the value of E_b is 250 volts?

Given: $P_o = 3.85$ watts   Find: Plate efficiency
$I_b = 60$ ma
$E_b = 250$ volts

Solution:

$$\text{Plate efficiency} = \frac{P_o}{E_b I_b} \times 100 = \frac{3.85 \times 100}{250 \times 0.06} = 25.6\%$$

Plate Dissipation. The heat given off at the plate of a tube as a result of electron bombardment is called the *plate dissipation*, and its symbol is P_p. This represents a loss in power and is equal to the difference between the power supplied to the plate of the tube and the a-c power delivered by the tube to its load.

Example 7-13.   What is the plate dissipation of the power-amplifier tube used in Examples 7-10, 7-11, and 7-12?

Given: P_o = 3.85 watts   Find: P_p
             E_b = 250 volts
             I_b = 60 ma

Solution:

$$P_p = P_i - P_o = (250 \times 0.06) - 3.85 = 11.15 \text{ watts}$$

The rated maximum plate dissipation of a power-amplifier tube can be obtained from a tube manual.   The maximum plate dissipation for the tube used in Example 7-13 is 15 watts.   This value is greater than the power dissipation calculated in Example 7-13; hence the tube is being operated within its rated plate-dissipation limit.

Power Sensitivity.   The ratio of the a-c power output to the square of the effective value of the input signal voltage is called the *power sensitivity.*   The basic unit of power sensitivity is the mho, but, because of the low values of power sensitivity usually obtained, the micromho (μmho) is commonly used.

$$\text{Power sensitivity} = \frac{P_o}{E_g^2} \times 10^6 \tag{7-15}$$

Power sensitivity is ordinarily only used in rating amplifier tubes that are operated so that no current flows in the grid circuit.   When this term is used in connection with an amplifier circuit in which the grid circuit consumes power, it refers to the entire amplifier section including the driver tube.   The power sensitivity of pentodes and beam power tubes is considerably greater than for triodes.

The term *power sensitivity* is also used in rating r-f power amplifiers but has an entirely different meaning than when used with a-f power amplifiers.   The power sensitivity of an r-f power amplifier expresses the ratio of the output power to the input power, or the power amplification of the circuit.

Example 7-14.   What is the power sensitivity of the tube and circuit used in Examples 7-10 to 7-13?

Given: P_o = 3.85 watts (Example 7-10)   Find: Power sensitivity
             E_g = 30.55 volts (Example 7-10)

Solution:

$$\text{Power sensitivity} = \frac{P_o}{E_g^2} \times 10^6 = \frac{3.85}{(30.55)^2} \times 10^6 = 4,125 \ \mu\text{mhos}$$

7-8. Class A$_2$ Operation.   Applying negative feedback to an amplifier circuit makes it possible to reduce the distortion produced in the amplifier.   When negative feedback is used, a tube can be operated as a Class A$_2$ amplifier without producing excessive distortion.

In a Class A_2 amplifier, the grid bias and the input signal voltage are such that the total instantaneous grid voltage e_c is driven positive during a portion of the input cycle.   Grid current will flow during the portion of the cycle in which the grid is positive.   Under this condition it is possible to obtain maximum plate current, at the positive peaks of the input signal, with a lower value of plate voltage.   The plate efficiency and power output is greater with Class A_2 operation than with Class A_1 and plate efficiencies of 30 to 40 per cent may be obtained.

The grid bias of a power tube operated as Class A_2 should be of such a value that rated plate current will flow when the rated plate voltage is applied.   The amount that the grid voltage may be driven positive during any portion of the input cycle will then depend upon the amount that the distortion is reduced by the addition of negative feedback.

7-9. Power Diagram for a Triode Power-amplifier Tube.   *Load Line and Power.*   The preceding discussion has shown that the load line can be used to determine the dynamic characteristics of a vacuum tube. These characteristics include the operating point, required amount of B supply voltage, maximum and minimum plate voltage, and the maximum and minimum plate current.   Other operating characteristics of an amplifier tube that can also be determined from the curves are the power lost in the load resistor, plate dissipation, power output, distortion, and voltage amplification.   Use of the load line in determining these operating characteristics is illustrated in Fig. 7-7.   Graphical determination of the power ratings of the typical triode power-amplifier tube of Fig. 7-4 and with the same operating voltages and load resistance as in Examples 7-10 to 7-13 is illustrated by the power diagram of Fig. 7-7.

Power Output.   The power diagram, Fig. 7-7, shows that with an input signal of 87 volts (peak-to-peak), the plate voltage will vary from 110 to 370 volts.   This peak-to-peak plate-voltage swing of 260 volts represents the useful signal voltage available to operate the following circuit.   For this same input signal, the plate current varies from 10 to 120 ma, or a peak-to-peak swing of 110 ma.

The a-c power output is represented on the power diagram by the area of the shaded triangle QNM.   The maximum power output of a Class A amplifier will occur when the area of the shaded triangle is QCH.   The power output would then be one-half of the plate power input or one-half of the area $IQCH$, and the plate efficiency would be 50 per cent.   In actual practice, it is impossible to obtain this condition since it requires that the instantaneous value of plate voltage reach zero and the instantaneous plate current reach its maximum value at full B supply voltage.

Example 7-15.   With the aid of the power diagram of Fig. 7-7, find: (*a*) the voltage amplification, (*b*) the a-c power output.   Also, (*c*) compare the results with the answer for VA in Example 7-10, and P_o in Example 7-6.

Given: Fig. **7-7**   Find: (*a*)   VA

(*b*) P_o

(*c*)   Comparison

Solution:

From Fig. **7-7**: $E_{M-D} = 370 - 110 = 260$ volts

$E_{E-D} = -87$ volts (input signal)

$E_{P-Q} = 250 - 110 = 140$ volts

$I_{P-M} = 60 - 10 = 50$ ma or 0.05 amp

(*a*) $\text{VA} = \dfrac{E_o}{E_i} = \dfrac{E_{M-D}}{E_{E-D}} = \dfrac{260}{87} = 2.99$

(*b*) $P_o = \text{area of } QNM = \dfrac{E_{PQ} \times I_{PM}}{2} = \dfrac{140 \times 0.05}{2} = 3.50$ watts

(*c*)   Comparison:

$\text{VA} = 2.99$ from power diagram

$= 3.15$ from Example 7-10

$P_o = 3.50$ from power diagram

$= 3.57$ from Example 7-6

Power Losses and Power Ratings.   The power consumed by the output resistor is equal to the product of the voltage drop across this resistor and the average plate current.   This power is represented by the area of the

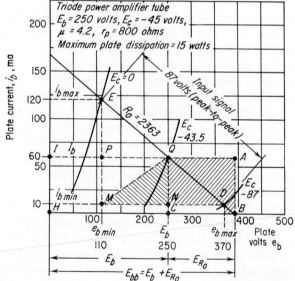

A–C power output ------------------- Area of the triangle QNM
Power lost in the output resistor ---------- Area of the rectangle QABC
Plate dissipation ------------------- Area of the rectangle IQCH
                        minus the area of the triangle QNM
Power supplied by the B power supply ------ Area of the rectangle IABH
Power supplied to the plate of the tube ----- Area of the rectangle IQCH

Fɪɢ. 7-7. Power diagram for a typical triode power amplifier tube (same tube as used in Fig. 7-4).

rectangle $QABC$. The power taken from the B supply is equal to the product of the B power-supply voltage and the average plate current and is represented by the area of the rectangle $IABH$. The input power to the plate circuit of the tube is equal to the product of the average plate voltage and the average plate current and is equal to the area of the rectangle $IQCH$. The plate dissipation of the tube is equal to the difference between the input power to the plate and the output power; it is represented by the area of the rectangle $IQCH$ minus the area of the triangle QNM.

Example 7-16. Using the power diagram of Fig. 7-7 and the value of P_o from Example 7-15, find: (a) the plate-circuit efficiency, (b) the plate efficiency, (c) the plate dissipation, and (d) the power lost in the output resistor.

Given: Power diagram, Fig. 7-7   Find: (a) Plate-circuit efficiency
$\quad\quad\quad\quad P_o = 3.5$ watts $\quad\quad\quad\quad\quad\quad\quad\quad$ (b) Plate efficiency
$\quad\quad\quad\quad\quad\quad\quad\quad\quad\quad\quad\quad\quad\quad\quad\quad\quad\quad$ (c) P_p
$\quad\quad\quad\quad\quad\quad\quad\quad\quad\quad\quad\quad\quad\quad\quad\quad\quad\quad$ (d) P_R

Solution:

From Fig. 7-7   . $\quad\quad\quad AB = CQ = 60$ ma $= 0.06$ amp
$\quad\quad\quad\quad\quad\quad\quad\quad\quad HB = 392$ volts
$\quad\quad\quad\quad\quad\quad\quad\quad\quad HC = 250$ volts
$\quad\quad\quad\quad\quad\quad\quad\quad\quad CB = 392 - 250 = 142$ volts

(a) Power of B supply $= P_B = HB \times AB = 392 \times 0.06 = 23.52$ watts

$\quad\quad$ Plate-circuit efficiency $= \dfrac{P_o}{P_B} \times 100 = \dfrac{3.5 \times 100}{23.52} = 14.9\%$

(b) $P_i = HC \times CQ = 250 \times 0.06 = 15$ watts

$\quad\quad$ Plate efficiency $= \dfrac{P_o}{P_i} \times 100 = \dfrac{3.5 \times 100}{15} = 23.3\%$

(c) $P_p = P_i - P_o = 15 - 3.5 = 11.5$ watts
(d) $P_R = CB \times AB = 142 \times 0.06 = 8.52$ watts

Comparison:

(a) Plate-circuit efficiency $= 14.9\%$ from power diagram
$\quad\quad\quad\quad\quad\quad\quad\quad\quad\quad\quad = 16.3\%$ from Example 7-11.
(b) Plate efficiency $= 23.3\%$ from power diagram
$\quad\quad\quad\quad\quad\quad\quad\quad = 25.6\%$ from Example 7-12.
(c) Plate dissipation $= 11.5$ watts from power diagram
$\quad\quad\quad\quad\quad\quad\quad\quad\quad = 11.15$ watts from Example 7-13.

Comparing the results of Example 7-16 with the results of Examples 7-11, 7-12, and 7-13 shows that the results obtained from the power diagram are fairly accurate and may be used for approximate values.

7-10. Dynamic Characteristics of a Pentode Vacuum-tube Amplifier Circuit. *Pentode and Beam Power Tube vs. the Triode.* The load line and operating characteristics for pentodes and beam power tubes are determined in much the same manner as for triodes. Figure 7-8 shows three load lines plotted on the static plate characteristic curves for a typical pentode power-amplifier tube. The recommended values for operating this tube Class A_1 are: $E_b = 250$ volts, $E_c = -16.5$ volts, $E_{g.max} = 16.5$ volts, $I_b = 34$ ma, and $R_o = 7,000$ ohms.

Load Line. The load line for the pentode serves the same purpose as for the triode and has been described in Art. 7-4.

Method of Plotting the Load Line for a Single-pentode Class A Power Amplifier with Resistive Load. The location of the load line will be governed by such factors as: (1) the value of the plate-load resistance, (2) the voltage of the B power supply, (3) the maximum plate-dissipation rating of the tube, (4) the desired power output, (5) the allowable per cent

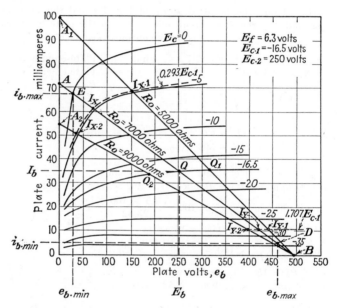

Fig. 7-8. Method of plotting the load line and of obtaining the dynamic operating characteristics for a pentode vacuum tube.

of distortion. For minimum distortion, the load line should be of such a value that $i_{b.max} - I_b$ is approximately equal to $I_b - i_{b.min}$. When operating on the 7,000-ohm load line, the difference between these two quantities is only 2 ma. Increasing the load resistance to 9,000 ohms increases this difference to 10 ma, and decreasing the load resistance to 5,000 ohms increases the difference to 14 ma.

The method of locating the load line on the static plate characteristic curves is shown in the following group of examples based on the curves of a typical power pentode shown in Fig. 7-8.

Zero-signal Bias. This should be the value of grid bias that for the specified operating plate voltage E_b will permit the greatest value of input signal voltage with the maximum acceptable amount of distortion. Examination of Fig. 7-8 will show that the recommended values of

$E_b = 250$ volts and $E_c = -16.5$ volts will best meet these conditions. With these values I_b is equal to 35 ma. The plate dissipation with $E_b = 250$ volts and $I_b = 35$ ma is 8.75 watts, which is well within the maximum rating of 11 watts.

Maximum-current Point. The maximum operating plate current should be located at or just below the knee of the curve for $E_c = 0$ in order to obtain maximum power output with minimum distortion. This is shown as point E on Fig. 7-8, and I_b is 67 ma for the location selected.

Plotting the Load Line. The load line, which is a straight line, can now be plotted from the available data which will locate two points on the line. The *quiescent* or zero-signal operating point Q is located along the curve for $E_c = -16.5$ volts where this curve crosses the plate-voltage value of $E_b = 250$ volts. Point E is located where the curve for $E_c = 0$ crosses the plate-current value of 67 ma. Points E and Q may now be connected by a straight line which is then extended to points A and B.

Operating Values Indicated by the Load Line. Operating the tube with a grid bias of 16.5 volts limits the maximum signal input voltage to 16.5 volts in order to avoid driving the grid positive during any part of the input signal cycle. A 16.5-volt input signal will cause the grid voltage to swing between zero and -33 volts. With the maximum positive input signal the tube will be operating at point E, and with the maximum negative input signal the tube will be operating at point D.

Example 7-17. A pentode power amplifier tube whose characteristics are shown in Fig. 7-8 is operated with a 16.5-volt input signal. What is the amount of (*a*) plate-current variation, (*b*) plate-voltage variation?

Given: Curves, Fig. 7-8   Find: (*a*) Plate-current variation
                                          (*b*) Plate-voltage variation

Solution:

From curves of Fig. 7-8:

At point E, $i_b = 67$ ma $e_b = 27$ volts
At point D, $i_b = 5$ ma $e_b = 460$ volts
(*a*)   Plate-current variation $i_{b.E} - i_{b.D} = 67 - 5 = 62$ ma
(*b*)   Plate-voltage variation $= e_{b.D} - e_{b.E} = 460 - 27 = 433$ volts

The phase relation among the voltages and current at the various electrodes of the tube are the same as described for a triode in Art. 7-4.

Value of the Load Resistance. The value of the plate-load resistance R_o required to produce the load line AQB in Fig. 7-8 is dependent upon the slope of the load line and is expressed by Eq. (7-4).

Example 7-18. What value of plate-load resistance is required to produce the circuit characteristics corresponding to the load line AQB on Fig. 7-8?

Given: Curves, Fig. 7-8   Find: R_o

Solution:

From Curves of Fig. 7-8:

At point E, $i_{b.\text{max}} = 67$ ma $e_{b.\text{min}} = 27$ volts
At point D, $i_{b.\text{min}} = 5$ ma $e_{b.\text{max}} = 460$ volts

$$R_o = \frac{e_{b.\text{max}} - e_{b.\text{min}}}{i_{b.\text{max}} - i_{b.\text{min}}} = \frac{460 - 27}{0.067 - 0.005} = 6,984 \cong 7,000 \text{ ohms}$$

Example 7-19. What value of B supply voltage will be required for the circuit whose characteristics are shown in Fig. 7-8 with a plate-load resistance as found in Example 7-18?

            Given: $E_b = 250$ volts   Find: E_{bb}
                     $I_b = 35$ ma
                     $R_o = 7,000$ ohms
Solution:

$$E_{bb} = E_b + I_b R_o = 250 + 0.035 \times 7,000 = 495 \text{ volts}$$

Power Output. Because of the relatively large amount of harmonic distortion in the output of pentodes and beam power tubes, the distortion is generally included in the power output. The total power output, taking the harmonic power into consideration, can be calculated by

$$P_o = \frac{[i_{b.\text{max}} - i_{b.\text{min}} + 1.41(I_x - I_y)]^2 R_o}{32} \qquad (7\text{-}16)$$

where I_x = plate current at $0.293E_c$
        I_y = plate current at $1.707E_c$

Example 7-20. What is the total a-c power output of the pentode power-amplifier tube represented by Fig. 7-8 with a 7,000-ohm load resistance?

           Given: $i_{b.\text{max}} = 67$ ma   Find: P_o
                    $i_{b.\text{min}} = 5$ ma
                    $I_x = 61$ ma
                    $I_y = 10.5$ ma
                    $R_o = 7,000$ ohms
Solution:

$$P_o = \frac{[i_{b.\text{max}} - i_{b.\text{min}} + 1.41(I_x - I_y)]^2 R_o}{32}$$

$$= \frac{[0.067 - 0.005 + 1.41(0.061 - 0.0105)]^2 \times 7,000}{32} = 3.88 \text{ watts}$$

Nonlinear Distortion. In addition to second harmonic distortion, third harmonic distortion is very pronounced in pentodes and beam power tubes. The reason for this high distortion can be explained by reference to the static plate characteristic curves for these two types of tubes (see Figs. 7-8 and 7-9). It can be seen from these curves that the distances between the grid-voltage lines are not uniform. The greatest distance occurs between zero grid bias and the adjacent grid-bias line. The distance between succeeding adjacent grid-bias lines gradually decreases, reaching a minimum between the maximum grid bias and its adjacent grid-bias line.   This nonuniform variation between the grid-voltage lines indicates that a change in grid voltage at the higher bias voltages will produce a smaller change in plate current than for the same grid-voltage change at the lower bias voltages.   The total distortion produced by pentodes and beam power tubes is therefore higher than the distortion obtained with triodes.

Effects of Harmonic Distortion.   The effects of harmonic distortion on a sinusoidal wave is shown in Fig. 7-10.   A second harmonic tends to

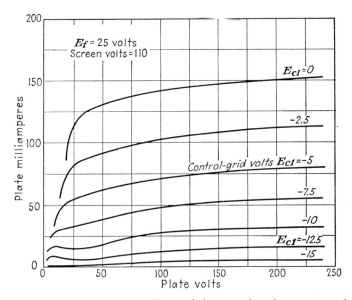

FIG. 7-9. Family of static plate characteristic curves for a beam power tube.

change the fundamental wave to a saw-tooth wave.   This type of distortion is characteristic of all even harmonics as the addition of the fundamental and all its even harmonics will tend to produce a saw-tooth wave. A third harmonic tends to change the fundamental wave to a square wave. This type of distortion is characteristic of all the odd harmonics, since the addition of the fundamental and all its odd harmonics will tend to produce a square wave.   The distortion produced by both the second and third harmonics results in a wave that is somewhat similar to the distorted saw-tooth wave produced by the second harmonic only.   However, the shape of this wave is also dependent upon the ratio of the maximum values of the second and third harmonics.   For the resultant wave, shown in Fig. 7-10c, the ratio is 1.

Calculation of the Amount of Distortion.   The per cent of harmonic distortion in pentodes and beam power tubes can be calculated by use of

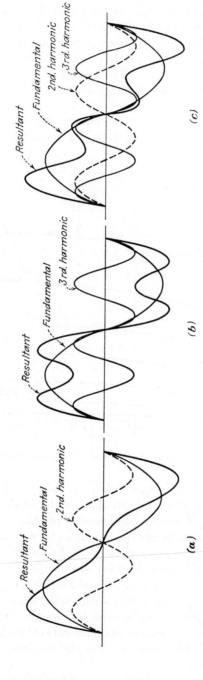

FIG. 7-10. Effects of harmonics on a sine wave. (a) Effect of a second harmonic. (b) Effect of a third harmonic. (c) Effect of a second and a third harmonic.

Eqs. (7-17), (7-18), and (7-19). As the derivations of these equations are rather complex, they are omitted in this text.

$$\% \text{ 2nd harmonic distortion} = \frac{i_{b.max} + i_{b.min} - 2I_b}{i_{b.max} - i_{b.min} + 1.41(I_x - I_y)} \times 100 \tag{7-17}$$

$$\% \text{ 3rd harmonic distortion} = \frac{i_{b.max} - i_{b.min} - 1.41(I_x - I_y)}{i_{b.max} - i_{b.min} + 1.41(I_x - I_y)} \times 100 \tag{7-18}$$

The per cent of the total harmonic distortion (second + third) is equal to the square root of the sum of the per cent of second harmonic distortion squared and the per cent of third harmonic distortion squared.

$$\% \text{ total (2nd + 3rd) harmonic distortion}$$
$$= \sqrt{(\% \text{ 2nd harmonic distortion})^2 + (\% \text{ 3rd harmonic distortion})^2} \tag{7-19}$$

Example 7-21. A pentode power tube whose characteristics are given in Fig. 7-8 is being operated with a 7,000-ohm load resistance and a grid bias of 16.5 volts. If an input signal with a peak value of 16.5 volts is applied, what is the per cent of (a) second harmonic distortion? (b) third harmonic distortion? (c) total second and third harmonic distortion?

Given: $E_c = -16.5$ volts   Find: (a) % 2nd harmonic distortion
            $E_{g.max} = 16.5$ volts   (b) % 3rd harmonic distortion
            $R_o = 7,000$ ohms   (c) % total (2nd + 3rd) harmonic
            Load line, Fig. 7-8   distortion

Solution:

From Fig. 7-8: $i_{b.max} = 67$ ma, $i_{b.min} = 5$ ma, $I_b = 35$ ma, $I_x = 61$ ma, $I_y = 10.5$ ma

(a)   % 2nd harmonic distortion $= \dfrac{i_{b.max} + i_{b.min} - 2I_b}{i_{b\ max} - i_{b.min} + 1.41(I_x - I_y)} \times 100$

$$= \frac{67 + 5 - 2 \times 35}{67 - 5 + 1.41(61 - 10.5)} \times 100 = 1.50$$

(b)   % 3rd harmonic distortion $= \dfrac{i_{b.max} - i_{b.min} - 1.41(I_x - I_y)}{i_{b.max} - i_{b.min} + 1.41(I_x - I_y)} \times 100$

$$= \frac{67 - 5 - 1.41(61 - 10.5)}{67 - 5 + 1.41(61 - 10.5)} \times 100 = 6.91$$

(c)   % total harmonic distortion $= \sqrt{(\% \text{ 2nd harmonic distortion})^2 +}$
$$\overline{(\% \text{ 3rd harmonic distortion})^2}$$
$$= \sqrt{1.50^2 + 6.91^2} = 7.07$$

The total harmonic distortion of pentodes and beam power tubes for normal operating conditions is usually listed in standard tube manuals. A total distortion of 8 per cent is listed for the typical pentode used in Example 7-21. The total distortion of 7.07 per cent as calculated in Example 7-21 compares favorably with the value of 8 per cent given in the tube manual. This example shows that the distortion is above the allowable maximum of 5 per cent and that most of the distortion is caused by the third harmonic. Consequently, operating this tube under these conditions would not permit its use in a good radio receiver or in a high-fidelity sound-reproducing system.   In order to reduce the amount of distortion, negative feedback may be applied to the circuits using these types

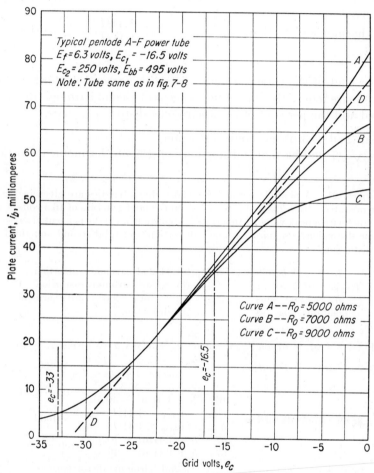

Fig. 7-11. Dynamic characteristics for a pentode power-amplifier tube (same tube as used in Fig. 7-8).

of tubes.   Beam power tubes have a lower percentage of distortion than pentodes and therefore are used more frequently as power-amplifier tubes.

7-11. Dynamic Characteristic Curves for a Pentode Class A Power-amplifier Tube.   Because the dynamic characteristic curve of a tube will vary with the value of the load impedance, the interpretation of the action of the tube can also be obtained from a family of dynamic characteristic curves.   Figure 7-11 shows the dynamic characteristics for a typical pentode power-amplifier tube with three values of plate resistance. These curves were plotted by using values obtained from the dynamic characteristic curves of Fig. 7-8.   For comparison, a straight line D–D was drawn along with the curves.   The curves of Fig. 7-11 show that for this particular tube type and operating voltages, the curves for the three values of R_o are nearly identical for values of e_c greater than -16.5 volts. However, for values of e_c between -16.5 volts and zero the curves spread out and their slopes vary considerably.   Comparing the curves A, B, and C with the straight line D–D for values of e_c between -16.5 volts and zero shows that with the 5,000-ohm load the slope of the curve increases as e_c approaches zero, while for the 7,000- and 9,000-ohm loads the slope decreases as e_c approaches zero.

The curves of Fig. 7-11 also provide a means of observing the relative amount of nonlinear distortion resulting with various values of R_o.   Minimum distortion will occur when the change in current for a change in values of e_c from -16.5 to -33 volts most nearly approaches the change in current for a change in values of e_c from -16.5 volts to zero.   For curve A, these values are $(37 - 5)$ or 32 ma and $(82 - 37)$ or 45 ma. For curve B, the values are $(35 - 5)$ or 30 ma and $(67 - 35)$ or 32 ma. For curve C, the values are $(35 - 5)$ or 30 ma and $(53 - 35)$ or 18 ma. These values show that with curve B, $R_o = 7,000$ ohms, the nonlinear distortion will be much lower than with either curves A and C.

7-12. Power Output of a Pentode Class A Power-output Tube. *Instantaneous Value.*   The varying component i_p of the plate current flowing through the load resistance R_o produces a varying component of voltage e_p across R_o.   The values of i_p and e_p are expressed by Eqs. (5-6) and (5-7).

$$i_p = g_m e_g \frac{r_p}{R_o + r_p} \tag{5-6}$$

$$e_p = g_m e_g \frac{r_p R_o}{R_o + r_p} \tag{5-7}$$

The instantaneous value of the a-c power output is

$$p_o = e_p i_p \tag{7-7}$$

Substituting Eqs. (5-6) and (5-7) for i_p and e_p in Eq. (7-7)

$$p_o = g_m e_g \frac{r_p R_o}{R_o + r_p} \times g_m e_g \frac{r_p}{R_o + r_p} = R_o (g_m e_g)^2 \left(\frac{r_p}{R_o + r_p}\right)^2 \tag{7-20}$$

A-C Power.   The a-c power output can best be determined from values taken from the characteristic curves as in Example 7-20.   However, if the nonlinear distortion is not too great, a reasonably approximate value of the a-c power output can be obtained by substituting the effective value of the signal voltage for the maximum value in Eq. (7-20).   Then

$$P_o = R_o(g_m E_g)^2 \left(\frac{r_p}{R_o + r_p}\right)^2 \tag{7-21}$$

Example 7-22.   The tube represented by Fig. 7-8 has a transconductance of 2,500 μmhos and a plate resistance of 80,000 ohms.   What is the a-c power output [use Eq. (7-21)] if the peak value of the input signal is 16.5 volts and R_o is 7,000 ohms?

Given: $g_m = 2,500$ μmhos   Find: P_o

$r_p = 80,000$ ohms

$e_{g \cdot max} = 16.5$ volts

$R_o = 7,000$ ohms

Solution:

$$P_o = R_o(g_m E_g)^2 \left(\frac{r_p}{R_o + r_p}\right)^2$$

$$= 7,000 \left(2,500 \times 10^{-6} \times \frac{16.5}{\sqrt{2}}\right)^2 \left(\frac{80,000}{7,000 + 80,000}\right)^2 \cong 5 \text{ watts}$$

Maximum Power Output.   The maximum power output will occur when $R_o = r_p$ (see Art. 7-6).   For this condition Eq. (7-20) becomes

$$p_{o \cdot max} = R_o(g_m e_g)^2 \left(\frac{r_p}{R_o + r_p}\right)^2 = \frac{R_o(g_m e_g)^2}{4} \tag{7-22}$$

If the effective value of the input signal is substituted for the instantaneous value in Eq. (7-22), the a-c power output of the circuit may then be found by

$$P_{o \cdot max} = \frac{R_o(g_m e_g)^2}{8} = \frac{R_o(g_m E_g)^2}{4} \tag{7-23}$$

Example 7-23.   (a) What value of plate load resistance would be required for the tube of Example 7-22 to produce maximum power output?   (b) What would be the maximum power output with an input signal having a peak value of 16.5 volts?

Given: $r_p = 80,000$ ohms   Find: (a) R_o

$g_m = 2,500$ μmhos   (b) P_o

$e_{g \cdot max} = 16.5$ volts

Solution:

(a) $R_o = r_p = 80,000$ ohms

(b) $P_{o \cdot max} = \dfrac{R_o(g_m e_{g \cdot max})^2}{8} = \dfrac{80,000(2,500 \times 10^{-6} \times 16.5)^2}{8} = 17 \text{ watts}$

Operating a tube at its maximum power output is impracticable in a-f amplifier circuits because: (1) its maximum plate-dissipation rating would be greatly exceeded, (2) the plate-supply voltage required would be too great, (3) the distortion would be excessive.

7-13. Ratings of Pentode and Beam Power Amplifiers.   *Power Amplification.*   The power amplification of a pentode or a beam power tube is found in the same manner as for the triode, Art. 7-7.   However, because pentodes require less driving voltage than triodes, the power amplification for pentodes will be greater than for triodes.

Example 7-24.   A pentode tube (Fig. 7-8) is being operated as a Class A_1 power amplifier with a 470,000-ohm grid input resistor R_g and a 7,000-ohm plate-load resistor R_o. The tube has a transconductance of 2,500 μmhos and a plate resistance of 80,000 ohms. In Example 7-20 the power output of this tube was found to be 3.88 watts. The input signal E_g causes a current of 25 μa (effective value) to flow in R_g. Find: (a) the input signal voltage, (b) the input signal power, (c) the power amplification, (d) the voltage amplification.

Given: $R_g = 470{,}000$ ohms   Find: (a) E_g
         $R_o = 7{,}000$ ohms   (b) P_i
         $g_m = 2{,}500$ μmhos   (c) PA
         $r_p = 80{,}000$ ohms   (d) VA
         $P_o = 3.88$ watts
         $I_g = 25$ μa

Solution:

(a) $E_g = I_g R_g = 25 \times 10^{-6} \times 0.47 \times 10^6 = 11.75$ volts
(b) $P_i = E_g I_g = 11.75 \times 25 \times 10^{-6} = 294$ μw
(c) $PA = \dfrac{P_o}{P_i} = \dfrac{3.88}{294 \times 10^{-6}} \cong 13{,}200$
(d) $VA = g_m \dfrac{r_p R_o}{R_o + r_p} = 2{,}500 \times 10^{-6} \dfrac{80{,}000 \times 7{,}000}{80{,}000 + 7{,}000} \cong 16$

Plate-circuit Efficiency.   Just as with the triode, the plate-circuit efficiency of the pentode and beam power tube expresses the ratio of the power output to the power taken from the B power supply. With pentodes and beam power tubes, the power taken by the screen grid must be included in the power provided by the B supply. The per cent efficiency of the plate circuit can then be expressed as

$$\text{Plate-circuit efficiency} = \frac{P_o}{E_{bb}(I_b + I_{g2})} \times 100 \qquad (7\text{-}24)$$

Example 7-25.   What is the plate-circuit efficiency of the power output tube of Examples 7-17 to 7-20 if the screen grid current is 6.5 ma? From preceding examples, $P_o = 3.88$ watts, $E_{bb} = 495$ volts, $I_b = 35$ ma.

Given: $P_o = 3.88$ watts   Find: Plate-circuit efficiency
         $E_{bb} = 495$ volts
         $I_b = 35$ ma
         $I_{g2} = 6.5$ ma

Solution:

$$\text{Plate-circuit efficiency} = \frac{P_o}{E_{bb}(I_b + I_{g2})} \times 100$$
$$= \frac{3.88 \times 100}{495(0.035 + 0.0065)} = 18.9\%$$

Plate Efficiency.   The plate efficiency of pentodes and beam power tubes expresses the ratio of the a-c power output to the power input based on the average values of plate voltage, plate current, and screen grid current with maximum signal input.

$$\text{Plate efficiency} = \frac{P_o}{E_{b \cdot \text{av}}(I_{b \cdot \text{av}} + I_{g2 \cdot \text{av}})} \times 100 \qquad (7\text{-}25)$$

Example 7-26. What is the plate efficiency of the tube used in Example 7-25 if $E_{b \cdot \text{av}} = 250$ volts, $I_{b \cdot \text{av}} = 35$ ma, and $I_{g2 \cdot \text{av}} = 10.5$ ma?

Given: $P_o = 3.88$ watts   Find: Plate efficiency
$E_{b \cdot \text{av}} = 250$ volts
$I_{b \cdot \text{av}} = 35$ ma
$I_{g2 \cdot \text{av}} = 10.5$ ma

Solution:

$$\text{Plate efficiency} = \frac{P_o}{E_{b \cdot \text{av}}(I_{b \cdot \text{av}} + I_{g2 \cdot \text{av}})} \times 100$$
$$= \frac{3.88 \times 100}{250(0.035 + 0.0105)} = 34\%$$

Plate Dissipation and Screen-grid Dissipation. The heat given off at the plate of a tube as the result of electron bombardment is called the *plate dissipation* and its symbol is P_p. This is a loss in power and is equal to the difference between the power supplied to the plate of the tube and the a-c power delivered by the tube to its load impedance. The heat given off at the screen grid of a tube as a result of electron bombardment is called the *screen-grid dissipation*. With pentodes and beam power tubes, the power dissipated in the screen grid must be added to the average d-c power supplied to the plate of the tube in order to obtain the total input to the tube from the B power supply.

Example 7-27. With the values given for the pentode in Example 7-26, find: (*a*) the plate dissipation, (*b*) the screen-grid dissipation, (*c*) the total dissipation of the plate and screen grid.

Given: $P_o = 3.88$ watts   Find: (*a*) P_p
$E_b = 250$ volts; $I_b = 35$ ma   (*b*) P_{g2}
$E_{g2} = 250$ volts; $I_{g2} = 10.5$ ma   (*c*) P_{total}

Solution:

(*a*) $P_p = P_i - P_o = 8.75 - 3.88 = 4.87$ watts
where $P_i = E_b I_b = 250 \times 0.035 = 8.75$ watts
(*b*) $P_{g2} = E_{g2} I_{g2} = 250 \times 0.0105 = 2.625$ watts
(*c*) $P_{\text{total}} = P_p + P_{g2} = 4.87 + 2.625 = 7.495$ watts

For the typical tube used in Example 7-27, the rated maximum plate dissipation is 11 watts and the rated maximum screen grid input is 3.75 watts. The values found in the example are both below these ratings; hence the tube is being operated within the maximum safe load conditions.

Power Sensitivity. The power sensitivity of pentodes and beam power tubes is found in the same manner as for the triode (Art. 7-7). However, because pentodes and beam power tubes require less driving voltage than triodes, the power sensitivity for pentodes and beam power tubes will be greater than for triodes.

Example 7-28.   What is the power sensitivity of the tube and circuit used in Example 7-20 if the effective value of the input signal is 11.65 volts?

$$\text{Given:} \quad P_o = 3.88 \text{ watts} \qquad \text{Find:} \quad \text{Power sensitivity}$$
$$E_g = 11.65 \text{ volts}$$

Solution:

$$\text{Power sensitivity} = \frac{P_o}{E_g{}^2} \times 10^6 = \frac{3.88 \times 10^6}{11.65^2} \cong 28{,}600 \ \mu\text{mhos}$$

7-14. Power Diagram for a Pentode. *Construction.* The power diagrams for pentodes and beam power tubes are constructed in the same manner as for triodes (Art. 7-9).   Figure 7-12 represents the power diagram for the typical pentode power-amplifier tube whose characteristics are shown in Figs. 7-8 and 7-11 and in Examples 7-17 to 7-28.   The same letter notations used in the triode power diagram (Fig. 7-7) are used in the pentode diagram.   The power ratings can therefore be found by following the procedure as explained for triodes.

Power Output. The power diagram (Fig. 7-12) shows that with an input signal of 33 volts (peak-to-peak), the plate voltage will vary from 27 to 460 volts.   This peak-to-peak plate voltage swing of 433 volts represents the useful signal voltage available to operate the following circuit. For this same input signal, the plate current varies from 5 to 67 ma, or a peak-to-peak swing of 62 ma.   The a-c power output is represented by the area of the shaded triangle QNM.

Example 7-29.   With the aid of the power diagram of Fig. 7-12, find (*a*) the voltage amplification and (*b*) the a-c power output.   Also, (*c*) compare the results with the answer for VA in Example 7-24, and P_o in Example 7-20.

$$\text{Given:} \quad \text{Fig. 7-12} \qquad \text{Find:} \quad (a) \ \text{VA}$$
$$(b) \ P_o$$
$$(c) \ \text{Comparison}$$

Solution:

From Fig. 7-12:
$$E_{M-D} = 460 - 27 = 433 \text{ volts}$$
$$E_{E-D} = 33 \text{ volts (input signal)}$$
$$E_{P-Q} = 250 - 27 = 223 \text{ volts}$$
$$I_{P-M} = 35 - 5 = 30 \text{ ma or } 0.03 \text{ amp}$$

(*a*) $\text{VA} = \dfrac{E_o}{E_i} = \dfrac{E_{M-D}}{E_{E-D}} = \dfrac{433}{33} = 13.1$

(*b*) $P_o = \text{area of } QNM = \dfrac{E_{PQ} \times I_{PM}}{2} = \dfrac{223 \times 0.03}{2} = 3.35 \text{ watts}$

(*c*)   Comparison:
$$\text{VA} = 13.1 \text{ from power diagram}$$
$$= 16 \text{ from Example 7-24}$$
$$P_o = 3.35 \text{ from power diagram}$$
$$= 3.88 \text{ from Example 7-20}$$

Power Losses and Power Ratings. The power consumed by the output resistor is represented by the area of the rectangle $QABC$. The power taken from the B supply by the plate circuit is represented by the area of

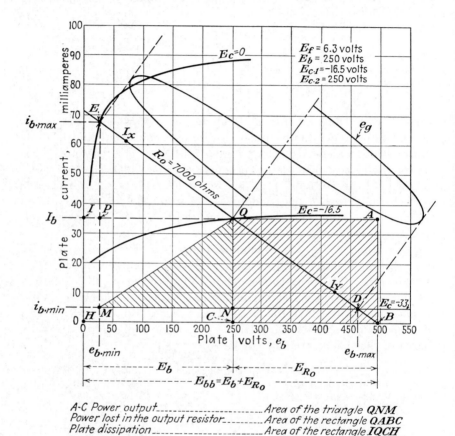

A-C Power output..Area of the triangle QNM
Power lost in the output resistor.........Area of the rectangle $QABC$
Plate dissipation..Area of the rectangle $IQCH$
                                         minus the area of the triangle QNM
Power supplied by the B power supply.....Area of the rectangle $IABH$
Power supplied to the plate of the tube.....Area of the rectangle $IQCH$

Fig. 7-12. Power diagram for a typical pentode power-amplifier tube (same tube as used in Fig. 7-8).

the rectangle $IABH$. The input power to the plate of the tube is represented by the area of the rectangle $IQCH$. The plate dissipation of the tube is represented by the area of the rectangle $IQCH$ minus the area of the triangle QNM.

Example 7-30. Using the power diagram of Fig. 7-12 and the value of P_o from Example 7-29, find: (*a*) the plate-circuit efficiency (use screen-grid dissipation as 2.625 watts from Example 7-27), (*b*) the plate efficiency, (*c*) the plate dissipation, and (*d*) the power lost in the output resistor.

Given: Power diagram, Fig. 7-12   Find: (a) Plate-circuit efficiency
$P_o = 3.35$ watts   (b) Plate efficiency
$P_{g2} = 2.625$ watts   (c) P_p
                                    (d) P_R

Solution:

From Fig. 7-12:

$$AB = CQ = 35 \text{ ma} = 0.035 \text{ amp}$$
$$HB = 495 \text{ volts}$$
$$HC = 250 \text{ volts}$$
$$CB = 495 - 250 = 245 \text{ volts}$$

(a) Power of B supply $= P_B = HB \times AB = 495 \times 0.035 = 17.325$ watts
Power to screen grid $= P_{g2} = 2.625$ watts

Plate-circuit efficiency $= \dfrac{P_o}{P_B + P_{g2}} \times 100 = \dfrac{3.35 \times 100}{17.325 + 2.625} = 16.8\%$

(b) $P_i = HC \times CQ = 250 \times 0.035 = 8.75$ watts

Plate efficiency $= \dfrac{P_o}{P_i + P_{g2}} \times 100 = \dfrac{3.35 \times 100}{8.75 + 2.625} = 29.4\%$

(c) $P_p = P_i - P_o = 8.75 - 3.35 = 5.4$ watts
(d) $P_R = CB \times AB = 245 \times 0.035 = 8.575$ watts

Comparison:

(a) Plate-circuit efficiency $= 16.8\%$ from power diagram
                                    $= 18.9\%$ from Example 7-25
(b) Plate efficiency $= 29.4\%$ from power diagram
                                 $= 34\%$ from Example 7-26
(c) Plate dissipation $= 5.4$ watts from power diagram
                                  $= 4.87$ watts from Example 7-27

7-15. Parallel Operation of Power Tubes. If the power output of a single tube is too small to produce a desired amount of volume, two or more power tubes may be connected in parallel. Operating tubes in parallel increases the power output, but the per cent of distortion remains the same as for a single tube. Two identical tubes connected in parallel will provide twice the output of a single tube for the same value of input signal voltage. The amount of distortion will also be doubled; but as the output and the distortion are both doubled, the per cent of distortion remains the same. Figure 7-13a illustrates the manner of connecting two triodes in parallel. The same procedure can be followed for pentodes.

Since the two plates are in parallel, the equivalent plate resistance will be one-half that of a single tube. The voltage generated by each tube will be equal to μe_g. The equivalent electrical circuit for two triodes in parallel can therefore be drawn as shown in Fig. 7-13b.

Example 7-31. A certain beam power tube is operated so that it produces an output of 6.5 watts. How many of these tubes must be connected in parallel, if it is desired to obtain a 35-db output from the same driving voltage?

Given: $P_{o.1} = 6.5$ watts   Find: Number of tubes in parallel
           Volume $= 35$ db

Solution:

$$P_{\bullet \cdot T} = P_R \times \text{antilog} \frac{db}{10} = 0.006 \times \text{antilog} \frac{35}{10} = 18.96 \text{ watts}$$

$$\text{Number of tubes} = \frac{P_{o \cdot T}}{P_{o \cdot 1}} = \frac{18.96}{6.5} = 2.92$$

Therefore, use three tubes.

7-16. Push-pull Amplifiers. *Push-pull Operation.* Push-pull operation of amplifier tubes is another method of obtaining a greater power

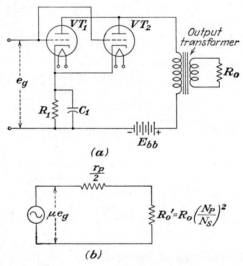

(a)

(b)

Fig. 7-13. Two triodes connected for parallel operation. (a) Actual circuit diagram. (b) Equivalent electrical circuit.

output than can be obtained from a single tube. A push-pull amplifier employs two identical tubes operating as a single stage of amplification. The grids and plates of the two tubes are connected respectively to opposite ends of the secondary of the input transformer and the primary of the output transformer (see Fig. 7-14a). A balanced circuit is obtained by connecting the cathode returns to center taps on the secondary and primary windings of the input and output transformers respectively. As a balanced circuit is necessary for push-pull operation, this system is also referred to as a *balanced amplifier*. A push-pull amplifier circuit may be either resistance-capacitance coupled or transformer coupled.

The varying current in the primary winding of the input transformer, which is actually the output current of the previous stage, induces a corresponding voltage in the secondary. At any instant the two ends of this secondary, 1 and 2 of Fig. 7-14a, are of opposite polarity. Thus the varying input voltages $e_{g \cdot A}$ and $e_{g \cdot B}$ will always be equal and 180 degrees out of phase with each other. Assuming the end of the secondary indicated as 1 to be positive, then the other end indicated as 2 will be negative. The grid of tube A will then become more positive, causing an increase in the plate current flowing through section 3-4 of the primary of the output transformer. The grid of tube B becomes more negative causing the plate current flowing in section 4-5 to decrease. As the two tubes are identical and the changes in their grid voltages are equal, the variation in plate current will also be equal but 180 degrees out of phase with each

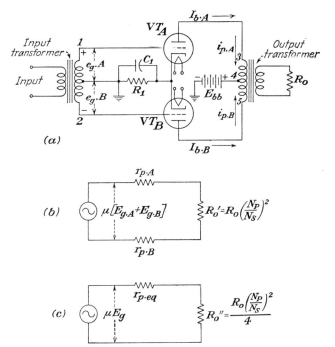

FIG. 7-14. Push-pull amplifier circuit. (*a*) Circuit diagram. (*b*) Equivalent electrical circuit. (*c*) Single-tube equivalent electrical circuit.

other. It is apparent that one tube pushes current through one-half of the primary winding of the output transformer while the second tube pulls an equal amount through the other half; hence the name *push-pull*. Push-pull operation is not limited to any particular type of tube; thus triodes, pentodes, or beam power tubes may be operated as balanced amplifiers.

Graphical Analysis of Push-pull Operation. A graphical analysis of push-pull operation for two tubes operating as a Class A_1 amplifier is illustrated in Fig. 7-15. The dynamic characteristics for tube B are plotted inverted with respect to the dynamic characteristics for tube A. The grid of each tube is biased to approximately one-half the cutoff value. With zero signal input, steady plate currents of $I_{b.A}$ and $I_{b.B}$ flow in their respective plate circuits. These two currents are equal and flow in opposite directions in each half of the primary winding of the output transformer.

When an alternating voltage is applied to the two grids, the plate current in one tube increases, while the plate current in the other tube decreases. The varying plate current of each tube is badly distorted, since both tubes operate over more than the linear portion of their

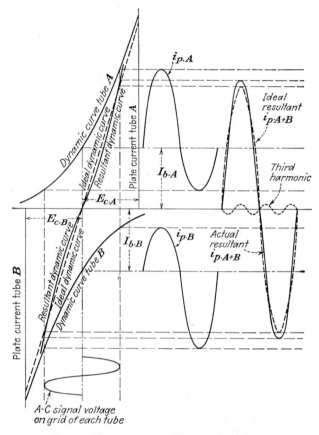

Fig. 7-15. Graphical analysis of Class A push-pull operation.

dynamic curves. The distortion is largely due to the second harmonic, since operating over the nonlinear portion of the characteristic curve produces this type of harmonic distortion. The phase relation of the second harmonic produced by each tube is such that they cancel each other in the output transformer. A graphical illustration of how the second harmonic is eliminated is seen in Fig. 7-16. This figure also shows that the combined output of tubes A and B is a sine wave devoid of any second harmonic and equal to twice that of either tube.

The net effect of the push-pull action of the varying plate current of each tube flowing through the primary winding of the output transformer is equivalent to an alternating current of twice the value of either plate current flowing through one-half of the primary winding. This effect is equivalent to that produced by an input signal voltage equal to the alternating voltage applied to the grid of either tube and operating on a dynamic curve that is the resultant of the dynamic curves of tubes A

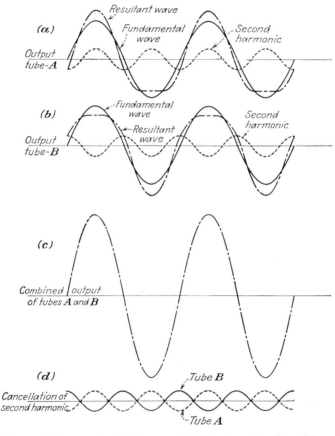

FIG. 7-16. Input and output waves of a Class A push-pull amplifier.

and B. This ideal dynamic curve is shown as a broken line in Fig. 7-15. The resultant plate current, also shown as a broken line, would be a sine wave. However, the actual resultant dynamic curve is not a straight line as is indicated by the ideal resultant curve, but is slightly curved. This curvature produces a third harmonic in the output; with an overload on the tube, a fifth harmonic will also be present.

7-17. Characteristics of Class A₁ Push-pull Amplifiers. Because the second harmonic distortion is balanced out in the output transformer, it is possible to use a load resistance whose value is equal to the plate resistance; thus maximum power output may be obtained. Also, greater values of input signal voltage can be applied to the grid of each tube, as its operation is not limited to the linear portion of its characteristic curve. It is therefore possible to obtain more than twice the power output of a single tube Class A_1 amplifier by operating two similar tubes in push-pull. However, the exciting voltage, measured between the two grids, must be twice that required for one tube.

The average plate currents $I_{b.A}$ and $I_{b.B}$ (Fig. 7-14a) flow in opposite directions in their respective halves of the primary winding of the output transformer. Thus the magnetizing effect of the direct currents on the iron core cancels out. Therefore, there can be no direct-current saturation in the core of the output transformer, regardless of how great the average plate currents may be. The incremental inductance will be higher and therefore will improve the low audio-frequency response. Large variable plate currents will produce proportionate changes in the magnetic flux rather than being distorted by the saturation bend in the magnetization curve of the iron.

At any instant, the resultant of the varying plate currents $i_{p.A}$ and $i_{p.B}$ flowing through the B power supply is zero. As there is no current of signal frequency flowing through the source of plate power, there will therefore be no regeneration.

Any alternating voltage that may be present in the plate power supply will also be balanced out in the primary winding of the output transformer. The a-c hum of a push-pull amplifier is therefore greatly reduced.

Because of these many advantages of balanced amplifiers, it is more advantageous to use two small tubes in push-pull rather than one large tube capable of producing the same amount of power output.

Although pentodes and beam power tubes may be operated in push-pull, little advantage is gained from the use of these tubes, because third harmonic distortion in pentodes is much higher than in triodes for comparable conditions. The odd harmonics are not balanced out in push-pull operation. In properly designed single-tube Class A pentode power amplifiers, the amount of second harmonic distortion is very low.

7-18. Equivalent Electrical Circuit of Push-pull Amplifiers. Equivalent electrical circuits for push-pull operation of two tubes are shown in Figs. 7-14b and 7-14c. In Fig. 7-14b, R_o' represents the plate-to-plate load impedance reflected to the full primary winding by the load connected to the secondary of the output transformer. In the circuit of Fig. 7-14c the two tubes are replaced by an equivalent single tube whose characteristics represent the resultant dynamic curve of Fig. 7-15. The load reflected to this single equivalent tube is designated as R_o'' and is one-fourth the value of R_o' because only one-half of the primary turns are effective with the single equivalent tube. While the operating characteristics of push-pull circuits may be observed from either of the equivalent circuits, the following discussion is based on the equivalent circuit of Fig. 7-14b.

It can be shown from Fig. 7-14b that the effective value of the varying plate current can be expressed as

$$I_p = \frac{\mu(E_{g \cdot A} + E_{g \cdot B})}{r_{p \cdot A} + r_{p \cdot B} + R_o'} \tag{7-26}$$

As $E_{g \cdot A}$ is equal to $E_{g \cdot B}$ and $r_{p \cdot A}$ equals $r_{p \cdot B}$, Eq. (7-26) can be expressed as

$$I_p = \frac{2\mu E_g}{2r_p + R_o'} \tag{7-26a}$$

where $E_g = E_{g \cdot A} = E_{g \cdot B}$

       $r_p = r_{p \cdot A} = r_{p \cdot B}$

(NOTE: E_g is equal to one-half of the voltage developed across terminals 1 and 2, Fig. 7-14a.)

For maximum power output R_o' should be equal to $2r_p$. For this condition, Eq. (7-26a) becomes

$$I_{p \cdot \max} = \frac{\mu E_g}{2r_p} \tag{7-27}$$

where $I_{p \cdot \max}$ = current at maximum power output

7-19. Power Output, Load Resistance, and Distortion for a Push-pull Amplifier Circuit. From the equivalent electrical circuit shown in Fig. 7-14b, the power output of a push-pull amplifier can be expressed as

$$P_o = \frac{4(\mu E_g)^2 R_o'}{(2r_p + R_o')^2} \tag{7-28}$$

The maximum power output is then equal to

$$P_{o \cdot \max} = \frac{(\mu E_{g \cdot \max})^2}{4r_p} \tag{7-29}$$

Example 7-32.   Two triode tubes, each having an amplification factor of 4.2 and a plate resistance of 800 ohms, are connected to operate as a Class A push-pull amplifier. The operating characteristics of each tube are: $E_b = 250$ volts, $E_c = -45$ volts, and $E_{g \cdot \max} = 45$ volts.   What is the maximum power output of the two tubes?

     Given: $\mu = 4.2$ Find: $P_{o \cdot \max}$

                 $r_p = 800$ ohms

                 $E_b = 250$ volts

                 $E_c = -45$ volts

            $E_{g \cdot \max} = 45$ volts

Solution:

$$P_{o \cdot \max} = \frac{(\mu E_{g \cdot \max})^2}{4r_p} = \frac{(4.2 \times 45)^2}{4 \times 800} = 11.16 \text{ watts}$$

The power output may also be determined by means of the tubes' plate characteristic curves and a load line.   To plot a load line, the desired operating plate voltage must first be known.   This value can usually be obtained from a tube manual.   A vertical line is drawn upward from a point on the abscissa of the plate characteristic curve equal to $0.6E_b$ (see Fig. 7-17).   The point of intersection of this vertical line and the zero grid-voltage curve represents the maximum plate current.   The load line

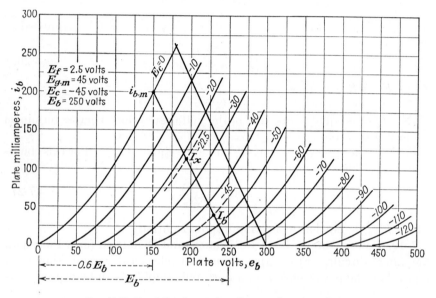

Fig. 7-17. Load line for push-pull operation of a triode.

is drawn from this point of intersection to a point on the abscissa representing the average plate voltage E_b.   The maximum power output for two tubes operating Class A push-pull will then be

$$P_{o\cdot\max} = \frac{4}{10} \times \frac{E_b}{\sqrt{2}} \times \frac{i_{b\cdot\max}}{\sqrt{2}} \tag{7-30}$$

or

$$P_{o\cdot\max} = \frac{E_b i_{b\cdot\max}}{5} \tag{7-30a}$$

This simple equation can be used for all triodes operated Class A push-pull.   The grid bias and maximum input signal voltage should be approximately equal to those specified for a single-tube Class A operation.

The resistance represented by the load line in Fig. 7-17 is equal to

$$R_o'' = \frac{E_b - 0.6E_b}{i_{b\cdot\max}} \tag{7-31}$$

or

$$R_o'' = \frac{0.4E_b}{i_{b\cdot\max}} \tag{7-31a}$$

The plate-to-plate load is then equal to

$$R'_o = \frac{1.6E_b}{i_{b\cdot\text{max}}} \qquad (7\text{-}32)$$

The distortion of push-pull amplifiers is generally very low. It can be shown that all even harmonics are eliminated; this is true for the same reason that the second harmonic is balanced out in the primary winding of the output transformer. Although the a-c hum from the plate power supply is balanced out, any hum induced in the input circuit of the amplifier will be amplified in the same manner as any other input signal. Because of the slight curvature of the resultant dynamic curve there will be some third harmonic distortion (see Fig. 7-15) and in some instances also a fifth harmonic distortion. The per cent of third harmonic distortion may be calculated by the equation

$$\% \text{ 3rd harmonic distortion} = \frac{i_{b\cdot\text{max}} - 2I_x}{2(i_{b\cdot\text{max}} + I_x)} \times 100 \qquad (7\text{-}33)$$

where I_x = plate current at $0.5E_{g\cdot\text{max}}$

Example 7-33. Using the plate characteristic curves for a tube represented by Fig. 7-17 find: (*a*) the maximum power output, (*b*) the plate-to-plate load resistance, (*c*) the per cent of third harmonic distortion for the push-pull circuit of Example 7-32.

Given: $E_b = 250$ volts   Find: (*a*) $P_{o\cdot\text{max}}$
 $E_c = -45$ volts   (*b*) R'_o
 $E_{g\cdot\text{max}} = 45$ volts   (*c*) % 3rd harmonic distortion

Solution:

From Fig. 7-17: $i_{b\cdot\text{max}} = 200$ ma, $I_x = 112$ ma

(*a*) $P_{o\cdot\text{max}} = \dfrac{E_b i_{b\cdot\text{max}}}{5} = \dfrac{250 \times 0.2}{5} = 10$ watts

(*b*) $R'_o = \dfrac{1.6E_b}{i_{b\cdot\text{max}}} = \dfrac{1.6 \times 250}{0.2} = 2,000$ ohms

(*c*) % 3rd harmonic distortion $= \dfrac{i_{b\cdot\text{max}} - 2I_x}{2(i_{b\cdot\text{max}} + I_x)} \times 100$

 $= \dfrac{0.2 - 2 \times 0.112}{2(0.2 + 0.112)} \times 100 = 3.84$

The values of power output of the push-pull Class A amplifier as found in Examples 7-32 and 7-33 compare favorably with each other. The maximum undistorted power output obtainable from a single tube operated Class A, and with the same electrode potentials and input signal voltage as used in Examples 7-32 and 7-33, is 3.5 watts. The power obtainable from push-pull operation, 10 watts (see Example 7-33), is considerably greater than twice this value. The per cent of distortion is less than the allowable maximum of 5 per cent. It may be noticed that there is a difference in the value of the plate-to-plate resistance used in Example 7-32 and the value calculated in Example 7-33. This difference is due to the fact that in Example 7-32 the plate resistance used is the value obtained for Class A operation at the electrode potentials employed, while in Example 7-33 the plate resistance used is the value obtained from operating on the resultant dynamic curve of the two tubes as indicated on Fig. 7-15.

7-20. Class AB Operation. The distortion caused by operating a tube on the lower bend of its characteristic curve, which is the region of high distortion, is eliminated by the push-pull circuit. It is then possible to increase the bias on the tubes used in a push-pull circuit so that they operate as Class AB. Increasing the grid bias of a tube decreases the value of the plate current with zero signal input. This decrease in the value of the operating plate current permits the use of higher screen-grid and plate voltages, and also increases the plate efficiency of the tube. Because of these factors, a greater power output can be obtained by operating two tubes as Class AB push-pull than by using the same tubes operated as Class A push-pull.

Class AB amplifiers may be operated with or without grid current flowing. As Class AB_1, the grid bias is always greater than the peak value of the input signal voltage applied to each tube. There will be no grid current in either tube, as the potential on their grids will not be positive during any part of the input cycle. In Class AB_2, the grid bias is always less than the peak value of the input signal voltage applied to each tube. There will, therefore, be some grid current in each tube during the portion of the input cycle that the grid is positive.

The general operating characteristics of Class AB_1 push-pull amplifiers are similar to those for Class A. The equations given for Class A operation are also applicable for Class AB_1. A higher power output and a higher plate efficiency (with a slight increase in the distortion) can be obtained from Class AB_1 than can be obtained from the same tubes operated Class A.

The grid current in a Class AB_2 amplifier represents a loss of power. This loss plus the power loss in the input transformer represents the total amount of driving power required by the grid circuit. In order to minimize the amount of distortion set up in the grid circuit, the power of the driving stage is generally made considerably higher than the minimum required amount. The input transformer of a Class AB_1 push-pull amplifier is usually a step-down transformer.

7-21. Class B Operation. The plate current with zero input signal voltage can be reduced to a minimum by adjusting the grid bias of a tube to approximately cutoff, which is Class B operation. Each tube of a Class B push-pull amplifier is operated in this manner. Referring to Fig. 7-14, it can be seen that when the grid of tube A is made more positive, the grid of tube B will be made more negative. During this half of the input cycle the plate current of tube B will be zero. During the next half-cycle the grid of tube A becomes more negative and the grid of tube B more positive. During this half-cycle the plate current of tube A will be zero. Thus, one tube amplifies the positive half-cycles and the other tube amplifies the negative half-cycles.

The two tubes alternately supply current to the primary winding of the output transformer. Each tube delivers power to one-half of the primary winding for one-half of the cycle. This is equivalent to one tube delivering power to one-half of the primary winding of the output transformer for an entire cycle. Under this condition, the effective load R_o'' is equal to one-fourth of the value of the impedance, R_o' reflected to the full primary by the load R_o connected to the secondary of the output transformer. The equivalent electrical circuit is similar to that for Class A and Class AB shown in Fig. 7-14. However, the values of μ and r_p are the values for Class B operation and are not equal to those values generally listed in a tube manual which are for Class A operation. The values of μ and r_p for Class B operation can be obtained from the plate characteristic curves. The power output of a Class B amplifier can be obtained by use of Eqs. (7-28) and (7-29) if the Class B operating values of μ and r_p are substituted in place of the Class A values.

As the fluctuations in plate current of a Class B amplifier will be higher than for Class AB, it is important that the power supply used should have good regulation. The remarks concerning the power supply used for Class AB operation therefore also apply to Class B operation.

To avoid the use of large fixed sources of biasing voltage, a number of tubes have been designed especially for Class B operation. These tubes have a high amplification factor and a very low plate current when the grid voltage is zero; they require no bias supply, as they can be operated as Class B at zero volts bias. It is also common practice to mount two triode units in one envelope so that only one tube is required for a Class B push-pull stage.

Because the plate current with zero input signal voltage will be practically zero, Class B amplifiers have a high plate efficiency. The grid is usually driven positive and the power output will be unusually high in proportion to the size of the tube. However, because the grids are driven positive and draw considerable power when operated as Class B, a high value of input signal power must be supplied from the driver stage to compensate for this loss of power.

7-22. Phase Inverter. The analysis of push-pull operation has shown that: (1) the input signal voltages on the grids of both tubes must be approximately equal in magnitude at all times, and (2) the two input signal voltages must be 180 degrees out of phase. With transformer coupling, the 180-degree phase difference between the two input voltages is obtained by means of an input transformer having a center-tapped secondary (see Fig. 7-14a). With resistance-capacitance coupling, the 180-

degree phase relation between the two input voltages is obtained by employing the inverter action of a vacuum tube. A tube used in this manner is called a *phase inverter.*

The circuit diagram for an *RC*-coupled push-pull amplifier is shown in Fig. 7-18. In this circuit, the driving voltage is obtained from tube 1 and the phase inversion from tube 2. Tubes *A* and *B* are being operated in push-pull and correspond to tubes *A* and *B* in Fig. 7-14*a*. When a varying input voltage is applied to the grid of tube 1, the varying output voltage of this tube will be applied to the grid of tube *A*. By means of the voltage divider R_1R_2, a portion of the output voltage of tube 1 is also

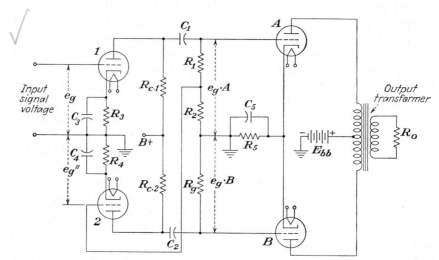

Fig. 7-18. Use of a phase-inverter tube in a resistance-capacitance-coupled push-pull amplifier circuit.

applied to the grid of tube 2. The output of the phase inverter (tube 2) is then applied to the grid of tube *B*. The action just described occurs practically instantaneously. Thus a positive output voltage from tube 1 causes the grids of tubes *A* and 2 to become more positive. The plate current of tube 2 increases and causes the grid of tube *B* to become more negative. In this manner, the input voltages at tubes *A* and *B* will always be 180 degrees out of phase. In order that the magnitude of the voltages applied to the grids of tubes *A* and *B* will always be equal, the voltage e_g'' applied to the grid of the phase inverter should always be equal in magnitude to the input voltage e_g. This is accomplished by making the ratio of $(R_1 + R_2)$ to R_2 equal to the voltage amplification of the circuit of tube 1. The characteristics of tubes 1 and 2 should be identical, and hence a twin triode is generally used for the driving tube and the phase inverter. As the output circuits of the two tubes should also be identical, the plate-coupling resistors $R_{c.1}$ and $R_{c.2}$ are of equal values, and the grid resistor R_g is equal to R_1 plus R_2.   The values of the resistors in the voltage-divider circuit can be calculated by

$$R_2 = \frac{R_g}{VA_1} \qquad (7\text{-}34)$$

$$R_1 = R_g - R_2 \qquad (7\text{-}35)$$

where VA_1 = voltage amplification from tube 1
R_1, R_2, and R_g = resistors as indicated on Fig. 7-18
The values to be used for the plate-coupling and grid resistors will depend upon the tube employed and can be obtained from a tube manual.

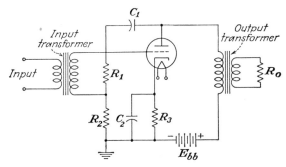

FIG. 7-19. A voltage-controlled feedback-amplifier circuit as applied to a single-tube power amplifier.

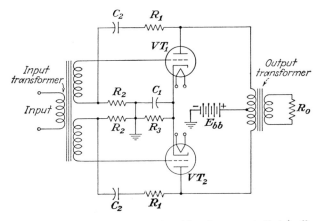

FIG. 7-20. A push-pull amplifier circuit with voltage-controlled feedback.

Another method of obtaining two signal voltages 180 degrees apart for driving a push-pull amplifier has been presented in Art. 6-13.

7-23. Negative Feedback.   The distortion developed in the power-amplifier circuit can be reduced to a negligible amount by the use of negative feedback.   The advantages, operation, limitations, and basic circuit connections for negative feedback as applied to power-amplifier circuits are similar to those for the audio-amplifier circuits as explained in Arts. 6-15 to 6-18. The equations used in the calculations of feedback in a-f amplifier circuits may also be used in the calculation of feedback for power-amplifier circuits. As the power output of most radio receivers exceeds the amount usually required for the average home receiver, the loss of power because of negative feedback is of little consequence.

Negative feedback as applied to a single stage of power amplification is shown in the circuit diagram of Fig. 7-19. In a push-pull amplifier circuit, negative feedback can be obtained by use of a separate voltage-divider network in the plate circuit of each tube as shown in Fig. 7-20.

QUESTIONS

1. (*a*) Explain the purpose of power amplifiers. (*b*) How are power amplifiers classified in terms of operating frequency? (*c*) How are power amplifiers classified in terms of tube-operating characteristics?

2. Compare voltage amplifiers and power amplifiers in terms of (*a*) plate-circuit impedance, (*b*) voltage amplification, (*c*) plate current.

3. (*a*) To which circuit or component in a radio receiver is the output of a power-amplifier tube generally fed? (*b*) What is the approximate value of the impedance into which the power output is coupled? (*c*) What component is generally used to couple the output of a power tube to its load?

4. (*a*) What important purpose does the output transformer serve? (*b*) Why is the output transformer generally required?

5. (*a*) How much power output is generally required of the average home receiver? (*b*) How does the power rating of power-output transformers compare with the amount of power used under average operating conditions?

6. In what manner does the output transformer of a push-pull amplifier differ from one used with a single-tube amplifier?

7. (*a*) Why is it possible to use the static characteristic curves of a tube for analyzing the operation of voltage amplifiers but not for the analysis of power amplifiers? (*b*) What types of characteristic curves are used in conjunction with power amplifiers?

8. (*a*) How can the dynamic characteristics of a tube be obtained from the static characteristic curves? (*b*) What form of load line is obtained when the plate load is resistive? (*c*) inductive? (*d*) capacitive?

9. (*a*) Why is it desirable to have the dynamic transfer characteristic curves approximately straight lines throughout the signal voltage range? (*b*) What factors affect the shape and location of the load line?

10. (*a*) What is the zero-signal bias and how is it determined? (*b*) How does the maximum plate-dissipation rating affect the choice of the zero-signal bias?

11. (*a*) What is the maximum-current point? (*b*) How is it determined?

12. (*a*) What is meant by the quiescent point? (*b*) What determines its location?

13. (*a*) What is the minimum number of points required in order to plot the load line? (*b*) Explain how the load line is then plotted.

14. What operating characteristics can be found by use of the load line?

15. Explain and establish proof of the phase relation between the following (*a*) e_g and i_p, (*b*) e_g and e_b, (*c*) e_g and e_p, (*d*) e_b and e_p, (*e*) e_b and i_p.

16. What information provided by the load line is used to calculate the value of the load resistance?

17. What factors determine the value of B supply voltage required?

18. What factors determined with the aid of the load line are used to determine the power output of the amplifier circuit?

19. (*a*) What is meant by distortion in an amplifier? (*b*) What is nonlinear distortion? (*c*) Why is distortion present in a Class A single-tube power amplifier?

20. What information obtained from the load line is used to calculate the nonlinear distortion?

21. How may a family of dynamic characteristic curves be obtained from the static characteristics?

22. Describe three effects on the dynamic characteristic curves when the value of the plate-load resistance is increased.

23. (*a*) What are the two components of the plate current of a tube when an a-c signal is applied to the grid? (*b*) What does each component contribute to the circuit?

24. How does the amount of nonlinear distortion affect the method of calculating the a-c power output of a power-amplifier circuit?

25. (*a*) State the principle of maximum power transfer. (*b*) Explain how the theory of this principle may be verified.

26. (*a*) Are power-amplifier circuits generally designed to produce maximum power output? (*b*) Explain why.

27. What factors affect the value of the maximum power output?

28. What is the relationship between the maximum power output and the amount of distortion?

29. What is the maximum per cent of harmonic distortion that is tolerated in (*a*) the average radio receiver? (*b*) high-fidelity sound equipment?

30. (*a*) What is meant by the maximum undistorted power output? (*b*) What per cent of distortion is tolerated in order to obtain maximum power output? (*c*) What value of load resistance is generally used in order to obtain maximum undistorted power output with a single-tube triode power amplifier?

31. (*a*) What is meant by power amplification? (*b*) What is meant by the driving power?

32. Does a power amplifier generally provide a large or small amount of voltage amplification? Explain.

33. (*a*) What is the plate-circuit efficiency? (*b*) What is the plate efficiency? (*c*) Which is more commonly used?

34. (*a*) What is the highest possible value of plate efficiency of a single-tube Class A amplifier? (*b*) Prove the answer given to (*a*). (*c*) What values of plate efficiency are ordinarily attained in practice?

35. (*a*) What is meant by the plate dissipation of a tube? (*b*) How may its rated value be obtained?

36. (*a*) What is meant by the power sensitivity of a power-amplifier tube? (*b*) When is this term used?

37. (*a*) What is meant by Class A_2 operation of an amplifier? (*b*) What circuit development has made Class A_2 operation practical? (*c*) What are the advantages of Class A_2 operation?

38. (*a*) What operating characteristics can be obtained from a power diagram similar to Fig. 7-7? (*b*) How is each of these characteristics represented on the power diagram?

39. (*a*) In what manner does the output power vary with a change in the value of the input signal voltage [see Eq. (7-9)]? (*b*) What limits the extent to which the power output may be increased by raising the value of the input signal voltage in Class A amplifiers? Explain.

40. (*a*) Do the operating characteristics of a pentode apply equally well to a beam power tube? (*b*) Why?

41. What factors govern the location of the load line for a single-pentode Class A power-amplifier tube with a resistive load?

42. With resistive load, what condition is necessary for minimum distortion?

43. In plotting the load line for a pentode (*a*) how is the zero-signal bias determined? (*b*) How is the maximum-current point determined?

44. (*a*) Does the pentode require a greater or lesser driving voltage than the triode power-amplifier tube? (*b*) Why?

45. (*a*) Can the output of a pentode power amplifier be determined by use of the same equation as for the triode? (*b*) Why?

46. (*a*) What harmonic distortion is present in amplifiers employing pentodes and beam power tubes? (*b*) What causes this distortion? (*c*) How does this differ from triodes?

47. What effect is produced on a sine wave by (*a*) a second harmonic? (*b*) A third harmonic? (*c*) Second and third harmonics?

48. How does the distortion in beam power tubes compare with that in pentodes of similar rating? (*b*) Which of these tube types is more commonly used? (*c*) What may be done to reduce the amount of distortion in circuits using these tubes?

49. What are the relative merits of amplifier circuits represented by curves *A*, *B*, and *C* of Fig. 7-11 in terms of (*a*) harmonic distortion? (*b*) power output?

50. What advantage does the pentode and beam power tube have over the triode in terms of (*a*) power amplification? (*b*) required driving voltage?

51. How does the power consumed by the screen grid enter into the calculation of the plate-circuit efficiency and plate efficiency with pentodes and beam power tubes?

52. (*a*) What is meant by the screen dissipation of a tube? (*b*) How may its rated value be obtained?

53. How does the power sensitivity of pentodes and beam power tubes compare with that of triodes?

54. (*a*) What operating characteristics can be obtained from a power diagram similar to Fig. 7-12? (*b*) How is each of these characteristics represented on the power diagram?

55. (*a*) What is the purpose of operating two or more tubes in parallel? (*b*) How does parallel operation affect the per cent of distortion?

56. (*a*) What is the purpose of operating two tubes in push-pull? (*b*) How are the grids and the plates of the tubes connected to the transformers? (*c*) Explain why center-tapped transformers are necessary.

57. Explain the importance of the phase relation of the various voltages in push-pull operation.

58. Show how second harmonic distortion is eliminated when two tubes are operated in push-pull.

59. (*a*) What is the cause of third and fifth harmonics in the output of a push-pull amplifier? (*b*) Why does not the push-pull operation reduce the effect of odd harmonics in addition to the even harmonics?

60. (*a*) What are the advantages of push-pull amplifiers? (*b*) Why are triodes frequently used in push-pull amplifiers?

61. Explain why the plate-to-plate load impedance should be four times the value of the plate resistance of a single equivalent tube when it is desired to obtain maximum power output from a push-pull amplifier?

62. Work out the derivation of Eq. (7-29) from Eq. (7-28), noting that maximum power output occurs when R_o' is equal to $2r_p$.

63. How does the amount of power output of two tubes operated in push-pull compare with the power output of the same tubes operated in parallel?

64. What are the advantages of Class AB push-pull operation over Class A push-pull?

65. What is the difference between Class AB_1 and Class AB_2 operation of push-pull amplifiers?

66. Why does a Class AB_2 amplifier require a greater amount of driving power than a single-tube Class A amplifier?

67. (a) What operating characteristic is required of the power supply for the plate circuit of Class AB power amplifiers? (b) Explain why this is necessary.

68. How is cathode bias usually obtained in a Class AB power amplifier?

69. Describe the operation of a Class B push-pull amplifier.

70. What are the operating characteristics of the Class B push-pull amplifier?

71. (a) What is a phase inverter? (b) What is its purpose? (c) When is it used?

72. Explain the operation of a phase inverter as used in conjunction with a push-pull amplifier.

73. (a) What advantage is gained by applying negative feedback to a power amplifier? (b) Why is it possible to apply negative feedback to some power-amplifier circuits without providing additional gain?

PROBLEMS

1. The output of a 50C5 beam-power amplifier tube is to be coupled to the voice coil of a loudspeaker by means of an output transformer. (a) What turns ratio should the output transformer have if the resistance of the voice coil is 6 ohms and the plate load should be 2,000 ohms? (b) What minimum current rating should the transformer have? (See a tube manual.)

2. A universal output transformer, that is, one with taps brought out from several points on its windings, is designed to provide a 2,500-ohm plate load from either a 2-, 4-, 8-, or 500-ohm load. What is the turns ratio if the output load tap used is (a) 2? (b) 4? (c) 8? (d) 500?

3. If an a-c signal whose maximum value is 23.5 volts is applied to the circuit represented by Figs. 7-3 and 7-4, what is the range of variation of the quantities: (a) e_c? (b) e_b? and (c) i_b?

4. If an a-c signal whose maximum value is 30 volts is applied to the circuit represented by Figs. 7-3 and 7-4, what is the range of variation of the quantities: (a) e_c? (b) e_b? and (c) i_b?

5. Draw a set of curves the same as in Fig. 7-4 for grid voltages of 0, -43.5, and -87 volts. Plot a load line for $R_o = 3,000$ ohms and Q to again appear at the point where $E_{b.o} = 250$ volts and $I_{b.o} = 60$ ma; the second point of the curve can be for E_{bb} at $i_b = 0$ as found in Example 7-5. If an input signal with a maximum value of 43.5 volts is applied, find (a) the grid-voltage variation, (b) the plate-voltage variation, and (c) the plate-current variation.

6. Draw a set of curves the same as in Fig. 7-4 for grid voltages of 0, -43.5, and -87 volts. Plot a load line for $R_o = 1,500$ ohms and Q to again appear at the point where $E_{b.o} = 250$ volts and $I_{b.o} = 60$ ma; the second point of the curve can be for E_{bb} at $i_b = 0$ as found in Example 7-5. If an input signal with a maximum value of 43.5 volts is applied, find (a) the grid-voltage variation, (b) the plate-voltage variation, and (c) the plate-current variation.

7. For the same conditions as in Prob. 5, find (a) the B supply voltage required, (b) the a-c power output, and (c) the per cent of second harmonic distortion.

8. For the same conditions as in Prob. 6, find (a) the B supply voltage required, (b) the a-c power output, and (c) the per cent of second harmonic distortion.

9. If the triode amplifier of Fig. 7-5 has an 800-ohm load resistance and a 43.5-volt input signal, find (a) $I_{b.o}$, (b) $i_{b.max}$, (c) $i_{b.min}$, and (d) per cent of second harmonic distortion.

10. If the triode amplifier of Fig. 7-5 has an 800-ohm load resistance and a 30-volt input signal, find (a) $I_{b.o}$, (b) $i_{b.max}$, (c) $i_{b.min}$, and (d) per cent of second harmonic distortion.

11. (a) Using Eq. (2-8) determine the maximum value of the varying component of the plate current for the triode of Fig. 7-4 with a load resistance of 2,363 ohms and an input signal with a maximum value of 43.5 volts. (b) Using Eq. (2-9) determine the maximum value of the varying component of the output signal voltage. (c) Compare the results of (a) and (b) with the answers to Example 7-3, bearing in mind that the answers in Example 7-3 are peak-to-peak values. (d) What are the effective values of the a-c component of the plate current and output signal voltage? (e) Find the a-c power output from the values found in (d) and compare with the results found in Example 7-6.

12. (a) By use of Eq. (2-8) determine the maximum value of the varying component of the plate current for the triode of Prob. 5. (b) By use of Eq. (2-9) determine the maximum value of the varying component of the output signal voltage. (c) Compare the results of (a) and (b) with the answers to Prob. 5, bearing in mind that the answers in Prob. 5 are peak-to-peak values. (d) What are the effective values of the a-c component of the plate current and output signal voltage? (e) Find the a-c power output from the values found in (d) and compare with the results found in Prob. 7.

13. By use of Eq. (7-9) determine the output of the amplifier of Prob. 11 and compare the results with those of Prob. 11 and Example 7-6.

14. By use of Eq. (7-9) determine the output of the amplifier of Prob. 12 and compare the results with those of Prob. 12 and Prob. 7.

15. By use of Eq. (7-11) or (7-12) determine the maximum power output of the amplifier of Prob. 11.

16. A certain triode power-amplifier tube having an amplification factor of 3.5 and a plate resistance of 1,650 ohms is being operated with 180 volts at its plate and -31.5 volts at its grid. The tube is to be used to amplify a signal with a peak value of 31.5 volts. (a) What value of load resistance is required to produce maximum power output? (b) What is the maximum power output? (c) What is the power output if the load resistance is increased to 2,700 ohms and the signal voltage is reduced to 30.6 volts (peak value)?

17. The power-amplifier tube of Prob. 11 is producing 3.92 watts output. If the input grid resistor is 220,000 ohms and has a current of 140 μa (effective value), find (a) the power input, (b) the power amplification.

18. The power-amplifier tube of Prob. 16 is producing 0.818 watt output. If the input grid resistor is 470,000 ohms and has a current of 47.5 μa (effective value), find (a) the power input, (b) the power amplification.

19. The amplifier of Prob. 5 and Prob. 7 produces 3.2 watts output with values of $E_{b.o} = 250$ volts, $I_{b.o} = 60$ ma, and $E_{bb} = 430$ volts. (a) What is its plate-circuit efficiency? (b) What is its plate efficiency?

20. The amplifier of Prob. 6 and Prob. 8 produces 2.05 watts output with values of $E_{b.o} = 250$ volts, $I_{b.o} = 60$ ma, and $E_{bb} = 340$ volts. (a) What is the plate-circuit efficiency? (b) What is its plate efficiency?

21. What is the plate dissipation of the power tube used in Prob. 19?

22. What is the plate dissipation of the power tube used in Prob. 20?

23. What is the power sensitivity of the tube used in Prob. 5 and Prob. 7? (NOTE: E_g is equal to 0.707 times the maximum value of the input signal.)

24. What is the power sensitivity of the tube used in Prob. 6 and Prob. 8? (NOTE: E_g is equal to 0.707 times the maximum value of the input signal.)

25. Using the curves of Prob. 5 as the basis of a power diagram, determine the operating characteristics of the circuit with a 3,000-ohm load resistance and a peak input signal of 43.5 volts. Find (a) the power output, (b) power in the load resistor, (c) power supplied by the B power supply, (d) plate-circuit efficiency, (e) plate efficiency.

26. Using the curves of Prob. 6 as the basis of a power diagram, determine the operating characteristics of the circuit with a 1,500-ohm load resistance and a peak input signal of 43.5 volts. Find (a) the power output, (b) power in the load resistor, (c) power supplied by the B power supply, (d) plate-circuit efficiency, (e) plate efficiency.

27. If the pentode represented by the curves of Fig. 7-8 is operated with a 5,000-ohm plate load, find (a) $i_{b.max}$, (b) $i_{b.min}$, (c) I_b, (d) I_x, (e) I_y, (f) E_b.

28. From the values obtained in Prob. 27, determine (a) the per cent of second harmonic distortion, (b) the per cent of third harmonic distortion, (c) the per cent of total harmonic distortion, (d) the power output, (e) the plate efficiency if the screen grid current is 10.5 ma.

29. If the pentode represented by the curves of Fig. 7-8 is operated with a 9,000-ohm plate load, find (a) $i_{b.max}$, (b) $i_{b.min}$, (c) I_b, (d) I_x, (e) I_y, (f) E_b.

30. From the values obtained in Prob. 29, determine (a) the per cent of second harmonic distortion, (b) the per cent of third harmonic distortion, (c) the per cent of total harmonic distortion, (d) the power output, (e) the plate efficiency if the screen grid current is 10.5 ma.

31. If the tube of Prob. 27 has a transconductance of 2,500 μmhos and a plate resistance of 80,000 ohms, find (a) the varying component of the plate current by Eq. (5-6), (b) the varying component of the plate voltage by Eq. (5-7).

32. If the tube of Prob. 29 has a transconductance of 2,500 μmhos and a plate resistance of 80,000 ohms, find (a) the varying component of the plate current by Eq. (5-6), (b) the varying component of the plate voltage by Eq. (5-7).

33. (a) By means of Eq. (7-21) calculate the power output of the tube used in Probs. 27, 28, and 31. (b) Compare the result with the answer to (d) of Prob. 28.

34. (a) By means of Eq. (7-21) calculate the power output of the tube used in Probs. 29, 30, and 32. (b) Compare the result with the answer to (d) of Prob. 30.

35. What is the power sensitivity of the pentode used in Probs. 27 and 28 if the input signal voltage is the result of 25μa flowing through a 470,000-ohm grid resistor?

36. What is the power sensitivity of the pentode used in Probs. 27 and 28 if the input signal voltage is the result of 53 μa flowing through a 220,000-ohm grid resistor?

37. What is the plate dissipation and the screen-grid dissipation for the pentode used in Probs. 27 and 28?

38. What is the plate dissipation and the screen-grid dissipation for the pentode used in Probs. 29 and 30?

39. What is the power sensitivity of the tube as operated in Probs. 27 and 28?

40. What is the power sensitivity of the tube as operated in Probs. 29 and 30?

41. From the curves of Fig. 7-8 determine the following values needed to plot a power diagram for the 5,000-ohm load line: (a) I_b, (b) $i_{b.min}$, (c) E_b, (d) E_{bb}, (e) $e_{b.min}$. (f) Plot the power diagram.

42. From the results of Prob. 41, find (a) the power output, (b) power in the load resistor, (c) power supplied to the screen grid, (d) power supplied to the plate circuit by the B power supply, (e) plate-circuit efficiency, (f) plate efficiency, (g) plate dissipation. (NOTE: The screen grid requires 10.5 ma at 250 volts.)

43. From the curves of Fig. 7-8 determine the following values needed to plot a power diagram for the 9,000-ohm load line: (a) I_b, (b) $i_{b.min}$, (c) E_b, (d) E_{bb}, (e) $e_{b.min}$. (f) Plot the power diagram.

44. From the results of Prob. 43, find (a) the power output, (b) power in the load resistor, (c) power supplied to the screen grid, (d) power supplied to the plate circuit by the B power supply, (*e*) plate-circuit efficiency, (*f*) plate efficiency, (*g*) plate dissipation. (NOTE: The screen grid requires 10.5 ma at 250 volts.)

45. A certain power-amplifier circuit employs four pentodes operated in parallel. Each tube delivers 3.4 watts.   (*a*) What is the power output of the circuit?   (*b*) What is the decibel rating of the circuit?

46. Five type 6F6 tubes are operated in parallel as Class A pentode amplifiers with the following circuit conditions: $E_b = 250$ volts, $E_{c.1} = -16.5$ volts, $E_{c.2} = 250$ volts, $R_o = 7,000$ ohms.   (*a*) From a tube manual determine the power output of a single tube.   (*b*) What is the power output of the amplifier circuit?   (*c*) What is the decibel rating of the circuit?

47. (*a*) What is the power output of each tube of Prob. 46 if the plate and screen voltages are increased to 285 volts and the grid bias is increased to 20 volts?   (*b*) What is the power output of the amplifier circuit?   (*c*) What is the decibel rating of the circuit?

48. Two triode tubes, represented by Fig. 7-17, are connected to operate as a Class AB$_1$ push-pull amplifier.   The operating characteristics of each tube are $E_b = 300$ volts, $E_c = -60$ volts, and $E_{g.m} = 60$ volts.   (*a*) What are the values of $i_{b.\max}$ and I_x?   (*b*) What is the maximum power output of the two tubes?   (*c*) What is the required value of plate-to-plate load resistance?   (*d*) What is the per cent of third harmonic distortion?

49. Two type 6L6 tubes are to be used in a Class A$_1$ push-pull amplifier.   The operating values of the tubes and circuit are $E_b = 250$ volts, $E_{c.1} = -16$ volts, $E_{c.2} = 250$ volts, $r_p = 24,500$ ohms, $g_m = 5,500$ μmhos, peak grid-to-grid signal voltage $= 32$ volts, plate-to-plate load resistance $= 5,000$ ohms.   What is the power output of the amplifier?   [NOTE: Use Eq. (7-28); let $\mu = g_m r_p$.]

50. Two type 6L6 tubes are to be used in a Class AB$_1$ push-pull amplifier.   The operating values of the tubes and circuit are $E_b = 360$ volts, $E_{c.1} = -22.5$ volts, $E_{c.2} = 270$ volts, $r_p = 28,000$ ohms, $g_m = 5,000$ μmhos, peak grid-to-grid signal voltage $= 45$ volts, plate-to-plate load resistance $= 4,300$ ohms.   What is the power output of the amplifier?   [NOTE: Use Eq. (7-28); let $\mu = g_m r_p$.]

51. Two type 6L6 tubes are to be used in a Class AB$_2$ push-pull amplifier.   The operating values of the tubes and circuit are $E_b = 360$ volts, $E_{c.1} = -22.5$ volts, $E_{c.2} = 270$ volts, $r_p = 28,000$ ohms, $g_m = 5,000$ μmhos, peak grid-to-grid signal voltage $= 72$ volts, plate-to-plate load resistance $= 3,800$ ohms.   What is the power output of the amplifier?   [NOTE: Use Eq. (7-28); let $\mu = g_m r_p$.]

52. A twin triode is to be used in a Class B push-pull amplifier.   The operating values of the tubes and circuit are $E_b = 300$ volts, $E_c = 0$ volts, $r_p = 13,000$ ohms, $\mu = 35$, peak grid-to-grid signal voltage $= 80$ volts, plate-to-plate load resistance $= 8,000$ ohms.   What is the power output of the amplifier?   (NOTE: The effective plate-to-plate load resistance is one-quarter of the actual value.)

53. A driving tube and a phase inverter, similar to tubes 1 and 2 of Fig. 7-18, are units of a 12AX7 twin-triode tube.   The plate power supply provides 90 volts, $R_{c.1} = 100,000$ ohms, C_1 and $C_2 = 0.007$ μf, and $R_g = 470,000$ ohms (see Fig. 7-18). (*a*) What value is recommended for $R_{c.2}$?   (*b*) What voltage amplification is produced by the driver stage?   (See resistance-coupled amplifier chart in a tube manual.) (*c*) What value is recommended for R_2?   (*d*) What value is recommended for R_1?

54. A push-pull amplifier is provided with negative feedback in the manner indicated in Fig. 7-20.   What is the per cent of feedback if the value of R_1 is 50,000 ohms and R_2 is 5,000 ohms?

55. What values would be required for R_1 and R_2 of Fig. 7-20 if it is desired to produce 10 per cent feedback and the total resistance of the feedback circuit is to be 100,000 ohms?

CHAPTER 8

VACUUM-TUBE OSCILLATOR CIRCUITS

The frequency of the varying currents or voltages associated with radio and electronic circuits may vary from a few cycles per second to millions of cycles per second. A vacuum tube, when used in conjunction with the proper combination of circuit elements, may be made to produce an alternating current having almost any value of frequency. The vacuum tube does not create any electrical energy; it merely changes one kind of current to another. The electrical circuit associated with a vacuum tube when used to produce an alternating current is called an *oscillator circuit.* In addition to having a wide frequency range, the frequency and amplitude of the output of vacuum-tube oscillator circuits are comparatively easy to control. The circuits used for oscillators are rather simple to construct and are relatively inexpensive. A vacuum-tube oscillator circuit can be designed so that its output is devoid of any harmonics or so that it is rich in harmonic content. Because of these favorable characteristics, vacuum-tube oscillator circuits form an important part of modern radio and electronic equipment.

8-1. Uses of Vacuum-tube Oscillator Circuits. *Use of Oscillators for Radio Circuits.* The waves sent out by a radio transmitter consist of electromagnetic and electrostatic fields. As these fields are caused by an alternating current, their frequencies are the same as that of the alternating current producing them. The frequency of the alternating currents used in communications ranges from hundreds of thousands of cycles per second to millions of cycles per second. Radio transmitters depend on vacuum-tube oscillators to produce their high-frequency carrier currents, and the vacuum-tube oscillators are as important to a radio transmitting station as the generator is to the power plant.

An oscillator circuit, called the *local oscillator*, is used in superheterodyne receivers to generate a stable r-f signal that is heterodyned with the incoming signal to produce an intermediate frequency.

Uses of Oscillators in Test Equipment. The oscillator circuit is also used in radio test equipment such as the signal generator. The r-f signal generator is very useful in aligning the r-f tuned circuits, the oscillator circuit, and the i-f amplifier circuits of a radio receiver. The a-f signal generator is very useful in locating the source of rattles and buzzes in radio equipment.

Use of Oscillators in Electronic and Industrial Equipment.   A sweep generator, which is a special form of oscillator circuit, is used to deflect a spot of light across the screen of a cathode-ray tube at a uniform rate. An oscillator circuit is therefore used in practically all types of electronic equipment employing a cathode-ray tube.   Among the list of apparatus using the cathode-ray tube are the oscilloscope, television, radar, and numerous types of industrial control equipment.

8-2. Types of Oscillator Circuits.

The types and kinds of oscillator circuits are many and varied.   It is beyond the scope of this text to discuss every type of oscillator circuit; however, a brief outline description of the most important types used in radio and electronics will be given. The general discussion of the principles and the operation of oscillator circuits in this chapter will be limited to those types which are frequently used in the production of a-f and r-f oscillations in radio equipment.

General Classifications of Oscillator Circuits.   Vacuum-tube oscillator circuits may be broadly divided into two groups: those circuits used to produce (1) *nonsinusoidal waves,* and (2) *sinusoidal waves.*

Oscillators Producing Nonsinusoidal Waves.   Oscillator circuits producing a nonsinusoidal wave are generally used as electronic timing and control circuits in television, radar, oscilloscope, and industrial control equipment.   Nonsinusoidal voltages are generally produced by some form of *relaxation oscillator circuit.*   In this type of oscillator circuit, one or more voltages or currents change abruptly one or more times during each cycle of oscillation.   The types of relaxation-oscillator circuit most commonly used are: (1) Van der Pol, (2) multivibrator, (3) glow-tube discharge, (4) arc-tube discharge, (5) saw-tooth-wave generators, (6) rectangular- or square-wave generators.

Oscillators Producing Sinusoidal Waves.   An oscillator circuit used to produce a sinusoidal voltage may come under any one of the following classifications: (1) negative resistance, (2) feedback, (3) heterodyne, (4) crystal, (5) magnetostriction, (6) ultrahigh frequency.

A circuit element is said to possess negative resistance when at some portion of its operating characteristic the current through the circuit decreases with an increase in the voltage and vice versa.   Any circuit having a negative-resistance characteristic can be used as an oscillator. Upon first thought, feedback oscillator circuits might be considered as negative-resistance oscillators; however, only those circuits having negative a-c resistance, even when not used in connection with an oscillatory circuit, are classified as negative-resistance oscillators.   The electric arc and the screen-grid tube possess this characteristic and therefore may be used in this type of circuit.   The four types of negative-resistance oscillator circuits most commonly used are (1) the *dynatron,* which makes use of the negative screen-grid resistance characteristic of a tetrode; (2) the *transitron,* which makes use of the negative transconductance characteristic of a pentode; (3) the push-pull circuit, which is a variation of the *Eccles-Jordan trigger circuit;* (4) the resistance-capacitance or *resistance-tuned circuit,* in which the frequency of oscillation is dependent upon the resistance and capacitance of the circuit.

A tuned oscillatory circuit combined with positive feedback is the type of oscillator circuit most commonly used in communications circuits. There are numerous circuits that may be used as feedback oscillators; however, for the purpose of classification, it is necessary to list only the basic circuits, as all others are merely modifications of these circuits. The basic feedback oscillator circuits may be considered as follows:

1. Tuned plate
2. Tuned grid
   (a) Inductive feedback
   (b) Capacitive feedback
   (c) Tickler feedback
3. Hartley
4. Colpitts
5. Complex types using more than one tuned circuit
   (a) Tuned-grid tuned-plate
   (b) Meissner
   (c) Tri-tet
   (d) Electron coupled
6. Resistance-capacitance tuned

The heterodyne or beat frequency oscillator consists of two vacuum-tube circuits that are made to oscillate at slightly different frequencies. The output of these oscillators is simultaneously applied to a common detector. By means of a filter, the higher r-f currents are removed, and the output will be of a frequency equal to the difference of the two original frequency values. The heterodyne oscillator circuit is a means of obtaining precise audio frequencies and is commonly used in test equipment operating at audio frequencies.

Crystal-controlled oscillators are used when the frequency of oscillation must be maintained at a fixed value. The crystal is not used to produce oscillations but controls the output frequency of the oscillator with which it is used.

The magnetostriction oscillator circuit is based on the principle that a change in magnetization will cause a magnetic material to expand or contract and, conversely, that a contraction or expansion of a magnetic material will cause a change in magnetization. A strong stable oscillation having a frequency of the order of 10,000 to 100,000 cps can be obtained from this type of oscillator circuit.

The types of oscillator circuits used to generate a-f or r-f currents or voltages cannot be used to produce ultrahigh-frequency currents or voltages. Various means are employed to obtain these ultrahigh frequencies, among which are (1) positive grid, (2) magnetron, (3) velocity modulation, (4) resonant cavities, (5) resonant lines. The theory and analysis of ultrahigh-frequency oscillators is beyond the scope of this text.

8-3. The Amplifier as an Oscillator. *Amplifier Action of the Oscillator.* The essential parts of a vacuum-tube oscillator are (1) the oscillatory or tank circuit, usually a parallel resonant circuit, (2) a vacuum-tube amplifier, (3) a feedback circuit. The vacuum-tube oscillator circuit is

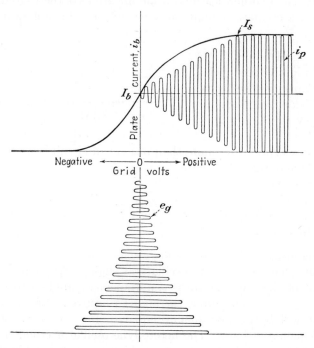

Fig. 8-1. Illustration of how the oscillations in the tube circuit build up until the saturation value of the plate current is reached.

so arranged that a portion of the energy developed in the output (or plate) circuit is returned in the proper phase relationship to the input (or grid) circuit. The energy returned is amplified by the tube, and a portion of the energy it develops in the plate circuit is then returned to the grid circuit. Each time some energy is taken from the output circuit it is regenerated in the input circuit. This cycle of operations is continually repeated. Because of the amplifying properties of the tube, the energy in the output circuit will increase with each cycle of operation until the maximum or saturation value of plate current has been reached (see Fig. 8-1). The value of the output plate current will depend upon the characteristics of the tube and the manner in which it is operated.

As the energy consumed by the input circuit of an amplifier is considerably less than that in its output circuit, it is possible to have an amplifier supply its own input. An amplifier operated in this manner will generate oscillations at a frequency that is determined by the electrical constants of the circuit. Furthermore, since the tube operates as an amplifier, the oscillator can be made to supply power to an external circuit in addition to supplying the circuit losses required to sustain oscillations. The tube thus acts as a power converter, changing the d-c power supplied to its plate circuit into a-c energy in the amplifier output circuit. It may thus be seen that a tube cannot produce oscillations by itself and also that the function of the oscillator is not to create energy but to change d-c energy to a-c energy. The combination of the vacuum tube and its associated circuits is called a *vacuum-tube oscillator.*

Classification of Tube Operation. The vacuum-tube oscillator can be considered as an amplifier in which part of the output voltage is returned to the input circuit in such a manner that the tube drives itself. The tube may be operated as a Class A, B, or C amplifier. Class A operation is generally used in high-quality a-f oscillators. Because high efficiency and low distortion are obtainable at high frequencies from Class C operation, r-f oscillators are usually operated Class C.

Oscillator Circuits. A number of circuits may be used to produce the required oscillations, each circuit having its own advantages and disadvantages. The circuit used will depend upon the frequency and power required and its application. The oscillatory circuit may be in the grid circuit, the plate circuit, or both. Energy can be applied to the input circuit by either capacitive or inductive coupling. Capacitive coupling can be accomplished by use of an external circuit or by use of the interelectrode capacitances within a tube. Inductive coupling can be accomplished only by means of an external circuit.

8-4. The Oscillatory Circuit. In the oscillatory circuit of the oscillator, electrical oscillations occur according to the fundamental laws governing capacitor and inductor actions. The oscillatory or alternating flow of electrons in the parallel resonant or tank circuit is caused by the repeated exchange of energy between the capacitor and inductor.

The operation of the oscillator circuit is explained with the aid of Fig. 8-2. It is assumed that the capacitor is fully charged (Fig. 8-2a) and just starting to discharge (Fig. 8-2b). As the capacitor discharges through the inductor, the flow of electrons (as indicated by the arrows on the wires connecting the capacitor and inductor) causes a magnetic field to be built up around the inductor. Electrons will continue to flow in this direction until the charges on the plates of the capacitor are equal to each other (Fig. 8-2c). At the instant that the rate of electron flow is zero, the magnetic field about the inductor will start to collapse (Fig. 8-2d). According to Lenz's law, the collapsing magnetic field causes electrons to flow in the same direction that produced the expanding field, thus causing the capacitor to become charged with a polarity opposite to its original charge. When the field about the inductor has completely collapsed, the energy that had been stored in its magnetic field will be transferred to the electrostatic field of the capacitor (Fig. 8-2e).

Electron flow ordinarily ceases when the charge on the two plates of the capacitor are equal. However, because of the effects of the collapsing magnetic field, electron flow continues past this neutral point. This

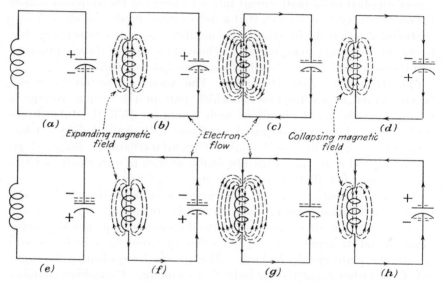

Fig. 8-2. Oscillatory action of a parallel resonant circuit.

action of a parallel resonant circuit is sometimes referred to as the *flywheel effect.*

The capacitor will now discharge, and electrons will flow in a direction opposite to that used to charge the capacitor. This flow of electrons produces an expanding field about the inductor (Fig. 8-2f), until the difference in charge between the two plates is zero (Fig. 8-2g). As before, the magnetic field collapses, thus causing the electrons to continue to flow in the same direction (Fig. 8-2h). When the energy stored in the electromagnetic field has been transferred to the electrostatic field, the capacitor becomes fully charged in the opposite polarity. It is thus restored to its original state as in Fig. 8-2a.

This cycle of operations is repeated at a frequency approximately equal to the frequency of resonance of the parallel circuit as expressed by

$$f_r = \frac{159}{\sqrt{LC}} \qquad (8\text{-}1)$$

where f_r = frequency of resonance, kc

   L = inductance, μh

   C = capacitance, μf

If the circuit is assumed to have zero resistance, each cycle of electron flow in the tank circuit will be similar to that shown in Fig. 8-3*a*. Thus, theoretically, a sustained alternating-current flow of constant magnitude is produced. As it is impossible to construct a circuit without some amount of resistance, some energy will be lost in the form of heat during

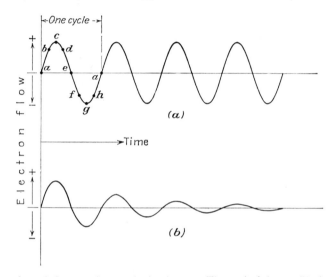

FIG. 8-3. Flow of electrons in a tank circuit.   (*a*) Theoretical flow.   (*b*) Actual flow.

each cycle. If no energy is supplied to replace this loss, the magnitude of each oscillation will diminish as shown in Fig. 8-3*b*. Therefore, a simple tank circuit by itself cannot produce an alternating current of constant magnitude.

8-5. Fundamental Oscillator Theory. In practical oscillator circuits a vacuum tube and a power supply are used to provide the energy required to overcome the losses caused by the resistance in the circuit, thus producing an alternating current of constant magnitude.

The fundamental oscillator circuit (Fig. 8-4) is capable of producing alternating currents of constant magnitude and constant frequency. The operation of this circuit is similar to most types of feedback oscillators. In order to simplify the description of the operation of this circuit, the action in each part will be presented individually. However, these actions are not independent of each other but are closely related, all occurring almost instantaneously.

Analysis of the Fundamental Oscillator Circuit. At the instant of closing the switch *S* (Fig. 8-4), the electrons being emitted from the cathode will be drawn to the plate. This causes a current to flow from the cathode to the plate, through the plate coil L_P, through the B power supply, and back to the cathode. As the current through L_P increases, a magnetic field builds up around the plate coil and, by mutual inductance, a voltage of increasing magnitude is induced in the grid coil L_G. The grid and plate coils are connected in their respective circuits so that the voltage induced in the grid coil will have a positive potential at the terminal connected to the grid of the tube when the current in L_P is increasing. Two immediate actions result from this positive voltage: (1) The voltage on the grid becomes positive, thereby increasing the amount of plate current. (2) The capacitor C in the tank circuit becomes charged. The

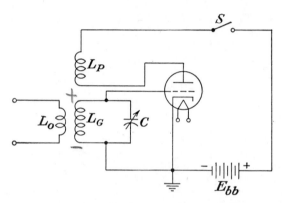

Fig. 8-4. A fundamental oscillator circuit.

increasing plate current produces an increase in the strength of the magnetic field about the plate coil, thus causing a greater voltage to be induced in the grid coil. This action makes the grid still more positive, thereby increasing the plate current still further. This action continues until saturation is reached (Fig. 8-5). The point of saturation will depend on the resistance of the circuit, the plate-supply voltage, and the characteristics of the tube. As soon as the plate current ceases to increase, the field about the plate coil ceases to expand and no longer induces a voltage in the grid coil. The tank capacitor C, having been charged to its maximum potential, starts to discharge through the grid coil. This decrease in voltage across the capacitor makes the voltage on the grid less positive, thus causing the plate current to decrease, which in turn causes the magnetic field about the plate coil to collapse. A voltage will again be induced in the grid coil but will be opposite in direction to that produced by an expanding field. The voltage on the grid thus becomes negative, thereby decreasing the plate current still further. This action continues until the grid voltage is such that zero plate current flows (point C on Fig. 8-5). During the time in which the plate current has decreased from saturation current to cutoff, the tank capacitor has lost its original charge and has again become charged to its maximum potential; the plates now have a polarity opposite to that of the previous charged condition. As the induced voltage in L_G is zero when cutoff is reached, the capacitor will now start to discharge through the grid coil. This decrease in voltage across the capacitor makes the voltage on the grid less negative, thus causing the plate current to increase and the magnetic field about the grid coil to expand. The voltage induced in the grid coil will make the grid more positive, thereby increasing the plate current still further until the saturation point is reached in the same manner as explained previously. This cycle of operations repeats itself

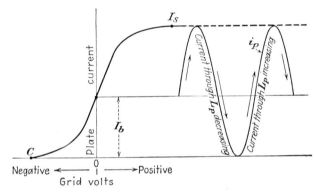

Fig. 8-5. Variation of plate current in the oscillator circuit of Fig. 8-4.

continuously as long as energy is supplied to overcome the losses in the circuit.

Output power can be obtained by inductively coupling the output coil L_O to the grid coil L_G (Fig. 8-4). Oscillating energy from the grid circuit can thus be transferred to the output circuit. In addition to supplying power to overcome the losses in the circuit, the B power supply must now also supply the power required by the output load.

8-6. Grid Bias. *Need for a Grid-leak Resistor and Grid Capacitor.* During part of each cycle of alternating current in the oscillator the grid is driven positive. In order to prevent the tube from drawing an excessive amount of plate current during this portion of the cycle, practically all oscillator circuits employ grid-leak bias. Two methods of connecting the grid resistor are shown in Fig. 8-6.

Operation of Oscillator Circuits with Grid-leak Resistor and Grid Capacitor. When the oscillator of Fig. 8-6a or 8-6b is started, the initial bias on the grid is zero. Thus, when positive voltage is applied to the grid, because of the expanding magnetic field about the plate coil, the grid will be driven positive. The flow of electrons in the grid circuit (caused by the positive grid) produces a voltage drop at the resistor R_g with point A negative and point B positive. The capacitor C_g will become charged to this polarity. The grid current flowing through the grid resistor thus causes the voltage on the grid to become negative. The plate current will then decrease, thus causing the magnetic field about the plate coil to collapse. The cycle then continues as explained for the fundamental oscillator circuit.

Graphical Representation of the Oscillator Circuit Action. The manner in which the oscillations are produced is illustrated in Fig. 8-7. In order to simplify the diagram, the dynamic characteristic curve of the oscillator tube, line AC, is assumed to be a straight line. Actually it is slightly curved. At the instant that plate voltage is applied, the plate

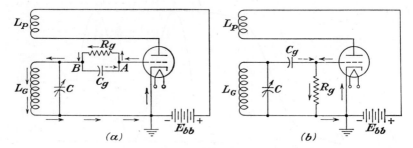

Fig. 8-6. Two methods of connecting the grid-leak resistor and grid capacitor.

current will start increasing at a rapid rate. However, the inductive reactance of the plate coil prevents the plate current from becoming maximum instantaneously. The tube is operated as a Class C amplifier with a grid bias whose value is twice the amount required to produce cutoff. The grid bias, being dependent upon the charge (and discharge) of the grid capacitor, will increase to its proper value according to the exponential curve xy, Fig. 8-7. Figure 8-7 shows that the grid is driven slightly positive for a small part of each cycle due to the flywheel effect of the tank circuit. During the interval that the grid is positive it acts as a diode plate; as a result grid current will flow, thereby charging the grid capacitor. The time required for the grid bias to attain its maximum value is determined by the time constant of the $R_g C_g$ circuit.

Values of C_g and R_g. Any alternating voltage across the grid capacitor will vary the grid excitation voltage. In order to limit this voltage to a minimum, the value of the capacitor should be as large as practical. The maximum value of capacitance is, however, limited by the time constant desired. The time constant should be small enough so that the bias voltage cannot attain a value high enough to stop oscillations, in which case it is said that the tube has become blocked. The grid-bias requirements for the particular tube used will determine the value of the grid resistor. Therefore, in order to reduce the time constant, it becomes necessary to use a smaller value of grid capacitance.

Effect of C_g and R_g upon Grid Current.   When the voltage on the grid is positive, current will flow in the grid circuit. This flow of current produces a power loss in the form of heat. As this power loss must be supplied from the power supply, the efficiency of the oscillator is decreased. Use of the grid resistor and grid capacitor limits the amount of grid current flow, thus making the oscillator circuit more efficient.

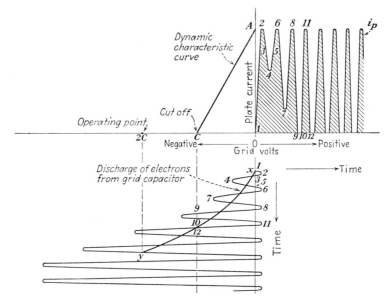

Fɪɢ. 8-7. Use of the dynamic characteristic curve to illustrate the manner in which oscillations are produced.

Grid-leak Bias vs. Fixed Bias.   In order to obtain high efficiency, the oscillator tube should be biased beyond the point of cutoff or operated as Class C.   If fixed bias of an amount sufficient to produce cutoff were used, the initial plate current would be zero and thus there would be no power input to the plate coil to start oscillations.   Using a grid-leak resistor and grid capacitor, the initial grid bias will be zero.   The plate current will immediately start oscillations and automatically build up the bias to its ultimate operating value.   Thus if the oscillator circuit is to be self-starting, it is essential that grid-leak bias be used.

8-7. Basic Oscillator Circuits. *The Hartley Oscillator.*   The Hartley oscillator (Fig. 8-8) is one of the simplest types of oscillator circuits. The operation of each circuit of Fig. 8-8 is identical except for the manner in which the d-c power is supplied to the plate.   As in the fundamental oscillator circuit just described, the amplified energy in the plate circuit is fed back to the grid circuit by means of inductive coupling. However, in the Hartley oscillator only one coil is used.   A portion L_P of this coil is in the plate circuit, and the remainder L_G is in the grid circuit.   The amount of inductive feedback between these two sections of the common coil will depend on the number of turns in sections L_P and L_G.

The alternating current in the plate section of the coil induces a voltage in the grid section.   This induced voltage is applied to the grid of the tube, is amplified, and again applied to the plate section.   The plate and grid voltages are 180 degrees out of phase with each other, since they are taken from opposite ends of the coil with respect to the common lead connected to the cathode.

In the circuit shown in Fig. 8-8a, the plate power supply is connected in series with the plate section and is called a *series-fed circuit.*   The

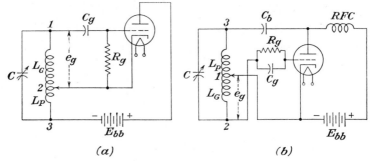

FIG. 8-8. Hartley oscillator circuits.   (a) A series-fed circuit.   (b) A parallel-fed circuit.

voltages across the tuning capacitor and the inductor (terminals 1 and 3) will be approximately equal to the plate-power-supply voltage.   In order to avoid the danger of high plate voltages appearing across these two circuit elements, parallel feed is generally used.

The circuit shown in Fig. 8-8b is called a *parallel* or *shunt-fed oscillator* because the plate circuit is divided into two parallel branches, one to provide a path for the direct current and the other for the alternating current.   The r-f choke coil keeps the alternating current out of the d-c circuit, and the blocking capacitor C_b keeps the direct current out of the a-c circuit.   In order that most of the r-f current may flow through the plate section L_P instead of through the plate power supply, the reactance of C_b at the resonant frequency should be small in comparison with the reactance of the choke coil at the same frequency.

The frequency of the oscillator output is equal to the frequency at which the voltage in the plate circuit is fed back to the grid circuit.   In oscillators employing a resonant circuit, the frequency is approximately equal to the resonant frequency of the tank circuit.   The frequency of oscillation of the circuits shown in Fig. 8-8 is thus equal to the resonant frequency of the parallel circuit formed by the tuning capacitor C and the entire inductor, L_G plus L_P.

Of the various types of self-excited oscillators, the Hartley oscillator is most generally used. However, this circuit is seldom used exactly as shown in Fig. 8-8, since many modifications (see Fig. 8-9) have been made in order to adapt the basic circuit to meet the individual requirements for specific applications. The circuit shown in Fig. 8-9a is used in some superheterodyne receivers. A circuit using mutual-inductance coupling

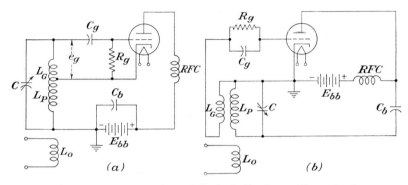

FIG. 8-9. Two modifications of the basic Hartley oscillator circuit.

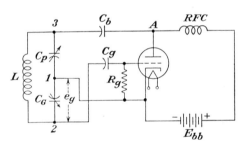

FIG. 8-10. The Colpitts oscillator circuit.

is shown in Fig. 8-9b. The excitation voltage for the grid is obtained by the mutual inductance between the grid and plate coils.

The Colpitts Oscillator. Another basic type of oscillator circuit is the Colpitts or capacitive-feedback oscillator (Fig. 8-10). In this circuit, the feedback of energy from the plate circuit to the grid circuit is accomplished by means of electrostatic coupling. Except for the manner in which feedback is obtained, this circuit is similar to the parallel-feed Hartley oscillator. In the Colpitts oscillator the tank voltage is divided into two parts by tapping the tank capacitor, or rather by connecting two capacitors in series, instead of the inductor. The grid excitation voltage is obtained from the grid tank capacitor C_G. The plate and grid voltages are 180 degrees out of phase with each other, since they are taken from opposite ends of the tank circuit with respect to the common lead connected to the cathode. Although the use of parallel feed is optional (and preferable) in the Hartley circuit, it is *necessary* in the Colpitts circuit because of the high voltage obtainable at the tank capacitor C_P.

The operation of this circuit is similar to that of the parallel-feed Hartley oscillator. As plate current starts to flow, the plate blocking capacitor C_b starts to charge, which in turn also causes the plate tank capacitor C_P to charge. This increase in voltage is transferred to the grid tank capacitor C_G, through the r-f choke, causing the terminal connected to the grid to become more negative. When the feedback voltage causes the plate current to decrease, less energy will be stored in the plate tank capacitor, thus reversing the direction of the feedback voltage. As this action continues, sustained oscillations will be produced.

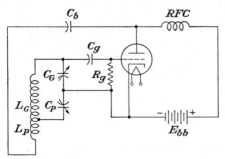

Fig. 8-11. A modified Colpitts oscillator circuit.

The capacitance of the tank circuit is equal to the resultant capacitance of C_P and C_G in series. By varying the capacitance of C_P and/or C_G, the voltage across the tank circuit may be divided to produce the voltage drop required across C_G for proper grid excitation. Increasing the reactance of the grid tank capacitor will increase the voltage across it.

A modified form of the basic Colpitts oscillator (Fig. 8-11) employs both capacitive and inductive feedback. The plate tank capacitor C_P provides capacitive feedback, and section L_P of the inductor provides inductive feedback. This oscillator can produce fairly uniform output over a wide frequency range by the proper selection of circuit elements.

Tuned-grid Tuned-plate Oscillator. Capacitive feedback is also used in the tuned-grid tuned-plate oscillator (Fig. 8-12a). Capacitive feedback is obtained by utilizing the grid-plate capacitance C_{gp} of the tube. Usually this capacitance is a nuisance, but in this circuit it is used to advantage. Two parallel resonant circuits are required, one in the grid circuit of the tube and the other in the plate circuit, and no inductive coupling should exist between the coils in these two circuits. The frequency of oscillation is dependent on the resonant frequency of each of the tuned circuits, therefore the name *tuned-grid tuned-plate oscillator.*

The operation of the tuned-grid tuned-plate oscillator may be more easily understood by rearranging the schematic diagram into the equivalent circuit form shown in Fig. 8-12b. The equivalent circuit is simplified by showing the interelectrode capacitance C_{gp} connected across the resonant circuit instead of across the grid and plate of the tube as shown in Fig. 8-12a. This simplification is possible because C_g and C_b act in series with the interelectrode capacitance C_{gp} to form the feedback circuit, and since C_g and C_b are much larger in value then C_{gp}, they are disregarded.

In order for the system to oscillate, the plate circuit should be tuned to a slightly lower frequency than the grid circuit. The resultant inductive reactance of the two tuned circuits will then resonate with the interelectrode capacitive reactance. The oscillator circuit thus formed is

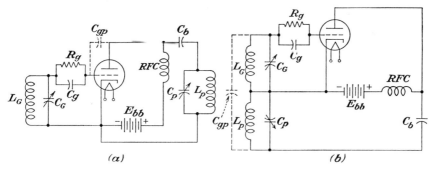

(a)   *(b)*

FIG. 8-12. Tuned-grid–tuned-plate parallel-feed oscillator circuit. (a) Schematic diagram. (b) Equivalent circuit.

similar in many respects to the Hartley circuit. The frequency of the tuned-grid tuned-plate oscillator is determined by the tuned circuit having the highest value of Q.

The frequency stability and the voltage amplification of the tuned-grid tuned-plate oscillator is comparatively high because of the high impedance of the parallel resonant plate load. Varying the resonant frequency of either of the tuned circuits will vary the amount of feedback, thus increasing or decreasing the amplitude of the oscillations. In practice, the frequency of oscillation is usually controlled by varying the resonant frequency of the plate-tuned circuit, and the amount of grid excitation is controlled by varying the resonant frequency of the grid-tuned circuit. The plate circuit is designed to deliver the required power output. The prop-

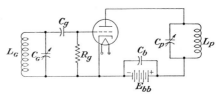

FIG. 8-13. Tuned-grid–tuned-plate series-feed oscillator circuit.

erties of the grid-tuned circuit are not too important, providing the desired frequency range can be obtained. The plate power supply of the tuned-grid tuned-plate oscillator may be either series fed or parallel fed. The tuned-grid tuned-plate circuit is used to some extent in high-frequency oscillators.

8-8. Circuit Considerations. *Effect of the Tank Circuit on the Amplitude Stability.* If the amplitude of the output of an oscillator is unstable the waveform will be distorted, thus producing undesirable harmonics. An oscillator tube generally operates as a Class C amplifier delivering power during less than one-half of the input cycle.   Under this condition, the plate current variation will not produce a sine wave, and a tank circuit (parallel resonant circuit) is added in order to obtain an approximate sine-wave output.   In order to have an output of approximately sine-wave form, it is necessary that the tank circuit store energy during the portion of the cycle in which the tube is delivering power and to deliver this stored energy to the load during the portion of the cycle in which no power is being delivered by the tube.

Effect of the Tank Circuit Q on the Waveshape.   The tank circuit consists of a capacitor and an inductor connected in parallel.   The resistance of the inductor acts as a resistance in series with the inductance and the Q of the tank circuit is equal to the ratio of the inductive reactance to the resistance.   When the tank circuit is supplying power to a load the effect is similar to increasing the series resistance of the tank circuit and consequently reduces the value of Q.   The Q of the circuit when supplying power is called the *effective Q*, designated Q_{eff}.   The factor Q_{eff} is also a measure of the ratio of the energy stored during each cycle to the energy dissipated during each cycle [see Eq. (8-21a)].   An increase in the value of Q_{eff} indicates an increase in the amount of energy stored. Furthermore, the ability to carry each cycle past its neutral point (flywheel effect) can be increased by increasing the effective Q of the tuned circuit.

In order to obtain a satisfactory waveshape at the output of an oscillator (or a Class C amplifier), it is necessary that the amount of energy stored during a portion of each cycle appreciably exceeds the amount of energy dissipated during the remainder of the cycle.   Experience has shown that the value of Q_{eff} should be in the order of 10 to 30.   Under this condition, the amount of energy stored per cycle will be approximately 2 to 5 times the amount of energy dissipated per cycle.   When Q_{eff} has a value of 12.5, the harmonic content of the output wave is approximately 3 per cent.   Decreasing the value of Q_{eff} causes an increase in the amount of distortion, while increasing the value of Q_{eff} produces a more nearly sinusoidal output wave.   However, a high value of Q_{eff} results in a high value of tank circuit current, thereby increasing the losses and reducing the efficiency of the tank circuit.

Efficiency of the Tank Circuit.   The efficiency of the tank circuit may be expressed as the ratio of the power delivered to the load to the power delivered to the tank circuit.   In terms of the circuit Q

$$\text{Efficiency} = \frac{Q - Q_{eff}}{Q} \times 100 \qquad (8\text{-}2)$$

where Q = tank circuit Q at no-load
     Q_{eff} = tank circuit Q with load

Example 8-1.   The tank circuit of an r-f amplifier, used in conjunction with the oscillator of a radio transmitter, has a circuit Q of 85 at no-load and an effective Q of 12 when load is applied.   What is the efficiency of the tank circuit?

Given: $Q = 85$ Find: Efficiency
         $Q_{eff} = 12$

Solution:

$$\text{Efficiency} = \frac{Q - Q_{eff}}{Q} \times 100 = \frac{85 - 12}{85} \times 100 = 85.8\%$$

Analysis of the Tank Circuit.   Important considerations in the study of the tank circuit are the frequency, impedance, current, power, and energy stored.

The tank circuit should be adjusted so that its resonant frequency corresponds to that of the oscillator.   For the values of Q commonly used with these circuits, the results obtained by use of Eq. (8-1) are sufficiently accurate for general purposes.   Accordingly, Eqs. (8-3) and (8-4) may be used for finding the values of the inductance and capacitance required in the tank circuit.

$$f_r = \frac{159}{\sqrt{LC}} \tag{8-1}$$

$$L = \frac{25{,}300}{f_r{}^2 C} \tag{8-3}$$

$$C = \frac{25{,}300}{f_r{}^2 L} \tag{8-4}$$

As the resistance of the tank circuit is small compared to the inductive reactance, a sufficiently accurate value of the impedance (at resonance) can be obtained by

$$Z_{t.r} = QX_L = \frac{X_L{}^2}{R} \tag{8-5}$$

As $X_L = X_C$ at resonance, then

$$Z_{t.r} = \frac{X^2}{R} = \frac{X_L X_C}{R} = \frac{2\pi f_r L}{2\pi f_r CR} = \frac{L}{CR} \tag{8-6}$$

The voltage at the tank circuit will be approximately equal to the alternating component of the plate voltage E_p.   It may also be expressed as

$$E_t = I_p Z_{t.r} \tag{8-7}$$

Correspondingly, the tank-circuit current may be expressed as

$$I_{t.r} \cong \frac{E_t}{X_L} \cong \frac{E_p}{X_L} \tag{8-8}$$

An approximate value of the tank-circuit current can also be found by

$$I_{t \cdot r} = I_L = I_C = QI_p \qquad (8\text{-}9)$$

At resonance the power delivered to the tank circuit (output load plus tank-circuit losses) may be expressed as

$$P_{t \cdot r} = I_p{}^2 Q X_L = \frac{E_t{}^2}{Q X_L} = \frac{E_p{}^2}{Q X_L} \qquad (8\text{-}10)$$

Example 8-2.   The tank circuit of the oscillator in a certain radio receiver has the following circuit values: $L = 80\ \mu h$, $C = 365\text{-}35\ \mu\mu f$, $Q_{\text{eff}} = 20$, $E_t = 10$ volts.   Find (a) the value of capacitance when the oscillator is adjusted to a resonant frequency of 1,000 kc, (b) the impedance of the tank circuit at resonance, (c) the plate current required to produce the desired voltage at the tank circuit, (d) the current in the tank circuit, (e) the power supplied to the tank circuit.

Given:	$L = 80\ \mu h$	Find:	(a) C
	$C = 365\text{-}35\ \mu\mu f$		(b) $Z_{t \cdot r}$
	$Q_{\text{eff}} = 20$		(c) I_p
	$E_t = 10$ volts		(d) I_t
	$f_r = 1,000$ kc		(e) P_t

Solution:

(a) $\quad C = \dfrac{25,300}{f_r{}^2 L} = \dfrac{25,300}{1,000^2 \times 80} = 0.000316\ \mu f$ or $316\ \mu\mu f$

(b) $\quad Z_{t \cdot r} = Q_{\text{eff}} X_L = 20 \times 6.28 \times 10^6 \times 80 \times 10^{-6} = 10,048$ ohms

(c) $\quad I_p = \dfrac{E_t}{Z_{t \cdot r}} = \dfrac{10}{10,048} = 0.000995$ amp or 0.995 ma

(d) $\quad I_t = Q_{\text{eff}} I_p = 20 \times 0.995 = 19.9$ ma

(e) $\quad P = E_p I_p = 10 \times 0.000995 = 0.00995$ watt

Energy Stored in the Tank Circuit.   An analysis of the tank circuit will show that the amount of energy stored will increase with an increase in the value of either the capacitance or the voltage.   The total energy stored in a tank circuit is equal to the sum of the energy stored in the inductor and in the capacitor.

$$w_t = \frac{L i_t{}^2}{2} + \frac{C e_t{}^2}{2} \qquad (8\text{-}11)$$

where w_t = total instantaneous energy stored in tank circuit, watt-sec

$\quad i_t$ = instantaneous value of current flowing in oscillatory or tank circuit, amp

$\quad e_t$ = instantaneous value of voltage across tank circuit, volts

$\quad L$ = inductance of tank coil, henrys

$\quad C$ = capacitance of tank capacitor, farads

For all practical purposes the voltage across the tank circuit can be considered as being equal to the alternating voltage output of the tube. Then

$$w_t = \frac{L i_t{}^2}{2} + \frac{C e_p{}^2}{2} \qquad (8\text{-}12)$$

When the energy stored in the tank circuit is sufficient to reduce the harmonic distortion to a negligible amount

$$e_p = E_{p \cdot m} \sin \theta \tag{8-13}$$

where $\theta = 2\pi ft$
t = time, sec

As the resistance of a well-designed tank circuit is very small in comparison to either of its reactances, it can generally be ignored in considering the impedance of either branch.   The current in the tank inductor lags the applied voltage and can be expressed as

$$i_t = -\frac{E_{p \cdot m}}{2\pi f_r L} \cos \theta \tag{8-14}$$

Substituting Eqs. (8-13) and (8-14) in Eq. (8-12)

$$w_t = \frac{L E_{p \cdot m}{}^2}{2(2\pi f_r L)^2} \cos^2 \theta + \frac{C E_{p \cdot m}{}^2}{2} \sin^2 \theta \tag{8-15}$$

At resonance

$$2\pi f_r L = \frac{1}{2\pi f_r C} \tag{8-16}$$

and

$$L = \frac{1}{(2\pi f_r)^2 C} \tag{8-17}$$

Substituting this value of L in Eq. (8-15)

$$w_t = \frac{C E_{p \cdot m}{}^2}{2} \cos^2 \theta + \frac{C E_{p \cdot m}{}^2}{2} \sin^2 \theta \tag{8-18}$$

or

$$w_t = \frac{C E_{p \cdot m}{}^2}{2} (\cos^2 \theta + \sin^2 \theta) \tag{8-18a}$$

However, from trigonometry, $\cos^2 \theta + \sin^2 \theta = 1$

and

$$W_t = \frac{C E_{p \cdot m}{}^2}{2} \tag{8-18b}$$

where W_t = energy stored in the tank circuit per cycle

Also $E_{p \cdot m} = \sqrt{2}\, E_p$ and $E_{p \cdot m}{}^2 = 2 E_p{}^2$
Thus $W_t = C E_p{}^2$ (8-18c)

Equation (8-18c) shows that the total energy stored in the tank circuit can be increased by using a higher value of capacitance.   In order to maintain a fixed value of frequency, an increase in the value of capacitance must be accompanied by a decrease in the value of inductance, thereby reducing the L/C ratio.   An increase in the value of the capacitance (and a decrease in the value of the inductance) will result in a lower value of reactance at the fixed value of resonant frequency. Equation (8-10) shows that the effective Q of the tank circuit increases with a decrease in the reactance; this is characteristic of a loaded tank circuit and should not be confused with the no-load value of Q, which decreases with a decrease in the reactance. Thus, an increase in the value of capacitance increases the amount of energy stored, decreases the L/C ratio of the tank circuit, and increases the effective Q of the tank circuit.

Increasing the effective Q of the tuned circuit increases its impedance [see Eq. (8-5)], thereby increasing the varying plate voltage output of the oscillator tube. This increase in plate voltage will further increase the total energy stored [Eq. (8-18c)]. However, an increase in the voltage applied to a tuned circuit increases the value of the circulating current, thus increasing the I^2R losses and reducing the over-all efficiency of the oscillator circuit. It is therefore good practice to limit the L/C ratio to the value required to produce a satisfactory sinusoidal output.

Relation of Q_{eff} to the Ratio of Energy Stored to Energy Dissipated. It has previously been stated that the value of Q_{eff} is also a measure of the ratio of the energy stored during each cycle to the energy dissipated during each cycle. This can now be shown as follows. In the same manner in which Eq. (8-18c) was developed, it can be shown that

$$W_t = LI_t^2 \tag{8-19}$$

The energy dissipated in the tank circuit per cycle may be expressed as

$$W_d = \frac{I_t^2 R_{eq}}{f_r} \tag{8-20}$$

where W_d = energy dissipated in tank circuit per cycle

R_{eq} = equivalent series resistance of tank circuit when loaded.

The ratio of energy stored per cycle to energy dissipated per cycle may then be expressed as

$$\frac{\text{Energy stored per cycle}}{\text{Energy dissipated per cycle}} = \frac{W_t}{W_d} = \frac{LI_t^2}{\dfrac{I_t^2 R_{eq}}{f_r}} = \frac{f_r L}{R_{eq}} \tag{8-21}$$

Multiplying both the numerator and the denominator by 2π

$$\frac{\text{Energy stored per cycle}}{\text{Energy dissipated per cycle}} = \frac{2\pi f_r L}{2\pi R_{eq}} = \frac{Q_{eff}}{2\pi} \tag{8-21a}$$

Example 8-3. (a) How much energy is stored per cycle in the tank circuit of Example 8-2? (b) Determine the value of R_{eq} from the values of Q_{eff} and X_L in Example 8-2. (c) How much power is dissipated by the tank circuit? (d) How much energy is dissipated per cycle? (e) What is the ratio of the energy stored per cycle to the energy dissipated per cycle? (f) Check the result of part (e) by use of Eq. (8-21a).

Given: $E_t = 10$ volts   Find: (a) W_t
         $C = 316$ $\mu\mu f$ (b) R_{eq}
         $Q_{eff} = 20$ (c) P_{dis}
         $L = 80$ μh (d) W_d
         $f_r = 1,000$ kc   (e) $\dfrac{W_t}{W_d}$
         $I_t = 19.9$ ma   (f) $\dfrac{W_t}{W_d}$

Solution:

(a) $W_t = CE_t^2 = 316 \times 10^{-12} \times 10^2 = 316 \times 10^{-10}$ watt per cycle

(b) $R_{eq} = \dfrac{X_L}{Q_{eff}} = \dfrac{6.28 \times 10^6 \times 80 \times 10^{-6}}{20} = 25.12$ ohms

(c) $P_{dis} = I_t^2 R_{eq} = (19.9 \times 10^{-3})^2 \times 25.12 = 9,947 \times 10^{-6}$ watt

(d) $W_d = \dfrac{P_{dis}}{f_r} = \dfrac{9,947 \times 10^{-6}}{10^6} = 99.47 \times 10^{-10}$ watt per cycle

(e) $\dfrac{W_t}{W_d} = \dfrac{316 \times 10^{-10}}{99.47 \times 10^{-10}} = 3.17$

(f) $\dfrac{W_t}{W_d} = \dfrac{Q_{eff}}{2\pi} = \dfrac{20}{6.28} = 3.18$

Excitation Voltage. The excitation voltage required to drive the oscillator tube depends upon the characteristics of the tube, the losses in the circuit, and the power consumed by the load. As this voltage is dependent upon the varying output plate voltage, it is also affected by changes in the operating potentials of the tube. Because the varying plate voltage is also dependent upon the effective Q of the tank circuit, any changes in the tuned circuit will also affect the excitation voltage.

If the excitation voltage is too low, the tube will not oscillate, and if it is too high, it will increase the amount of positive grid. The power consumed by the grid will then be excessive and will decrease the over-all efficiency of the oscillator.

Power Output. In radio receivers oscillator tubes operate as voltage amplifiers and the power output and efficiency are not important. Oscillator tubes in radio transmitters operate as Class C power amplifiers. The operating circuit considerations for voltage and power amplifiers previously studied will also apply to oscillators.

The power output of an oscillator is the useful a-c power consumed by its load. This load may be coupled by means of capacitive, inductive, or electron coupling. The frequency and amplitude stability of an oscillator will be affected by changes in the power taken by the load. In order to maintain a high degree of stability, oscillator circuits are seldom designed to deliver large amounts of power. When large amounts of power are required the oscillator circuit is used to drive a power amplifier, which in turn produces the desired amount of power output. Most oscillator circuits are therefore used as frequency-controlling devices delivering a small amount of power at a comparatively high voltage.

The maximum power output of an oscillator will occur when the effective resistance of the plate load, usually the tuned circuit, is equal to the plate resistance of the tube. However, as oscillator tubes are generally operated as Class C amplifiers, the plate resistance will vary appreciably over the entire cycle. When the plate voltage is high, the plate resistance is low, and when the plate voltage is low, the plate resistance will be high. The equation for maximum power transfer is quite complex, thus making it rather difficult to calculate exact values of the circuit constants required to obtain maximum power transfer.

Efficiency. The plate efficiency of the oscillator tube depends upon its operating characteristics such as the load, plate resistance, excitation voltage, etc. Operating the tube as a Class C amplifier increases its plate efficiency and also increases the over-all circuit efficiency.

With Class C operation, a higher plate voltage can be used and a greater power output can be obtained. Increasing the plate voltage also increases the plate current, thus increasing the plate dissipation. However, higher efficiency and higher power output is obtained as the power output increases at a greater rate than the plate dissipation. Furthermore, it is possible to limit the plate voltage so that the allowable plate dissipation of the tube is not exceeded and still obtain higher efficiency and power output.

The power lost in the grid will vary, the loss being between 10 and 20 per cent of the output power. As this loss must be supplied by the oscillator's power supply, the over-all efficiency will be less than if the tube were being operated as an amplifier.

8-9. Frequency Stability. *Factors Affecting the Frequency Stability.* The ability of an oscillator to maintain a constant frequency in the presence of variable operating conditions is referred to as its *frequency stability.* Because of the small difference in frequency of adjacent radio broadcasting channels, the carrier frequency of these stations must be held to a very close tolerance. The allowable variation in frequency for broadcasting stations is ± 20 cycles. In a radio receiver, any variation in the output frequency of its local oscillator will vary the beat frequency that is applied to the i-f amplifier. In order to obtain the maximum voltage amplification with a minimum of distortion the beat frequency and the frequency to which the i-f circuit is tuned should be identical. The factors affecting the frequency stability of an oscillator are (1) plate voltage, (2) output load, (3) temperature, (4) mechanical variation of the circuit elements.

Plate Voltage and Output Load. Any variation in the operating voltages applied to the tube will vary the operating characteristics, thus causing a shift in the output frequency of the oscillator. This type of frequency change is referred to as *dynamic instability.* The operating voltages can be stabilized by the use of a regulated power supply.

Any variation in the plate load will affect the frequency of an oscillator in a similar manner as changes in plate voltage. Dynamic instability caused by a variable load can be reduced by the use of (1) electron coupling, (2) a buffer amplifier, (3) a tank circuit having a high effective Q. Since the tube and load represent a comparatively low resistance in parallel with the tank circuit, the dynamic instability can be reduced by using a tuned circuit having a high effective Q and a low L/C ratio. Dynamic instability can also be reduced by increasing the effective resistance that the tube reflects to the tank circuit. This can be accomplished by using a higher value of grid-leak resistor, thus increasing the grid bias.

Temperature and Mechanical Variations. The expansion and contraction of the elements in a tube due to temperature changes will cause the interelectrode capacitances to vary. As these capacitances are part of the tuned circuit, the frequency of oscillation will be affected.

Temperature changes will also slightly affect the values of the tank coil and capacitor, causing an additional shift in the resonant frequency. Because the effects of temperature change are comparatively slow in operation, the frequency change is referred to as *drift*. The obvious means of minimizing the amount of drift are by the use of (1) adequate ventilation, (2) a coil wound with large wire, (3) a low direct input voltage. Use of any of these methods will reduce the temperature of the oscillator unit, thus reducing the amount of drift. The variations in the interelectrode capacitances of the tube will have very little effect when they are shunted by a large value of capacitance. It is thus possible to reduce the amount of drift by using a large value of capacitance in the tank circuit, thus producing a high C/L ratio.

Mechanical vibration of the circuit elements such as tubes, capacitors, and inductors also causes their values to vary. These changes in the values of inductance and capacitance will cause the resonant frequency to vary with the mechanical vibration. Instability due to mechanical vibration can be minimized by isolating the oscillator from the source of mechanical vibration.

8-10. Crystals. *Uses of Crystals.* The frequency of oscillation of the oscillator circuits previously discussed is controlled by the electrical constants of the circuit. Such oscillators are called *self-controlled oscillators*. Because the values of the circuit elements will be affected by the operating conditions, the frequency of oscillation of all self-controlled oscillators has a tendency to drift. There are a number of applications of oscillator circuits where the frequency of oscillation must be maintained at a fixed frequency or at several definite frequencies. Some of these applications are (1) transmitters, (2) time-signal receivers, (3) police-car radio receivers, (4) military, navigation, and aircraft communication apparatus, (5) test equipment used for calibration purposes. In order to maintain the output frequency of an oscillator at a constant value, a crystal may be used to control the frequency of oscillation. This type of oscillator circuit is called a *crystal-controlled oscillator*.

Characteristics of Crystals. When certain crystalline materials are placed under a mechanical strain, such as compression or expansion, an electrical difference of potential will be developed across opposite faces of the crystal.   This action is called the *piezoelectric effect*.   Conversely, when a voltage is impressed across opposite faces of this type of crystal, it will cause the crystal to expand or contract.   If the voltage applied is alternating, the crystal will be set into vibration.   The frequency of

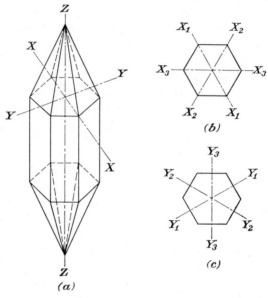

FIG. 8-14. Quartz-crystal axes.   (a) Symmetrical quartz crystal.   (b) Electrical axes. (c) Mechanical axes.

vibration will be equal to the resonant frequency of the crystal as determined by its structural characteristics.   When the frequency of the applied voltage is equal to the resonant frequency of the crystal, the amplitude of vibration will be maximum.   Because of its piezoelectric effect, a high-quality crystal can maintain the frequency of oscillation so that the variation in frequency will be less than one part in a million.

Piezoelectric effects may be obtained from rochelle salts, tourmaline, and quartz; for practical reasons quartz is most commonly used.   The voltage set up in a rochelle salt crystal is much greater than would be set up in a quartz crystal; however, it is much weaker mechanically and is more likely to break.   Tourmaline is mechanically stronger than quartz, but because it is much less sensitive electrically it is only used for very high-frequency applications where quartz crystals must be ground so thin as to make its use impractical.   Other advantages of quartz are its low cost and its availability in comparatively large quantities.

Characteristics of Quartz Crystals.   Quartz crystals in their natural state assume the general form of hexagonal prisms with each end surmounted by a hexagonal pyramid (Fig. 8-14a).   Each crystal has three principal axes: (1) Z or optical axis, (2) X or electrical axis, (3) Y or mechanical axis.   The axis joining the two points at the ends of the crystal is called the Z or *optical axis.*   No piezoelectric effects are produced when electrostatic charges are applied in this direction.   The X axes join opposite points of the hexagonal prism, and each axis is parallel to one pair of sides of the hexagon and perpendicular to the optical axis.   The X axes are called the *electrical axes* because the greatest

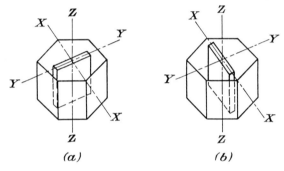

FIG. 8-15. Methods of cutting a quartz crystal.   (a) X cut.   (b) Y cut.

piezoelectric effect is produced along this axis.   The Y axes join opposite sides of the hexagonal prism and are perpendicular to the Z axis and the flat sides that it joins.   Each Y axis is also perpendicular to an X axis. The Y axis is called the *mechanical axis.*   Each crystal therefore has one Z axis, three X axes, and three Y axes (Fig. 8-14).

Crystals used for electronic purposes are cut into thin plates.   When a plate is cut so that the flat surfaces are perpendicular to an X axis (Fig. 8-15a), it is called an *X-cut crystal.*   A plate cut so that its flat surfaces are perpendicular to a Y axis (Fig. 8-15b) is called a *Y-cut crystal.*

Temperature Coefficient.   The resonant frequency of a crystal will vary with temperature changes.   The number of cycles change per million cycles for a 1°C change in temperature is called the *temperature coefficient.*   An X-cut crystal has a negative temperature coefficient, and thus an increase in temperature will cause the resonant frequency to decrease.   A Y-cut crystal has a positive temperature coefficient, and thus an increase in temperature will cause the resonant frequency to increase.

Two methods are employed to stabilize the resonant frequency of a crystal: (1) enclosing the crystal in a thermostatically controlled container, (2) varying the angle of cut of the crystal in order to obtain a zero temperature coefficient.   As an *X*-cut crystal has a negative temperature coefficient and a *Y*-cut crystal a positive temperature coefficient, a zero temperature coefficient can be obtained by cutting the crystal at the proper angle between the *X* and *Y* axes.   This type of crystal is called an *XY cut*.   Because of the inherent irregularities in the natural structure of a crystal, the lengths of all the *X* axes, or *Y* axes, are not equal to each other.   In an *XY*-cut crystal it is thus possible to have two resonant frequencies very close to each other.   This type of crystal is not very practical, as it has the tendency to change its frequency abruptly.

Practical crystals with zero temperature coefficient have been obtained by rotating the cutting plane about the *X* axis and cutting the crystal at

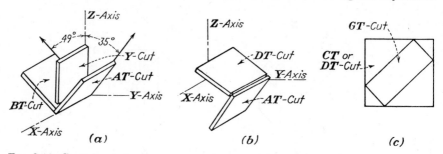

Fig. 8-16. Crystal cuts having zero temperature coefficient.   (*a*) *AT* and *BT* cuts.   (*b*) *DT* cut.   (*c*) *GT* cut.

an angle to the *Z* axis.   Crystals obtained by cutting at an angle to the *Z* axis are called *A T, BT, CT, DT*, and *GT cuts*.   The *A T* cut is obtained by rotating the cutting plane about 35 degrees in a clockwise direction, and the *BT* cut by a rotation of about 49 degrees in a counterclockwise direction (see Fig. 8-16*a*).   The *CT* cut is made at approximately right angles to the *BT* cut, and the *DT* cut is made at approximately right angles to the *A T* cut (see Fig. 8-16*b*).   These four types of crystal cuts have zero temperature coefficient at only one temperature.   A crystal having zero temperature coefficient over a wide temperature range is obtained by rotating the principal axes of a *CT* or *DT* crystal by 45 degrees (Fig. 8-16*c*) and is called a *GT cut*.

Frequency-Thickness Ratio.   The length and width of a crystal is of relatively minor importance, as crystals are generally cut so that the resonant frequency is determined mainly by its thickness.   The frequency of vibration will vary inversely with the thickness as expressed by

$$f = \frac{k}{t} \tag{8-22}$$

where f = frequency, kc
    t = thickness, inches
    k = 112.6 for *X* cut, 77 for *Y* cut, 66.2 for *A T* cut

Example 8-4.   What is the thickness of a crystal whose resonant frequency is 1,000 kc, when (*a*) *X*-cut? (*b*) *Y*-cut? (*c*) *AT*-cut?

Given: $f = 1,000$ kc   Find: (*a*) *t* for *X* cut
                                             (*b*) *t* for *Y* cut
                                             (*c*) *t* for *AT* cut

Solution:

(*a*) $t = \dfrac{k}{f} = \dfrac{112.6}{1,000} = 0.1126$ inch

(*b*) $t = \dfrac{k}{f} = \dfrac{77}{1,000} = 0.077$ inch

(*c*) $t = \dfrac{k}{f} = \dfrac{66.2}{1,000} = 0.0662$ inch

Example 8-4 shows that for a given frequency the *X*-cut crystal is thicker and thus mechanically stronger than *Y*- or *AT*-cut crystals.

At the very high frequencies, the crystal required becomes too thin for practical use.   Crystal control at these frequencies is obtained by the use of harmonic crystals, which will be discussed later in this chapter.

Power Limitations.   The amount of current that can safely pass through a crystal ranges from 50 to 200 ma.   When the rated current is exceeded, the amplitude of mechanical vibration becomes too great and may crack the crystal.   Overloading the crystal affects the frequency of vibration, because the power dissipation and crystal temperature will increase with the amount of load current.

8-11. Crystal Oscillator Circuits.   *Equivalent Electrical Circuit of a Crystal.*   In order to make electrical connections to the crystal, it is usually mounted horizontally between two metal plates (Fig. 8-17*a*). As the crystal must vibrate to produce oscillations, it must be mounted between these two plates in such a manner as to allow for the required amount of mechanical vibration.

As vibration of the crystal will induce electrical charges on the two plates, it is thus possible to consider the crystal and its mountings as an electrical resonant circuit (Fig. 8-17*b*).   In this circuit, *C* represents the elasticity of the crystal, *L* represents its mass, and *R* represents the resistance offered to the vibration by its internal friction. C_M represents the capacitor formed by the two metal plates of the holder separated by the crystal dielectric.

The reactances of *L* and *C* will be numerically equal to each other at the resonant frequency of the crystal.   Since the crystal forms a series resonant circuit, maximum current will flow through the circuit at its resonant frequency, thus causing the magnitude of the crystal's vibrations to be maximum at this frequency.   When the crystal vibrates at its resonant frequency, the voltage that it generates will be maximum and will also be of the same value of frequency.

At some frequency slightly higher than the resonant frequency, the combined effective reactance of L and C is inductive and will be numerically equal to the reactance of the capacitor C_M.   At this frequency the crystal circuit acts as a parallel resonant circuit, and its impedance is maximum.   The circulating current in the circuit $CLRC_M$ is maximum and therefore the crystal vibrations will also be maximum.   Thus a parallel resonant tank circuit can be obtained by operating the crystal at the resonant frequency of the parallel tuned circuit formed by $CLRC_M$.

The equivalent inductance of a crystal is very high in comparison with either its equivalent capacitance or its equivalent resistance.   Because of

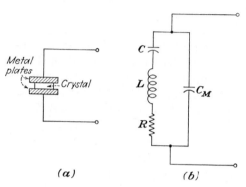

(a) (b)

Fig. 8-17. Crystal characteristics. (*a*) Crystal and mounting plates. (*b*) Equivalent electrical circuit.

this high L/R ratio, the Q of a crystal circuit is many times greater than can be obtained from an electric circuit, Example 8-5.   Greater frequency stability and frequency discrimination are obtained because of the high Q and high L/C ratio of the series resonant circuit CLR.

Example 8-5.   A certain X-cut crystal is resonant at 450 kc.   For this frequency, its equivalent inductance is 3.65 henrys, its equivalent capacitance is 0.0342 $\mu\mu$f, and its equivalent resistance is 9,040 ohms.   What is the Q of the crystal?

Given: $L = 3.65$ henrys   Find: Q
          $R = 9,040$ ohms
          $f = 450$ kc

Solution:

$$Q = \frac{2\pi fL}{R} = \frac{6.28 \times 450,000 \times 3.65}{9,040} = 1,141$$

Simple Crystal Circuit.   Since a properly cut quartz crystal has the same characteristics as a high-Q tuned circuit, it can be used to control the frequency in an oscillator circuit.   A simple crystal oscillator circuit is shown in Fig. 8-18.   This circuit is similar to the tuned-grid tuned-plate oscillator except for the substitution of the crystal for the parallel tuned-grid circuit.

The voltage returned to the grid circuit by the grid-to-plate capacitance of the tube is applied to the crystal and causes it to vibrate.   The voltage set up by the crystal vibrations is applied to the grid of the tube, thereby controlling the amount of energy released in the plate circuit.

The voltage returned to the grid circuit will be maximum when the impedance of the crystal is maximum.   Maximum impedance occurs at the parallel resonant frequency of the crystal circuit.   At this frequency the crystal vibrations will be maximum and thus the voltage generated by the crystal will also be maximum.   The parallel resonant frequency of

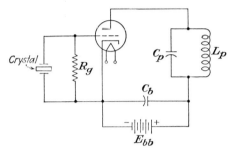

Fig. 8-18. A simple crystal oscillator circuit.

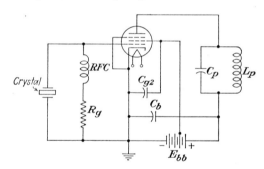

Fig. 8-19. A crystal oscillator circuit using a pentode.

the crystal circuit thus determines the frequency of oscillation of a crystal-controlled oscillator.   As with a tuned-grid tuned-plate oscillator, the resonant frequency of the tuned-plate circuit should be slightly greater than the parallel resonant frequency of the crystal circuit.

Pentode Crystal Oscillator Circuit.   There are many practical variations of the fundamental crystal-controlled oscillator circuit.   Generally, the higher the amplification factor of a tube, the smaller will be the amount of driving voltage required to produce the desired output.   Crystals used in circuits employing tubes having a high amplification factor require low values of crystal current, and hence the magnitude of the crystal vibrations is also low.   Because of their high amplification factor, pentodes and beam power tubes are ideal for use in crystal-controlled oscillator circuits.   A pentode crystal-oscillator circuit is shown in Fig. 8-19.

8-12. Electron Coupling. *Tetrode Electron-coupled Oscillator.* The tank circuit of an oscillator can be isolated from its load by using electron coupling between the oscillator and the load. The impedance of the tank circuit of this type of oscillator will be practically constant, since it will not be affected by changes in plate load. As the oscillator frequency depends upon the impedance of the tank circuit, it will also be practically constant. Thus electron coupling is an effective means of making the frequency of an oscillator independent of variations in load.

Figure 8-20 illustrates two oscillator circuits employing electron coupling. A shunt-fed Hartley circuit is formed by the cathode, control

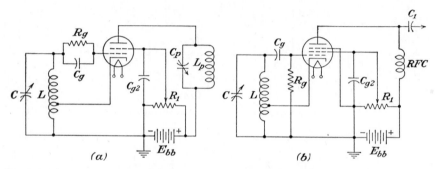

Fig. 8-20. Electron-coupled oscillator circuits. (*a*) Tetrode oscillator circuit with a tuned output load. (*b*) Pentode oscillator circuit with an untuned output load.

grid, screen grid, and LC tank circuit. Although the oscillator portion of both of these circuits is of the Hartley type, the Colpitts circuit may also be used. In this type of circuit the screen grid serves as a plate and the r-f path to the inductor L in the tank circuit is completed through $C_{g.2}$. The plate of the tube serves only as an output electrode. Since $C_{g.2}$ blocks the direct voltage and passes the high-frequency alternating voltage, the screen grid is in effect grounded for the r-f voltages. The plate is thus shielded from the oscillatory section of the tube, thereby preventing the load impedance from reacting on the oscillator.

Since the screen grid is constructed of mesh or fine wire, some of the electrons drawn to it will pass through. As the plate is maintained at a higher potential than the screen grid these electrons will be drawn to the plate. The frequency of the a-c component of the plate current is therefore the same as the oscillator frequency, and thus energy is delivered to the output load through an electron stream. Because the coupling medium is an electron stream, the circuit is called an *electron-coupled oscillator.*

Pentode Electron-coupled Oscillator. Because of the interelectrode capacitance between the plate and control grid of a tetrode, a small amount of feedback will always be present. The frequency of electron-coupled oscillator circuits using tetrodes is therefore not completely independent of load variations.   By substituting a pentode for the tetrode, feedback due to interelectrode capacitance can be eliminated.   The frequency stability of electron-coupled oscillator circuits is thereby further improved by using a pentode.

An electron-coupled oscillator using a pentode is shown in Fig. 8-20b. The connections and circuit actions are the same as for the tetrode oscillator except for the suppressor grid.   This grid is connected to either the cathode or to ground potential and acts as an electrostatic screen to shield the plate from the screen grid and the control grid.   For practical purposes, the capacitances between the tube elements are thereby eliminated.   Since the electron stream flows in only one direction and the interelectrode capacitances are eliminated, the variations in load will have no effect on the frequency of oscillation.

Plate Voltage Supply.   Increasing the plate voltage of an electron-coupled oscillator will cause the frequency of oscillation to change. Increasing the screen-grid voltage will also change the frequency of oscillation but in an opposite direction to that caused by a plate-voltage increase.   If the voltage on the screen grid is made variable (Fig. 8-20), the screen-grid voltage can be adjusted so that these two actions balance each other.   The frequency of oscillation will then be practically independent of variations in the supply voltages.

8-13. Frequency Multiplication.   *Harmonic Generation.*   Because the thickness of a crystal varies inversely as its frequency, at very high frequencies the crystal would have to be very thin and might easily be broken.   The frequency at which a crystal becomes too thin to be practical will vary with the crystal material and the type of cut.   The practical limit for quartz crystals is approximately 11 mc; however, it is possible to grind a quartz plate to operate as high as 20 mc.   In order to obtain crystal control of an oscillator at the high frequencies, frequency multiplication is employed.   This is accomplished by using a crystal-controlled oscillator of a comparatively low resonant frequency and using a harmonic of the oscillator output to drive a power amplifier.

The output current of a highly biased Class C operated tube is not a sine wave but is a wave made up of a fundamental and a number of its harmonics.   Increasing the bias on the tube will increase the intensity of the higher order of harmonics.   Knowing the frequency of a crystal, the harmonic components of the output of the tube can be used as a frequency standard over a wide frequency range.   Harmonics of a high order can be obtained from such a circuit and it is possible that as high as a fiftieth harmonic may be used as a means of checking and calibrating high frequencies.   Thus, if the fundamental frequency of an oscillator is 1,000 kc, the twentieth harmonic is 20,000 kc or 20 mc, and the fiftieth harmonic is 50,000 kc or 50 mc.   The crystal used to control these circuits would be required to have a resonant frequency of only 1,000 kc.

A frequency doubler is commonly used in connection with crystal oscillators to obtain a frequency higher than that for which the crystal is ground. A crystal-controlled oscillator circuit which possesses good frequency stability is practical at only relatively low radio frequencies. In order to obtain good frequency stability at the very high and ultrahigh radio frequencies it is necessary to obtain the frequency by multiplication or harmonic generation. An oscillator circuit that gives a harmonic output is called a *frequency multiplier*.

Tri-tet Oscillator. The frequency-load stability of electron-coupled oscillators is excellent; however, the output waveform contains many

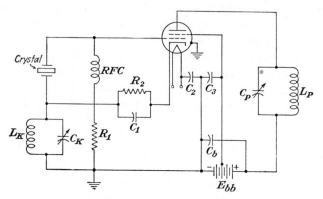

Fig. 8-21. A tri-tet oscillator circuit.

harmonics. These harmonics may be advantageous, since they can be used to produce frequencies that are multiples of either the tank-circuit frequency or the crystal frequency. The oscillator circuit illustrated in Fig. 8-21 uses a single tube as both a triode and tetrode, hence the name *tri-tet oscillator*. This circuit is a combination of a triode crystal oscillator and an electron-coupled oscillator. The oscillator circuit is electron-coupled to the output circuit $L_P C_P$ and is thus electrostatically shielded from the output by the screen grid. A pentode-type tube can also be used if the connection to the suppressor grid can be made externally. The suppressor grid should be connected either directly to ground or may be operated at approximately 50 volts positive. Operating the suppressor grid slightly positive produces a higher power output.

The tank circuit $L_K C_K$ in the cathode circuit is always tuned to a frequency somewhat higher than that of the crystal. The tank circuit $L_P C_P$ in the output circuit is tuned to either the fundamental crystal frequency or one of its multiples.

Two outstanding features of the tri-tet oscillator are (1) its ability to operate as a frequency multiplier, and (2) the buffer action between the oscillator and output circuits, that is, isolating the output circuit from the crystal oscillator circuit.

QUESTIONS

1. (*a*) Define the oscillator circuit. (*b*) What is the function of a vacuum tube in an oscillator circuit?

2. (*a*) Why are vacuum-tube oscillators necessary in radio equipment? (*b*) What is their purpose in transmitters? (*c*) What is their purpose in receivers?

3. Name some applications of vacuum-tube oscillators other than radio transmitters and receivers.

4. (*a*) Into what two classifications may vacuum-tube oscillator circuits be divided? (*b*) Give a further subdivision of each of these general classifications.

5. What is meant by (*a*) negative resistance? (*b*) Dynatron? (*c*) Transitron?

6. How may feedback oscillator circuits be classified?

7. (*a*) What is the principle of the heterodyne oscillator? (*b*) What are its uses?

8. (*a*) When are crystal-controlled oscillators used? (*b*) What purpose does the crystal serve?

9. (*a*) What is the principle of the magnetostriction oscillator? (*b*) What are its uses?

10. Can any oscillator circuit be used to produce ultrahigh-frequency currents? Explain.

11. Explain the amplifier action of an oscillator circuit.

12. (*a*) What are the essential parts of a vacuum-tube oscillator? (*b*) What is the purpose of each part?

13. Why are r-f oscillator tubes generally operated as Class C?

14. Explain the oscillatory action that takes place in the tank circuit of an oscillator.

15. What is the relation of the factors that affect the frequency of oscillations in a tank circuit?

16. (*a*) What is the waveform of the current in a tank circuit if no energy is supplied to the circuit to compensate for the I^2R loss? (*b*) What waveform will be obtained if sufficient energy is supplied to the circuit to overcome its losses?

17. Describe the operation of the fundamental oscillator circuit.

18. How may energy from the oscillator circuit be applied to the load?

19. Explain the purpose and operation of the grid-leak resistor and grid capacitor in an oscillator circuit.

20. (*a*) What determines the value of the time constant recommended for the grid-bias circuit of an oscillator? (*b*) How are the values of R_g and C_g determined?

21. What is the objection to using a fixed bias with an oscillator whose tube is being operated Class C?

22. (*a*) What are the characteristics of the Hartley circuit? (*b*) Describe the operation of the Hartley circuit. (*c*) What determines its frequency?

23. Compare the characteristics of series feed and parallel feed for the Hartley oscillator.

24. (*a*) What are the characteristics of the Colpitts circuit? (*b*) Describe the operation of the Colpitts circuit. (*c*) What determines its frequency?

25. Why is parallel feed necessary with the Colpitts oscillator?

26. (*a*) What are the characteristics of the tuned-grid tuned-plate oscillator? (*b*) Describe its operation. (*c*) What determines its frequency?

27. (*a*) What is meant by amplitude stability? (*b*) What factors affect the amplitude stability?

28. How does the tank circuit Q affect the waveshape of the oscillator output?

29. What factors affect the amount of energy stored in the tank circuit?

30. (*a*) What factors affect the value of excitation voltage required to drive the oscillator tube? (*b*) What is the effect of too high an excitation voltage? (*c*) What is the effect of too low an excitation voltage?

31. (a) Why are oscillator circuits seldom required to deliver large amounts of power?   (b) How is a large power output generally obtained?

32. What factors affect the plate efficiency of an oscillator tube?

33. (a) What is meant by frequency stability?   (b) What factors affect the frequency stability?

34. (a) What is meant by dynamic instability?   (b) What is its cause?

35. (a) What is meant by frequency drift?   (b) What is its cause?

36. (a) What is meant by a crystal-controlled oscillator?   (b) What are some applications of the crystal-controlled oscillator?

37. (a) Describe the piezoelectric effect of the crystal.   (b) What materials can be used for crystals and what is the advantage or disadvantage of each?

38. (a) What are the three axes of a quartz crystal?   (b) Describe the electrical axes.

39. (a) What is meant by the temperature coefficient of a crystal?   (b) What is its importance?

40. Describe and give the temperature characteristics of the following types of crystal cuts; (a) X cut, (b) Y cut, (c) XY cut, (d) CT cut, (e) DT cut, (f) GT cut.

41. (a) What is the range of current rating of the average crystal?   (b) What is the possible effect of overloading a crystal?

42. Describe the equivalent electrical circuit of a crystal.

43. Compare the crystal-controlled oscillator circuit with the tuned-grid tuned-plate oscillator.

44. Explain the operation of a simple crystal-controlled oscillator.

45. (a) Explain the principle of electron coupling.   (b) What are the advantages of electron-coupled oscillators?

46. (a) Explain the principle of frequency multiplication.   (b) When is frequency multiplication used in oscillator circuits?

47. (a) Explain the principle of the tri-tet oscillator.   (b) What are its advantages?

PROBLEMS

1. The tank section of an oscillator circuit used in a superheterodyne receiver contains an inductance of 90 μh shunted by a variable capacitor of 365–35 $\mu\mu$f.   What is the frequency range of the oscillator circuit if the distributed capacitance of the coil and wiring is disregarded?

2. What is the frequency range of the oscillator circuit of Prob. 1 if the distributed capacitance of the coil and wiring is known to be 15 $\mu\mu$f?

3. What is the frequency range of the oscillator circuit of Prob. 1 if a padding capacitor of 0.001 μf is connected in series with the tuning capacitor?   The distributed capacitance of the coil and wiring is 15 $\mu\mu$f.

4. If the oscillator of Prob. 3 produces a frequency 465 kc higher than the frequency of the r-f circuit throughout its working range, what is the tuning range of the receiver?

5. The oscillator section of a certain radio receiver obtains its grid bias by use of a 50,000-ohm grid-leak resistor and a 50-$\mu\mu$f grid capacitor.   (a) What is the time constant of the circuit?   (b) What is the time required to complete one-half of an output cycle if the frequency of the oscillator is 2,200 kc?   (c) Is the time constant of the $R_g C_g$ combination long or short compared to the time of one-half of a cycle of the oscillator output?

6. The oscillator section of a certain radio receiver uses a type 12BE6 tube.   A grid bias of 10 volts is to be obtained by means of a grid-leak resistor.   (a) What value of resistance is required if the grid current is 0.5 ma?   (b) What value of grid capacitor should be used if the time constant of the R_gC_g combination should be approximately 1 μsec?

7. A parallel-fed Hartley oscillator circuit similar to Fig. 8-8b uses an r-f choke of 2.5 mh and a blocking capacitor of 0.005 μf. (*a*) What is the reactance of the choke and of the capacitor at 2,200 kc? (*b*) What is the ratio of X_L to X_C at this frequency? (*c*) What is the reactance of the choke and of the capacitor at 1,000 kc? (*d*) What is the ratio of X_L to X_C at this frequency?

8. What is the frequency of a Hartley oscillator circuit similar to Fig. 8-8a if the total inductance of L_P and L_G is 80 μh and the total capacitance of the tuning circuit is 180 $\mu\mu$f?

9. What is the frequency of a Colpitts oscillator circuit similar to Fig. 8-10 if the value of the inductor L is 140 μh, C_P is 200 $\mu\mu$f, and C_G is 300 $\mu\mu$f?

10. The inductor of a certain tank circuit has an inductance of 100 μh and a resistance of 10 ohms at 1,600 kc. What is the efficiency of the tank circuit at 1,600 kc if the Q of the circuit when loaded is 15?

11. The efficiency of a tank circuit is 90 per cent and its no-load value of Q is 150. What is the effective Q of the circuit when loaded?

12. The effective Q of a certain tank circuit is 12.5. What is the no-load value of Q if the efficiency of the circuit is 92 per cent?

13. The oscillator section of a certain radio receiver has a 100-μh inductor and a 350–40-$\mu\mu$f tuning capacitor for its tank circuit. The Q of the circuit when loaded is 30 and the alternating voltage across the tank circuit is 15 volts (rms). Find (*a*) the value of capacitance when the oscillator is adjusted to a resonant frequency of 1,500 kc, (*b*) the current in the tank circuit, (*c*) the power supplied to the tank circuit, (*d*) the amount of energy stored in the tank circuit during each cycle, (*e*) the ratio of energy stored per cycle to the energy dissipated per cycle.

14. An r-f Class C power amplifier is to supply 60 watts at a frequency of 2 mc. The value of E_p (and E_t) is 1,000 volts and in order to obtain satisfactory waveform the effective Q of the plate tank circuit should not be less than 15. (*a*) What value of inductance should be used in the tank circuit? (*b*) What value of capacitance should be used? (*c*) What is the current in the tank circuit? (*d*) What is the plate circuit current?

15. The tank circuit of a certain Class C power amplifier is operated at a frequency of 5 mc. The power supplied to the tank circuit is 500 watts and the value of the alternating component of the plate voltage (also E_t) is 1,500 volts. It is desired that the effective Q of the tank circuit be 12.5. Find (*a*) the value of inductance required, (*b*) the value of capacitance required, (*c*) the current in the tank circuit, (*d*) the current supplied to the tank circuit, (*e*) the amount of energy stored in the tank circuit per cycle.

16. Repeat Prob. 15 for a value of Q equal to 50.

17. A Hartley oscillator, similar to Fig. 8-9a, is to be operated so that the tube will deliver 2.5 watts of power to the tank circuit (plate section) at a resonant frequency of 1,500 kc. The alternating component of the plate voltage is to be 140 volts, and the voltage at L_G is to be 20 volts. In order to obtain the desired waveform at the output, the effective Q of the tank circuit (plate section L_P) is to have a value of approximately 20. Find (*a*) the inductance required for section L_P of the tank circuit, (*b*) the inductance required for section L_G assuming 100 per cent coupling between L_P and L_G, (*c*) the value of capacitance required for the tank circuit, (*d*) the amount of energy stored in the tank circuit per cycle.

18. A Colpitts oscillator, similar to Fig. 8-10, is to be operated so that the tube will deliver 2.5 watts of power to the tank circuit (plate section) at a resonant frequency of 1,500 kc. The alternating component of the plate voltage is to be 140 volts and the voltage at C_G is to be 20 volts. In order to obtain the desired waveform at the output, the effective Q of the tank circuit (plate section C_P) is to have a value of approximately 20. Find (a) the capacitance required for the capacitor C_P, (b) the capacitance for the capacitor C_G, (c) the value of inductance required for the tank circuit. HINT: For solution of part (a) use Eq. (8-20), substituting X_C in place of X_L.

19. A certain X-cut quartz crystal has a resonant frequency of 3 mc at 20°C. The crystal has a negative temperature coefficient of 25 cycles per mc per degree C. How much will the frequency vary if the temperature changes to (a) 30°C? (b) 15°C? (c) 10°C?

20. A certain Y-cut quartz crystal has a resonant frequency of 3 mc at 20°C. The crystal has a positive temperature coefficient of 80 cycles per mc per degree C. How much will the frequency change if the temperature changes to (a) 30°C? (b) 15°C? (c) 10°C?

21. A certain crystal whose resonant frequency is 450 kc has an inductance of 3 henrys, and its effective series resistance is 2,000 ohms. What is the value of Q for this crystal?

22. If the crystal of Prob. 21 is represented by the equivalent circuit of Fig. 8-17b, what is the equivalent capacitance C?

23. What is the thickness of an X-cut crystal whose resonant frequency is (a) 250 kc? (b) 2,500 kc? (c) 10,000 kc?

24. What is the thickness of a Y-cut crystal whose resonant frequency is (a) 200 kc? (b) 2,000 kc? (c) 8,000 kc?

25. What is the resonant frequency of an AT-cut crystal if its thickness is (a) 0.1655 inch? (b) 0.0301 inch? (c) 0.01655 inch?

CHAPTER 9

POWER-SUPPLY CIRCUITS

The *power-supply* unit is an essential part of every type of radio and electronic equipment, since it supplies the proper voltages and currents to the filaments (or heaters), plates, and grids of the various tubes used. The general requirements of a power supply are: (1) the output voltages should be of the correct values for the apparatus used; (2) the variation in the output voltage between no-load and full-load conditions should be as small as is economically practical; (3) the output voltage should be an unvarying voltage or as nearly constant in value as is economically practical.

9-1. Sources of Power Supply. *Batteries.* In the early stages of radio development batteries were used to supply the power for practically all equipment. The use of batteries is generally divided into three classifications: (1) *A power supply*, used to supply power to the heaters or filaments; (2) *B power supply*, used to supply power to the plate and screen-grid circuits; (3) *C power supply*, used to supply voltage for the grid bias. Some types of radio equipment still are operated by batteries. Batteries are used to supply the power required to operate (1) portable receivers, (2) transceivers (transmitter and receiver in one unit), and (3) radio equipment in automobiles and aircraft.

The current from a battery is smooth, as it is devoid of any ripple or other variation. Uniform flow of power is one of the desirable characteristics of batteries. However, batteries have a limited amount of energy and must be either recharged or replaced periodically. As the available energy decreases, the voltage delivered also decreases and eventually causes unsatisfactory operation of the equipment. Other disadvantages of batteries are their weight, bulk, and cost.

Electromechanical Systems. The power required to operate certain types of radio transmitters, radio receivers, and electronic equipment is obtained from electromechanical systems. This type of power equipment may be in either of two forms: (1) generator systems, (2) vibrator systems.

Generator systems appear in various forms such as electric-motor-driven generators, gasoline-engine-driven generators, hand-driven generators, dynamotors, and converters. The type of mechanical equipment used will depend upon the kind and amount of energy available as a prime mover. The output of the d-c generator is not a continuous current but is a pulsating current. The pulsations are caused by the commutator and are generally referred to as *commutator ripple*. As one of the requirements of power supplies is to provide a current with a minimum amount of pulsations, it is necessary to use filter circuits with this type of equipment in order to smooth out the commutator ripple.

The type of power supply commonly used for automobile and aircraft radio equipment is a storage battery used in conjunction with a vibrator unit. Power for the filaments (or heaters) is obtained directly from the battery. Any high voltage required for the plates and screen grids is obtained by means of a vibrator unit (Art. 9-13).

Power Lines. Power obtained from an a-c or d-c power line is the least expensive and the most convenient source of power. Whenever practical, it is desirable to use power lines to supply all the voltages required by electronic devices such as radio receivers and transmitters. As the high operating voltages required for the plates and screen grids of electronic tubes cannot be taken directly from the power lines, suitable equipment must be used to change the input voltage to the desired values of voltage. Since a-c power is almost universally used for lighting circuits, it is also the source of power for most stationary types of electronic equipment.

The power-supply unit used in radio equipment to convert alternating current to direct current consists of four parts: (1) the power transformer, (2) the rectifier, (3) the filter, (4) the voltage divider. The power transformer increases the line voltage to a value high enough to obtain the high voltage required at the output of the power-supply unit. The rectifier allows current to flow in only one direction and therefore converts the alternating current to a pulsating unidirectional current. The filter removes the ripples from the pulsating current so that the output of the power supply is practically a continuous current. The voltage divider, as its name implies, divides the output voltage of the filter into the several values of voltage required by the plates and grids of the tubes.

Types of Power Supplies. Power-supply units in radio receivers operated from a-c power lines may be divided into four basic types: (1) the half-wave rectifier, (2) the full-wave rectifier, (3) the bridge-type rectifier, (4) the voltage-doubler-type rectifier. Each type of power supply has different operating characteristics. The choice of power supply will depend upon the operating voltages and currents desired. The operating characteristics of a power supply will also be affected by the type of filter circuit used and the values of the circuit elements. The output requirements of a power supply determine the type of rectifier and filter circuit to be used. In analyzing the requirements of the power supply, the characteristics to be taken into consideration are (1) the required output voltage, (2) the required output current, (3) the allowable peak voltage, (4) the ripple voltage, (5) the voltage regulation.

9-2. The Power Transformer. *Requirements of the Power Transformer.* The main purpose of the power transformer is to increase the line voltage so that the power-supply unit will be able to furnish the high operating voltages required by the plates and screen grids of the vacuum tubes in the equipment being supplied.   The power transformer may also be required to furnish the low voltages for the filaments or heaters of the tubes.   In this case, one or more low-voltage secondary windings are provided on the power transformer.

Where a comparatively high voltage is required by the plate circuits, the heaters are usually operated from a separate transformer.   A transformer used to supply only the plate power requires a primary winding and only a single secondary winding.   A transformer of this type, used

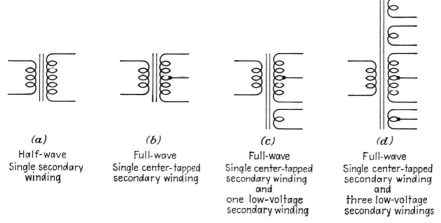

(a)	*(b)*	*(c)*	*(d)*
Half-wave	Full-wave	Full-wave	Full-wave
Single secondary winding	Single center-tapped secondary winding	Single center-tapped secondary winding and one low-voltage secondary winding	Single center-tapped secondary winding and three low-voltage secondary windings

FIG. 9-1. Power transformer circuits.

in a half-wave rectifier circuit, is shown in Fig. 9-1*a*.   In a full-wave rectifier circuit the secondary is usually center-tapped (Fig. 9-1*b*).

When the output of the power supply is comparatively low, the plate and heater transformers are combined into one unit, Figs. 9-1*c* and 9-1*d*. In this type of power transformer, the line voltage will also have to be decreased to the value, or values, required by the heaters of the tubes to be supplied.   As it is sometimes necessary to provide separate heater voltage sources for the rectifier tube, the r-f tubes, and the a-f tubes, the number of heater windings on the power transformer is determined by the requirements of the particular radio receiver.   All the heater windings of a power transformer are secondary windings and, together with the secondary for the plate supply, operate from a common primary winding (Fig. 9-1).

Ratings of Power Transformers.   Power transformers are usually rated in secondary volts and milliamperes output (d-c) at full load.   The secondary volts is measured across the full high-voltage winding, or from plate-to-plate of the rectifier tube.   This should be the voltage at full load and hence should be measured with the rated current being drawn from the rectifier.   Thus a transformer rated 350-0-350 secondary volts and 120 ma output current should indicate 700 volts across the two ends of the high-voltage secondary winding and 350 volts from either end to the center tap when 120 ma is being drawn from the rectifier.

The rms value of the high-voltage output of the power transformer should be slightly higher than the sum of the required output voltage and the voltage drops at the filter and the rectifier tube.   Transformers used with full-wave center-tapped rectifiers should produce this amount of voltage from the center tap to each side of the secondary winding.   The output current rating should be approximately 10 per cent greater than the sum of the currents taken by the various tubes and the bleeder resistor.

In a full-wave center-tapped rectifier circuit, the load is alternately transferred electronically from one-half of the secondary winding to the other, and thus only one-half of the secondary winding is used at a time. Each half of the secondary should therefore be capable of delivering the required voltage and current.

Voltage Regulation.   The voltage delivered by the secondary winding of a power transformer will decrease as the current taken by the load is increased.   Power transformers used in radio receivers are designed to operate continuously under full-load condition with excellent regulation. The variation in output voltage from no-load to full-load for this type of transformer is small and usually can be ignored.

9-3. Rectifiers of Alternating-current Power.   *Purpose of the Rectifier.* The purpose of the rectifier is to change alternating current to unidirectional current.   The action of a rectifier that permits current to flow through it more easily in one direction than the other makes possible its use for changing alternating current to direct current.

The Diode Rectifier.   The principle of operation of a rectifier may be either mechanical, thermal, chemical, or electronic.   Of these, the electronic principle, as applied to a diode, is the one most generally used in the rectifier of d-c power supplies associated with electronic equipment.

Current will flow in the plate-cathode circuit of a tube only when the plate is positive with respect to the cathode.   If an alternating current is applied between the plate and cathode of a diode as in Fig. 9-2, current will flow only during that portion of the cycle when the plate is positive. During the half-cycle in which the plate is negative, no current will flow. The diode rectifier is thus alternately a conductor and an insulator.

The rectifier tube may be either of the high vacuum type or the gaseous type.   The high vacuum tube is generally used for high-voltage and low-current outputs.   The gaseous-type rectifier tube is generally used for high-current outputs.

Tube Voltage Drop.   During the portion of the cycle in which the tube is conducting, voltage drops will be produced at the tube and the secondary winding of the transformer.   Since these voltage drops reduce the voltage available across the output load they should be kept low.

The internal voltage drop of a high-vacuum-type rectifier tube will vary in almost direct proportion to its load current (see Fig. 9-3).   A varying load current will cause the voltage drop across the tube to vary, thus also causing a variation in the voltage across the output load.   The voltage regulation of high-vacuum-type rectifier tubes is therefore very

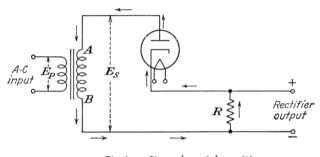

→ *Electron flow when* **A** *is positive*

Fig. 9-2. A single-diode rectifier circuit.

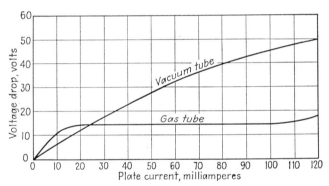

Fig. 9-3. Internal voltage drop between cathode and plate for comparable gas and vacuum tubes.

poor.   The internal voltage drop of directly heated (filament-type) high-vacuum rectifier tubes is high, in some cases as much as 50 to 60 volts. The internal voltage drop for a specific rectifier tube can be obtained from a tube manual.

As the internal voltage drop in a tube is affected by its space charge, the voltage drop can be decreased by reducing the effect of the space charge.   In some high-vacuum-type rectifier tubes, such as the 35W4, the space charge is reduced by decreasing the spacing between the cathode and plate.   The internal voltage drop of these tubes at rated operating values may be in the order of 20 volts.

The space charge in a tube can also be reduced by ionization. This method is used in the mercury-vapor rectifier tube (type 83) and also in the ionic-heated cathode rectifier tube (type 0Z4). The internal voltage drop of mercury-vapor tubes is very low, approximately 15 volts. The voltage regulation of mercury-vapor tubes is very good, since the internal voltage drop is practically independent of the load current (see Fig. 9-3).

The cathode of the ionic-heated rectifier tube is heated by the bombardment of the cathode by the ions from within the tube. As no external current is required to heat the cathode, it is also called a *cold-cathode* rectifier tube. As energy is taken from the ionization discharge to heat the cathode, the internal voltage drop of this tube is slightly higher than the hot-cathode mercury-vapor rectifier tube. The internal voltage drop of the 0Z4 at rated operating conditions is 24 volts.

Ratings of Rectifier Tubes. Rectifier tubes are generally rated according to (1) the alternating voltage per plate, (2) the peak inverse voltage, (3) the peak plate current, (4) the load current. These ratings are not fixed values but vary with the rectifier tube and filter circuits with which the tube is associated.

The *alternating voltage per plate* is the highest rms value of voltage that can safely be applied between the plate and cathode of the tube. In a half-wave rectifier circuit the open circuit voltage across the secondary winding of the power transformer should not exceed this value. In a full-wave rectifier circuit the open circuit voltage between either end of the secondary winding of the power transformer and its center tap should not exceed this value.

The *peak inverse voltage* rating is the maximum value of voltage that a rectifier tube can safely withstand between its plate and cathode when the tube is not conducting. During the portion of the a-c input cycle when the plate is negative with respect to the cathode, no voltage drops will exist in the rectifier circuit, and hence the full secondary voltage will be impressed between the plate and cathode. For normal operating conditions this voltage is equal to the peak value of the transformer secondary voltage. For transient conditions this peak value of voltage may be greatly exceeded.

The *peak plate current* represents the maximum amount of electron emission that the cathode can supply. It is the maximum instantaneous value of current that can safely flow through the rectifier tube.

The *load current* is the maximum safe value of direct current that the tube can deliver. Since current flows through a plate circuit of a rectifier tube during only half of the input cycle, the average value of its d-c output will be less than one-half of its peak plate current.

Metallic Rectifiers. A metal disk or plate that is held in contact, under pressure, with another substance is also used as a rectifier of alternating current. This type of unit is called a *metallic* or *contact rectifier*. The resistance to electron flow from the metal to the substance with which it is in contact is very low. The resistance to electron flow in the reverse direction is very high. Because of this characteristic the metallic rectifier can be used to rectify alternating current. Among the types of metallic rectifiers are the selenium, copper oxide, and copper sulfide. The selenium rectifier, which is used frequently in radio receivers, consists of an aluminum plate or disk that has one of its surfaces coated with selenium. Electrical contact with the selenium-coated surface is made directly with the uncoated surface of the adjacent plate or disk.

Semiconductor Rectifiers. The development of semiconductors introduced a new group of rectifiers including selenium, silicon, and germanium diodes. The selenium diode rectifier is a junction diode consisting of thin sections of P-type and N-type selenium joined together (see also Art. 13-4). These semiconductors have very low resistance to the flow of current in one direction (low forward resistance) and very high resistance to the flow of current in the opposite direction (high reverse resistance). A selenium diode for radio-receiver application may have a forward resistance in the order of 1 ohm and a reverse resistance as high as 100 megohms. Other advantages of semiconductor rectifiers are (1) high efficiency, (2) low forward voltage drop, (3) high peak-inverse-voltage ratings, (4) wide range of current ratings, (5) low heat radiation, (6) small size, (7) long life.

9-4. Fundamental Rectifier Circuits. *Half-wave Rectifier Circuit.* In a half-wave rectifier circuit a single-diode vacuum or gas tube is used and may be connected as shown in Fig. 9-4a. When the tube is conducting, electrons flow from the cathode to the plate, through the secondary of the power transformer, through the output circuit, and back to the cathode, as indicated by the arrows. When the plate is positive with respect to the cathode, the tube acts as a conductor, and current flows in the output circuit. When the plate is negative with respect to the cathode, the tube acts as an insulator, and no current flows in the output circuit. The relation between the input voltage and the output current for the single-diode rectifier is shown in Fig. 9-4. Because output current flows during only one-half of the input cycle, the single diode tube is also referred to as a *half-wave rectifier.*

The inverse peak voltage is equal to the maximum voltage of the transformer secondary. For a sine-wave input this peak value is equal to 1.41 times the rms value of the secondary output voltage, or 1.41 times E_S.

Full-wave Rectifier Circuit Using a Center-tapped Transformer. The operating characteristics of a rectifier will be improved if current can be made to flow in the output circuit for the entire period of the input cycle. This may be accomplished by using two single-diode tubes or a duplex diode (two single diodes in one envelope) and a center-tapped power transformer (Fig. 9-5). Because the two plates of the rectifier tube are connected to opposite ends of the secondary winding, their polarity with respect to the cathode will always be opposite to each other. Thus during one-half of the input cycle one plate will be conducting, and the other will not be conducting, and during the second half-cycle the second plate will be conducting, and the first will not be conducting. Under this condition, current will flow in the output circuit during both halves of the input

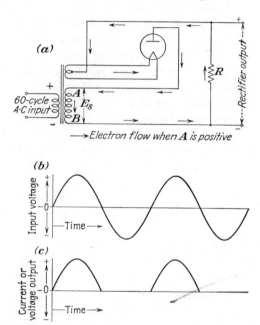

FIG. 9-4. A half-wave diode rectifier. (a) The circuit. (b) Waveform of the input voltage. (c) Waveform of the output current or voltage.

cycle. Figure 9-5 shows that both halves of the input cycle have been rectified and hence the name *full-wave rectifier*. Since each section supplies energy for one-half of the cycle, the plates of a full-wave rectifier will carry only one-half the amount of plate current required by a half-wave rectifier for equal amounts of load current.

The action of this type of full-wave rectifier circuit may be explained in the following manner. During one-half of the input cycle, plate 1 will be positive and plate 2 negative. During this period electrons will flow from cathode to plate 1, through one-half of the secondary winding (terminal 1 to the center tap), to the output circuit, and back to the cathode. The waveform of the output current for this half-cycle is indicated by the sections labeled 1 on Fig. 9-5c. During the other half of the input cycle, plate 2 becomes positive and plate 1 negative. Electrons will now flow from the cathode to plate 2, through the other half of the secondary winding (terminal 2 to the center tap), to the output circuit, and back to the cathode.   The waveform of the output current for this half-cycle is indicated by the sections labeled 2 on Fig. 9-5c.

The full-wave center-tapped transformer rectifier circuit is used mostly in transformer-type power supplies.   While one section of the rectifier is conducting, the peak inverse voltage across the other section will be 1.41 times the rms value of the full secondary voltage, or 1.41 E_S.

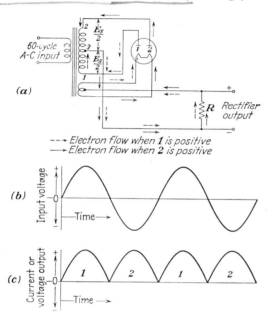

FIG. 9-5. A full-wave diode rectifier using a center-tapped transformer.   (a) The circuit. (b) Waveform of the input voltage.   (c) Waveform of the output current or voltage.

Output Voltage.   The output voltage of a power supply unit whose rectifier output is applied directly to a resistance load will be a pulsating unidirectional voltage as represented by Figs. 9-4c and 9-5c. (The output voltage as recorded with a d-c voltmeter would be the average value of the pulsations.) For sine-wave input voltages and resistance loads, the output voltage for full-wave rectification will be

$$E_o = \frac{0.637}{0.707} E_{\text{a-c}} = 0.9 E_{\text{a-c}} \tag{9-1}$$

For sine-wave input voltages and resistance loads, the output voltage for half-wave rectification will be

$(.637 \times 1.41 \times E_{\text{a-c}})$

$$E_o = \frac{0.637}{2 \times 0.707} E_{\text{a-c}} = 0.45 E_{\text{a-c}} \tag{9-2}$$

where $E_{\text{a-c}}$ is the rms value of alternating voltage per plate.

These two equations are based on the assumption that there is no voltage drop at the transformer secondaries or the rectifier tubes.   With the same value of alternating volts per plate, the output voltage with half-wave rectification is one-half the value of that with full-wave rectification, because the half-wave rectifier supplies current and voltage for only one-half the amount of time that a full-wave rectifier supplies current and voltage to the load.

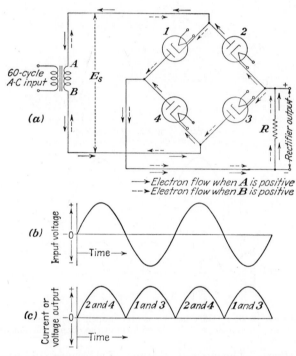

Fig. 9-6. A full-wave rectifier using four diodes.   (*a*) Simple bridge-type rectifier circuit. (*b*) Waveform of the input voltage.   (*c*) Waveform of the output voltage or current.

When filter circuits are used in conjunction with rectifiers, it will be found that a full-wave rectifier will require a transformer with twice the value of secondary voltage that is required with a half-wave rectifier in order to obtain the same output voltage.

9-5. Bridge Rectifier Circuits.   *Full-wave Bridge Rectifier Circuits.* The bridge-type full-wave rectifier circuit (Fig. 9-6) is another method of obtaining a continuous flow of rectified current in the output circuit of a rectifier.   During the half of the input cycle that terminal *A* is positive, electrons will flow from terminal *B*, through tube 4, the output circuit, tube 2, and back to terminal *A*.   The waveform of the output current for this half-cycle is indicated by the sections labeled 2 and 4 on Fig. 9-6*c*. During the other half of the input cycle terminal *B* is positive, and electrons will flow from terminal A, through tube 1, the output circuit, tube 3, and back to terminal B (Fig. 9-6a). The waveform of the output current for this half-cycle is indicated by the sections labeled 1 and 3 on Fig. 9-6c. (The transformer used with a bridge-type rectifier does not require a center-tapped secondary winding.) As this secondary supplies plate voltage for two tubes in series, its voltage is made twice the plate voltage required for each tube. This circuit utilizes the complete secondary winding for the entire period of each cycle. The output voltage for the bridge-type rectifier circuit will thus be twice that obtainable from a full-wave center-tapped circuit using a similar transformer. The inverse peak voltage is 1.41 E_S.

Another form of bridge-type rectifier (Fig. 9-7) operates directly from the power source without a transformer and uses two duplex diodes of the

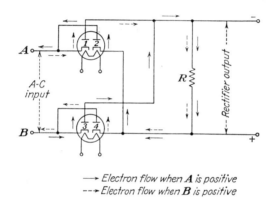

→ *Electron flow when **A** is positive*
--→ *Electron flow when **B** is positive*

FIG. 9-7. A full-wave duplex-diode bridge-type rectifier circuit.

heater-type cathode. Contact rectifiers may be used in bridge circuits instead of vacuum tubes, as shown in Fig. 9-8. The basic circuits of Figs. 9-6, 9-7, and 9-8 are all the same. The operation and the output waveform of the duplex-diode bridge rectifier circuit and the contact rectifier circuit will therefore be the same as for the circuit using four single diodes.

Important characteristics of the bridge-type rectifier circuit are its low plate voltage, low power output, compactness, and economy. Because of these features, it is generally used for supplying power to radio test equipment such as test oscillators and vacuum-tube voltmeters.

9-6. Other Rectifier Circuits. *Full-wave Voltage Doubler.* The voltage multiplier type of rectifier circuit makes it possible to obtain a d-c output whose voltage is equal to some multiple of the alternating voltage applied to the plates of the rectifier tube. The most common type of voltage multiplier circuit is the voltage doubler. This circuit is used to obtain an output voltage equal to approximately twice the alternating input voltage.

Figure 9-9 shows the circuit connections for a full-wave voltage-doubler circuit which operates in the following manner.   During a portion of the half-cycle that terminal A is positive, electrons will flow from terminal B to C_1, from C_1 through tube 1 and back to terminal A; C_1 will become charged with the polarities indicated.   During a portion of the next half-cycle terminal B is positive, and electrons will flow from terminal A through tube 2, to C_2, and from C_2 to terminal B; C_2 will become charged with the polarities indicated.   The positive (+) terminal of C_2 is connected to the negative (−) terminal of C_1, and if neither capacitor is discharged through the output circuit, the voltage across the two capacitors

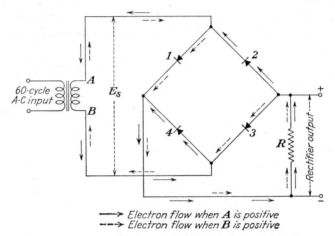

→ *Electron flow when A is positive*
--→ *Electron flow when B is positive*

FIG. 9-8. A full-wave contact-type rectifier bridge circuit.

in series will be double the peak plate voltage less any voltage drops in the tubes; hence the name *voltage doubler*.   However, the capacitors do discharge through the output circuit, C_1 discharging while C_2 is charging and vice versa.   If a comparatively large time constant is provided, these capacitors will lose only a small portion of their charge in the short interval required for the line voltage to reverse its polarity.   Therefore, during the half-cycle that one tube is conducting, the capacitor in its output circuit will not start charging until the instantaneous value of the line voltage (less the tube drop) exceeds the capacitor terminal voltage. The waveform of the output voltage illustrating the charge and discharge actions of the two capacitors is shown in Fig. 9-9c.

Because of the high values of capacitance used at C_1 and C_2, the output of a full-wave voltage-doubler circuit is nearly uniform.   For best results, these capacitors should be of equal value and not less than 20 μf.   The regulation of the voltage-doubler circuit is inherently poor and may be improved by using higher values of capacitance for C_1 and C_2.   As increasing the capacitance of C_1 and C_2 will increase the peak plate current of the rectifier tubes, the maximum values of these capacitors will be limited by the peak current rating of the tubes.   Each capacitor is charged by a separate diode and thus the voltage across either capacitor will never be greater than the peak value of the plate voltage.   Because the size of the capacitors required would make the cost prohibitive, this circuit is not practical for rectifying large amounts of current.

Half-wave Voltage Doubler.   The half-wave voltage doubler shown in Fig. 9-10 is another form of voltage-doubler circuit.   This rectifier

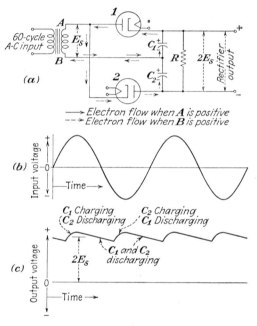

Fig. 9-9. A full-wave voltage-doubler circuit.   (*a*) The circuit.   (*b*) Waveform of the input voltage.   (*c*) Waveform of the output voltage.

operates in a somewhat different manner from the full-wave voltage doubler.   During the half-cycle that terminal A is positive, electrons will flow from terminal B to C_1, from C_1 through tube 1 to terminal A; C_1 will become charged to peak line voltage (less any drop in tube 1) with the polarities indicated.   During the next half-cycle terminal B will be positive, and electrons will flow from terminal A to C_2, from C_2 through tube 2 to C_1, from C_1 to terminal B; C_2 will become charged to a voltage equal to the peak line voltage plus the voltage across C_1 or double the peak line voltage (less any drop in the tubes).   However, capacitor C_2 would only attain this full charge when no load is being drawn from the power-supply unit.   When the unit is supplying power to a load, capacitor C_2 will charge and partially discharge during alternate halves of the input cycle.

Capacitor C_1, likewise, will charge and discharge during alternate halves of the input cycle, but during opposite times that C_2 is charging and discharging. Thus capacitor C_2 will be charging when tube 2 is conducting, the energy being obtained from capacitor C_1, which is then discharging to C_2 and the load. When tube 2 is not conducting, capacitor C_2 is discharging through the load. Tube 1 is then conducting, and capacitor C_1, which acts as a reservoir for C_2, is being recharged, so that the cycle

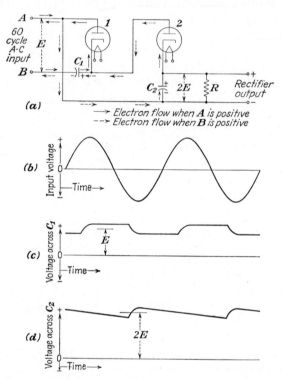

FIG. 9-10. A half-wave voltage-doubler circuit. (a) The circuit. (b) Waveform of the input voltage. (c) Waveform of the voltage across C_1. (d) Waveform of the voltage across C_2 or the output voltage.

repeats itself continually. The waveform of the voltages across capacitors C_1 and C_2, shown in Figs. 9-10(c) and (d) respectively, illustrates the charging and discharging periods of these two capacitors.

The regulation of the half-wave voltage-doubler circuit is not as good as that of the full-wave voltage-doubler circuit. Another disadvantage is that the voltage rating of capacitor C_2 must be twice as high as that of the capacitors in a full-wave circuit of the same rating.

An advantage of the circuit of Fig. 9-10a is that one side of the power line and the negative terminal of the output capacitor C_2 may be connected to a common terminal, usually the chassis. This makes it possible to use a series-connected heater circuit arranged so that the heaters of the high-gain r-f tubes will be at practically ground potential.   Thus, by keeping the voltage difference between the heater and cathode of the high-gain tubes at a low value, the possibility of cathode-to-heater leakage and its resulting hum is reduced.

Other Types of Voltage-multiplier Circuits.   The principles of the half-wave voltage-doubler circuit of Fig. 9-10a are utilized to obtain a rectified output voltage that may be any multiple of the input voltage.   In the voltage-multiplier circuit of Fig. 9-11, capacitors C_1 and C_2 operate in the same manner as capacitors C_1 and C_2 of Fig. 9-10.   Capacitor C_2 then adds its voltage (now double the line voltage) to the line voltage when tubes 1 and 3 are conducting.   This action then continues for the remaining capacitors and tubes.   After steady operating condition has been obtained, current will flow from the individual rectifiers for only that portion of the cycle necessary to replace the amount of charge lost by the capacitors in the previous half-cycles.   Thus, after the steady-state operating condition has been reached, C_1 will be charged to approximately the peak line voltage, C_2 to approximately twice the peak line voltage, C_3 to approximately three times the peak line voltage, etc.

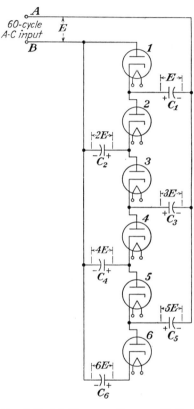

Fig. 9-11. A half-wave voltage-multiplier circuit.

The voltage-multiplier circuit is very useful for obtaining high output voltages.   However, the voltage regulation of the multiplier circuit is poor.   Furthermore, unless adequate provisions are made, the difference in voltage between the cathode and heater of the tube (or tubes) at the high-voltage terminal of a series heater string may greatly exceed the maximum voltage rating of the tube and thus cause it to break down.

Parallel Operation of Rectifiers.   One method of obtaining a higher output current from a rectifier is to connect two or more rectifier units in parallel.   The circuit connections for parallel operation of four vacuum tubes to produce full-wave rectification is shown in Fig. 9-12.   From this figure it can be seen that two plates are connected together, and two cathodes are connected together so that the circuit operates in a manner similar to the single-tube full-wave rectifier circuit of Fig. 9-5.

When gas tubes are operated in parallel, a slight difference in the operating characteristics of the tubes may cause one tube to ionize at a lower voltage than the other. The tube ionizing first will carry the entire load, since the voltage drop in this tube will decrease the input voltage to the second tube, thus preventing it from becoming ionized. In order to correct this condition, a resistor is connected in series with the plate of each tube. Then, if one tube ionizes before the other, the voltage at the second tube will still remain high enough to cause it to ionize, thus ensuring successful parallel operation of the two tubes.

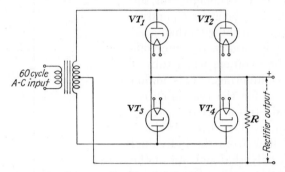

FIG. 9-12. Parallel operation of two full-wave rectifier circuits.

Selenium-rectifier Circuits. Selenium rectifiers may be obtained for a wide range of ratings and applications. Standard sizes range from 50 ma to 2,000 amp and from 6 to 30,000 volts. For average radio-receiver applications, current ratings may range from 75 to 300 ma and 156 volts (maximum value of a-c input). Compared with the vacuum-tube recti-fier, the selenium rectifier has many advantages, among which are (1) smaller size, (2) greater durability, (3) cooler operation, (4) less fragile, (5) longer life, (6) only two required connections, therefore easy installa-tion, (7) lower voltage drop at the rectifier, (8) better voltage regulation, (9) lower cost of rectifier unit since no tube socket or special mounting is required.

A selenium rectifier has two terminals, one positive and the other nega-tive. These terminals correspond to the plate and cathode of a vacuum tube. The positive side is usually indicated by a red dot or a " + " sign. The negative side is indicated by a yellow dot or a " − " sign, or it may be left blank. Rectifier circuits employing selenium-rectifier units as a half-wave rectifier, a full-wave rectifier, and a voltage-doubler circuit are shown in Fig. 9-13.

9-7. Filters. *Purpose of the Filter.* Although the output of a rectifier circuit is unidirectional, this output is not steady but is pulsating.

Because of the variations in magnitude, called *ripples*, of the current of a rectifier circuit, the current cannot be used in this form for radio applications. [A filter must be used in conjunction with a rectifier to smooth out the ripples in order that the output of the power-supply unit will become practically a steady direct current.]

Ripple Voltage. The unidirectional output voltage of a power-supply unit may be considered as a steady voltage having a pulsating voltage

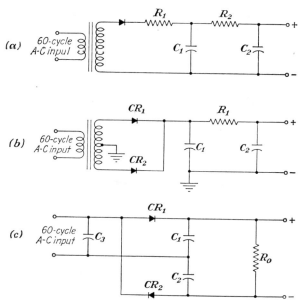

FIG. 9-13. Several applications of contact rectifiers (selenium or silicon) to rectifier circuits. (*a*) Half-wave rectifier. (*b*) Full-wave rectifier. (*c*) Voltage doubler.

superimposed upon it. The pulsating component of the output voltage is referred to as the *ripple voltage*. The frequency of the ripple voltage will depend upon the frequency of the input voltage and the type of rectifier. Since the ripple voltage does not vary in the same manner as a perfect sine-wave voltage, it may be considered as consisting of a fundamental and a series of harmonics. In general, the relative effect of the harmonics is negligible as compared to the fundamental and the harmonics can usually be ignored. The fundamental frequency of the ripple voltage is equal to the input frequency for half-wave rectifiers and twice the input frequency for full-wave rectifiers.

The effectiveness of a filter is measured by the ratio of the effective (rms) value of the fundamental component of the ripple voltage to the output voltage. This ratio is called the *ripple factor*.

$$k_r = \frac{E_r}{E_{d\text{-}c}} \qquad\qquad (9\text{-}3)$$

where k_r = ripple factor

$\quad\quad E_r$ = rms value of fundamental component of ripple voltage

$\quad\quad E_{\text{d-c}}$ = average value of output voltage

The ripple voltage is often expressed in terms of its percentage of the output voltage, as

$$\text{Per cent } E_r = \frac{E_r}{E_{\text{d-c}}} \times 100 \quad\quad\quad (9\text{-}4)$$

The type of service for which a power supply is to be used determines its allowable value of ripple voltage. For the plate-supply voltages of the average radio receiver, a ripple voltage of 0.25 per cent or less is required in order to reduce the hum to a negligible amount. The ripple voltage for the microphone circuit in a radio transmitter should be less than 0.003 per cent. In cathode-ray oscilloscopes, a ripple voltage as high as 1 per cent is sometimes permitted.

Example 9-1. The output voltage of a power-supply unit is 300 volts and the rms value of the ripple voltage is 0.6 volt. What is the per cent of ripple voltage?

$\quad\quad$ Given: $\quad E_{\text{d-c}}$ = 300 volts $\quad\quad\quad$ Find: $\quad$ Per cent E_r

$\quad\quad\quad\quad\quad\quad E_r$ = 0.6 volt

Solution:

$$\text{Per cent } E_r = \frac{E_r}{E_{\text{d-c}}} \times 100 = \frac{0.6}{300} \times 100 = 0.2$$

Operation of the Filter Circuit. Filter circuits associated with rectifier units use the energy-storing properties of capacitors and inductors to smooth out the ripple in the rectified output. The capacitor smooths out the voltage variations and also increases the value of the output voltage. The inductor smooths out the variations in current. The capacitor will store electrons during a portion of each cycle that the voltage increases, indicated as 1 to 2, 3 to 4, and 5 to 6 on Fig. 9-14*a*. During the portion of the cycle that the voltage decreases (2 to 3, 4 to 5, and 6 to 7 on Fig. 9-14*a*), the capacitor will slowly discharge some of its stored electrons. The voltage across the capacitor is thus made more uniform as indicated by Fig. 9-14*b*. Because electrolytic capacitors provide high voltage and high capacitance ratings in comparatively small-size units, they are generally used in power-supply filter circuits associated with radio receivers. The capacitance of electrolytic capacitors used for this purpose generally ranges from 10 to 100 μf, and the d-c voltage rating may be as high as 800 volts.

A characteristic of inductors is that they oppose any change in the amount of current that flows through them. Thus, when the output current of a rectifier flows through an inductor, the variations in current (both increases and decreases) will be opposed by the action of the inductor. The output will thus be more uniform as indicated in Fig. 9-14*c*. Inductors used in power-supply filter circuits are called *filter chokes* and are wound on a soft-iron core. In order to maintain a high value of inductance for a wide variation in current flow, some chokes use an iron core with a small air gap to prevent saturation. The inductance of the filter chokes used in power-supply units ranges from 10 to 30 henrys.

Resistors may be used with a capacitor to form an *RC* filter circuit. The time constant of *RC* filters must be large compared to the time of one cycle of the lowest frequency to be attenuated. Because of this, the d-c resistance of this type of filter is comparatively high and thus the voltage drop, voltage regulation, and heat dissipation are great. The development of electrolytic capacitors having a high capacitance makes it possible to use lower values of resistance with this type of circuit. *RC* filters

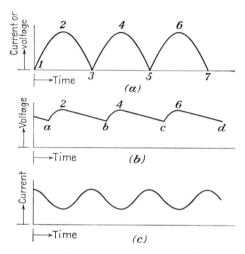

FIG. 9-14. Filter action of a capacitor and an inductor. (*a*) Current or voltage output of a full-wave rectifier. (*b*) Output from a capacitor. (*c*) Output from an inductor.

are used when the requirements of low cost and compactness outweigh the desirability of a high degree of filtering.

Types of Filter Circuits. Power-supply filter circuits are of the low-pass type, using one or more series inductors and one or more shunt capacitors. These filter circuits are usually referred to as being *choke input* or *capacitor input* filters depending upon whether an inductor or a capacitor is the first element in the filter network. A number of different types of filter circuits are shown in Fig. 9-15, the choke-input filters being represented by (*a*), (*b*), and (*c*), and the capacitor-input filters by (*d*), (*e*), (*f*), and (*g*). These filter circuits may be further classified as single-section filters represented by (*a*), (*b*), and (*d*); two-section filters as in (*c*), (*e*), and (*g*); and three-section filters as in (*f*). The number of sections of filtering that is required will depend upon the rectifier, the type of filter circuit, and the allowable ripple factor.

9-8. Capacitor-input Filter.   *Theory of Operation.*   A capacitor-input filter is a filter circuit in which the first element is a capacitor connected in parallel with the input from the rectifier.   During the time that the rectifier tube is conducting, energy will be stored in the capacitor, and when the rectifier tube is not conducting, part of the stored energy will be discharged through the filter network to the load.   The capacitor increases the average value of the output voltage.   The waveform of the capacitor voltage shown in Fig. 9-14b may be considered as consisting of two parts: (1) the portion during the charging period represented

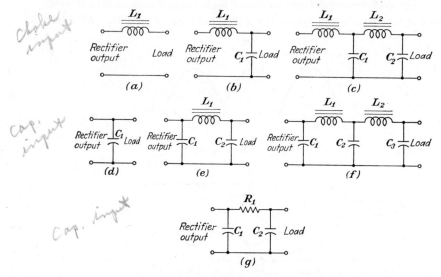

FIG. 9-15. Types of filter circuits used with rectifiers.

by *a* to 2, *b* to 4, *c* to 6, etc.; (2) the portion during the discharge period represented by 2 to *b*, 4 to *c*, 6 to *d*, etc.   If the capacitor were to be discharged directly through a resistor, the discharge portion of the curve would decrease exponentially.   If the capacitor is discharged through an inductor, which helps further to smooth out the output current, the discharge portion of the curve will decrease linearly, and the resultant voltage wave that is applied to the inductor will be a wave with a saw-tooth characteristic.   The ripple component of the voltage across the input capacitor is prevented from reaching the output circuit by the combined actions of any inductors and capacitors that follow the input capacitor in the filter circuit.

Ripple Voltage.   The per cent of ripple voltage developed across the input capacitor will vary inversely with the frequency of the rectified output, the effective load resistance, and the capacitance of the input capacitor.   Thus, increasing any of these factors in a filter circuit will decrease the per cent of ripple voltage.   The effect of a variation in any of these factors on the d-c output can be seen by observation of Fig. 9-14b.

Increasing the frequency decreases the time that the input capacitor is permitted to discharge. The capacitor will lose less of its charge, thus maintaining the voltage across it more nearly uniform. The extent to which the voltage across the input capacitor drops off is also affected by the time constant of the RC circuit, consisting of the input capacitor C_1 and the effective load resistance R_o. Increasing either of these two values will increase the time constant of the R_oC_1 circuit, thereby decreasing the rate of discharge. The voltage across the input capacitor will thus be maintained more nearly uniform.

Although it is difficult to obtain accurate calculations of the per cent of ripple voltage, the following equations will provide reasonable results for ripple voltages of 10 per cent or less; beyond this amount the accuracy decreases. The per cent of ripple voltage at the output of a single-section capacitor-input filter as shown in Fig. 9-15d is

$$\text{Per cent } E_{r \cdot 1} \cong \frac{10^8 \sqrt{2}}{2\pi f_r R_o C_1} \cong \frac{2{,}245 \times 10^4}{f_r R_o C_1} \tag{9-5}$$

where f_r = frequency of ripple voltage, cps
$\quad R_o$ = resistance of load, ohms
$\quad C_1$ = capacitance of input filter capacitor, μf

This equation can also be used to calculate the per cent of ripple voltage at the output of the first section (per cent of ripple voltage at C_1) for any multisection capacitor-input filter, examples of which are shown as (e), (f), and (g) of Fig. 9-15.

Example 9-2. A power-supply unit has a 60-cycle input to its rectifier, uses a single-section capacitor-input filter (Fig. 9-15d), and delivers 40 ma direct current at 320 volts to the load. Determine the per cent of ripple voltage for (a) half-wave rectification and a 10-μf filter capacitor, (b) full-wave rectification and a 10-μf filter capacitor, (c) half-wave rectification and a 20-μf filter capacitor, (d) full-wave rectification and a 20-μf filter capacitor.

Given: $E_{d\text{-}c}$ = 320 volts   Find: (a) Per cent $E_{r \cdot 1}$, half-wave
           $I_{d\text{-}c}$ = 40 ma   (b) Per cent $E_{r \cdot 1}$, full-wave
           C_1 = 10 μf (a), (b)   (c) Per cent $E_{r \cdot 1}$, half-wave
           C_1 = 20 μf (c), (d)   (d) Per cent $E_{r \cdot 1}$, full-wave
           f_r = 60 (a), (c)
           f_r = 120 (b), (d)

Solution:

$$R_o = \frac{E_{d\text{-}c}}{I_{d\text{-}c}} = \frac{320}{40 \times 10^{-3}} = 8{,}000 \text{ ohms}$$

(a)   Per cent $E_{r \cdot 1} \cong \dfrac{2{,}245 \times 10^4}{f_r R_o C_1} = \dfrac{2{,}245 \times 10^4}{60 \times 8{,}000 \times 10} = 4.67$

(b)   Per cent $E_{r \cdot 1} \cong \dfrac{2{,}245 \times 10^4}{f_r R_o C_1} = \dfrac{2{,}245 \times 10^4}{120 \times 8{,}000 \times 10} = 2.34$

(c)   Per cent $E_{r \cdot 1} \cong \dfrac{2{,}245 \times 10^4}{f_r R_o C_1} = \dfrac{2{,}245 \times 10^4}{60 \times 8{,}000 \times 20} = 2.34$

(d)   Per cent $E_{r \cdot 1} \cong \dfrac{2{,}245 \times 10^4}{f_r R_o C_1} = \dfrac{2{,}245 \times 10^4}{120 \times 8{,}000 \times 20} = 1.17$

The results of Example 9-2 show that the per cent of ripple voltage for a single-section filter is considerably higher than is usually acceptable for radio applications. The additional filtering necessary to reduce the ripple voltage to an acceptable value can be obtained by providing additional filter sections, Figs. 9-15e, 9-15f, and 9-15g. The per cent of ripple voltage at the output of a two-section capacitor-input filter similar to that of Fig. 9-15e can be found by use of Eq. (9-6). The results obtained with this equation are not extremely accurate but provide reasonable accuracy for low percentages of ripple voltage.

$$\text{Per cent } E_{r\cdot2} \cong \frac{\% \, E_{r\cdot1}}{[10^{-6}(2\pi f_r)^2 L_1 C_2] - 1} \tag{9-6}$$

where $E_{r\cdot2}$ = ripple voltage at capacitor C_2 (Fig. 9-15e)
$\%E_{r\cdot1}$ = obtained by use of Eq. (9-5), which is the ripple at capacitor C_1 (Fig. 9-15e)
f_r = frequency of the ripple voltage, cps
L_1 = inductance of L_1 (Fig. 9-15e), henrys
C_2 = capacitance of C_2 (Fig. 9-15e), μf

Example 9-3. A power-supply unit using a full-wave rectifier and a filter circuit similar to Fig. 9-15e has a 60-cycle input and delivers 40 ma direct current at 320 volts to the load. Capacitors C_1 and C_2 are each 20 μf, and the inductance of L_1 is 15 henrys. What is the per cent of ripple voltage at (a) C_1? (b) C_2?

Given: f_r = 120 cycles   Find: (a) Per cent E_r at C_1
                L_1 = 15 henrys   (b) Per cent E_r at C_2
          C_1, C_2 = 20 μf

Solution:

(a)   Per cent $E_{r\cdot1} \cong 1.17$ [Same as Example 9-2(d).]

(b)   Per cent $E_{r\cdot2} \cong \dfrac{\% \, E_{r\cdot1}}{[10^{-6}(2\pi f_r)^2 L_1 C_2] - 1}$

$$= \frac{1.17}{[(2 \times 3.14 \times 120)^2 15 \times 20 \times 10^{-6}] - 1} \cong 0.007$$

An approximate value of the per cent of ripple voltage $E_{r\cdot3}$ at the output of a three-section capacitor-input filter similar to that of Fig. 9-15f can be found by modifying Eq. (9-6).

$$\text{Per cent } E_{r\cdot3} \cong \frac{\% \, E_{r\cdot2}}{[10^{-6}(2\pi f_r)^2 L_2 C_3] - 1} \tag{9-6a}$$

In a similar manner, Eq. (9-6) can be modified to determine the per cent of ripple voltage at the output of any number of succeeding sections of a multisection capacitor-input filter of this type.

For a two-section RC filter similar to that of Fig. 9-15g, an approximate value of the per cent of ripple voltage at the output can be found as in Example 9-3 by use of Eqs. (9-5) and (9-7). Equation (9-7) can also be modified in the same manner as for Eq. (9-6) to determine the per cent of ripple voltage at the output of succeeding sections of multisection RC filters.

$$\text{Per cent } E_{r \cdot 2} \cong \frac{\% \ E_{r \cdot 1} \times 10^6}{2\pi f_r C_2 R_1} \tag{9-7}$$

where C_2 = capacitance, μf

R_1 = resistance, ohms

The results obtained with this equation are not extremely accurate but provide reasonable accuracy when the product of $C_2 R_1$ is 10,000 or more.

Output Voltage. The output voltage of a capacitor-input filter circuit will vary with the capacitance and with changes in the effective load resistance. At no-load, or with a comparatively light load, the effective load resistance is comparatively high. The time constant of $R_o C_1$ will also be high and the output voltage will approach the peak value (also called *crest value*) of the alternating voltage being rectified. As the load current increases, the effective load resistance will decrease, thus also decreasing the time constant of the circuit. The resulting increase in the rate of discharge of the input capacitor will lower the average value of the voltage across this capacitor. When the output current is high, the effective load resistance is low, thus causing a considerable decrease in the output voltage. For average applications of capacitor-input filters, the output voltage at full-load will be approximately equal to the effective value of the alternating voltage being rectified.

The variation in output voltage with changes in current is called the *voltage regulation* of the circuit and expressed as a percentage is

$$\text{Voltage regulation} = \frac{E_{NL} - E_L}{E_L} \times 100 \tag{9-8}$$

where E_{NL} = no-load voltage

E_L = full-load voltage

Since variations in the load current of a capacitor-input filter circuit result in a wide range of output voltage, the voltage regulation of this type of circuit is very poor.

Example 9-4. It is desired to have a power-supply unit that will provide 300 volts at the output terminals when supplying its rated full-load current. What is the per cent of regulation if (a) the unit employs a full-wave rectifier and a capacitor-input filter, and the voltage with no-load rises to 426 volts? (b) The unit employs a half-wave rectifier and a capacitor-input filter, and the voltage with no-load rises to 480 volts?

Given: $E_L = 300$ volts   Find: (a) Per cent of regulation
        $E_{NL} = 426$ volts   (b) Per cent of regulation
        $E_{NL} = 480$ volts

Solution:

(a) Per cent of regulation $= \dfrac{E_{NL} - E_L}{E_L} \times 100 = \dfrac{426 - 300}{300} \times 100 = 42$

(b) Per cent of regulation $= \dfrac{E_{NL} - E_L}{E_L} \times 100 = \dfrac{480 - 300}{300} \times 100 = 60$

Characteristics of the Capacitor-input Filter. Compared to the choke-input filter, the capacitor-input filter will deliver a higher voltage at light loads, has a slightly better filtering characteristic, but has poorer voltage regulation. The current in the rectifier associated with a capacitor-input filter circuit does not flow uniformly but flows in pulses; hence the ratio of peak rectifier current to average current will be higher than in the choke-input system. The d-c voltage rating of the input capacitor should never be less than the peak input voltage, since at light loads the output voltage approaches this value. In order to provide a safety factor, it is usually desirable to use a capacitor whose working voltage rating is somewhat higher than this value. Capacitor-input filter circuits are generally used in power-supply units that are required to deliver only small amounts of power, such as radio receivers, public-address systems, and testing apparatus.

9-9. Choke-input Filter. *Theory of Operation.* In a choke-input filter, the first element is an inductor connected in series with the input from the rectifier. The filtering action of the series input inductor of the choke-input filter circuit (Fig. 9-15) is explained in the following manner. During the portion of the rectified output cycle in which the current increases, the strength of the magnetic field about the inductor will increase, and energy will be stored in the field. The inductor also opposes the increase in current. During the portion of the rectified output cycle in which the current decreases, the magnetic field about the inductor collapses, returning part of its stored energy to the circuit, and also opposes the decrease in current. These actions of an inductor smooth out the ripple in the rectified output, as shown in Fig. 9-14c. This figure shows that the current through the inductor is made up of an a-c component and a d-c component. The capacitor following the input inductor will tend to short-circuit the a-c component, thus producing a practically smooth voltage at its output terminals.

If the ripple voltage from a single-section filter circuit exceeds the allowable percentage, it can be further reduced by using additional filter sections as shown in Fig. 9-15c. The first inductor is called the *input choke* and the second inductor is called the *smoothing choke.*

Ripple Voltage. The per cent of ripple voltage developed across the first capacitor C_1 (Fig. 9-15b and 9-15c) will vary inversely with the capacitance of C_1 and the inductance of the input choke. For most practical purposes an approximate value of the per cent of ripple voltage for a single-section filter can be obtained by the equation

$$\text{Per cent } E_{r \cdot 1} \cong \frac{144 \times 10^4}{f_r^2 L_1 C_1} \qquad (9\text{-}9)$$

where f_r = ripple frequency, cps
L_1 = inductance of input choke, henrys
C_1 = capacitance of first capacitor, μf

This equation may be further simplified for determining the approximate per cent of ripple voltage of a single-section filter whose ripple frequency is 120 cycles, as is the case when the input to the filter is obtained from a full-wave rectifier operated from a 60-cycle power source. Equation (9-9) may then be simplified to

$$\text{Per cent } E_{r \cdot 1} \cong \frac{100}{L_1 C_1} \qquad (9\text{-}9a)$$

Example 9-5. Determine the approximate per cent of ripple voltage at the output of a single-section choke-input filter circuit using a 15-henry choke and a 10-μf capacitor. The input to the filter is obtained from a full-wave rectifier operated from a 60-cycle power source.

Given: $L_1 = 15$ henrys Find: Per cent $E_{r \cdot 1}$
$\quad\quad\quad C_1 = 10 \ \mu$f

Solution:

$$\text{Per cent } E_{r \cdot 1} \cong \frac{100}{L_1 C_1} = \frac{100}{15 \times 10} = 0.66$$

An approximate value of the per cent of ripple voltage at the output of a two-section choke-input filter can be obtained by the equation

$$\text{Per cent } E_{r \cdot 2} \cong \frac{1,350 \times 10^8}{f_r{}^4 L_1 L_2 (C_1 + C_2)^2} \qquad (9\text{-}10)$$

where L_2 = inductance of smoothing choke, henrys
$\quad\quad C_2$ = capacitance of second capacitor, μf
This equation may be further simplified for determining the approximate per cent of ripple voltage of a two-section filter whose ripple frequency is 120 cycles, as is the case when the input to the filter is obtained from a full-wave rectifier operated from a 60-cycle power source. Equation (9-10) may then be simplified to

$$\text{Per cent } E_{r'2} \cong \frac{650}{L_1 L_2 (C_1 + C_2)^2} \qquad (9\text{-}10a)$$

Example 9-6. Determine the approximate per cent of ripple voltage at the output of a two-section choke-input filter circuit using two 15-henry chokes and two 10-μf capacitors. The input to the filter circuit is obtained from a full-wave rectifier operated from a 60-cycle power source.

Given: $L_1 = L_2 = 15$ henrys Find: Per cent $E_{r \cdot 2}$
$\quad\quad\quad C_1 = C_2 = 10 \ \mu$f

Solution:

$$\text{Per cent } E_{r \cdot 2} \cong \frac{650}{L_1 L_2 (C_1 + C_2)^2} = \frac{650}{15 \times 15(10 + 10)^2} \cong 0.0072$$

Example 9-7. Determine the approximate per cent of ripple voltage at the output of the single-section choke-input filter circuit used in Example 9-5, if the input is obtained from a full-wave rectifier circuit having a 25-cycle input.

Given: $L_1 = 15$ henrys Find: Per cent $E_{r \cdot 1}$
$\quad\quad\quad C_1 = 10 \ \mu$f
$\quad\quad\quad f_{in} = 25$ cycles

Solution:

$$f_r = 2f_{in} = 2 \times 25 = 50 \text{ cycles}$$

$$\text{Per cent } E_{r.1} \cong \frac{144 \times 10^4}{f_r^2 L_1 C_1} = \frac{144 \times 10^4}{50 \times 50 \times 15 \times 10} \cong 3.8$$

The Input Choke. The input choke of a filter circuit serves two functions: (1) to maintain a continuous flow of current from the rectifier, (2) to prevent the output voltage from increasing above the average value of the alternating voltage applied to the rectifier. The output voltage and the peak plate current of the rectifier are both dependent upon the inductance of the input choke and the d-c resistance of the load. The minimum value of inductance required to maintain the output voltage at the average value of the alternating voltage being rectified is called the *critical inductance.* For a rectified output having a 120-cycle ripple frequency, an approximate value of the critical inductance may be obtained by use of the following equation.

$$L_c = \frac{R_o}{1,000} \tag{9-11}$$

where L_c = critical value of inductance, henrys

R_o = output load resistance, ohms

If the inductance of the input choke is less than its critical value, its impedance to the varying component of the rectified output will be so small that the filter circuit will tend to operate as a capacitor-input filter. Increasing the inductance of the input choke to more than its critical value will further decrease the ratio of peak to average plate current, thus maintaining a more nearly uniform flow of current through the inductor. Increasing the value of the inductance beyond twice the critical value does not correspondingly improve the operating characteristics of the filter. The *optimum* value of inductance is thus equal to twice the critical value of inductance.

Example 9-8. Determine the optimum value of inductance for the input choke of a filter circuit having a d-c load resistance of 4,000 ohms. The frequency of the rectified input is 120 cycles.

Given: R_o = 4,000 ohms   Find: L_o

     f_r = 120 cycles

Solution:

$$L_c = \frac{R_o}{1,000} = \frac{4,000}{1,000} = 4 \text{ henrys}$$

$$L_o = 2L_c = 2 \times 4 = 8 \text{ henrys}$$

Swinging Choke. Equation (9-11) shows that the value of the inductance required for the input choke will vary directly with the effective load resistance and inversely with the load current. Thus, if the load current varies over a wide range, some means must be provided for preventing the ratio of peak to average plate current from becoming excessive.   The inductance of the choke coil will vary inversely with the value of the direct current flowing through it.   A choke coil having an inductance of 10 henrys with 100 ma flowing through it may have an inductance of 15 henrys when the current flow is reduced to practically zero. A choke designed to have a critical value of inductance at full load and an optimum value of inductance at no-load is called a *swinging choke.*

Output Voltage.   The average value of the output voltage at full-load of a choke-input filter is in the order of 65 to 75 per cent of the rms volts per plate at the rectifier.   It should be observed that the output voltage with a choke-input filter is lower than that with a capacitor-input filter supplied with the same value of rms volts per plate.   The decrease in output voltage is due to the effect of the inductance being introduced into the circuit and to the voltage drop at the choke due to its d-c resistance.   In order to reduce the drop in voltage at the filter chokes, their d-c resistance should be kept as low as possible.

Characteristics of the Choke-input Filter.   Although the choke-input filter circuit delivers a lower output voltage than the capacitor-input filter circuit, its voltage regulation is much better.   Another advantage of the choke-input filter is that the input choke prevents high instantaneous peak currents, thus protecting the rectifier tube from being damaged.

The input inductor and the first capacitor of a choke-input filter form a series resonant circuit.   If the values of these two circuit elements make the circuit resonant to the ripple frequency, high values of ripple voltage will be produced. (It is therefore important that the values of inductance and capacitance used do not form a series resonant circuit that is tuned to the ripple frequency.)

Choke-input filter circuits are generally used where the output current is large or where the voltage regulation must be fairly good.   Because choke-input filters operate best when the current flow is sustained over the complete cycle, they are usually used only with full-wave rectifiers.

9-10. The Voltage Divider.   *Bleeder Resistor.*   Removing the external load from a power-supply unit causes a high voltage to be developed across the filter capacitors.   If the voltage becomes too high, it will cause a breakdown of the insulation in these capacitors.   The voltage across the terminals of the filter capacitors at zero load can be reduced to a safe value by connecting a fixed resistor, called a *bleeder resistor*, across the output terminals of the filter circuit.   The amount of bleeder current varies with the requirements of the individual power-supply unit and generally ranges from 10 to 25 per cent of the total current drawn from the rectifier.

As a capacitor will retain its charge for a considerable length of time, the bleeder resistor also provides a path through which the filter capacitors will discharge when the power is turned off.   This eliminates the danger of a high-voltage shock when occasion arises to repair the power-supply unit.

Since the bleeder resistor draws a fixed amount of current continuously from the power-supply unit, it reduces the value of the output voltage at no-load. The bleeder resistor thus also serves to improve the voltage regulation of the power-supply unit by reducing the difference in voltage between that obtained at no-load and that obtained at full-load.

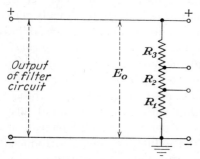

The Voltage Divider. When the elements of the various tubes in a circuit require different amounts of voltage, the power-supply unit generally delivers more than one value of voltage. The various voltages may be obtained by connecting a tapped resistor, called

Fig. 9-16. Connection of a voltage divider used in a power-supply unit.

a *voltage divider*, across the output terminals of the filter circuit (see Fig. 9-16). The voltage and current requirements of the power-supply unit will determine the number of taps and the value of resistance between each pair of taps. The voltage divider also serves as a bleeder resistor.

In addition to supplying the high operating voltages required by the plates and screen grids of electronic tubes, the power-supply unit may also

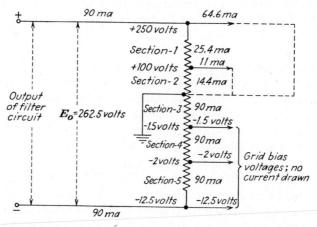

Fig. 9-17. A voltage-divider circuit used for providing plate and grid-bias voltages.

be required to supply the necessary grid-bias voltages. This is accomplished by connecting the voltage divider as shown in Fig. 9-17.

Calculation of a Voltage-divider Circuit. The resistance values and the power rating of the voltage divider may be calculated by use of Ohm's law. The following procedure should be followed.

1. Determine the voltage required at each tap and the current to be drawn from it.

2. Determine the amount of bleeder current. This is the difference between the current required by the tubes and the current necessary to operate the power supply at approximately 90 per cent of its rated value.

3. Determine the current in each section of the voltage divider.

4. Calculate the resistance of one section at a time.

5. Determine the power rating of the voltage divider.

Example 9-9. Determine the resistance values of a voltage divider for a small superheterodyne receiver that employs a 6BE6 oscillator-mixer tube, a 6BA6 i-f amplifier tube, a 6AV6 detector amplifier tube, and a 6AQ5-A power-output tube. The operating voltages and currents are to be obtained from a tube manual for plate voltages of 250 volts. A power transformer rated at 100 ma is to be operated at 90 per cent of its rated value.

Given: Tubes: 6BE6, 6BA6, 6AV6, 6AQ5-A   Find: R of each section
$I_T = 100$ ma
$I_S = 0.9 I_T$

Solution:

See Table 9-1.

TABLE 9-1. VALUES FROM A TUBE MANUAL

Tube	6BE6	6BA6	6AV6	6AQ5-A
E_b, volts	250	250	250	250
$E_{c\cdot2}$, volts	100	100	. . .	250
$E_{c\cdot1}$, volts	. . . .	. . .	−2	−12.5
$E_{g\cdot3}$, volts	−1.5			
I_b, ma	2.9	11	1.2	45
$I_{c\cdot2}$, ma	6.8	4.2	. . .	4.5

Using the values of voltage and current listed in Table 9-1, the voltage and current at each tap of the voltage divider will be as shown in Fig. 9-17. The total voltage required from the power-supply unit will be the sum of the highest plate voltage and the highest grid-bias voltage.

$E_o = 250 + 12.5 = 262.5$ volts
$I_{250} = 2.9 + 11 + 1.2 + 45 + 4.5 = 64.6$ ma
$I_{100} = 6.8 + 4.2 = 11$ ma
$I_o = 64.6 + 11 = 75.6$ ma
$I_{bleeder} = 0.9 I_T - I_o = (0.9 \times 100) - 75.6 = 14.4$ ma

Solution:

$$R_{sec\cdot1} = \frac{e_1}{i_1} = \frac{250 - 100}{25.4 \times 10^{-3}} = 5,905 \text{ ohms}$$

$$R_{sec\cdot2} = \frac{e_2}{i_2} = \frac{100 - 0}{14.4 \times 10^{-3}} = 6,944 \text{ ohms}$$

$$R_{sec\cdot3} = \frac{e_3}{i_3} = \frac{1.5 - 0}{90 \times 10^{-3}} = 16.7 \text{ ohms}$$

$$R_{sec\cdot4} = \frac{e_4}{i_4} = \frac{2 - 1.5}{90 \times 10^{-3}} = 5.5 \text{ ohms}$$

$$R_{sec\cdot5} = \frac{e_5}{i_5} = \frac{12.5 - 2}{90 \times 10^{-3}} = 117 \text{ ohms}$$

Power Rating of the Voltage Divider. If a single resistor of uniform wire size is to be used, the voltage divider will have a uniform power rating. For this type of voltage divider, the highest current flowing through any part of the voltage divider must be used in determining the power rating. However, a separate resistor can be used for each section of the voltage divider, in which case the power rating of each section or resistor is determined by its own current and resistance ratings.

As the voltage divider is usually mounted under the chassis of the radio receiver and therefore does not have much ventilation, it is recommended that its power rating be approximately double that of the load it is to carry.

Example 9-10.   Determine the power rating of the voltage divider used in Example 9-9 for (*a*) a single resistor having four taps, (*b*) five separate resistors.

Given: Resistance and current values deter-   Find: (*a*) P_R, single resistor
          mined in Example 9-9   (*b*) P_R, separate resistors

Solution:

(*a*) $R_T = R_{\text{sec·1}} + R_{\text{sec·2}} + R_{\text{sec·3}} + R_{\text{sec·4}} + R_{\text{sec·5}}$
        $= 5,905 + 6,944 + 16.7 + 5.5 + 117 = 12,988$ ohms
        $P_R = 2I_{\text{max}}^2 R_T = 2(90 \times 10^{-3})^2 \times 12,988 = 210$ watts

(*b*) $P_{R\text{·sec·1}} = 2i_1^2 R_1 = 2(25.4 \times 10^{-3})^2 \times 5,905 = 7.62$ watts
        $P_{R\text{·sec·2}} = 2i_2^2 R_2 = 2(14.4 \times 10^{-3})^2 \times 6,944 = 2.87$ watts
        $P_{R\text{·sec·3}} = 2i_3^2 R_3 = 2(90 \times 10^{-3})^2 \times 16.7 = 0.27$ watt
        $P_{R\text{·sec·4}} = 2i_4^2 R_4 = 2(90 \times 10^{-3})^2 \times 5.5 = 0.088$ watt
        $P_{R\text{·sec·5}} = 2i_5^2 R_5 = 2(90 \times 10^{-3})^2 \times 117 = 1.88$ watts

Thus, if a single resistor having four taps is to be used, the approximate power rating of the voltage divider of Example 9-9 would be 210 watts. Using five separate resistors, each designed to carry the amount of current actually flowing through it, the total power rating of the five resistors would be 12.7 watts. As the five-section voltage divider has a much lower power rating, it is less expensive and also more practical.

If one or more sections of a voltage divider breaks down, it is not necessary to replace the entire divider. Resistors of the correct values may be substituted in place of the defective sections after they have been disconnected from the circuit.

9-11. The Power-supply Unit.   *Components of the Power-supply Unit.* In the preceding discussion, the four components of the power-supply unit, namely, the transformer, rectifier, filter, and voltage divider, have been considered as separate units. These four units are closely associated with each other and may be connected to one another to form the complete power-supply unit as shown in Fig. 9-18. The rating of the transformer and the type of rectifier tube to be used will be determined by the values of output current and voltage required. The per cent of allowable ripple voltage will determine the type of filter circuit, the number of sections required, and the values of its components.   The position of the taps on the voltage divider will be determined by the value of the plate and screen-grid voltages of the tubes to be supplied.

Output Voltage. Because of the resistance of the tube and the filter chokes, there will be a drop in voltage at each of these circuit elements.

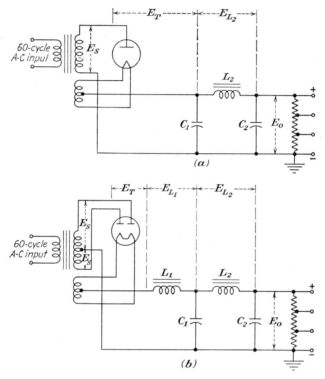

FIG. 9-18. Power-supply circuits.   (a) Circuit with a half-wave rectifier and a capacitor-input filter.   (b) Circuit with a full-wave rectifier and a choke-input filter.

These voltage drops must be added to the output voltage in order to determine the rms value of the voltage required at the secondary of the power transformer.   The output voltage at normal load for the full-wave-rectifier choke-input-filter power-supply unit of Fig. 9-18b can be expressed as

$$E_o = 0.9E_S - E_T - E_{L.1} - E_{L.2} \qquad (9\text{-}12)$$

For the half-wave-rectifier capacitor-input-filter power-supply unit of Fig. 9-18a

$$E_o = E_S - E_T - E_{L.2} \qquad (9\text{-}13)$$

where E_o = output voltage of power-supply unit

$\quad E_S$ = rms volts per plate at rectifier

$\quad E_T$ = voltage drop at rectifier tube

$\quad E_{L \cdot 1}$ = voltage drop at input choke

$\quad E_{L \cdot 2}$ = voltage drop at smoothing choke

Example 9-11.   A choke-input filter circuit similar to Fig. 9-18*b* is to be used in a power supply that is to provide 100 ma at 285 volts.   The resistance of the input choke is 80 ohms, the resistance of the smoothing choke is 400 ohms, and the internal voltage drop at the rectifier tube is 30 volts when the tube is supplying full-load.   What voltage is required at the plates of the tube when operating as a full-wave rectifier?

Given: E_o = 285 volts   Find: E_S
$\quad\quad\quad E_T$ = 30 volts
$\quad\quad\quad R_{L \cdot 1}$ = 80 ohms
$\quad\quad\quad R_{L \cdot 2}$ = 400 ohms
$\quad\quad\quad I_S$ = 100 ma

Solution:

$$E_{L \cdot 1} = I_S R_{L \cdot 1} = 100 \times 10^{-3} \times 80 = 8 \text{ volts}$$
$$E_{L \cdot 2} = I_S R_{L \cdot 2} = 100 \times 10^{-3} \times 400 = 40 \text{ volts}$$
$$E_S = \frac{E_o + E_T + E_{L \cdot 1} + E_{L \cdot 2}}{0.9} = \frac{285 + 30 + 8 + 40}{0.9} = 403 \text{ volts}$$

Voltage Regulation.   Example 9-4 shows that the regulation of a power-supply unit that does not use a bleeder resistor is poor.   When a bleeder resistor (which may also be serving as a voltage divider) is used, disconnecting the external load does not cause the no-load current to drop to zero ma.   Instead, the no-load current now has an appreciable value that may be determined by Ohm's law.   Under this condition, the no-load voltage will be considerably lower than if the current had dropped to zero.   Equation (9-8) shows that a reduction in the no-load voltage will result in a smaller variation between the no-load and full-load voltages, thus improving the regulation of the power-supply unit. The curves of Fig. 9-19 show the operating characteristics of several power-supply systems.

Example 9-12.   A power-supply unit represented by the curves of Fig. 9-19 is to supply 300 volts at the input to the filter at 100 ma, and its no-load current is 20 ma. What is the per cent of regulation for (*a*) the capacitor-input filter circuit?   (*b*) the choke-input filter circuit?

Given: E_L = 300 volts   Find: (*a*) Per cent of regulation
$\quad\quad\quad I_L$ = 100 ma   (*b*) Per cent of regulation
$\quad\quad\quad I_{NL}$ = 20 ma

Solution:

(*a*)   From Fig. 9-19*b*, E_{NL} = 380 volts

$\quad$ Per cent regulation = $\dfrac{E_{NL} - E_L}{E_L} \times 100 = \dfrac{380 - 300}{300} \times 100 = 26.6$

(*b*)   From Fig. 9-19*a*, E_{NL} = 330 volts

$\quad$ Per cent regulation = $\dfrac{E_{NL} - E_L}{E_L} \times 100 = \dfrac{330 - 300}{300} \times 100 = 10$

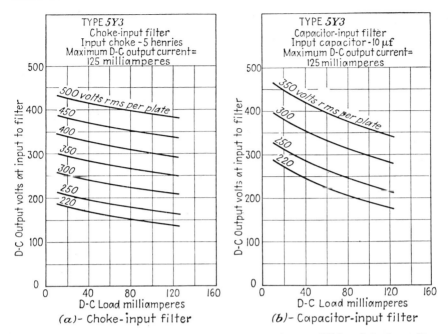

FIG. 9-19. Operating characteristics of a power-supply unit. (*a*) With a choke-input filter. (*b*) With a capacitor-input filter.

The curves of Fig. 9-19 show the voltage at the input to the filter. The voltage at the output terminals of the filter will be less than this amount because of the voltage drop at the filter chokes.

Example 9-13. What is the per cent of regulation on the basis of the voltage at the terminals of the voltage divider for the conditions of Example 9-12 with a capacitor-input filter whose smoothing choke has a resistance of 335 ohms?

Given: Voltages from Example 9-12*a*   Find: Per cent of regulation
I_L = 100 ma
I_{NL} = 20 ma
R_L = 335 ohms

Solution:

$$E_{o \cdot L} = E - IR_L = 300 - 100 \times 10^{-3} \times 335 = 266.5 \text{ volts}$$
$$E_{o \cdot NL} = E - IR_L = 380 - 20 \times 10^{-3} \times 335 = 373.3 \text{ volts}$$

$$\text{Per cent regulation} = \frac{E_{NL} - E_L}{E_L} \times 100 = \frac{373.3 - 266.5}{266.5} \times 100 = 40$$

9-12. Transformerless Power-supply Units. *A-C–D-C Power Supplies.* The voltage from an a-c power source may be rectified by applying this voltage directly to the plate-cathode circuit as in Fig. 9-20. The output voltage of this power-supply unit will be approximately equal to the alternating voltage input to the rectifier. If a higher output voltage is required, a voltage-multiplier rectifier circuit is used (see Fig.

9-21 and Art. 9-6).   Since a transformer is no longer required to step up the line voltage, this type of power-supply unit can be operated directly from either an a-c or a d-c power line.

Another advantage of the line rectifier type of power supply is its compactness, which has made it possible to manufacture the small a-c–d-c radio receivers.   In addition to radio-receiver applications, the transformerless power supply is also used in test equipment and electronic apparatus where a limited amount of space is available.

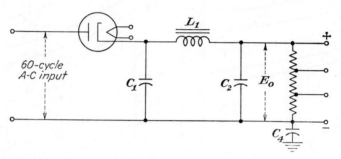

Fig. 9-20. A transformerless power-supply unit employing a half-wave rectifier operated directly from the power line.

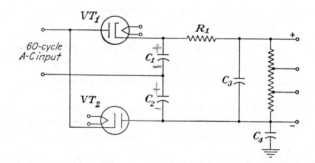

Fig. 9-21. A transformerless power-supply unit employing a voltage-doubler rectifier.

A resistor is generally substituted for the smoothing choke (Fig. 9-21), thus making the power-supply unit still more compact by eliminating the choke coil.   The cost of this type of power-supply unit is comparatively low, as both the power transformer and the filter choke are eliminated.

In some types of electronic equipment, the magnetic fields set up about the power transformer and filter chokes are undesirable.   The line rectifier type of power supply has an advantage in this kind of application since both units can be eliminated.

Rectifier Tubes.   In order to eliminate the filament transformer, the heater of a line-voltage rectifier is connected in series with the heaters of the other tubes in the receiver (or other electronic device) and the power line. If the sum of the heater voltages of all the tubes is less than the line voltage, a suitable resistor is connected in series with the heaters and the power line to provide the required voltage drop (see Art. 2-3). Since the energy consumed by the dropping resistor represents a loss, high-voltage heater tubes are generally used with transformerless power supplies in order to minimize or totally eliminate this loss. All the high-voltage heater tubes designed expressly for use with transformerless power supplies are of small size. Rectifier tubes for this type of service may be half-wave, full-wave, or of the voltage-multiplier type. The heater voltages range from 12 to 117 volts.

Characteristics of the Transformerless Power-supply Unit. From a study of the principle of operation of the line-voltage rectifier, it can be seen that a capacitor-input filter circuit should be used with this type of rectifier. The voltage regulation is comparatively poor and is dependent upon the capacitance values of the filter capacitors. The higher the capacitance, the better will be the voltage regulation. Filter capacitors used with transformerless power supplies may range from 20 to 80 μf.

Because of the comparatively low output current of the rectifier tube, the output current of the power-supply unit is also low.

In the transformerless power-supply units of Figs. 9-10, 9-11, and 9-20 one side of the power line is connected directly to the negative terminal of the output of the power-supply unit, generally referred to as B—. Essentially the same condition exists in the case of the circuit of Fig. 9-21, since the B— terminal is connected to one side of the power line through capacitor C_2. The high value of capacitance normally used at C_2 results in a low impedance, and hence the line may be considered as being conductively connected to the B— terminal. If this negative terminal is connected directly to the chassis, as is commonly done with transformer-type power-supply units, it introduces two possible sources of danger, namely, electric shock and fire hazard. This danger is present when one side of the power line is grounded at the service entrance, as is done in most communities. Since either side of the power line may be connected to the chassis, depending upon the direction in which the attachment plug is inserted in the power outlet, it will be possible to connect either the grounded or the ungrounded side of the power line to the chassis. When the ungrounded side is connected to the chassis, touching any metal part of the receiver may result in an electric shock. Also, an accidental short circuit of the B power leads or a breakdown of a capacitor will cause a virtual short circuit of the power lines with its attendant fire hazard. Grounding the radio receiver to a radiator, water pipe, etc., whether intentionally or not, will also be a possible source of danger.

The danger of electric shock, and the danger caused by grounding the chassis to water pipes, etc., can be eliminated by using a bus bar or conductor that is insulated from the chassis as the B— line. Under this condition, the chassis is usually connected to the B— line through a capacitor of about 0.1 μf or less. This capacitor, C_4 in Figs 9-20 and 9-21, provides a low-impedance path for r-f grounding yet presents a high enough impedance at 60 cycles to practically eliminate the afore-mentioned dangers. The danger of fire hazard is generally reduced by enclosing in metal containers those parts connected in the circuit in such a manner that a breakdown may present fire hazards.

9-13. Vibrator Power-supply Units. *Types of Vibrator.* When a radio receiver or other type of electronic equipment is to be operated from a low-voltage d-c power supply such as a storage battery, the con-ventional power transformer and rectifier tube circuit cannot be used for obtaining the high operating voltages required for the plate and screen-grid circuits. The d-c power from the battery can be changed to alter-nating current by use of an electromagnetic device that reverses the direction of current flow in the power transformer during each vibration of its vibrating armature. This device, called a *vibrator*, is used in some power supplies of battery-operated equipment to convert the low direct voltage of the battery to an alternating voltage that can be increased to any desired value by means of a transformer.

Basically, there are two types of vibrators: (1) the *synchronous* vibrator, (2) the *nonsynchronous* vibrator. The nonsynchronous vibra-tor interrupts the d-c circuit at a frequency that is unrelated to the other circuit constants. Since the high alternating voltage output of the power transformer is then normally rectified by means of a rectifier tube, this type of vibrator may also be called a *tube-type* vibrator. A synchronous vibrator, in addition to changing the low direct voltage to an alternating voltage, simultaneously rectifies the high alternating voltage output at the secondary of the power transformer. Rectification is accomplished by employing an additional set of contacts, thereby eliminating the need of a rectifier tube.

Nonsynchronous-vibrator Power-supply Unit. A circuit diagram of a nonsynchronous-vibrator power-supply unit is shown in Fig. 9-22. At the instant that the switch S_1 is closed, current will flow from the battery through coil L_2, section 1-2 of the primary winding of the transformer, coil L_1, and back to the battery. Coil L_2 is wound on a soft-iron core, which becomes magnetized when current flows through the coil. The vibrating reed R is so constructed that when a current flows through L_2, the reed is attracted toward the magnet. As the reed approaches the magnet it makes contact with point A, thus short-circuiting the coil L_2. Since the iron core is then no longer magnetized, the vibrating reed is released and tends to return to its normal position. However, because of the force present when it is released, the reed moves past its normal position and makes contact with point B and then returns to its normal position. At the instant that contact is being made at point B, current will flow through section 1-3 of the primary winding of the transformer. Current will flow in this circuit for only a very short period of time because contact is made at point B only instantaneously during the forward swing of the vibrating reed after it is released from point A. When the reed returns to its normal position, current will again flow through coil L_2 and the cycle of operations will be repeated. These operations occur very rapidly, and the complete cycle of operations is repeated many times per second.

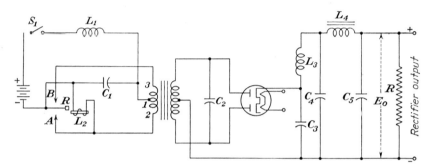

FIG. 9-22. Circuit diagram of a power-supply unit using a nonsynchronous vibrator.

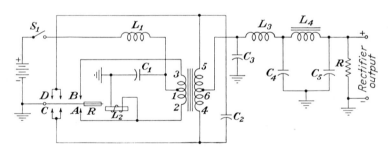

FIG. 9-23. Circuit diagram of a power-supply unit using a synchronous or self-rectifying vibrator.

It can be seen that for each cycle of operations the current is caused to flow in opposite directions through each half of the primary winding of the transformer. Since this flow of current is essentially the same as an alternating current, the voltage at the secondary terminals can now be increased to any desired value by increasing the ratio of secondary to primary turns of the transformer. The output of the secondary winding is then rectified by a tube and filtered in the usual manner.

Synchronous-vibrator Power-supply Unit. The output from the secondary of the power transformer can be rectified by adding another set of points to the vibrator, thereby eliminating the need of a rectifier tube. These points are connected to opposite ends of the secondary winding of the transformer as shown in Fig. 9-23. In this circuit, when the vibrating reed makes contact with point A, it also makes contact with point C, thus grounding terminal 2 of the primary winding and terminal 4 of the secondary winding of the transformer. In a similar manner, when the reed makes contact with point B it also makes contact with point D, thus grounding terminals 3 and 5 of the transformer. Terminal 6, which is one terminal of the output circuit, is thus always positive, and terminals 4 and 5 are alternately connected to ground during opposite halves of the cycle. It can thus be seen that the output of the trans-

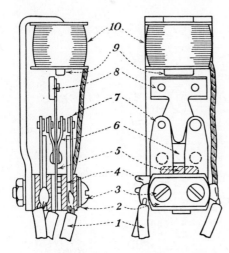

former is rectified since current flows in opposite directions in each half of the secondary winding during alternate halves of the cycle.

Filters. A surge of current will occur each time the primary winding is connected to or disconnected from the d-c power source. Connecting a capacitor across the primary winding will absorb this surge of current, thus preventing the contact points or rectifier tube from being damaged. The amount of capacitance required to absorb this surge will decrease with an increase in the applied voltage. It is thus more economical to connect a capacitor across the secondary winding of the power transformer (C_2 in Figs. 9-22 and 9-23) because the higher voltage at the secondary windings makes it possible to use a lower

1. Stranded leads and soft rubber tubing
2. Spring washer-plate
3. Stack clamping screws
4. Stops and solder lugs
5. Reed slot for starting
6. Reed contact arms
7. Outer contact arms
8. Reed and armature
9. Pole-piece integral with frame
10. Coil

Fig. 9-24. Construction of a vibrator unit employing four sets of contactors. (*Courtesy of P. R. Mallory and Co., Inc.*)

value of capacitance. This capacitor is called a *buffer capacitor*. As the capacitance reflected to the primary from the secondary increases as the square of the secondary to primary turns ratio, it produces substantially the same result as connecting a high value of capacitance in the primary circuit.

Each time the contacts are opened, sparking will take place at the contacts. This sparking produces r-f transients, thus causing interference if the power supply is used with a radio receiver. This interference is commonly referred to as *hash* and may be minimized by use of filter circuits and by shielding the entire vibrator unit. A hash-filter circuit is usually connected in the battery circuit and consists of an r-f choke coil and a 0.5-μf to 1-μf capacitor as represented by L_1 and C_1 in Figs.

9-22 and 9-23. Another hash-filter circuit is connected in the output circuit and consists of an r-f choke coil and a 0.01-μf to 0.1-μf capacitor as represented by L_3 and C_3 in Figs. 9-22 and 9-23.

Characteristics of the Vibrator Power-supply Unit. The vibrator power-supply unit represents an inexpensive and compact means of obtaining the high operating voltage for the plate and screen-grid circuits from a low-voltage battery. It can thus be very effectively used in all types of portable electronic equipment that is to be operated from low-

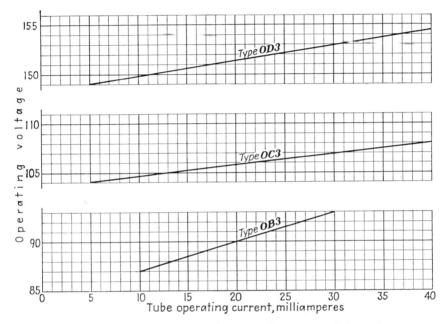

Fig. 9-25. Operating characteristics of typical voltage-regulating tubes.

voltage batteries. However, the vibrator has a limited life, and its associated circuits require a complex filtering and shielding system.

The type of vibrator to be used will depend upon the power-supply requirements. In addition to the choice of either the synchronous or nonsynchronous type of vibrator, the current output and the type of interrupter and rectifier circuits must also be considered. Figure 9-24 shows the construction of an eight-contact vibrator. This unit may be connected to operate simply as an interrupter or as a self-rectifier. In either circuit, the vibrator is capable of carrying the current required to produce 30 watts of output power.

9-14. Voltage Regulation. *Gaseous Regulator Tubes.* Because of various circuit conditions the output voltage will vary inversely with the output load. Variations in the voltage of the a-c input to the power supply will also cause the output voltage to vary. There are a number of power-supply applications where the voltage applied to the load must be maintained practically constant regardless of the voltage regulation of the power-supply unit. For low current applications, a cold-cathode gaseous diode having a practically constant internal voltage drop can be used as a voltage regulator. Typical voltage-regulator tubes are the 0B3/VR90, 0C3/VR105, and the 0D3/VR150. The voltage-regulating characteristics of these three tubes are illustrated by the curves of Fig. 9-25. These curves show that the variation in operating voltage from no-load to full-load for these tubes is approximately only 5 volts.

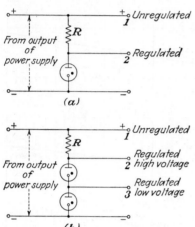

A voltage slightly higher than the operating voltage must be used to ionize the gas inside the tube in order to have it start operating. Once the tube is started, it will continue to operate at some value of voltage within its operating range.

Voltage-regulator Tube Circuits. The voltage-regulator tube is connected to the output of the power supply as shown in Fig. 9-26. In order to limit the current flowing through the regulator tube to a safe value, a resistor should always be connected in series with the tube and the power-supply unit. The value of this resistor will depend upon the output voltage of the power supply and the operating voltage of the regulator tube. When it is required to provide a regulated voltage that is higher than that obtained from one tube, two or more similar voltage-regulator tubes may be connected in series, as shown in Fig. 9-26b. The voltage at terminal 1 of either Fig. 9-26a or 9-26b will be approximately equal to the output voltage of the power-supply unit. This voltage is unregulated, and its variation will be dependent upon load conditions. The voltage obtained at terminal 2 or 3 is regulated, and its value will be dependent upon the voltage rating of the tube or tubes used.

FIG. 9-26. Voltage-regulating circuits using gaseous regulator tubes. (*a*) Circuit providing one value of regulated voltage. (*b*) Circuit providing two values of regulated voltage.

QUESTIONS

1. (*a*) What is the purpose of the power supply? (*b*) What are the general requirements of a power supply?

2. What is the purpose of (*a*) the A power supply? (*b*) The B power supply? (*c*) The C power supply?

3. (*a*) What are the advantages of battery power supplies? (*b*) What are the disadvantages of battery power supplies?

4. (*a*) What method is commonly used to obtain d-c power at high voltages from a low-voltage battery? (*b*) Where is this method generally used?

5. (*a*) Why is the power line the most desirable source of power for stationary electronic equipment? (*b*) Explain why the voltages required to operate electronic tubes cannot be taken directly from the power line.

6. (*a*) Name the four parts of a power-supply unit used to convert alternating current to direct current. (*b*) Explain the function of each of these parts.

7. (*a*) Name the four basic types of power supplies used in radio receivers that operate from a-c power lines. (*b*) What factors determine the type of power supply to be used? (*c*) What factors must be taken into consideration when determining the requirements of a power supply?

8. (*a*) What is the main purpose of the power transformer? (*b*) For what additional purpose is it sometimes used?

9. How are power transformers usually rated?

10. Explain how alternating current may be rectified by use of a diode.

11. (*a*) For what type of service are high-vacuum-type tubes used? (*b*) For what type of service are gaseous-type tubes used?

12. How are the following types of rectifier tubes classified in terms of their voltage regulation: (*a*) high-vacuum-type rectifier tubes? (*b*) mercury-vapor rectifier tubes? (*c*) ionic-heated cathode rectifier tubes?

13. (*a*) What four factors are generally used in rating rectifier tubes? (*b*) Are these ratings fixed values? (*c*) Explain your answer to part (*b*).

14. Define (*a*) alternating voltage per plate, (*b*) peak inverse voltage, (*c*) peak plate current, (*d*) load current.

15. (*a*) Describe the construction of the contact-type rectifier. (*b*) Explain the principle of operation of the contact-type rectifier.

16. What are the applications of selenium rectifiers?

17. (*a*) Explain the principle of operation of a half-wave rectifier circuit using a single diode. (*b*) What are the output current and inverse peak voltage characteristics of this type circuit?

18. (*a*) Explain the principle of operation of a full-wave rectifier circuit using two diodes and a power transformer having a center-tapped secondary. (*b*) What are the output current and inverse peak voltage characteristics of this type circuit?

19. Compare the output voltage of a half-wave and a full-wave rectifier for a sine-wave input. Assume that the output is applied directly to a resistive load and that the voltage drops at the transformer secondaries and the rectifier tube are negligible.

20. (*a*) Explain the principle of operation of a full-wave bridge rectifier circuit. (*b*) What are the output current and inverse peak voltage characteristics of this type of circuit?

21. (*a*) What are the important characteristics of bridge-type rectifier circuits? (*b*) Where is this type of circuit generally used?

22. Explain the principle of operation of a full-wave voltage-doubler circuit.

23. (*a*) How may the voltage regulation of the voltage-doubler circuit be improved? (*b*) What is the limiting factor that must be taken into consideration when using this method?

24. (*a*) Explain the principle of operation of a half-wave voltage-doubler circuit. (*b*) What are the advantages and disadvantages of this type of circuit?

25. Explain the principle of operation of a half-wave voltage multiplier circuit.

26. What provisions should be made in voltage-multiplier circuits to prevent the rectifier tubes from being damaged?

27. (*a*) What is the purpose of parallel operation of rectifier tubes? (*b*) What safety provisions should be made when operating gas tubes in parallel?

28. (*a*) What sizes (current and voltage ratings) of selenium rectifiers are available? (*b*) What sizes are commonly used in radio receiver circuits?

29. What are some of the advantages of the selenium-type contact rectifier over the vacuum-tube rectifier?

30. Define: (*a*) ripple voltage, (*b*) ripple factor, (*c*) per cent of ripple voltage.

31. What is the function of (*a*) the filter circuit? (*b*) the filter capacitor? (*c*) the filter choke?

32. Explain the filtering action of (*a*) the filter capacitor, (*b*) the filter choke.

33. (*a*) What are the operating requirements and characteristics of an *RC* filter circuit? (*b*) Where is this type of filter circuit used?

34. What is meant by (*a*) a low-pass filter? (*b*) a choke-input filter? (*c*) a capacitor-input filter? (*d*) a single-section filter? (*e*) a multisection filter?

35. Explain the principle of operation of a single-section capacitor-input filter circuit.

36. Explain how the per cent of ripple voltage is affected by changes in (*a*) the frequency of the rectified output, (*b*) the effective load resistance, (*c*) the capacitance of the input capacitor.

37. Why does a capacitor-input filter circuit have poor voltage regulation?

38. How do the operating characteristics of a capacitor-input filter circuit compare with those of a choke-input filter circuit?

39. For what type of service are capacitor-input filter circuits generally used?

40. What is the purpose of connecting a low value of resistance directly in the rectifier tube circuit of a power-supply unit employing a capacitor-input filter?

41. Explain the principle of operation of a single-section choke-input filter circuit.

42. Explain how the per cent of ripple voltage of a choke-input filter circuit is affected by (*a*) the input choke, (*b*) the first capacitor, (*c*) the smoothing choke, (*d*) the second capacitor.

43. Give two functions of the input choke.

44. What is meant by (*a*) the critical value of inductance? (*b*) the optimum value of inductance?

45. How does the value of inductance of the input choke affect the operation of the filter circuit?

46. (*a*) What is meant by a swinging choke? (*b*) What is its function?

47. What are the characteristics of a choke-input filter circuit?

48. For what type of service are choke-input filter circuits generally used?

49. What are the functions of the bleeder resistor?

50. What are the functions of the voltage divider?

51. What are the advantages of using a separate resistor for each section of a voltage divider rather than a single resistor having the necessary taps?

52. Explain how the use of a bleeder resistor improves the voltage regulation of a power-supply unit.

53. Describe four advantages of the transformerless type of power supply.

54. Why do the rectifier tubes used in transformerless power supplies have high-voltage heaters?

55. Why are capacitor-input filter circuits used with transformerless power-supply units?

56. What are the characteristics of transformerless power supplies?

57. (*a*) Explain how the negative terminal of a transformerless power-supply unit introduces two possible sources of danger. (*b*) Describe a few of the methods that can be used to eliminate these sources of danger.

58. (*a*) What is a vibrator type of power supply unit? (*b*) What is the purpose of the vibrator? (*c*) Where are vibrator power supplies used?

59. What is meant by (*a*) a synchronous vibrator? (*b*) a nonsynchronous vibrator?

60. Describe the principle of operation of a nonsynchronous-vibrator power-supply circuit.

61. Describe the principle of operation of a synchronous-vibrator power-supply circuit.

62. (*a*) What is the purpose of the buffer capacitor?   (*b*) How is the buffer capacitor connected in the circuit?

63. (*a*) What is meant by hash?   (*b*) How is hash eliminated?

64. What are the advantages and disadvantages of vibrator power-supply units?

65. For what types of service are gaseous voltage-regulator tubes used?

66. (*a*) How are voltage-regulator tubes connected in the power-supply circuit? (*b*) Why should a resistor always be connected in series with the regulator tube and the power supply?

67. Explain how it is possible to obtain voltage regulation in a power-supply circuit whose regulated output voltage is higher than the voltage rating of the standard regulating tubes.

PROBLEMS

1. A certain power-supply unit using a single diode has its output connected directly to a resistance load. The voltage at the secondary terminals of the power transformer is 300 volts.   (*a*) What is the output voltage of the power-supply unit if the voltage drops at the tube and the transformer secondary are neglected?   (*b*) Draw a sketch of the waveform of the output voltage.   (*c*) What is the output voltage if the drop at the tube is 30 volts and the drop at the transformer secondary is 15 volts? (*d*) What is the peak inverse voltage?

2. A power-supply unit using a duodiode as a full-wave rectifier has its output connected directly to a resistance load. The secondary of the power transformer is center-tapped and has a voltage of 300 volts from each side to the center tap.   (*a*) What is the output voltage of the power-supply unit if the voltage drops at the tube and the transformer secondary are neglected?   (*b*) Draw a sketch of the waveform of the output voltage.   (*c*) What is the output voltage if the drop from each plate to cathode is 25 volts and the drop from the center tap to the outside terminals of the transformer secondary is 15 volts?   (*d*) What is the peak inverse voltage?

3. A power-supply unit using four diodes in a full-wave bridge rectifier circuit has its output connected directly to a resistance load. The voltage across the secondary terminals of the power transformer is 450 volts.   (*a*) What is the output voltage of the power-supply unit if the voltage drops at the tubes and the transformer secondary are neglected?   (*b*) Draw a sketch of the waveform of the output voltage.   (*c*) What is the output voltage if the drop at each tube is 30 volts and the drop from the center tap to the outside terminals of the transformer secondary is 15 volts?   (*d*) What is the peak inverse voltage?

4. A full-wave voltage-doubler power-supply circuit similar to Fig. 9-9*a* is being supplied with a secondary voltage of 300 volts. What is the output voltage if the capacitors used are capable of maintaining an average value of voltage equal to (*a*) the rms value of the voltage at the secondary terminals?   (*b*) 90 per cent of the rms voltage at the secondary terminals?   (*c*) 80 per cent of the rms voltage at the secondary terminals?

5. A half-wave voltage-doubler circuit similar to Fig. 9-10 is being operated from a 120-volt a-c power line.   (*a*) What is the peak voltage possible at capacitor C_1? (*b*) What is the peak voltage possible at capacitor C_2?   (*c*) What is the peak inverse voltage at tube 1?   (*d*) What is the peak inverse voltage at tube 2?

6. A half-wave voltage-multiplier circuit similar to Fig. 9-11 is being operated from a 120-volt a-c power line.   (*a*) What is the peak voltage possible at each capacitor?   (*b*) What is the peak inverse voltage at each tube?

7. (*a*) Draw a diagram of a half-wave voltage-multiplier circuit using three tubes and three capacitors to provide an output voltage of approximately 350 volts when operated from a 117-volt a-c power line. (*b*) What is the minimum standard voltage rating recommended for each capacitor? (*c*) What standard voltage rating capacitors would you recommend if it is desired to allow for a reasonable safety factor? (*d*) What is the peak inverse voltage at each tube?

8. (*a*) Draw a diagram of a half-wave voltage-multiplier circuit using four tubes and four capacitors to provide an output voltage of approximately 450 volts when operated from a 115-volt a-c power line. (*b*) What is the minimum standard voltage rating recommended for each capacitor? (*c*) What standard voltage rating capacitors would you recommend if it is desired to allow for a reasonable safety factor? (*d*) What is the peak inverse voltage at each tube?

9. If the rectifier circuit of Fig. 9-12 employs four 35W4 tubes, what is the maximum load current that the circuit can supply? Obtain current rating of the tubes (without panel light) from a tube manual.

10. A power-supply circuit similar to Fig. 9-12 employs four duodiodes with their plates connected in parallel so that each tube acts as a diode. The maximum d-c output per plate of a tube connected in this manner is 225 ma. What is the maximum load current that the circuit can supply?

11. A certain half-wave power-supply unit is being operated with 260 volts and 60 cycles at the input. It is being operated without a filter circuit; hence the output is fed directly to its resistance load. The rms value of the fundamental component of the ripple voltage is 130 volts. (*a*) What is the average value of the output voltage? (*b*) What is the per cent of ripple voltage? (*c*) What is the frequency of the ripple voltage?

12. A certain full-wave power-supply unit is being operated with 300 volts at 60 cycles applied between the center tap and each side of the secondary of its power transformer. It is being operated without a filter circuit; hence the output of the rectifier is fed directly to its resistance load. The rms value of the fundamental component of the ripple voltage is 127 volts. (*a*) What is the average value of the output voltage? (*b*) What is the per cent of ripple voltage? (*c*) What is the frequency of the ripple voltage?

13. The plate circuit of a certain a-f amplifier is to be operated at 180 volts and the per cent of ripple is not to exceed 0.05 per cent for operation without noticeable hum. What is the highest rms value of ripple voltage permitted?

14. The amplifier of Prob. 13, if operated from a power supply with a ripple voltage of 1 per cent, will produce a slight amount of hum in the output, which may not be objectionable for some applications. What is the highest rms value of ripple voltage permitted under this condition?

15. A simple power-supply unit using a single half-wave rectifier tube is to be operated from a 60-cycle power line. What is the approximate per cent of ripple voltage at the first capacitor of its capacitor-input filter circuit if the effective load resistance is 5,000 ohms and the value of capacitance is 20 μf?

16. A simple power-supply unit using a full-wave rectifier tube is to be operated from a 60-cycle power line. (*a*) What is the approximate per cent of ripple voltage at the first capacitor of its capacitor-input filter circuit if the effective load resistance is 3,000 ohms and the value of capacitance is 30 μf? (*b*) What is the magnitude of the ripple voltage if the output is 250 volts?

17. It is desired to determine whether it would be practical to use a 35W4 rectifier tube and a single-unit filter, consisting of a single capacitor, to provide a 60-ma load at 90 volts with a maximum ripple of 0.25 per cent. The peak plate current of the 35W4 is 600 ma. (*a*) What is the effective load resistance? (*b*) What value of capacitance is required? (*c*) What is the capacitive reactance of the required capacitor to 60-cycle current? (*d*) If the peak value of the 60-cycle input voltage to the rectifier is 165 volts, what is the magnitude of the rectifier current during the first cycle after the power is applied? (*e*) What effect will the current during this first cycle have upon the tube? (*f*) Is this circuit design practical?

18. A power-supply unit being operated from a 60-cycle power line is supplying a load with 100 ma at 250 volts. A full-wave rectifier is used and the filter circuit is similar to that of Fig. 9-15*e*. The values of C_1 and C_2 are 10 μf each and L_1 is 20 henrys. What is the per cent of ripple voltage at (*a*) C_1? (*b*) C_2?

19. A power-supply unit being operated from a 60-cycle power line is supplying a load with 120 ma at 300 volts. A half-wave rectifier is used and the filter circuit is similar to that of Fig. 9-15*e*. The values of C_1 and C_2 are 20 μf each and L_1 is 20 henrys. What is the per cent of ripple voltage at (*a*) C_1? (*b*) C_2?

20. A power-supply unit being operated from a 60-cycle power line is supplying a load with 120 ma at 360 volts. A full-wave rectifier is used and the filter circuit is similar to that of Fig. 9-15*f*. The values of C_1, C_2, and C_3 are 10 μf each; L_1 is 20 henrys and L_2 is 15 henrys. What is the per cent of ripple voltage at (*a*) C_1? (*b*) C_2? (*c*) C_3?

21. A power-supply unit being operated from a 60-cycle power line is supplying a load with 120 ma at 360 volts. A half-wave rectifier is used and the filter circuit is similar to that of Fig. 9-15*f*. The values of C_1, C_2, and C_3 are 10 μf each; L_1 is 20 henrys and L_2 is 15 henrys. What is the per cent of ripple voltage at (*a*) C_1? (*b*) C_2? (*c*) C_3?

22. A power-supply unit being operated from a 60-cycle power line is supplying a load with 50 ma at 100 volts. A half-wave rectifier is used and the filter circuit is similar to that of Fig. 9-15*g*. The values of C_1 and C_2 are 40 μf each and R_1 is 1,000 ohms. What is the per cent of ripple voltage at (*a*) C_1? (*b*) C_2?

23. A power-supply unit being operated from a 60-cycle power line is supplying a load with 50 ma at 80 volts. A half-wave rectifier is used and the filter circuit is similar to that of Fig. 9-15*g*. The value of C_1 is 30 μf, C_2 is 50 μf, and R_1 is 800 ohms. What is the per cent of ripple voltage at (*a*) C_1? (*b*) C_2?

24. A power-supply unit being operated from a 60-cycle power line is supplying a load with 14 ma at 80 volts. A half-wave rectifier is used and the filter circuit is similar to that of Fig. 9-15*g*. The current for the power tube is taken off at capacitor C_1, and hence this current does not flow through resistor R_1. The value of C_1 is 40 μf, C_2 is 20 μf, and R_1 is 1,200 ohms. What is the per cent of ripple voltage at (*a*) C_1? (*b*) C_2?

25. What is the per cent of regulation of the power-supply unit of Prob. 18 if its output voltage is 250 volts at full-load and 350 volts at no-load?

26. What is the per cent of regulation of the power-supply unit of Prob. 19 if its output voltage is 300 volts at full-load and 450 volts at no-load?

27. What is the per cent of regulation of the power-supply unit of Prob. 23 if the output voltage is 80 volts at full-load and 150 volts at no-load?

28. If it is desired to limit the regulation of the power-supply unit of Prob. 23 to 20 per cent by adding a bleeder resistor to the unit, to what value of no-load voltage must the circuit be limited?

29. A power-supply unit being operated from a 60-cycle power line is supplying a load with 125 ma at 300 volts. A full-wave rectifier is used and the filter circuit is similar to that of Fig. 9-15*b*. The value of L_1 is 5 henrys and C_1 is 10 μf. What is the per cent of ripple voltage at the output?

30. A power-supply unit being operated from a 60-cycle power line is supplying a load with 150 ma at 450 volts. A full-wave rectifier is used and the filter circuit is similar to that of Fig. 9-15*b*. The value of L_1 is 5 henrys and C_1 is 20 μf. What is the per cent of ripple voltage at the output?

31. A power-supply unit being operated from a 60-cycle power line is supplying a load with 170 ma at 425 volts. A full-wave rectifier is used and the filter circuit is similar to that of Fig. 9-15c. The values of L_1 and L_2 are 10 henrys each, and C_1 and C_2 are 10 μf each. What is the per cent of ripple voltage at (a) C_1? (b) C_2?

32. A power-supply unit being operated from a 60-cycle power line is supplying a load with 200 ma at 450 volts. A full-wave rectifier is used and the filter circuit is similar to that of Fig. 9-15c. The value of L_1 is 10 henrys, L_2 is 5 henrys, C_1 is 10 μf, and C_2 is 20 μf. What is the per cent of ripple voltage at (a) C_1? (b) C_2?

33. Determine the per cent of ripple voltage at the output of the power-supply unit of Prob. 29 if it is to be operated from a 25-cycle power line.

34. Determine the per cent of ripple voltage at the output of the power-supply unit of Prob. 30 if it is to be operated from a 40-cycle power line.

35. Determine the per cent of ripple voltage at the output of the power-supply unit of Prob. 31 if it is to be operated from a 50-cycle power line.

36. Determine the optimum value of inductance for the input choke of the power-supply unit in Prob. 29.

37. Determine the optimum value of inductance for the input choke of the power-supply unit in Prob. 30.

38. Determine the resistance values of a voltage divider for a superheterodyne receiver that employs a 6SA7 converter, a 6SG7 i-f amplifier tube, a 6SQ7 detector-amplifier tube, and a 6K6-GT/G power-output tube. The operating voltages and currents are to be obtained from a tube manual for plate voltages of 250 volts. (The screen grid of the 6SG7 is to be operated at 150 volts.) The voltage divider should have taps to supply all plate, screen-grid, and grid-bias voltages. Its power transformer, rated at 90 ma, is to be operated at 90 per cent of its rated current.

39. Determine the power rating of the voltage divider of Prob. 38 for (a) a single resistor having five taps, (b) six separate resistors.

40. Determine the resistance values of a voltage divider for a superheterodyne receiver that employs a 7B8 converter, a 7A7 i-f amplifier tube, a 7B6 detector-amplifier tube, and a 7B5 power output tube. The operating voltages and currents are to be obtained from a tube manual for plate voltages of 250 volts. All control-grid bias voltages are to be obtained by separate cathode-bias resistors, and hence these voltages are not to be provided by the voltage divider. Its power transformer, rated at 90 ma, is to be operated at 90 per cent of its rated current.

41. Determine the power rating of the voltage divider of Prob. 40 for (a) a single-tapped resistor, (b) two separate resistors.

42. In the receiver circuit shown in Fig. 11-13, the voltage divider is replaced by the 12,000-ohm resistor R_{18}. (a) What purpose does this resistor serve? (b) What voltage is applied to the grids of VT_1, VT_2, and VT_3 if 12.5 ma flows through this resistor and the voltage at the junction of R_{18}, R_{19}, and C_{26} is 250 volts? (c) How much power is dissipated by this resistor?

43. What rms value of alternating voltage must be supplied by the power source to the plate of the rectifier tube of Prob. 19 if the resistance of L_1 is 150 ohms and the internal tube drop is 25 volts? Assume that the average value of the rectifier output voltage is equal to the rms value of the applied voltage when the tube drop is neglected.

44. What rms value of alternating voltage must be supplied by the power source to the plate of the rectifier tube of Prob. 20 if the resistance of L_1 is 150 ohms, L_2 is 100 ohms, and the internal drop of the tube is 30 volts? Assume that the average value of the rectifier output voltage is equal to the rms value of the applied voltage when the tube drop is neglected.

45. What rms value of alternating voltage must be supplied by the power source to the plate of the rectifier tube of Prob. 24 if the internal drop of the tube is 20 volts? Assume that the average value of the rectifier output voltage is equal to the rms value of the applied voltage when the tube drop is neglected.

CHAPTER 10

AUDIO UNITS AND
HIGH-FIDELITY REPRODUCTION

The sound waves produced at a sending station may be reproduced practically instantaneously at numerous receiving stations situated at either short or long distances from the transmitter. Sound waves are capable of traveling only comparatively short distances and travel at a speed of approximately 1,130 ft per sec. The instantaneous reproduction of sound waves over long distances therefore cannot be accomplished by the direct transmission of the sound waves, but requires the transmission of electrical waves whose frequencies are a faithful reproduction of the frequencies of the sound waves to be reproduced. For complete radio communication it is therefore necessary that some device for changing the sound waves to electrical waves be situated at the sending station and another device for changing the electrical waves back to sound waves be situated at the receiving station. The audio unit used at the sending station for converting sound waves to electrical waves is called a *microphone*. At the receiving station, an audio unit called the *loudspeaker*, commonly referred to as simply the *speaker*, is used to convert the electrical waves back to sound waves.

Two other audio units used in radio communication are (1) a device called a *phonograph pickup*, which is used for converting sound on records to electric waves, (2) a device called a *magnetic reproducer*, which is used for converting sound on tape to electric waves.

10-1. Microphones. *Air-pressure Type.* The most commonly used microphone is the *air-pressure type* also called *sound-wave type*. This instrument converts the mechanical variations in air pressure caused by the human voice or musical instrument to an equivalent electrical wave of similar frequency. Because of the many fields of audio applications a number of variations of the basic air-pressure microphone have been developed. Maximum performance can only be obtained by using the instrument designed for the specific application. Several sound-wave type microphones are illustrated in Fig. 10-1.

Contact Type. An instrument used to convert the mechanical vibrations of a medium other than air into electrical waves is called a *contact microphone*. This type of microphone is fastened in direct contact with that part of an object or person's body whose vibration it is designed to pick up. Contact microphones have a number of fields of application

369

and hence are made in a variety of forms, each to suit a particular need (see Fig. 10-2).

This type of instrument can be used to pick up the mechanical vibrations of wind, string, or percussion-type musical instruments. Such a unit is small and compact and is provided with a spring mounting clamp that permits it to be fastened with comparative ease to various types of musical instruments. The contact microphone is also used for industrial purposes. An analysis of the vibration of a machine or any of its parts can easily be obtained by mounting a contact microphone directly to that part of the machine to be studied. Contact microphones are very useful for applications where the surrounding noise level is very high.

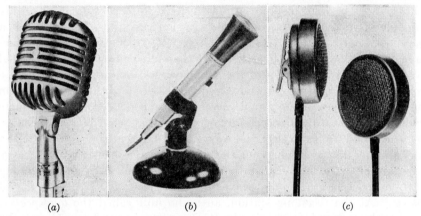

(a)   (b)   (c)

Fig. 10-1. Sound-wave-type microphones. (a) Floor model. (b) Desk-stand type. (c) Lapel type. (*Courtesy of Shure Brothers, Inc.*)

Two types of units used in this field of application are the *throat microphone* and the *lip microphone*. These microphones are strapped to the throat or lip and react directly to the vibration of these organs rather than to the sound waves they produce. The extraneous noises in the vicinity of the microphone produce only variations in air pressure and thus are not picked up.

10-2. Requirements of the Microphone. *Frequency Response.* The sounds produced by the loudspeaker of a radio receiver should be essentially the same as those produced by the artists in the studio of the transmitting station. The intensity characteristics of audible sounds as presented in Art. 6-1 for the audio amplifier will therefore also apply to the microphone. Since the fidelity of reception can only approach but never equal the fidelity of transmission, the microphone should have a more uniform response over a wider range than the a-f components in the radio receivers. The frequency range of microphones used at a transmitter usually is from 30 to 10,000 cycles as compared to a range of 50 to 5,000 cycles for the average home receiver.

Sensitivity. Sound waves are the mechanical vibration of air in space at a-f rates.   These vibrations cause the air pressure to vary above and below its normal pressure.   The variations in air pressure on the ear-drum produce the sensation of sound in the brain.   A sound that is just barely audible increases the normal air pressure by only about one-millionth of 1 per cent.   A sound that is so loud that it causes pain increases the normal air pressure by only about one-tenth of 1 per cent.   For perfect fidelity, a microphone should be sensitive to these small variations in air pressure for the entire range of hearing.

Since the intensity of a sound wave decreases as the distance from its source increases, so, too, the sensitivity of a microphone will decrease as the distance between the microphone and the source of the sound wave is

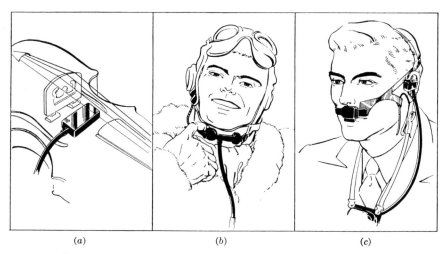

        (a)   (b)   (c)

Fig. 10-2. Contact-type microphones.   (a) For musical instruments.   (b) Throat microphone.   (c) Lip microphone.

increased.   Maximum response is thus obtained when the distance between the microphone and the source of the sound is at a minimum.

Range of Pickup.   Microphones may be classified according to their directional characteristics as: (1) *cardioid*, (2) *bidirectional*, and (3) *all-directional*.   A cardioid microphone, also called *unidirectional*, has a heart-shaped response pattern (Fig. 10-3).   The ratio of front-to-back pickup of random sound energy is 7 to 1 for the cardioid, and 14 to 1 for the supercardioid.   This high front-to-back sensitivity ratio permits this type of microphone to be operated at a higher volume level than the other two types before acoustic feedback occurs.   It is therefore ideal for use in applications requiring unidirectional response characteristics, such as public-address systems.   The bidirectional microphone has a figure-eight response pattern, thus permitting sound waves to be picked up equally as well from the front or rear.   The sound waves on either side of the microphone will be rejected.   This type of microphone is generally used in radio and recording studios, and other types of fixed installations. The all-directional microphone has a 360-degree circular response pattern, thus permitting the sound waves to be picked up equally from all directions in a plane perpendicular to the axis of the microphone. This type of microphone is ideal for use with large groups such as orchestras, choirs, and stage productions. Its small diameter makes it especially suitable for use where the microphone must be held in the hand or suspended from the neck.

10-3. Microphone Ratings. *Pressure Rating.* As the output of a microphone varies with the pressure of the sound waves, it is necessary

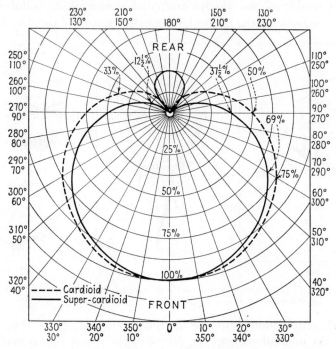

Fig. 10-3. Range of pickup for cardioid- and supercardioid-type microphones. (*Courtesy of Shure Brothers, Inc.*)

to rate the output of a microphone for some definite unit of pressure. The unit of pressure most generally used is the *bar*. The term *bar* actually means a unit of atmospheric pressure, which is equal to a pressure of approximately 1 million dynes per sq cm. However, in the rating of microphones many manufacturers consider the bar as the cgs (centimeter-gram-second) absolute unit of pressure which is equal to a pressure of 1 dyne per sq cm. The ASA (American Standards Association) recommends the use of the phrase *dynes per square centimeter* as the unit of sound pressure.

Impedance Classification. Microphones may be classified according to their impedance as being either low-impedance or high-impedance microphones. Carbon, velocity, and dynamic type microphones have a low impedance, and the crystal microphone has a high impedance. The velocity and dynamic microphones can be obtained with a self-contained high-impedance transformer that enables these two types of microphones to be used as high-impedance instruments.

Low-impedance Microphone Ratings. One of the important factors in the choice of low-impedance microphones is their power output. Low-impedance microphones are therefore rated on the basis of the decibels below a zero power level for a zero reference pressure level. These microphones are usually rated with a zero power reference level of 6 mw, although a 1-mw zero reference level is also used. The zero reference pressure level is generally one bar, although a ten-bar zero reference level is also used. Because of the lack of consistent use of a single reference level, it is very important to express and interpret carefully the manner in which a microphone is rated.

The power in a microphone circuit can readily be determined from its decibel rating by the methods presented in Art. 6-3. However, before determining the power output of a microphone from its decibel rating, it is necessary that the decibel be expressed in terms of its reference pressure level. The effect of variations in pressure upon the decibel rating is expressed by the equation

$$\text{Rating} = 20 \log \frac{F}{F_R} \tag{10-1}$$

where Rating = output rating at higher pressure, db

F = higher pressure, bars

F_R = reference pressure level, bars

The following example illustrates the interpretation of low-impedance microphone ratings.

Example 10-1. A certain low-impedance microphone is rated at 62.8 db below 6 mw per 10-bar signal. What is the power ouput for (*a*) a 10-bar signal? (*b*) A 300-bar signal?

Given: $P_R = 6$ mw   Find: (*a*) P_o @ 10 bars
   Rating $= -62.8$ db   (*b*) P_o @ 300 bars
   $F_R = 10$ bars

Solution:

(*a*) db $= 10 \log \dfrac{P_R}{P_o}$

$P_o = \dfrac{P_R}{\text{antilog} \dfrac{\text{db}}{10}} = \dfrac{0.006}{\text{antilog} \dfrac{62.8}{10}} = \dfrac{0.006}{1.906 \times 10^6}$

$= 0.0031 \times 10^{-6}$ watt, or 0.0031 μw

(*b*) Output rating $= 20 \log \dfrac{F}{F_R} = 20 \log \dfrac{300}{10} = 20 \times 1.4771 = 29.54$ db

Output $= -62.8 + 29.54 = -33.26$ db

$P_o = \dfrac{P_R}{\text{antilog} \dfrac{\text{db}}{10}} = \dfrac{0.006}{\text{antilog} \dfrac{33.26}{10}} = \dfrac{0.006}{2.12 \times 10^3} = 2.83 \times 10^{-6}$ watt, or 2.83 μw

High-impedance Microphone Ratings. One of the important factors in the choice of high-impedance microphones is their voltage output. High-impedance microphones are therefore rated in decibels below 1 volt per bar. This means that an alternating sound pressure of 1 dyne per sq cm rms will produce an output voltage, across an open circuit grid, of the rated decibels below 1 volt.

Example 10-2. A certain high-impedance microphone is rated at 55 db below 1 volt per bar. What is the voltage output for (a) a signal of 1 bar? (b) A 300-bar signal?

Given: Rating $= -55$ db   Find: (a) e_o @ 1 bar
             $F_R =$ one bar   (b) e_o @ 300 bars
Solution:

(a) $\text{db} = 20 \log \dfrac{e_R}{e_o}$

   $e_o = \dfrac{e_R}{\text{antilog } \dfrac{\text{db}}{20}} = \dfrac{1}{\text{antilog } \dfrac{55}{20}} = \dfrac{1}{562.3} = 0.00177 \text{ volt}$

(b)   Output rating $= 20 \log \dfrac{F}{F_R} = 20 \times \log \dfrac{300}{1} = 20 \times 2.4771 = 49.54$ db

   Output $= -55 + 49.54 = -5.46$ db

   $e_o = \dfrac{e_R}{\text{antilog } \dfrac{\text{db}}{20}} = \dfrac{1}{\text{antilog } \dfrac{5.46}{20}} = \dfrac{1}{1.875} = 0.533 \text{ volt}$

Impedance Matching. In order to obtain maximum power transfer it is necessary to match the microphone impedance with its load. Thus, a low-impedance microphone should feed into a low load impedance, and a high-impedance microphone should feed into a high load impedance.

Transformers are generally used to couple low-impedance microphones to their amplifier circuits. The turns ratio of the microphone transformer should be of such a value that the load reflected to the primary circuit by the loaded secondary will be equal to the impedance of the microphone, and may be expressed as

$$n = \sqrt{\frac{Z_o}{Z_M}} \tag{10-2}$$

where $n =$ turns ratio (secondary to primary)
   $Z_o =$ output impedance, ohms
   $Z_M =$ microphone impedance, ohms

Example 10-3. A certain low-impedance microphone has a resistance of 50 ohms. The secondary of the microphone transformer feeds into a 450-ohm line. Determine the transformer turns ratio.

Given: $Z_M = 50$ ohms   Find: n
             $Z_o = 450$ ohms
Solution:

$$n = \sqrt{\frac{Z_o}{Z_M}} = \sqrt{\frac{450}{50}} = 3$$

Resistance-capacitance coupling is generally used to couple high-impedance microphones to their amplifier circuits. Maximum power transfer is obtained by using the proper value of plate resistor.

Effect of the Microphone Connecting Line. The distributed capacitance of the line connecting the microphone to its amplifier may be considered as a capacitor of equivalent value connected across the load. The impedance of this capacitor is dependent upon the frequency of the a-f signal and the length of the connecting line. In a low-impedance microphone circuit the lowest impedance of this capacitor, which occurs at the highest audio frequency, will be many times greater than the microphone impedance. The shunting effect of this capacitance is thus very small and can usually be ignored. Increasing the length of the connecting line increases its distributed capacitance, which decreases its impedance and thereby increases the shunting effect of the line. However, the shunting effect of even long lines may still be ignored when used with low-impedance microphones.

If the connecting line used with a high-impedance microphone is too long, the microphone impedance may be many times greater than the highest value of shunting impedance owing to the distributed capacitance of the line. Under this condition, the distributed capacitance of the line will have considerable effect on both the frequency response and the energy output of a high-impedance microphone circuit and therefore the length of the connecting line is an important factor. The ratings of high-impedance microphones are usually listed by the manufacturers for definite lengths of connecting lines.

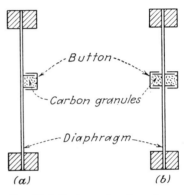

FIG. 10-4. Basic construction of a carbon microphone. (*a*) Single-button type. (*b*) Double-button type.

10-4. Carbon Microphones. *Operation of Carbon Microphones.* The operation of the carbon microphone is based upon the principle that the resistance of a pile of carbon granules will vary as the pressure exerted upon it is varied. The basic construction of a carbon microphone is illustrated in Fig. 10-4. The carbon granules are loosely piled in an insulated cup called the *button.* This button is mounted so that it is always in direct contact with the diaphragm. Any movement of the diaphragm will vary the amount of pressure exerted on the carbon granules and will vary the resistance of the carbon pile. In the circuit shown in Fig. 10-5a the microphone button is connected in series with a battery and the primary winding of the microphone transformer. With no movement of the diaphragm the resistance of the carbon pile remains constant and a steady amount of direct current will flow through the circuit. When sound waves strike the diaphragm they cause it to vibrate. Vibrations of the diaphragm will cause variations in the amount of pressure being exerted on the pile of carbon granules. Increasing the pressure packs the carbon granules more closely, thus decreasing the amount of contact resistance between granules and thereby decreasing the resistance of the entire carbon pile. Just as the variations in pressure on the diaphragm are controlled by the intensity and frequency of the sound waves that strike it, so, too, the variations in resistance of the carbon pile will also be dependent on the intensity and frequency of the sound waves. The variations in resistance of the carbon pile will cause the current flowing in this circuit to vary with the intensity and

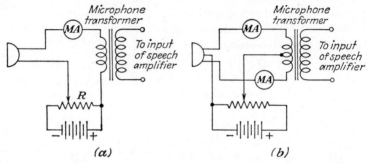

FIG. 10-5. Carbon microphone circuits.   (a) Single-button type.   (b) Double-button type.

frequency of the sound waves. The output at the secondary of the microphone transformer will therefore be an alternating voltage of magnitude and frequency corresponding to the sound waves.

Any distortion that may be caused by the nonlinear response of the carbon microphone can be reduced by use of the double-button microphone circuit shown in Fig. 10-5b. The two buttons in conjunction with the center-tapped primary form a push-pull arrangement that tends to cancel out all the even-order harmonics.

Characteristics of Carbon Microphones. The response of the carbon microphone is practically independent of frequency up to the mechanical resonant frequency of the diaphragm. The response decreases rapidly for those sound waves having a frequency above this value. This type microphone has certain advantages that make it important in its field. It is the only microphone that is also an amplifier in that its electrical energy output is greater than the energy required to produce the vibrations of the diaphragm. Other advantages are its light weight, low initial cost, rugged construction, and the fact that it is portable. It is therefore used where high sensitivity is an important factor or where voice reproduction rather than musical entertainment is the primary object.   Carbon microphones are used for amateur, police, and military work and also in some contact microphones.

10-5. Crystal Microphones. *Operation of Crystal Microphones.*   The crystal microphone makes use of the piezoelectric characteristics of certain crystalline materials (see Art. 8-10).   Because the voltage output of a rochelle salt crystal is much greater than for other crystalline materials, it is the one most generally used for microphone applications.   The basic crystal sound cell, sometimes called a *bimorph cell*, is made by clamping two thin crystal slabs together, as is shown in Fig. 10-6.   Electrical contact is made with each crystal by cementing a sheet of tin foil on both of its faces.   The microphone can be constructed so that the movement of the crystal is actuated directly by the sound waves.   However, this

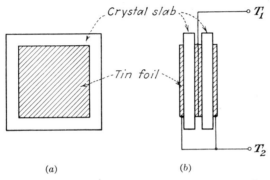

(a)   (b)

Fig. 10-6. Basic construction of a crystal sound cell.   (a) Single crystal.   (b) Bimorph cell.

method is not very effective, and in the more sensitive types of microphones the crystal is fastened directly to the diaphragm.   Vibration of the diaphragm by sound waves will cause the crystal to vibrate, and because of its piezoelectric action an alternating voltage having a frequency and intensity corresponding to the sound waves will be produced at the terminals of the crystal.   Because the output voltage of a single cell is very small, several cells are usually connected in series in order to obtain a higher voltage.   The crystal microphone requires no separate source of current or voltage and hence its output is applied directly to the input circuit of a speech amplifier.   Because of the comparatively low-voltage output of the crystal microphone several stages of high-gain speech amplification are required.   Since transformer coupling may pick up an excessive amount of hum, the coupling between stages should be of the resistance-capacitance type.

Characteristics of Crystal Microphones.   Crystal microphones are purely pressure operated and do not possess any cutoff effect at the low audio frequencies.   Its frequency response is comparatively flat over the entire a-f range.   It is also entirely nondirectional and therefore may be used at any angle. Other advantages of the crystal microphone are its light weight, comparative ruggedness, ease of maintenance, and the fact that no external power source is required.

Crystal microphones are used in high-quality public-address systems, broadcasting stations, and recording equipment. Because exposure to high temperatures (above 125°F) may permanently damage a crystal, precautions should be taken when using crystal microphones to keep the crystal from being exposed to high temperatures such as rays from the sun, radiators, etc. If the seal on a crystal microphone is broken, the crystal may absorb moisture from the air and become useless. Another precaution that should be taken when using crystal

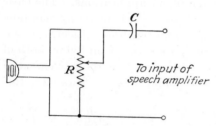

Fig. 10-7. A crystal microphone circuit.

microphones is to make sure that no voltage is applied to it. The battery voltage of an ohmmeter can destroy a crystal microphone.

10-6. Dynamic Microphones. *Operation of Dynamic Microphones.* The dynamic microphone makes use of a nonrigid unstretched diaphragm. A coil of wire that is rigidly attached to the back of the diaphragm is so arranged that it is free to move back and forth in the strong magnetic field produced by a permanent magnet (see Fig. 10-8). When sound waves strike the diaphragm, the coil will move back and forth in a radial

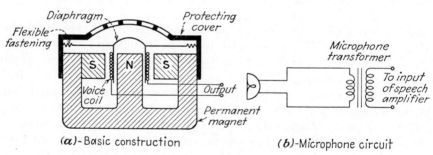

(*a*)-Basic construction   (*b*)-Microphone circuit

Fig. 10-8. A dynamic microphone and its circuit.

magnetic field, thus inducing in the coil a voltage that is a faithful electrical reproduction of the sound waves. A number of flexible circular corrugations on the diaphragm permit it to have a large amount of displacement. The response of a dynamic microphone at the very low audio frequencies is therefore very good. If the microphone is designed so that the voltage induced in the moving coil is exactly proportional to the pressure of the sound waves striking the diaphragm, the response of the dynamic microphone can be made practically independent of the frequency over a range of 30 to 18,000 cycles.   Because the coil of a dynamic microphone moves with the diaphragm it is also called a *moving-coil microphone.*

Characteristics of Dynamic Microphones.   The sensitivity of the moving-coil microphone is higher than other types of nonamplifying microphones.   It is light in weight, small in size, and requires no external source of power.   It is unusually rugged, since it is practically immune to the effects of moisture, temperature, and mechanical vibration.   It can therefore be used outdoors even in a strong wind.   The impedance of the dynamic microphone is normally very low, thus making it possible to use long connecting lines between the microphone and the first stage

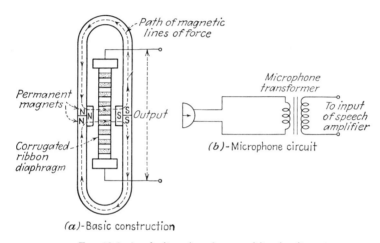

(a)-Basic construction

Fig. 10-9. A velocity microphone and its circuit.

of amplification.   However, it is also possible to obtain a dynamic microphone having a high impedance, which may be used in conjunction with a crystal microphone amplifier circuit or any other amplifier having an input impedance of 100,000 ohms or more.

10-7. Velocity Microphones.   *Operation of Velocity Microphones.*   The velocity microphone is a variation of the moving-coil microphone.   In place of the moving coil, a strip of metal is caused to vibrate in a magnetic field.   As only a single conductor cuts the magnetic lines of force, the velocity microphone is not so sensitive as the dynamic microphone.   The basic construction of this type of microphone is shown in Fig. 10-9.   A thin, lightweight, flexible corrugated metal strip (usually aluminum or duralumin) is suspended between the poles of a permanent magnet.   Because this metal strip resembles a ribbon, this type of microphone is called a *ribbon microphone.*   The ribbon will vibrate in accordance with the sound waves that strike it.   As the conductor cuts the magnetic lines of force, a voltage proportional to the frequency and pressure of the sound waves will be induced in the ribbon.   The force exerted on the ribbon by a sound wave is equal to the difference between the pressures on the front and back of the ribbon.   The resulting force is proportional to the velocity of the particles of air set in motion by the sound wave and hence the name *velocity microphone.*

Characteristics of Velocity Microphones.   The velocity microphone has an extremely good frequency response.   It is comparatively rugged and requires no external source of power.   Its response is very directional, and thus it will only reproduce those sound waves made directly in front of the microphone.   However, the ribbon is very delicate and must be protected from strong drafts or winds.

10-8. Dynamic Loudspeakers.   *Principle of the Dynamic Loudspeaker.* In the dynamic loudspeaker (Fig. 10-10) the a-f current flows through a coil that is mechanically coupled to a paper cone.   Since only a-f currents flow through this coil it is generally referred to as the *voice coil.* The interaction between the varying magnetic field set up about the voice coil by the varying a-f current and the constant magnetic field of the stationary magnet causes the voice coil to move back and forth in the constant magnetic field.   The paper cone being mechanically coupled to the voice coil will move back and forth at a frequency and intensity that are determined by the frequency and intensity of the a-f current.   The motion of the paper cone will thus set up sound waves corresponding to the a-f currents in the voice coil.   Because of the large surface area of the paper cone, a large quantity of air is set into motion, thus creating a loud sound.   Freedom of movement of the paper cone is obtained by the use of a flexible fastening between the large end of the cone and the surface to which it is connected.   Because of the movement of the voice coil this type of loudspeaker is also called a *moving-coil loudspeaker.*   The constant stationary magnetic field may be produced either by a permanent magnet, as in the case of the permanent-magnet dynamic loudspeaker, or by an electromagnet, as in the case of the electrodynamic loudspeaker.   The dynamic loudspeaker is the type of loudspeaker most commonly used in modern receivers.   The response of the dynamic loudspeaker is fairly uniform over a frequency range that is adequate for the ordinary a-m broadcast radio receiver, namely, 40 to 5,000 cycles.   For high-fidelity reproduction, a high-quality dynamic loudspeaker can be obtained with fairly uniform response from 30 to 18,000 cycles.

Permanent-magnet Dynamic Loudspeaker.   The basic construction of a permanent-magnet dynamic loudspeaker, commonly referred to as a *p-m dynamic loudspeaker,* is shown in Fig. 10-10.   The fixed magnetic field is produced by a permanent magnet.   The voice coil is wound on a thin bakelite tube that is mounted in a manner which permits it to move back and forth along the core of the permanent magnet.   The voice coil is centered between the poles of the permanent magnet by means of a thin springy sheet of bakelite or metal called a *spider*.   The spider permits the coil to move down along the core, and because of its spring action it also forces the coil to move back when the pull on the coil is reduced or eliminated.

The movement of the coil and the cone to which it is attached is directly dependent upon the strength of both the fixed and the varying magnetic fields.   Thus the greater the strength of the fixed magnetic field, the greater will be the movement of the paper cone.   Also, the greater the current flowing through the voice coil, the greater will be the movement of the cone.   The strength of the fixed magnetic field is limited

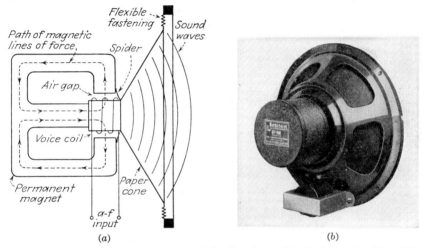

FIG. 10-10. A permanent-magnet dynamic loudspeaker. (a) Basic construction, (b) a commercial loudspeaker. (*Courtesy of Jensen Radio Manufacturing Co.*)

by the size and the material of which the permanent magnet is made. Advantages of the p-m dynamic loudspeaker are its light weight, compactness, low cost, and the fact that it requires no external source of power. Because of these advantages this type of loudspeaker is used extensively.

Permanent-magnet dynamic loudspeakers are usually rated as to their power output, impedance of the voice coil, and the outside diameter of the cone.   The power output ranges from 1 to 30 watts.   The impedance of its voice coil has a range of approximately 3 to 15 ohms.   The outside diameter of the cone may be as small as 2 inches or as large as 16 inches.

In order to obtain the maximum transfer of energy from the a-f signal in a high-impedance plate circuit of an output stage to the low-impedance voice coil, it is necessary to use an output transformer, as described in Art. 7-2.   It is not at all uncommon to find the output transformer mounted on the framework of the loudspeaker.

10-9. Horns. Because of the poor coupling between the air and the vibrating cone (also called a *direct radiator*), the efficiency of cone-type loudspeakers in transforming electrical energy to sound energy is very low. The efficiency of the average loudspeaker is less than 5 per cent. Because the average radio receiver has a much greater power output than the relatively small amount required by the loudspeaker, the poor efficiency of transformation is of little consequence and is usually ignored. Where a large amount of power output is required, such as in public-address systems, the efficiency of transformation becomes an important factor.

A higher efficiency of transformation can be obtained by mechanically coupling a dynamic loudspeaker, called the *driver unit*, to a horn, as shown

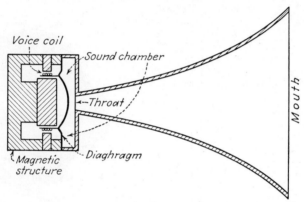

FIG. 10-11. Basic structure of horn and driver unit. (*Courtesy of Jensen Radio Manufacturing Co.*)

in Fig. 10-11. The small end of the horn is called the *throat*, and the large end is called the *mouth*. The purpose of the horn is to transform sound energy having a high pressure and low velocity to sound energy having a low pressure and high velocity. A horn may be considered as a matching device for coupling the heavy vibrating surface at the throat of the horn to a relatively light medium, which is the air, at the mouth of the horn. The shape of the horn will depend upon the rate of increase of the cross-sectional area as shown in Fig. 10-12. The function of the horn contour is to produce a smooth and continuous increase in the cross-sectional area. The efficiency of transmission along the length of the horn is controlled by the rate of increase in cross-sectional area, also called the *taper*. High audio frequencies are transmitted quite well by all horns. At the low audio frequencies, the efficiency decreases for both the conical and the parabolic tapers, is fairly uniform for the exponential taper, and is comparatively uniform for the hypex. In order to make the horn more compact, it is generally folded upon itself one or more times, as shown in Fig. 10-13.

10-10. Recording and Reproduction of Sound. A permanent record of any audible program or event can be made with the aid of modern recording apparatus. It is also possible to reproduce readily and faithfully the audible sounds from this record with the aid of modern reproducing

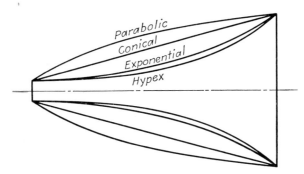

FIG. 10-12. Comparative shapes of different types of horns with the same throat and mouth diameters. (*Courtesy of Jensen Radio Manufacturing Co.*)

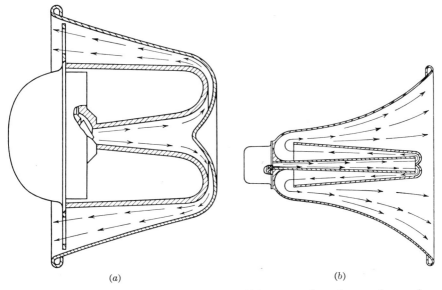

(a)   (b)

FIG. 10-13. Horn-type reproducers. (a) Single-fold horn reproducer for speech reproduction. (b) Two-fold horn reproducer for general reproduction of speech and music. (*Courtesy of Jensen Radio Manufacturing Co.*)

apparatus. It thus becomes possible for a broadcasting station to record a program at any convenient time and to broadcast this program at the most advantageous time for that station. If proper equipment is employed, it is almost impossible for the average listener to detect whether a program is originating directly from the studio or from a recording.

Because of this flexibility in broadcasting, practically all broadcasting stations have apparatus capable of accurately recording and reproducing such program transcriptions.

Entertainment to suit the mood or occasion can be obtained by the average person with the aid of a record player or tape recorder. These units may be obtained separately or in combination with a radio receiver.

The various methods of making a record of sound are: (1) mechanical, (2) magnetic, and (3) light. In the mechanical method a thin disc made of wax, aluminum, acetate-coated cardboard, or a form of plastic such as vinylite, may be used. A continuous groove is cut in the disc, starting near the outside and ending near the center. Two methods of cutting the groove, the lateral cut and the vertical cut, are employed. In the *lateral-cut* method, the depth of the groove is kept constant and the sides are undulated according to the variations in frequency and amplitude of the program being recorded. In the *vertical-cut* method, the sides of the groove are kept uniform and the depth of the cut is varied. The lateral cut is the one most commonly used. In the magnetic method, sound is recorded on a plastic and iron-oxide tape, or a thin wire, by magnetizing the metal according to the variations in frequency and amplitude. The principle of light variations is used in sound on film recording. The variations in light are controlled by a photoelectric cell that changes the electrical equivalent of the sound waves to light waves.

10-11. Mechanical Sound Recorders and Reproducers. *Principles of Mechanical Sound Recorders and Reproducers.* The mechanical method is used in radio broadcasting stations, and in home recording and reproduction apparatus. In recording apparatus the variations in electrical energy are transformed to an equivalent mechanical energy by cutting a groove in a disc, also called a *record*. The unit used for this purpose is commonly referred to as a *cutting head*. In reproduction apparatus a unit called a *pickup head*, or *pickup cartridge*, transforms the variations in mechanical energy obtained from the grooves in a record to an equivalent electrical energy. Mechanical energy may be transformed to electrical energy by the pickup cartridge, or vice versa by the cutting head, using any one of the following three methods: (1) magnetic, (2) piezoelectric, and (3) capacitance. Three basic principles are used in magnetic heads: (1) a moving armature in a magnetic field, referred to as *variable reluctance*, (2) a moving coil in a magnetic field, referred to as *dynamic*, and (3) a moving magnet in a stationary coil, referred to as *Dynetic*.

By means of a suitable switching apparatus a single amplifier circuit can be used for both recording and reproducing. When the unit is recording, sound energy is transformed to an equivalent electrical energy by a microphone and then strengthened by a speech amplifier. The output of the amplifier is fed into the cutting head and thus controls the manner in which the groove is cut.   When the unit is reproducing, the vibration of the needle, caused by the undulations in the groove of the record, is changed to an equivalent electrical energy by the pickup cartridge.   This electrical energy is strengthened by the speech amplifier, whose output is then fed into a loudspeaker.   A sharp point, called a *stylus*, is used for cutting the groove in a record.   Points used for this purpose are made of hard steel, a sapphire, or a diamond.   A duller point, also called a *stylus* or *needle*, is used for reproducing the audible sounds from a record. Points used for this purpose are also made of hard steel, a sapphire, or a diamond.

Variable-reluctance Cartridge.   The basic circuit of a variable-reluctance cartridge is shown in Fig. 10-14.   The needle used in conjunction with a phonograph pickup, when traveling in the groove of a record, is caused to move from side to side because of the lateral-cut groove. An iron core is mounted in the center of the coil and also on a pivot so that it is free to move between two poles of a permanent magnet.   A needle is mechanically coupled to this armature. In its normal position the needle is at point A and the top of the armature is at A'.   When the needle moves to point B the top of the armature moves to B', and when the needle moves to C the top of the armature moves to C'.   As the

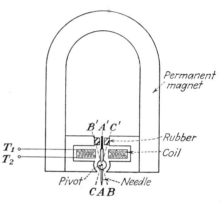

Fig. 10-14. A variable-reluctance cartridge.

armature moves back and forth the magnetic lines of force linking the coil are varied, thus causing an a-f voltage to be induced in the coil.   The output from this coil is taken from terminals T_1 and T_2 and is fed into a speech amplifier.

When the magnetic unit is used in conjunction with a cutting head, the a-f current from a speech amplifier is fed into the coil through terminals T_1 and T_2.   The varying current flowing in the coil sets up a varying magnetic field that causes the armature to move from side to side.   This movement of the armature also moves the cutting stylus, thus causing a lateral-cut groove to be made in the record.   The undulations in the sides of the groove will thus be made in accordance with the frequency and amplitude variations of the a-f current.

Modern versions of the variable-reluctance cartridge are popular for high-fidelity use because of their low tracking force, high compliance, and good frequency response.   However, because of its relatively low output, additional stages of amplification are required with this type of pickup.

Dynamic Cartridge. The basic circuit of a dynamic cartridge is shown in Fig. 10-15. A magnetic field is set up by the strong permanent magnet. Movement of the stylus causes a small coil of wire to move through the lines of force created by the magnetic field. A small signal voltage is thus produced in the coil, and its output is fed into a speech amplifier from

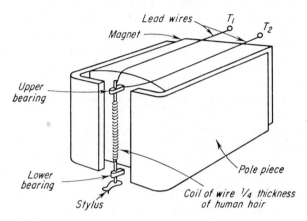

Fig. 10-15. A dynamic (moving-coil magnetic) cartridge.

terminals T_1 and T_2. The wire used in the moving coil is of microscopic size, being approximately one-fourth the diameter of human hair. The dynamic pickup has an accurate frequency response with low distortion, especially at the low frequencies. Its output is very low, being less than the variable-reluctance type. In addition to requiring additional amplification, a high-ratio booster transformer is generally used with this type of pickup.

Dynetic Cartridge. The basic circuit of a Dynetic cartridge is shown in Fig. 10-16. Movement of the stylus causes the permanent magnet to move through the fixed coil. A relatively low signal voltage will be induced in the coil, and is fed to the speech amplifier from terminals T_1 and T_2. The effective mass of the

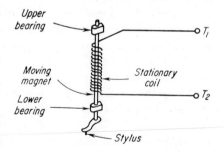

Fig. 10-16. A Dynetic cartridge.

stylus magnet is much lower than the other two types of magnetic pickups, and its tracking force is very low, being approximately 1 gram. Its frequency response over the entire audio range is extremely accurate with a minimum amount of distortion.

Crystal Cartridge. The basic circuit of a crystal cartridge is shown in Fig. 10-17. When used as a recording pickup, the vibration of the needle in the record groove distorts the crystal element. Because of the piezo-electric effect, a voltage will be generated across the faces of the crystal. This voltage will vary with the amplitude and frequency at which the crystal is distorted. The a-f output of this unit is obtained from terminals T_1 and T_2 and is fed to a speech amplifier. When used in conjunction with a cutting head, the operation is reversed. The a-f output of a speech amplifier is applied across the faces of the crystal, thus causing the crystal to be distorted because of the piezoelectric action. These distortions will then cause the stylus, to which it is mechanically coupled, to cut a groove whose undulations conform to the variations in the amplitude and frequency of the a-f current. Rochelle salt and ammonium dihydrogen phosphate crystals are generally used for this purpose.

The crystal pickup produces approximately six times as much voltage as the best magnetic pickup, and may be used without a preamplifier. It is impervious to stray magnetism and therefore does not require shielding. The crystal cartridge distorts the recorded signal as the strength of the output signal varies inversely with frequency. The lower

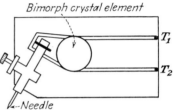

Fig. 10-17. A crystal cartridge.

the frequency the stronger the output voltage, and the higher the frequency the weaker the output voltage. It will not respond to the entire audio range as the upper limit for commercial crystals is approximately 14,000 cycles. The effective mass of the crystal stylus is higher than a magnetic stylus in that it must bend a crystal as compared to moving a light coil of wire, or an equally light metal tube.

Ceramic Cartridge. A piezoelectric effect can be made to occur in a ceramic unit in the same manner as a crystal unit. The ceramic unit is made by mixing a solution of barium and calcium titanate and allowing it to harden in a mold at normal temperature. It is then exposed to an intense baking heat, after which it is allowed to cool while under the influence of a very strong electric field. After it has cooled, the ceramic unit will exhibit piezoelectric effects when bent or twisted. Its structure and principle of operation are similar to the crystal cartridge. Ceramic cartridges are not affected by humidity or temperature, and therefore no special protective coating is required. They have a relatively smooth, distortion-free response at a high output signal level over a 50- to 10,000-cycle range.

Capacitance Cartridge. The capacitance cartridge consists of a fixed metal plate and a floating metal plate. The two plates are placed parallel to each other and separated by a tiny air gap. A very rapidly oscillating charge is fed onto the fixed plate, and the floating plate is attached to the stylus.   As the stylus moves through the grooves in a record, its vibration, produced by the undulations in the grooves, causes the floating plate to flutter toward and away from the charged plate. The expansion and contraction of the air gap causes the oscillating voltage to be modulated by the frequency of the vibration of the stylus.   Since the resultant signal is frequency modulated, this type of cartridge is also referred to as an *f-m pickup*.   An f-m detector must be used to separate the audio signal from the f-m signal in the same manner as used in an f-m radio receiver.   This type of cartridge therefore cannot be used alone, as it requires an additional unit containing specially designed oscillator and f-m detector circuits.   The less work the stylus must do, the more accurate the pickup.   In magnetic and piezoelectric pickups the electric signal is generated by the movement of the stylus.   In the capacitance

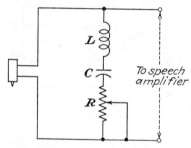

pickup the stylus does the least amount of work, as it is not required to produce the output voltage but is used to merely modify it.   The frequency response of an f-m cartridge is practically unlimited, and units have been made having a response as high as 30,000 cycles. Record and stylus wear are very slow with this type of unit, since tracking is obtained with a pressure of only 1 gram. The unit is very fragile, and if it is not

FIG. 10-18. A scratch filter circuit.

handled carefully, it may become damaged and will then produce intermodulation distortion.

Scratch Filter.   Surface noise or needle scratch is produced when the needle passes over the minute irregularities in the surface of the record groove.   This surface noise can be reduced by use of a scratch filter. Where maximum fidelity is not important the level of the audible needle scratch can be practically eliminated by use of the scratch filter circuit shown in Fig. 10-18.   The circuit is essentially a series resonant circuit with the values of L and C adjusted to resonate at the frequency of the scratch, which is approximately 5,000 cycles.   The degree of attenuation can be controlled by varying the rheostat R, which is connected in series with the resonant circuit.

10-12. Magnetic Sound Recorders and Reproducers.   *Principles of Operation*.   Magnetic recordings can be made on either a plastic and iron-oxide tape or on a wire of small diameter.   As tape is used in practically all modern magnetic recording equipment, this method of magnetic transcription will be described.   However, the basic principles of operation are the same for both types of materials.   The principle of operation of magnetic recording is illustrated in Fig. 10-19 and is based on the fact that the magnetic characteristics of a material can be varied. This characteristic is also referred to as the *coercivity* of a material and represents the resistance a magnetized metal offers to changes to its magnetic state. The two most important materials used in magnetic recorders are: (1) the laminations for constructing the recording or playback head, and (2) the metal oxides used on the tape. The metal used in the construction of the head should be capable of having its magnetic characteristics changed thousands of times per second, and it should therefore have a low coercivity. The metal oxide used on

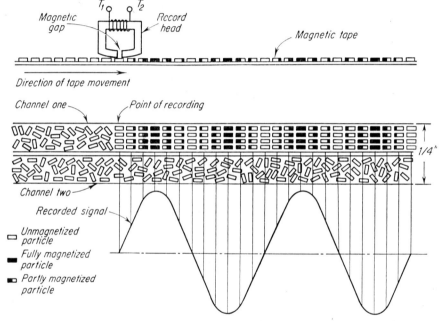

FIG. 10-19. Effect produced on the magnetic particles in a recording tape by applying a signal to the tape.

the tape should have a high coercivity, as once it is magnetized it should resist any effort to demagnetize it (unless passed through a special demagnetizing head). The tape is approximately one-fourth inch wide, and is divided into two equal sections along its length so that each half can be used for making a separate recording.

A single unit can be used as a recorder and also as a reproducer. To record, the sound impulse is changed to an electrical impulse by the microphone. This electrical impulse is strengthened by an amplifier, and it is then fed into a recording head. This head basically consists of a coil of wire that is wound on a soft-iron core having a small air gap. As the tape passes over the air gap, the magnetic lines of force pass through the section of tape that is over the air gap. A magnetic pattern of the sound fed into the microphone is induced onto the tape (Fig. 10-19). When used as a reproducer, the above operations are reversed. As the magnetized tape passes over the air gap, a pulsating magnetic field is produced in the recording head. The pulsating lines of force cut the coil and thus induce electrical impulses in the coil. The electrical impulses are then strengthened by the amplifier and transformed into sound waves by a loudspeaker.

Recorder-Reproducer Head. Although the general construction of the recorder-reproducer head varies with different manufacturers, their basic operations are the same. The basic construction of a combination

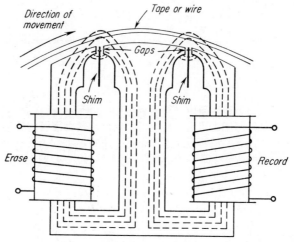

FIG. 10-20. A combination recording and erasing head.

record-erase head is shown in Fig. 10-20. As the tape moves through the head structure it first passes over the erase section. The purpose of this portion of the head is to remove any audible signals that may be on the tape. This may be accomplished by passing a direct current, of proper polarity and value, through the erase coil. A more effective method is to pass a high-frequency signal (40 to 80 kc) through the erase coil. As the tape passes over the air gap in the erase section, the magnetic particles are magnetized according to the strength and character of the inaudible high-frequency erase signal. For all practical purposes the tape is now devoid of any audible signal. Because of the hysteresis effect of the magnetic material on the tape, a preconditioning signal called *bias* is applied at the same time as the recording signal. The purpose of this bias is to establish linearity by applying the audio signal on the linear portion of the hysteresis loop. Either a direct-current or a high-frequency signal may be used as bias. The high-frequency signal is preferred, as using the d-c bias will produce distortion during playback. The high-frequency signal is taken from the same oscillator circuit that produced the erase signal, and it is mixed with the audio signal being applied to the recording head. As the tape passes over the air gap in the record section, the magnetic particles are energized by both the audio signal and the inaudible bias signal. The audio signal provided by a recording amplifier has now been impressed on the recording tape. During playback neither the erase or bias signal is applied. The magnetic impulses on the tape produce signal voltages across the record coil, now operating as a playback coil. The small signal voltages on the playback coil are amplified by voltage and power amplifiers and are then fed to a loudspeaker system.

In some of the better-quality tape recorders a separate playback head is provided for monitoring the signal on the tape during recording, and also for later playback service. This head is similar to the recording head in structure, except for a smaller gap in the pole pieces. The magnetic pattern on the tape causes a signal to be produced in the associated head circuit. This signal is amplified by a separate amplifier system, and is then fed to earphones or a loudspeaker.

Tape Transport Mechanism. The functions of the tape transport system are: (1) to remove the tape from the supply reel, (2) to move the tape past the head structure at a uniform speed, and (3) to wind the tape on the take-up reel. Variations in the speed of the tape, or its tension, will produce audible changes in the sound or music being recorded, or being played back. These variations in sound are called *flutter* and *wow*, and they may be minimized by using smooth-operating hysteresis synchronous motors.

Two tape speeds are usually provided, 3.75 inches per sec and 7.5 inches per sec. This permits the recording of one hour of high-fidelity music, or two hours of speech, from 1,200 feet of tape (a 7-inch reel). The recorders are generally dual-track mechanisms, in which only one-half of the tape width is used during recording or playback. The two tracks are used for separate recordings.

Characteristics of Tape and Wire Recorders. As there is no wear on either the wire or tape in the recording or playback operations, they can be used almost indefinitely. If the wire should break, the two ends can be tied together. If the tape should break, the two ends can be joined by the use of a plastic tape. The use of magnetic tape has added great flexibility to recording operations. If a mistake is made, a correction can be repeated. The section of the tape containing the error is removed, and the corrected version is applied in its place. Many unique effects can be accomplished before the recording is complete: (1) two or more recordings can be superimposed on the same tape, (2) the range of certain sections can be increased or decreased, and (3) echo and other effects can be added. If the recording is no longer needed, the sounds can be removed and the tape or wire reused.

Some of the other advantages of the magnetic recorder are: (1) excellent fidelity can be obtained, (2) longer recording time, (3) the practically indestructible and compact record simplifies storage and handling, (4) it provides instantaneous playback without processing, (5) it has a minimum of background noise, (6) it is affected very little by external vibrations, and (7) it is completely portable. A disadvantage of the magnetic recorder is that regular preventive maintenance must be performed. In order to prevent deterioration of the recording on the tape, all parts of the equipment touching the oxide-side of the tape—such as tape guides, capstan and roller, and head surfaces—should be cleaned regularly with grain alcohol. Also, during recording and/or playback the head may become slightly magnetized. This small amount of magnetization may (1) add noise to the recording, (2) cause part of the signal to be erased, and (3) lower the signal-to-noise ratio. It is therefore of prime importance that the heads be demagnetized regularly, using a device designed for this purpose.

10-13. Sound Amplifiers. *Requirements of the Sound Amplifier.* Sound amplifiers are used to increase the voltage and current output of a microphone, record pickup, magnetic recorder, or radio receiver. Since this amplifier is used to increase the voltage and power of a-f signals, the explanation of the principles, operation, and characteristics of the a-f voltage and power amplifiers described in Chaps. 6 and 7 will also apply to sound amplifiers. Because of the number of different applications of the sound amplifier, it is referred to by various names such as the microphone amplifier, preamplifier, speech amplifier, phonograph amplifier, record amplifier, playback amplifier, public-address amplifier, etc. The power-output, frequency-response, and distortion requirements of a sound amplifier will depend upon its use.

The purpose of the voltage amplifier is to increase the input voltage to an amount high enough to operate a power amplifier that will produce the desired amount of power output. The power output of sound amplifiers ranges from a few watts to several hundred watts. The amount of power needed will depend upon the area to be covered, and whether the amplifier is to be used indoors or outdoors. The driver-voltage requirements of the power-amplifier stage will determine the number of stages of voltage amplification that must be used. The over-all voltage amplification is equal to the product of the amplifications of each stage of voltage amplification used. In order to reproduce speech with a fair degree of fidelity, the variation in the over-all gain of the amplifier over a range of 100 to 4,000 cycles should not exceed 1 db. Because of the much wider frequency range required for the reproduction of music, the frequency response of an amplifier used for general reception should be fairly uniform over a range of 50 to 8,000 cycles, and for high-fidelity reception the range is increased to 30 to 20,000 cycles.

Public-address Systems. A public-address unit is usually entirely self-contained. The complete unit consists of a power supply, voltage amplifier, power amplifier, one or more microphones, and a loudspeaker system. As some public-address applications require that more than one microphone be available, the input circuit is usually provided with more than one channel. Thus the input circuit may have one or more low-impedance channels for carbon, dynamic, or velocity microphones; and one or more high-impedance channels for crystal or magnetic pickups. Provision is also usually made for operating more than one loudspeaker from the output circuit. In order to facilitate impedance matching to the voice coil, the output transformer is usually tapped at various voice-coil impedances such as 2, 4, 6, 8, 15, etc., ohms.

Example 10-4. A microphone whose output is 60 db below 6 mw is feeding an amplifier that delivers 15 watts of undistorted output power. What is the over-all gain from the microphone to the loudspeaker?

Given: Input = 60 db below 6 mw   Find: Over-all gain
P_o = 15 watts

Solution:

$$\text{Output} = 10 \log \frac{P_o}{P_r}$$

$$= 10 \log \frac{15}{0.006} = 10 \log 2,500 = 10 \times 3.3979 \cong 34 \text{ db}$$

Over-all gain = db output − db input = 34 − (−60) = 94 db

Sound Amplifiers for Reproducing Music. This type of amplifier is used for reproducing the sounds from (1) musical instruments, such as a guitar, accordion, etc., and (2) recording and reproduction equipment, such as record players, tape recorders, and f-m tuners. Basically the function and circuit operation for an amplifier used for reproducing music are the same as those for reproducing speech. However, because of the wider frequency response and the higher fidelity requirements, this type of amplifier uses a number of circuits that are not generally used in speech amplifiers. The additional circuits may include (1) negative feedback, (2) low- and high-frequency peaking, (3) preemphasis and deemphasis, (4) harmonic-, intermodulation-, frequency-, phase-, and transient-distortion elimination, (5) equalization control, (6) loudness control, (7) treble control, and (8) bass control. The requirements for the equipment in which the amplifier is to be used will determine which of these circuits are to be included in the over-all amplifier circuit. The requirements of a high-fidelity amplifier, and an analysis of the operation of the additional circuits listed above, will be taken up in Arts. 10-15 and 10-16.

10-14. High-fidelity Equipment. *High-fidelity System.* The best sound reproduction is that which offers the listener as nearly as possible an exact replica of the original sound. A *distortion* is any difference between the original and the reproduced sound, and it may take various forms, as frequency, harmonic, etc. The term *high fidelity* refers to sound reproduction in which the various distortions are kept below the audible limits of the majority of listeners. A high-fidelity system is used to reproduce sounds with a minimum of distortion, and may include: (1) a microphone, (2) a tuner, (3) a record player, (4) a tape recorder and reproducer, (5) a preamplifier and control unit, (6) voltage and power amplifiers, (7) a loudspeaker system, and (8) a power supply. The over-all fidelity of the sound reproduction of the entire system can never be better than that of any of its components. For example, if the micro-phone does not pick up a low-frequency signal, it cannot be reproduced by the loudspeaker; or if the loudspeaker cuts off at 10,000 cps, a 15,000-cps signal reaching the input terminals of the loudspeaker will not be repro-duced. Because of the high amplification used in high-fidelity systems, the slightest noise or distortion produced by a single circuit element or a vibrating unit may be amplified millions of times and will be heard through the loudspeaker. Therefore, in order to obtain high-fidelity sounds from a reproduction system all the parts in each of the components used must be capable of high-fidelity reproduction.

Microphones. The important characteristics of microphones are (1) directivity, (2) frequency range, (3) output voltage, (4) output level, decibels, and (5) output impedance. The characteristics of the various types of microphones have been taken up in detail in Arts. 10-1 through Art. 10-7.

Tuners. A tuner may be designed for either a-m or f-m reception. Since a-m broadcast is limited to a channel width of 10 kc, the highest audio frequency that can be transmitted is 5,000 cycles. Using an a-m tuner with a high-fidelity system will produce only a slight increase in fidelity over that obtained with an ordinary system. The high-fidelity reproduction of the f-m tuner has many advantages over the a-m tuner, as, (1) it can receive audio frequencies up to 15,000 cycles, (2) it has the ability to discriminate against noise, and (3) it has less signal interference. The important characteristics of an f-m tuner are (1) sensitivity, (2) dis-tortion, and (3) i-f bandwidth. The circuit operations and character-istics of a-m tuners are taken up in Chap. 4, and f-m tuners are described in Chap. 12.

Record Players. A record player consists of (1) a turntable, (2) a tone arm, (3) a pickup cartridge and stylus, and (4) a record changer.

Turntables. Turntables are generally made to rotate at four dif-ferent speeds, $16\frac{2}{3}$, $33\frac{1}{3}$, 45, and 78 rpm. Distortion may be produced in a turntable by (1) a rumble in the drive system, (2) a variation in speed, and (3) a vibration of any component. The most widely used method for obtaining rotation employs an induction motor and a rim-drive system of power transmission. Drive power and speed reduction are provided through a series of rubber-tired wheels, one of which is driven by the motor armature. This wheel, in turn, drives a second wheel that is placed in contact with the rim of the turntable. The advantages of this method are: (1) Speed reduction is easily obtained. (2) Vibration and rumble are reduced. (3) While the turntable is being driven at the rim, the center hub can be used during change cycles to power a mechanical disc changer. (4) It is a relatively small unit. The rubber wheels are usually disengaged when not in use, so as to prevent a *flat* or *indentation* from being formed on them during periods of idleness.

The power from a motor may be coupled to the center shaft of a turntable by means of a flexible shaft or belt. In these two methods, the full power of the motor can be transmitted to the turntable without vibration, as the motor can be placed on its own shock mounting at a distance from the turntable. Speed changes are accomplished by (1) using a series of nylon or Teflon gears, or (2) shifting the belt to motor pulleys of different diameters. A direct-drive system is mainly used in professional equipment, as the cost of a low-speed motor is relatively high.

Tone Arms. The tone arm is a moving device that holds the pickup cartridge and stylus. Its functions are (1) to move freely over the horizontal surface of a disc record, (2) to allow the stylus to be easily set down on a record, or to be easily removed from it, and (3) to approach accurate tangential tracking for each successive groove as it moves toward the center of the record. The quality of a tone arm is determined by the degree to which it accomplishes these functions. A high-fidelity tone arm may incorporate one or more adjustments to improve its quality, namely (1) stylus weight, (2) height relationship to turntable, (3) length of tone arm, and (4) tracking angle.

Pickup Cartridge and Stylus. The general requirements for a high-fidelity phonograph pickup cartridge and stylus are: (1) widest possible frequency response, (2) high lateral compliance, (3) high signal output, (4) low stylus pressure, (5) minimum stylus vibration, (6) no signal to be produced from a vertical movement of the stylus, (7) low distortion, and (8) no noticeable hum at normal or low-signal levels. The characteristics of the various types of pickup cartridges and styluses have been taken up in detail in Arts. 10-10 and 10-11.

Record Changers. The automatic record changer makes provision for stacking a number of records to provide up to 2 hours of continuous playing with 78-rpm records, and up to 20 hours of continuous programing with long playing records. The method used by many manufacturers is the *drop system*. In this method a series of records are stacked above the rotating table on a bent spindle. After the completion of one side of the record on the turntable, and after the tone arm has been moved safely out of the way, the bottommost record in the stack drops onto the record on the turntable. A balanced tone arm starts picking up the sounds from the rotating record by automatically placing the stylus in its outermost groove.

The motor of a single-disc turntable has only the job of rotating the table. The motor of a record changer in addition to turning the table must (1) move the tone arm out of harm's way, and (2) drop records one on top of the other. The basic job of the motor, to rotate the table, is therefore done less efficiently in a record changer. The tone arm for a single installation holds the stylus in the groove and moves in as the record plays. The tone arm for a record changer, in addition to these two functions, must also trip a switch to start the changing cycle. As the tone arm presses against the switch, toward the end of the recording, it may drag the stylus against the outside edge of the groove. This dragging action may distort the sound, and cause excessive wear to the grooves. Record changers do not employ heavily-weighted turntables and therefore they lack the flywheel effect which helps to maintain a constant speed. Pickups are made to perform most accurately when the stylus is directly perpendicular to the flat record. A tone arm can be adjusted to hold the stylus in this position if there is to be only one record on the turntable. A record changer plays a stack of records, and the tone arm will hold the stylus perpendicular to only one record. The weight on the turntable increases with the number of records on it, and as the weight increases, the speed of the turntable may decrease. Because of these disadvantages, and the fact that the quality of a recording is impaired by dropping it into the grooves of a rotating record, the single-disc turntable is used with high-fidelity systems rather than the automatic record changer.

Tape Recorders and Reproducers. The high-fidelity requirements and characteristics of tape recorders and reproducers have been taken up in Art. 10-12.

Preamplifiers and Control Units. A preamplifier is basically an a-f voltage amplifier; the basic high-fidelity preamplifier circuit and its circuit operations are similar to the a-f voltage-amplifier circuits described in Chap. 6. Although the basic principles of the control circuits in a high-fidelity preamplifier are the same as those used for controlling the output of a voltage amplifier in a radio receiver, their requirements are more exacting. Preamplifiers and their control circuits will be taken up in Art. 10-15.

Voltage and Power Amplifiers. The basic voltage- and power-amplifier circuits used in high-fidelity systems are similar to the a-f voltage-amplifier circuits described in Chap. 6, and the a-f power-amplifier circuits described in Chap. 7. The application of these circuits to high-fidelity systems will be taken up in Art. 10-15.

Loudspeaker System with an Enclosure. The basic operation of, and the characteristics of, the various types of loudspeakers has been taken up in Arts. 10-8 and 10-9. The fidelity of sound reproduction can be improved by the use of (1) a specially designed single loudspeaker, (2) two or more loudspeakers, (3) a mechanical crossover system, (4) an electrical crossover network, and (5) a speaker enclosure. The application of these features to produce high-fidelity sound reproduction will be taken up in Art. 10-16.

Power Supply. The basic power-supply circuits used to provide the direct voltages to a high-fidelity system generally employ vacuum-tube rectification, and are similar to the power-supply circuits described in Chap. 9. However, the performance of any amplifier depends directly upon the quality of its power supply, and for high-fidelity applications the power supply should have (1) sufficient power output, (2) minimum distortion, and (3) good regulation. To minimize a-c hum the heaters of the vacuum tubes in the early stages of the preamplifier may be powered with direct current rather than with alternating current. Metallic-disk-type rectifiers are generally used to provide the low-voltage and high-current power required for this purpose.

10-15. Electronic Channel. *Requirements.* The major component of the audio reproduction system is the *audio amplifier section*, also called the *electronic channel.* The functions of this section are: (1) to accept the a-f electrical impulses from a transducer (a microphone, a phonograph pickup, a magnetic recorder head, or the output of an f-m or a-m detector), (2) to select the desired input signal, (3) to amplify the impulses to the required signal strength, (4) to select and apply the required amount of frequency correction or equalization, (5) to reduce noise and other types of distortion to a minimum, and (6) to supply sufficient power to produce the sound intensity required from the loudspeaker system. The electronic channel section may contain all the components and circuits required to perform these functions in one or more units. It may include (1) a preamplifier having provisions for accepting various types of input signals, (2) an input-signal selector switch, (3) volume and loudness controls, (4) tone controls, (5) an equalization control, (6) a voltage amplifier, and (7) a power amplifier. The amount of power output required from the audio amplifier will depend on (1) the size of the room in which it is to be used, and (2) the acoustic efficiency of the loudspeaker.

Example 10-5. An acoustic power output of 0.3 watt is required for loud listening in a room having a volume of approximately 2,000 cu ft. Determine the undistorted power output required from an audio-amplifier system when (*a*) the acoustic efficiency of the loudspeaker is 5 per cent, (*b*) the loudspeaker is mounted in an enclosure that increases its acoustic efficiency to 15 per cent.

$$\text{Given:} \qquad P_{o \cdot s} = 0.3 \text{ watt} \qquad \text{Find:} \quad (a)\ P_{o \cdot a}$$
$$(a)\ \text{Efficiency} = 5 \text{ per cent} \qquad \qquad (b)\ P_{o \cdot a}$$
$$(b)\ \text{Efficiency} = 15 \text{ per cent}$$

Solution:

(a) $\quad P_{o \cdot a} = \dfrac{P_{o \cdot s} \times 100}{\% \text{ efficiency}} = \dfrac{0.3 \times 100}{5} = 6$ watts

(b) $\quad P_{o \cdot a} = \dfrac{0.3 \times 100}{15} = 2$ watts

✓ *Preamplifiers.* A preamplifier is a device used to increase the voltage level of the input signal sufficiently to be used for mixing, equalization, or further amplification. The progress of the a-f signal through a typical preamplifier unit is illustrated in Fig. 10-21. Although there are seven input circuits, there is only one output circuit. Each of the amplified input signals is fed through the same output circuit to the input of the power amplifier. The path taken by an input signal depends upon the position of the *input-selector switch.* High-level input signals are fed directly, through the input-selector switch and the volume control circuit, to the input of the first stage of voltage amplification. Low-level input signals are first amplified by a *preamplifier stage,* and are then sent through the input selector switch and the volume-control circuit to the input of the first amplifier stage. The strength of the input signal can be varied by adjusting the separate *input-level controls* in each of the input circuits.

An *equalization circuit* is required in the magnetic phonograph-pickup input circuit to compensate for the *preemphasis* and *deemphasis* of the frequency response performed in making a recording. When a recording is made, the low frequencies are deemphasized (cut) to prevent over-cutting of the record grooves, and the high frequencies are preemphasized (reinforced) to permit a better balance between the high-frequency tones and the surface noise, thus producing a quieter recording. In order to obtain an approximately flat frequency response during the playing of a recording, an equalizer circuit is used to preemphasize the low-frequency signals and to deemphasize the high-frequency signals. The various disc recording companies have not standardized on the amount of compensation to be used. Hence, four equalizer circuits are generally provided to compensate for the amount used by the different companies. The *equalizer switch* selects the proper circuit to be used. The general classification of the positions on the switch are (1) AES—for modern 78-rpm recordings, and records recorded under the NARTB curve, (2) EUR—for European LP's and old 78-rpm recordings, (3) RIAA—for modern 33⅓- and 45-rpm recordings, and records recorded under the ortho curve, and (4) FFRR—a special curve for full-frequency-range recordings.

The human hearing system has the faculty of losing sensitivity to both bass and treble frequencies as the intensity level of the sound is reduced. A volume control is used to compensate for this hearing deficiency. However, since everyone does not hear all audio frequencies equally well, a *level control,* also called a *loudness control,* is added to the output circuit

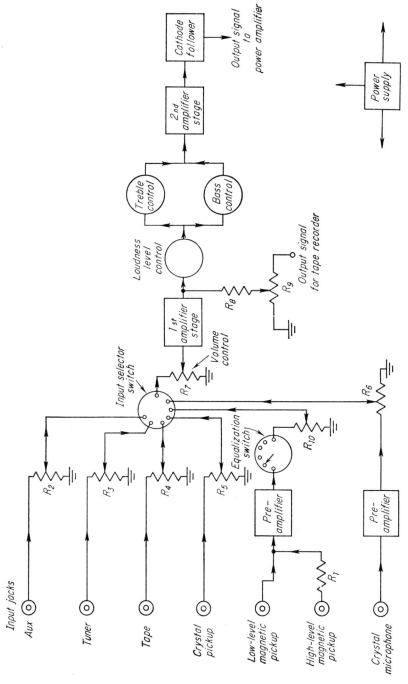

Fig. 10-21. Block diagram of a preamplifier.

of the first voltage-amplifier stage to provide adjustable compensation for individual differences in hearing. The output of the level control is fed to the input of the second voltage-amplifier stage through separate *bass* and *treble tone-control circuits*. The output of this stage is then coupled through a cathode-follower circuit to the input of the power amplifier. The preamplifier is also called a *control unit* because of all the selector and control circuits in this section of the high-fidelity system.

Power Amplifiers. The fundamental job of a power amplifier is to increase the power level of the output signal from the preamplifier sufficiently to operate a loudspeaker at a desired volume level, and with a minimum of distortion. The fundamental characteristics and circuit operations of the various types of power-amplifier circuits have been taken up in Chap. 7. A high-fidelity power amplifier may use (1) a pentode-amplifier circuit, (2) a beam-power-tube amplifier circuit, or (3) a variation of the basic Williamson amplifier circuit. The fundamental characteristics of the Williamson circuit are the use of (1) a push-pull output, (2) a specially designed output transformer, (3) two beam power tubes connected to operate as triodes, and (4) a negative-feedback voltage. One modification of the basic Williamson amplifier circuit uses an *ultralinear* output transformer. Each half-section of the primary winding of this transformer is tapped at approximately 20 per cent of its total turns. The screen-grids of the power-output tubes are connected to these two taps, and as a result partial triode and partial pentode operation occurs. The frequency response of an amplifier circuit using this modification may vary as little as ±1 db from 2 cps to 200 kc, with no peaking or raggedness. Intermodulation distortion can be reduced to less than 1 per cent, and its rated output can be fed to the loudspeaker system practically undistorted, from 20 cps to 20 kc.

The progress of the a-f signal through a typical power-amplifier unit is illustrated in Fig. 10-22. The input stage is a basic voltage-amplifier circuit. In some units the over-all maximum volume of the power amplifier is controlled by connecting a *gain control* in the input circuit of this stage. This control protects the loudspeaker, by limiting the output of the amplifier to the power rating of the loudspeaker. The output of the voltage amplifier is fed through a phase inverter stage to the input grids of the push-pull power-output tubes. Phase-inversion may be provided by (1) a center-tapped secondary interstage transformer, (2) a single-tube cathode-follower phase inverter, or (3) a dual-triode phase inverter. The output of the push-pull amplifier tubes is fed through a specially designed output transformer to the loudspeaker system. High-fidelity output transformers use balanced coils rather than single coils, and special methods of winding the coils on laminated iron cores, to operate at their rated power input with (1) a frequency range of 10 to 40,000 cycles, (2) a minimum of frequency distortion (produced by low-winding inductance, leakage reactance, or resonance), and (3) a minimum of intermodulation or harmonic distortion (produced by overloading the primary winding during low-frequency signals). Negative feedback is almost universally used, because its corrective capabilities can (1) reduce harmonic, intermodulation, and phase distortion, (2) correct for a poor output transformer, and (3) stabilize the entire amplifier circuit.

10-16. Corrective Networks. *Equalizer and Tone-control Circuits.* An equalizer, or tone-control, circuit is any network whose response varies in some desired manner over a given frequency range. Therefore,

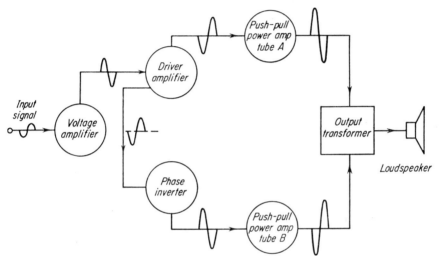

Fig. 10-22. Block diagram of a power amplifier.

the basic principles used for both these types of corrective networks are the same. The preemphasis or deemphasis circuits used to compensate for (1) the inherent nature of magnetic recording, and/or (2) the planned changes made to increase the frequency range, or to reduce the extraneous noise signals, may use *RL* or *RC* networks. The basic *RL* and *RC* equalizer circuits are shown in Fig. 10-23. In the simple series-attenuator circuit shown in Fig. 10-23*a*, the two impedances are pure resistance. The output level of this circuit does not vary with frequency changes, because the impedance of the resistors does not vary with frequency changes. When either of these two resistors is replaced by a capacitor or an inductor, the output level of the circuit will vary according to (1) the circuit element used, (2) the manner in which it is connected, and (3) the frequency of the a-f signal. When the series impedance is a capacitor, or the shunt impedance is an inductor, as shown in Fig. 10-23*c* and *e*, the response level increases at high frequencies. This type circuit is referred to as a *treble-boost* or *bass-attenuation network*. When the series impedance is an inductor, or the shunt impedance is a capacitor, as shown in Fig. 10-23*d* and *b*, the response level decreases at high frequencies. This type of circuit is referred to as a *bass-boost* or *treble-attenuation network*. Any number of different combinations of these three circuit elements, such as a series resonant or parallel resonant circuit placed in either the shunt or series arm, can be used to vary the slope of the attenuation curve, or to cause this curve to peak or dip at a specific frequency.

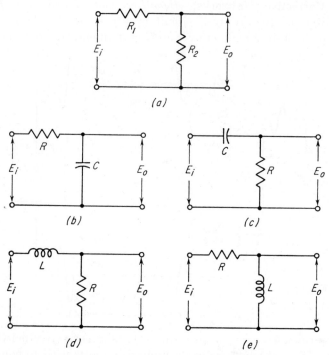

Fig. 10-23. Basic *RC* and *RL* equalizer circuits. (*a*) Flat frequency response. (*b*) High-frequency deemphasis–low-frequency preemphasis, or treble-attenuation–bass-boost. (*c*) Low-frequency deemphasis–high-frequency preemphasis, or bass-attenuation–treble-boost. (*d*) same as part (*b*), (*e*) same as part (*c*).

An equalizer, or tone-control, circuit that uses only resistors, inductors, and capacitors as circuit elements is called a *passive network*. This type circuit does not contain an amplifier or a source of voltage and is the most commonly used type of equalizer or tone-control circuit. A passive network cannot produce an output signal that is larger than its input signal. In order to obtain a desired frequency response, these circuits operate on the principle of attenuating the signals at certain frequencies. For example, a circuit used to obtain a treble boost of 10 db actually attenuates all the low- and medium-frequency signals by 10 db and leaves the level of the high-frequency signals unchanged. The *RC* network is the most widely used type of equalizer circuit employed for home reproduction applications. Inductors are bulky and expensive, and therefore are not generally used.

Cutoff and Turnover Frequencies. Since preemphasis and deemphasis concern the shape of the audio-response curve, it is convenient to have definite reference frequencies for which the response relative to some given frequency will be automatically known and which may be expressed in voltage, current, or power ratios. Usually three frequencies are used: (1) a reference frequency (approximately 900 cps) f_o, (2) a high frequency (approximately 1,100 cps) f_h, and (3) a low frequency (approximately 500 cps) f_l. The response at the reference frequency is considered to be unity, while the high- and low-frequency references are taken at points where the power has dropped to one-half its value at f_o, or has increased to twice its value at f_o. Since a power ratio of 1 to 2 or 2 to 1 is a 3-db change, the points f_h and f_l may be called *half-power points, double-power points, 3-db points, turnover frequencies,* or *rolloff frequencies.* In audio amplifiers, low-frequency compensation is obtained through circuits connected to the selector-equalizer switch and is called *turnover equalization.* High-frequency compensation is obtained through circuits connected to the rolloff switch, and is called *rolloff equalization.*

Example 10-6. What is the decibel change of an amplifier that has a power output of 10 watts at f_o and 5 watts at f_l and f_h?

Given: P_o = 10 watts   Find: Decibel change
$P_l = P_h$ = 5 watts

Solution:

$$\text{Change} = 10 \log \frac{P_h}{P_o} = 10 \log \frac{5}{10} \cong 10(9.70 - 10) \cong -3 \text{ db}$$

Note 1: log 5 = 0.699 ≅ 0.70. Therefore log 0.5 ≅ 9.70 − 10 ≅ −0.3.
Note 2: The minus sign indicates that the response is less at f_l and f_h than at f_o.

Example 10-7. What is the decibel change of an amplifier that has a power output of 5 watts at f_o and 10 watts at f_l and f_h?

Given: P_o = 5 watts   Find: Decibel change
$P_l = P_h$ = 10 watts

Solution:

$$\text{Change} = 10 \log \frac{P_h}{P_o} = 10 \log \frac{10}{5} = 10 \times 0.301 \cong 3 \text{ db}$$

Deemphasis Circuits. In order to produce deemphasis, a circuit must offer greater impedance to some signal frequencies than it does to others. All equalizer circuits are basically alternating voltage-divider networks. In the high-frequency deemphasis circuit of Fig. 10-23b, a higher voltage will be developed across C for low frequencies than for high frequencies. The voltage E_o developed across C will approach the value of E_i as the frequency is lowered. As the frequency is increased, the reactance of C drops progressively to lower values and approaches a short circuit at extremely high frequencies. When the reactance of C approaches zero, the voltage developed across it also approaches zero. If the voltage E_o is fed into an amplifier, the high-frequency components of E_o will not drive the amplifier as much as the low-frequency components will. The output of the amplifier will have a response curve shape similar to the high-frequency section of curve B of Fig. 10-24.

Interchanging the positions of C and R converts the high-frequency deemphasis circuit of Fig. 10-23b into the low-frequency deemphasis circuit shown in Fig. 10-23c. At high frequencies, the reactance of C is so low that only a small voltage drop will exist across C. Practically the full signal input will appear across R; hence, E_o will be very nearly equal to E_i. At low frequencies, the reactance of C becomes very great,

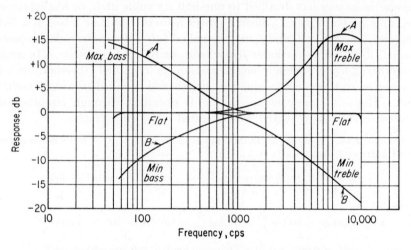

Fig. 10-24. Frequency-response curves of a tone-control circuit.

so that as the frequency changes from high to low values, more of the signal voltage will exist as a voltage drop across C, and less will be available across R. Thus as the frequency is lowered, E_o becomes progressively smaller. This action will affect the response of the amplifier to produce low-frequency deemphasis, and the response curve shape for the circuit is similar to the low-frequency section of curve B of Fig. 10-24.

The response of high-frequency deemphasis circuits may be expressed in terms of the time constant of the circuit used, as

$$R_{f \cdot h} = 20 \log \frac{1}{\sqrt{(2\pi f_h)^2 t^2 + 1}} \tag{10-3}$$

For low-frequency deemphasis circuits,

$$R_{f \cdot l} = 20 \log \frac{1}{\sqrt{\dfrac{1}{(2\pi f_l)^2 t^2} + 1}} \tag{10-4}$$

where $R_{f \cdot h}$ = high-frequency response, db
$\quad\ R_{f \cdot l}$ = low-frequency response, db
$\qquad\ t$ = time constant of circuit (RC), μsec
$\qquad f_h$ = rolloff frequency, cps
$\qquad f_l$ = turnover frequency, cps

The rolloff and turnover frequencies will occur at the 3-db points, and in terms of voltage and impedance occur when $X_c = R$. Consequently f_l and f_h are equal to $\dfrac{1}{2\pi t}$.

Example 10-8. If the RC circuit of Fig. 10-23b has a time constant of 150 μsec and the resistance of R is 500,000 ohms, find (a) the value of C, (b) the rolloff frequency, (c) the response in decibels at the rolloff frequency and whether it is rising or falling, (d) the reactance of C at the rolloff frequency, (e) the response at $2f_h$.

Given: $t = 150\ \mu$sec $\qquad\qquad$ Find: (a) C
$\qquad\qquad R = 500{,}000$ ohms $\qquad\qquad\qquad$ (b) f_h
$\qquad\qquad$ Circuit as in Fig. 10-23b $\qquad\qquad$ (c) $R_{f \cdot h}$
$\qquad\qquad\qquad\qquad\qquad\qquad\qquad\qquad\qquad$ (d) X_c
$\qquad\qquad\qquad\qquad\qquad\qquad\qquad\qquad\qquad$ (e) $R_{2f \cdot h}$

Solution:

(a) $C = \dfrac{t}{R} = \dfrac{150 \times 10^{-6}}{5 \times 10^5} = 3 \times 10^{-10}$ farad $= 300\ \mu\mu$f

(b) $f_h = \dfrac{1}{2\pi t} = \dfrac{1}{6.28 \times 150 \times 10^{-6}} \cong 1{,}060$ cps

(c) $R_{f \cdot h} = 20\ \log\ \dfrac{1}{\sqrt{(2\pi f_h)^2 t^2 + 1}} = 20 \times \log 0.707$

$\qquad\qquad = 20(9.85 - 10) \cong -3$ db (falling)

$\qquad$ where $(2\pi f_h)^2 t^2 = 6.28^2 \times 1{,}060^2 (150 \times 10^{-6})^2 \cong 1$

(d) $X_c = \dfrac{1}{2\pi f_h C} = \dfrac{10^6}{6.28 \times 1{,}060 \times 0.0003} \cong 500{,}000$ ohms

(e) $R_{2f \cdot h} = 20\ \log\ \dfrac{1}{\sqrt{(2\pi 2f_h)^2 t^2 + 1}} = 20 \times \log 0.446 = 20(9.649 - 10) \cong -7$ db

$\qquad$ where $(2\pi 2f_h)^2 t^2 = 6.28^2 \times 2{,}120^2 (150 \times 10^{-6})^2 \cong 4$

Example 10-9. If the RC circuit of Fig. 10-23c has a capacitance of 0.0068 μf and a resistance of 50,000 ohms, find (a) the time constant, (b) the turnover frequency, (c) the response in decibels at the turnover frequency and whether it is rising or falling, (d) the reactance of C at the turnover frequency, (e) the response at $0.5 f_l$.

Given: $R = 50{,}000$ ohms $\qquad\qquad$ Find: (a) t
$\qquad\qquad C = 6{,}800\ \mu\mu$f $\qquad\qquad\qquad\qquad\ $ (b) f_l
$\qquad\qquad\qquad\qquad\qquad\qquad\qquad\qquad\qquad$ (c) $R_{f \cdot l}$
$\qquad\qquad\qquad\qquad\qquad\qquad\qquad\qquad\qquad$ (d) X_c
$\qquad\qquad\qquad\qquad\qquad\qquad\qquad\qquad\qquad$ (e) $R_{0.5f \cdot l}$

Solution:

(a) $t = RC = 50 \times 10^3 \times 6{,}800 \times 10^{-12} = 340\ \mu$sec

(b) $f_l = \dfrac{1}{2\pi t} = \dfrac{10^6}{6.28 \times 340} \cong 470$ cps

(c) $R_{f\cdot l} = 20 \log \dfrac{1}{\sqrt{\dfrac{1}{(2\pi f_l)^2 t^2} + 1}} = 20 \times \log 0.707$

$\qquad = 20(9.85 - 10) \cong -3$ db (falling)

where $\dfrac{1}{(2\pi f_l)^2 t^2} = \dfrac{1}{6.28^2 \times 470^2 (340 \times 10^{-6})^2} \cong 1$

(d) $X_c = \dfrac{1}{2\pi f_l C} = \dfrac{10^6}{6.28 \times 470 \times 0.0068} \cong 50{,}000$ ohms

(e) $R_{0.5f\cdot l} = 20 \log \dfrac{1}{\sqrt{\dfrac{1}{(2\pi 0.5 f_l)^2 t^2} + 1}} = 20 \times \log 0.446 = 20(9.649 - 10) \cong -7$ db

where $\dfrac{1}{(2\pi 0.5 f_l)^2 t^2} = \dfrac{1}{6.28^2 \times 235^2 (340 \times 10^{-6})^2} \cong 4$

Preemphasis Circuits. In the high-frequency deemphasis circuit of Fig. 10-23b, the voltage E_o developed across the capacitor C approaches the value of E_i as the frequency is decreased. Hence, this circuit can also be used as a low-frequency preemphasis circuit. In the low-frequency deemphasis circuit of Fig. 10-23c, the voltage E_o developed across the resistor R approaches the value of E_i as the frequency is increased. Hence, this circuit can also be used as a high-frequency preemphasis circuit. The response of high-frequency preemphasis circuits may be expressed in terms of the time constant of the circuit used, as

$$R_{f\cdot h} = 20 \log \sqrt{(2\pi f_h)^2 t^2 + 1} \qquad (10\text{-}5)$$

For low-frequency preemphasis circuits

$$R_{f\cdot l} = 20 \log \sqrt{\dfrac{1}{(2\pi f_l)^2 t^2} + 1} \qquad (10\text{-}6)$$

Example 10-10. If the RC circuit of Fig. 10-23c has a time constant of 150 μsec and the resistance of R is 500,000 ohms, find (a) the value of C, (b) the rolloff frequency, (c) the response in decibels at the rolloff frequency and whether it is rising or falling, (d) the reactance of C at the rolloff frequency, (e) the response at $2f_h$.

Given: $t = 150\ \mu$sec   Find: (a) C
$\qquad\qquad R = 500{,}000$ ohms   (b) f_h
$\qquad\qquad\qquad\qquad\qquad\qquad\qquad\qquad$ (c) $R_{f\cdot h}$
$\qquad\qquad\qquad\qquad\qquad\qquad\qquad\qquad$ (d) X_c
$\qquad\qquad\qquad\qquad\qquad\qquad\qquad\qquad$ (e) $R_{2f\cdot h}$

Solution:

(a) $C = \dfrac{t}{R} = \dfrac{150 \times 10^{-6}}{5 \times 10^5} \cong 3 \times 10^{-10}$ farad $\cong 300\ \mu\mu f$

(b) $f_h = \dfrac{1}{2\pi t} = \dfrac{1}{6.28 \times 150 \times 10^{-6}} \cong 1{,}060$ cps

(c) $R_{f\cdot h} = 20 \log \sqrt{(2\pi f_h)^2 t^2 + 1} = 20 \log 1.414$
$\qquad = 20 \times 0.15 \cong +3$ db (rising)
where $(2\pi f_h)^2 t^2 = 6.28^2 \times 1{,}060^2 (150 \times 10^{-6})^2 \cong 1$

(d) $X_c = \dfrac{1}{2\pi f_h C} = \dfrac{10^6}{6.28 \times 1{,}060 \times 0.0003} \cong 500{,}000$ ohms

(e) $R_{2f\cdot h} = 20 \log \sqrt{(2\pi 2f_h)^2 t^2 + 1} = 20 \times \log 2.24 = 20 \times 0.35 \cong +7$ db
where $(2\pi 2f_h)^2 t^2 = 6.28^2 \times 2{,}120^2 (150 \times 10^{-6})^2 \cong 4$

Example 10-11.   If the RC circuit of Fig. 10-23b has a capacitance of 0.0068 μf and a resistance of 50,000 ohms, find (a) the time constant, (b) the turnover frequency, (c) the response in decibels at the turnover frequency and whether it is rising or falling, (d) the reactance of C at the turnover frequency, (e) the response at $0.5f_l$.

$$\text{Given:}\quad \begin{aligned} R &= 50,000 \text{ ohms} \\ C &= 6,800 \ \mu\mu\text{f} \end{aligned} \qquad \text{Find:}\quad \begin{aligned} &(a)\ t \\ &(b)\ f_l \\ &(c)\ R_{f\cdot l} \\ &(d)\ X_c \\ &(e)\ R_{0.5f\cdot l} \end{aligned}$$

Solution:

(a) $t = RC = 50 \times 10^3 \times 6,800 \times 10^{-12} = 340 \ \mu$sec

(b) $f_l = \dfrac{1}{2\pi t} = \dfrac{10^6}{6.28 \times 340} \cong 470$ cps

(c) $R_{f\cdot l} = 20 \log \sqrt{\dfrac{1}{(2\pi f_l)^2 t^2} + 1} = 20 \log 1.414$

$\qquad\qquad = 20 \times 0.15 \cong +3$ db (rising)

$\qquad$ where $\dfrac{1}{(2\pi f_l)^2 t^2} = \dfrac{1}{6.28^2 \times 470^2 (340 \times 10^{-6})^2} \cong 1$

(d) $X_c = \dfrac{1}{2\pi f_l C} = \dfrac{10^6}{6.28 \times 470 \times 0.0068} \cong 50,000$ ohms

(e) $R_{0.5f\cdot l} = 20 \log \sqrt{\dfrac{1}{(2\pi 0.5 f_l)^2} + 1} = 20 \times \log 2.24 = 20 \times 0.35 \cong +7$ db

$\qquad$ where $\dfrac{1}{(2\pi 0.5 f_l)^2 t^2} = \dfrac{1}{6.28^2 \times 235^2 (340 \times 10^{-6})^2} \cong 4$

Practical Tone-control Circuits.   Two practical applications of the basic preemphasis and deemphasis circuits are shown in Figs. 10-25 and 10-26.   In the tone-control circuit of Fig. 10-25a the value of C_1 is selected so as to be effectively a very high impedance compared with R_1 at f_o.   At this frequency C_2 is selected to practically provide a short circuit.   Thus, at the reference frequency the circuit is essentially a resistive voltage-divider network consisting of R_1 and R_2.   The maximum amount of frequency correction, and the insertion loss of the network, is determined by the output of this voltage-divider circuit.   As the frequency is increased above the reference frequency, the reactance of C_1 decreases until it effectively short-circuits R_1, and the output voltage E_o will approach the input voltage E_i.   The response curve for this circuit is similar to the high-frequency section of curve A of Fig. 10-24.

As the frequency is decreased below the reference frequency, the impedance of the series arm $R_1 C_1$ remains practically constant (approximately equal to R_1).   However, the reactance of C_2 increases with frequency decrease, thus increasing the impedance of section $R_2 R_3 C_2$.   Hence, at the low-frequency range the impedance of the shunt arm of the voltage divider becomes much greater than the impedance of the series arm, and the output voltage E_o will approach the input voltage E_i.   The response curve for this circuit is similar to the low-frequency section of curve A of Fig. 10-24.

In the circuit of Fig. 10-25a the amount of high-frequency and low-frequency boost are determined by the relative values of R_1, R_2, and R_3. The frequency at which turnover occurs is determined by the value of $R_2R_3C_2$, and the frequency at which rolloff occurs is determined by the value of R_1C_1. A variable amount of equalization can be obtained by replacing R_1 and R_3 with variable resistors, as shown in Fig. 10-25b. In this circuit, the low- or high-frequency equalization circuits may be varied independently of each other. A variable boost can thus be obtained at either or both ends of the a-f band, while the response at the reference frequency remains constant.

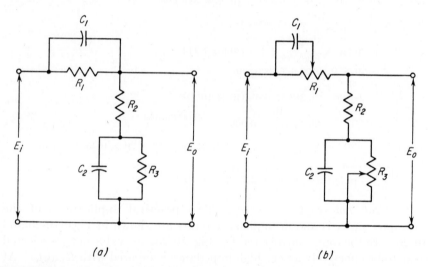

Fig. 10-25. Low-frequency and high-frequency boost circuits. (a) Fixed low and high frequency. (b) Variable low and high frequency.

Completely separate bass and treble control can be obtained from the tone-control network shown in Fig. 10-26a. Simplified diagrams of the bass and treble control circuits for the extreme positions of the controls R_3 and R_4 are shown in Figs. 10-26b, 10-26c, 10-26d, and 10-26e. The signal loss for any frequency and any position of the controls R_3 and R_4 may be expressed as

$$\text{Loss} = 20 \log \frac{E_o}{E_i} \tag{10-7}$$

When the potentiometer R_3 is in its extreme boost position, C_1 is short-circuited, and C_2 is thereby connected across R_3. The resultant voltage-divider network consists of the series circuit formed by R_1, R_2, and the parallel combination of R_3 and C_2 (Fig. 10-26b). At low frequencies the reactance of C_2 is much greater than the resistance of R_3 and the resultant output impedance (approximately equal to $R_3 + R_2$)

is a high percentage of the total input impedance. At high frequencies the reactance of C_2 is much smaller than the resistance of R_3, and the resultant output impedance (approximately equal to $\sqrt{R_2^2 + X_{c.2}^2}$) is a lower percentage of the total input impedance. Since the input and output voltages vary in direct proportion to their impedances, the ratio of the output voltage to the input voltage for maximum bass boost may be expressed as

$$\frac{E_o}{E_i} = \frac{\sqrt{R_o^2 + X_s^2}}{\sqrt{R_i^2 + X_s^2}} \tag{10-8}$$

where $R_o = R_2 + R_s$
$\quad R_i = R_1 + R_2 + R_s$
$\quad R_s = \dfrac{R_3 X_{c.2}^2}{R_3^2 + X_{c.2}^2}$
$\quad X_s = \dfrac{R_3^2 X_{c.2}}{R_3^2 + X_{c.2}^2}$

In the above equations R_s and X_s are the equivalent values of resistance and reactance that could be connected in series with R_1 and R_2 to produce the same effect in the circuit as is produced by the parallel combination of R_3 and C_2. The solutions of the following examples may then be simplified by considering them as series circuits.

Example 10-12. In the tone-control circuit of Fig. 10-26a the bass control is adjusted for maximum boost. Find the signal loss at 100 cycles.

Given: Circuit and values Fig. 10-26a   Find: Signal loss
$\qquad\qquad R_3$ in maximum-boost position

Solution:

NOTE: For ease of mathematical computation all values of resistance and reactance are expressed in kilo-ohms (k).

$$\text{Loss} = 20 \log \frac{E_o}{E_i}$$

$$\frac{E_o}{E_i} = \frac{\sqrt{R_o^2 + X_s^2}}{\sqrt{R_i^2 + X_s^2}} = \frac{\sqrt{68.5^2 + 145^2}}{\sqrt{288.5^2 + 145^2}} \cong 0.497$$

$$\text{Loss} = 20 \log 0.497 = 20(9.696 - 10) \cong -6.1 \text{ db}$$

where $R_s = \dfrac{R_3 X_{c.2}^2}{R_3^2 + X_{c.2}^2} = \dfrac{500 \times 160^2}{500^2 + 160^2} \cong 46.5 \text{ k}$

$\qquad X_s = \dfrac{R_3^2 X_{c.2}}{R_3^2 + X_{c.2}^2} = \dfrac{500^2 \times 160}{500^2 + 160^2} \cong 145 \text{ k}$

$\qquad X_{c.2} = \dfrac{1}{2\pi f C_2} = \dfrac{10^3}{6.28 \times 100 \times 0.01} \cong 160 \text{ k}$

$\qquad R_o = R_2 + R_s = 22 + 46.5 \cong 68.5 \text{ k}$

$\qquad R_i = R_1 + R_2 + R_s = 220 + 22 + 46.5 \cong 288.5 \text{ k}$

When the potentiometer R_3 is in its extreme cut position, C_2 is short-circuited, and C_1 is thereby connected across R_3. The resultant voltage-divider network consists of the series circuit formed by R_1, R_2, and the parallel combination of R_3 and C_1 (Fig. 10-26c). At low frequencies the reactance of C_1 is much greater than the resistance of R_3, and the output resistance R_2 is a low percentage of the resultant input impedance (approximately equal to $R_1 + R_2 + R_3$). At high frequencies the reactance of C_1 is much smaller than the resistance of R_3, and the output resistance is a higher percentage of the resultant input impedance (approximately equal to $\sqrt{(R_1 + R_2)^2 + X_{c.1}{}^2}$). The ratio of the output voltage to the input voltage for maximum bass cut may be expressed by

$$\frac{E_o}{E_i} = \frac{R_2}{\sqrt{R_i{}^2 + X_s{}^2}} \qquad (10\text{-}9)$$

Example 10-13. In the tone-control circuit of Fig. 10-26a the bass control is adjusted for maximum cut. Find the signal loss at 100 cycles.

Given: Circuit and values, Fig. 10-26a   Find: Signal loss
         R_3 in maximum cut position
Solution:

$$\text{Loss} = 20 \log \frac{E_o}{E_i}$$

$$\frac{E_o}{E_i} = \frac{R_2}{\sqrt{R_i{}^2 + X_s{}^2}} = \frac{22}{\sqrt{697^2 + 142^2}} \cong 0.031$$

$$\text{Loss} = 20 \log 0.031 = 20(8.49 - 10) \cong -30.2 \text{ db}$$

$$\text{where } R_s = \frac{R_3 X_{c.1}{}^2}{R_3{}^2 + X_{c.1}{}^2} = \frac{500 \times 1,600^2}{500^2 + 1,600^2} \cong 455 \text{ k}$$

$$X_s = \frac{R_3{}^2 X_{c.1}}{R_3{}^2 + X_{c.1}{}^2} = \frac{500^2 \times 1,600}{500^2 + 1,600^2} \cong 142 \text{ k}$$

$$X_{c.1} = \frac{1}{2\pi f C_1} = \frac{10^3}{6.28 \times 100 \times 0.001} \cong 1,600 \text{ k}$$

$$R_i = R_1 + R_2 + R_s = 220 + 22 + 455 = 697 \text{ k}$$

Example 10-14. Assuming that the preemphasis and deemphasis response curves for the tone-control circuit of Examples 10-12 and 10-13 are symmetrical, find for the mid-frequency setting: (a) the amount of gain or loss in decibels at 100 cycles, (b) the decibel output at the tone control.

Given: Results of Examples 10-12   Find: (a) Decibel gain or loss
         and 10-13   (b) Decibel output
Solution:

(a) $\text{Gain or loss} = \dfrac{\text{decibel cut at 100 cycles} - \text{decibel boost at 100 cycles}}{2}$

$$= \frac{-30.2 - (-6.1)}{2} \cong \pm 12 \text{ db}$$

(b)   Output = decibel cut + decibel gain = $-30.2 + 12 \cong -18.2$ db

In the treble control circuit of Fig. 10-26a the voltage-divider network consists of R_4, C_3, and C_4. When the potentiometer R_4 is in its extreme-boost position, the output impedance consists of the series circuit formed by R_4 and C_4 (Fig. 10-26d). At high frequencies the reactances of C_3 and C_4 are low, and the output impedance (approximately equal to R_4)

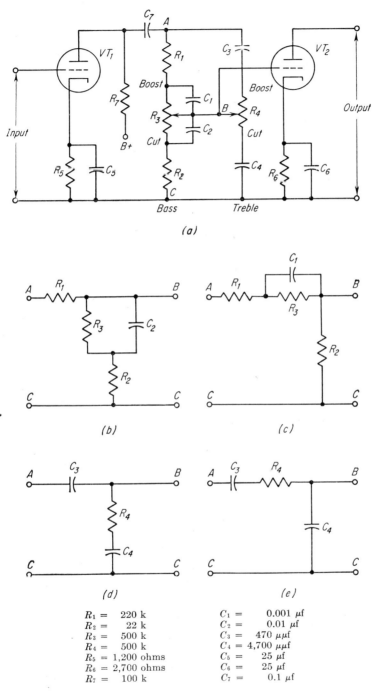

$R_1 =$	220 k	
$R_2 =$	22 k	
$R_3 =$	500 k	
$R_4 =$	500 k	
$R_5 =$	1,200 ohms	
$R_6 =$	2,700 ohms	
$R_7 =$	100 k	

$C_1 =$	0.001 μf
$C_2 =$	0.01 μf
$C_3 =$	470 $\mu\mu$f
$C_4 =$	4,700 $\mu\mu$f
$C_5 =$	25 μf
$C_6 =$	25 μf
$C_7 =$	0.1 μf

FIG. 10-26. (*a*) A bass and treble tone-control circuit. (*b*) Bass boost. (*c*) Bass cut. (*d*) Treble boost. (*e*) Treble cut.

is a high percentage of the total input impedance. At low frequencies the reactances of C_3 and C_4 are high, and the output impedance (approximately equal to $\sqrt{R_4{}^2 + X_{c\cdot4}{}^2}$) is a lower percentage of the total input impedance. The ratio of the output voltage to the input voltage for maximum treble boost may be expressed as

$$\frac{E_o}{E_i} = \frac{\sqrt{R_4{}^2 + X_{c\cdot4}{}^2}}{\sqrt{R_4{}^2 + (X_{c\cdot3} + X_{c\cdot4})^2}} \tag{10-10}$$

Example 10-15. In the tone-control circuit of Fig. 10-26a the treble control is adjusted for maximum boost. Find the signal loss at 4,000 cycles.

Given: Circuit and values Fig. 10-26a   Find: Signal loss
             R_4 in maximum-boost position
Solution:

$$\text{Loss} = 20 \log \frac{E_o}{E_i}$$

$$\frac{E_o}{E_i} = \frac{\sqrt{R_4{}^2 + X_{c\cdot4}{}^2}}{\sqrt{R_4{}^2 + (X_{c\cdot3} + X_{c\cdot4})^2}} = \frac{\sqrt{500^2 + 8.5^2}}{\sqrt{500^2 + (85 + 8.5)^2}} \cong 0.985$$

$$\text{Loss} = 20 \log 0.985 = 20(9.994 - 10) \cong -0.12 \text{ db}$$

where $X_{c\cdot3} = \dfrac{1}{2\pi f C_3} = \dfrac{10^9}{6.28 \times 4{,}000 \times 470} \cong 85 \text{ k}$

$\qquad\quad X_{c\cdot4} = \dfrac{1}{2\pi f C_4} = \dfrac{10^9}{6.28 \times 4{,}000 \times 4{,}700} \cong 8.5 \text{ k}$

When the potentiometer R_4 is in its extreme-cut position, the output impedance consists only of C_4 (Fig. 10-26e). At high frequencies the reactances of C_3 and C_4 are low, and the resultant output impedance is a low percentage of the input impedance. At low frequencies the reactances of C_3 and C_4 are high, and the resultant output impedance is a higher percentage of the input impedance. The ratio of the output voltage to the input voltage for maximum treble cut may be expressed as

$$\frac{E_o}{E_i} = \frac{X_{c\cdot4}}{\sqrt{R_4{}^2 + (X_{c\cdot3} + X_{c\cdot4})^2}} \tag{10-11}$$

Example 10-16. In the tone-control circuit of Fig. 10-26a the treble control is adjusted for maximum cut. Find the signal loss at 4,000 cycles.

Given: Circuit and values Fig. 10-26a   Find: Signal loss
            R_4 in maximum cut position
Solution:

$$\text{Loss} = 20 \log \frac{E_o}{E_i}$$

$$\frac{E_o}{E_i} = \frac{X_{c\cdot4}}{\sqrt{R_4{}^2 + (X_{c\cdot3} + X_{c\cdot4})^2}} = \frac{8.5}{\sqrt{500^2 + (85 + 8.5)^2}} \cong 0.0167$$

$$\text{Loss} = 20 \log 0.0167 = 20(8.223 - 10) = -35.5 \text{ db}$$

where $X_{c\cdot3} = 85 \text{ k}$
$\qquad\quad X_{c\cdot4} = 8.5 \text{ k (from Example 10-15)}$

Example 10-17. Assuming that the preemphasis and deemphasis response curves for the tone-control circuit of Examples 10-15 and 10-16 are symmetrical, find for the mid-frequency setting: (*a*) the amount of gain or loss in decibels at 4,000 cycles, (*b*) the decibel output at the tone control.

Given: Results of Examples 10-15   Find: (*a*) Decibel gain or loss
and 10-16   (*b*) Decibel output

Solution:

(*a*) Gain or loss $= \dfrac{\text{decibel cut at 4,000 cycles } - \text{ decibel boost at 4,000 cycles}}{2}$

$$= \frac{-35.5 - (-0.12)}{2} = \pm 17.7 \text{ db}$$

(*b*) Output $=$ decibel cut $+$ decibel gain $= -35.5 + 17.7 = -17.8$ db

10-17. Practical Audio-amplifier Circuit. *Specifications.* The circuit diagram for a commercial high-fidelity audio amplifier is shown in Fig. 10-27. This amplifier contains a number of the circuits and features described in the previous articles. The specifications for this unit are similar to those for most high-fidelity amplifiers; however, the values presented will vary with power output and quality.

Power output. 18 watts, sine wave, into a resistive load

Harmonic distortion. less than 0.5 per cent at 18 watts, at 1 kc through the tuner input, tone-control settings at normal

Hum and noise. Better than 50 db below 18 watts at low-level inputs referred to 10-mv input at "General Electric;" better than 60 db below 18 watts at high-level inputs referred to 0.8-volt input at "tape"

Input voltages for 18 watts output. See Table 10-1

TABLE 10-1

Low level		High level	
Input	Input, mv	Input	Input, volts
Tape head	5	Auxiliary (Aux.)	1.8
General Electric (GE)	5	Ceramic Phono (Cer.)	0.5
Pickering (Pic.)	12	Tape preamplifier (Tape)	0.4
Microphone (Mic.)	35	Tuner	1.8

Feedback. 18-db negative feedback

Frequency response. ± 1 db, 20 to 30,000 cps at half power

Tone controls. 12 db of bass boost or bass cut at 20 cps, 12 db of treble boost or treble cut at 20,000 cps

Output impedances. 4, 8, and 16 ohms

Power consumption. 90 watts at 117/120 volts a-c, 60 cps

Circuit Operation. The input signals from a phonograph pickup or a tape head are fed to the input grid of the first stage of preampification, first half of VT_1, through a double-pole slide switch S_1. The otlher input signals are fed to the input grid of the second stage of preamplification, second half of VT_1, through the selector switch S_{2D}. A flat frequency response is obtained by connecting the proper equalizer circuit, by means of switches S_{2A}, S_{2B}, and S_{2C}, to correct the input signal from (1)

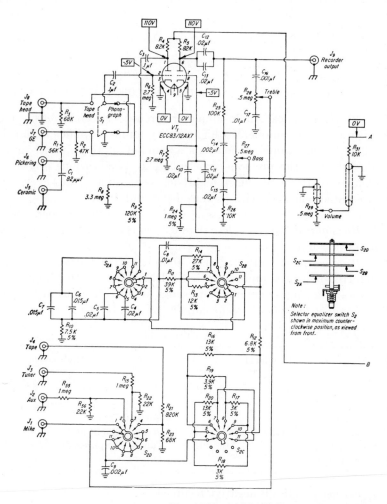

Fig. 10-27. A high-fidelity amplifier circuit.

disc recordings, (2) tape recordings, and (3) an f-m tuner. Bass and treble adjustments are made by varying the output from the preamplifier by means of a tone-control circuit that is similar to Fig. 10-26a. The bass-control circuit consists of R_{25}, R_{26}, R_{27}, C_{14}, and C_{15}. The treble-control circuit consists of R_{28}, C_{16}, and C_{17}. The output from the tone-control circuit is fed through a volume control R_{29} to the input grid of the driver stage, VT_2; and its output is fed through the grounded-grid phase inverter, VT_{3A} and VT_{3B}, to the push-pull amplifiers VT_4 and VT_5. Negative feedback is provided by R_{40} and C_{25}, which are connected from a tap on the secondary winding of the output transformer T_2 to the

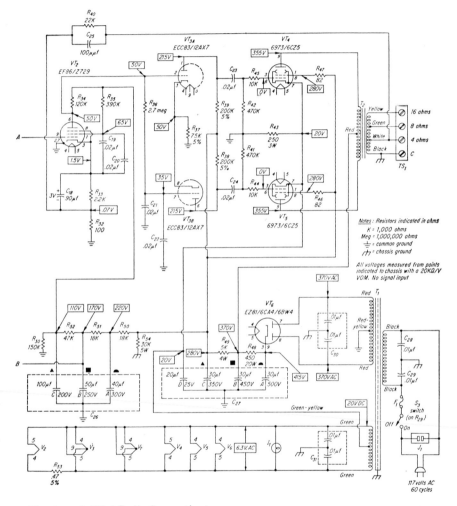

(*Courtesy of Allied Radio Corporation.*)

cathode of the driver-stage tube VT_2. The power supply uses a step-up transformer T_1, and a full-wave rectifier VT_6. The a-c hum interference is minimized by the filter circuits connected in (*a*) the power line, (*b*) the heater circuits, and (*c*) the rectified a-c output circuit.

Since the grounded-grid phase inverter has not been described previously, a brief description of its circuit operation will be presented. The output of the driver stage VT_2 is fed to the grid of VT_{3A}, which is one-half of a grounded-grid phase inverter. Because the grid of VT_{3A} is also connected to B+, the bottom end of the grid resistor R_{36} and the grid of VT_{3B} are effectively grounded to a-f signals by C_{21} and C_{22}. Phase inversion is obtained as follows: (1) A positive signal at the grid of VT_{3A} will produce a negative signal at the output of VT_{3A} and the input of VT_4. (2) A positive signal at the grid of VT_{3A} will cause the plate current in VT_{3A} to increase, therefore also increasing the current through the cathode resistor R_{37}. (3) As there is no bypass capacitor across R_{37}, a positive signal will be developed at the cathode of VT_{3B}. (4) With the grid of VT_{3B} grounded, the positive signal at its cathode will produce the same effect as if the grid was receiving a negative signal. (5) The output of VT_{3B}, and therefore also the input of VT_5, will have a positive signal. (6) The input signals at VT_4 and VT_5 are of opposite polarities, and phase reversal has been accomplished.

10-18. High-fidelity Reproduction of Sound. *Full-range Single-unit Loudspeaker.* Because of the inertia of a loudspeaker cone, the portion nearest the rim is unable to keep pace with the rapid movement of the portion nearest the voice coil. Hence, large-diameter cones will reproduce low-frequency sounds the best, and small-diameter cones will reproduce high-frequency sounds the best. Single-unit loudspeakers compensate for the inertia-caused disadvantages by utilizing the central portion of the cone for high frequencies, and the peripheral portion for low frequencies. This type of speaker may use a large cone having a hardened-central area, or an aluminum dome over its voice coil, to radiate high-frequency sounds. To reproduce the low-frequency sounds from the peripheral portion the cone is made large, using a stiff lightweight material having (1) annular embossed rim corrugations to lower the mass of the diaphragm at high frequencies, and (2) variations in the shape and depth of the cone.

Full-range Multiunit Loudspeaker Systems. Because of the limitations of a full-range single-unit loudspeaker, high-fidelity loudspeaker systems may use two, three, or four loudspeakers to cover the entire a-f range. Each of these loudspeakers is designed (1) to operate over a narrow range of frequencies, and (2) to overlap in a manner to smoothly cover the complete audio range. A loudspeaker designed to reproduce low-frequency sounds is generally referred to as a *woofer*. In a two-loudspeaker system the woofer is usually designed to cover a frequency range of 35 to 1,200 cps, while in a three- or four-loudspeaker system the range is only 35 to 600 cps (or lower). The general requirements of a woofer are: (1) good low-frequency performance, obtained by using a moderately stiff large-diameter cone and a high-efficiency large permanent magnet; (2) a highly compliant suspension system that permits unrestricted movement at high driving power; (3) a heavy frame to prevent distortion as a result of warping of the mounting structure, (4) a deep-cone diaphragm for use in direct-radiator enclosures, having a resonant frequency that is less than the lowest frequency to be reproduced, and (5) a proper crossover frequency in relation to the other loudspeakers in the system.

A loudspeaker designed to reproduce high-frequency sounds is called a *tweeter*. This type of loudspeaker may use (1) a relatively hard cone, (2) a small-diameter diaphragm, or (3) a hard-shelled dome diaphragm to reproduce high-frequency sounds with high efficiency and good fidelity. These diaphragm-driver units use a relatively large-diameter voice coil in comparison with their small-diameter cones or domes. Tweeter units usually operate with a horn having a small apex, or throat opening, in comparison with the size of the dome. The dispersion of high-frequency sounds may be accomplished by using (1) a flared horn, (2) acoustic lenses (multicellular horn unit), and (3) a dispersion plug at the mouth of the unit. The frequency range of a tweeter depends on (1) the crossover frequency of the midrange loudspeaker, and (2) the quality of reproduction required; and it may vary from 1,500 to 40,000 cps.

A loudspeaker designed to reproduce middle-frequency sounds is called a *driver*. Cone-type loudspeakers are often used to reproduce the middle-frequency sounds, and the general requirements for this type of loudspeaker are similar to those for any other narrow-range unit. A horn-type driver using a tweeter unit is also used because of the high efficiency and good fidelity obtained in the reproduction of middle-frequency and high-frequency sounds. The frequency range of a driver unit depends on the crossover frequencies of the other loudspeakers in the system and may range from 300 to 5,000 cycles.

Coaxial and Triaxial Loudspeakers. When two separate loudspeakers are mounted together on the same axis the unit is called a *coaxial loudspeaker*. The small treble loudspeaker is mounted inside the big bass loudspeaker so that a line running back from the midpoint at the mouth of the loudspeaker unit will pass through the center of both loudspeakers. When three separate loudspeakers are mounted together on the same axis, the unit is called a *triaxial loudspeaker*. The specific construction of coaxial and triaxial loudspeakers varies with the manufacturer, and are also called *two-way loudspeaker, two-way diffaxial loudspeaker, three-way loudspeaker,* and *three-way diffaxial loudspeaker.* In the triaxial loudspeaker of Fig. 10-28 the mid-range driver horn passes through the voice-coil form of the low-frequency driver. The high-frequency driver is mounted off-center within the low-frequency cone, so as not to obstruct the mid-frequency sound waves coming through the center of the woofer. Some of the advantages of coaxial and triaxial loudspeakers are (1) they occupy less space than multiple units, (2) the sound waves come from a single source, and (3) it is possible to use a lower crossover frequency in the bass unit because of the efficient self-powered mid-range driver.

Loudspeaker Phasing.   As the cone in a loudspeaker moves in and out, it creates pressure and rarification waves, thus producing the sound waves that impinge upon the eardrums of a listener.   When two loudspeakers are used, the sound intensity is maximum when the loudspeakers

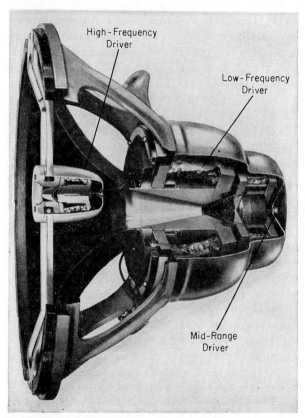

High-Frequency Driver

Low-Frequency Driver

Mid-Range Driver

Fig. 10-28. Driver units in a triaxial loudspeaker.   (*Courtesy of Jensen Radio Manufacturing Co.*)

are *in phase*, that is when the two cones move in the same direction at the same time.   When two loudspeakers are connected *out of phase*, one of the loudspeaker cones will move out while the other cone is moving in. This out-of-phase movement may result in (1) distortion of the signal, and (2) a reduction of the intensity of the sound wave.   This phasing effect is of greatest importance when the distance between the two loudspeakers is less than a few wavelengths.   For example; at 3,000 cycles one wavelength is approximately 4.5 inches, at 300 cycles it is approximately 45 inches, and at 30 cycles it is approximately 450 inches.

Thus, it would seldom make much difference whether or not two high-frequency loudspeakers were in phase.   However, it is of great importance when two low-frequency loudspeakers are (1) placed in the same cabinet, or (2) placed on opposite ends of one side of an average room.

The polarity of a loudspeaker may be determined by using a single 1.5-volt cell in the following manner: (1) The cell is connected across the voice coil, and the movement of the cone noted.   (2) At the instant the cell is connected the cone will be drawn in or pushed out, depending upon the manner in which the cell is connected.   (3) Mark with a " | " sign the terminal of the voice coil that is connected to the positive terminal of the cell when the cone is pushed out.   (4) Repeat this test

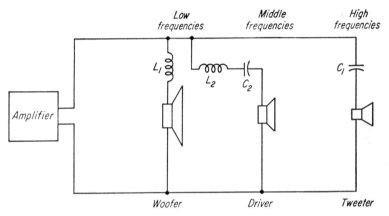

Fig. 10-29. A simple crossover system.

for the other loudspeakers to be used in the same system.   To obtain proper phasing all the terminals marked + should be connected together, and all the terminals having no marks should be connected together.

Crossover Networks.   When two or more loudspeakers are used to reproduce the entire range of a-f sounds, the frequency below which the lower-frequency loudspeaker receives its electrical signals and above which the higher-frequency loudspeaker receives its electrical signals is called the *crossover frequency.*   The audio signal can be separated into two or more frequency ranges by the use of a *crossover network.*   The simplest crossover network consists of (1) a low-pass filter using an inductor that is connected in series with the low-frequency loudspeaker, and (2) a high-pass filter using a capacitor that is connected in series with the high-frequency loudspeaker.   If three loudspeakers are used, a bandpass filter using a series LC circuit is connected in series with the mid-frequency loudspeaker (Fig. 10-29).   In order to obtain maximum efficiency, the crossover network should present a constant impedance of a proper value to match the impedance of the loudspeaker used.   A good impedance match cannot be obtained from the simple crossover circuits of Fig. 10-29. A better impedance match, plus a greater attenuation at the crossover frequency, can be obtained by using the parallel- or series-crossover network shown in Fig. 10-30. An attenuation of approximately 15 db for the first octave away from the crossover frequency can be obtained by using the following equations:

$$L_1 = \frac{R_o}{2\pi f_c} \qquad (10\text{-}12a)$$

$$L_2 = 1.6L_1 \qquad (10\text{-}12b)$$
$$L_3 = 0.625L_1 \qquad (10\text{-}12c)$$

$$C_1 = \frac{1}{2\pi f_c R_o} \qquad (10\text{-}13a)$$

$$C_2 = 0.625C_1 \qquad (10\text{-}13b)$$
$$C_3 = 1.6C_1 \qquad (10\text{-}13c)$$

where R_o = impedance of voice coil, ohms
$\quad\ f_c$ = crossover frequency, cps

Example 10-18.   A low- and high-frequency loudspeaker are connected as shown in Fig. 10-30c. The impedance of their voice coils is the same and is equal to 8 ohms. Find the values of capacitance and inductance required for a crossover frequency of 1,250 cycles.

Given: Circuit of Fig. 10-30c   Find: L_1, L_2
$\qquad\qquad R_o$ = 8 ohms C_1, C_2
$\qquad\qquad f_c$ = 1,250 cycles

Solution:

$$L_1 = \frac{R_o}{2\pi f_c} = \frac{8}{6.28 \times 1,250} \cong 0.001 \text{ henry} \cong 1 \text{ mh}$$

$$L_2 = 1.6L_1 = 1.6 \times 1 \cong 1.6 \text{ mh}$$

$$C_1 = \frac{1}{2\pi f_c R_o} = \frac{1}{6.28 \times 1,250 \times 8} \cong 0.000016 \text{ farad} \cong 16 \text{ }\mu f$$

$$C_2 = 0.625C_1 = 0.625 \times 16 = 10 \text{ }\mu f$$

Since the two loudspeakers may not have been designed to be used together, the efficiency of one may be greater than the other, and therefore they will not produce equal sound outputs over the two frequency ranges. The differences in the efficiencies of the two loudspeakers may be compensated for by inserting a constant-resistance attenuator pad, R_1, R_2, and R_3 of Fig. 10-30c, in the circuit of the more efficient loudspeaker. This attenuator pad is usually connected in the circuit of the high-frequency loudspeaker, and is used to balance the response over the two a-f ranges. The general requirements for the attenuator pad are (1) to maintain a constant input impedance for any attenuator setting, and (2) to dissipate the power that is not applied to the loudspeaker.

10-19. Loudspeaker Enclosures and Baffles. *Need for a Baffle.* If a loudspeaker were suspended in the open air, all the high-frequency sounds would be reproduced with a minimum loss in volume, while practically all the low-frequency sounds would be lost. This happens for two reasons: (1) high-frequency sound waves tend to move out in a forward direction along a straight line that is a continuation of the center line of the loudspeaker. (Because of this feature tweeter horns are usually designed to distribute their sound waves over a wide hemisphere.) (2) Low-frequency sound waves do not have a directional bias

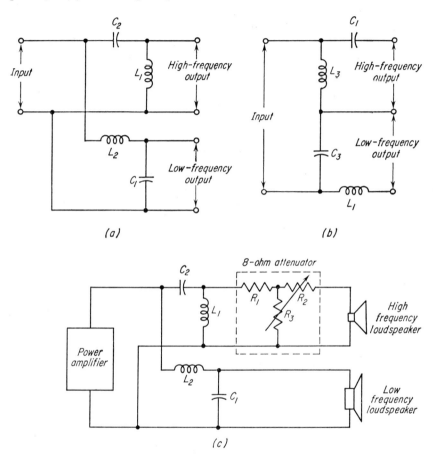

FIG. 10-30. Loudspeaker crossover networks. (a) A parallel network. (b) A series network. (c) A practical application of a parallel network.

and therefore spread out in all directions. As a result of the in-and-out movement of the diaphragm, or cone, the sound waves in the back of the loudspeaker will be 180 degrees out of phase with the sound waves in front of the loudspeaker. Should these two sound waves meet, because of reflection or other means, they would cancel each other, and there would be no sound. A loudspeaker mounting that is used to prevent the sound waves from the rear of a loudspeaker from interfering with the sound waves in front of the loudspeaker is called a *baffle,* or *loudspeaker enclosure.*

Infinite Baffle. If a loudspeaker is mounted on a board, with a hole cut out to the diameter of the cone, the sound waves reproduced from the rear of the loudspeaker will be baffled in their attempt to come around to the front of the loudspeaker. If a wall in a room was substituted for the board, it would be impossible for any of the sound waves from the rear of the loudspeaker to get around to the front of the loudspeaker, and the wall is called an *infinite baffle.* With the same reasoning a totally enclosed box, with just one hole for the loudspeaker cone, is also an infinite baffle. However, the difficulty of using a totally enclosed box is that it may reenforce the natural resonance of the loudspeaker causing the response to rise or drop sharply at this frequency.

Bass-reflex Baffle. A bass-reflex baffle is an enclosed cabinet having two openings at the front (Fig. 10-31a)—the top opening is for the loudspeaker cone, and the bottom opening is called the *reflex port,* or *port.* With an enclosure of sufficient volume, and the port opening of the correct size in relation to the loudspeaker, it is possible to reverse the phase of the sound energy coming from the back of the loudspeaker. This reversal occurs near the resonant frequency of the system. The size of the port is made just large enough to cause the resonant frequency of the column of air inside the box to be slightly lower than the fundamental resonant frequency of the loudspeaker. If the port opening is of the proper size the low-frequency sound waves will emerge from the port opening in phase with the sound waves from the front of the loudspeaker, thus aiding the loudspeaker response at the low frequencies. The back, the top, and one side of the inside of the enclosure are covered with an absorbent material, such as Ozite, Celotex, or Fiberglas, to absorb all the back-pulsing high-frequency sound waves.

In order to obtain a relatively low resonant frequency from the air inside a bass-reflex enclosure the box must be large, at least 6 cu ft for a 12-inch loudspeaker, and 8 cu ft for a 15-inch loudspeaker. The outstanding feature of this type of enclosure is that it can be used with almost any loudspeaker merely by changing the size of the port opening. However, since sound waves vary in length, the mid-range frequencies will tend to emerge from the port out-of-phase with the sound waves emanating from the front of the loudspeaker. Some distortion is thus produced at the middle frequencies with this type enclosure.

Acoustic Labyrinth. An acoustic labyrinth is a modification of the bass-reflex enclosure in which the back pulsations are bounced back and forth through a maze of small internal boxes (Fig. 10-31b) before emerging from the port. This type of enclosure causes the sound waves from the back of the loudspeaker to travel through a longer path before reaching the port opening than in the bass-reflex enclosure. Since the resonant frequency decreases as the column of air increases in length, the acoustic labyrinth lowers the bass response of a loudspeaker.   The length of the air column should be one-fourth the wavelength of the resonant frequency of the loudspeaker.   As with the bass-reflex enclosure, the high-frequency reflections are absorbed by covering the inside surfaces of the path with a sound-absorption material.

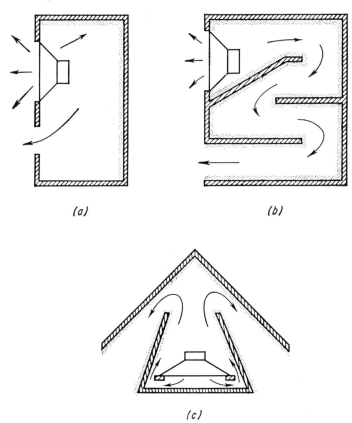

(a)   *(b)*

(c)

Fig. 10-31. Loudspeaker enclosures. (*a*) Bass-reflex. (*b*) Acoustic-labyrinth. (*c*) Klipschorn.

Klipschorn Enclosure.   The Klipschorn enclosure loads the front of the loudspeaker by having it face a closed board (Fig. 10-31*c*) instead of facing out into the room through a hole in the front of the enclosure. The sound waves leaving the front of the loudspeaker are forced to move around to join the sound waves leaving the rear of the loudspeaker. The two sound waves travel together through a folded exponential horn and emerge from both sides of the enclosure into a corner of a room. Since the largest part of the horn is the enormous mouth, formed by the corner of the room, the enclosure can be made relatively small.

This type of enclosure sets up a great air pressure against both the front and back of the loudspeaker cone, thus lowering the fundamental resonant frequency of the loudspeaker. Approximately one octave of bass response is added to a woofer by this pressure. Two disadvantages of the Klipschorn enclosure are (1) it can only hold a woofer, as high-frequency waves cannot efficiently follow the path of a folded horn, (2) it is too efficient, as the amplification of the bass response is so great that it cannot be used beyond a loudspeaker's inefficient low-frequency range (approximately 100 cycles).

10-20. Stereophonic Sound. *Description.* The purpose of stereophonic sound is to present a true sense of the relative positions of the sounds being produced. Although a person hears a sound with both ears, the sound heard by each ear is slightly different. The brain utilizes this difference to determine, among other things, the direction of the sound. It is this feature of hearing that gives a sound depth as well as direction. Theoretically, the ultimate in stereophonic recording of the sounds produced by a symphony orchestra would be to use a vast array of microphones, and have each one feed into an individual recorder. The reproduction of this recorded sound would require a similar array of an equal number of loudspeakers with each one being fed the recorded information from the microphone that was located in the same relative position. Practical systems of stereophonic recording and reproduction use two or three channels.

Two-channel System. A satisfactory stereophonic sound system is obtained by using only two channels, as illustrated in Fig. 10-32. Two identical directional microphones are placed approximately 10 ft apart and directly in front of the orchestra. One microphone will receive the sounds from predominantly one side of the orchestra, while the other microphone will receive sounds predominantly from the other side. The signals from the two microphones are fed through two identical amplifiers to a stereophonic recording head. The sound signals from a disc or tape recording are transferred by means of a stereophonic pickup to two identical amplifiers. The output from each amplifier is fed to one of the two identical loudspeakers that are placed approximately 10 ft apart.

The two-channel system provides the listener with a psychophysical sensation of being present at a live performance. The term *psychophysical* is used because there is also a mental interpretation of the two sounds in addition to their physical sensation. Thus, the sound seems to come from the entire area around and in between the two sound sources, rather than from one or two specific points. The two-channel system therefore provides the listener with sound feelings of volume, tone, depth, and direction.

In order to obtain a true stereophonic effect it is necessary that the two amplifiers used in recording the sounds be identical, and that the two amplifiers used in reproducing the sounds also be identical.   If the number of stages in each of the two amplifiers are not equal, an undesirable phase difference may be produced between the two channels. Incorrect phasing produces a loss of the stereophonic effect in the area between the two loudspeakers.   Incorrect phasing may also be produced by stereophonic recordings as (1) the phase relationship between the

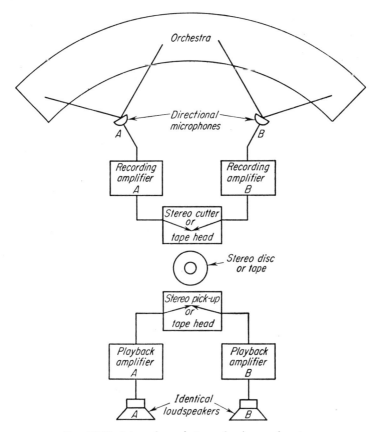

FIG. 10-32. A two-channel stereophonic sound system.

two channels at the end of a recording is not always the same as at the beginning, and (2) the phase relationship between the two channels may vary with the manufacturer.

Three-channel System.   To compensate for phase differences a third channel is added, as shown in Fig. 10-33, to produce a *curtain of sound.* Three nondirectional microphones are used to provide the signals for two channels of stereophonic sound.   All the sounds from the orchestra are heard by each microphone, but with different sound levels and from different directions.   The third, or center, channel is blended by the recording engineer into the two outside channels to provide two separate signals for the stereo cutter. This method of recording has greatly improved the quality of stereophonic recordings, even though this quality is dependent to a large extent on the skill of the recording engineer. In the reproduction of stereophonic recordings it has been determined that sounds having frequencies below 300 cycles contribute very

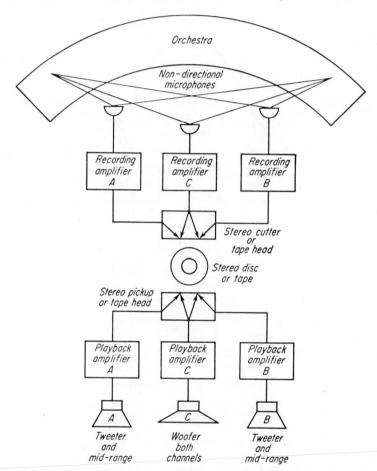

Fig. 10-33. A three-channel stereophonic sound system.

little to the over-all stereophonic effect. The wavelengths of sounds at these frequencies are so long that the relatively small distance between a person's ears is not an appreciable part of a wavelength. Hence, it is difficult to determine directivity at these low frequencies, as the head does not mask the sounds between the ears. This principle is utilized in the arrangement of the loudspeakers for the reproduction of three-channel stereophonic-sound. In Fig. 10-33 two sets of matched tweeter and mid-frequency loudspeakers are placed approximately 10 ft apart to reproduce sounds from these two frequency ranges from each of the two channels.   The two woofer loudspeakers are placed in the center and reproduce the low-frequency sounds from both channels.   This loudspeaker arrangement produces a greater area of sound source, which permits a greater separation between the two outside sets of loudspeakers. Since the sounds from the woofer loudspeakers fill any holes that may appear in the center of the curtain of sound, the phasing between the two channels is not too important.

The loudspeaker arrangement for the reproduction of three-channel stereophonic sound shown in Fig. 10-34 utilizes a potentiometer between

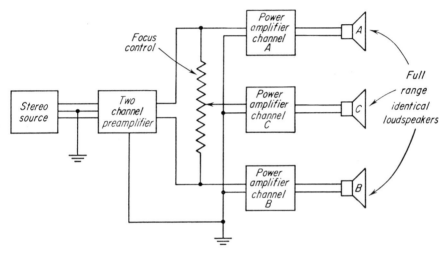

FIG. 10-34. A method of obtaining a third channel from a two-channel stereophonic sound system.

the two channels to provide the signals for a third power amplifier.   The signals from the three identical amplifiers are fed to three identical full-range loudspeakers.   Because a variation in the position of the movable contact appears to move the center of the orchestra, the potentiometer is called a *focus* or *stereo-centering control.*

Stereophonic Equipment.   The requirements for the pickup head, the amplifiers, and the loudspeakers used in the reproduction of stereophonic sounds are the same as for those used in high-fidelity reproduction. However, in addition to using identical units in the two channels, the controls (tone, volume, and equalizer) for both amplifiers should be mechanically ganged to provide identical variations in the control settings. Stereophonic tapes use the full width of the tape for a single recording, in which the top half is used for one channel and the bottom half for the other.   An in-line reproducer head is used whereby the signals for both channels are recorded, and reproduced, on the same vertical axis. Stereophonic discs use the *"45-45"* or *Westrex system* in which both channels are recorded at a 90-degree angle to each other. The axes of modulation are placed at an angle of 45 degrees to both the horizontal and vertical planes. The groove produced not only moves from side to side (lateral recording) but also varies in width. The variation in the width of the groove is caused by the triangular cutting stylus moving in a vertical direction. The "45-45" system reproduces a mono recording equally on both channels, since the lateral movement of the mono-groove would be split into the two 45-degree axes.

QUESTIONS

1. What is the purpose of each of the following audio units: (*a*) the microphone? (*b*) the loudspeaker? (*c*) the phonograph pickup? (*d*) the magnetic reproducer?

2. Define (*a*) air-pressure-type microphones, (*b*) contact-type microphones.

3. Name and describe the applications of two types of air-pressure microphones.

4. Name and describe the applications of three types of contact microphones.

5. Describe the range-of-pickup characteristics and the general applications of the following types of microphones: (*a*) unidirectional, (*b*) bidirectional, (*c*) all-directional.

6. (*a*) Why is it necessary to rate a microphone for some definite unit of pressure? (*b*) What is the unit of pressure most generally used?

7. (*a*) What does the term *bar* actually mean? (*b*) In rating microphones, what do many manufacturers consider the term *bar* to represent?

8. (*a*) What is meant by a low-impedance microphone? (*b*) Name three types of low-impedance microphones.

9. (*a*) What is meant by a high-impedance microphone? (*b*) Name one type of high-impedance microphone.

10. (*a*) How are low-impedance microphones usually rated? (*b*) Why is this type of rating used? (*c*) What zero power reference level is usually used? (*d*) What zero pressure reference level is usually used?

11. (*a*) How are high-impedance microphones usually rated? (*b*) Why is this type of rating used?

12. What method is used to couple the amplifier circuit to (*a*) a low-impedance microphone? (*b*) a high-impedance microphone?

13. How is maximum power transfer obtained between the microphone and amplifier circuit (*a*) with a low-impedance microphone? (*b*) with a high-impedance microphone?

14. Explain the effect of the length of line connecting the microphone to the amplifier for (*a*) a low-impedance microphone, (*b*) a high-impedance microphone.

15. (*a*) Describe the construction and principle of operation of a carbon microphone. (*b*) What are the advantages and disadvantages of carbon microphones?

16. (*a*) Describe the construction and principle of operation of a crystal microphone. (*b*) What are the advantages and disadvantages of crystal microphones?

17. (*a*) Describe the construction and principle of operation of a dynamic microphone. (*b*) What are the advantages and disadvantages of moving-coil-type microphones?

18. (*a*) Describe the construction and principle of operation of a velocity microphone. (*b*) What are the advantages and disadvantages of ribbon-type microphones?

19. (a) Describe the construction and principle of operation of a permanent-magnet dynamic loudspeaker. (b) What are the advantages of a p-m dynamic loudspeaker?

20. Explain why it is necessary to use an output transformer to couple the power-output stage of an audio amplifier to a dynamic-type loudspeaker.

21. In horn-type loudspeakers what is meant by (a) the driver unit? (b) the throat? (c) the mouth?

22. How does the efficiency of transmission for the audio-frequency range of sounds vary for a horn with (a) a conical taper? (b) a parabolic taper? (c) an exponential taper? (d) a hyperbolic taper?

23. (a) How is sound recorded mechanically on discs? (b) What is meant by a vertical-cut record? (c) What is meant by a lateral-cut record?

24. Describe the basic operations of the mechanical method of sound recording and reproduction.

25. (a) What is the purpose of the stylus? (b) What materials are used in making a stylus?

26. Describe the construction, principle of operation, and operating characteristics of a variable-reluctance cartridge.

27. Describe the construction, principle of operation, and operating characteristics of a dynamic cartridge.

28. Describe the construction, principle of operation, and operating characteristics of a Dynetic cartridge.

29. Describe the construction, principle of operation, and operating characteristics of a crystal cartridge.

30. Describe the construction, principle of operation, and operating characteristics of a ceramic cartridge.

31. Describe the construction, principle of operation, and operating characteristics of a capacitance cartridge.

32. (a) What is meant by surface noise? (b) Explain the operation of the circuit used to minimize surface noise.

33. Explain the principle of operation of a magnetic sound recorder and reproducer.

34. Describe the coercivity characteristic of (a) the laminations used in the recorder head, (b) the metal oxides used on the recording tape.

35. Describe the construction and principle of operation of a recorder-reproducer head.

36. With magnetic reproduction, describe the purpose of and method of obtaining (a) a d-c bias, (b) a high-frequency preconditioning signal, (c) an erase signal.

37. Describe the functions of the tape transport system.

38. Describe the advantages and disadvantages of magnetic recording.

39. Describe the basic requirements of a sound amplifier used for reproducing speech.

40. (a) Describe the basic units required for a complete public-address system. (b) Why is it desirable for the input circuit to have more than one channel? (c) What provision is made for operating more than one loudspeaker from the output circuit?

41. Describe the basic requirements of a sound amplifier used for reproducing music.

42. Describe the basic circuit requirements for a complete high-fidelity system.

43. Describe the factors affecting the over-all fidelity of sound reproduction.

44. (a) What are the advantages of an f-m tuner as compared to an a-m tuner? (b) What are the important characteristics of f-m tuners?

45. (a) How may distortion be produced in a turntable? (b) What methods are used for minimizing this distortion?

46. (a) Describe the rim-drive system of power transmission. (b) What are the advantages and disadvantages of this system?

47. (*a*) What are the functions of a tone arm? (*b*) What adjustments are incorporated in a high-fidelity tone arm to improve its quality of reproduction?

48. What are the requirements for a high-fidelity pickup cartridge and stylus?

49. (*a*) Describe the drop system of automatic record changing. (*b*) Describe some of the disadvantages of this method as used for high-fidelity reproduction.

50. What are the requirements of a power supply used for high-fidelity applications?

51. (*a*) Describe the functions of the electronic channel. (*b*) Name the units that may be included in this channel.

52. How are the following types of input signals fed to the first stage of voltage amplification (*a*) low-level? (*b*) high-level?

53. Describe the following terms: (*a*) preemphasis, (*b*) deemphasis, (*c*) equalization circuit.

54. Describe the following terms: (*a*) volume control, (*b*) loudness control, (*c*) treble control, (*d*) bass control.

55. (*a*) What are the fundamental characteristics of the Williamson amplifier circuit? (*b*) What advantages are obtained by using an ultralinear output transformer with this circuit?

56. (*a*) Describe the progress of the a-f signal through the typical power-amplifier unit illustrated in Fig. 10-22. (*b*) What methods are used for minimizing distortion in a power amplifier?

57. Describe the operation of a basic (*a*) treble-boost or bass-attenuation circuit, (*b*) bass-boost or treble-attenuation circuit.

58. Explain how a corrective gain or loss is obtained by use of a passive network.

59. What is meant by (*a*) reference frequency? (*b*) turnover frequency? (*c*) roll-off frequency?

60. Describe the operation of the deemphasis circuit of (*a*) Fig. 10-23*b*, (*b*) Fig. 10-23*c*.

61. Describe the operation of the preemphasis circuit of (*a*) Fig. 10-23*b*, (*b*) Fig. 10-23*c*.

62. Describe the operation of the tone-control circuit of Fig. 10-25*a* for (*a*) medium frequencies, (*b*) low frequencies, (*c*) high frequencies.

63. What is the purpose of making R_1 and R_3 of Fig. 10-25*b* variable resistors?

64. Explain the operation of the bass section of the tone-control circuit of Fig. 10-26*a* when (*a*) R_3 is in its extreme-boost position, (*b*) R_3 is in its extreme-cut position.

65. Explain the operation of the treble section of the tone-control circuit of Fig. 10-26*a* when (*a*) R_4 is in its extreme-boost position, (*b*) R_4 is in its extreme-cut position.

66. In the high-fidelity amplifier circuit of Fig. 10-27 describe the circuit operation (*a*) from the input circuit to the input of the grounded-grid phase inverter, (*b*) from the input of the grounded-grid phase inverter to the input of the loudspeaker.

67. Describe the construction of a full-range single-unit loudspeaker.

68. In a multiunit loudspeaker system describe the construction and operating characteristics of (*a*) a tweeter, (*b*) a woofer, (*c*) a driver.

69. What is meant by (*a*) a coaxial loudspeaker? (*b*) a triaxial loudspeaker?

70. (*a*) What is meant by loudspeaker phasing? (*b*) What are the effects of incorrect loudspeaker phasing?

71. (*a*) What is the purpose of a loudspeaker crossover network? (*b*) What is meant by crossover frequency? (*c*) What is the purpose of a constant-resistance attenuator pad as used in a multiunit loudspeaker system?

72. (*a*) Why is it necessary to use a baffle? (*b*) What is meant by an infinite baffle?

73. Describe the construction and operating characteristics of a bass-reflex baffle.

74. Describe the construction and operating characteristics of an acoustic labyrinth.

75. Describe the construction and operating characteristics of a Klipschorn enclosure.

76. What is the purpose of stereophonic sound?

77. Describe the operation of a two-channel stereophonic sound system (*a*) at the transmitter, (*b*) at the receiver.

78. Describe the operation of a three-channel stereophonic sound system (*a*) at the transmitter, (*b*) at the receiver.

79. Describe the requirements of the equipment used for stereophonic sound reproduction.

80. Describe the Westrex system of stereophonic sound recording.

PROBLEMS

1. A certain single-button carbon microphone has an output power rating of 12 db below 6 mw for a 100-bar signal. What is the power output with an input signal of (*a*) 100 bars? (*b*) 10 bars? (*c*) 10 dynes per sq cm? (*d*) 200 bars?

2. What is the power output of a carbon microphone with an input signal of 100 dynes per sq cm if the output level rating of the microphone is 27 db below 6 mw for a pressure of 10 dynes per sq cm?

3. If the microphone of Prob. 2 has a voltage rating of 50 db below 1 volt per bar, what is the output voltage with a pressure of 100 dynes per sq cm?

4. What is the power output of a double-button carbon microphone whose output level rating is 55 db below 6 mw for a 10-bar speech signal when the input signal is (*a*) 10 bars? (*b*) 100 bars?

5. What is the power output of a low-impedance (250 ohms) dynamic-type microphone whose output level rating is 63.8 db below 6 mw, when feeding into a 250-ohm impedance, for a 10-bar signal when the input signal is (*a*) 10 bars? (*b*) 50 bars?

6. A certain dynamic-type microphone has a self-contained transformer, which increases its output impedance to 35,000 ohms. If its output level rating is 55 db below 1 volt per bar, what is the output voltage for (*a*) a 10-bar signal? (*b*) A 100-bar signal?

7. (*a*) A certain low-impedance (50 ohms) dynamic-type microphone has a voltage output level rating of 86 db below 1 volt per bar. What is the output voltage for a 1-bar signal? (*b*) When the same microphone is equipped with a transformer that raises the impedance at its output terminals to 38,000 ohms, the voltage output level is 54 db below 1 volt per bar. What is the output voltage for a 1-bar signal?

8. What is the output voltage of a velocity-type microphone equipped with a transformer that produces an output impedance of 35,000 ohms and then has a voltage output level rating of 65 db below 1 volt per bar, for an input signal of (*a*) 1 bar? (*b*) 10 bars?

9. What is the voltage output of a crystal-type microphone that has a voltage output level rating of 48 db below 1 volt per bar for an input signal of (*a*) 1 bar? (*b*) 10 bars?

10. A certain crystal-type microphone has a voltage output level rating of 58 db below 1 volt per bar at the microphone terminals and 61.5 db below 1 volt per bar at the end of a 25-ft cable. For an input signal of 10 bars, find (*a*) the output voltage at the microphone terminals, (*b*) the output voltage at the end of the 25-ft cable, (*c*) the voltage drop in the cable, (*d*) the per cent of voltage loss in the cable.

11. What turns ratio is required of a transformer used to couple a single-button carbon microphone whose impedance is 100 ohms to a grid circuit whose input impedance should be 60,000 ohms?

12. A certain transformer designed for coupling the output of a dynamic microphone to the grid circuit of a tube provides a choice of 7.5 ohms or 30 ohms primary impedance and 50,000 ohms secondary impedance.   What is the turns ratio for (a) the 7.5/50,000-ohm connection?   (b) The 30/50,000-ohm connection?

13. A microphone transformer used for matching the output of a crystal microphone to its connecting cable has a number of secondary terminals so that it can be operated with lines of various values of impedance.   The rated primary impedance is 100,000 ohms and the secondary is rated at 50, 200, and 500 ohms.   What is the turns ratio for (a) the 100,000/50-ohm connection?   (b) The 100,000/200-ohm connection? (c) The 100,000/500-ohm connection?

14. The microphone transformer used to couple a certain velocity-type microphone to its line has an impedance ratio of 0.2/500 ohms.   What is its turns ratio?

15. A general-purpose transformer designed for matching low-impedance sources to a grid circuit has an impedance ratio of 200/50 ohms primary to 500,000 ohms secondary.   What is the turns ratio for (a) 200/500,000 ohms?   (b) 50/500,000 ohms?

16. A general-purpose output transformer that can be used to match the output of the average output tube of a small radio receiver to a line or to a voice coil of the loudspeaker has a primary impedance of 5,000 ohms and by choice of terminals secondary impedance values of 500, 200, 50, 15, 8, 5, 3, and 1.5 ohms can be obtained.   What is the turns ratio for each of these impedance ratios?

17. (a) If the scratch filter circuit of Fig. 10-18 uses a 160-mh choke, what value of capacitance should be used to make the filter most effective for 5,000-cycle currents? (b) What is the impedance of the filter circuit to 5,000-cycle currents if the value of the resistance is 100 ohms?   (c) What is the impedance to 3,000-cycle currents for the same circuit values as in (b)?   (d) What is the impedance to 7,000-cycle currents for the same circuit values as in (b)?

18. (a) If the scratch filter circuit of Fig. 10-18 uses a 220-mh choke, what value of capacitance should be used to make the filter most effective for 5,000-cycle currents? (b) What is the impedance of the filter circuit to 5,000-cycle currents if the value of the resistance is 100 ohms?   (c) What is the impedance to 3,000-cycle currents for the same circuit values as in (b)?   (d) What is the impedance to 7,000-cycle currents for the same circuit values as in (b)?

19. A public-address sound system that is capable of delivering 35 watts of undistorted output power is being used with a microphone whose power output level rating is 72 db below 6 mw.   What is the over-all decibel gain from the microphone to the loudspeaker when the system is delivering its rated output of 35 watts?

20. A certain public-address sound system is rated to produce an output of 15 watts.   The gain of the system for the microphone input channel is 113 db and for the phonograph input channel it is 72 db.   (a) What is the minimum output that the microphone may supply in order to produce the rated output?   (b) What is the minimum output that the phonograph pickup unit must supply in order to produce the rated output?

21. A room having a volume of 4,000 cu ft requires an acoustic power output of 0.6 watt.   Determine the undistorted power output required from an audio-amplifier system if the efficiency of the loudspeaker system is (a) 10 per cent, (b) 20 per cent.

22. An auditorium having a volume of 100,000 cu ft requires 10 watts of acoustic power.   Determine the acoustic efficiency of the loudspeaker system when the required power output of the audio-amplifier system is (a) 50 watts, (b) 40 watts.

23. What is the decibel change of an amplifier that has a power output of 6 watts at f_o and 3 watts at f_l and f_h?

24. What is the decibel change of an amplifier that has a power output of 4 watts at f_o and 8 watts at f_l and f_h?

25. In the high-frequency deemphasis circuit of Fig. 10-23b, $R = 470,000$ ohms and

$C = 270 \ \mu\mu f$. Find (a) the rolloff frequency, (b) the response in decibels at the rolloff frequency, and whether it is rising or falling, (c) the response in decibels at twice the rolloff frequency.

26. In the high-frequency deemphasis circuit of Fig. 10-23b, $t = 155 \ \mu sec$ and $R = 470,000$ ohms. Find (a) the value of C, (b) the rolloff frequency, (c) the response in decibels at the rolloff frequency, (d) the reactance of C at the rolloff frequency.

27. In the high-frequency deemphasis circuit of Prob. 10-25, determine the response in decibels at (a) 1,500 cycles, (b) 3,000 cycles.

28. In the high-frequency deemphasis circuit of Prob. 10-26, determine the response in decibels at (a) 1,600 cycles, (b) 6,400 cycles.

29. In the low-frequency deemphasis circuit of Fig. 10-23c, $t = 350 \ \mu sec$ and $R = 47,000$ ohms. Find (a) the value of C, (b) the turnover frequency, (c) the response in decibels at the turnover frequency, (d) the reactance of C at the turnover frequency.

30. In the low-frequency deemphasis circuit of Fig. 10-23c, $R = 47,000$ ohms and $C = 5,600 \ \mu\mu f$. Find (a) the turnover frequency, (b) the response in decibels at the turnover frequency, and whether it is rising or falling, (c) the response in decibels at one-half the turnover frequency.

31. In the low-frequency deemphasis circuit of Prob. 10-29, determine the response in decibels at (a) 400 cycles, (b) 100 cycles.

32. In the low-frequency deemphasis circuit of Prob. 10-30, determine the response in decibels at (a) 450 cycles, (b) 150 cycles.

33. In the high-frequency preemphasis circuit of Fig. 10-23c, $R = 470,000$ ohms and $C = 270 \ \mu\mu f$. Find (a) the rolloff frequency, (b) the response in decibels at the rolloff frequency, and whether it is rising or falling, (c) the response in decibels at twice the rolloff frequency.

34. In the high-frequency preemphasis circuit of Fig. 10-23c, $t = 155 \ \mu sec$ and $R = 470,000$ ohms. Find (a) the value of C, (b) the rolloff frequency, (c) the response in decibels at the rolloff frequency, (d) the reactance of C at the rolloff frequency.

35. In the high-frequency preemphasis circuit of Prob. 10-33, determine the response in decibels at (a) 1,500 cycles, (b) 3,000 cycles.

36. In the high-frequency preemphasis circuit of Prob. 10-34, determine the response in decibels at (a) 1,600 cycles, (b) 6,400 cycles.

37. In the low-frequency preemphasis circuit of Fig. 10-23b, $t = 350 \ \mu sec$ and $R = 47,000$ ohms. Find (a) the value of C, (b) the turnover frequency, (c) the response in decibels at the turnover frequency, (d) the reactance of C at the turnover frequency.

38. In the low-frequency preemphasis circuit of Fig. 10-23b, $R = 47,000$ ohms and $C = 5,600 \ \mu\mu f$. Find (a) the turnover frequency, (b) the response in decibels at the turnover frequency, and whether it is rising or falling, (c) the response in decibels at one-half the turnover frequency.

39. In the low-frequency preemphasis circuit of Prob. 10-37, determine the response in decibels at (a) 400 cycles, (b) 100 cycles.

40. In the low-frequency preemphasis circuit of Prob. 10-38, determine the response in decibels at (a) 450 cycles, (b) 150 cycles.

41. In the tone-control circuit of Fig. 10-26a the bass control is adjusted for maximum boost. Find the signal loss at 75 cycles.

42. In the tone-control circuit of Fig. 10-26a the bass control is adjusted for maximum boost. Find the signal loss at 200 cycles.

43. In the tone-control circuit of Fig. 10-26a the bass control is adjusted for maximum cut. Find the signal loss at 75 cycles.

44. In the tone-control circuit of Fig. 10-26a the bass control is adjusted for maximum cut. Find the signal loss at 200 cycles.

45. Assuming that the preemphasis and deemphasis response curves for the tone-control circuit of Probs. 10-41 and 10-43 are symmetrical, find for the mid-frequency setting (a) the amount of gain or loss in decibels at 75 cycles, (b) the decibel output at the tone control.

46. Assuming the preemphasis and deemphasis response curves for the tone-control circuit of Probs. 10-42 and 10-44 are symmetrical, find for the mid-frequency setting (a) the amount of gain or loss in decibels at 200 cycles, (b) the decibel output at the tone control.

47. In the tone-control circuit of Fig. 10-26a the treble control is adjusted for maximum boost. Find the signal loss at 2,000 cycles.

48. In the tone-control circuit of Fig. 10-26a the treble control is adjusted for maximum boost. Find the signal loss at 6,000 cycles.

49. In the tone-control circuit of Fig. 10-26a the treble control is adjusted for maximum cut. Find the signal loss at 2,000 cycles.

50. In the tone-control circuit of Fig. 10-26a the treble control is adjusted for maximum cut. Find the signal loss at 6,000 cycles.

51. Assuming that the preemphasis and deemphasis response curves for the tone-control circuit of Probs. 10-47 and 10-49 are symmetrical, find for the mid-frequency setting (a) the amount of gain or loss in decibels at 2,000 cycles, (b) the decibel output at the tone control.

52. Assuming that the preemphasis and deemphasis response curves for the tone-control circuit of Probs. 10-48 and 10-50 are symmetrical, find for the mid-frequency setting (a) the amount of gain or loss in decibels at 6,000 cycles, (b) the decibel output at the tone control.

53. A low- and high-frequency loudspeaker are connected as shown in Fig. 10-30a. The impedance of their voice coils is the same, and is equal to 8 ohms. Find the values of capacitance and inductance required for a crossover frequency of 1,100 cycles.

54. A low- and high-frequency loudspeaker are connected as shown in Fig. 10-30a. The impedance of their voice coils is the same, and is equal to 16 ohms. Find the values of capacitance and inductance required for a crossover frequency of 1,330 cycles.

55. A low- and high-frequency loudspeaker are connected as shown in Fig. 10-30b. The impedance of their voice coils is the same, and is equal to 8 ohms. Find the values of capacitance and inductance required for a crossover frequency of 1,100 cycles.

56. A low- and high-frequency loudspeaker are connected as shown in Fig. 10-30b. The impedance of their voice coils is the same, and is equal to 16 ohms. Find the values of capacitance and inductance required for a crossover frequency of 1,330 cycles.

CHAPTER 11

AMPLITUDE-MODULATION RECEIVER CIRCUITS

The operation of simple receiving circuits capable of reproducing a-m broadcast signals without the use of vacuum tubes was presented in Chap. 1. While these simple circuits do not have much practical value, they were presented to acquaint the reader with the fundamental functions of a receiving system before introducing the theory of vacuum tubes and their associated circuits. In the chapters that followed, a detailed study of vacuum tubes and their uses as amplifiers, detectors, oscillators, and rectifiers was presented. With this knowledge of vacuum-tube applications, it is now possible to understand the purpose and operation of all the components in the average receiver.

11-1. Characteristics of Receivers. *Definitions.* A radio receiver is a device for converting radio waves into perceptible signals. How well a receiver accomplishes its purpose is generally determined by investigation of its characteristics, the most important of which are the sensitivity, selectivity, fidelity, stability, and signal-to-noise ratio (see also Art. 4-1).

Sensitivity. The IRE definition states that the sensitivity of a radio receiver is that characteristic which determines the minimum strength of signal input capable of causing a desired value of signal output.

Selectivity. The IRE definition states that the selectivity of a radio receiver is that characteristic which determines the extent to which it is capable of differentiating between the desired signal and disturbances of other frequencies.

Fidelity. The IRE definition states that fidelity is the degree with which a system, or a portion of a system, accurately reproduces at its output the essential characteristics of the signal which is impressed upon its input.

Stability. Stability may be defined as a measure of the ability of a radio receiver to deliver a constant amount of output for a given period of time when the receiver is supplied with a signal of constant amplitude and frequency. Factors affecting the stability of a receiver are the variations in output voltage of the power-supply unit, temperature variations, and occasionally features of mechanical construction. Instability of a receiver, in addition to affecting its fidelity, may also cause it to break into oscillations that produce a whistle or a howling noise at the loudspeaker.

Signal-to-noise Ratio. The signal-to-noise ratio of a radio receiver is one of the important operating characteristics of the receiver. Various definitions have been presented, depending upon the source of noise to be considered. In terms of the receiver itself, the signal-to-noise ratio is the ratio of the signal power output to the noise power output at a specified value of modulated carrier voltage applied to the input terminals.

Kinds and Sources of Noise. The source of noise may be in the receiver itself or may be due to external causes such as static, background noises from other stations, or noises originating in the transmitter. The noises produced within the receiver may be divided into four classifications, namely (1) thermal agitation, (2) shot effect, (3) microphonics, (4) hum from the a-c power source. *Thermal agitation* may be defined as an irregular random movement of the free electrons in a conductor that is carrying a normal flow of electron current. The random motion of the free electrons produces minute currents, which upon being amplified result in noise at the loudspeaker. The magnitude of these minute currents increases with the temperature. *Shot effect* is

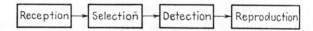

Fig. 11-1. Block diagram showing the essential functions of a receiver.

caused by the small irregularities in plate current, which normally exist as a result of the individual electrons striking the plate, and produces noise at the loudspeaker. It is generally present only in high-gain amplifiers operated with a low input signal. *Microphonics* is the name generally applied to the noise present in the loudspeaker as a result of mechanical vibration at one or more points in a radio receiver, which causes corresponding variations in the a-f currents. Although microphonic noises are most generally due to vibration of the elements of amplifier tubes, they may also be caused by the vibration of other circuit elements such as capacitors and coils. *Hum from the a-c power source* is usually picked up in the a-f section of the receiver, although it may also be picked up at other points. Among the causes of hum are (1) operating the heaters or filament circuits of the tubes on alternating current, (2) insufficient filtering of cathode-bias resistors, (3) insufficient filtering in the B power-supply unit, (4) stray magnetic and electrostatic fields near the circuit elements of the receiver.

11-2. Fundamental Principles of Amplitude-modulation Receivers. *Essential Functions.* The minimum essential functions of a radio receiver are: (1) reception, (2) selection, (3) detection, (4) reproduction. These functions are performed in the order shown on the block diagram of Fig. 11-1. A simple discussion of these functions was presented in Chap. 1, and a simple receiving circuit is shown in Fig. 1-9.

The simple receiving circuit of Fig. 1-9 has many disadvantages such as poor sensitivity, poor selectivity, and possible loss of the signal when the crystal is jarred. Some of these disadvantages may be lessened or overcome by substituting a vacuum tube for the crystal detector. Two single-tube receiving circuits are shown in Fig. 11-2. Both of these circuits will provide increased signal strength (Arts. 3-5 and 3-8), but still only sufficient to operate earphones.

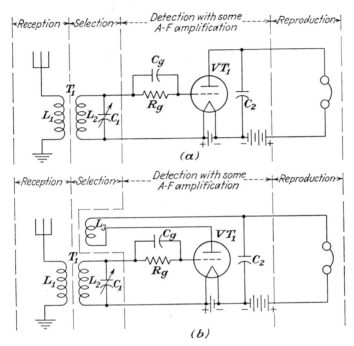

FIG. 11-2. Two single-tube receiver circuits. (*a*) Simple receiver circuit with a grid-leak detector. (*b*) Simple receiver circuit with a regenerative detector.

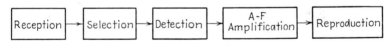

FIG. 11-3. Block diagram of a simple receiver with the addition of a-f amplification.

Addition of A-F Amplification. In order to obtain sufficient signal strength to operate a loudspeaker, one or more stages of a-f amplification may be added, as shown in the block diagram of Fig. 11-3. The a-f amplifier stages may be either of the transformer-coupled or RC-coupled type. Figure 11-4*a* illustrates the addition of one stage of a-f amplification to the simple regenerative receiver of Fig. 11-2*b*. Figure 11-4*b* illustrates the addition of two stages of a-f amplification to the single-tube receiving circuit of Fig. 11-2*a*. In the circuit of Fig. 11-4*b* the first amplifier stage

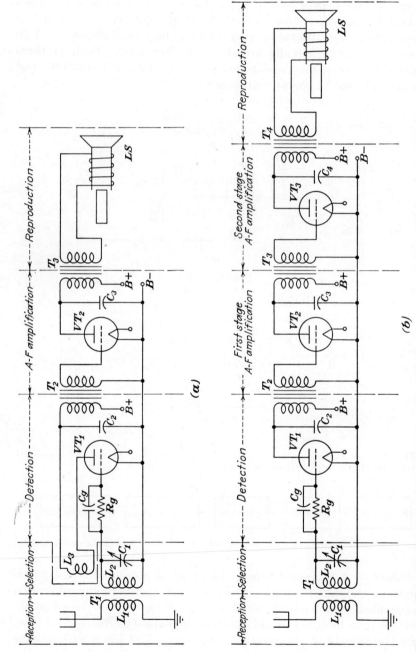

Fig. 11-4. Two simple receiver circuits with a-f amplification. (a) Receiver with a-f amplification. (b) Receiver with a grid-leak detector and two stages of a-f amplification.

provides voltage amplification and the second stage provides power amplification.

The circuits of Fig. 11-4 do not have very great practical value because of the poor selectivity of the circuit.   The a-f amplifier stages increase the strength of all signals passed on to them from the detector and consequently the interstation background noises are amplified to the same degree as the signal of the desired station.   This usually results in poor over-all operation of the receiving circuit.

Addition of R-F Amplification.   The selectivity of a receiver may be improved by adding one or more additional tuning stages.   As an additional tube is required for each stage added, amplification also takes place with the introduction of each new stage.   As both tuning and amplification take place within each added stage, the process is commonly referred to as *tuned-radio-frequency amplification*, or it is said that one or more *trf* stages have been added.   The block diagram of a receiver employing trf amplification is shown in Fig. 11-5.   The circuit of Fig. 11-6 illustrates the

Fig. 11-5. Block diagram of a simple receiver with the addition of both r-f and a-f amplification.

addition of a single stage of trf amplification and one stage of a-f amplification to the simple receiving circuit of Fig. 11-2a.   The manner in which the additional stages of tuning improve the selectivity has been discussed in Art. 4-6, and trf amplification was presented in Art. 5-5.   If too many tuned stages are added, the circuit may become too selective and adversely affect the fidelity of the receiver.   Another disadvantage of trf amplification is that the selectivity of the circuit varies with the frequency of the received signal, decreasing with an increase in frequency, as indicated by Fig. 11-7.

Addition of I-F Amplification.   The disadvantages of the trf circuit can be overcome to a large extent by reducing the received frequency of the selected station to a lower fixed value of radio frequency (called the intermediate frequency) and providing further tuning and amplification at this intermediate frequency. A receiver employing this principle is called a *superheterodyne receiver*.   The portion of the receiver in which this tuning and fixed frequency amplification take place is called the *i-f amplifier*.   Because the i-f amplifier operates at a fixed frequency, greater selectivity, sensitivity, stability, and fidelity can be obtained than with an amplifier that must operate over the entire frequency range of the receiver.

The intermediate frequency is obtained by beating the frequency of

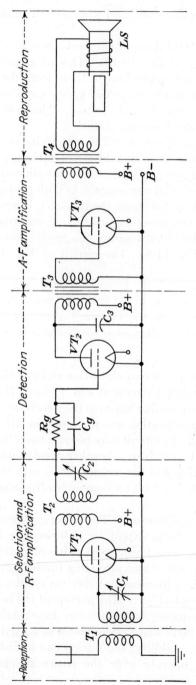

FIG. 11-6. A simple receiving circuit with both r-f and a-f amplification.

a separate oscillator circuit that is added to the superheterodyne receiver with the modulated frequency of the desired station.   In order to maintain a constant value of intermediate frequency for all stations to be received, the frequency of the oscillator circuit must be varied whenever a new station is selected.   The block diagram of a superheterodyne receiver is shown in Fig. 11-8.   A more complete analysis of the superheterodyne receiver is presented later in this chapter.

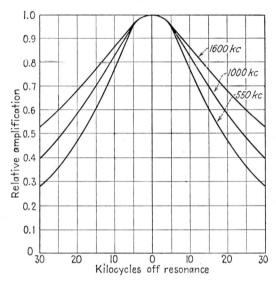

FIG. 11-7. Curves showing the effect of frequency upon the selectivity of a typical trf amplifier stage.

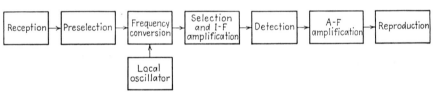

FIG. 11-8. Block diagram of a simple superheterodyne receiver.

Other Additions to Receiving Circuits.   Numerous improvements and additional operating features have been introduced for radio receiver circuits.   They include such features as short-wave reception, all-wave reception, automatic volume control, automatic frequency control, tone control, noise-suppression circuits, pushbutton control, tuning indicators, band switching, bandspread tuning, and preselection.   The purpose of each is indicated by its name.

11-3. The Tuned-radio-frequency (TRF) Receiver.   *The TRF circuit.* A trf receiver is one in which the incoming signal is passed through one or more stages of trf amplification and then applied to the detector with the same frequency and waveform at which it was received. The circuit diagram of a five-tube, a-c operated, trf receiver is shown in Fig. 11-9. The circuit uses two stages of tuned r-f amplification, a biased detector, one stage of a-f voltage amplification, and one stage of a-f power amplification. It is designed to provide (1) sufficient selectivity for the reception of all local stations and some distant stations, (2) ample volume for average home use, (3) good fidelity, (4) high signal-to-noise ratio.

Volume Control. Control of the volume for the circuit of Fig. 11-9 is obtained by means of the potentiometer R_1. As the movable arm C approaches A a greater portion of the signal input is shunted to ground, thereby decreasing the signal strength in the primary of the antenna coil T_1. As the arm C approaches B, less current is shunted to ground and the signal strength in the primary of T_1 increases. Also, advancing the movable arm C toward point A increases the cathode-bias resistance (R_2 plus the resistance in section B-C of R_1), thereby reducing the amplification produced at VT_1 and VT_2, and thus decreasing the volume. When the movable arm C is advanced toward B, the cathode bias of VT_1 and VT_2 is decreased, and the volume will be increased.

TABLE 11-1. VALUES OF THE CIRCUIT ELEMENTS OF FIG. 11-9

Resistors, ohms	Capacitors		Vacuum tubes	Other elements
$R_1 = 10,000$	C_1		VT_1	T_1 = antenna coil
$R_2 = 250$	C_2 $\}$ = 10–365 $\mu\mu$f		VT_2 $\}$ = 6SK7	T_2
$R_3 = 50,000$	C_3		VT_3 = 6SJ7	T_3 $\}$ = r-f transformer
$R_4 = 68,000$	C_4		VT_4 = 6F6	
$R_5 = 33,000$	C_5 $\}$ = 4–30 $\mu\mu$f		VT_5 = 5Y3	T_4 = output transformer
$R_6 = 1$ megohm	C_6			T_5 = power transformer
$R_7 = 220,000$	C_7			LS = loudspeaker
$R_8 = 470,000$	C_8 $\}$ = 0.1 μf			L_1 = field winding of speaker
$R_9 = 400$	C_9			S_1 = spst switch
$R_{10} = 75,000$	C_{10}			
	$C_{11} = 220$ $\mu\mu$f			
	$C_{12} = 0.05$ μf			
	$C_{13} = 10$ μf			
	$C_{14} = 0.1$ μf			
	$C_{15} = 0.01$ μf			
	C_{16} $\}$ = 20 μf			
	C_{17}			

Screen-grid Voltage-dropping Resistor. Some pentode tubes are operated with a lower voltage at the screen grids than at the plates. In order to avoid the use of high-wattage resistors necessary at a voltage-divider network, the screen grids are usually fed through a comparatively low-wattage carbon resistor directly from the high-voltage B power supply source. Resistors R_3 and R_6 of Fig. 11-9 are used for this purpose.

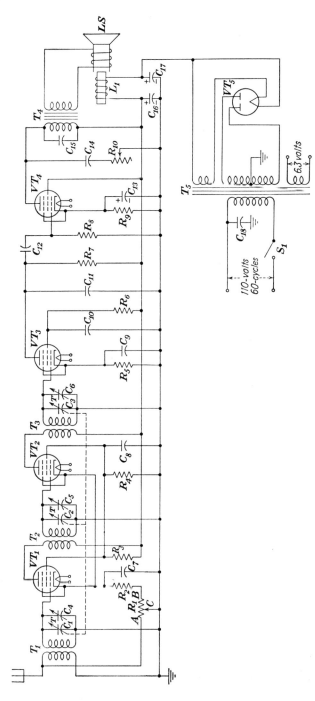

Fig. 11-9. Circuit diagram of a five-tube, a-c operated, trf receiver.   (See also Table 11-1.)

Screen-grid R-F Bypass Capacitors. Capacitors C_8 and C_{10} of Fig. 11-9 bypass the r-f currents to ground instead of permitting these currents to flow through the screen-grid voltage-dropping resistors and on through the B power supply. Since the power supply acts as a common coupling impedance for all plate and screen-grid currents, any r-f currents flowing through the power source may cause trouble. Capacitors C_8 and C_{10} are also called *decoupling capacitors;* and the combination of R_4 and C_8 is often referred to as a *decoupling circuit.*

Tone Control. The potentiometer R_{10} and the capacitor C_{14} in Fig. 11-9 serve as a tone-control circuit. These two circuit elements are connected in series, and they provide a path to ground for a portion of the a-f current. The value of C_{14} is such that only the higher values of the audio frequencies are shunted to ground. The value of R_{10} determines the percentage of the a-f current that is shunted to ground instead of being permitted to flow through the primary of the output transformer. Thus, when the value of R_{10} is made very low, only the currents of the low and medium audio frequencies will pass through the output transformer and hence the loudspeaker accentuates the bass notes. By increasing the value of R_{10}, the higher frequency notes approach their normal strength, and the bass is no longer accentuated.

Line Filter Capacitor. In receivers operated from power lines, it is common practice to connect a capacitor from one input line wire to ground as indicated by capacitor C_{18} in Fig. 11-9. The purpose of this capacitor is to bypass any line voltage disturbances to ground, so that they will have little or no effect upon the operation of the receiver.

11-4. The Superheterodyne Receiver. *Definitions.* From the IRE definitions of terms: (1) *superheterodyne reception* is a form of heterodyne reception in which one or more frequency changes take place before detection, (2) *heterodyne reception* (beat reception) is the process of operation on radio waves to obtain similarly modulated waves of different frequency; in general, this process includes the use of a locally generated wave, which determines the change of frequency.

A *superheterodyne receiver* may thus be defined as one in which one or more changes of frequency are produced before the a-f signal is extracted from the modulated wave. However, the name superheterodyne is generally applied to receivers in which only one frequency change is made before a-f detection takes place, while a receiver using two intermediate frequencies is usually called a *double superheterodyne receiver.* As the average a-m broadcast receiver is required to operate at only comparatively low frequencies (550 to 1,600 kc), there is no advantage of practical value in having two or more r-f frequency changes.

Advantages of the Superheterodyne Receiver. In the development of radio receivers, two types of receiving circuits have had outstanding use. They are the trf receiver and the superheterodyne receiver. The superheterodyne receiver has a number of advantages that have resulted in almost universal use of the superheterodyne circuit for average broadcast receivers. Its advantages over the trf receiver are (1) improved selectivity in terms of stations on adjacent channels, (2) more uniform selectivity over the broadcast band, (3) improved stability of operation, (4) a large portion of its amplification is obtained at a single (i-f) frequency instead of over the entire r-f range of the receiver, (5) higher gain per stage due to obtaining the amplification at the lower frequency value of the i-f stages.

The fundamental principle of the superheterodyne receiver has been stated in Art. 11-2, and Fig. 11-8 shows the various functions performed in the receiver. The functions not previously described will now be studied.

11-5. Reception and Preselection. *Reception.* The function of reception is performed at the antenna. Because superheterodyne receivers are very sensitive and can thus operate with weak signals, it is possible to obtain satisfactory reception of all local stations and some distant stations with a loop antenna. The loop antenna can be made small enough to fit inside the receiver cabinet even with small portable receivers.

There are several types of loop-antenna designs. In one type, the loop antenna contains both primary and secondary windings, which are connected in the same manner as an ordinary r-f antenna coil. The primary winding has a low number of turns and is made as large as the cabinet of the receiver will permit. The secondary has a greater number of turns, usually of a smaller size of wire, and may or may not be of the same over-all dimensions as the primary. In most cases only the secondary circuit is tuned by use of a variable capacitor. Another type of loop antenna has only a single coil, which is connected in a manner similar to that of the secondary winding of the ordinary r-f antenna transformer. This circuit is tuned by use of a variable capacitor, and if a fairly large size of wire is used, the coil resistance can be kept low, thereby producing a higher value of Q and obtaining more signal voltage from the desired station.

The *loopstick* and the *ferrite-rod* antennas have their coils wound over a specially processed piece of iron that is suitable for use over the frequency range of the a-m broadcast band. The advantages of these antennas are (1) very small size, (2) high values of Q (in the order of 400), (3) high signal pickup, (4) good image rejection.

Preselection. In superheterodyne receivers, any tuning circuits located before frequency conversion takes place are generally referred to as *preselectors.* In the circuit of Fig. 11-10 preselection takes place at the tuned circuit formed by the secondary of T_1 and capacitor C_1. Most receivers have at least one preselection stage. In the more expensive broadcast receivers, and in some short-wave receivers, additional preselection stages (actually transformer-coupled r-f amplifiers) are used (see Fig. 11-13). These r-f amplifiers are generally operated with untuned primaries and tuned secondaries and usually employ pentode tubes. The advantages derived by the use of preselection are (1) improved selectivity, (2) improved image suppression (Art. 11-9), (3) improved signal-to-noise ratio.

11-6. Frequency Conversion. *Need for the Frequency Converter.* The frequency converter is the heart of the superheterodyne receiver since the advantages of the superheterodyne receiver are gained by reducing the frequency of the various r-f input signals to a constant i-f signal. This change in frequency is accomplished at the *frequency converter*, sometimes called the *mixer* or *first detector*.

Methods of Obtaining Frequency Conversion. The process of obtaining frequency conversion requires three fundamental functions, namely, oscillation, mixing, and detection. An oscillator circuit, generally called the *local oscillator*, is required to set up a frequency differing in value from the signal frequency in order to produce heterodyne action. A *mixer* is required to obtain a new frequency by combining the signal frequency with that of the local oscillator through heterodyne action. As detector action is required to extract the beat frequencies obtained by the heterodyne action, the mixer is also called the *first detector*.

Explanation of the Fundamental Functions. The local oscillator circuits are either similar to or are merely variations of the fundamental oscillator circuits. The oscillator may employ a separate vacuum tube or it may use a portion of the converter tube. In order that the oscillator circuit will sustain oscillation over the entire frequency range, it must (1) avoid driving the grid of the mixer positive at the high frequencies when the oscillator output increases, and (2) avoid having the oscillator cease functioning at the lower frequencies if the feedback of the oscillator drops too low.

The function of mixing is accomplished by applying both the modulated r-f signal and the unmodulated local oscillator output to the mixer tube. There are several methods of feeding these two voltages to the mixer. In general, when two signal voltages of different frequencies are applied to the mixer, the current in the plate circuit will contain many frequencies, namely, (1) the original signal frequency, (2) the local oscillator frequency, (3) the sum of the signal and oscillator frequencies, (4) the difference of the signal and oscillator frequencies, (5) numerous other frequencies produced by combinations of the fundamentals and harmonics of the signal and oscillator frequencies. Of these, the signal, the local oscillator, and the sum and difference frequencies will be strongest. The sum and difference frequencies are the result of heterodyne action (Art. 3-7) and of these only the difference frequency is used in a-m broadcast-band super-heterodyne receivers.

Applying the r-f signal and the local oscillator output voltages to a tube does not necessarily result in a beat frequency output.   Although only the two original frequencies are applied, the envelopes formed by combining these frequencies will be of new values equal to the sum and difference of the two.   The detector action of the converter tube makes it possible to obtain the envelope frequencies in the output circuit.   The first detector tube is usually operated on a nonlinear portion of its characteristic curve and operates as a plate detector.   The output of the detector will contain the same a-f signal modulation that was present in the original r-f signal input.

Conversion Gain.   In addition to the functions of mixing and detection, a certain amount of i-f amplification also takes place during the process of frequency conversion.   The ratio of the i-f voltage developed in the output of the converter to the r-f signal input voltage is called the *conversion gain* or *conversion efficiency.*   The voltage amplification obtained at the converter varies from about 0.3 to 0.5 times the value which would be obtained with a similar tube operated as an i-f amplifier.   For example, a tube that can produce a voltage amplification of 100 when operated as an i-f amplifier will only provide a voltage amplification of 30 to 50 when operated as a frequency converter.

11-7. Frequency-converter Circuits.   In the development of superheterodyne receivers, numerous circuits designed to obtain frequency conversion have been introduced.   Some circuits require the use of two tubes, while others accomplish the purpose with a single tube.   Only the fundamental and the commonly used circuits will be presented.

Simple Converter Circuits.   The first superheterodyne receivers used separate tubes for the oscillator and the mixer as shown in Fig. 11-10; although only triodes are shown, tetrodes and pentodes may also be used. The circuits in both figures are identical except for the manner in which the oscillator output is coupled to the mixer.   In Fig. 11-10a the oscillator output is inductively coupled to the mixer at T_4 and in Fig. 11-10b the output is capacitively coupled by means of capacitor C_7.

As both the r-f signal and the oscillator output are applied to the control grid, the electron stream flowing from the cathode to the plate will be affected by both voltages.   With the proper amount of grid bias, VT_1 can be operated as a plate detector.   The current in the plate circuit of VT_1 is composed of a number of signals of different frequencies, and as only the signal of the difference frequency is desired, the double-tuned transformer T_3 is tuned to the desired i-f value.   Because of the excellent selectivity of the double-tuned i-f transformer, only the desired frequency is passed on to the succeeding circuits, all other frequencies being bypassed through capacitor C_4.   Disadvantages of this circuit are: (1) the tuning of one circuit affects the tuning of the second circuit, which may cause the two circuits to lock together and thereby make it difficult to track the two

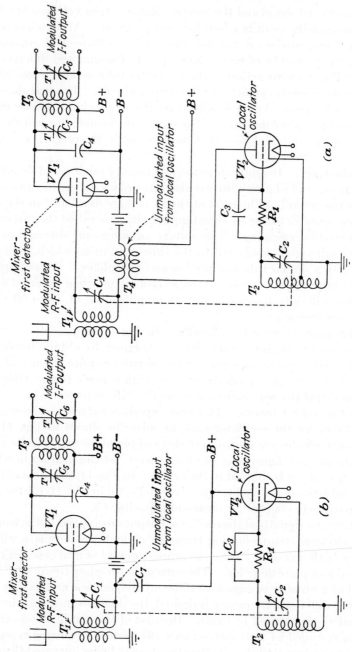

Fig. 11-10. Simple frequency-converter circuits. (a) Oscillator output inductively coupled. (b) Oscillator output capacitively coupled.

circuits; (2) the increase of the oscillator output voltage with an increase in the oscillator frequency may drive the grid of VT_1 positive and thereby cause unsatisfactory operation of the circuit.

Among the methods used to avoid the difficulties present when both the r-f signal and the oscillator voltage are applied to the same grid as in Fig. 11-10 are (1) using a tetrode for the mixer and applying the r-f signal to grid 1 and the oscillator output to grid 2; (2) using a pentode for the mixer and applying the r-f signal to grid 1, the oscillator output to grid 3, and connecting grid 2 to B+ to provide shielding action between the r-f signal input at grid 1 and the oscillator input at grid 3.

Pentagrid Converters.   The functions of oscillation and mixing can be accomplished in a single tube by providing a cathode, a plate, and five grids in a single envelope.   Such a tube is called a *pentagrid converter*.   There are two fundamental types of pentagrid converters differing chiefly in the order in which the grids are used for their various functions; of these, the type described in Art. 11-8 is used most frequently.

11-8. Pentagrid Converter. *Tube Electrodes.* In the type 6BE6 pentagrid converter tube the order of the grids numbered from the cathode out are (1) oscillator grid, (2) inner screen grid, (3) r-f signal control grid, (4) outer screen grid, (5) suppressor grid.   Grids 2 and 4 are connected together internally and a single lead is brought out to one of the base pins.

Operation of the Tube.   The diagram of a converter circuit using this type of tube is illustrated in Fig. 11-11.   The tube operates in the following manner.   The cathode, grid 1, and grids 2 and 4 are connected to the external circuit in such a manner that they operate as a triode oscillator, with grid 1 serving as the oscillator grid and grids 2 and 4 serving as a composite anode for the oscillator.   The stream of electrons leaving the cathode and flowing toward the plate will be modulated by the oscillator voltage at grid 1.   Some of the electrons upon passing grid 1 will return to the cathode by way of grid 2.   The remainder of the electrons will be drawn on toward grid 4 because of their acceleration and because of the positive potential of grid 4.   However, before reaching grid 4, the electron stream will be further modulated by the r-f signal voltage at grid 3. As grid 3 is generally biased negatively, some electrons will be driven back toward the cathode.   If these electrons succeed in reaching the space-charge area about the cathode, they will cause an undesirable change in the amount of cathode current.   However, these electrons are prevented from reaching the cathode space-charge area because of the shielding effect produced by placing grid 3 between grids 2 and 4.   Thus far it has been shown that the electron stream in its travel from the cathode toward grid 4 has been modulated by both the oscillator and r-f signal voltages.   Of those electrons which pass grid 3, some will return to the cathode by way of grid 4 and the remainder by way of the plate circuit.   The r-f signal voltage at grid 3 produces variations in the plate current, which is of course essential to the operation of the converter circuit.   As the signal voltage at grid 3 also produces an approximately equal change in the current of grid 4 but opposite in direction to the plate current change, the cathode current will not be affected to any appreciable extent by changes in the r-f signal voltage.   Grid 5 acts as a suppressor grid and may be connected directly to the cathode or to ground.   Its function is similar to the suppressor grid in a pentode.

Operation of the Circuit.   The circuit of Fig. 11-11 operates in the following manner.   The r-f signals are received at the primary of the r-f transformer T_1 and selection of the desired station's signal is made by

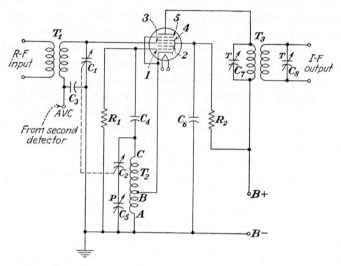

Fig. 11-11. Frequency-converter circuit using a pentagrid converter tube similar to the type 6BE6.

means of the tuning circuit consisting of the secondary of T_1 and capacitor C_1.   The r-f signal voltage of the selected station is applied to grid 3, where it modulates the electron stream flowing from the cathode to the plate of the tube.   The local oscillator is a variation of the Hartley circuit. The oscillator coil T_2 is an autotransformer connected so that any variations in the cathode current flowing through section AB will induce a voltage in section BC.   This induced voltage is applied to the oscillator grid and provides the feedback necessary to produce and sustain oscillations.   The oscillator frequency is determined by the tuned circuit formed by the oscillator coil T_2 and capacitors C_2 and C_5.   The oscillator grid bias is obtained by means of resistor R_1 and capacitor C_4. Capacitor C_3 is used to prevent short-circuiting the avc voltage to ground. Resistor R_2 is a screen-grid voltage-dropping resistor and together with capacitor C_6 also serves as a screen-grid filter circuit.   The plate current, from which the output of the converter is obtained, is fed through the primary of the doubly tuned i-f transformer T_3. Because of the bandpass tuning characteristics of the i-f transformer, only the desired frequency appears at the output terminals of T_3.

Characteristics of the Circuit. The advantages of this type of converter circuit are: (1) the r-f signal voltage has practically no effect upon the cathode current; hence variations in the avc bias applied to the r-f input will not cause any detuning of the oscillator; (2) the use of a suppressor grid increases the plate resistance and thereby increases the conversion efficiency of the tube and circuit. This type of tube and its associated circuit are used extensively with a m broadcast-band receivers.

11-9. The Intermediate-frequency Characteristics. *Frequencies Present at the Converter.* When the r-f signal and the oscillator output are combined, the output current of the first detector will contain (1) the r-f signal frequency, (2) the oscillator frequency, (3) the sum of the r-f signal and oscillator frequencies, (4) the difference of the r-f signal and oscillator frequencies, (5) numerous other frequencies produced by combinations of the fundamentals and harmonics of the r-f signal and oscillator frequencies. For example, if a receiver has its r-f circuit tuned to 550 kc and its oscillator is tuned to 1,015 kc, the frequencies appearing at the first detector output will be 550, 1,015, 1,565, and 465 kc and harmonics of these frequencies. Of these, only the difference frequency of 465 kc is desired. As the output of the first detector is fed directly to the highly selective i-f amplifier, tuning this amplifier to 465 kc will result in acceptance of the 465-kc currents and rejection of the currents of all other frequencies.

Image-frequency Signals. It is possible that with a certain value of oscillator frequency the desired value of intermediate frequency can be obtained from two different carrier frequencies at the same time. For example, an r-f signal of 550 kc and an oscillator frequency of 1,015 kc will produce an i-f signal of 465 kc. It is also possible to obtain a 465-kc i-f signal with the same oscillator frequency if an r-f signal of 1,480 kc reaches the first detector. Of these two 465-kc i-f signals only one is desired; the undesired signal is called the *image-frequency signal.* The effect of image-frequency signals may be minimized or eliminated by providing one or more stages of r-f tuning or preselection. The ratio of the output from the desired r-f signal to that from the undesired r-f signal is called the *signal-to-image ratio* or simply the *image ratio.* The effect of image-frequency signals is present mostly in short-wave receivers operating at high frequencies and in receivers designed to operate at a low i-f value.

Choice of the I-F Value. Throughout the development of superheterodyne receivers a wide range of values have been used for the intermediate frequency. At the start, values as low as 50 kc were used, and in short-wave receivers values of several megacycles may be used. In receivers intended for use on the broadcast band only, the i-f values range from approximately 130 to 485 kc, the value varying with the manufacturer and the design of the receiver.   Among the values of intermediate frequency that have been used for broadcast receivers are 130, 175, 262, 345, 450, 455, 456, 460, 465, 470, and 485 kc.   The majority of home receivers use either 455, 456, or 465 kc, while many receivers designed for use in automobiles employ an i-f value of either 175 or 262 kc.

The choice of frequency for the i-f is affected by a number of factors. Two important factors are (1) the tuning ratio, and (2) selectivity.

Tuning Ratio.   The purpose of the converter is to combine the variable r-f signal input with the variable oscillator frequency and thereby obtain a constant i-f output from the converter.   This can best be accomplished by increasing the oscillator frequency by the same amount that the frequency of the preselector circuit is increased.   Under this condition the difference in frequency between the r-f input and the oscillator can be maintained practically constant.   The intermediate frequency can be obtained by making the oscillator frequency either higher or lower than the r-f signal by the amount of the intermediate frequency desired.   In broadcast receivers it is almost universal practice to make the oscillator frequency higher than the r-f signal.   In short-wave receivers the oscillator frequency may be made either higher or lower than the r-f signal.   By making the oscillator frequency of a broadcast receiver higher than the r-f signal it becomes possible to obtain a more desirable frequency range for the oscillator, as is illustrated by the following example.

If it is desired to have a receiver operate over an r-f range of 550 to 1,600 kc and maintain a 465-kc intermediate frequency, the oscillator will have to be tuned from 85 to 1,135 kc if the oscillator frequency is to be lower than the r-f signal or from 1,015 to 2,065 kc if the oscillator frequency is to be higher than the r-f signal.   With the lower oscillator frequency the required tuning ratio is approximately 13 to 1, while for the high frequency the ratio is approximately only 2 to 1.   It is thus obvious why broadcast receivers are designed to have the oscillator frequency higher than the r-f signal input.

Selectivity.   The use of low i-f values, that is, in the order of 130, 175, and 262 kc, has the advantage of improved selectivity of stations on adjacent broadcast channels.   This becomes apparent when the difference in frequency between stations on adjacent channels is expressed as a percentage of the frequency at which tuning and amplification take place. As adjacent broadcast transmitter frequencies may be as little as 10 kc apart, this percentage for an i-f of 175 kc is 5.7 per cent, for 262 kc it is 3.8 per cent, and for 465 kc it is 2.1 per cent.   Incidentally, this shows one advantage of the superheterodyne over the trf receiver where the percentage is 1.8 at the lower frequency limit of 550 kc and only 0.625 per cent at the upper frequency limit of 1,600 kc.   Use of low i-f values, however, results in greater possibility of image-frequency interference because the image frequency gets closer to the desired station frequency as the i-f value is reduced.   Accordingly, most manufacturers of broadcast receivers now use i-f values of approximately 450 to 470 kc.   It is interesting to note that most i-f values are odd numbers such as 455, 465, 472.5, etc., in preference to such values as 450, 460, and 470.   Use of the odd-number values reduces the possibility of two transmitter carrier frequencies heterodyning with one another and producing a signal of the same value as the intermediate frequency, which, of course, would cause interference with the signal of the desired carrier wave.

Spurious Responses.   There are some additional sources of interference sometimes present in superheterodyne receivers.   These are usually due to the effects of harmonics generated in some portion of the receiver, which find their way to the i-f input.   These interferences are generally referred to as *spurious responses.*   Among the possible causes of spurious responses are: (1) harmonics of the intermediate frequency generated by the a-f detector (sometimes called the second detector) that may find their way back to the r-f circuit or to the i-f input through stray coupling or feedback; (2) harmonics of an r-f signal generated by the first detector, particularly the second harmonics of stations with carrier frequencies of from 550 to 800 kc; (3) harmonics of the oscillator, which may beat with the signal from a short-wave station and thus be received by the broadcast receiver; (4) interference caused by the beat frequency of two r-f signals differing in frequency by an amount equal to the i-f value; (5) reception of the r-f signal of a transmitter operating on a frequency equal to that of the i-f of the receiver.   These various interferences are not present in a well-designed and carefully adjusted receiver, since they can be avoided by providing sufficient preselection and by shielding of wires, coils, and tubes to eliminate stray pickup of any undesired frequencies.

11-10. Typical Superheterodyne Receiver Circuit.   The circuit diagram of a five-tube a-c–d-c superheterodyne receiver is shown in Fig. 11-12.   The r-f signals reaching the antenna are acted upon directly by the tuning circuit T_1, C_1, C_2.   The signals of the station selected by this circuit are applied to the r-f signal control grid (grid 3) of the pentagrid converter tube VT_1.   The local oscillator circuit is formed by the cathode, grid 1, and grids 2 and 4 of VT_1 and the circuit elements R_1, C_6, T_2, C_3, C_4.   Tuning capacitors C_1 and C_3 are controlled by a common shaft and tuning dial, and the two tuning circuits are adjusted so that the oscillator frequency will always be 455 kc above the resonant frequency of the antenna tuning circuit.   Trimmer capacitors C_2 and C_4 are used to make the two tuning circuits track evenly with one another over the complete a-m broadcast band.

The plate circuit output of VT_1, whose current contains components of

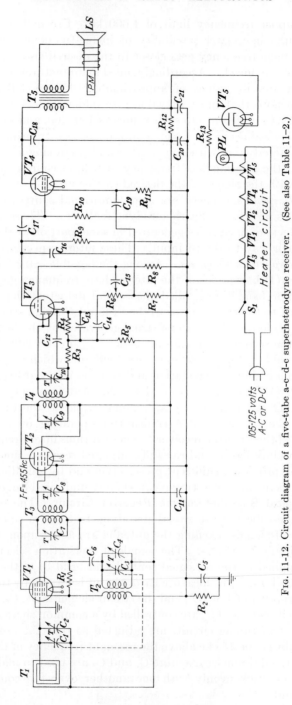

FIG. 11-12. Circuit diagram of a five-tube a-c-d-c superheterodyne receiver. (See also Table 11-2.)

TABLE 11-2. CIRCUIT ELEMENTS SHOWN IN FIG. 11-12

Part	Name	Function	Approximate value
T_1	Antenna coil	Reception and preselection	
T_2	Oscillator coil	Part of local oscillator	
T_3, T_4	I-f transformer	Bandpass tuning	
T_5	Output transformer	Impedance matching	
VT_1	Pentagrid converter	Frequency conversion	12BE6
VT_2	Pentode	I-f amplification	12BA6
VT_3	Duodiode-triode	Detection, a-f voltage amplification, and automatic volume control	12AV6
VT_4	Beam power tube	A-f power amplification	50C5
VT_5	Half-wave rectifier	B power supply	35W4
C_1	Variable capacitor	Tuning antenna circuit	11–390 $\mu\mu$f
C_2	Trimmer capacitor	Alignment	2–14 $\mu\mu$f
C_3	Variable capacitor	Tuning oscillator circuit	7–115 $\mu\mu$f
C_4	Trimmer capacitor	Alignment	2–14 $\mu\mu$f
C_5	Paper capacitor	Chassis isolation	0.1 μf, 400 volts
C_6	Ceramic capacitor	Oscillator bias	56 $\mu\mu$f, 500 volts
C_7, C_8, C_9, C_{10}	Adjustable capacitor	I-f bandpass adjustment	130–170 $\mu\mu$f
C_{11}	Paper capacitor	Avc circuit	0.047 μf, 400 volts
C_{12}	Ceramic capacitor	I-f bypass	220 $\mu\mu$f
C_{13}	Ceramic capacitor	I-f filter circuit	220 $\mu\mu$f
C_{14}	Paper capacitor	A-f coupling	0.005 μf, 400 volts
C_{15}	Paper capacitor	A-f coupling	0.01 μf, 400 volts
C_{16}	Mica capacitor	R-f bypass	330 $\mu\mu$f
C_{17}	Paper capacitor	Coupling and d-c blocking	0.005 μf, 400 volts
C_{18}	Paper capacitor	High a-f attenuation	0.01 μf, 400 volts
C_{19}	Paper capacitor	Couple negative feedback	0.1 μf, 400 volts
C_{20}, C_{21}	Electrolytic capacitor	B supply filter	50–50 μf, 150–150 volts
R_1	Carbon resistor	Oscillator bias	33,000 ohms, $\frac{1}{2}$ watt
R_2	Carbon resistor	Chassis isolation	220,000 ohms, $\frac{1}{2}$ watt
R_3	Carbon resistor	Portion of diode load	47,000 ohms, $\frac{1}{2}$ watt
R_4	Carbon resistor	Portion of diode load	220,000 ohms, $\frac{1}{2}$ watt
R_5	Carbon resistor	Avc circuit	3.3 megohms, $\frac{1}{2}$ watt
R_6	Potentiometer	Volume control	2 megohms
R_7	Carbon resistor	Negative feedback	1,000 ohms, $\frac{1}{2}$ watt
R_8	Carbon resistor	Grid-load resistor	5.6 megohms, $\frac{1}{2}$ watt
R_9	Carbon resistor	Plate-load resistor	470,000 ohms, $\frac{1}{2}$ watt
R_{10}	Carbon resistor	Grid-load resistor	470,000 ohms, $\frac{1}{2}$ watt
R_{11}	Carbon resistor	Provides cathode bias	150 ohms, $\frac{1}{2}$ watt
R_{12}	Carbon resistor	B supply filter	1,200 ohms, 1 watt
R_{13}	Carbon resistor	Surge protection	18 ohms, $\frac{1}{2}$ watt
LS	Loudspeaker	Reproduction of sound	
S_1	Spst switch	"On and off" control	
PL	Pilot light	Illumination of dial	

a number of frequencies (Art. 11-9), is fed to the primary of the first i-f transformer T_3. Because this transformer tunes very sharply, only the 455-kc component is passed on to the control grid of the i-f voltage-amplifier tube VT_2. The output of VT_2 is fed to the primary of the second i-f transformer T_4, which provides additional tuning. The band-pass characteristics of the i-f amplifier stage are controlled by the adjustable capacitors C_7, C_8, C_9, C_{10}.

Another function of T_4 is to couple the i-f signal from VT_2 to the diode detector (second or a-f detector) in VT_3. The detector load resistance consists of R_3 and R_4, and the detector i-f bypass capacitor is C_{12}. The combination of R_3, R_4, $R_6 + R_7$, and C_{12}, C_{13}, C_{14} forms an r-f filter network (Art. 3-2, Fig. 3-5). Potentiometer R_6 provides the means of controlling the volume output of the receiver. R_5 and C_{11} produce the avc voltage which is applied to the control grids of VT_1 and VT_2.

The triode section of VT_3 operates as an a-f voltage amplifier and has the variable a-f signal obtained at R_6 coupled to its grid by C_{15}. R_8 is the grid-load resistor, R_9 is the plate-load resistor, and C_{16} is an r-f bypass capacitor. The output of VT_3 is coupled to the grid of a beam power tube VT_4 by C_{17}. R_{10} is the grid-load resistor and R_{11} provides the cathode bias for VT_4. The signal voltage appearing at R_{11} is coupled through C_{19} back to R_7, thereby providing negative feedback to VT_3. The output of VT_4 is fed to the output transformer T_5, which provides the proper impedance match to efficiently couple the a-f signal to the voice coil of the loudspeaker.

The pulsating voltage output of the half-wave rectifier tube VT_5 is fed through the capacitor-input type filter circuit consisting of C_{21}, R_{12}, C_{20} to the B+ supply line. R_{13} is called a *surge resistor* and protects VT_5 from being damaged by a possible high current surge at the instant the power switch S_1 is closed. The heaters of the five tubes are connected in series and may be connected directly across the power lines as the sum of the rated heater voltages is approximately equal to the line voltage. R_2 and C_5 are used to isolate, in effect, the chassis from the B+ line and the power line.

A list of the parts, their function, and their values for the receiver shown in Fig. 11-12 is given in Table 11-2.

11-11. Nine-tube Superheterodyne Receiver. Figure 11-13 is a schematic diagram showing the type of circuit that might be used in a nine-tube a-c operated superheterodyne receiver; it is not being presented as being a specific commercial receiver. Among the features that distinguish it from the typical five-tube superheterodyne receiver of Art. 11-10 are the following: (1) It can be operated only from an a-c power source. (2) A three-gang variable capacitor is required to control the three tuned circuits simultaneously. (3) It has one stage of trf amplification preceding the converter stage. (4) It uses a two-tube push-pull a-f output

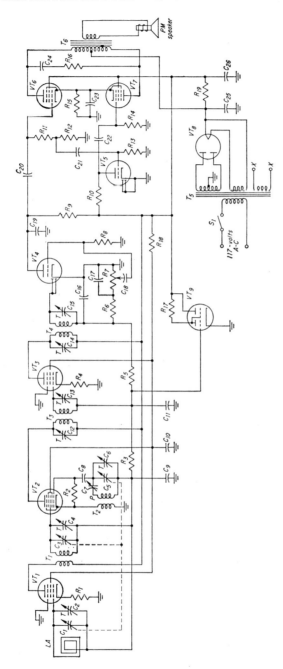

Fig. 11-13. Circuit diagram of a nine-tube a-c operated superheterodyne receiver.   (See also Table 11-3.)

stage.   (5) A phase inverter is required for the push-pull amplifier stage. (6) An electron-ray tube is included to assure accurate tuning.   (7) A dual avc circuit is used.   (8) A full-wave rectifier is used in the B power-supply circuit.

The r-f signals reaching the loop antenna are acted upon directly by the tuning circuit LA, C_1, C_2 and amplified by the pentode voltage amplifier VT_1. This trf amplifier stage will improve both the sensitivity and selectivity over that of the receiver of Fig. 11-12.

TABLE 11-3. VALUES OF THE CIRCUIT ELEMENTS OF FIG. 11-13

Resistors			Capacitors		Tubes		Other elements	
Part	Ohms	Watts	Part	Value	Part	Type	Part	Description
R_1	180	½	C_1	10–365 μμf	VT_1	6BA6	LA	Loop antenna
R_2	22,000	½	C_2	4–30 μμf	VT_2	6BE6	LS	Loudspeaker
R_3	2.2 megohms	½	C_3	10–365 μμf	VT_3	6BA6	T_1	R-f transformer
R_4	180	½	C_4	4–30 μμf	VT_4		T_2	Oscillator transformer
R_5	2.2 megohms	½	C_5	10–365 μμf	VT_5	6AV6	T_3	I-f transformers
R_6	100,000	½	C_6	4–30 μμf	VT_6		T_4	
R_7	1-megohm		C_7	depends on T_2	VT_7	6AQ5	T_5	Power transformer
	potentiometer	...	C_8	56 μμf, mica	VT_8	5Y3GT	T_6	Output transformer
R_8	10 megohms	½	C_9	0.05 μf, 50 volts	VT_9	6E5		
R_9	220,000	½	C_{10}	0.05 μf, 300 volts				
R_{10}	220,000	½	C_{11}	0.05 μf, 50 volts				
R_{11}	470,000	½	C_{12}	Part of T_3				
R_{12}	8,200	½	C_{13}					
R_{13}	10 megohms	½	C_{14}	Part of T_4				
R_{14}	470,000	½	C_{15}					
R_{15}	270	5	C_{16}	180 μμf, mica				
R_{16}	15,000	1	C_{17}					
R_{17}	1 megohm	½	C_{18}	0.01 μf, 50 volts				
R_{18}	12,000	2	C_{19}	120 μμf, mica				
R_{19}	1,800	2	C_{20}	0.02 μf, 50 volts				
			C_{21}	0.01 μf, 50 volts				
			C_{22}	0.02 μf, 50 volts				
			C_{23}	20 μf, 50 volts, electrolytic				
			C_{24}	0.05 μf, 600 volts				
			C_{25}	20 μf, 450 volts, electrolytic				
			C_{26}					

The converter and i-f amplifier stages are similar to those explained in analyzing the circuit of Fig. 11-12 except for capacitor C_7.   Because the oscillator frequencies are higher than the r-f signal frequencies, a lower value of capacitance is used in the oscillator tuning circuit than in the r-f tuning circuits.   For the circuit of Fig. 11-12, the capacitance of C_3 is made lower than that of C_1 by reducing the area of the capacitor plates at C_3.   For the circuit of Fig. 11-13 and Table 11-3, tuning capacitors C_1, C_3, C_5 are all of the same size and the capacitance of the oscillator circuit, tuned by C_5, is reduced to the required value by the addition of the series-connected capacitor C_7.

The detector, r-f filter network, and the a-f voltage amplifier circuits (VT_4) are the same as those explained in analyzing the circuit of Fig. 11-12. The avc system contains two sections: the first section R_5C_{11} provides the avc bias for VT_3, and the second section R_3C_9 provides the avc bias for VT_1 and VT_2.

The a-f power output is obtained from two beam power tubes VT_6 and VT_7 connected as a push-pull amplifier. As a push-pull amplifier requires that the signal inputs to the two tubes have a 180-degree phase difference, the input signal for VT_6 is taken from the output of the a-f voltage amplifier VT_4, and the input signal for VT_7 is taken from the output of the phase-inverter stage VT_5 (Art. 7-22).

An electron-ray tube VT_9 is used to indicate when the receiver is properly tuned to the signal of the station selected (Art. 11-12).

11-12. Tuning Indicators. *Need for Tuning Indicators.* In the operation of a highly selective receiver (Fig. 11-13), the receiver should be carefully tuned so that the carrier frequency of the desired station is at the center of the response band. If the receiver is not properly tuned, the output at the loudspeaker may be badly distorted. The *electron-ray tube*, also called a *magic eye* or a *cathode-ray indicator tube*, may be used to indicate whether a receiver is properly tuned to the desired station.

Electron-ray Tube. The electron ray tuning indicator shown in Fig. 11-14 contains two sections: (1) a triode consisting of a cathode, a control grid, and a plate which together function as a d-c amplifier; (2) the fluorescent-coated target and the ray-control electrode. The circuit connections for this type of tube are illustrated in Figs. 11-13 and 11-14*d*. Electrons from the cathode have two paths, one to the triode plate and the other to the fluorescent target. The triode plate current is controlled by the voltage of the triode grid, and the target current is controlled by the voltage of the ray-control electrode. When electrons from the cathode strike the target, they cause the coating on the target to fluoresce and give off a faint green light. When the voltage of the ray-control electrode and the target are of approximately equal value, the target will be illuminated evenly or may have only a very small shaded area (Fig. 11-14*b*). When the voltage of the ray-control electrode is less than the target voltage, so that it is negative with respect to the target, fewer electrons reach the target and the fluorescent area decreases (Fig. 11-14*c*).

The tuning indicator tube is mounted in the receiver in such a manner that the fluorescent-coated target is visible to the person tuning the receiver. When the receiver is being tuned to a station, the operator should observe the action taking place at this tube. The receiver is properly tuned to a station when the shadow appearing on the target covers a minimum amount of area (Fig. 11-14*b*). Improper tuning causes the shadow to cover a greater area, as is indicated in Fig. 11-14*c*.

Electron-ray Indicator Circuit. In analyzing the action of the tuning indicator in the circuit of Fig. 11-13, it should be observed that the triode grid is connected to the avc line.   When the receiver is not tuned to a station, the avc voltage will be at its minimum value and the plate current of the triode section of the tuning indicator will be at its highest value.   An appreciable voltage drop will be present at the 1-megohm resistor R_{17}, and the ray-control electrode voltage will be considerably lower than the voltage at the fluorescent target.   This will result in the target having a large shaded area.   As the receiver tunes in a station, the negative voltage at the avc line increases.   This negative voltage is

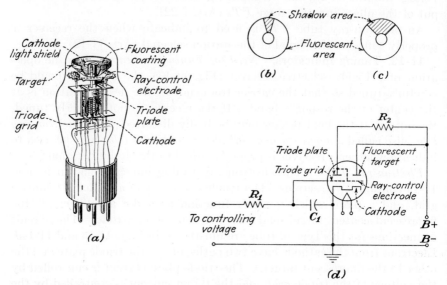

Fɪɢ. 11-14. Tuning indicator.   (*a*) Construction of the electron-ray tube.   (*b*) Indication when the receiver is properly tuned.   (*c*) Indication when the receiver is improperly tuned.   (*d*) Circuit connections.   (*Courtesy of RCA.*)

applied to the triode grid of the tuning indicator and causes a reduction in the triode-plate current, which in turn reduces the voltage drop at R_{17} and thereby increases the voltage at the ray-control electrode.   This in turn increases the current in the fluorescent target, thereby increasing the fluorescent area and decreasing the shaded area.   As the avc voltage reaches its highest negative value when a station is properly tuned, the diode-detector current through R_6 and R_7 will then be at its highest value; the shaded area of the tuning indicator will be at its minimum value when the receiver is properly tuned to a desired station.

11-13. Tracking and Alignment of Receivers. *Need for Alignment.* Modern receivers are constructed so that station selection can be obtained by turning a single dial.   In order to accomplish this, all the tuning circuits must be adjusted simultaneously.   This is ordinarily accomplished by use of ganged capacitors. In the trf receiver of Fig. 11-9, there are three tuning circuits, and thus a three-gang variable capacitor is used to obtain single-dial tuning. In the superheterodyne receiver of Fig. 11-13, two preselector circuits and the oscillator are tuned simultaneously by means of a three-gang capacitor. In order to have a receiver properly tuned to a station, each of its tuned circuits must be adjusted so that they are resonant to the proper value of frequency. When ganged tuning is employed, it is difficult to have two or more tuned circuits have their correct resonant frequencies at all points of the dial, even with modern precision-manufacturing methods. It thus becomes necessary that the tuning circuits be provided with means of adjustment so that alignment of the circuits will be as nearly uniform as is practically possible over the entire frequency range of the receiver. When this is accomplished, the circuits are said to be *tracking* each other.

There are two reasons why a receiver should be carefully aligned. First, correct alignment is necessary in order to obtain the best possible performance of the receiver. Second, proper alignment is required in order to calibrate the receiver dial, that is, to have it receive a desired station at its designated position on the dial.

Method of Aligning a Receiver. Accurate alignment can best be accomplished by use of a signal generator and a meter or other device that will indicate the strength of the output signal of the receiver. The signal generator should have a calibrated variable r-f output of sufficient frequency range to check all the frequencies at which the receiver is to operate. The signal generator should also be provided with a means of modulating its r-f output with an a-f signal of constant value, preferably in the order of 400 cycles. As the procedure in aligning a receiver is based on obtaining the maximum output for a given setting of the receiver, the device for indicating the output need only indicate whether maximum output is being obtained. Among the devices used for obtaining an indication of the relative value of the output are (1) an output meter, (2) a magic-eye tuning indicator, especially if it is already a part of the receiver circuit, (3) a vacuum-tube voltmeter, (4) a cathode-ray oscilloscope.

When the output of a signal generator is applied to the antenna and ground terminals of a receiver for the purpose of aligning the receiver, it is usually desirable to connect a small capacitor between the high-side terminal of the signal generator and the antenna terminal of the receiver. This capacitor is used to make allowance for the antenna that is normally connected to the receiver and is therefore called a *dummy antenna;* for broadcast receivers a 200-$\mu\mu$f capacitor is recommended. If the receiver has a loop antenna, the signal generator output may be coupled to the receiver by connecting a piece of wire about 2 ft in length to the high-side terminal of the signal generator. This wire is not connected to the receiver but acts as an antenna and should be kept about 2 ft from the receiver's loop antenna.

In general, the procedure in aligning a receiver is to apply the modulated r-f output of a signal generator to the receiver through a dummy antenna and to connect an output meter at the loudspeaker. With the receiver and the signal generator tuned to the same frequency, the tuning circuits are adjusted so that maximum output is obtained.

11-14. Aligning the Superheterodyne Receiver. *Procedure.* In aligning superheterodyne receivers, the following procedure may be used. First adjust the tuned circuit nearest the a-f output of the receiver and then work back toward the antenna. In the average superheterodyne receiver, the order in which the circuits are aligned is as follows: (1) the i-f amplifier circuit, (2) the oscillator circuit, (3) the r-f or preselector circuits.

Aligning the I-F Amplifier Circuits. When tuning the i-f circuits, the local oscillator and the automatic volume control should be made inoperative. In the circuit of Fig. 11-13 this may be done by (1) short-circuiting the oscillator capacitors C_5 and C_7 and (2) short-circuiting the avc capacitors C_9 and C_{11}. An output meter is connected across the secondary terminals of the output transformer T_6 and can remain there throughout the entire alignment and tracking procedure. A signal generator is connected with its ground terminal to the chassis of the receiver and its high-side terminal connected through a 0.01-μf capacitor to the control grid of VT_3. The receiver volume control should be turned on fully, the signal generator frequency set to correspond with the i-f value recommended for the receiver, and the signal generator output turned up just enough to produce an indication on the output meter. The trimmer capacitors C_{14} and C_{15} are then adjusted to obtain maximum indication on the output meter. When the i-f transformer T_4 is properly adjusted, the next step is to adjust i-f transformer T_3. This may be done by applying the output of the signal generator to the r-f control grid of VT_2 and then proceeding to adjust the trimmer capacitors C_{12} and C_{13} until maximum indication is obtained on the output meter. If the output indication becomes too high for the meter, it should be reduced by decreasing the amplitude of the output of the signal generator. In order to obtain satisfactory tracking and over-all operation of the receiver, the frequency to which the i-f circuits are adjusted must correspond with the value recommended by the manufacturer of the receiver. The correct i-f value for a receiver may be obtained from some markings or diagram supplied with the receiver or from a service manual.

Aligning the Oscillator and R-F Circuits. In order to align the oscillator, it is necessary to remove the short circuit at the oscillator tuning capacitors C_5 and C_7, which had been placed there when adjusting the i-f circuits. With the signal generator still connected to the r-f control grid of VT_2, set the receiver dial and the signal generator frequency each to a value near 1,400 kc, but not to a point at which a station is received, and adjust the oscillator trimmer capacitor C_6 to obtain maximum indication on the output meter.   With the tuning dial of the receiver and the signal generator frequency at the same settings, apply the signal generator output to the antenna terminal of the receiver through a 200-$\mu\mu$f capacitor and adjust trimmer capacitors C_4 and C_2 until maximum indication is obtained on the output meter.   It would also have been possible to adjust the local oscillator and the two preselector circuits with the signal generator output connected to the antenna terminal through a 200-$\mu\mu$f capacitor and, with the receiver and signal generator set near 1,400 kc, then progressively adjust the trimmer capacitors C_6, C_4, and C_2 to obtain maximum indication on the output meter.

The local oscillator must also be adjusted for some value at the low-frequency end of the dial.   It is generally desirable to make the low-frequency adjustment at about 10 per cent above the minimum frequency of the receiver; in fact, most manufacturers recommend a specific value at which the low-frequency-range tracking should be adjusted.   This information is usually obtained from a service manual.   As no specific value is given for the circuit of Fig. 11-13, 600 kc will be a convenient value to use. Accordingly, the low-frequency alignment may be made by setting the receiver dial and the signal generator frequency at 600 kc. The oscillator padding capacitor C_7 is then adjusted to obtain maximum indication on the output meter, while gently rocking the ganged capacitors back and forth to keep the signal tuned in.   The low-frequency adjustment should be made only at the padder capacitor, and no adjusting is to be done at the trimmer capacitors.

It is good practice to recheck the high-frequency setting after completing the low-frequency adjustments and make any corrections necessary. Although the use of tuning capacitors with slotted rotor plates makes it possible to make an additional adjustment at the middle of the frequency range, most superheterodyne receivers will track satisfactorily if they are carefully aligned at the high- and low-frequency ends.

Methods of Aligning the Oscillator and R-F Circuits.   There are two commonly used methods of making the oscillator circuit of the super-heterodyne receiver track with the preselector circuits.   One method is to use a padder capacitor as in the circuit of Fig. 11-13.   In the second method, sometimes called the *cut-plate method*, the variable capacitor used to tune the oscillator circuit has a lower value of capacitance than the capacitor used in the r-f tuning circuits (see C_1 and C_3 in Table 11-2).

In the method that employs a padder capacitor, all the ganged tuning capacitors are of the same construction.   Because the oscillator is tuned to a higher frequency, the ratio of high- to low-frequency values is less for the oscillator than for the r-f circuits.   For example, to cover the broadcast band from 550 to 1,650 kc, the frequency range of the r-f circuits will be 3 to 1. For a 465-kc i-f value, the oscillator frequency must range from 1,015 to 2,115 kc or approximately 2 to 1. Consequently, the capacitors used to tune the r-f circuits must have a greater range than the capacitors used to tune the oscillator. In order to use identical capacitors for all the tuning circuits, another capacitor, called a *padder capacitor*, is connected in series with the original tuning capacitor to reduce the ratio of the maximum to minimum values of the series combination. This may be verified by referring to Examples 4-9 and 4-10 in Art. 4-5. Radio receivers may be designed so that exact tracking will be obtained at three frequencies in the band and satisfactory tracking also provided over the remainder of the band. In order to provide a means of making tracking adjustments, the padding capacitor is made adjustable. The capacitance required for the padder is in the order of 350 $\mu\mu$f for a 465-kc intermediate frequency, and nearly 1,000 $\mu\mu$f for an intermediate frequency of 175 kc. When a high value of capacitance is required for the padder and only a small amount of variation is needed, some receivers use a fixed capacitor shunted by a suitable small trimmer for this purpose.

In the cut-plate method, the shape of the rotor plates of the oscillator tuning capacitor is different from those of the r-f tuning capacitors. By properly designing the shapes of the rotor plates, it is possible to achieve proper tracking of the oscillator circuit without the need of a padding capacitor. The disadvantages of this method are: (1) The cost of manufacture is higher. (2) It is not satisfactory for use in multiband receivers. (3) The capacitors can usually only be used for receivers with the same i-f value. (4) In some instances the capacitors can only be used for one circuit design. (5) It is more difficult to make service adjustments than with the padder-capacitor method.

QUESTIONS

1. Define the following terms: (*a*) sensitivity, (*b*) selectivity, (*c*) fidelity, (*d*) stability, (*e*) signal-to-noise ratio.

2. Name and describe four types of noise that may occur in a radio receiver.

3. Name four essential functions of a radio receiver.

4. What are the disadvantages of the simple receiving circuit of (*a*) Fig. 1-9? (*b*) Fig. 11-2*a*? (*c*) Fig. 11-2*b*?

5. (*a*) What is the purpose of adding one or more stages of a-f amplification? (*b*) Why is it generally undesirable to have more than two stages of a-f amplification in a radio receiver?

6. (*a*) What is the purpose of adding one or more stages of r-f amplification? (*b*) What is meant by a trf stage of amplification? (*c*) Describe two advantages of trf amplification.

7. (*a*) What is meant by an i-f amplifier? (*b*) What is the purpose of an i-f amplifier? (*c*) What advantages are gained in using i-f amplification?

8. With the aid of a block diagram explain the operation of a trf receiver.

9. Describe the operation of the tuning section of the trf receiver of Fig. 11-9.

10. Describe the operation of the detector section of the trf receiver of Fig. 11-9.

11. Describe the operation of the a-f section of the trf receiver of Fig. 11-9.

12. Describe the operation of the power-supply section of the trf receiver of Fig. 11-9.

13. Draw a circuit diagram and explain the operation of (a) a volume-control circuit, (b) a tone-control circuit.

14. With the aid of circuit diagrams explain the operation of (a) a voltage-dropping resistor, (b) an r-f bypass capacitor, (c) a decoupling network.

15. Define (a) superheterodyne receiver, (b) double superheterodyne receiver.

16. Explain five advantages of the superheterodyne receiver.

17. With the aid of a block diagram explain the operation of a superheterodyne receiver.

18. (a) Describe several types of loop antennas. (b) What are the advantages of the loop antenna? (c) What are the disadvantages of the loop antenna?

19. (a) What is meant by a preselector? (b) Describe the operation of a preselector circuit. (c) What advantages are obtained by the use of preselection?

20. (a) Name the three essential functions required to obtain frequency conversion. (b) Explain the purpose of each function.

21. Explain two factors that must be taken into consideration in the design of a local oscillator that is to be used with a superheterodyne receiver.

22. (a) What five frequencies are present in the output circuit of a mixer tube when two signal voltages of different frequencies are applied to its input circuits? (b) Which of these frequencies is used in a-m broadcast-band superheterodyne receivers?

23. (a) What is meant by conversion gain? (b) How does the voltage amplification of a tube used as a converter compare with the voltage amplification of the same tube when used as an i-f amplifier?

24. (a) Describe the operation of a converter circuit that uses separate tubes for the oscillator and mixer. (b) What are the advantages of this type of circuit?

25. (a) Describe the operation of a pentagrid converter tube of the 6BE6 type. (b) Describe the operation of a converter circuit using a pentagrid converter tube of the 6BE6 type. (c) What are the characteristics of this type of converter circuit?

26. What is meant by (a) image-frequency signal? (b) image ratio?

27. (a) Name two factors that affect the choice of i-f value selected for a superheterodyne receiver. (b) What i-f values are commonly used?

28. (a) Explain the relation between the tuning ratio, the oscillator frequency, and the intermediate frequency. (b) In broadcast receivers, why is it desirable to have the oscillator frequency higher than the r-f signal input?

29. (a) Explain the relation between the value of the intermediate frequency and the selectivity of the receiver. (b) Why are odd number values of intermediate frequency generally used?

30. (a) What is meant by spurious responses? (b) Name five possible causes of spurious responses. (c) What precautions are used in superheterodyne receivers to minimize the interference caused by spurious responses?

31. (a) Draw a diagram for the r-f and oscillator-mixer portions (including the primary of the input i-f transformer) for the superheterodyne receiver circuit of Fig. 11-12. (b) Explain the operation of these portions of the receiver. (c) Prepare a list of the parts used, and state the function of each.

32. (a) Draw a diagram for the i-f and detector portions of the superheterodyne receiver circuit of Fig. 11-12. (b) Explain the operation of these portions of the receiver. (c) Prepare a list of the parts used and state the function of each.

33. (a) Draw a diagram for the a-f portion of the superheterodyne receiver circuit of Fig. 11-12. (b) Explain the operation of this portion of the receiver. (c) Prepare a list of the parts used and state the function of each.

34. (*a*) Draw a diagram for the heater and B power supply portions of the super-heterodyne receiver circuit of Fig. 11-12.   (*b*) Explain the operation of these portions of the receiver.   (*c*) Prepare a list of the parts used and state the function of each.

35. (*a*) In what manner does the r-f section of the receiver of Fig. 11-13 differ from that of Fig. 11-12?   (*b*) What are the advantages and disadvantages of each?

36. (*a*) In what manner does the a-f detector circuit of the receiver of Fig. 11-13 differ from that of Fig. 11-12?   (*b*) What are the advantages and disadvantages of each?

37. (*a*) In what manner does the a-f section of the receiver of Fig. 11-13 differ from that of Fig. 11-12?   (*b*) What are the advantages and disadvantages of each?

38. (*a*) In what manner does the heater and B power supply circuits of the receiver of Fig. 11-13 differ from those of Fig. 11-12?   (*b*) What are the advantages and disadvantages of each?

39. (*a*) Describe the principle of operation of an electron-ray indicator tube. (*b*) Describe the operation of an electron-ray indicator circuit.

40. (*a*) Explain two reasons why a radio receiver should be properly aligned. (*b*) What instruments should be used to properly align a receiver?

41. In what order should the various tuned circuits in a superheterodyne receiver be aligned?

42. Describe the procedure for aligning an i-f amplifier circuit.

43. Describe the procedure for aligning the oscillator and r-f circuits.

44. Explain the use of padders and cut-plate capacitors in aligning superheterodyne receivers.

CHAPTER 12

FREQUENCY-MODULATION RECEIVER CIRCUITS

Radio development started with the a-m system and grew to a tremendous industry in a relatively few years. The f-m radio system is the outgrowth of research in quest of a means of radio communication that would overcome the disadvantages encountered in a-m broadcasting. By the time that the f-m system was discovered and developed to a commercially acceptable degree, the a-m system had become firmly entrenched as our major system of radio broadcasting. However, after a slow start, the use of f-m broadcasting has increased rapidly toward its proper place in the field of radio broadcasting.

12-1. Frequency-modulation vs. Amplitude-modulation Reception. *Frequency Bands.* A-m broadcasting stations are operated on a band of frequencies (assigned by the FCC) ranging from 540 to 1,600 kc, and f-m stations operate on assigned frequencies ranging from 88 to 108 mc.

Noise Reduction. The outstanding advantage of f-m over a-m reception is the great reduction in undesired external noises. Numerous types of external and internal noises may be present in the sounds produced by the conventional a-m receiver. External noises, generally called *static*, may be caused by nature, as in the case of lightning, northern lights, sunspots, etc., or it may be man-made static as produced by vacuum cleaners, electric shavers, neon signs, electric elevators, electric vibrator mechanisms, etc. As these disturbances affect the amplitude of the received signal, they introduce undesired noises in the output of a-m receivers. In some instances the noise produced is so severe that it completely ruins the reception of the desired program.

In f-m reception, amplitude disturbances due to static do not affect the transmitted signal frequencies. The variations in amplitude are smoothed out by the limiter circuit in the f-m receiver, and hence the output of the receiver is practically free from noises due to static.

Fidelity. Potentially, the f-m receiver is superior to the a-m receiver in the fidelity of reproduction. With the f-m system it is possible to reproduce sounds at frequencies up to 15,000 cycles as compared with only 5,000 cycles for the a-m system. Thus an f-m receiver, particularly in the case of a symphonic muscial program, will provide a more nearly exact reproduction of the original sounds than an a-m receiver.

Interstation Interference. With a-m reception it is possible to be tuned to a desired station and to have another station interfering in the background.   In some instances the interference may be sufficiently disturbing to spoil the program of the desired station, even though the signal strength of the interference is only 1 or 2 per cent of that of the desired station.   With f-m reception, it is possible to successfully separate the signals of two stations operating on the same frequency if the strength of the interfering signal is not more than 50 per cent of that of the desired station.   Thus, it is less likely to have distant stations interfere with local stations with f-m receivers than with a-m receivers.

Image Interference.   Image interference can only occur from signals differing by twice the value .of the frequency of the i-f section of the receiver.   With the 10.7-mc i-f value used in f-m receivers, double the i-f value (21.4 mc) exceeds the complete span of the f-m band of 88 to 108 mc, and hence there can be no image interference among f-m stations. However, this does not exclude the possibility of image interference from other signal sources outside of the f-m service band.

Selectivity.   The f-m receiver must have better adjacent channel selectivity than an a-m receiver because the ratio of channel width to

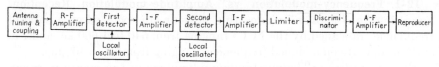

Fig. 12-1. Block diagram showing the sequence of the functions performed in an f-m receiver.

carrier frequency is greater for f-m receivers.   For example, an f-m station operating on 100 mc with a 200-kc channel separation has a ratio of 2 to 1,000, while an a-m station operating on 1,000 kc with a 10-kc channel separation has a ratio of 10 to 1,000.   Another way of stating this is that for the 100-mc f-m station the channel separation is 0.2 per cent of the carrier frequency while for a 1,000-kc a-m station it is 1 per cent, or five times as great.

12-2. Comparison of F-M and A-M Receivers. *Similarities.*   The f-m receiver employs the superheterodyne principle and is similar in many respects to the a-m superheterodyne receiver but its circuits are a little more complex.   In addition to performing the same functions as the a-m receiver, the f-m receiver has two additional functions, namely, amplitude limiting and frequency discrimination.   The limiter circuit is largely responsible for the noise-free reception obtained with the f-m receiver, and the discriminator serves as the detector to convert the frequency modulations into the audio signal.   A block diagram of an f-m receiver is shown in Fig. 12-1.

Differences.   Specifically, the f-m receiver differs from the a-m receiver in the following respects: (1) The signal frequencies in the r-f and i-f circuits are much higher.   (2) An antenna coupling network is used.

(3) At least one stage of r-f amplification is used.   (4) The i-f amplifier must pass a wider band of frequencies.   (5) Two or three stages of i-f amplification are generally used.   (6) A voltage-limiter circuit is provided between the i-f amplifier and the discriminator.   (7) The f-m detector is different from an a-m detector.   (8) The a-f amplifier is designed to provide a wider range of frequency response.   (9) A loudspeaker capable of reproducing a wider range of frequencies is used.

Because of the high frequencies present in the f-m receiver, some of the parts used in these receivers may have a different physical appearance than the parts used to perform similar functions in an a-m receiver.   Also, because of the high frequencies involved, the placement of the components and the length and location of the connecting wires must conform strictly with the specified locations and lengths as prescribed for each particular receiver design.

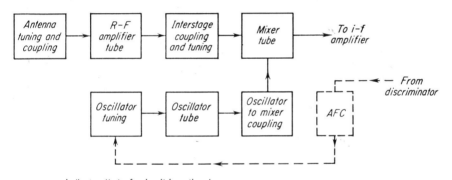

— — — — indicates that afc circuit is optional

Fig. 12-2. Block diagram of the basic circuits in the front end of an f-m receiver.

12-3. The Front End. *Functions.*   The section of the receiver which first acts upon the received signals is called the *front end,* and its primary functions are (1) to select the signal of the desired station, (2) to convert the selected r-f signal into a 10.7-mc i-f signal, and (3) to attenuate all other signals.   In the block diagram of Fig. 12-1 these functions are performed by (1) the antenna tuning and coupling unit, (2) the r-f amplifier, (3) the first detector, and (4) the local oscillator.   A more complete block diagram showing the basic circuits and the paths of the signals for the front end in commercial receivers is shown in Fig. 12-2.

Electrical Requirements.   The front end of an f-m receiver has more exacting operational requirements than an a-m receiver for these reasons: (1) It operates at much higher frequencies.   (2) It requires a greater bandwidth.   (3) The input circuit must be matched to the antenna.   (4) The signal-to-noise ratio must be high.   (5) Image rejection must be high. (6) Signals from the local oscillator must be attenuated to prevent feedback to the antenna.   (7) The local oscillator should not be subject to any appreciable frequency drift or microphonic noise.

Physical Requirements.   The circuits performing the functions of the front end are placed physically very close to one another (frequently mounted on a separate subchassis) because: (1) at their high operating frequencies the required values of inductance and capacitance are very low; (2) unless the connecting leads are kept very short and are carefully positioned, the stray inductances and capacitances may become high enough to seriously affect the proper functioning of these circuits; (3) errors of a fraction of an inch in the length of a lead or the positioning of a lead can cause improper operation of the receiver.

12-4. The Antenna and its Coupling Unit.   *Antenna Requirements.* The type of antenna required for use with an f-m receiver will vary with the location of the receiver with respect to the broadcasting stations.   If the receiver is used in a strong signal area a self-contained type of antenna will probably provide satisfactory results.   One commonly used type of self-contained antenna employs the inherent ability of the power lines to pick up radio signals and consequently is called a *power-line* or *power-cord antenna*.   In areas of only moderate signal strength satisfactory reception may be obtained from either (1) a power-cord antenna, or (2) a twin-lead antenna made from a 56-inch length of 300-ohm television wire with the two wires connected together at each end and with one wire opened at the mid-point for connecting the antenna to the antenna-input terminals of the receiver.   In weak-signal areas a commercial type of outdoor antenna must be used.   Because f-m signals do not follow the curvature of the earth, the antenna must be placed high enough to be in a straight line with the antenna of the transmitting station.   Both the simple twin-lead antenna and most commercial antennas possess strong directional characteristics; hence it is important to carefully position such antennas.

Antenna Tuning Unit and its Functions.   The *antenna coupling unit*, also called the *antenna coil* or *antenna input transformer*, is a network of inductance, capacitance, and resistance located between the input of the first tube and the signal-input terminals of the receiver.   This unit must properly couple the transmission line to the receiver in order to (1) develop maximum signal voltage at the input of the first tube, (2) terminate the line properly, (3) improve the selectivity, and (4) balance the line to ground when required.   The correct termination is a resistive load of the same magnitude as the characteristic impedance of the line, in most cases 75 or 300 ohms.   The impedance at the input terminals of the receiver also should be purely resistive and equal to the characteristic impedance of the line.

Classification of Antenna Couplers.   Antenna-coupler networks may be divided into two main groups, namely, balanced and unbalanced.   Each group may be further subdivided as being tuned or untuned.   A coupler which maintains each side of a transmission line at the same impedance above ground is called a *balanced antenna coupler*. When the impedance above ground differs for each side of a transmission line, it is called an *unbalanced antenna coupler*. If the coupler is also tuned on either the primary or secondary side, it may be either a tuned balanced or tuned unbalanced coupler. If both the primary and secondary are untuned, it may be either an untuned balanced or untuned unbalanced coupler.

Usually balanced couplers are used with twin-lead transmission lines, and unbalanced couplers with shielded or coaxial lines. To function properly, each conductor of a twin-lead transmission line must be at the same impedance above ground, and the line is then balanced to ground.

12-5. Tubes Used in the Front End of F-M Receivers. *Character-istics of Triodes and Pentodes.* Either a triode or a pentode may be used in the high-frequency circuits of the f-m receiver, and each has some advantages and disadvantages. Important requirements of an r-f

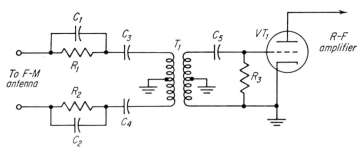

Fig. 12-3. Antenna coupling circuit with balanced input.

amplifier tube are (1) low noise figure, (2) low interelectrode capacitances, (3) high transconductance, and (4) high input resistance that does not vary with changes in grid voltage. Triodes have a much lower noise figure than pentodes. However, because of their low transconductance they have relatively low gain. Also the input impedance and the grid-to-plate capacitance of triodes are much higher than of pentodes. When triodes are used as r-f amplifiers, provision must be made for minimizing feedback, such as (1) using neutralizing circuits, or (2) using the tube in a grounded-grid or cathode-driven type of circuit. Although the pentode has a higher noise figure than the triode, it has the advantage of a high transconductance and generally does not require neutralization. Tubes having a remote-cutoff characteristic reduce cross modulation; however, sharp-cutoff tubes have a lower input impedance.

Miniature tubes are used in f-m receivers because of their (1) compactness, (2) low interelectrode capacitance, (3) high transconductance, (4) low tube-lead inductance, and (5) very low loss in the tube base material. In miniature tubes the elements are connected directly to the rigid base pins sealed into the bottom of the glass envelope.

Factors Affecting the Input Resistance of a Tube. At frequencies of 88 to 108 mc the input resistance of a tube will be much lower than at the frequencies used in a-m receivers. This is due largely to the electron transit time, which is the time required for electrons to move from the cathode to the plate. At very high frequencies, especially at 100 mc and above, the transit time causes power to be consumed at the grid, even though the bias is higher than the peak positive swing of the r-f input signal voltage. The power loss is proportional to the equivalent resistance, which is represented by

$$R_{eq \cdot g} = \frac{1}{k g_m f^2 t^2} \tag{12-1}$$

where $R_{eq \cdot g}$ = equivalent grid resistance due to transit time, ohms
$\quad\quad k$ = constant depending upon (1) plate voltage, (2) grid voltage, (3) ratio of time required for electrons to flow from cathode to grid to time required to flow from grid to plate
$\quad\quad g_m$ = transconductance of tube, μmhos
$\quad\quad f$ = frequency, cps
$\quad\quad t$ = time for electron to travel from cathode to grid, sec

This equation shows that the input resistance varies inversely with the square of the frequency. For example, a pentode with an input resistance of 20 megohms at 1,000 kc will have a resistance of 22,000 ohms at 30 mc and only 2,000 ohms at 100 mc. The input resistance is also directly proportional to the square root of the electrode voltages and inversely proportional to the square of the linear dimensions of the tube; consequently, improved operation can be obtained with smaller tubes. The input resistance of a miniature tube is about 20 times higher than that of a standard-size tube.

Variations in Input Capacitance Due to Automatic Volume Control. The input capacitance of a vacuum tube decreases as the grid bias becomes more negative. For the tuned circuits of very-high-frequency receivers the capacitance may be composed principally of the input capacitance of the tube, and consequently a small change in this capacitance may cause an appreciable change in the resonant frequency and the bandpass of the amplifier. A 1- to 3-$\mu\mu$f change in the grid capacitance of the tube may be caused by (1) a shift in the space charge as the grid bias is varied, and (2) feedback through the grid-to-plate capacitance. The effect of such changes can be decreased by use of the circuit shown in Fig. 12-4. The unbypassed resistor R_n produces negative feedback, which reduces the changes in input capacitance, since the bias voltage due to avc action varies with the strength of the r-f input signal.

Connections of R-F Amplifier Tubes. The use of a pentode does not always ensure stability for a circuit and eliminate the need for neutralization. If a pentode tube with its suppressor grid connected internally to the cathode is used in a high-frequency amplifier stage, the circuit might be as unstable as with an unneutralized triode and under certain conditions might oscillate.   For this reason, it is better to use a pentode with the suppressor grid brought out to a separate pin which is then connected to ground as close as possible to the tube socket.   To minimize coupling to other stages and also to reduce the impedance of ground paths, the various bypass capacitors should be connected to ground near the socket of the associated stage and as close as possible to the ground end of the cathode. The lead between cathode and ground introduces inductance into the cathode circuit, and if this lead becomes too long, the resultant degeneration may cause too great a decrease in the grid input resistance.   A long lead in the screen-grid circuit will introduce regeneration and cause negative resistance to be reflected in the grid circuit which decreases the grid losses and may cause oscillation.   Regeneration may also be caused by (1) an inductive plate load, (2) a capacitive cathode circuit, or (3) a capacitive plate or screen-grid load.

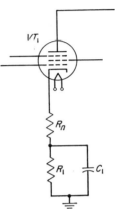

Fig. 12-4. Method of decreasing the effect of variations in the input capacitance.

12-6. Noise. *Types of Noise.* A certain amount of noise is generated in an f-m receiver in addition to the noise that may be picked up by the antenna.   The four types of disturbance that may produce noise in a receiver are: (1) shot effect, (2) thermal agitation, (3) microphonics, and (4) a-c hum (Art. 11-1).   If the noise voltages approach the magnitude of the signal voltage, the signal will be rendered useless for enjoyable listening.

Shot Effect.   The amount of noise due to shot effect increases with the number of elements in a tube.   Thus, a pentode being used as an r-f amplifier may produce several times as much noise as a triode.   A tube used as a mixer or converter produces more noise than when it is used as an amplifier.   By using a tube having a low noise factor and high amplification characteristics, an r-f amplifier can increase the signal voltage enough to override the noise produced in the mixer.

Noise Figure.   The weakest signal that a receiver is capable of receiving successfully is determined by the amount of noise generated in the circuits of the front end.   A quantity called the *noise figure* is used to compare the quality, as far as noise is concerned, of various receivers.   The noise figure is the ratio of the signal-to-noise power at the antenna terminals of the receiver to the signal-to-noise power at the loudspeaker.   This ratio is expressed in decibels and is compared with the signal-to-noise power ratio of an assumed perfect receiver, which has a noise figure of 1, or zero db.

12-7. The Radio-frequency Amplifier. *Advantages.* The r-f amplifier stage increases the signal strength at the same frequency at which it is received.   In strong signal areas, the principal advantage of having an r-f amplifier stage is in the improved selectivity of the receiver.   In congested areas, receivers without an r-f stage may cause mutual interference among the receivers because of oscillator voltage getting into the antenna system.   The receiver may then act as a miniature transmitter and radiate disturbing signals in the area.   However, the addition of an r-f stage to the receiver will reduce the possibility of radiating a signal from the receiver.   In weak signal areas, the avc action may cause a receiver to operate at its full sensitivity, whereupon the amplification approaches the noise limit because of the shot-effect and thermal-agitation noise.   When the signal is very weak, an r-f stage may strengthen the signal to a level that permits operation at less than full sensitivity.

Basic Circuit Principles.   Most f-m receivers use one stage of r-f amplification.   This amplifier requires an antenna coupling circuit (Art. 12-4) to couple the r-f signal from the antenna to the input of the r-f amplifier tube and an interstage coupling circuit (Art. 5-5) to couple the output signal of this tube to the input of the mixer tube.   (The performance characteristics of r-f amplifier circuits have been presented in Chap. 5.)   The r-f amplifiers used in f-m receivers are usually broadband amplifiers designed to have a wideband frequency response at the antenna coupler and a relatively narrow-band response at the interstage coupler. The bandwidth of the input circuit must be sufficient to cover the f-m band (88 to 108 mc), while the bandwidth of the output circuit should be in the order of 200 to 400 kc, that is, sufficient to pass the signal of one, and only one, station.   None of the antenna couplers shown in this chapter indicate any means of varying their resonant frequency; hence they must have sufficient bandwidth to cover the entire f-m band.   In order to get the desired bandwidth, these circuits use coils having relatively low values of Q; in some cases, the desired values of Q are obtained by adding a loading resistor.

Three types of amplifier circuits are in common use for the r-f amplifier stage; they are (1) grounded-cathode amplifier, (2) grounded-grid amplifier, and (3) cascode amplifier.

Grounded-cathode Amplifier.   This is the type of amplifier commonly used in a-m receiver circuits.   The characteristics of this circuit are (1) high gain, (2) high input impedance, and (3) minimum loading of the driving circuit.   Because of its high gain and low effective plate-to-grid capacitance, the pentode may be used in a grounded-cathode r-f amplifier circuit.   Figure 12-10 illustrates the use of a grounded-cathode r-f amplifier in an f-m receiver.

Grounded-grid Amplifier.   Neutralization is unnecessary with a grounded-grid amplifier circuit (Fig. 12-21).   In this circuit, the grid of

VT_1 is grounded and thereby serves as a shield or screen between the plate and cathode. Regeneration can now be caused only by the interelectrode capacitance between the plate and the cathode, since the input is now to the cathode. Grounding the grid effectively shields the input circuit to the cathode and prevents it from receiving any feedback. Grounding the grid and introducing the signal between the cathode and ground produce about the same amount of amplification as applying the signal directly to the grid with the cathode grounded. The tube behaves as if the amplification factor was $\mu + 1$. However, the input resistance R_i is much lower and is approximately equal to

$$R_i \simeq \frac{1}{g_m} \tag{12-2}$$

This equation shows that a tube having a transconductance of 5,000 μmhos would have an input impedance of only 200 ohms when used in a grounded-grid circuit.

The characteristics of grounded-grid amplifier circuits are as follows: (1) They have high gain when the source resistance is low. (2) Alternating signal plate current flows through the source of the input signal. (3) The output and input circuits are isolated in so far as stray capacitance is concerned. (4) Triodes can be used in low-noise, simple, inexpensive circuits.

Example 12-1. One section of a twin triode being used as a grounded-grid r-f amplifier has a transconductance of 5,300 μmhos and a plate resistance of 7,100 ohms. If the required shunt resistance of the output circuit is 2,660 ohms, what is the voltage amplification of this circuit?

Given: $g_m = 5,300$ μmhos   Find: VA
$r_p = 7,100$ ohms
$R = 2,660$ ohms

Solution:

$$VA = g_m \frac{r_p R}{r_p + R} = 5,300 \times 10^{-6} \frac{7,100 \times 2,660}{7,100 + 2,660} = 10.27$$

Cascode Amplifier. The circuit of Fig. 12-5, called a *cascode amplifier*, combines the gain of a pentode and the low noise factor of the triode by using the two sections of a twin triode connected in cascode. The first section VT_1 is a grounded-cathode amplifier with the input signal applied between the grid and the ground. The second section VT_2 is a grounded-grid amplifier with the output of the first stage applied between cathode and ground. VT_1 is essentially an impedance-matching device which matches the relatively high impedance of the input tuned circuit to the low impedance of the cathode circuit of VT_2. Coupling between the two stages is provided by a single-tuned r-f coil L_2, which resonates with the stray capacitance at the desired frequency. Although the impedance of the coupling circuit is infinite at resonance, it is loaded down by the grounded-grid stage to an extent that makes this impedance unimportant. The r-f choke L_3 increases the impedance of the output circuit to the value required for proper bandwidth. Although the tube is completely stable, a neutralizing coil L_1 is connected between the plate and grid of VT_1 to resonate with the grid-to-plate capacitance and decrease the effect of this capacitance to zero. Because of the relatively low input impedance of VT_2, VT_1 effectively operates as a power amplifier with a voltage amplification of 1. Since VT_2 is operated as a grounded-grid amplifier, it

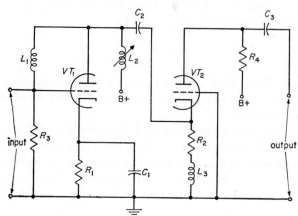

Fig. 12-5. Basic cascode amplifier circuit.

does not require neutralization and has the normal gain of such an amplifier. The over-all voltage amplification of the circuit equals that of a well-designed pentode, with the noise factor of a triode; its amplification may be expressed as

$$VA = \frac{\mu(1 + \mu)}{1 + \frac{r_p}{R}(2 + \mu)} \tag{12-3}$$

Example 12-2. A twin triode is connected as a cascode r-f amplifier and is operated so that its transconductance is 6,800 μmhos and its plate resistance is 5,600 ohms. If the required shunt resistance of the output circuit is 2,660 ohms, what is the voltage amplification of this circuit?

Given: $g_m = 6,800$ μmhos   Find: VA
         $r_p = 5,600$ ohms
         $R = 2,660$ ohms

Solution:

$$\mu = g_m r_p = 6,800 \times 10^{-6} \times 5,600 = 38.1$$

$$VA = \frac{\mu(1 + \mu)}{1 + \frac{r_p}{R}(2 + \mu)} = \frac{38.1(1 + 38.1)}{1 + \frac{5,600}{2,660}(2 + 38.1)} = 17.45$$

The circuit of Fig. 12-11 illustrates the use of a cascode r-f amplifier in an f-m receiver. The circuit and its operation are basically the same as that of Fig. 12-5. The left-hand portion of the twin triode VT_{1A} operates as a grounded-cathode amplifier with the 10-ohm resistor R_2 added to obtain the desired bandwidth. The plate of this triode is connected to B+ through the cathode-to-plate path of the right-hand portion of the twin triode VT_{1B}. This section of the tube is being operated as a grounded-grid amplifier because the grid is effectively at ground level to the r-f signals due to the 0.01-μf capacitor C_4. Because VT_{1B} is connected in series with the plate of VT_{1A}, the cathode of VT_{1B} is being operated at a high direct voltage, actually 92 volts in this circuit. Resistors R_4 and R_5 form a voltage divider to make the voltage at the grid of VT_{1B} +92 volts and hence this tube operates at zero bias. Inductor L_3 serves the same function as L_2 in Fig. 12-5. This circuit therefore operates as a twin-triode broadband r-f cascode amplifier.

Noise Figure. The equivalent noise resistance for r-f amplifier tubes may be found by using these equations:

For triodes
$$R_n \cong \frac{2.5}{g_m} \tag{12-4}$$

For pentodes
$$R_n \cong \frac{I_b}{I_b + I_{c2}} \left(\frac{2.5}{g_m} + \frac{20 I_{c2}}{g_m{}^2} \right) \tag{12-5}$$

where R_n = equivalent noise resistance, ohms
 I_b = plate current
 I_{c2} = screen-grid current

Example 12-3.   What is the equivalent noise resistance of an r-f amplifier triode whose transconductance is 6,250 μmhos?

Given: $g_m = 6{,}250$ μmhos   Find: R_n

Solution:

$$R_n \cong \frac{2.5}{g_m} \cong \frac{2.5}{6{,}250 \times 10^{-6}} \cong 400 \text{ ohms}$$

If the input signal-to-noise ratio is unity, the noise figure of an r-f amplifier circuit may be expressed as

$$F_n = 20 \log \sqrt{\frac{2R_i + 4R_n}{R_i}} \tag{12-6}$$

where F_n = noise figure, decibels
 R_i = input resistance, ohms
 R_n = equivalent noise resistance, ohms

Example 12-4.   The first section of a twin triode connected as a cascode r-f amplifier has an input resistance of 1,200 ohms and an equivalent noise resistance of 400 ohms. What is the noise figure of this section of the r-f amplifier?

Given: $R_i = 1{,}200$ ohms   Find: F_n
     $R_n = 400$ ohms

Solution:

$$F_n = 20 \log \sqrt{\frac{2R_i + 4R_n}{R_i}} = 20 \log \sqrt{\frac{2 \times 1{,}200 + 4 \times 400}{1{,}200}} = 5.2 \text{ db}$$

Example 12-5. A certain pentode being used in an r-f amplifier circuit has an equivalent noise resistance of 1,432 ohms. What is the noise figure of this circuit when the input is a 300-ohm line?

$$\text{Given: } R_n = 1{,}432 \text{ ohms} \qquad \text{Find: } F_n$$
$$R_i = 300 \text{ ohms}$$

Solution:

$$F_n = 20 \log \sqrt{\frac{2R_i + 4R_n}{R_i}} = 20 \log \sqrt{\frac{2 \times 300 + 4 \times 1{,}432}{300}} = 13.25 \text{ db}$$

The solutions of Examples 12-4 and 12-5 show the advantage of using a triode over the pentode in terms of the noise figure.

12-8. The Local Oscillator. *Requirements.* The local oscillator must supply an r-f voltage of correct frequency to the mixer so that when it is combined with the signal from the desired station, the output of the mixer will contain a signal of the required i-f value, namely 10.7 mc. The output voltage of the oscillator must be high enough so that the amplitude of the i-f output of the mixer is dependent mainly on the strength of the r-f signal being received. As station selection requires varying both the r-f and local oscillator signals being fed to the mixer, the oscillator frequency must be easily adjusted to the exact frequency needed to produce the desired i-f value. When the receiver is tuned to the signal of a particular station, the oscillator frequency should remain steady and not be subject to variations resulting from such causes as (1) changes in interelectrode or circuit capacitances caused by changes in temperature as the tube and circuit elements warm up, and (2) changes in the power-supply voltages such as the B+ and heater voltages.

Since the sound signal is frequency-modulated, any circuit disturbances which tend to cause the local oscillator to be frequency-modulated at an audio rate will actuate the f-m detector and will appear as noise in the loudspeaker. Such noises may be caused by (1) hum voltages in the B supply, and (2) microphonic tube elements and/or circuit components.

Frequency Drift. Undesired changes in the oscillator frequency, generally due to changes in the temperature of the tube and its associated circuit components, are called *frequency drift*. A change of as little as 0.1 to 0.5 per cent in the oscillator frequency may shift the i-f value far enough to cause the detector to operate on an unfavorable portion of its characteristic curve and cause distortion. Although a detuned oscillator may not cause complete loss of the signal, it will cause a loss in the volume and will also affect the quality of the sound reproduction. An automatic means of compensating for a slow frequency drift may be obtained through the use of an afc (automatic-frequency-control) circuit.

Such a circuit will maintain an oscillator constantly on frequency and will even pull an oscillator into tune if it is adjusted to within approximately 300 kc of the correct value.

12-9. Automatic Frequency Control. *Purpose.* Many f-m receivers use an afc circuit to hold the oscillator frequency constant and thereby eliminate the undesirable effect of frequency drift. A number of circuits have been developed for this purpose, some employing a vacuum tube and others a crystal diode.

Principle of Operation. The afc circuit is essentially a variable reactance associated with the oscillator tuning circuit. As the oscillator frequency drifts, the reactance of the afc circuit changes until the oscillator is returned to its correct resonant frequency. In a tuning circuit either or both the inductance or capacitance may be varied; consequently, the afc circuit may behave as either an inductance or a capacitance. However, in order to maintain the L/C ratio constant, which will also keep the oscillator voltage constant, the afc circuit should present the same kind of reactance as the variable element it is to correct.

One type of afc circuit frequently used employs a triode tube which is generally one-half of a twin triode, the other half being used as the oscillator tube. Whether a circuit element is inductive or capacitive may be determined by the phase angle between the current in the element and the voltage across it. If the current lags the voltage, the element is behaving as an inductance, and the smaller the current, the larger is the apparent inductance. If the current leads the voltage, the element is behaving as a capacitance, and the greater the current, the larger is the apparent capacitance. Since plate current is controlled by the grid bias, the amount of reactance that a tube introduces into the oscillator circuit may be varied by changing the voltage on the grid. The grid-bias voltage is generally taken from the f-m detector. This detector produces a direct voltage which will vary as the i-f signal shifts away from its center frequency because of the frequency drifts in the oscillator circuit.

Capacitive-reactance Tube Circuit. A reactance tube circuit of the variable-capacitance type is shown in Fig. 12-6a. The reactance of C_1, which is connected from plate to grid, is much higher than the resistance of R_1, which is connected from grid to cathode. The blocking capacitor C_2 prevents the direct current from entering the oscillator tank circuit, while L_1 keeps the r-f voltage from the B power supply. The bias voltage applied to the grid from the C supply is controlled by the potentiometer R_2. The tuned circuit of the oscillator is represented by the r-f oscillator tank. Since the reactance of C_1 is much higher than the resistance of R_1, the path C_1R_1 is predominantly capacitive. The voltage E_t from the oscillator tank circuit causes a current to flow in the C_1R_1 path, and this current will lead the voltage E_t by 90 degrees. The current in R_1 will produce a voltage e_g across this resistor; the voltage e_g will be in phase with the current in the C_1R_1 path.   Since the current leads the voltage E_t by 90 degrees, the voltage e_g is also 90 degrees ahead of the voltage E_t. This leading voltage e_g is applied across the grid and cathode of VT_1, thus causing the plate current to lead the r-f voltage E_t by 90 degrees. Since the voltage E_t is also connected across the plate and cathode, the

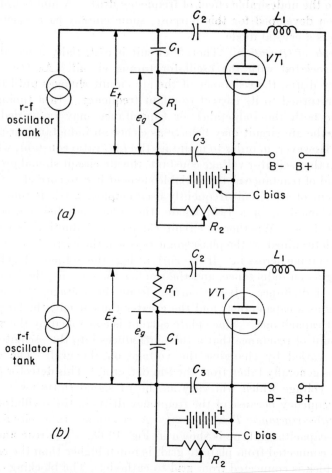

Fig.   12-6.   Automatic-frequency-control   circuits.   (a) Variable-capacitance-reactance tube.   (b) Variable-inductance-reactance tube.

leading plate current caused by e_g will also lead the voltage E_t.   The tube and its circuit will therefore appear as a capacitance to the oscillator. The voltage from the oscillator will have very little effect on the plate current because the tube is more sensitive to the voltage applied to its grid; in fact, it is μ times as sensitive to a grid-voltage change as it is to a similar plate-voltage change.   For a tube having an amplification factor of 20, a change of 1 volt on the grid will produce the same effect as a change of 20 volts on the plate.   As the plate current is dependent upon the amount of grid-bias voltage, the apparent capacitive reactance can be controlled by varying the amount of lagging current by adjusting the potentiometer R_2.   In a practical circuit, the control-grid bias is obtained from the detector; hence, the potentiometer R_2 and the C battery are not actually used.

Inductive-reactance Tube Circuit.   A reactance tube circuit of the variable-inductance type is shown in Fig. 12-6*b*.   In this circuit, the resistance of R_1 is made much higher than the reactance of C_1. R_1 is connected from the plate to the grid and C_1 from the grid to the cathode, which is just the reverse of the connections for R_1 and C_1 in the capacitive-reactance tube circuit of Fig. 12-6*a*.   As the series combination of R_1 and C_1 is predominantly resistive, the current in the circuit is substantially in phase with the voltage E_t.   The voltage e_g across C_1 lags the current

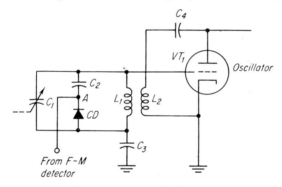

FIG. 12-7. Automatic-frequency-control circuit using a crystal diode.

through it by 90 degrees, and the resultant plate current will also lag this current by 90 degrees.   As the plate current will also lag the voltage E_t across the plate and cathode by 90 degrees, the tube acts as an inductive reactance.   The magnitude of this reactance can be controlled by the bias voltage developed at the detector.

Practical AFC Circuits Using Vacuum Tubes.   In the circuit of Fig. 12-10, VT_{3B} acts as a variable reactance in an afc circuit.   Any frequency drift at the oscillator tube VT_{3A} will cause a change in the output voltage at the detector.   This voltage change is coupled to the grid of VT_{3B} which then causes a change in the reactance of this tube.   As C_{13} connects the reactance of this tube to the tank circuit of the oscillator, the change in reactance will automatically bring the resonant frequency of the oscillator back to the value required for maximum output at the f-m detector.

In the circuit of Fig. 12-11, VT_3 operates as a triode oscillator and an afc triode in the same manner as described for the circuit of Fig. 12-10.

Variable-reactance Circuit Using a Diode.   The circuit of Fig. 12-7 illustrates the basic principles of an afc system using a crystal diode. Any change in the center frequency of the i-f signal at the f-m detector produces a change in the direct voltage obtained from the detector circuit which is then applied to point A in the afc circuit. When this direct voltage varies, a positive- or negative-going change in voltage occurs at point A. These voltage changes will increase or decrease the normal conduction of the crystal diode CD and decrease or increase the effective resistance of the diode. Capacitor C_2 is therefore connected in series with a variable resistance, and as the value of the resistance changes, the impedance of the C_2CD path will also change. When the resistance decreases, the capacitance effect of the circuit will increase, and when the resistance increases, the capacitance effect will decrease. The variation in the effective capacitance changes the resonant frequency of the oscillator tank circuit and restores the i-f signal to its proper center-frequency value.

12-10. The Mixer Stage. *Mixers and Converters.* The section of the front end in which the r-f signal is combined with the local oscillator signal is called either the *mixer* or *converter*. When the oscillator and mixing functions are performed in a single tube, it is called a *converter*. When the oscillator function is performed in a separate tube, the tube in which the mixing function is performed is called a *mixer*. In either case, it is in this stage that the r-f signal voltage is heterodyned with the oscillator voltage to produce the desired i-f signal voltage.

Basic Converter Circuit. A basic circuit for a frequency converter is shown in Fig. 12-8. This converter consists of two distinct circuits (1) the oscillator circuit, and (2) the mixer circuit. The value of R_1 is made high enough to operate VT_1 near cutoff so that this tube has nonlinear characteristics and is therefore suitable for use as a mixer. The circuit of Fig. 12-8 contains the fundamental components of a frequency converter, namely: (1) a supply of local r-f voltage of controllable frequency to develop the heterodyning voltage, (2) an input source of r-f signal voltage, (3) a nonlinear circuit element or mixer, (4) a means of coupling the oscillator voltage to the mixer, and (5) a means for extracting the desired new i-f signal from the plate circuit. The relative positions of L_2, L_3, and L_4 are such that magnetic coupling exists between these circuits. The coupling between L_3 and L_4 produces the oscillations, and the coupling between L_2 and L_3 provides a means for introducing the oscillator voltage into the grid of VT_1. The oscillator and signal voltages are in series, as both voltages are induced into L_2 by the currents in L_3 and L_1 respectively. Because the oscillator signal is made the stronger of the two and is also of sufficient strength to swing the tube from cutoff to near grid conduction, the transconductance of VT_1 will periodically vary over a wide range. Thus at any instant, the plate current due to signal voltage is dependent upon the instantaneous value of the transconductance.

The resulting plate current will therefore be complex and will contain the new sum and difference frequency signals. The tuned circuit in the mixer-plate section separates the desired i-f signal from all others.

Characteristics of Suitable Mixing Devices. Crystal detectors, diodes, triodes, and pentodes can perform satisfactorily as mixers. It is important that the mixer produce as little noise as possible, since any noise voltage produced in the front end will be amplified by the following i-f stage and other amplifying circuits. Crystals and diode mixers are characteristically low noise producers but they do not provide any

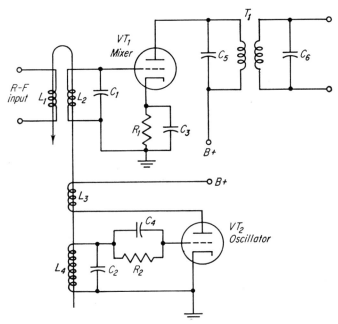

Fig. 12-8. Basic frequency-converter circuit using separate tubes for oscillator and mixing functions.

amplification of the signal; therefore triodes and pentodes are used whenever amplification is desired and the operating frequency and noise requirements permit their use. Generally, the tube having the fewest electrodes will produce the least noise. Thus, a triode is rated better than a pentode for noise qualities; furthermore, a triode can produce almost as much gain as a pentode at the high frequencies used in f-m broadcasting. The relatively low impedance of the triode makes it possible more nearly to load the tube suitably. Consequently, a greater portion of the amplified signal is produced across the useful load instead of being expended in the internal impedance of the tube.

Oscillator Voltage Injection. The heterodyning voltage generated by the oscillator must be injected into the mixer circuit so that it can vary the transconductance at the oscillator frequency. The oscillator voltage can be applied to any of the electrodes of the mixer tube but for a given value of voltage change applying the voltage to the control grid will produce the greatest variation. Because of the greater sensitivity of the control grid, the oscillator signal voltage is usually fed or injected into the control grid. The oscillator signal could also be fed into either the cathode or screen-grid circuits. However the low impedance of the cathode seriously loads the oscillator, making it more difficult to obtain sufficient oscillator voltage that is also stable in amplitude and frequency over the full f-m band. The screen grid does not load the oscillator so much as does the cathode. However, it is not nearly so sensitive as the control grid, and therefore a higher value of oscillator voltage is required for injection into the screen grid in order to produce the same results.

Practical Mixer Circuits. The circuit of Fig. 12-10 illustrates the use of a triode mixer tube VT_2 with the oscillator voltage injected into the grid circuit; the oscillator voltage is capacitively coupled to the mixer by C_9. In the circuit of Fig. 12-11, the oscillator voltage from the triode oscillator VT_{3A} is capacitively coupled by C_{10} to grid 1 of the pentagrid mixer tube VT_2.

12-11. Single-tube Frequency Converters. *Pentagrid Converter.* A pentagrid converter tube of the type 6BE6 may be used to produce the desired i-f signal. The circuit and principles of operation are similar to those of the pentagrid converter applications in a-m receivers (Art. 11-8). Most of the disadvantages of using the pentagrid converter in f-m receivers result from the high frequencies at which the f-m system operates. Among the disadvantages of the pentagrid converter at high operating frequencies are (1) poor signal-to-noise ratio, (2) reduced conversion gain, (3) interaction between signals on the oscillator and r-f signal grids, (4) loading effect due to decreased effective resistance, and (5) difficulty in producing sufficiently strong and uniform oscillator signal voltage.

Autodyne Converter. A method frequently used to perform both the oscillator and mixer functions in a single tube employs the autodyne converter. With this system all of the tube elements used in producing the oscillator signal voltage are also used in the operation of the mixing function. The circuit of Fig. 12-9 illustrates the basic principles of the autodyne converter. The r-f signal received at the antenna is coupled to the cathode of VT_1 and amplified by VT_1. After additional tuning at L_1C_2, the r-f signal is capacitively coupled to the grid of VT_2 by capacitor C_3.

One function of the triode VT_2 is to operate as a tuned-grid oscillator. Variations in the plate current are passed through L_3 while the steady direct current is blocked by C_6. The current variations in L_3 induce voltages in L_2 which are fed back to the grid of VT_2 in the proper phase to sustain oscillations. The tuning circuit L_2C_5 controls the frequency of the oscillator signal.

Another function of VT_2 is to operate as a mixer. The voltages of both the r-f and oscillator signals are applied to the grid of this tube. As with other methods of frequency conversion, the oscillator voltage is made much higher than the r-f signal voltage. The relatively large oscillator signal voltage will cause plate-current variations between zero and maximum plate current which will cause corresponding variations in the transconductance of the tube. Because of the relatively high oscillator signal voltage, the plate-current and transconductance variations become a series of pulses whose peaks will be modulated by the relatively small r-f signal. The plate current therefore contains pulses occurring at the oscillator frequency rate, and also sidebands above and below the

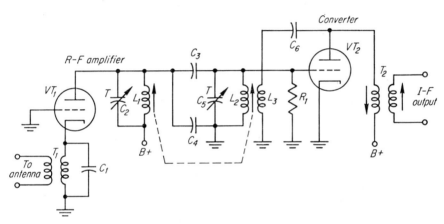

Fig. 12-9. Circuit illustrating the basic principles of the autodyne converter.

oscillator signal frequency; the sidebands still contain the r-f signal modulations.

The transformer T_2 is tuned to select the lower sideband frequency from among the various signals flowing in the plate circuit of VT_2. This frequency is equal to the difference between the oscillator and r-f signal frequencies and in f-m receivers is made to be 10.7 mc.

Applying an avc (automatic-volume-control) voltage to the simple autodyne converter of Fig. 12-9 will cause unsatisfactory operation of the circuit in the form of a series of pulses called *motorboating*. This difficulty may be overcome by using a tube containing a diode plate in addition to the triode and incorporating certain circuit modifications.

Advantages of the autodyne converter are (1) high gain, (2) less noise, (3) simpler circuit, and (4) good image rejection. Applications of the autodyne converter may be observed in the circuits of Figs. 12-21 and 12-23.

12-12. Front-end Circuits of Practical Receivers.   The block diagram
of Fig. 12-2 illustrates the sequence of the functions performed in the
front-end portion of typical f-m receivers.   The manner in which each
of these functions is performed, together with their circuit characteristics, has been described in the preceding articles.   Combinations of
these functions and circuits into a composite front end will now be
presented.

The front end represented by Fig. 12-10 employs three tubes: (1) a
pentode VT_1, used as an r-f broadband grounded-cathode amplifier, (2)
a triode mixer VT_2, and (3) a twin triode VT_3, of which one section is

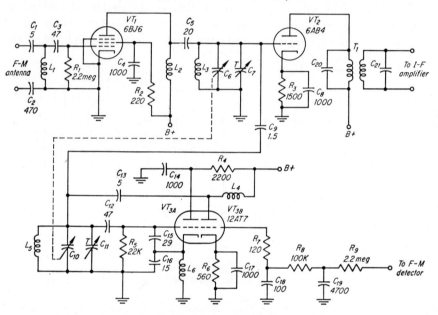

Fig. 12-10. Front-end circuits of an f-m receiver.

used as a tuned-grid oscillator and the other section as a variable-reactance tube for the afc circuit.   The antenna coupler $C_1C_2L_1$ is of the
unbalanced type, and its r-f signals are coupled to VT_1 by C_3 and R_1.
The r-f amplifier has a wide-band response and increases the strength of
all signals in the f-m band; its output is coupled by C_5 to the tuning circuit $L_3C_6C_7$.   The r-f tuning circuit selects the signal of one station and
rejects all others; the signal of the selected station is fed to the grid of the
mixer tube VT_2. VT_{3A} operates as a tuned-grid triode oscillator that is
tuned by $L_5C_{10}C_{11}$.   The oscillator signal is coupled to the grid of the
mixer tube by C_9.   Capacitors C_6 and C_{10} are mechanically controlled by
a single tuning dial. VT_{3B} is operated as a variable-reactance tube
which is coupled to the oscillator tuning circuit by C_{13} and keeps the
oscillator frequency constant.

The front end represented by Fig. 12-11 employs three tubes: (1) a twin triode VT_1, used as a cascode r-f amplifier, (2) a pentagrid mixer VT_2, and (3) a twin triode VT_3, of which one section is used as a tuned-grid oscillator and the other section as a variable-reactance tube for the afc circuit. The antenna coupling unit consisting of L_1, L_2, C_2, and R_2 has a wide-band response and couples the r-f signals to the cascode amplifier tube VT_1 (see Art. 12-7). The r-f output of VT_{1B} is coupled by C_5 to the r-f tuning section $L_4C_7C_8$, whose output is capacitively coupled by C_6 to grid 3 of the pentagrid converter VT_2. The tuned-grid oscillator circuit uses the triode VT_{3A} with its feedback produced by C_{15}, L_6, L_5, R_{14}, and C_{19}; the oscillator is tuned by L_5, C_{16}, C_{17}, C_{18}. Automatic frequency control is provided by the variable-reactance triode VT_{3B}, with the controlling reactance coupled to the oscillator tuning circuit by C_{20}.

In the f-m receiver circuit diagram shown in Fig. 12-21, the front-end functions are performed by a single twin-triode tube VT_1. The r-f signals are picked up by a line-cord type of antenna and fed to the unbalanced antenna coupling circuit consisting of T_1, C_1, C_2, C_3, R_1, and R_2. The antenna-coupling circuit is broadly tuned to cover the entire f-m band. The triode VT_{1A} is operated as a grounded-grid r-f amplifier with the r-f input signal coupled to the cathode. Additional tuning takes place in the plate circuit of the tube by means of the variable tuning circuit L_1C_4, in which the main tuning is performed by the movable core in L_1; C_4 is a trimmer capacitor for adjusting the receiver alignment. The selected r-f signal is then coupled by C_7 to the grid of the autodyne converter VT_{1B}. The triode VT_{1B}, together with T_2, C_8, C_{12}, and R_4, forms a tuned-grid oscillator whose frequency is varied by the movable core in T_2; C_8 is a trimmer capacitor for adjusting the receiver alignment. The movable cores of L_1 and T_2 are operated by a common tuning dial to achieve station selection. As both the r-f and oscillator signals are applied between the grid and cathode of the triode VT_{1B}, this tube is operating as an autodyne converter. The i-f transformer T_3 is tuned to the 10.7-mc i-f value and selects this signal from among those present in the plate circuit of VT_{1B}. The arrows shown at T_3 indicate that the frequency response of this transformer may be adjusted by two movable cores, one accessible at the top and the other at the bottom of the transformer. R_3, R_5, C_5, and C_{11} form the decoupling units for the B power supply.

12-13. The Intermediate-frequency Amplifier Section. *Requirements.* Except for the higher operating frequency and wider bandpass, an i-f amplifier stage used in an f-m receiver is similar in many respects to its counterpart in an a-m broadcast receiver. Although early f-m receivers used various i-f values, such as 2.1, 4.3, and 8.3 mc, the standard i-f value is now 10.7 mc. The ideal over-all response for the i-f section of an

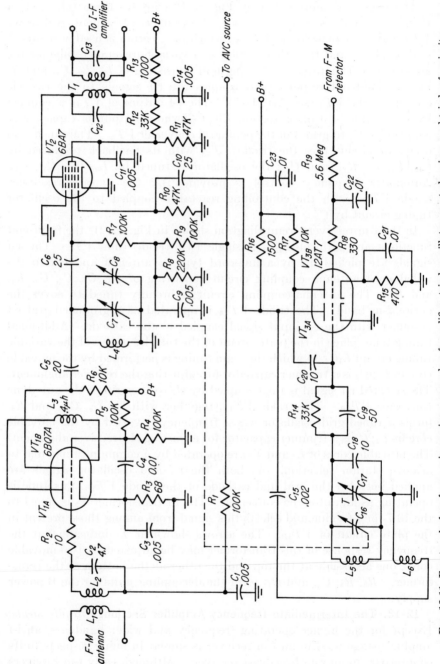

FIG. 12-11. Front end of an f-m receiver using a cascode r-f amplifier, triode oscillator, pentode mixer, and triode afc circuits.

f-m receiver should be flat over a 200-kc band and drop sharply on either side of the response curve; a good practical i-f response curve is shown in Fig. 12-12. The i-f bandwidth should not be less than 150 kc at the 50 per cent power points. The response curve should be flat, or at least nearly flat, as any excessive amount of depression in the center (see Fig. 12-13) can result in slope detection of the signal which will introduce new amplitude modulations to the f-m signal.

Because of the increase in circuit losses and input loading at the higher frequencies, the gain per stage will be lower for an f-m amplifier than for an a-m amplifier. This decrease in gain per stage,

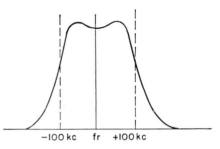

FIG. 12-12. A symmetrical i-f response curve.

together with the loss introduced by the limiter stage (if a limiter is used), makes it necessary to use two or three stages of i-f amplification in an f-m receiver.

Methods of Obtaining the Desired Bandpass Characteristics. The shape of the response curve for an i-f amplifier stage can be varied by

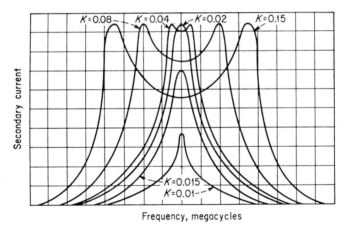

FIG. 12-13. Effect of coupling on the frequency-response characteristics of tuned circuits. Critical coupling for the circuit represented by these curves occurs at $K = 0.02$, loose coupling at values of K less than 0.02, and tight coupling at values of K greater than 0.02.

changing the amount of coupling between the primary and secondary coils of the i-f transformer as is shown in Fig. 12-13. Three methods may be used to obtain the desired over-all response of an i-f system. For convenience, a three-stage transformer-coupled i-f system is assumed. In one method the three stages are each adjusted to the same resonant frequency with slightly less than critical coupling. The over-all response will be as shown in Fig. 12-14a. In a second method, all three stages are adjusted for the same resonant frequency with two of the stages loosely coupled and one tightly coupled. Generally the first and third stages are loosely coupled, and the second stage is tightly coupled. The over-all response is shown in Fig. 12-14b. In the third method, each

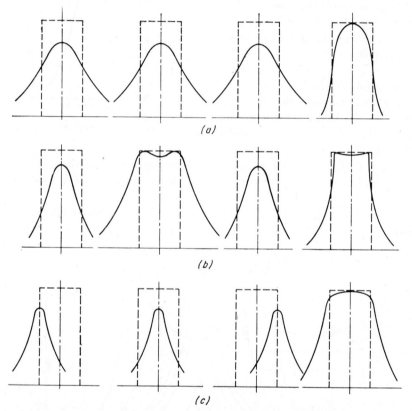

FIG. 12-14. Over-all response curves of a three-stage transformer-coupled i-f system. (a) Three transformers tuned to the same frequency and all near critical coupling. (b) Three transformers all tuned to the same frequency with number 1 and 3 loosely coupled and number 2 tightly coupled. (c) Three transformers all loosely coupled but each at a slightly different frequency.

stage is loosely coupled, but each is tuned to a slightly different frequency. This method is called *stagger tuning*. The over-all response is shown in Fig. 12-14c.

I-F Transformers. Because of the high operating frequency, the values of capacitance and inductance required in the tuned circuits will be low; consequently, the transformer windings will have very few turns and the capacitances used will be of low values. In general, i-f transformers may be tuned by varying either the capacitance or the inductance;

however at the higher frequencies it is not very practical to tune the circuits by varying the capacitance. Tuning is usually accomplished by varying the position of a core—made of molded powdered-iron dust and mounted on an adjusting screw—in the coils' magnetic influence.

12-14. Limiters. *Principle of the Limiter.* An important factor in the noise-free operation of the f-m system is that the transmitted f-m waves are of constant amplitude. However, even though the transmitter provides an output wave train of constant amplitude, the waveform at the i-f section of the receiver may contain variations in amplitude due to noise or other causes picked up in transmission. (To eliminate amplitude variations, a limiter circuit is placed between the last i-f amplifier stage and the detector.)

The limiter is essentially an i-f amplifier operated with such (low) values of voltage at its electrodes that the tube reaches saturation for

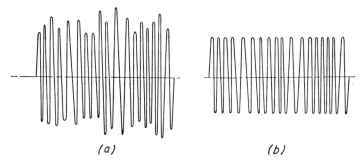

(a)   *(b)*

Fig. 12-15. Effect of the limiter circuit on the waveform of a varying amplitude f-m signal. (*a*) Signal input to the limiter. (*b*) Signal output of the limiter.

very low signal voltages. Under this condition, the portion of each input cycle above a certain value is cut off, and hence the amplitude of each cycle is limited to a fixed amount. The i-f output should be of sufficient strength so that even the weakest signal will cause the limiter to saturate. The waveshapes of the input and output signals of a limiter are shown in Fig. 12-15. The distorted flat-topped wave at the output of the limiter is no problem for three reasons: (1) The relative frequencies present in the frequency deviation are not changed. (2) The frequency of the harmonics introduced by the flat-topped waves is outside the range of the resonant plate circuit of the limiter. (3) Bypassing the output of the limiter through a tuned transformer reduces to a negligible value the harmonics resulting from the flat-topped waves.

Limiting action may be obtained by use of grid limiting and/or plate limiting. Grid limiting is achieved by using sufficient grid-leak bias to limit the plate current on both the positive and negative peaks of the input signal. Plate limiting is obtained by using values of plate and screen-grid voltages so low that reasonably strong signals will saturate the tube.

Grid Limiters. The principle of operation of a grid-limiting circuit is similar to the operation of the grid-leak-detector circuit. A simple grid-limiter circuit is shown in Fig. 12-16a. When the signal input is zero,

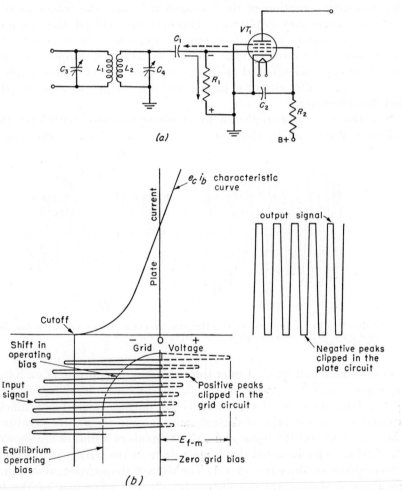

Fig. 12-16. Grid limiting. (a) Grid-limiter circuit. (b) Curves showing the clipping action in the grid and plate circuits.

the voltages on the grid and cathode are zero. During the initial positive half-cycle of an input signal, the grid is driven positive, thereby causing electrons to flow in the grid circuit and charge C_1 to E_{f-m} volts. During the following negative half-cycle, C_1 will start discharging through the grid resistor R_1 and develop a voltage across this resistor with the polarity indicated in Fig. 12-16a. Because of the time constant R_1C_1, some charge will still remain on C_1 at the time that the second input cycle is about to start. Electrons will again flow in the grid circuit and recharge C_1 during that portion of the cycle when the positive input voltage exceeds the charge present on the capacitor. Since during each cycle only a portion of the charge on C_1 discharges through R_1, the charge remaining after each cycle will be cumulative until a point of equilibrium is reached and the voltage across R_1 remains constant (see Fig. 12-16b).

The point of equilibrium establishes the operating bias about which the input f-m voltage fluctuates. The operating bias should be slightly less than the peak value of its maximum positive signal swing. If the grid bias is too large, limiting will not occur for small amplitude variations, thus causing interfering a-m signals to be heard in the audio output. The exact value of the operating bias is dependent upon (1) the voltages at the electrodes of the tube, (2) the voltage of the input signal, (3) the values of C_1 and R_1.

Plate Limiters. With plate limiting, the limiter tube is operated at very low plate and screen-grid voltages in order that effective saturation may be obtained with small values of input signal. In general, when the operating potentials of a voltage amplifier pentode are so low that they cause the tube to operate below the knee of its plate-characteristic curve, the output plate current becomes independent of the input grid voltage over a wide range.

To obtain a limited i-f output, the limiter tube must operate at cutoff on a portion of each negative half-cycle. Thus the sensitivity of a plate limiter is determined by the amount of voltage required to drive the tube to cutoff. For optimum sensitivity and limiting performance, a sharp-cutoff pentode is generally used. With abnormally low plate and screen-grid voltages, a relatively small input signal will drive the tube to cutoff on the negative peaks and to saturation on the positive peaks.

Double Limiters. Although satisfactory limiting may be obtained by use of either grid or plate limiters, a combination of both may be used. By adding grid-leak bias to a tube operating as a plate limiter, it is possible to increase the operating voltages on the plate and screen grid and thus obtain a higher gain while still providing good limiting action. When a tube is operated as both a grid and plate limiter, it is called a *double limiter*.

Limiter Stages. A substantial reduction in amplitude variations of an f-m signal can be obtained with a single limiter stage. Although the reduction in a-m interference obtained with one limiter stage is usually sufficient, the output is not perfectly constant. Almost perfect limiting action can be obtained by using two limiters. The first limiter stage usually has a short time constant to counteract amplitude variations due to sharp impulses. Better regulation and higher gain are obtained in the second limiter stage by using a longer time constant. Resistance-capacitance, impedance, or transformer coupling may be used to connect two or more limiter stages in cascade.

12-15. Frequency-modulation Detector Circuits. *Function.* The f-m detector converts the frequency variations of an f-m wave to electrical signal variations that are proportional to the original audio signal. The ideal response curve of an f-m detector (Fig. 12-17) shows that each frequency variation produces a definite value of voltage and also that a linear relationship should exist between the frequency and voltage. This linear relationship is essential in order to produce distortionless conversion.

Types of F-M Detector Circuits. Many types of f-m detectors have been developed and used, some for only a short time. Among the various types of f-m detectors are (1) the slope detector, (2) the Travis f-m discriminator, (3) the Foster-Seeley discriminator, (4) the ratio

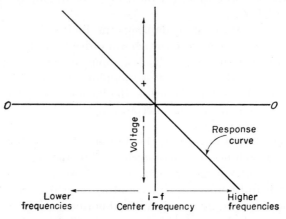

Fig. 12-17. Ideal response curve for an f-m detector.

detector, (5) the locked-in oscillator detector, (6) the Fremodyne super-regenerative detector, (7) the gated-beam detector, (8) the quadrature-grid f-m detector. Most modern f-m receivers use either (1) the discriminator detector (Foster-Seeley) or (2) the ratio detector; hence only these two types will be described.

12-16. The Foster-Seeley Discriminator. *Requirements.* In modern f-m receivers, the Foster-Seeley discriminator, also called a *phase discriminator*, is probably the most frequently used type of f-m detector circuit. Because it is also sensitive to amplitude modulations, it must be preceded by a limiter stage. This type of circuit requires a transformer with a center-tapped secondary, generally called the *discriminator transformer*, two diodes, and a number of resistors and capacitors. Figure 12-18 shows the circuit of a basic Foster-Seeley discriminator.

Circuit Analysis. The output voltage E_p of the limiter stage is fed to the primary L_1 of the discriminator transformer and is also capacitively coupled by C_3 to the center tap of the secondary winding L_2L_3.   The primary tuned circuit L_1C_1 and the secondary tuned circuit $L_2L_3C_2$ are both tuned to the center i-f value of 10.7 mc.   Center-tapping the secondary winding produces two equal voltages, $E_{s.1}$ and $E_{s.2}$, 180 degrees out-of-phase across terminals 3 and 4 and 4 and 5.   This circuit also shifts the relative phase between the voltages E_p and $E_{s.1}$, $E_{s.2}$ as the frequency modulation produces deviations of the frequency away from the center i-f value.   Since E_p is applied between the center tap of the secondary and ground, the voltages $E_{s.1}$ and $E_{s.2}$ must act in series with the primary voltage E_p in order to be effective across their respective diodes.   The reactances of C_3, C_4, C_5, C_6, and C_7 are very low at the operating frequency of the discriminator transformer and high at audio

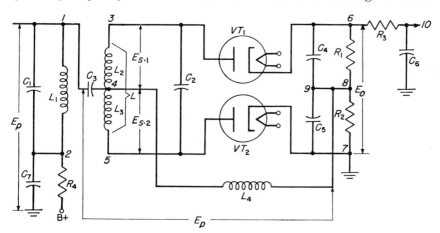

Fig. 12-18. Basic Foster-Seeley discriminator circuit.

frequencies.   Therefore terminal 1 of the primary winding L_1 is effectively connected to the cathodes of VT_1 and VT_2 and also to ground for i-f signals.   The voltage across VT_1 is the vector sum of $E_{s.1}$ and E_p and the voltage across VT_2 is the vector sum of $E_{s.2}$ and E_p.   Inductor L_4 provides a d-c return path for the diode plate current; in some practical circuits this inductor is replaced by a resistor or is omitted.

Behavior of the Circuit for Center Frequency Signals.   When the center (or unmodulated) frequency of the signal coincides with the resonant frequencies of the transformer primary and secondary circuits $E_{s.1}$ and $E_{s.2}$ will be equal and 180 degrees apart and also 90 degrees out-of-phase with E_p (Fig. 12-19).   The resultant voltages E_1 and E_2 will also be equal in magnitude.   Because of the rectifying action of the diodes, two direct voltages of equal magnitude will be developed across R_1 and R_2.   By tracing the direction of electron flow in the circuits of VT_1 and VT_2, it will be found that the polarities of the voltage across R_1 will be negative at point 8 and positive at point 6; the polarities across R_2 will be negative at point 8 and positive at point 7.   As R_1 and R_2 are made of equal value, the voltages $E_{R.1}$ and $E_{R.2}$ will be of equal magnitude but of opposite polarities with respect to ground.   The output voltage E_o is equal to the sum of $E_{R.1}$ and $E_{R.2}$, and for the above condition will be zero.   The voltages E_1 and E_2 may vary in value, but as long as they remain equal the output voltage of the discriminator will be zero.

Behavior of the Circuit with Signals above the Center Frequency.   When the modulations produce i-f values higher than the center value, the phase angle between $E_{s.1}$ and E_p will get larger than 90 degrees and the phase angle between $E_{s.2}$ and E_p will get smaller than 90 degrees.   Figure 12-19b shows that for this condition E_1 and E_2 will no longer be equal, even though the values of $E_{s.1}$, $E_{s.2}$, and E_p have not changed.   As E_2 becomes greater than E_1, $E_{R.2}$ will be greater than $E_{R.1}$, and E_o will have

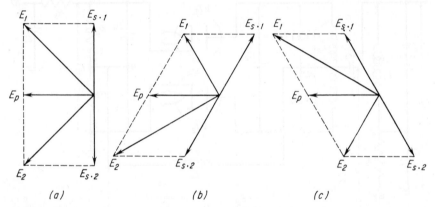

(a)   (b)   (c)

Fig. 12-19. Phase relation of voltages in a discriminator circuit.   (a) For center-frequency signal.   (b) For signals above the center frequency.   (c) For signals below the center frequency.

a significant value with the output at point 6 being negative with respect to ground.   When the deviation to a higher frequency increases, the phase shift of $E_{s.1}$ and $E_{s.2}$ will also vary, and the value of E_o will increase.

Behavior of the Circuit with Signals below the Center Frequency.   When the modulations produce i-f values lower than the center value, the phase angle between $E_{s.1}$ and E_p will get smaller than 90 degrees and the phase angle between $E_{s.2}$ and E_p will get larger than 90 degrees. Figure 12-19c shows that for this condition E_1 becomes greater than E_2. Also $E_{R.1}$ will be greater than $E_{R.2}$ and E_o will have a significant value with the output at point 6 being positive with respect to ground.   When the deviation toward a lower frequency increases, the difference between $E_{R.1}$ and $E_{R.2}$ becomes greater and results in a higher output voltage.

Behavior of the Circuit with an F-M Signal.   When the i-f signal applied to the discriminator is frequency-modulated, deviations in frequency above and below the center i-f value will take place continually.

From the preceding explanations, it is evident that the output signal E_o will be an alternating voltage whose value is dependent upon the frequency modulations. The rate of voltage changes will be the same as the a-f signals that produced the frequency modulations at the transmitting station.

Control Voltage for Automatic Frequency Control. When an afc circuit is used in the receiver, the control voltage for the variable-reactance tube can be obtained from the output of the discriminator. However, the output contains an a-f signal component which must be removed so that a d-c control voltage is available. This may be accomplished with an RC filter such as R_3C_6 in Fig. 12-18. The time constant of R_3C_6 must be large enough to prevent any fluttering of the signal due to rapid changes of the afc control bias, which may cause the reactance tube to vary the oscillator frequency at or near an audio rate. In the circuit of Fig. 12-10 this filter action is produced by R_8, R_9, C_{18}, and C_{19}.

12-17. The Ratio Detector. *Comparison of the Foster-Seeley and Ratio Detectors.* The ratio detector shown in Fig. 12-20, which is a modification of the basic Foster-Seeley circuit, is inherently insensitive to amplitude modulations. Because noiseless operation can be obtained without limiter stages, the ratio detector is frequently used in f-m receivers. The ratio detector differs from the Foster-Seeley detector mainly in that (1) diode VT_1 is reversed, (2) a large capacitor (electrolytic of 5 to 15 μf) is used across the load resistors R_1 and R_2, (3) the high side of the audio signal is taken from the junction of C_4, C_5, and L_4, (4) the low side of the audio signal is taken from the junction of R_1 and R_2, (5) the d-c return path is completed through the series connection of the diodes, (6) inductor L_4 supplies the charging current for capacitors C_4 and C_5.

Basic Principles. The ratio detector derives its name from the fact that the ratio of the voltages across the diodes varies only with f-m signals. The voltages e_1 and e_2 (Fig. 12-20) are developed at the discriminator transformer in the same manner as described for the Foster-Seeley circuit. The magnitudes of e_1 and e_2 depend on the relative phase and magnitude of E_p, $E_{s.1}$, and $E_{s.2}$; however, the ratio of e_1 to e_2 will remain constant as long as the phase angle remains unchanged. This ratio is independent of the magnitudes of E_p, $E_{s.1}$, and $E_{s.2}$ and of the resultant voltages e_1 and e_2. The ratio varies only when the phase angle varies, and the phase angle varies only with the f-m signals and in proportion to the percentage of frequency modulation. Thus, amplitude modulations will vary with the magnitude of the vectors E_p, $E_{s.1}$, $E_{s.2}$, e_1, and e_2; but as the angle will vary only with frequency modulations, the ratio of e_1 to e_2 will not be affected by amplitude modulations.

Circuit Actions for Amplitude Variations. The voltages e_1 and e_2 obtained from the discriminator transformer are applied across the diodes VT_1 and VT_2 and cause a rectified current to flow through the series circuit VT_1, R_1, R_2, and VT_2.   (VT_1 and VT_2 are effectively connected in series by C_3 as far as the r-f voltages e_1 and e_2 are concerned.)   Direct-voltage drops E_1 and E_2 are produced across R_1 and R_2 and are of such polarities that they are series aiding.   Capacitor C_6 is charged to the voltage E_3 which is approximately equal to the sum of the peak values that e_1 and e_2 would assume for the unmodulated carrier of the signal being received.   Since E_3 acts as a bias voltage on the diodes, only very small amounts of rectified current will flow to C_6 after it has reached the operating value for a given carrier signal strength.   The time constant $C_6R_1R_2$ is of the order of 0.2 sec, which is long compared with the time

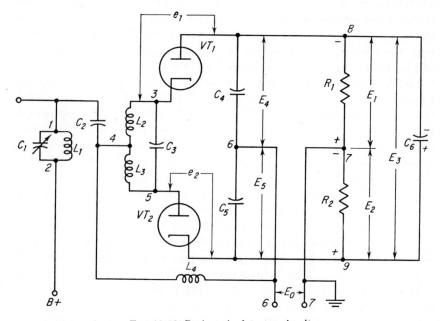

FIG. 12-20. Basic ratio-detector circuit.

required for amplitude variations at a-f rates.   A short rapid decrease in carrier signal strength produces a condition where E_3 is greater than the signal voltage, and the diode current is thus cut off.   The discharge of C_6 through R_1R_2 is so small that E_3 remains practically constant.   A short rapid increase in carrier signal strength will also fail to produce any appreciable change in the voltage E_3, since a considerable amount of current is required to change the charge on so large a capacitor as C_6. Therefore, no change is produced in E_3 by rapid fluctuations in the amplitude of the f-m signal such as may occur due to noise and other disturbances.

Behavior of the Circuit for Center Frequency Signals.   When an unmodulated f-m signal is being received and it produces an i-f signal of the same value as the resonant frequency of the discriminator transformer, e_1 and e_2 are equal, and hence the ratio of e_1 to e_2 is unity. For this condition the voltages E_4 and E_5 are equal, and the voltages E_1 and E_2 are equal. Furthermore, for this condition E_1, E_2, E_4, and E_5 are all of the same magnitude. The output voltage E_o, which is equal to $E_2 - E_5$ (or also $-E_1 + E_4$), will be zero.

Behavior of the Circuit with Signals above the Center Frequency. When the modulations produce i-f values higher than the center value, the phase angle will vary in the same manner as with the Foster-Seeley circuit. The voltage e_1 will decrease, and e_2 will increase, and similar changes will occur in E_4 and E_5. The voltages E_1 and E_2, however, will remain equal to each other. Under this condition the output voltage will now have a significant value with the voltage at terminal 6 being negative with respect to ground.

Behavior of the Circuit with Signals below the Center Frequency. When the modulations produce i-f values lower than the center value, the phase angle will vary in the same manner as with the Foster-Seeley circuit. The voltage e_1 will increase and e_2 will decrease, and similar changes will occur in E_4 and E_5. The voltages E_1 and E_2, however, will remain equal to each other. Under this condition the output voltage will have a significant value with the voltage at terminal 6 being positive with respect to ground.

Behavior of the Circuit with an F-M Signal. When the i-f signal applied to the discriminator is frequency modulated, deviations in frequency above and below the center i-f value will take place continually. From the preceding explanations, it is evident that the output signal E_o will be an alternating voltage whose value is dependent upon the frequency modulations. The rate of the voltage changes will be the same as the a-f signals that produced the frequency modulations at the transmitting station. The a-f output signal will not be affected by changes in amplitude of the f-m signal applied to the detector; hence the ratio detector can be used without a limiter.

Control Voltage for Automatic Frequency Control. A d-c component is present at terminals 6 and 7 with f-m signals whenever the center frequency of the i-f signal differs from the center frequency of the discriminator transformer. The value of this direct voltage depends on the amount of difference between the two frequencies, while the polarity depends on whether the signal is above or below the center frequency of the discriminator. This direct voltage can be used to operate a variable-reactance tube to obtain automatic frequency control of the oscillator.

12-18. Preemphasis and Deemphasis. *Purpose.* From the time a-f signals are picked up at the microphone in a studio, worked through the transmitter, broadcast along the air waves, picked up by the receiving antenna, worked through the receiver, and reproduced at the loudspeaker, extraneous noise signals are added to the original audio signals. The disturbances added to the a-f signal are more predominant at the higher audio frequencies.   In general, the higher a-f signals of normal program material have lower amplitudes than the lower a-f signals.   These two effects cause a relatively low signal-to-noise ratio for the higher a-f signals.   To counteract this condition, high-frequency preemphasis and deemphasis are introduced intentionally in modern f-m broadcast transmitters and receivers (see also Art. 10-16).

Preemphasis.   In all radio broadcast f-m transmitters, emphasis is intentionally added to the higher frequencies in accordance with a standard preemphasis curve.   This curve shows that for a-f signals up to 500 cycles no preemphasis is employed; after 500 cycles the signal strength is gradually boosted until at 1 kc it is up 1 db, at 5 kc it is up 8 db, at 10 kc it is up 13.5 db, and at 15 kc it is up 17 db.   This procedure results in (1) a better signal-to-noise ratio, and (2) better-quality high a-f signals at the receiver's loudspeaker.

Deemphasis.   A disadvantage of adding emphasis to the higher audio frequencies at the transmitter is that the signals reproduced by the loudspeaker at the receiver will not have the same ratio of signal strength for the low and high frequencies as prevailed at the microphone in the broadcast studio.   To correct this condition a high-frequency deemphasis circuit is added to the receiver.   The deemphasis effect should be just the opposite of the preemphasis, namely it should introduce a loss of 17 db at 15 kc, 13.5 db at 10 kc, 8 db at 5 kc, 1 db at 1 kc, and produce a flat response with zero-db loss below 500 cycles.

Deemphasis Circuits.   The deemphasis circuit is generally connected at the a-f output of the f-m detector.   The deemphasis network used in many f-m receivers is simply a series-connected resistor followed by a parallel-connected capacitor as, for example, R_{16} and C_{25} in Fig. 12-21.

Circuit Component Values.   According to broadcasting practices, the standard time constant for the preemphasis circuit is 75 μsec.   In order to restore the original relative strength to the high-frequency signals, the deemphasis circuit in the receiver should also have a time constant of 75 μsec.   It will be found that not all receivers employ a time constant of exactly 75 μsec; however they are in the general order of this value. In the circuit of Fig. 12-21, the deemphasis network uses a 100,000-ohm resistor and a 0.001-μf capacitor which produces a time constant of 100 μsec.   The deemphasis circuit in Fig. 12-23 consists of $R_{12}C_{11}$ and produces a 68-μsec time constant.

12-19. The Audio-frequency Section.   The a-f section of an f-m receiver is very similar to that of an a-m receiver.   The main difference is that the f-m receiver can reproduce audio frequencies up to 15,000 cycles, which is three times as high as the a-f range required of ordinary a-m broadcast receivers.   In order to utilize this high-fidelity reproduction, the a-f circuits should be designed to provide satisfactory response over the high-frequency range. This also requires the use of a suitable high-quality loudspeaker. All of these requirements and how they are fulfilled have been fully treated in the preceding chapters.

12-20. The Frequency-modulation Receiver. The circuit diagram of a six-tube (plus rectifier) f-m receiver is shown in Fig. 12-21. Actually this circuit is only part of a complete seven-tube (plus rectifier) f-m/a-m receiver. In order to simplify tracing and explaining the circuit, it was divided into its f-m and a-m operation. Figure 12-21 shows the f-m circuits, and Fig. 12-22 shows the a-m circuits. The f-m operation will be described here, and the a-m operation is described in the following article. Most f-m receivers presently being manufactured are combination f-m/a-m receivers.

The operation of the front end of this circuit (Fig. 12-21) has already been described in Art. 12-12. The i-f output of the front end is coupled by T_3 to the input of the first i-f amplifier tube VT_2 whose output is then transformer-coupled (T_4) to the second i-f amplifier tube VT_3. The output of VT_3 is impedance-coupled ($L_2C_{18}C_{19}$) to the i-f amplifier-limiter tube VT_4. The voltage-dropping resistor R_{13} provides the reduced voltage (approximately 60 volts) required to operate VT_4 as a limiter. The constant-amplitude i-f signal is coupled by T_5C_{22} to the Foster-Seeley-type f-m discriminator circuit. The operation of this type of discriminator was fully described in Art. 12-16. R_{16} and C_{25} form the deemphasis network; there is no automatic volume control used by this receiver on f-m operation. VT_5 is shown connected for operation as (1) a twin-diode f-m detector, (2) a triode a-f amplifier; the diode plate shown without a connection is used for the a-m detector (see Fig. 12-22).

The a-f section is similar to that used in a-m broadcast receivers. The a-f output from the discriminator is fed to the volume control R_{17} and is coupled from there by C_{26} to the input of the triode a-f voltage amplifier section of VT_5. The output of VT_5 is RC coupled to the beam power-output tube VT_6 whose output is transformer coupled (T_6) to the loudspeaker.

The power-supply circuits are similar to those used in a-m receivers. However, a number of variations may be observed: (1) the heater circuit uses a number of filter elements, namely, L_4, L_5, C_{33}, C_{34}, C_{35}, C_{36}, and C_{37}; (2) a selenium rectifier is used; (3) a choke L_3 is used in the B supply filter network.

12-21. Amplitude-modulation Functions of an A-M–F-M Receiver. Many of the circuits and components of the f-m receiver described in Art. 12-20 are used for both f-m and a-m operation. Those parts used only for a-m operation were omitted from Fig. 12-21 and are shown separately on Fig. 12-22.

For a-m operation an f-m–a-m function switch (not shown on Figs.

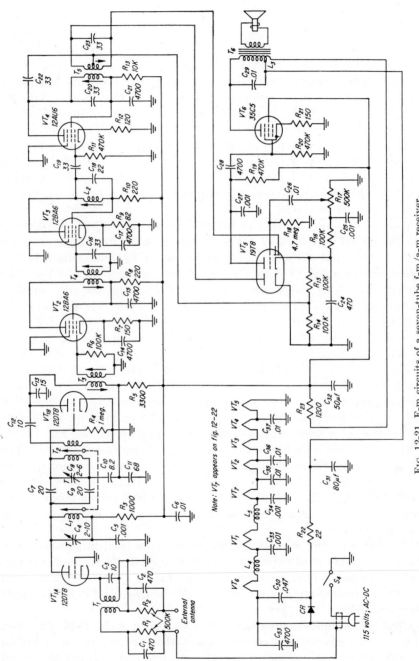

Fig. 12-21. F-m circuits of a seven-tube f-m/a-m receiver.

12-21 or 12-22) performs the following operations: (1) removes the B supply from VT_1, VT_2, and VT_4 of Fig. 12-21, thereby disabling the f-m tuner and i-f circuits, (2) disconnects the a-f output of the f-m receiver from the top terminal of the volume control (R_{17} in Fig. 12-21), (3) connects the plate circuit of VT_7 to the B power supply which is similar to closing S_1 of Fig. 12-22, (4) opens the grounding switch shown as S_2 on Fig. 12-22, and (5) connects the a-m detector output to the top terminal of the volume control R_{17} which is similar to closing S_3 of Fig. 12-22.

The receiver will operate as an a-m receiver in the following manner. Signals of 540 to 1,600 kc will be picked up by the antenna tuning circuit L_6, C_{38}, C_{39} and fed to the conventional type of pentagrid converter VT_7. The output of the converter is fed to the i-f transformer T_8. The output of T_8 is fed to the control grid of VT_3 whose plate circuit is connected to the primary of the second i-f transformer T_9. It should be observed that VT_3 is used as an i-f amplifier for both f-m and a-m operation. The output of T_9 is fed to the diode plate of VT_5; thus this tube is used for both f-m and a-m detection. The a-f output of VT_5 is fed to the volume control R_{17} and then through the triode amplifier section of VT_5, the beam power amplifier VT_6, and on to the loudspeaker, just as with f-m operation. The power-supply circuit is not changed in switching from f-m to a-m operation. Resistor R_{26} and capacitors C_{50}, C_{51} form an avc network that supplies avc bias to VT_3 and VT_7 when the receiver is connected for a-m operation.

12-22. A-M–F-M Receivers. The receiver described in Arts. 12-20 and 12-21 represents just one of many methods of combining f-m and a-m receiver functions into a single receiver. Reducing the number of tubes and components by using them for both f-m and a-m reception requires additional switching each time another tube is used for dual purpose. Eliminating switching from the high-frequency circuits reduces the possibilities of malfunctioning of these sensitive circuits but requires the use of one or more additional tubes.

12-23. A-M–F-M Tuner. *Application.* The electronic industry has experienced a phenomenal growth in the field of music reproduction which resulted in the production of many expensive high-quality amplifier and loudspeaker systems. In many cases the equipment provided only for reproducing music from records and/or tape. In order to use the existing high-fidelity amplifiers and reproducing systems for radio listening, auxiliary tuning units were introduced. Tuners are available for (1) only f-m reception, (2) a-m and f-m reception.

Because the tuner is generally used with an already available audio amplifying unit, the tuner includes (1) the front end, (2) the i-f section, (3) the detector, (4) the power supply; it may also include (5) a tuning eye indicator, (6) one a-f stage, (7) an output cable.

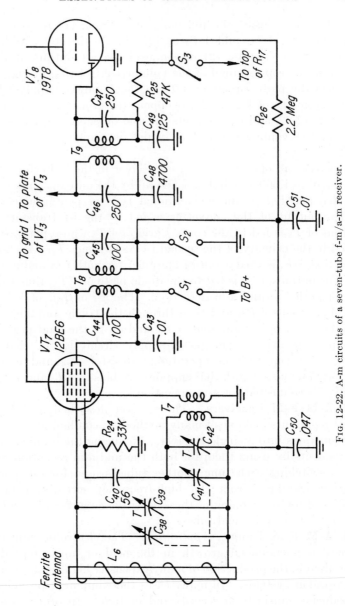

Fig. 12-22. A-m circuits of a seven-tube f-m/a-m receiver.

An A-M–F-M Tuner Circuit. The tuner, whose circuit diagram is shown in Fig. 12-23, is an eight-tube a-m–f-m tuner designed for use with an external two-channel stereo amplifier. The tuner has completely separate a-m and f-m channels, which may be used independently for regular broadcast reception, or simultaneously for reception of a-m–f-m stereophonic broadcasts. There are three controls (1) a-m tuning, (2) f-m tuning, and (3) a three-position function switch for a-m, f-m, and a-m–f-m stereo. A neon tuning indicator and an f-m multiplex jack are also included.

The F-M Circuit. The f-m channel, which uses five tubes plus the rectifier, consists of an r-f amplifier, an autodyne converter, two i-f amplifier stages, a limiter, and a discriminator stage; it may use either a built-in line-cord antenna or an external antenna.

The f-m signals at the antenna are coupled to the r-f amplifier tube V_{101A} through C_{32} and the unbalanced antenna coupling transformer T_{102}. The transformer has fixed broadband tuning characteristics produced by connecting C_{101} across its secondary. V_{101A}, which is one-half of a twin triode tube, is used as a grounded-grid r-f amplifier. Tuning of the r-f amplifier stage for station selection is performed by the variable slug-tuned coil L_{102A} and the alignment trimmer capacitor C_{104}. The r-f signal is coupled to the converter tube by C_{105}. The autodyne converter V_{101B} is the same as is used by the circuit described in Art. 12-11. The oscillator and r-f tuning circuits are varied simultaneously by controlling the movement of the tuning slugs in L_{102A} and L_{102B} with a common dial. Inductors L_{101} and L_{103} and capacitors C_{102} and C_{111} are used as filters to isolate the effects of heater currents from the cathodes of the high-frequency tubes.

The i-f section contains two i-f amplifier stages and a limiter stage. The i-f output of the converter is transformer-coupled (T_{101}) to the first i-f tube (V_3) which is then transformer coupled (T_2) to the second i-f tube (V_4) and then impedance coupled (T_3) to the limiter tube (V_5). The limiter tube, whose plate and screen-grid voltages are considerably lower than for the preceding i-f tubes, maintains a constant amplitude for the i-f signals that are transformer coupled (T_4) to the Foster-Seeley type discriminator detector (V_{6A}). High-frequency deemphasis is provided by R_{12} and C_{11}. The a-f signal from the discriminator is fed (1) through switch S_{1A} to the a-f output plug P_3, and (2) to the f-m multiplex output jack J_2.

The A-M Circuit. The a-m channel, which uses three tubes plus the rectifier, consists of a pentagrid converter, an i-f amplifier stage, and a detector stage. It uses a *Filteramic* ferrite antenna. The circuits for the converter (V_2), the i-f amplifier (V_7), and the a-m detector (V_{6B}) are the same as for the conventional a-m broadcast receiver (Chap. 11). The avc bias is produced by R_{18} and C_{16}. The a-f signal from the detector circuit is fed through switch S_{1A} to the a-f output plug (P_3).

The Tuning Indicator.   The triode section of V_{6B} is used as a tuning amplifier and is the only feature of the tuner not previously explained. A neon lamp (D_{S3}) is connected in series with the plate of the triode so that the plate current must flow through the lamp.   The grid of the triode is connected (1) through R_{22} to the avc line of the a-m tuner, and (2) through R_{23} to the grid of the f-m i-f amplifier-limiter tube (V_5). With zero station-signal condition, the plate current will be sufficient

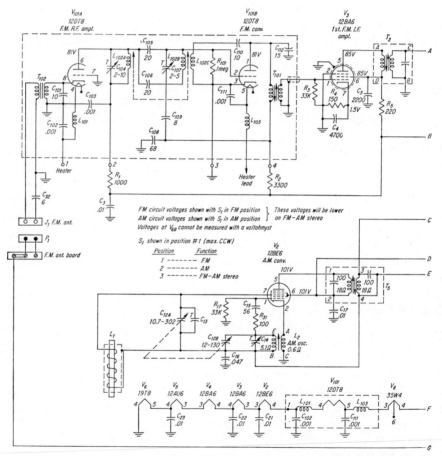

FIG. 12-23. Circuit diagram of

to cause the neon lamp to glow brightly.   As either an a-m or f-m signal is tuned in, a bias voltage is added to the grid of the triode which decreases the plate current and also decreases the illumination of the neon lamp.   A station is properly tuned in when the illumination of the neon lamp is minimum. R_{16} causes a bleeder current to flow through the neon lamp sufficient to maintain at least a slight glow at all times.   With a-m–f-m stereo operation, the tuning indicator circuit provides some degree of automatic stereo balance.

Function Switch. The two-deck function switch, S_{1A} rear and S_{1A} front, is shown in the position for f-m reception. In this position the a-m tuner is disabled by (1) removing the B supply of V_2 as contacts 8 and 10 of S_{1A} front are kept open, and (2) attenuating any a-f output by shorting out R_{14} as contacts 3 and 5 on S_{1A} rear are closed. The f-m tuner is utilized by (1) feeding the B supply to V_{101}, V_3, and V_5 by closing contacts

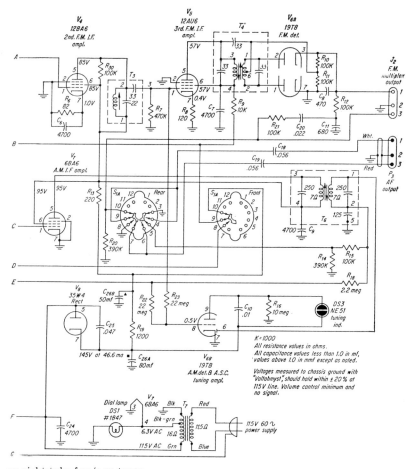

an eight-tube f-m/a-m tuner.

9 and 10 on S_{1A} front, and (2) connecting the a-f signal to pins 1 and 3 of the plug P_3 by closing contacts 7, 9, and 10 on S_{1A} rear.

When the switch is rotated one notch clockwise into position 2, the tuner is set for a-m reception. The f-m circuit is disabled by (1) shorting out R_{20} by closing contacts 8 and 10 on S_{1A} rear, (2) removing the B supply of V_{101}, V_3, and V_5 by disconnecting contacts 9 and 10 on S_{1A} front. The a-m circuit is utilized by (1) feeding a B supply to V_2, which is done by closing contacts 8 and 10 of S_{1A} front, and (2) connecting the a-f output to pins 1 and 3 of plug P_3 by connecting together contacts 4, 5, and 6 of S_{1A} rear.

When the switch is advanced to position 3, the tuner is set for a-m–f-m stereo operation. All of the tubes receive B voltage as contacts 8, 9, and 10 on S_{1A} front are all connected together. The a-f output of the f-m tuner is applied to pin 1 of plug P_3 by joining contacts 9 and 10 on S_{1A} rear. The a-f output of the a-m tuner is applied to pin 3 of plug P_3 by joining contacts 5 and 7 on S_{1A} rear. For this type of operation the audio amplifier system used with this tuner must have two separate audio channels, one for the a-m signals and the other for the f-m signals.

QUESTIONS

1. Compare the f-m and a-m systems with respect to their (*a*) operating frequencies, (*b*) a-f range, (*c*) selectivity.

2. Compare the f-m and a-m systems with respect to their (*a*) noise reduction, (*b*) interstation interference, (*c*) image interference.

3. In what respects does the f-m receiver differ from an a-m receiver?

4. (*a*) What is the *front end* of an f-m receiver? (*b*) What functions are performed in the front end?

5. In what manner do the components in the front end of an f-m receiver differ physically from their counterparts in an a-m receiver?

6. What is the purpose of the antenna-coupling unit?

7. Name four classifications of antenna couplers.

8. How does a balanced antenna coupler differ from an unbalanced antenna coupler?

9. (*a*) What are the advantages of using a triode in the high-frequency circuits of a receiver? (*b*) What are the advantages of using a pentode in the high-frequency circuits of a receiver?

10. What are the advantages of using miniature-type tubes in the high-frequency circuits?

11. What factors affect the equivalent input resistance of a tube?

12. (*a*) What precautions should be observed in positioning the ground connections of tube circuits? (*b*) What effect does the length of the cathode connecting lead have in the operation of a high-frequency circuit?

13. (*a*) Name four causes of noise in an f-m receiver. (*b*) Define *noise figure*.

14. Why is an r-f amplifier stage used more frequently in f-m receivers than in a-m receivers?

15. (*a*) Name three types of r-f amplifier circuits used in radio receivers. (*b*) Which type (or types) is used most frequently in f-m receivers? (*c*) Give the reason for your answer to (*b*).

16. What are the advantages of a grounded-grid amplifier?

17. What are the advantages of a cascode amplifier?

18. How does the noise figure of a triode compare with that of a pentode?

19. Why is it important that the oscillator frequency of a receiver remain steady?

20. (*a*) What is meant by frequency drift? (*b*) What undesirable effect does it cause?

21. (*a*) What is the purpose of an afc circuit? (*b*) What is the principle of operation of an afc circuit?

22. (*a*) What is a capacitive-reactance tube afc circuit? (*b*) What is an inductive-reactance tube afc circuit? (*c*) When is each type used?

23. Explain the principle of operation of an afc circuit using a crystal diode.

24. (*a*) What are the distinguishing operating characteristics of a converter and of a mixer?   (*b*) What basic function is performed by each?

25. (*a*) What types of devices can be used as mixers and/or converters?   (*b*) May this component have either linear or nonlinear characteristics?   (*c*) Explain your answer to (*b*).

26. (*a*) Give a definition for an autodyne converter.   (*b*) Explain the principle of operation of the autodyne converter.   (*c*) What are the advantages of the autodyne converter?

27. (*a*) How do the requirements of the i-f section of an f-m receiver differ from those in an a-m receiver?   (*b*) Why do f-m receivers frequently use two or three i-f stages?

28. Describe the shape of the response curve recommended for the i-f section of an f-m receiver.

29. (*a*) Name three methods used to obtain the desired over-all response of an i-f system.   (*b*) Explain briefly how each achieves its purpose.

30. (*a*) What is the function of the limiter?   (*b*) Explain briefly the principle of operation of the limiter stage.

31. (*a*) What is a grid limiter?   (*b*) What is a plate limiter?   (*c*) What is a double limiter?

32. (*a*) Name eight types of f-m detector circuits.   (*b*) Which type (or types) is commonly used in modern f-m receivers?

33. Describe the principle of operation of the Foster-Seeley type of f-m detector circuit.

34. Describe the action of the Foster-Seeley circuit for the following signal conditions: (*a*) center-frequency signal, (*b*) signals above the center frequency, (*c*) signals below the center frequency, (*d*) an f-m signal.

35. Describe the principle of operation of the ratio detector.

36. Describe the action of the ratio detector for the following signal conditions: (*a*) center-frequency signal, (*b*) signals above the center frequency, (*c*) signals below the center frequency, (*d*) an f-m signal.

37. (*a*) Why are preemphasis and deemphasis circuits used?   (*b*) Where is each located?

38. (*a*) What are the response characteristics of the preemphasis circuits?   (*b*) How should the response characteristics of the deemphasis circuits compare with those of the preemphasis circuits?

39. (*a*) What circuit components are used to provide deemphasis?   (*b*) What determines the values used for these components?

40. How do the requirements of the a-f section of an f-m receiver differ from those in an a-m receiver?

PROBLEMS

1. The r-f section of a certain f-m receiver is tuned by varying the inductance (Fig. 12-21). If the trimmer and stray capacitances are 6- and 4-$\mu\mu$f respectively, what range of inductance values is required to tune the circuit between 88 and 108 mc?

2. The oscillator section of a certain f-m receiver is tuned by varying the inductance (Fig. 12-21). If the trimmer and stray capacitances are each 4 $\mu\mu$f, what range of inductance values is required to tune the oscillator between 98.7 and 118.7 mc?

3. The r-f section of a certain f-m receiver is tuned by varying the capacitance (Fig. 12-10). (*a*) If the inductance of the circuit is 0.2 μh, what are the maximum and minimum values of total circuit capacitance required to tune the circuit between 88 and 108 mc?   (*b*) If the stray capacitance of the circuit is 4.2 $\mu\mu$f and the trimmer capacitor is adjusted to 2.1 $\mu\mu$f, what must be the range of the tuning capacitor?

4. The oscillator section of a certain f-m receiver is tuned by varying the capacitance (Fig. 12-10). (a) If the inductance of the circuit is 0.2 μh, what are the maximum and minimum values of total circuit capacitance required to tune the circuit between 98.7 and 118.7 mc? (b) If the stray capacitance of the circuit is 4 $\mu\mu$f and the trimmer capacitor is adjusted to 2 $\mu\mu$f, what must be the range of the tuning capacitor?

5. A certain high-frequency tube has an input resistance of 1,400 ohms at 100 mc. What is its input resistance at (a) 108 mc, (b) 88 mc, (c) 20 mc, (d) 1 mc?

6. A certain high-frequency tube has an input resistance of 8,000 ohms at 50 mc. What is its input resistance at (a) 108 mc, (b) 88 mc, (c) 25 mc, (d) 1 mc?

7. A certain triode being operated as a grounded-grid r-f amplifier has a transconductance of 3,000 μmhos and a plate resistance of 6,000 ohms. (a) What is the approximate input resistance? (b) What is the voltage amplification of the circuit if the shunt resistance of the output circuit is 3,000 ohms?

8. One section of a certain twin-triode being used as a grounded-grid r-f amplifier has a transconductance of 9,000 μmhos and a plate resistance of 4,500 ohms. (a) What is the approximate input resistance? (b) What is the voltage amplification of the circuit if the shunt resistance of the output circuit is 3,000 ohms?

9. A certain twin triode being used in a cascode amplifier stage has a transconductance of 3,000 μmhos and a plate resistance of 6,000 ohms. What is the voltage amplification of the stage if the shunt resistance of the output circuit is 3,000 ohms?

10. A certain twin triode being used in a cascode amplifier stage has a transconductance of 9,000 μmhos and a plate resistance of 4,500 ohms. What is the voltage amplification of the stage if the shunt resistance of the output circuit is 3,000 ohms?

11. What is the approximate equivalent noise resistance of an r-f amplifier triode whose transconductance is (a) 5,000 μmhos, (b) 3,000 μmhos?

12. What is the approximate equivalent noise resistance of an r-f amplifier triode whose transconductance is (a) 4,000 μmhos, (b) 3,333 μmhos?

13. What is the approximate equivalent noise resistance of an r-f amplifier pentode when $g_m = 4,000$ μmhos, $I_b = 5$ ma, and $I_{c.2} = 2$ ma?

14. What is the approximate equivalent noise resistance of an r-f amplifier pentode when $g_m = 8,000$ μmhos, $I_b = 13$ ma, and $I_{c.2} = 3.7$ ma?

15. What is the noise figure of the triode of Prob. 11a ($R_n = 500$ ohms) if the input resistance of the amplifier circuit is 300 ohms?

16. What is the noise figure of the pentode of Prob. 13 ($R_n = 2,230$ ohms) if the input resistance of the amplifier circuit is 300 ohms?

17. What is the circuit Q of an r-f amplifier stage that is to pass a 20-mc band (88 to 108 mc) using the mid-frequency value of 98 mc as the reference resonant frequency? (Note: The circuit Q is equal to the ratio of the resonant frequency to the bandwidth being passed.)

18. What is the circuit Q of an i-f amplifier stage that is to pass a 200-kc band if the resonant frequency is 10.7 mc? (Note: The circuit Q is equal to the ratio of the resonant frequency to the bandwidth being passed.)

19. In the receiver circuit of Fig. 12-21 C_{19} and R_{11} correspond to C_1 and R_1 in the limiter circuit of Fig. 12-16a. (a) What is the time constant of this circuit? (b) What is the time required to complete 1 cycle of the 10.7-mc i-f signal? (c) Is the time constant of the RC circuit long or short compared with the time of one input cycle?

20. The limiter circuit (Fig. 12-16a) of a certain f-m receiver uses a 33-$\mu\mu$f capacitor and a 100-k resistor. (a) What is the time constant of this circuit? (b) What is the time required to complete 1 cycle of the 10.7-mc i-f signal? (c) Is the time constant of the RC circuit long or short compared with the time of one input cycle?

CHAPTER 13

TRANSISTORS

A transistor is basically a resistor that amplifies electrical impulses as they are transferred through it from its input to its output terminals. The name *transistor* is derived from the words *transfer* and *resistor*. Transistors may be used to fulfill most of the functions performed by vacuum tubes and with greater efficiency. The basic material in most transistors is either germanium or silicon, and the devices are usually made in two types, *point contact* and *junction*. The transistor has many advantages among which are its: (1) low current requirement, (2) small size, (3) light weight, (4) long operating and shelf life, (5) elimination of warm-up time, (6) mechanical strength, and (7) photosensitivity. Because of these features it can replace the vacuum tube in many electronic-circuit applications.

13-1. Vacuum Tubes and Transistors. *Vacuum Tubes.* The operation of a vacuum tube involves two fundamental actions: (1) liberation of electrons from a solid, and (2) control of electrons in a vacuum. A vacuum tube contains many separate components such as filament wire, coated cathode cylinder, one or more grid assemblies, carefully shaped plate assembly, getter, spacers, shield pieces, support rods, insulators, connecting strips, leads, envelope, and base. Each part must be made to precise tolerances, from different materials, and all assembled with extreme care.

Transistors. The operation of a transistor, or other related solid-state electronic device, depends on the flow of electric charges carried within the solid. Two fundamental actions are involved: (1) generation of carriers within the solid, and (2) control of these carriers within the solid.

A triode junction transistor consists of a single crystal, the leads, and an envelope. The crystal is essentially a three-layer unit. However, the layers are not separate pieces but are areas within the crystal which have slightly different electrical characteristics. Transistors are made from a *semiconductor* whose electrical properties are midway between that of a conductor such as silver, and an insulator such as porcelain. A semiconductor under one condition may act as a conductor allowing an easy flow of current, while under another condition it may act as an insulator and virtually block the flow of current. The amount of current will depend on various physical influences, such as (1) electric fields, (2) heat, and (3) light.

511

In a conductor, an electric current is thought of only as the movement of free electrons along the conductor.   In a semiconductor, current consists of the movement of free electrons and holes.   When an electron leaves an atom the resulting charge on the atom is positive.   The absence of an electron from the atomic structure in a crystal is referred to as a *hole.*   The application of an electric field will cause electrons and holes to drift in opposite directions.   The holes being positively charged will drift toward the negative terminal, while the electrons being negatively charged will drift toward the positive terminal.

13-2. Physical Concepts of Solids.   *Conductors.*   Solids having a low electrical resistivity at room temperature are called *conductors.*   The solids in this group include most of the common metals such as copper, aluminum, and silver.   The resistivity of copper is of the order of 1.6×10^{-6} ohm per centimeter cube, and increases approximately linearly with temperature.

Insulators.   Solids having a high electrical resistivity at room temperature are called *insulators.*   Their resistivity is of the order of 10^9 to 10^{18} ohms per centimeter cube.   Examples of solids in this group are porcelain, quartz, glass, and mica.

Semiconductors.   Solids having a value of resistivity midway between that of conductors and insulators are called *semiconductors.*   Their resistivity is of the order of 0.1 to 50 ohms per centimeter cube at room temperature.   Their resistivity varies nonlinearly with temperature changes, and these solids may possess either a positive or negative temperature coefficient.   The resistance characteristics depend largely upon the amount of impurities in the material.   Examples of solids in this group are elements such as germanium and silicon, and compounds such as zinc oxide and copper oxide.

When the particle concentration in a given semiconductor consists of both electrons and holes in approximately equal numbers, the material is called an *intrinsic semiconductor.*   When the holes predominate, the material is called a *positive* or *P-type semiconductor.*   When the electrons predominate the material is called a *negative* or *N-type semiconductor.*   An intrinsic semiconductor may be given P or N type characteristics by adding various substances, called *impurities,* to the base material.

Energy Levels.   In order for conduction to take place there must be a movement of electrons.   In atomic theory, the electrical properties of the elements are explained by the concept of energy bands.   The electrons in the outer orbit of an atom can be moved with the least amount of energy and are called *valence electrons.*   Valence electrons have definite energy levels or bands (Fig. 13-1), and the conductivity of an element is determined by the energy required to move its valence electrons from their normal energy level, or *valence band,* to its highest energy level, the *conduction* or *energy band.*   The distance a valence electron moves in its travel from the valence band to the energy band varies with each type of atom. The energy gap separating the valence and conduction bands in an insulator is extremely large and it is very difficult for a valence electron to reach the energy band. In a conductor the valence and energy bands overlap, and the valence electrons are available for conduction. In a semiconductor the energy gap is very small, and the thermal energy of the valence electrons at room temperature is sufficient to permit an appreciable amount of conduction. Because an electron cannot remain in the space between the valence and energy bands, this region is sometimes referred to as the *forbidden zone*.

13-3. Conduction in Crystals. *Conduction.* In a semiconductor the valence electrons of one atom bond to the valence electrons of adjacent atoms to form a lattice structure called a *crystal*. This *covalent bond* causes the crystal unit to possess various energy bands. The outer

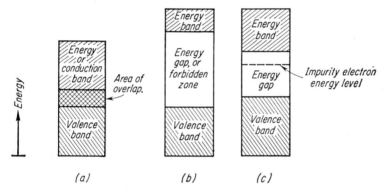

FIG. 13-1. Energy and valence bands. (*a*) Conductor. (*b*) Insulator. (*c*) Semiconductor.

energy band provides conduction when only a few electrons are present. Conduction is thus obtained through the use of a deficiency of electrons, or the presence of holes. A hole is sometimes referred to as a *positive charge* in terms of polarity; however its mass and mobility are different from those of an electron. In negative conduction the electrons flow toward the positive terminal, while in positive conduction the holes flow toward the negative terminal.

Crystalline Structure. The crystals of germanium and silicon are tetrahedral in shape, with each of the four valence electrons of an atom forming a covalent bond with one of the four adjacent atoms (Fig. 13-2). Thus, valence electron *a* from atom *A* and valence electron *b* from atom *B* form a covalent bond between atoms *A* and *B*. In a similar manner covalent bonds are formed between atoms *A* and *D*, *F* and *D*, *F* and *G*, etc. The resulting lattice structure is in equilibrium, as there are no excess positive or negative charges, and at low temperatures it is an insulator.

Impurity Conduction. Pure germanium and pure silicon have some conductivity as electrons are released in these crystals at room temperature. However, for transistor operation, this normal conductivity must be increased. The conductivity of germanium or silicon can be greatly increased by the addition of certain impurities. The conductivity of the resulting crystal will depend on the amount and type of impurity that is added. The added impurity may be either a donor or an acceptor. The following explanation of impurity addition will consider only germanium; however, the same principles also apply to silicon.

The amount of an impurity that is added to pure germanium is relatively very small and very critical, as is shown by the following example. If the impurity is added in the ratio of one atom of the impurity for every 100 million atoms of germanium, the resistivity of the germanium drops

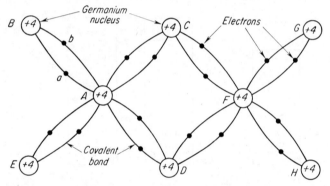

Fig. 13-2. Lattice structure of a germanium or a silicon crystal illustrating the covalent bonds between adjacent atoms.

from 60 to 3.8 ohms per centimeter cube. This value is satisfactory for transistor use. However, if the ratio is one impurity atom for every 10 million germanium atoms, the resistivity drops to 0.38 ohm per centimeter cube. This value is too low for transistor use.

Donor Impurities. If an impurity having five valence electrons is added to pure germanium, each impurity atom will take the place of a germanium atom. Four of the five valence electrons will form covalent bonds with four valence electrons from four germanium atoms (Fig. 13-3). The fifth valence electron is free to wander about the crystal and contributes to the conductivity of the crystal in a manner similar to the movement of free electrons in a metallic conductor. As more impurity atoms of this type are added, more electrons will be released to wander about the crystal, and therefore the conductivity of the germanium increases, and its resistivity decreases. Germanium having an excess of electrons is called *N-type germanium.* When the impurity donates electrons to the crystal conductivity, it is called a *donor impurity.* Arsenic and antimony are typical donor impurities that may be used to make N-type germanium.

Acceptor Impurities.　If an impurity having three valence electrons is added to pure germanium, each impurity atom will take the place of a germanium atom.　The three valence electrons will form covalent bonds with three valence electrons from three germanium atoms (Fig. 13-4).

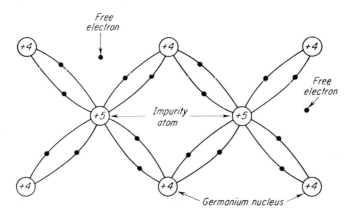

FIG. 13-3. Effect of adding a donor impurity to pure germanium.

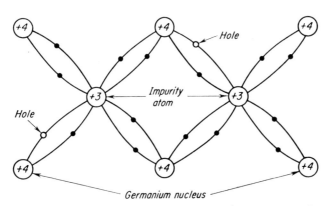

FIG. 13-4. Effect of adding an acceptor impurity to pure germanium.

There is a deficiency of one electron, and this deficiency is called a *hole.* Under the influence of an electric field, an electron from an electron-pair bond somewhere in the crystal will move into the hole, thus leaving a net positive charge in the half-empty bond.　Holes contribute to the conductivity of a crystal in much the same manner as do electrons, as they also move from atom to atom.　As more impurity atoms of this type are added, more holes are formed and the conductivity of the crystal increases.　Germanium having an excess of holes is called *P-type germanium.*　When the impurity accepts electrons in order to increase crystal conductivity, it is called an *acceptor impurity.* Aluminum, boron, gallium, and indium are typical acceptor impurities that may be used to make P-type germanium.

13-4. The PN Junction or Diode. *Characteristics.* A PN junction is formed when similar sections of P-type and N-type germanium are joined together to produce a *germanium diode* (Fig. 13-5a). The circles with negative symbols in the *P-type germanium* represent acceptor atoms. The negative symbol is used because of the presence of the additional electron which was obtained from an adjacent electron-pair bond. The hole left by the removed electron is represented by a small positive symbol. The circles with positive symbols in the *N-type germanium* repre-

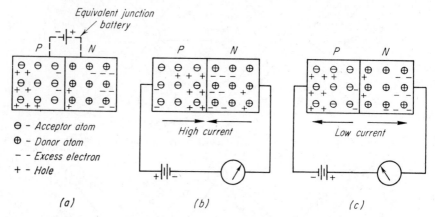

FIG. 13-5. Effect of junction bias voltage. (*a*) Neutral or zero bias. (*b*) Forward bias. (*c*) Reverse bias.

sent donor atoms. The positive symbol is used because the fifth electron has been removed, leaving the atom with a resultant positive charge. The free electron is represented by a small negative symbol.

At first glance it would seem that the excess electrons in the N-type germanium would immediately cross the junction and combine with the holes in the P-type germanium. This action does not occur because of the repelling force set up by the negatively charged acceptor atoms. However, a relatively small number of electrons will attain sufficient thermal energy to overcome this negative potential and cross through the junction. In a similar manner the holes in the P-type germanium are prevented from crossing the junction by the positive repelling force set up by the donor atoms. A relatively small number of holes will, however, attain sufficient energy to overcome this positive potential and cross through the junction. As a result the net current through the junction is zero, as indicated by the current vectors in Fig. 13-6a. The opposition to the movement of electrons and holes from crossing the junction is called a *potential hill,* and its effect is the same as that produced by a battery connected across the junction as shown in Fig. 13-5a. The PN junction may be used as a rectifier by connecting an external battery across it to aid or oppose the potential of the equivalent junction battery.

Forward Bias. In order to obtain a current through the junction, the potential hill across the junction has to be neutralized. This is done by applying an external potential across the two sections of germanium so that it opposes the junction potential (Fig. 13-5b). The voltage applied in this manner is called a *forward potential* or *forward bias.* Free electrons in the N section will be repelled by the negative force set up by the power source and will move toward the junction. At the same time the holes in the P section will be repelled by the positive force set up by the power source and will also move toward the junction. The voltage

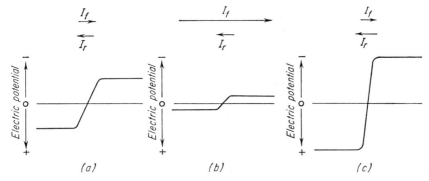

Fig. 13-6. Variation of potential at PN junction. (a) Zero bias. (b) Forward bias. (c) Reverse bias.

applied should be high enough to impart sufficient energy to these carriers to overcome the potential barrier at the junction, and enable them to cross through it. Once the junction is crossed, the free electrons from the N section will combine with the holes in the P section, and the holes from the P section will combine with the free electrons in the N section. This action decreases the potential barrier at the junction. For each hole in the P section that combines with an electron from the N section, an electron from an electron-pair bond leaves the crystal and enters the positive terminal of the battery. This action creates a new hole that is forced to move toward the junction because of the electric field produced by the battery. For each electron in the N section that combines with a hole from the P section, an electron enters the crystal from the negative terminal of the battery. This constant movement of electrons toward the positive terminal, and the holes toward the negative terminal, produces a high *forward current* I_f (Fig. 13-6b). A small number of newly released electrons from the P section and a similar number of newly created holes from the N section will cross through the junction to produce a *reverse current* I_r. This current is very small being only a few microamperes, while the forward current is measured in milliamperes (Fig. 13-7a). The net current $I_f - I_r$ increases exponentially as the forward potential is increased.

Reverse Bias. If the applied voltage is reversed so that the external potential aids the junction potential (Fig. 13-5c), a *reverse potential* or *reverse bias* is obtained. The excess holes in the P section will be attracted by the negative force of the power source and will move away from the junction. At the same time the free electrons in the N section will be attracted by the positive force of the power source and will also

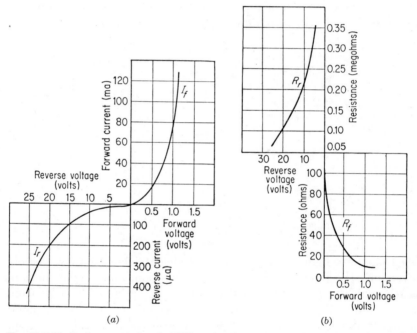

FIG. 13-7. Static characteristics of a PN junction diode. (a) Volt–Ampere. (b) Volt–Ohm.

move away from the junction. The net current approaches a constant value as the forward current reduces to zero, and the reverse current remains constant (Fig. 13-6c).

Static Characteristics. The PN junction possesses the property of a rectifier since the current resulting from a voltage applied in one direction across the junction is different from the current resulting from the same amount of voltage applied in the opposite direction across the junction. The static volt-ampere characteristic curve for a typical germanium diode is shown in Fig. 13-7a. The forward and reverse current curves are drawn to different voltage and current scales, and both are nonlinear over a considerable portion of their ranges. The forward-current curve has a sharp upward swing at a relatively low value of voltage. The reverse current curve indicates that a relatively high voltage is required to produce even a very low value of current. As the applied reverse voltage is increased, a point is reached where the valence bonds start to break up, thus releasing a large number of holes and electrons. This point is indicated on the curve by a sharp increase in current, and the voltage causing this breakdown of the crystalline structure is called the *Zener voltage*. This voltage compares to the maximum inverse voltage rating of a diode vacuum tube. The Zener voltage will therefore indicate the maximum reverse voltage that can be applied to a semiconductor without

producing an excessive reverse current. The Zener voltage of the diode junction used for obtaining the characteristic curve of Fig. 13-7a is 15 volts. From this curve it can be seen that the initial increase of 15 volts reverse voltage (0 to 15) produces an increase of 100 μa in the reverse current, while an additional increase of only 5 volts produces the same amount of increase, and the next 5-volt increase produces an increase of 200 μa.

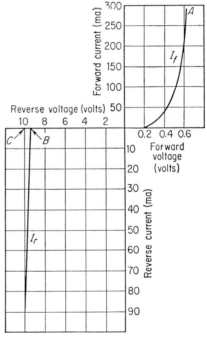

The static voltage-resistance curve for a typical germanium diode is shown in Fig. 13-7b. This curve was obtained by dividing the voltage at a number of points on the curve of Fig. 13-7a by the current at these voltages. The forward and reverse resistance curves are drawn to different voltage and resistance scales. Increasing the forward voltage decreases the resistance to a low value,

Fig. 13-8. Static conduction characteristics of a Zener diode.

usually less than 100 ohms. Decreasing the forward voltage increases the resistance, and at zero voltage the resistance is in the order of hundreds of thousands of ohms. As the reverse voltage is increased, the resistance increases to a peak value, then starts to decrease sharply.

Zener Diode. The Zener voltage of a silicon junction diode may be used to obtain voltage regulation, signal limiting, or other similar circuit applications. A silicon junction diode that is designed to place the Zener point at a definite voltage level is called a *Zener diode*. The static conduction characteristics of a typical silicon junction Zener diode are shown in Fig. 13-8. A relatively low forward resistance is indicated by the steep slope of the current curve between zero voltage and the forward voltage A. A very high inverse resistance is indicated by the very slight slope of the curve between zero voltage and the reverse voltage B. As the reverse voltage is increased beyond point B, the reverse current increases very sharply for a very small increase in reverse voltage, B to C. Within a few tenths of a volt the back resistance decreases very sharply from several megohms to a few ohms. For this same voltage change the current increases from a few microamperes to many milliamperes. The high conduction interval BC is nondestructive providing the maximum allowable power dissipation of the diode is not exceeded. The voltage at point B is called the *Zener voltage*, and its value may be held to a tolerance of approximately ± 1 per cent. The important features of Zener diodes are their reverse-conduction characteristics, namely: (1) the sharp-break-down characteristic, (2) the high ratio of the resistance between zero volts and voltage B to the resistance between voltages B and C, and (3) the low resistance of the diode between voltages B and C.

13-5. Types of Junction. *Methods of Construction.* The characteristics of a semiconductor diode, triode, or tetrode will depend on the material and type PN junction used. Germanium is usually better than silicon at high frequencies, while at high power levels silicon is generally better than germanium. The PN junction may be made in various ways; among these are: (1) point contact, (2) grown, (3) diffused, (4) recrystallized, (5) alloyed, and (6) surface barrier (Fig. 13-9). There are many variations of these six methods, and in the commercial manufacture of transistors more than one method is generally used.

Point-contact Junction. Basically the point-contact type of junction (Fig. 13-9a) consists of a fine pointed wire that makes pressure contact with the face of a N-type germanium wafer. The PN junction is formed by applying a relatively high current momentarily across the pointed wire and the crystal face. The heat generated during this interval drives some of the electrons from the atoms in a small region around the point of contact, leaving holes. This region in the N-type germanium is thus converted to P-type germanium.

Grown Junction. There are several methods of growing crystals; one of the methods generally used is the Czochralski technique. In this method a single crystal is slowly pulled out of a vat of molten germanium. The melted germanium crystallizes in exactly the right order on the single crystal, or seed. If a donor impurity is added to the molten germanium the grown crystal will be N type, and if an acceptor impurity is added the grown crystal will be P type. Using this idea, a grown junction is made by starting with an N-type material. At a certain point, a small pellet of an acceptor impurity is dropped into the melt, and a P-type germanium crystal starts to grow. After a definite period, the crystal is removed and the junction located so that the crystal can be cut into thin wafers. Each wafer possesses the properties of a PN junction (Fig. 13-9b).

Diffused Junction. The diffused junction (Fig. 13-9c) is made by placing a small pellet of an acceptor impurity on one face of an N-type wafer. The combination is then heated in order to melt the impurity. A portion of the impurity will diffuse a short distance into the wafer, creating a P-type region that is in close contact with an N-type region.

Recrystallized Junction. The recrystallized junction (Fig. 13-9d) is made by placing a piece of N-type germanium between hot and cold temperature areas. The germanium will melt back partway so that it is

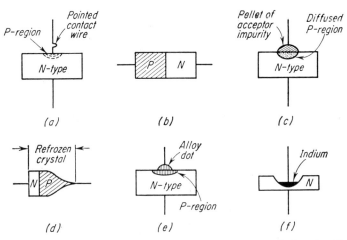

Fig. 13-9. PN junctions. (a) Point contact. (b) Grown. (c) Diffused. (d) Recrystallized. (e) Alloy technique. (f) Surface barrier.

part molten and part solid. The temperature is then reduced, and the germanium refreezes into a single crystal having a P region and a N region in close contact.

Alloyed-technique Junction. The alloyed junction is the one most commonly used. It is known by various names, such as *fused junction*, *fusion alloy*, and *diffused alloyed*. In its simplest form, a small dot of an indium alloy is placed on one face of an N-type germanium wafer. The combination is heated until the alloy melts and dissolves some of the germanium. The temperature is then lowered, and the germanium refreezes to form a single crystal having a PN junction (Fig. 13-9e).

Surface-barrier Junction. The structure of the surface barrier junction is similar to the alloyed junction. Instead of alloying a dot deep into the base material to obtain a narrow base width, a well is electrochemically etched into the base material until the thickness of the base is only a few tenths of a mil. Indium is then plated to the well, and a PN junction is formed (Fig. 13-9f). Microetching techniques as used with surface-barrier junctions are also used to produce bonded-barrier and microalloy junctions.

13-6. Junction Transistors.　*The Triode.*　By joining a layer of N-type germanium to a PN junction an NPN junction is formed (Fig. 13-10*a*); and joining a layer of P-type germanium to a PN junction, a PNP junction is formed (Fig. 13-11*a*).　Either of these two junctions can be used as a triode junction transistor, and are generally referred to as a *NPN*

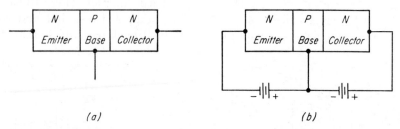

(a)　　　　　　　　　　(b)

Fig. 13-10. NPN junction transistor.　(*a*) The junction and the name of each section. (*b*) Biasing potentials.

junction transistor or a *PNP junction transistor.*　The center section is called the *base,* one of the outside sections the *emitter,* and the other outside section the *collector.*

Transistor Biasing.　In order to use the NPN transistor as an amplifier, the emitter-base junction is biased with a forward voltage, and the base-collector junction with a reverse voltage (Fig. 13-10*b*).　A relatively high current will flow through the emitter-base junction, and a relatively low

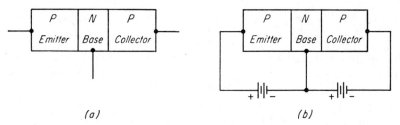

(a)　　　　　　　　　　(b)

Fig. 13-11. PNP junction transistor.　(*a*) The junction and the name of each section. (*b*) Biasing potentials.

current through the base-collector junction.　If the base is made relatively wide, practically all the current will be between the emitter and base.　The base is usually made very narrow so that its width is less than the average length of the path taken by an electron in the base before combining with a hole.　Most of the electrons crossing the emitter-base junction will therefore diffuse through the base to the base-collector junction, where the applied positive voltage will draw them to the collector terminal.　The base current is very small, as only a small number of emitter electrons (usually less than 5 per cent) will combine with the base holes. Since the remaining electrons flow toward the collector, the emitter-collector current is large.

The operation of a PNP transistor is similar to the operation of an NPN transistor except that the bias voltage polarities are reversed and the current carriers are holes instead of electrons. The emitter-base junction is biased with a forward voltage, and the base-collector junction

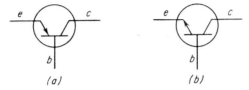

FIG. 13-12. Symbol identification for triode transistors: (*a*) PNP. (*b*) NPN.

with a reverse voltage (Fig. 13-11*b*). The holes from the emitter will diffuse through the base to the base-collector junction, where the applied negative voltage will draw them to the collector terminal.

Symbol and Lead Identification. The symbols for triode transistors are shown in Fig. 13-12. Although there are several variations of symbols in use, these are frequently used and have been adopted by the IRE. The horizontal line represents the base, and the two angular lines represent the emitter and collector. The arrowhead drawn on the emitter

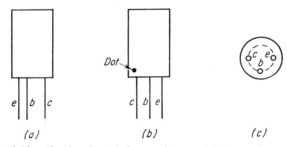

FIG. 13-13. Lead identification for triode transistors. (*a*) Unevenly spaced leads. (*b*) Evenly spaced leads. (*c*) Symmetrically spaced leads.

indicates the direction of current. In the PNP type, the arrow is drawn pointing toward the base (Fig. 13-12*a*), and in the NPN type the arrow is drawn pointing away from the base (Fig. 13-12*b*).

The lead identification of a triode transistor varies with the manufacturer. Three systems in general use are shown in Fig. 13-13 When the leads of a transistor are in the same plane but unevenly spaced (Fig. 13-13*a*), they are identified by the position and spacing of the leads. The center lead is the base lead, and the emitter and collector leads are on either side of this lead. The collector lead is identified by the larger spacing existing between it and the base lead. When the leads on a transistor are in the same plane but evenly spaced (Fig. 13-13*b*), the center lead is the base, the lead identified by a dot the collector, and the remaining lead the emitter. When the leads are spaced around the circumference of a circle as shown in Fig. 13-13c, the center lead is the base lead. The collector and emitter leads are identified by their relative position to the base lead, when viewed looking into the base of the transistor. Some manufacturers add a fourth lead that is electrically connected to the mounting base, or case, of the transistor. Other manufacturers connect the common electrode to the base, or case, and use only two leads.

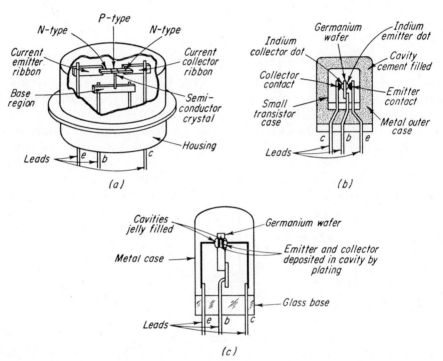

FIG. 13-14. Cutaway view of commercial transistors.   (a) Grown junction.   (b) Diffused junction.   (c) Surface barrier.

13-7. Point-contact Transistors. *The Triode.* Point-contact transistors are usually of the PNP-type junction, and are made by pressure-mounting two closely spaced finely pointed electrodes against one face of a thin wafer of N-type germanium (Fig. 13-15a). The PN junctions are formed by applying a high current momentarily across each pointed wire and the crystal face. A small region around each contact is converted to P-type germanium in a manner similar to the formation of a point-contact diode. The germanium wafer is called the *base*, one electrode the *emitter*, and the other electrode the *collector*.

Transistor Biasing. In order to use the PNP point-contact transistor as an amplifier, the emitter-base junction is biased with a forward voltage, and the collector-base junction with a reverse voltage (Fig. 13-15b). A relatively high current will flow through the emitter-base junction, and a relatively low current through the base-collector junction. The high forward current consists of holes moving from the emitter to the base. Because of the proximity of the two junctions, the applied negative voltage on the collector will draw a large number of holes across the base-collector junction to the collector terminal.

The proximity of the two junctions produces other actions between the holes and the electrons that increase the collector current to a value that is two to three times that of the emitter current. There are several theories for the explanation of this current amplification. One is that when the emitter holes travel toward the collector they form a positive space charge that attracts electrons from other sections of the germanium

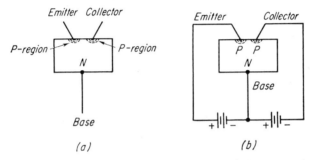

(a)   *(b)*

Fig. 13-15. Point-contact transistor. (*a*) The junction and the name of each section. (*b*) Biasing potentials.

crystal, and causes these electrons to add to the collector current. Another theory is that the movement of holes in the closely spaced junction reduces the resistance of the collector junction, thus permitting a greater current to flow for the same applied voltage. The input resistance is approximately 300 ohms, while the output resistance is approximately 20,000 ohms. The transistor voltage gain is equal to the current gain times the resistance gain. Thus the resistance gain adds to the net voltage gain of a transistor. The voltage gains obtainable from point-contact transistors are comparable to those obtained from high-mu vacuum tubes.

13-8. Transistor Characteristics. *Methods of Operation.* The input and output impedances of a transistor vary with the manner in which it is connected into a circuit. Correct impedance match can be made by selecting the proper method of connection. The characteristics of a transistor will therefore depend upon its method of connection. Basically there are three methods of connection: (1) *common base*, (2) *common emitter*, (3) *common collector*. The transistor may be considered similar to a vacuum tube in the following respects, (1) the emitter and cathode both serve as the source of electron flow, (2) the base and control grid each serves to control the electron flow through the unit, (3) the collector and plate are normally part of the output circuit.

Because of the comparable relation between the components of a transistor and those of a vacuum tube, the basic circuit operation of transistors may be compared with a comparable basic circuit of a vacuum tube (Fig. 13-16). However, in these comparable circuit operations it should be noted that (1) the transistor is current controlled [its collector characteristics are plotted with the control current I_b as the parameter (Fig. 13-18)] and (2) the vacuum tube is voltage controlled [its plate characteristics are plotted with the control voltage E_c as the parameter (Fig.

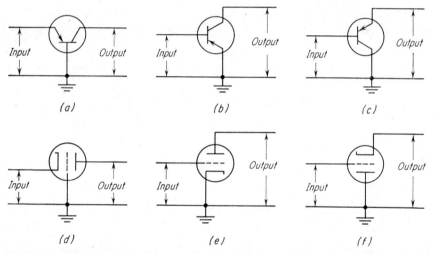

Fig. 13-16. Comparison of transistor and vacuum-tube operation. (*a*) Common base. (*b*) Common emitter. (*c*) Common collector. (*d*) Grounded grid. (*e*) Grounded cathode. (*f*) Grounded plate.

2-24)]. In transistor operation, the term *common* is used to denote the electrode that is common to both the input and output circuits. Since the common electrode is usually grounded, it may also be referred to as a *grounded base, grounded emitter*, or *grounded collector*. The common-base type is similar to a vacuum tube operated as a grounded-grid amplifier. The advantage of using a transistor in this manner is that it has a low input impedance and a high output impedance. The common-emitter type is similar to a grounded-cathode vacuum-tube circuit. The input impedance is low to medium, and the output impedance medium to high. The common-collector type is similar to a vacuum tube used as a cathode follower. The input impedance is high, and the output impedance low.

Static Characteristics. The static characteristics of a transistor may be plotted in many ways. The two families of curves generally used are:

(1) collector current versus collector voltage for constant values of emitter current (Fig. 13-17) and (2) collector current versus collector voltage for constant values of base current (Fig. 13-18).   The curves of Fig. 13-17 are generally used for grounded-base connected transistors, and those of Fig. 13-18 for grounded-emitter connected transistors.   The shapes of the characteristic curves for transistors are similar to those for a pentode vacuum tube (Fig. 2-24).   The plate current in a pentode is relatively independent of plate voltage, and primarily dependent on the grid voltage.   The collector current in a grounded-base connected transistor is

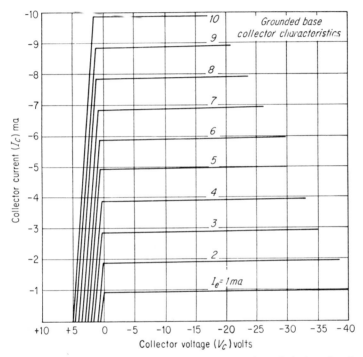

FIG. 13-17. Static common-base collector characteristics of a triode junction transistor.

relatively independent of collector voltage, and primarily dependent on the emitter current.   A reversal of the collector voltage causes the collector current to drop sharply to zero for very low values of collector voltage.   If the voltage is increased beyond these values, the collector current rapidly reverses itself and starts increasing in a forward direction. The resulting excessive current may cause overheating and possible permanent damage to the transistor.   When the transistor is connected as a grounded emitter, the collector voltage has very little effect on the collector current for low values of base current.   However, as the base current increases, the effect of the collector voltage on the collector current also increases.

High-frequency Effects. At high frequencies, the operation of transistors will be adversely affected by: (1) charge-carrier diffusion, (2)
phase shift, and (3) base-collector junction capacitance. *Charge-carrier
diffusion* is caused by the differences in the length of the paths taken by
the emitter current carriers in traveling through the base from the emitter
to the collector.   The resulting variations in traveling time cause some of
the current carriers that left the emitter at the same instant to arrive at
the collector at different times.   When the differences in the arrival time
represent a large portion of the input signal cycle (for example 180 degrees)
some of the current carriers will cancel each other, thus causing distortion.

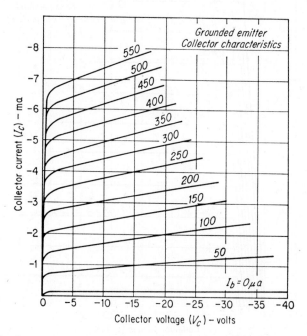

Fig. 13-18. Static common-emitter collector characteristics of a triode junction transistor.

Since there is no potential across the base to accelerate the current carriers from the emitter junction to the collector junction, they diffuse
relatively slowly through the base to the collector. This action results
in a time delay between the emitter and collector currents, which represents a phase shift between the two currents at the signal frequency.
The effects of charge-carrier diffusion and phase shift are less for point-
contact transistors than for junction transistors because (1) there is a
shorter space between the emitter and collector, and (2) the electric field
from the collector extends into the base.
    The reverse voltage at the base-collector junction causes current carriers
of both the base and collector to be repelled from the junction.   The

charge at the junction will then be the result of the donor and acceptor atoms on each side of the junction. The density of the charge carriers increases with the distance from the junction, and decreases with an increase in voltage. Since the charge at the junction is a function of the voltage, the base-collector junction indicates the presence of capacitance. The capacitance of the junction varies with: (1) voltage, (2) current, (3) area, and (4) type of junction. The effects of base-collector junction capacitance are less for point-contact transistors than for junction transistors because of their (1) lower collector resistance, and (2) smaller collector junction area.

Frequency Cutoff. The upper frequency limit of a transistor is determined by its transit time, which is the time taken by the electrons or holes to pass from the emitter to the collector. Transistors are rated according to their *alpha cutoff frequency*, which is the frequency at which the grounded-base current gain has decreased to 0.707 of its low-frequency reference value (usually 1,000 cps). To obtain transistor performance that is equivalent to vacuum-tube performance, the operating frequency should be approximately 20 per cent of the alpha cutoff frequency. Using this rule, the minimum cutoff frequency for transistors used in a-m broadcast-band receivers should be approximately 5×455 kc or 2.2 mc for the i-f section, 5×2.0 mc or 10.0 mc for the oscillator section, and 5×20 kc or 100 kc for the audio section.

Voltage Gain and Power Gain. The direct voltage gain in a transistor can be expressed as the product of the direct current gain and resistance gain, or

$$VG = IG \times RG \tag{13-1}$$

In a junction transistor, the collector current is equal to the difference between the emitter and base currents, and the resulting current represents a loss rather than a gain. The voltage gain is obtained because of the high ratio between the collector-base resistance and the emitter-base resistance. This high ratio of resistance can also be used to produce a direct power gain. A small amount of power in the input circuit (emitter-base) controls a much larger amount of power in the output circuit (collector-base).

$$PG = (IG)^2 \times RG \tag{13-2}$$

Example 13-1. The collector current in a junction transistor is 96 per cent of the emitter current, the collector-base resistance is 750,000 ohms, and the emitter-base resistance is 500 ohms. (*a*) What is the direct voltage gain of the transistor? (*b*) What is the direct power gain of the transistor?

Given: $IG = 0.96$ Find: (*a*) VG
$r_{cb} = 750{,}000$ ohms   (*b*) PG
$r_{eb} = 500$ ohms

Solution:

(a) $VG = IG \times RG = 0.96 \times \dfrac{750,000}{500} = 1,440$

(b) $PG = (IG)^2 \times RG = 0.96 \times 0.96 \times 1,500 = 1,382$

The voltage and power gains obtained in Example 13-1 can be obtained only if the transistor is operated into a very-high-impedance circuit. In practical amplifier circuits, the voltage and power gains obtained are much less than these values because of the impedances that must be used to match the high output impedance of a prior stage with the low input impedance of the following stage.

The collector current in point-contact transistors is generally two to three times the value of the emitter current. In spite of its greater current gain, its relatively lower resistance gain makes its voltage and power gains lower than those of junction transistors.

Example 13-2.   The collector current in a point-contact transistor is 2.6 times its emitter current, the collector-base resistance is 22,000 ohms, and the emitter-base resistance is 250 ohms.   (a) What is the direct voltage gain of the transistor?   (b) What is the direct power gain of the transistor?

Given: $IG = 2.6$ Find: (a) VG
            $r_{cb} = 22,000$ ohms   (b) PG
            $r_{eb} = 250$ ohms

Solution:

(a) $VG = IG \times RG = 2.6 \times \dfrac{22,000}{250} = 228.8$

(b) $PG = (IG)^2 \times RG = 2.6 \times 2.6 \times 88 = 594.8$

13-9. Transistor Specifications. *Electrical Characteristics.* Although many properties of a transistor may be specified, this section will describe only those specifications usually listed on manufacturers' transistor characteristic charts.   The values are listed under three general headings: (1) general data, (2) maximum ratings, and (3) typical operation.   Under general data are listed the following: (1) application or class of service, (2) type, (3) outline dimensions, and (4) lead arrangement.   The maximum ratings are the direct voltage and current values that must not be exceeded in the operation of the unit.   Maximum ratings usually include the direct: (1) collector-base voltage, (2) emitter-base voltage, (3) collector current, (4) emitter current, and (5) collector power dissipation. With point-contact transistors, the maximum peak-inverse emitter voltage may also be included.   The typical operating values are presented only as a guide, since the values vary widely, as they are dependent on the operating voltages and which electrode is used as the common circuit element.   The values listed may include the: (1) common electrode, (2) collector-emitter volts, (3) collector current, (4) current-transfer ratio, (5) input resistance, (6) load resistance, (7) power gain, (8) noise factor, and (9) alpha cutoff frequency.   Some manufacturers also list for the common emitter circuit the small-signal: (1) hybrid-π parameters, (2) H parameters, and (3) T parameters.

Letter Symbols. Maximum, average (d-c), and effective (root-mean-square) values are represented by the upper-case letter of the proper symbol.

$$I = \text{current}, \quad V = \text{voltage}, \quad R = \text{resistance}, \quad P = \text{power}$$

Instantaneous values that vary with time are represented by the lower-case letter of the proper symbol.

$$i = \text{current}, \quad v = \text{voltage}, \quad r = \text{resistance}, \quad p = \text{power}$$

D-c values and instantaneous total values are indicated by upper-case subscripts.

$$I_C, \; V_{EB}, \; P_C, \; i_C, \; v_{EB}, \; p_C$$

Varying component values are indicated by lower-case subscripts

$$I_c, \; V_{eb}, \; P_c, \; i_c, \; v_{eb}, \; p_c$$

To distinguish among maximum, average, and effective values the maximum value is represented by the addition of the subscript m or M, and the average value by the addition of the subscript av or AV.

$$i_{c \cdot m}, \; I_{c \cdot m}, \; I_{C \cdot M}, \; I_{C \cdot AV}, \; i_{C \cdot AV}$$

Other abbreviations used as subscripts are

E, e = emitter electrode
B, b = base electrode
C, c = collector electrode

J, j = electrode, general
X, x = circuit node
Q = average (d-c) value with signal applied

The first subscript designates the electrode at which the current is measured, or where the electrode potential is measured with respect to the reference electrode or circuit node, which is designated by the second subscript. When the reference electrode or circuit node is understood, the second subscript may be omitted where its use is not required to preserve the meaning of the symbol.

Supply voltages may be indicated by repeating the electrode subscript. The reference electrode may then be designated by the third subscript.

$$V_{EE}, \; V_{CC}, \; V_{BB}, \; V_{EED}, \; V_{CCB}, \; V_{BBC}$$

In devices having more than one electrode of the same type, the electrode subscripts are modified by adding a number following the subscript, and on the same line.

$$V_{B1}, \; V_{b2}, \; I_{B2}$$

In multiple-unit devices, the electrode subscripts are modified by a number preceding the electrode subscript, as 1b, 2B, 1c, 2C.

Current Gain. When a transistor circuit uses a common base, its current gain is the ratio of the change in collector current to the change in emitter current for a constant value of collector voltage. When the current gain is specified for static or relatively large values of collector and emitter currents, it is called the *alpha direct-current gain.* It is represented by the symbol α, or the H-parameter symbol h_{FB}, and may be expressed as

$$\alpha = h_{FB} = \frac{\text{change in collector current}}{\text{change in emitter current}} = \frac{dI_C}{dI_E} \qquad (V_C \text{ is constant}) \quad (13\text{-}3)$$

The value of alpha for junction transistors is always less than 1, as the collector current can never exceed the emitter current (Fig. 13-17). Its value is in the range of 0.94 to 0.99. The value of alpha for point-contact transistors ranges from 2 to 3.

When a transistor circuit uses a common emitter, its current gain is the ratio of the change in collector current to the change in base current for a constant value of collector voltage. When the current gain is specified for static or relatively large values of collector and base currents, it is called the *beta direct-current gain.* It is represented by the symbol β, or the H-parameter symbol h_{FE}, and may be expressed as

$$\beta = h_{FE} = \frac{\text{change in collector current}}{\text{change in base current}} = \frac{dI_C}{dI_B} \qquad (V_C \text{ is constant}) \quad (13\text{-}4)$$

The value of beta is always greater than unity and can be obtained by the use of a transistor's common-emitter $I_C V_{CE}$ characteristic curves (Fig. 13-18).

Example 13-3. A junction transistor having the characteristics shown in Fig. 13-18 is operated as a grounded emitter with a collector-emitter potential of -7 volts. What is the current-amplification factor when the base current is changed from 50 to 100 μa?

$$\text{Given:} \quad \begin{aligned} V_{ce} &= -7 \text{ volts} \\ I_{B1} &= 50 \ \mu\text{a} \\ I_{B2} &= 100 \ \mu\text{a} \end{aligned} \qquad\qquad \text{Find:} \quad \beta$$

Solution:

From Fig. 13-18 it can be noted that a change in base current from 50 to 100 μa with $V_{ce} = -7$ volts produces a change in the collector current from 0.8 to 1.6 ma.

$$\beta = \frac{dI_C}{dI_B} = \frac{1.6 \text{ ma} - 0.8 \text{ ma}}{100 \ \mu\text{a} - 50 \ \mu\text{a}} = \frac{0.8 \times 10^{-3} \text{ amp}}{50 \times 10^{-6} \text{ amp}} = 16$$

For a grounded emitter, the relation between β and α is

$$\beta = \frac{\alpha}{1 - \alpha} \qquad\qquad (13\text{-}5)$$

This equation shows that β increases as α approaches unity.

Example 13-4. A junction transistor operated as a grounded emitter is biased so that α is 0.941. What is β for these conditions?

Given: $\alpha = 0.941$ Find: β

Solution:

$$\beta = \frac{\alpha}{1 - \alpha} = \frac{0.941}{1 - 0.941} \cong 16$$

The current gain is the most important property of a transistor in determining the gain of an amplifier. Since beta is partially dependent on frequency, its value is usually specified for a definite frequency, and for transistors used at audio and low radio frequencies, it is generally specified at 1 kc.

Current-transfer Ratio. When a transistor is connected with its base grounded, the current-transfer ratio is the ratio of the a-c variation in the collector current to the a-c variation in the emitter current for a constant value of collector voltage. This ratio is called the *forward-current-transfer ratio*, and is represented by the symbol h_{fb}.

When a transistor is connected with its emitter grounded, the current-transfer ratio is the ratio of the a-c variation in the collector current to the a-c variation in the

Fig. 13-19. Input and output voltage and current measurements for any network.

base current for a constant value of collector voltage. This ratio is called the *forward-current-transfer ratio* and is represented by the symbol h_{fe}.

H Parameters. A *parameter*, also called a *network constant*, is a constant that enters into a functional equation, and corresponds to some characteristic of a circuit such as resistance, inductance, capacitance, or any other property value in a network. The small signal parameters of transistors are usually specified in terms of the H parameters. These parameters are defined for any network (Fig. 13-19) by the following equations.

$$v_i = h_i i_i + h_r v_o \qquad\qquad (13\text{-}6)$$
$$i_o = h_o v_o + h_f i_i \qquad\qquad (13\text{-}7)$$

where v_i = input voltage, volts
         v_o = output voltage, volts
         i_i = input current, amp
         i_o = output current, amp
         h_i = input impedance, ohms (output circuit shorted)
         h_o = output admittance, μmhos (input circuit open)
         h_f = forward-current-transfer ratio (output circuit shorted)
         h_r = reverse-voltage-transfer ratio (input circuit open)

When H parameters are used for a transistor circuit, a second subscript is added to indicate the terminal that is grounded. For example: h_{FE} would be the forward-current-transfer ratio of a grounded-emitter circuit, and h_{IB} the input impedance of a grounded-base circuit.

Power Gain. The power gain P_g of a transistor is expressed in decibels. Since its value is dependent on the transistor (1) circuit, (2) input resistance, and (3) load resistance, these values are usually listed in the characteristic chart for the power gain specified. The variations of the power gain of a transistor for different circuits and values are listed in Table 13-1. This table shows (1) that the greatest power gain is obtained

TABLE 13-1. POWER GAIN OF A TRANSISTOR

Common electrode	Input resistance, ohms	Load resistance, ohms	Power gain, db
E	1,980	100,000	44.1
E	2,670	2,670	34.5
C	500,000	10,000	17.0
B	215	500,000	32.5

by using the transistor with a grounded emitter and a high-load resistance, (2) that the least power gain is obtained with the transistor connected with a grounded collector, and (3) that a low-load resistance with a grounded emitter produces a power gain less than that obtained with a high-load resistance but greater than that obtained with either of the other two connections. These conditions are typical for all transistors.

Noise Figure. The primary source of internal noise in a transistor is the molecular agitation resulting from the movement of electrons and holes through a semiconductor. The magnitude of the noise currents in a transistor is dependent upon the: (1) average current, (2) frequency, and (3) bandwidth. With zero bias current, the noise output of a transistor is equal to the thermal noise obtained from a resistor of equivalent resistance. A quantity called the *noise figure* is used to evaluate the amount of noise produced by a transistor. It is generally specified in decibels above the thermal input power for a 1-cycle bandwidth at 1,000 cycles, or

$$F_n = 10 \log \frac{P_{no}}{P_{ni}} \qquad (13\text{-}8)$$

where F_n = noise figure, db

P_{no} = output power noise, watts

P_{ni} = thermal input power noise, watts

The output power noise per 1-cycle bandwidth varies approximately inversely with frequency to about 100 kc, where it attains a value several times that of its equivalent resistor thermal noise, and then remains relatively constant for the higher frequencies. An approximate noise figure for reference frequencies other than 1,000 cycles can be obtained by use of the equation

$$F_{nf} = F_n + 10 \log \frac{1,000}{f} \qquad (13\text{-}9)$$

where F_{nf} = noise figure for 1-cycle bandwidth at frequency f, db
  f = new frequency, cps

Example 13-5. The specified noise figure for a certain transistor at 1,000 cycles and a bandwidth of 1 cycle is 23 db. What is its approximate noise figure at (a) 3,000 cycles? (b) 455 kc?

Given: F_n = 23 db   Find: (a) F_{na}
     f_a = 3,000 cycles   (b) F_{nb}
     f_b = 455 kc

Solution:

$$F_{nf} = F_n + 10 \log \frac{1,000}{f}$$

(a) $F_{na} = 23 + 10 \log \dfrac{1,000}{3,000} = 23 + 10(9.5224 - 10) = 18.22$ db

Since the noise figure is relatively constant above 100 kc, this value can be used for f_b in place of 455 kc.

(b) $F_{nb} = 23 + 10 \log \dfrac{1,000}{100,000} = 23 + 10(-2) = 3$ db

Although the noise figure for junction transistors is much lower than for point-contact transistors, it is much higher than for a vacuum tube operating at the same frequency. In the a-f range, a vacuum tube has a noise figure of the order of 3 db, a junction transistor a noise figure of 9 to 20 db, and a point-contact transistor a noise figure of 20 to 40 db. Transistors designed for high and very high radio frequencies have noise figures of the order of 6 db.

13-10. Tetrodes. To improve the high-frequency operation of a transistor, a fourth lead is added to an NPN transistor. The additional lead $b2$ is attached to the base region at a position that is on the side opposite to the original base connection $b1$ (Fig. 13-20a). This type of transistor is called a *tetrode*, and its schematic symbol is shown in Fig. 13-20b. In the same manner as with an NPN triode transistor, the emitter is biased in the direction of greatest electron flow, and the collector is biased in the direction of least electron flow. The additional lead is biased with a negative potential that is considerably greater than the normal emitter-base voltage. This negative potential restricts the electrons flowing through the base, and causes them to flow through a relatively narrow area of the base region (Fig. 13-20a). The improvement in high-frequency operation is obtained because: (1) the reduced effective area of each region adjacent to the base decreases the emitter and collector capacitances, (2) the shorter path taken by the base current decreases the base resistance.   The base resistance of a tetrode may be decreased in the order of 10 to 1 over that of a triode.

In forcing the electrons through a narrow channel in the base region, the collector-current capabilities of the transistor are reduced, thus also reducing its power-handling capabilities.   The tetrode is primarily intended for high-frequency use as an r-f amplifier, i-f amplifier, mixer,

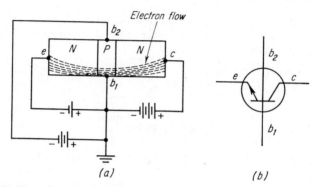

FIG. 13-20. Tetrode transistor.   (*a*) D-c biasing voltages.   (*b*) Schematic symbol.

and oscillator.   The second base connection may be used for automatic gain control, as this connection will cause very little detuning of the collector circuit.

QUESTIONS

1. Compare the fundamental actions involved in the operation of a vacuum tube with those involved in the operation of a transistor.

2. Compare the general construction of a vacuum tube with that of a transistor.

3. Explain the movement of electrons and holes in a semiconductor.

4. Define the following terms: (*a*) conductor, (*b*) insulator, (*c*) semiconductor.

5. What is meant by (*a*) intrinsic semiconductor? (*b*) P-type semiconductor? (*c*) N-type semiconductor?

6. Define the following terms (*a*) valence electron, (*b*) valence band, (*c*) conduction band, (*d*) energy band, (*e*) forbidden zone.

7. Explain the lattice structure of a silicon or germanium crystal.

8. (*a*) What is meant by impurity conduction?   (*b*) How does the relative amount of impurity addition affect the resistance of a semiconductor?

9. Explain the effect on a semiconductor of adding (*a*) a donor impurity, (*b*) an acceptor impurity.

10. Describe the characteristics of a PN junction.

11. Describe how a high forward current is obtained from a PN junction.

12. What is meant by (*a*) reverse bias? (*b*) reverse current? (*c*) potential hill?

13. Compare the forward and reverse current static volt-ampere characteristics of a semiconductor diode.

14. What is meant by (*a*) Zener voltage? (*b*) Zener diode?

15. (*a*) Describe the operating characteristics of a Zener diode.   (*b*) What are some of the applications of **Zener** diodes?

16. Describe the construction of the following types of junctions: (*a*) point-contact, (*b*) grown, (*c*) surface-barrier.

17. Describe the construction of the following types of junctions: (*a*) recrystallized, (*b*) diffused, (*c*) alloyed.

18. Describe the construction of the following types of transistors: (*a*) NPN, (*b*) PNP.

19. Describe the method used for biasing the following types of transistors for use as amplifiers: (*a*) NPN, (*b*) PNP.

20. What methods are used for identifying the leads of a triode transistor?

21. Describe the construction of a point-contact transistor.

22. Explain two theories for the high current amplification of point-contact transistors.

23. Compare the operating characteristics of the following transistor connections with those of a comparable vacuum-tube connection: (*a*) common base, (*b*) common emitter, (*c*) common collector.

24. In comparing transistor and vacuum-tube circuit operation what important factors must be considered?

25. Describe the static characteristics of transistors for (*a*) collector current versus collector voltage for a constant value of emitter current, (*b*) collector current versus collector voltage for a constant value of base current.

26. Describe the high-frequency effects of transistor operation as caused by (*a*) charge-carrier diffusion, (*b*) phase shift, (*c*) base-collector junction capacitance.

27. (*a*) What is meant by the alpha-cutoff frequency?   (*b*) What is the relation between the alpha-cutoff frequency and the operating frequency?

28. Explain the following terms (*a*) direct voltage gain, (*b*) direct current gain, (*c*) direct resistance gain, (*d*) direct power gain.

29. What are the maximum ratings of a transistor that are generally included in manufacturers' specifications?

30. What are the operating values of a transistor that are generally included in manufacturers' specifications?

31. What is meant by (*a*) alpha direct-current gain? (*b*) beta direct-current gain? (*c*) current-transfer ratio $h_{f \cdot b}$? (*d*) current-transfer ratio $h_{f \cdot e}$?

32. How are transistor small-signal parameters usually specified?

33. How is the power gain of a transistor affected by (*a*) its method of connection? (*b*) its input resistance? (*c*) its load resistance?

34. (*a*) What is the primary source of internal noise in a transistor?   (*b*) What factors affect the magnitude of the noise currents in a transistor?

35. (*a*) Explain the specifications used to evaluate the amount of noise produced by a transistor.   (*b*) How does the noise figure for the a-f range of a transistor compare to that of vacuum tubes?

36. (*a*) Describe the construction and operation of a tetrode transistor.   (*b*) What are the primary applications of tetrode transistors?

37. Describe how the tetrode improves the high-frequency operation of transistors.

PROBLEMS

1. The collector current in a junction transistor is 95 per cent of the emitter current, the collector-base resistance is 1,200,000 ohms and the emitter-base resistance is 600 ohms.   (*a*) What is the direct voltage gain of the transistor?   (*b*) What is the direct power gain of the transistor?

2. A junction transistor has an input resistance of 1,400 ohms, an output resistance of 3.5 megohms, and $\alpha = 0.97$. Find (a) the direct voltage gain of the transistor, (b) the direct power gain of the transistor.

3. A point-contact transistor has an input resistance of 310 ohms, an output resistance of 22,000 ohms, and $\alpha = 2.7$. Find (a) the direct voltage gain of the transistor, (b) the direct power gain of the transistor.

4. The collector current in a point-contact transistor is 2.4 times its emitter current, the collector-base resistance is 21,000 ohms, and the emitter-base resistance 270 ohms. (a) What is the direct voltage gain of the transistor? (b) What is the direct power gain of the transistor?

5. A junction transistor having the characteristics shown in Fig. 13-17 is operated with a grounded base with a collector-base potential of -5 volts. What is the alpha direct current gain when the emitter current changes from 1 to 2 ma?

6. A junction transistor having the characteristics shown in Fig. 13-17 is operated with a grounded base with a collector-base potential of -10 volts. What is the alpha direct current gain when the emitter current changes from 2 to 3 ma?

7. A junction transistor having the characteristics shown in Fig. 13-18 is operated with a grounded emitter with a collector-emitter potential of -10 volts. What is the beta direct current gain when the base current is changed from 100 to 150 μa?

8. A junction transistor having the characteristics shown in Fig. 13-18 is operated with a grounded emitter with a collector-emitter potential of -5 volts. What is the current amplification factor when the base current is changed from 50 to 100 μa?

9. A junction transistor operated with a grounded emitter is biased so that $\beta = 38$. What is α for these conditions?

10. A junction transistor operated with a grounded emitter is biased so that $\alpha = 0.97$. What is β for these conditions?

11. The specified noise figure for a certain transistor at 1,000 cycles and a bandwidth of 1 cycle is 16 db. What is the approximate noise figure at (a) 4,000 cycles? (b) 455 kc?

12. The specified noise figure for a certain transistor at 1,000 cycles and a bandwidth of one cycle is 28 db. What is the approximate noise figure at (a) 2,000 cycles? (b) 10.7 mc?

CHAPTER 14

TRANSISTOR AMPLIFIER CIRCUITS

Amplification is the process whereby small amounts of voltage, current, and/or power at the input side of a circuit or system can produce larger amounts of voltage, current, and/or power at the output side of the circuit or system. A transistor, in some respects, may be compared to the vacuum tube; similar circuit arrangements are used for both vacuum-tube and transistor amplifiers. However, it is incorrect to consider the transistor as a direct substitute for a vacuum tube, as the similarity applies only to the circuit arrangement; in fact, the differences in their operation far outnumber their similarities.

14-1. Basic Amplifiers. *Transistor vs. Vacuum-tube Amplification.* A vacuum tube is a voltage-operated device in which the input voltage controls the output current and/or voltage. That is, the a-c voltage applied to the grid of a vacuum tube will control the current and/or voltage in the plate circuit. This type of amplifier works best with a constant-voltage power source. A transistor is a current-operated device in which the input current controls the output current. For example, the input current in the emitter-base circuit controls the output current in the collector circuit. This type of amplifier works best with a constant-current power source. Unlike the vacuum tube, the transistor does not provide isolation between its input and output circuits. Because of this, the output circuit parameters affect the input circuit parameters, and vice versa. Most vacuum-tube circuits have high input and high output impedances. Transistor circuits generally have a low-to-medium input impedance and a moderate-to-high output impedance. The current flowing through these impedances determines the voltage or power gain of a transistor amplifier circuit. The gain of transistor circuits is usually specified in decibels of power rather than multiplications of voltage.

There are three basic types of transistor-amplifier circuits: (1) common or grounded emitter, (2) common or grounded base, (3) common or grounded collector. The term *grounded* will generally be used in this text, as its use in the field of electronics is far more frequent than the term *common*, although an electrode so specified does not necessarily have to be connected to ground.

Transistor Circuit Parameters. In the analysis of transistor circuits, the principal transistor parameters are

α = emitter-collector current-amplification factor
β = base-collector current-amplification factor
r_b = base resistance
r_e = emitter-junction resistance
r_c = collector-junction resistance
V_B = base voltage
V_E = emitter voltage
V_C = collector voltage
I_b = base current
I_e = emitter current
I_c = collector current
P_e = emitter power
P_c = collector power
P_g = power gain
I_{co} = collector cutoff current

The external circuit into which the transistor operates may contain

R_B = base resistance
R_E = emitter resistance
R_C = collector resistance
R_L = output load resistance
R_i = input circuit resistance
R_o = output circuit resistance
V_{BB} = base-bias supply voltage
V_{EE} = emitter-bias supply voltage
V_{CC} = collector-bias supply voltage

Grounded Emitter. The configuration of the basic grounded-emitter transistor amplifier circuit is similar to the basic grounded-cathode vacuum-tube amplifier circuit (Fig. 14-1). In Fig. 14-1a the grid is the driven element, and the input signal is applied between the grid and cathode; the output signal is produced between the plate and cathode. For all practical purposes the resistance of the power supplies E_{cc} and E_{bb} may be considered as being zero. Thus, the cathode is common to both input and output circuits. The average grid bias is determined by the biasing voltage E_{cc}. The average plate voltage, with reference to the cathode, is determined by the fixed d-c power source E_{bb} and the voltage drop across R_L. The input and output impedances of the tube are relatively high, and the input and output signals are 180 degrees out of phase with each other.

The corresponding PNP transistor amplifier circuit (Fig. 14-1b) shows that the base is the driven element and the input signal is applied between the base and emitter; the output signal is taken off between the collector and the emitter. For all practical purposes the resistance of the power supplies V_{EE} and V_{CC} may be considered as being zero. Thus, the emitter is common to both input and output circuits.   The emitter is biased in a forward direction by the voltage V_{EE}, and the collector is biased in a reverse direction by the voltage V_{CC}.   The input resistance of the transistor is relatively low, usually in the order of 1,000 to 2,000 ohms.   The output resistance of the transistor is relatively high, usually

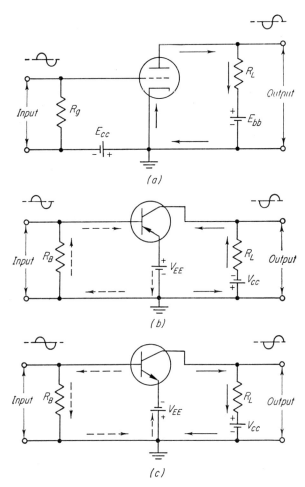

(a)

(b)

(c)

Fig. 14-1. Basic grounded-emitter transistor amplifier circuits and the analogous vacuum-tube circuit.   (a) Grounded cathode.   (b) PNP grounded emitter.   (c) NPN grounded emitter.

in the order of 50,000 ohms.   The input and output signals are 180 degrees out of phase with each other.

When a signal is applied to the input circuit (Fig. 14-1b), the base current will vary about its average value.   A corresponding variation will take place in the collector current but with a much greater amplitude.

The base-emitter circuit is biased with the negative side of V_{EE} connected to the base, and the positive side connected to the emitter. When the input signal is positive, its polarity will be opposite that of V_{EE}; consequently the bias voltage V_E will decrease and cause a decrease in the emitter and base currents. The decrease in the emitter and base currents produces a decrease in the collector current with a corresponding decrease in the output voltage developed across R_L. As a result the potential at the collector with reference to ground will become more negative. By the same reasoning, a negative signal at the input will produce an increase in the output voltage across R_L, and as a result the potential at the collector will become more positive. Thus, a positive-going input signal produces a negative-going output signal, and vice versa. As these signals are 180 degrees out of phase, phase reversal has taken place.

The characteristic curves of Fig. 13-18 show that a small change in base current produces a relatively large change in collector current. This large change in collector current, plus the high output resistance, produces a large voltage gain, in the order of 1,500. As this circuit produces both voltage and current gains, it also produces high power gains. Power gains of approximately 10,000 times (40 db) can be obtained with this type of circuit. Although the common-emitter type of amplifier has the highest power gain, some other transistor-amplifier circuits are capable of handling greater amounts of power. Its stability with temperature change is slightly less than the common-base type of amplifier. The common-emitter amplifier circuit is popular because it produces: (1) a large current gain, (2) a large voltage gain, and (3) the highest power gain. This type of amplifier circuit is frequently used in radio receiver circuits, especially with RC coupling.

A common-emitter circuit using an NPN transistor is shown in Fig. 14-1c. The operation and characteristics of this circuit are the same as for the PNP transistor. However, it should be noted that the polarities of the biasing voltages are opposite of those for the PNP transistor. When connecting transistor circuits, it is extremely important that the proper d-c polarities be observed, since application of the wrong polarity to a transistor may cause instantaneous damage to it.

Grounded Base. The basic grounded-base transistor amplifier circuits (Figs. 14-2b and 14-2c) are similar to the basic grounded-grid vacuum-tube amplifier circuit shown in Fig. 14-2a. In this circuit the cathode is the driven element and the input signal is applied between the cathode and grid; the output signal is produced between the plate and grid. Thus, the grid is common to both input and output circuits. The average cathode bias is determined by the biasing voltage E_{cc}, and the average plate voltage, with reference to the grid, is determined by the fixed d-c power source E_{bb} and the voltage drop across R_L. The input impedance of the tube is relatively low, and its output impedance relatively high.   There is no phase reversal between the input and output signals.

The corresponding PNP transistor amplifier circuit is shown in Fig. 14-2b.   In this circuit the emitter is the driven element and the input signal is applied between the emitter and base; the output signal is

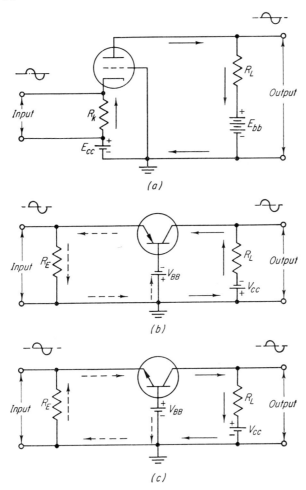

(a)

(b)

(c)

Fig. 14-2. Basic grounded base transistor amplifier circuits and the analogous vacuum-tube circuit.   (a) Grounded grid.   (b) PNP grounded base.   (c) NPN grounded base.

produced between the collector and base.   Thus, the base is common to both the input and output circuits.   The emitter is biased in a forward direction by the voltage V_{BB}, and the collector is biased in a reverse direction by the voltage V_{CC}.   The input resistance of the transistor is very low, and is in the order of 20 to 50 ohms.   The output resistance of the transistor is very high, and is in the order of 1 to 2 megohms.

When a signal is applied to the input circuit, the emitter current will vary about its average value. A corresponding variation will take place in the collector current but with a slightly smaller amplitude. When the input signal is positive, its polarity is the same as V_{BB}; consequently the bias voltage V_B will increase and cause an increase in the emitter and base currents. The increase in the emitter and base currents produces an increase in the collector current, with a corresponding increase in the output voltage developed across R_L. As a result the potential at the collector with reference to ground will become less negative, or more positive. By the same reasoning, a negative signal at the input will produce a decrease in the output voltage across R_L, and as a result the potential at the collector will become more negative. Thus, a positive-going input signal produces a positive-going output signal, and vice versa. As these signals are in phase, no phase reversal has taken place.

The characteristic curves of Fig. 13-17 show that a change in emitter current produces practically the same amount of change in collector current. The resistance gain between the base and the collector is very high, and with relatively equal current values a voltage gain is produced for the circuit in the order of 1,500. Since no current gain is produced, the resultant power gain of this type of circuit is less than can be obtained with a common-emitter type of amplifier. The stability of this circuit with temperature change is slightly better than with the common-emitter type of amplifier. The grounded-base amplifier is useful in circuit applications requiring a very low input impedance and a very high output impedance. However, it is more generally used with point-contact transistors than with junction transistors.

A common-base amplifier circuit using an NPN transistor is shown in Fig. 14-2c. The operation and characteristics of this circuit are the same as for the PNP transistor. It should be noted that the polarities of the biasing voltages are opposite of those for the PNP transistor.

Grounded Collector. The basic grounded-collector transistor amplifier circuit is similar to the basic grounded-plate, or cathode follower, vacuum-tube amplifier circuit shown in Fig. 14-3a. The plate of the vacuum tube is not connected directly to ground, but is at a-c ground by virtue of the capacitor C connected between plate and ground. The plate requires a positive potential relative to the cathode in order to attract the electrons; hence it cannot be at d-c ground. In this circuit the grid is the driven element and the input signal is applied between the grid and the plate, and the output signal is produced between the cathode and plate. Thus, the plate is common to both input and output circuits for a-c signals. The average grid bias is determined by the fixed power source E_{cc} and the average plate current flowing through R_L; and the average plate voltage, with reference to the cathode, is determined by the fixed power source E_{bb} and the voltage drop across R_L. The input impedance of the tube is very high, and the output impedance is very low. There is no phase reversal between the input and output signals.

The corresponding PNP transistor amplifier circuit (Fig. 14-3*b*) shows that the base is the driven element and the input signal is applied between

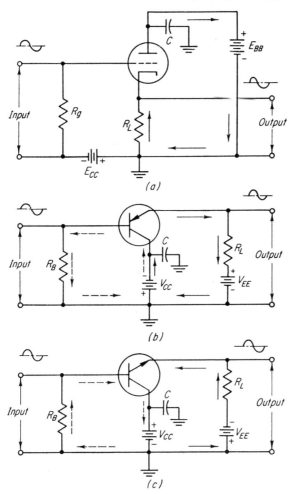

Fig. 14-3. Basic grounded-collector transistor amplifier circuits and the analogous vacuum-tube circuit. (*a*) Grounded plate. (*b*) PNP grounded collector. (*c*) NPN grounded collector.

the base and collector; the output signal is taken off between the emitter and collector. The collector is not d-c grounded, but is at a-c ground by virtue of capacitor C connected between the collector and ground. Thus, the collector is common to both input and output circuits for a-c signals. The emitter is biased in a forward direction by the voltage

V_{EE}, and the collector is biased in a reverse direction by the voltage V_{CC}. The input resistance of the transistor is very high and may be in the order of 150,000 to 600,000 ohms. The output resistance of the transistor is very low and may be in the order of 100 to 1,000 ohms. When a signal is applied to the input circuit, the base current will vary about its average value. A similar variation will take place in the emitter current but with a much greater amplitude. The base-collector circuit is biased with the positive side of V_{CC} connected to the base, and the negative side connected to the collector. When the input signal is positive, its polarity will be the same as that of V_{CC}; consequently the bias voltage V_C will increase and cause a decrease in the emitter and base currents. The decrease in emitter current produces a decrease in the output voltage developed across R_L. As a result the potential at the emitter with reference to the collector will become more positive. By the same reasoning, a negative signal at the input will produce an increase in the output voltage across R_L, and as a result the potential at the emitter will become less positive, or more negative. Thus, a positive-going input signal produces a positive-going output signal, and vice versa. As these signals are in phase, no phase reversal has taken place.

The load impedance required by the grounded-collector type of amplifier is dependent on the input impedance. Because of the low value of the required output impedance, the voltage gain is always less than unity. The current gain is high but the resultant power gain is less than can be obtained from the other two types of amplifier circuits. Because of the high input impedance and low output impedance of a grounded-collector circuit, the circuit is generally used as an impedance-matching device. The grounded-collector circuit can pass a signal in either direction, and the circuit can therefore be used in a two-way amplifier.

14-2. Biasing Methods. *Methods Used.* In order that a transistor may function satisfactorily as an amplifier, it should be operated with proper d-c bias conditions. A constant current should be provided on: (1) the emitter side of a grounded base, (2) the base side of a grounded emitter, or (3) the base side of a grounded collector. The methods used for providing a constant current bias may be classified as: (1) fixed bias, (2) self-bias, and (3) stabilized bias. The collector should be connected to a d-c source of several volts. This voltage is usually applied through an output resistor, or the primary winding of an interstage transformer. The voltage of the collector supply should be equal to the sum of the voltage drop across the resistor or primary winding, and the voltage required at the collector

$$V_{CC} = V_L + V_C \qquad (14\text{-}1)$$

Fixed Bias. Fixed bias may be obtained by use of a direct voltage supply connected in series with a relatively large resistance (Figs. 14-1, 14-2, and 14-3). In the grounded-emitter circuit of Fig. 14-1

$$I_b = \frac{V_{EE}}{R_B + r_e + r_b} \qquad (14\text{-}2)$$

As $r_e + r_b$ is much smaller than R_B,

$$I_b \cong \frac{V_{EE}}{R_B} \qquad (14\text{-}2a)$$

In the grounded-base circuit of Fig. 14-2

$$I_e = \frac{V_{BB}}{R_E + r_e + r_b} \qquad (14\text{-}3)$$

$$I_e \cong \frac{V_{BB}}{R_E} \qquad (14\text{-}3a)$$

In the grounded-emitter circuit of Fig. 14-1b, both the base and collector are negative with respect to the emitter. The bias arrangement may be simplified by eliminating the power supply V_{EE}, as shown in Fig. 14-4. The required base current can be obtained by using a resistor of the proper value in the base circuit, or

$$R_B = \frac{V_{CC}}{I_b} - (r_b + r_e) \qquad (14\text{-}4)$$

However, since the internal base-emitter resistance is usually only a few hundred ohms, and the external base resistance is generally in the hundreds of kilo-ohms, the internal resistance can normally be neglected, and

$$R_B \cong \frac{V_{CC}}{I_b} \qquad (14\text{-}4a)$$

Example 14-1.   A junction transistor is connected as shown in Fig. 14-4, with $V_{CC} = 6$ volts. What approximate value of base resistance is required to operate the transistor with a base current of: (a) 30 μa?   (b) 60 μa?

Given: $V_{CC} = 6$ volts   Find: (a) R_B
          (a) $I_b = 30$ μa   (b) R_B
          (b) $I_b = 60$ μa

Solution:

$$R_B \cong \frac{V_{CC}}{I_b}$$

(a) $R_B \cong \dfrac{6}{30 \times 10^{-6}} \cong 200{,}000$ ohms

(b) $R_B \cong \dfrac{6}{60 \times 10^{-6}} \cong 100{,}000$ ohms

The bias arrangement for the grounded-base amplifier circuit of Fig. 14-2*b* can be simplified by eliminating the power supply V_{BB} and connecting a biasing resistor in the base lead, as shown in Fig. 14-5. The current flowing through this resistor will bias the emitter of the PNP transistor positive with respect to its base, and the emitter of the NPN transistor negative with respect to its base.

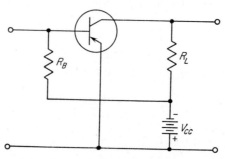

Fig. 14-4. Grounded-emitter amplifier circuit using a single power source.

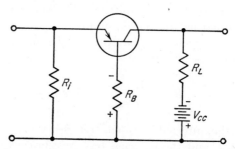

Fig. 14-5. Grounded-base amplifier circuit using a single power source.

Example 14-2.   The junction transistor shown in Fig. 14-5 has a base current of 0.05 ma.   What value of base resistance is required to operate the transistor with an emitter bias of: (*a*) 1.5 volts?   (*b*) 3.0 volts?

Given: $I_b = 0.05$ ma   Find: (*a*) R_B
          (*a*) $V_E = 1.5$ volts   (*b*) R_B
          (*b*) $V_E = 3.0$ volts
Solution:

$$R_B \cong \frac{V_E}{I_b}$$

(*a*) $R_B \cong \dfrac{1.5}{0.05 \times 10^{-3}} \cong 30{,}000$ ohms

(*b*) $R_B \cong \dfrac{3.0}{0.05 \times 10^{-3}} \cong 60{,}000$ ohms

Need For Stabilization.   The current flowing through the collector-base junction consists of two parts.   One part is the collector current, which is due to normal transistor action and is proportional to the emitter current and the current amplification factor.   The other part is the current that flows through the junction when the emitter current is zero; it is called the *collector cutoff current.*   This cutoff current also consists of two parts. One part is the reverse leakage current which is dependent on the collector voltage.   The other part is a current that varies exponentially with temperature and is practically independent of the collector-base voltage.

In a common-base junction transistor circuit, the total collector current is determined by the emitter current, the collector cutoff current, and the current amplification factor α, or

$$I_c = \alpha I_e + I_{co} \qquad (14\text{-}5)$$

Since the collector current is usually in milliamperes and the collector cutoff current in microamperes, any normal increase in temperature will not increase the collector cutoff current to a value that would seriously affect the total collector current.   For this reason the problem of stabilization of the collector current because of temperature change is not an important problem in grounded-base junction transistor circuits.

In a common-emitter junction transistor circuit the total collector current is determined by the base current, the collector cutoff current, and the current amplification factor β, or

$$I_c = \beta I_b + (1 + \beta)I_{co} \qquad (14\text{-}6)$$

This equation shows that any increase in the collector cutoff current will be amplified $(1 + \beta)$ times.   Since β is of the order of 30 to 60, and both the base and collector cutoff currents are in microamperes, a slight increase in the cutoff current will have a significant effect on the total collector current.   An increase in collector current increases the collector power dissipation, thus also increasing the operating temperature, which will cause an increase in the collector cutoff current, which will cause a further increase in the collector current.   If this process is permitted to continue, the total collector current can increase beyond the safe operating value of the transistor, causing it to become damaged.

Self-bias.   It is not generally desirable to operate a transistor with a fixed bias, since it is difficult to maintain a critical base current.   There are two reasons for this difficulty: (1) the variation in transistor units and (2) their sensitivity to temperature change.   One method of partially overcoming this problem in the common-emitter circuit is to connect the base resistor directly to the collector rather than to the supply voltage (Fig. 14-6).   The value of the required base resistance can be determined by substituting the collector voltage for the supply voltage in Eq. (14-4a)

$$R_B \cong \frac{V_c}{I_b} \qquad (14\text{-}7)$$

If the collector current increases because of an increase in junction temperature, the voltage drop across the collector load resistor will also increase, thus decreasing the voltage at the collector.   This decrease in collector voltage causes the base current to decrease, thereby also decreasing the collector current and counteracting its original increase.   This method of obtaining self-bias produces a negative feedback of the a-c

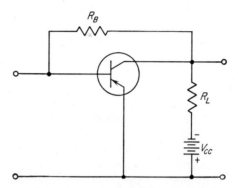

FIG. 14-6. Grounded-emitter amplifier circuit using self-bias.

signal and reduces the effective gain of the amplifier.   Because of this degenerative action this method is not generally used.

Self-bias may also be obtained by connecting a resistor and capacitor between the emitter and ground, as shown in Fig. 14-7.   Capacitor C_E is connected across R_E to prevent degeneration and its resultant loss of gain.

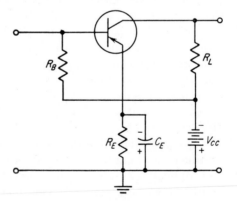

FIG. 14-7. Grounded-emitter amplifier circuit with self-bias, using emitter bias.

This method of obtaining self-bias is called *emitter bias,* and C_E and R_E operate in a similar manner to the capacitor and resistor used in the cathode-bias circuit of a vacuum tube.   The value of R_E is usually less than one-tenth the value of R_B.   The value of C_E will depend on the value of R_E and the lowest frequency to be amplified.   Because of the large ratio between the resistance of R_E and R_B practically all of the collector current will flow through the emitter resistor. This current produces a voltage drop across R_E that serves to make the emitter negative with respect to ground. Since the base is also negative with respect to ground, the base-emitter voltage will be approximately equal to the difference between the battery voltage drop across R_B and the small reverse voltage produced across R_E. If the collector current increases because of an increase in the collector-base junction temperature, the voltage across R_E will also increase, thereby decreasing the base-emitter voltage. This decrease in the forward bias of the base-emitter circuit causes a decrease in both the base and emitter currents, thereby counteracting the increase in collector current. This method of biasing stabilizes the circuit by reducing the effects of temperature drift and also compensates for differences in transistor units.

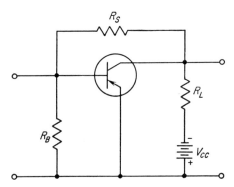

Fig. 14-8. Grounded-emitter amplifier circuit using a voltage divider to obtain both fixed bias and self-bias.

Fixed Bias and Self-bias. An increase in circuit stability with a minimum loss of gain may be obtained by using both fixed and self-bias. A circuit using a voltage divider to obtain both fixed and self-bias is shown in Fig. 14-8. The voltage divider R_SR_B biases the base negative with respect to the emitter. The bleeder current flowing through this voltage divider fixes a bias at the base, with the amount of bias current being determined by the value of R_B, R_S, and V_C. An increase in collector current will decrease the collector voltage, thereby also decreasing the bias current and the emitter-to-base bias. This decrease in the forward bias will cause a decrease in both the emitter and base currents, thus counteracting the original increase in collector current. As with the circuit of Fig. 14-6, this method of obtaining a stabilized bias also produces a degenerative signal that decreases the effective gain of the amplifier.

A more effective means of obtaining stabilization with a minimum loss of gain is shown in Fig. 14-9. The series resistor R_S is returned to the negative terminal of V_{CC}, and the voltage applied to the base is reduced by the voltage-divider network R_BR_S. Connecting a resistor R_E in series with the emitter limits the base current to the desired bias value. Capacitor C_E is connected across R_E to bypass the a-c signals so as to minimize the effects of degeneration. Any increase in collector current also increases the base-emitter voltage, thus reducing the base current. The amount of d-c variation will depend on the value of R_E, a higher value giving the better stabilization. However, the feedback also depends on how constant the base potential can be maintained during changes; and low values of R_S and R_B will therefore also improve stability. When R_E is large and R_S and R_B are so small that the potential at the base is practically equivalent to an effective zero circuit resistance, the circuit is equivalent to the grounded-base arrangement with its inherent high stability. When R_E is small and R_S and R_B very large, the circuit is equivalent to the grounded-emitter arrangement with its low inherent stability. The circuit is very flexible and any desired amount of stability

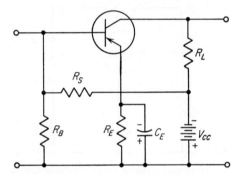

Fig. 14-9. Grounded-emitter amplifier circuit using a voltage divider and an emitter resistor to obtain both fixed bias and self-bias.

can be obtained between these two extremes. A disadvantage of this arrangement is that a resistance equal to the parallel combination of $R_S R_B$ (assuming the resistance of the power source V_{CC} to be zero) is added to the base return circuit, thereby increasing the variation in collector current with changes in I_{co}.

Another method used for obtaining stabilization with a minimum loss of gain is shown in Fig. 14-10. This circuit is a variation of Fig. 14-8. The d-c return resistor R_S is divided into two resistors R_1 and R_2, and all a-c variations are bypassed by capacitor C. The value of R_2 is usually five to ten times that of R_1. Using the principles described for obtaining stabilization with a minimum amount of degeneration, many other circuit arrangements can be made. The voltage gain and circuit stability are usually the determining factors in selecting the circuit to be used.

14-3. Equivalent Circuits. The transistor is an active resistance network that may be represented by three resistors connected in either a three- or four-terminal network. In solving for the current, voltage, or power amplification of a transistor-amplifier circuit, it is important to know the following: (1) input impedance (Z_i), (2) output impedance (Z_o), (3) current-transfer ratio (α, β) and (4) active mutual characteristics of the network (r_m).   The equations used in solving for these parameters in either a three- or four-terminal network are rather complex, and their derivations are beyond the scope of this text.   In addition, a change in the input impedance is reflected into the output circuit, and a change in the load impedance is reflected into the input circuit.   For example, as the load impedance increases, the input impedance decreases, and vice versa.   These equations also vary with frequency, signal input, and type of impedance used in the input and output circuits.

In this chapter two sets of equations will be presented, (1) a complex equation whose number will contain the suffix a, and (2) a simplified

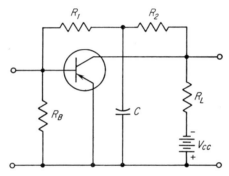

Fig. 14-10. Grounded-emitter amplifier circuit using a divided d-c return and a capacitor for obtaining self-bias and fixed bias.

equation whose number will contain the suffix b.   The complex equations are for a small-signal, low-audio-frequency, four-terminal network having resistive input and output impedances, in which

$$r_m = \alpha\, r_c \tag{14-8}$$

where α is the current-transfer ratio when $R_L = 0$.

$$\alpha_L = \frac{\beta}{1 + \beta} \tag{14-9}$$

where α_L is the current-transfer ratio when R_L has a significant value. The simplified equations assume that: (1) $r_e + r_b$ is much smaller than R_L, (2) R_L is much smaller than r_m, and (3) R_G is much smaller than r_c. The current-transfer ratio α as listed in transistor manuals is for a short-circuited output, which is a usual mode of operation; thus $R_L = 0$.   In solving the complex equations that do not contain the term R_L, this value of α is used.   In solving the complex equations containing the term R_L or the simplified equation in which the term R_L has been dropped, the current-transfer ratio must be calculated to include the effects of this load.

In Figs. 14-11, 14-12, and 14-13, which are the equivalent circuit diagrams for the basic circuits of Figs. 14-1, 14-2, and 14-3, the bias supplies have been omitted; V_G is the source generator having an internal resistance of R_G, and R_L is the load resistance.   The arrows on these diagrams indicate the direction of electron flow in the input and output circuits due to the fixed-bias supplies which are shown in Figs. 14-1, 14-2, and 14-3, but omitted in Figs. 14-11, 14-12, and 14-13 to avoid confusion. As the input signal is a varying voltage, the polarity of V_G has been arbitrarily chosen.   The polarities of V_G should not be associated with the directions of electron flow indicated on the diagrams.   The voltage of

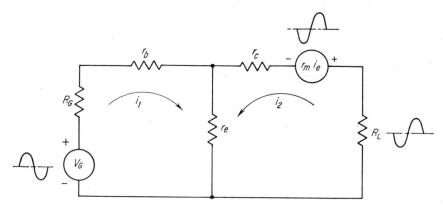

Fig. 14-11. Equivalent four-terminal network for a PNP transistor grounded-emitter amplifier.

the output generator is equal to $r_m i_e$, which is analogous to μe_g in a vacuum-tube circuit, and the polarity shown indicates the phase relationship between the input and output signals.

14-4. Equivalent Circuit for a Grounded-emitter Amplifier.   A four-terminal equivalent circuit for the grounded-emitter amplifier circuit of Fig. 14-1b is shown in Fig. 14-11.   The equations for determining the input resistance, output resistance, current gain, voltage gain, and power gain are

Input resistance.

$$R_i = r_e + r_b + \frac{r_e(r_m - r_e)}{R_L + r_e + r_c - r_m} \qquad (14\text{-}10a)$$

$$R_i \cong r_b + \frac{r_e}{1 - \alpha_L} \qquad (14\text{-}10b)$$

Output resistance.

$$R_o = r_e + r_c - r_m + \frac{r_e(r_m - r_e)}{R_G + r_b + r_e} \qquad (14\text{-}11a)$$

$$R_o \cong r_c(1 - \alpha) + \frac{r_e r_m}{R_G + r_b} \qquad (14\text{-}11b)$$

Current gain.

$$I_g = \frac{r_e - r_m}{R_L + r_e + r_c - r_m} \tag{14-12a}$$

$$I_g \cong \frac{r_m}{R_L + r_c - r_m} \tag{14-12b}$$

Voltage gain.

$$V_g = \frac{R_L(r_m - r_e)}{(r_b + r_e)(R_L + r_c) - r_b(r_m - r_e)} \tag{14-13a}$$

$$V_g \cong \frac{\alpha_L R_L}{r_e + r_b(1 - \alpha_L)} \tag{14-13b}$$

Power gain.

$$P_g = \frac{V_g^2 R_i}{R_L} \tag{14-14a}$$

$$P_g \cong \frac{\alpha_L^2 R_L}{(1 - \alpha_L)[r_e + r_b(1 - \alpha_L)]} \tag{14-14b}$$

Example 14-3.   A certain PNP junction transistor when connected as a grounded emitter has the following circuit parameters: $r_e = 23$ ohms, $r_b = 1,430$ ohms, $r_c = 3.93$ megohms, $R_L = 100,000$ ohms, $\alpha = 0.982$.   Find the input resistance using (*a*) the complex equation, (*b*) the simplified equation.

Given: $r_e = 23$ ohms   Find: (*a*) R_i
              $r_b = 1,430$ ohms   (*b*) R_i
              $r_c = 3.93$ megohms
              $R_L = 100,000$ ohms
              $\alpha = 0.982$

Solution:

$$r_m = \alpha r_c = 0.982 \times 3.93 = 3.86 \text{ megohms}$$

$$I_g = \frac{r_e - r_m}{R_L + r_e + r_c - r_m}$$

$$= \frac{23 - 3.86 \times 10^6}{100,000 + 23 + 3.93 \times 10^6 - 3.86 \times 10^6} = 22.7$$

$$\alpha_L = \frac{\beta}{1 + \beta} = \frac{22.7}{1 + 22.7} = 0.957$$

(*a*) $R_i = r_e + r_b + \dfrac{r_e(r_m - r_e)}{R_L + r_e + r_c - r_m}$

$$= 23 + 1,430 + \frac{23(3.86 \times 10^6 - 23)}{100,000 + 23 + 3.93 \times 10^6 - 3.86 \times 10^6} \cong 1,970 \text{ ohms}$$

(*b*) $R_i \cong r_b + \dfrac{r_e}{1 - \alpha_L} \cong 1,430 + \dfrac{23}{1 - 0.957} \cong 1,965 \text{ ohms}$

Example 14-4.   The internal resistance of the source generator of Example 14-3 is 100 ohms.   Find the output resistance using (*a*) the complex equation, (*b*) the simplified equation.

Given: $R_G = 100$ ohms   Find: (*a*) R_o
        Transistor and circuit of Example 14-3   (*b*) R_o

Solution:

(a) $R_o = r_e^{\cdot} + r_c - r_m + \dfrac{r_e(r_m - r_e)}{R_G + r_b + r_e}$

$= 23 + 3.93 \times 10^6 - 3.86 \times 10^6 + \dfrac{23(3.86 \times 10^6 - 23)}{100 + 1{,}430 + 23} \cong 127{,}000$ ohms

(b) $R_o \cong r_c(1 - \alpha) + \dfrac{r_e r_m}{R_G + r_b}$

$\cong 3.93 \times 10^6(1 - 0.982) + \dfrac{23 \times 3.86 \times 10^6}{100 + 1{,}430} \cong 128{,}000$ ohms

Example 14-5.   Find the voltage gain of the transistor circuit of Example 14-3 using (a) the complex equation, (b) the simplified equation.

Given: Transistor and circuit of Example 14-3   Find: (a) V_g
                                                                            (b) V_g

Solution:

(a) $V_g = \dfrac{R_L(r_m - r_e)}{(r_b + r_e)(R_L + r_c) - r_b(r_m - r_e)}$

$= \dfrac{100{,}000(3.86 \times 10^6 - 23)}{(1{,}430 + 23)(100{,}000 + 3.93 \times 10^6) - 1{,}430(3.86 \times 10^6 - 23)} = 1{,}170$

(b) $V_g \cong \dfrac{\alpha_L R_L}{r_e + r_b(1 - \alpha_L)} \cong \dfrac{0.957 \times 100{,}000}{23 + 1{,}430(1 - 0.957)} \cong 1{,}135$

Example 14-6.   Find the power gain (in decibels) of the transistor circuit of Example 14-3 using (a) the complex equation, (b) the simplified equation.

Given: Transistor and circuit of Example 14-3   Find: (a) P_g
             $V_g = 1{,}170$ (b) P_g
             $R_i = 1{,}970$ ohms

Solution:

(a) $P_g = \dfrac{V_g^2 R_i}{R_L} = \dfrac{1{,}170 \times 1{,}170 \times 1{,}970}{100{,}000} \cong 27{,}000$

$P_g = 10 \times \log P_g = 10 \times \log 27{,}000 = 44.31$ db

(b) $P_g \cong \dfrac{\alpha_L^2 R_L}{(1 - \alpha_L)[r_e + r_b(1 - \alpha_L)]}$

$\cong \dfrac{0.957 \times 0.957 \times 100{,}000}{(1 - 0.957)[23 + 1{,}430(1 - 0.957)]} \cong 25{,}200$

$P_g \cong 10 \times \log P_g \cong 10 \times \log 25{,}200 \cong 44$ db

14-5. Equivalent Circuit for a Grounded-base Amplifier. A four-terminal equivalent circuit for the grounded-base amplifier circuit of Fig. 14-2b is shown in Fig. 14-12.   The equations for determining the input resistance, output resistance, current gain, voltage gain, and power gain are as follows:

Input resistance.

$$R_i = r_b + r_e - \frac{r_b(r_b + r_m)}{R_L + r_b + r_c} \qquad (14\text{-}15a)$$

$$R_i \cong r_e + r_b(1 - \alpha_L) \qquad (14\text{-}15b)$$

where $\alpha_L = I_g$ (for grounded-base amplifier)

Output resistance.

$$R_o = r_b + r_c - \frac{r_b(r_b + r_m)}{R_G + r_b + r_e} \qquad (14\text{-}16a)$$

$$R_o \cong r_c - \frac{r_b r_m}{R_G + r_b + r_e} \qquad (14\text{-}16b)$$

Current gain.

$$I_g = \frac{r_b + r_m}{r_b + r_c + R_L} \qquad (14\text{-}17a)$$

$$I_g \cong \frac{r_m}{r_c + R_L} \qquad (14\text{-}17b)$$

Voltage gain.

$$V_g = \frac{R_L(r_b + r_m)}{(R_G + r_b + r_e)(R_L + r_b + r_c) - r_b(r_b + r_m)} \qquad (14\text{-}18a)$$

$$V_g \cong \frac{\alpha_L R_L}{R_G + r_e + r_b(1 - \alpha_L)} \qquad (14\text{-}18b)$$

Power gain.

$$P_g = \frac{V_g{}^2 R_i}{R_L} \qquad (14\text{-}19a)$$

$$P_g \cong \frac{\alpha_L{}^2 R_L}{R_G + r_e + r_b(1 - \alpha_L)} \qquad (14\text{-}19b)$$

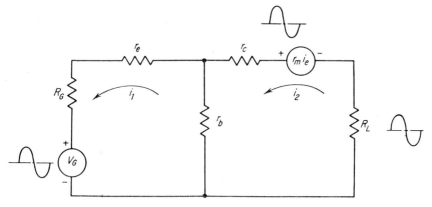

FIG. 14-12. Equivalent four-terminal network for a PNP transistor grounded-base amplifier.

Example 14-7.   The PNP junction transistor used in Example 14-3 is connected with a grounded base and an output load resistance of 500,000 ohms.   Find the input resistance using (*a*) complex equation, (*b*) simplified equation.

Given: $R_L = 500,000$ ohms   Find: (*a*) R_i
         Transistor of Example 14-3   (*b*) R_i

Solution:

$$I_g = \frac{r_b + r_m}{r_b + r_c + R_L} = \frac{1,430 + 3.86 \times 10^6}{1,430 + 3.93 \times 10^6 + 500,000} = 0.871$$

(a) $\quad R_i = r_b + r_e - \dfrac{r_b(r_b + r_m)}{R_L + r_b + r_c}$

$\qquad = 1{,}430 + 23 - \dfrac{1{,}430(1{,}430 + 3.86 \times 10^6)}{5 \times 10^5 + 1{,}430 + 3.93 \times 10^6} = 206$ ohms

(b) $\quad R_i \cong r_e + r_b(1 - \alpha_L) \cong 23 + 1{,}430(1 - 0.871) \cong 207$ ohms

Example 14-8. The internal resistance of the source generator of Example 14-7 is 50 ohms. Find the output resistance using (a) the complex equation, (b) the simplified equation.

Given: Transistor and circuit of Example 14-7   Find: (a) R_o
$\qquad\qquad\quad R_G = 50$ ohms   (b) R_o

Solution:

(a) $\quad R_o = r_b + r_c - \dfrac{r_b(r_b + r_m)}{R_G + r_b + r_e}$

$\qquad = 1{,}430 + 3.93 \times 10^6 - \dfrac{1{,}430(1{,}430 + 3.86 \times 10^6)}{50 + 1{,}430 + 23} \cong 260{,}000$ ohms

(b) $\quad R_o \cong r_c - \dfrac{r_b r_m}{R_G + r_b + r_e}$

$\qquad \cong 3.93 \times 10^6 - \dfrac{1{,}430 \times 3.86 \times 10^6}{50 + 1{,}430 + 23} \cong 260{,}000$ ohms

Example 14-9. Find the voltage gain of the junction transistor circuit of Example 14-7 using (a) complex equation, (b) simplified equation.

Given: Transistor and circuit of Example 14-7   Find: (a) V_g
                                                                  (b) V_g

Solution:

(a) $\quad V_g = \dfrac{R_L(r_b + r_m)}{(R_G + r_b + r_e)(R_L + r_b + r_c) - r_b(r_b + r_m)}$

$\qquad = \dfrac{500{,}000(1{,}430 + 3.86 \times 10^6)}{(50+1{,}430+23)(500{,}000 + 1{,}430 + 3.93\times10^6) - 1{,}430(1{,}430+3.86\times10^6)}$

$\qquad \cong 1{,}700$

(b) $\quad V_g \cong \dfrac{\alpha_L R_L}{R_G + r_e + r_b(1 - \alpha_L)}$

$\qquad \cong \dfrac{0.871 \times 500{,}000}{50 + 23 + 1{,}430(1 - 0.871)} \cong 1{,}690$

Example 14-10. Find the power gain (in decibels) of the junction-transistor circuit of Example 14-7 using (a) the complex equation, (b) the simplified equation.

Given: Transistor and circuit of Example 14-7   Find: (a) P_g
$\qquad\qquad\quad V_g = 1{,}700$ (b) P_g
$\qquad\qquad\quad R_i = 206$

Solution:

(a) $\quad P_g = \dfrac{V_g{}^2 R_i}{R_L} = \dfrac{1{,}700 \times 1{,}700 \times 206}{500{,}000} \cong 1{,}190$

$\qquad P_g = 10 \times \log P_g \cong 10 \times \log 1{,}190 \cong 30.75$ db

(b) $\quad P_g \cong \dfrac{\alpha_L{}^2 R_L}{R_G + r_e + r_b(1 - \alpha_L)}$

$\qquad \cong \dfrac{0.871 \times 0.871 \times 500{,}000}{50 + 23 + 1{,}430(1 - 0.871)} \cong 1{,}470$

$\qquad P_g \cong 10 \times \log P_g \cong 10 \times \log 1{,}470 \cong 31.67$ db

14-6. Equivalent Circuit for a Grounded-collector Amplifier. A four-terminal equivalent circuit for the grounded-collector amplifier circuit of Fig. 14-3b is shown in Fig. 14-13. The equations for determining the

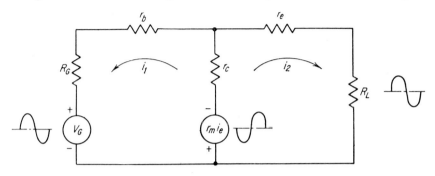

Fig. 14-13. Equivalent four-terminal network for a PNP transistor grounded-collector amplifier.

input resistance, output resistance, current gain, voltage gain, and power gain are as follows:

Input resistance.

$$R_i = r_b + r_c - \frac{r_c(r_c - r_m)}{R_L + r_e + r_c - r_m} \qquad (14\text{-}20a)$$

$$R_i \cong \frac{r_b + R_L}{1 - \alpha_L} \qquad (14\text{-}20b)$$

Output resistance.

$$R_o = r_e + \frac{(r_b + R_G)(r_c - r_m)}{R_G + r_b + r_c} \qquad (14\text{-}21a)$$

$$R_o \cong r_e + (r_b + R_G)(1 - \alpha) \qquad (14\text{-}21b)$$

Current gain.

$$I_g = \frac{r_c}{R_L + r_e + r_c - r_m} \qquad (14\text{-}22a)$$

$$I_g \cong \frac{r_c}{R_L + r_c - r_m} \qquad (14\text{-}22b)$$

Voltage gain.

$$V_g = \frac{r_c R_L}{(R_G + r_b + r_c)(R_L + r_e + r_c - r_m) - r_c(r_c - r_m)} \qquad (14\text{-}23a)$$

$$V_g \cong 1 \qquad (14\text{-}23b)$$

Power gain.

$$P_g = \frac{V_g{}^2 R_i}{R_L} \qquad (14\text{-}24a)$$

$$P_g \cong \frac{1}{1 - \alpha_L} \qquad (14\text{-}24b)$$

Example 14-11. The PNP junction transistor used in Example 14-3 is connected with a grounded collector, and an output load resistance of 10,000 ohms. Find the input resistance using (*a*) the complex equation, (*b*) the simplified equation.

Given: Transistor of Example 14-3   Find: (*a*) R_i
$\qquad\qquad R_L = 10,000$ ohms $\qquad\qquad\qquad\qquad$ (*b*) R_i

Solution:

$$I_g = \frac{r_c}{R_L + r_e + r_c - r_m}$$

$$= \frac{3.93 \times 10^6}{10,000 + 23 + 3.93 \times 10^6 - 3.86 \times 10^6} \cong 49$$

$$\alpha_L = \frac{\beta}{\beta + 1} \cong \frac{49}{49 + 1} \cong 0.98$$

(*a*) $R_i = r_b + r_c - \dfrac{r_c(r_c - r_m)}{R_L + r_e + r_c - r_m}$

$$= 1,430 + 3.93 \times 10^6 - \frac{3.93 \times 10^6 (3.93 \times 10^6 - 3.86 \times 10^6)}{10,000 + 23 + 3.93 \times 10^6 - 3.86 \times 10^6}$$

$$\cong 490,000 \text{ ohms}$$

(*b*) $R_i = \dfrac{r_b + R_L}{1 - \alpha_L} = \dfrac{1,430 + 10,000}{1 - 0.98} = 571,500$ ohms

Example 14-12. The internal resistance of the source generator of Example 14-11 is 100 ohms. Find the output resistance using (*a*) the complex equation, (*b*) the simplified equation.

Given: Transistor and circuit of Example 14-11   Find: (*a*) R_o
$\qquad\qquad\qquad\qquad\qquad\qquad\qquad\qquad\qquad\qquad\qquad\qquad$ (*b*) R_o

Solution:

(*a*) $R_o = r_e + \dfrac{(r_b + R_G)(r_c - r_m)}{R_G + r_b + r_c}$

$$= 23 + \frac{(1,430 + 100)(3.93 \times 10^6 - 3.86 \times 10^6)}{100 + 1,430 + 3.93 \times 10^6} \cong 50 \text{ ohms}$$

(*b*) $R_o \cong r_e + (r_b + R_G)(1 - \alpha) \cong 23 + (1,430 + 100)(1 - 0.982) \cong 50.5$ ohms

Example 14-13. Find the voltage gain of the junction transistor circuit of Example 14-11 using (*a*) the complex equation, (*b*) the simplified equation.

Given: Transistor and circuit of Example 14-11   Find: (*a*) V_g
$\qquad\qquad\qquad\qquad\qquad\qquad\qquad\qquad\qquad\qquad\qquad\qquad$ (*b*) V_g

Solution:

(*a*) $V_g = \dfrac{r_c R_L}{(R_G + r_b + r_c)(R_L + r_e + r_c - r_m) - r_c(r_c - r_m)}$

$$= \frac{3.93 \times 10^6 \times 10,000}{(100 + 1,430 + 3.93 \times 10^6)(10,000 + 23 + 3.93 \times 10^6 - 3.86 \times 10^6) - 3.93 \times 10^6 (3.93 \times 10^6 - 3.86 \times 10^6)}$$

$$\cong 1$$

(*b*) $V_g \cong 1$

Example 14-14. Find the power gain (in decibels) of the junction transistor circuit of Example 14-11 using (*a*) the complex equation, (*b*) the simplified equation.

Given: Transistor and circuit of Example 14-11   Find: (*a*) P_g
$\qquad\qquad V_g = 1 \qquad\qquad\qquad\qquad\qquad\qquad\qquad\qquad$ (*b*) P_g
$\qquad\qquad R_i = 490,000$ ohms

Solution:

(a) $P_g = \dfrac{V_g^2 R_i}{R_L} = \dfrac{1 \times 1 \times 490,000}{10,000} = 49$

   $P_g = 10 \times \log P_g = 10 \times \log 49 = 16.9 \text{ db}$

(b) $P_g \cong \dfrac{1}{1 - \alpha_L} = \dfrac{1}{1 - 0.98} = 50$

   $P_g \cong 10 \times \log P_g \cong 10 \times \log 50 \cong 17 \text{ db}$

14-7. Methods of Coupling. *Methods Used.* Two transistor amplifier stages may be coupled to one another by using: (1) transformer coupling, (2) resistance-capacitance (*RC*) coupling, (3) impedance coupling, or (4) direct coupling. These methods of coupling transistor amplifier stages are basically similar to the methods used for coupling vacuum-tube-amplifier stages. A major difference is that, compared to vacuum tubes,

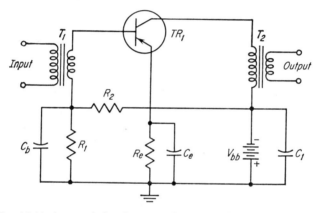

Fig. 14-14. A grounded-emitter transformer-coupled amplifier circuit.

the input impedance varies widely with (1) the type of transistor, (2) the type of circuit, and (3) the circuit parameters.

Transformer Coupling. A grounded-emitter transformer-coupled amplifier circuit is shown in Fig. 14-14. With this method of coupling both the input and output impedances of the transistor can be matched for maximum power gain. The collector-emitter circuit of the preceding stage is coupled to the base-emitter circuit of this stage by means of the step-down transformer T_1. Although the voltage across the secondary of this transformer indicates a loss in voltage, the step-down transformer provides an excellent means for obtaining maximum power transfer. Since the transistor is a current-operated device, a power gain is produced across the primary of transformer T_2.

The proper bias is obtained by the voltage divider $R_1 R_2$, which is bypassed by capacitor C_b to minimize signal attenuation. Variations in the transistor, or its circuit elements, are absorbed automatically without any adverse effects by means of the stabilizing resistor R_e. This resistor is bypassed by capacitor C_e to prevent degeneration and its resulting loss of gain. Feedback of a-c signals through the common power supply is reduced to a minimum by the bypass capacitor C_1 connected across V_{bb}.

Transformer coupling may be used where good selectivity is required, as in r-f and i-f amplifiers. They may also be used in large-signal audio amplifiers for the driver and output stages. With low-level audio amplifiers RC coupling is generally used, as transformers are more likely to pick up hum.

RC Coupling. A ground-emitter RC-coupled amplifier circuit is shown in Fig. 14-15. Proper bias is obtained by resistor R_b, stabilization by resistor R_e, prevention of degeneration by capacitor C_e, and reduction of signal feedback by capacitor C_1, in a manner similar to the transformer-coupled amplifier circuit. The additional components are the load

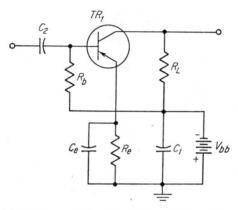

FIG. 14-15. A grounded-emitter RC-coupled amplifier circuit.

resistor R_L and coupling capacitor C_2. The value of R_L varies with the transistor used and the impedance match required to produce the desired gain. Because the output and input resistances are quite low in value, capacitances in the order of 2 to 10 μf are required. Electrolytic capacitors are commonly used to provide the relatively high values of capacitance required for the coupling capacitors in transistor audio-amplifier circuits. Electrolytic capacitors may be used in transistor applications because transistors are not as sensitive to leakage currents as are electron tubes.

Impedance Coupling. A grounded-emitter impedance-coupled amplifier circuit is shown in Fig. 14-16. This circuit is similar to the RC-coupled amplifier circuit except that an inductor L is used in place of the load resistor R_L. This type of coupling is used for frequencies above the a-f range, as the load inductance then permits a fairly good impedance match to be obtained at these frequencies. Series and shunt peaking coils may be used with an impedance-coupled circuit in the same manner as they are used in electron-tube circuits.

Direct Coupling.   In some applications it becomes desirable to connect a load device directly in series with the output electrode of the transistor (Fig. 14-17a).   The load device may be (1) a headphone, (2) a d-c meter, (3) a d-c relay, (4) a high-impedance loudspeaker, or (5) a neon lamp. This method of connection is permissible (1) when the d-c component of the output current does not interfere with the normal operation of the

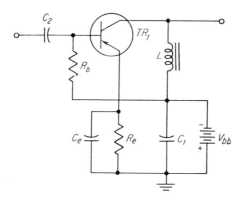

Fig. 14-16. A grounded-emitter impedance-coupled amplifier circuit.

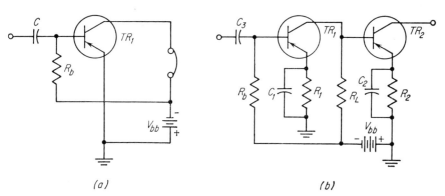

(a)   (b)

Fig. 14-17. Direct-coupled amplifier circuits.   (a) Simple circuit.   (b) Common resistor circuit.

device and (2) when the resistance of the device is not high enough to appreciably reduce the electrode voltage.

Direct coupling is also used with d-c amplifiers, or where inductors or capacitors cannot be used as the coupling element.   In the circuit of Fig. 14-17b, the resistor R_L serves as both the collector load for TR_1 and the bias resistor for TR_2.   In direct-coupled circuits the magnitude of the source and load impedances should be comparable to the input and output impedances of the transistor.   That is, the resistance of the source generator should be equal to, or less than, the input resistance of the transistor, and the resistance of the load should be equal to, or higher than, the output resistance of the transistor.

An NPN transistor may be coupled to a similar PNP transistor (or vice versa) by means of direct coupling (Fig. 14-18). This method of coupling takes advantage of the fact that the two types of transistors are symmetrical counterparts of each other. For each operating point in an NPN transistor there is an equivalent operating point in a similar PNP transistor. The polarity of the input signal required to increase conduction in an NPN transistor is opposite to that required to increase conduction in a PNP transistor. Therefore, a circuit using this method of coupling is called a *complementary symmetrical circuit*. In Fig. 14-18, the circuit parameters in the first stage are connected so that the collector

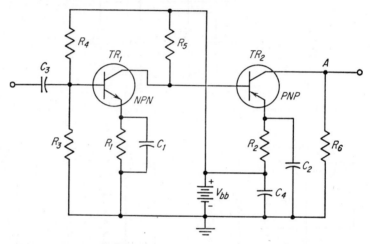

Fig. 14-18. Direct-coupled amplifier circuit using complementary symmetry.

current from TR_1 flowing through the load resistor R_5 develops the amount of voltage required to make the base of TR_2 negative with respect to its emitter. The emitter-base circuit of TR_2 will then be biased in a forward direction. When a positive signal is applied at the base of TR_1, an amplified negative signal will be developed across R_5, as the collector terminal of this resistor is negative with respect to its battery terminal. This increase in negative voltage across R_5 causes the forward bias of TR_2 to increase, thus causing an increase in the collector current flowing through load resistor R_6. The direction of electron flow through this resistor is such that terminal A is positive with respect to ground. Thus, when a signal is applied to the input of a complementary symmetrical circuit, it will appear at the output of the second stage in an amplified form and of the same polarity.

14-8. Multistage Amplifiers. *Requirements.* Transistor amplifier stages may be cascaded following in general the same principles employed in multistage electron-tube amplifier circuits. Any desirable combination of grounded-emitter, grounded-base, or grounded-collector stages may be used to increase the voltage and/or power gain. The maximum power gain of a transistor amplifier is equal to the product of the maximum current gain and maximum voltage gain. The maximum theoretical current gain is obtained when $R_L = 0$, and the maximum theoretical voltage gain is obtained when $R_L = \infty$. Since these two conditions oppose one another, the maximum power gain will occur when (1) the input impedance of the transistor circuit equals the source impedance, $R_i = R_G$ and (2) the output impedance of the transistor circuit equals the load impedance, $R_o = R_L$.

The relative impedance match of the output circuit of a transistor with the input circuit of another transistor for various common electrode connections is indicated in Table 14-1. A good match is obtained when

TABLE 14-1. RELATIVE IMPEDANCE MATCHING OF TRANSISTORS

From	To		
	Common emitter	Common base	Common collector
Common emitter	Fair	Poor	Good
Common base	Poor to fair	Poor	Good
Common collector	Good	Good	Poor

the two impedances are of the same order of magnitude. From the impedance-matching characteristics listed in this table the following conclusions may be reached: (1) Where the impedance mismatch is poor, it is necessary to use an impedance-matching device, such as a transformer. (2) Where the impedance match is fair, acceptable performance may be obtained without the aid of an impedance-matching device. (3) Where the impedance match is good, direct coupling may be used to produce good results. Thus, in practical RC-coupled amplifier circuit applications the maximum value of power gain cannot be achieved. The decrease in gain obtained with RC coupling can be made up by using additional stages to produce the desired over-all gain. RC coupling is generally used in a-f amplifier circuits because of: (1) the simplicity of an RC network, (2) the flat over-all response that can be obtained over a wide band of frequencies, and (3) its low cost.

The high output impedance of a preceding stage can be matched to the low input impedance of a transistor circuit by using special transformers having the required step-down impedance ratio. Interstage transformers are generally used in the r-f and i-f stages of radio receivers in order to meet the tuning and bandwidth requirements of these stages in addition to their impedance requirements. A transformer is also used to match the high impedance of the output circuit of the last audio stage with the low impedance of the voice coil in the loudspeaker.

When connecting RC-coupled transistor amplifier circuits in cascade the following general rules may be used to determine the type of common electrode amplifier circuit to use for the various stages.

First stage. (1) When R_{G1} (resistance of source generator to first stage) is less than 500 ohms, use either the grounded-emitter or grounded-base circuit. (2) When R_{G1} ranges between 500 and 1,500 ohms, use the grounded-emitter circuit. (3) When R_{G1} has a value greater than 1,500 ohms use the grounded-collector circuit.

Intermediate stages. A grounded-emitter circuit is generally used for these stages.

Final stage. (1) When R_{Lf} (load resistance of the final stage) is less than 10,000 ohms, use a grounded-collector circuit. (2) When R_{Lf} ranges between 10,000 and 500,000 ohms, use a grounded-emitter circuit. (3) When R_{Lf} has a value greater than 500,000 ohms, use a grounded-base circuit.

Over-all Gain. The over-all, or total, current gain of a multistage transistor amplifier is equal to the product of the current gain for each stage, or

$$I_{gt} = I_{g1} \times I_{g2} \times I_{g3} \times \cdots \tag{14-25}$$

The over-all, or total, power gain may then be expressed as

$$P_{gt} = \frac{4R_{G1}R_{Lf}I_{gt}^2}{(R_{G1} + R_{i1})^2} \tag{14-26}$$

where R_{i1} is the input resistance of the first stage.

The procedure to be followed in calculating the over-all power gain of a multistage transistor RC-coupled amplifier circuit can best be explained by an example.

Example 14-15. Three transistors, each one similar to the transistors used in Examples 14-3 to 14-14, are connected as shown in Fig. 14-19. Find the over-all power gain (in decibels) of this three-stage amplifier circuit.

Given: $r_e = 23$ ohms   Find: P_{gt}
$r_b = 1,430$ ohms
$r_c = 3.93$ megohms
$r_m = 3.86$ megohms
$R_{Lf} = 10,000$ ohms
$R_{G1} = 100$ ohms

Solution:

As only the value of over-all power gain for the three-stage transistor RC-coupled amplifier is desired, the biasing arrangements, coupling networks, etc., have been omitted from Fig. 14-19.

Since the input resistance of a transistor stage is the output resistance of the previous stage it is necessary to start the calculations for the over-all gain with the final stage.

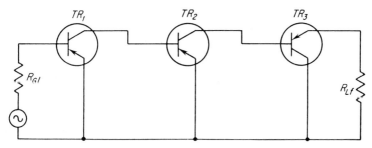

FIG. 14-19. Three-stage transistor amplifier.

This stage is connected as a grounded-collector amplifier circuit, and both the current gain and input resistance will be the same as those obtained for this transistor in Example 14-11.

$$I_{g3} = 49 \qquad R_{i3} = 490,000 \text{ ohms}$$

The input resistance of the third stage is the output resistance of the second stage; therefore $R_{L2} = 490,000$ ohms. The second stage is connected as a grounded-emitter amplifier, and

$$I_{g2} = \frac{r_e - r_m}{R_{L2} + r_e + r_c - r_m}$$

$$= \frac{23 - 3.86 \times 10^6}{490,000 + 23 + 3.93 \times 10^6 - 3.86 \times 10^6} = 6.9$$

$$R_{i2} = r_e + r_b + \frac{r_e(r_m - r_e)}{R_{L2} + r_e + r_c - r_m}$$

$$= 23 + 1,430 + \frac{23(3.86 \times 10^6 - 23)}{490,000 + 23 + 3.93 \times 10^6 - 3.86 \times 10^6}$$

$$= 1,611 \text{ ohms}$$

The input resistance of the second stage is the output resistance of the first stage; therefore $R_{L1} = 1,611$ ohms. The first stage is connected as a grounded-emitter amplifier, and

$$I_{g1} = \frac{23 - 3.86 \times 10^6}{1,611 + 23 + 3.93 \times 10^6 - 3.86 \times 10^6} = 54$$

$$R_{i1} = 23 + 1,430 + \frac{23(3.86 \times 10^6 - 23)}{1,611 + 23 + 3.93 \times 10^6 - 3.86 \times 10^6} = 2,695 \text{ ohms}$$

The over-all current gain is then equal to

$$I_{gt} = I_{g1} \times I_{g2} \times I_{g3} = 54 \times 6.9 \times 49 \cong 18,300$$

The over-all power gain is then equal to

$$P_{gt} = \frac{4R_{G1}R_{Lf}I_{gt}^2}{(R_{G1} + R_{i1})^2} = \frac{4 \times 100 \times 10,000 \times 18,300 \times 18,300}{(100 + 2,695)^2} \cong 172,000,000$$

$$P_{gt} = 10 \times \log P_{gt} = 10 \times \log 172,000,000 = 10 \times 8.2355 \cong 82.3 \text{ db}$$

Decoupling Network. When RC coupling is used in multistage amplifiers, it is necessary to decouple one or more stages in order to prevent feedback (Art. 6-19). One of the methods used is the decoupling network (Fig. 14-20), consisting of the series resistor R_7 and the bypass capacitor C_4. The values of R_7 and C_4 are determined by the time constant required to satisfactorily bypass the lowest frequency to be amplified. The value of R_7 should be low in order to prevent any appreciable decrease in the voltage supplied to the preceding stages. The value of

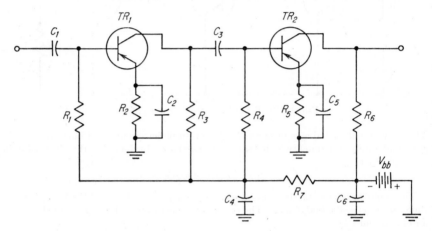

Fig. 14-20. Two-stage RC-coupled grounded-emitter amplifier.

C_4 is therefore made large, and may be in the order of 100 μf. The time constant of the decoupling network is generally made equal to the reciprocal of the lowest frequency to be amplified; thus

$$R_D C_D = \frac{10^6}{f_l} \qquad (14\text{-}27)$$

Example 14-16. Two similar transistors each having a base bias current of 250 μa are connected as shown in Fig. 14-20. The maximum amount of voltage that can be lost across the decoupling resistor is 0.25 volt. Find the value of decoupling capacitance required to pass (*a*) 20 cycles, (*b*) 100 cycles.

Given:　　$V_D = 0.25$ volt　　　　Find:　(*a*) C_D
　　　　　　$I_b = 250$ μa　　　　　　　　(*b*) C_D
　　　　(*a*) $f_l = 20$ cycles
　　　　(*b*) $f_l = 100$ cycles

Solution:

$$R_D = \frac{V_D}{2I_b} = \frac{0.25}{2 \times 250 \times 10^{-6}} = 500 \text{ ohms}$$

(*a*)　$C_D = \dfrac{10^6}{f_l R_D} = \dfrac{10^6}{20 \times 500} = 100$ μf

(*b*)　$C_D = \dfrac{10^6}{f_l R_D} = \dfrac{10^6}{100 \times 500} = 20$ μf

Frequency Response of an RC-coupled Amplifier. The frequency response of an *RC*-coupled amplifier can be analyzed from its equivalent circuits for the low, medium, and high audio frequencies. In the mid-frequency range, the reactance of the coupling capacitor is very low compared with the input resistance to the next stage, and therefore its

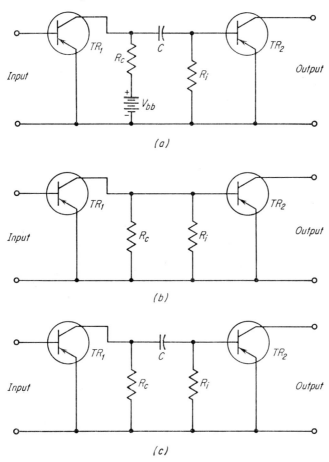

(a)

(b)

(c)

Fig. 14-21. Frequency response circuits. (*a*) Basic *RC*-coupled grounded-emitter amplifier circuit. (*b*) Equivalent circuit for intermediate audio frequencies. (*c*) Equivalent circuit for low audio frequencies.

effects can be neglected for this frequency range. The equivalent circuit for the medium frequency range for a grounded-emitter amplifier may then be drawn as shown in Fig. 14-21*b*. The output current from TR_1 divides between R_c and R_i, thus reducing the over-all current gain at the medium frequencies by the factor

$$K_M = \frac{R_c}{R_c + R_i} \qquad (14\text{-}28)$$

where R_c = collector load resistance, ohms
$\quad\quad R_i$ = input resistance to TR_2, ohms

Example 14-17. Find the over-all current gain at the medium frequencies for the first stage of amplification in Example 14-15. The collector load resistance is 4,000 ohms.

Given: R_c = 4,000 ohms   Find: I_{gM}
$\quad\quad TR_1$ of Example 14-15
$\quad\quad I_{g1}$ = 54
$\quad\quad R_{i2}$ = 1,611 ohms

Solution:

$$I_{gM} = I_{g1}\frac{R_c}{R_c + R_{i2}} = 54\,\frac{4,000}{4,000 + 1,611} \cong 38.5$$

Theoretically the current gain at the low audio frequencies will be lower than for the intermediate-frequency range because of the reactance introduced by the coupling capacitor at these frequencies (Fig. 14-21c). The output current from TR_1 will be reduced at the low audio frequencies by the factor

$$K_L = \frac{R_c}{\sqrt{(R_c + R_i)^2 + X_c^2}} \tag{14-29}$$

Much higher values of capacitance are used for coupling transistor a-f amplifiers than for coupling vacuum-tube a-f amplifiers. Because coupling capacitors are of the order of 10 μf, the current loss is very small, and for most practical purposes the low frequency gain can be considered as being equal to the current gain obtained at the intermediate audio frequencies.

Example 14-18. When the coupling capacitance is equal to 10 μf, find the over-all current gain of the transistor and circuit of Example 14-17 at (a) 100 cycles, (b) 20 cycles.

Given: Transistor and circuit Example 14-17   Find: (a) I_{gL}
$\quad\quad C$ = 10 μf   (b) I_{gL}
$\quad\quad$(a) f = 100 cycles
$\quad\quad$(b) f = 20 cycles

Solution:

$$I_{gL} = I_{g1}\frac{R_c}{\sqrt{(R_c + R_{i2})^2 + X_c^2}}$$

(a) $X_c = \dfrac{10^6}{2\pi fC} = \dfrac{10^6}{6.28 \times 100 \times 10} = 159$ ohms

$\quad I_{gL} = 54\,\dfrac{4,000}{\sqrt{(4,000 + 1,611)^2 + 159^2}} = 38.4$

(b) $X_c = \dfrac{10^6}{6.28 \times 20 \times 10} = 795$ ohms

$\quad I_{gL} = 54\,\dfrac{4,000}{\sqrt{(4,000 + 1,611)^2 + 795^2}} = 38.1$

The frequency response at the high audio frequencies is complicated by the internal reactances and diffusion times of the transistors, and hence an exact equivalent circuit cannot be readily drawn. Because of these complex factors, the calculations for the frequency response at the high audio frequencies is beyond the scope of this text. In general, the response at the high audio frequencies is sufficiently uniform to permit the use of RC-coupled transistor amplifier circuits in audio-amplifier circuit applications.

14-9. Power Amplifiers. *Classifications.* In a manner similar to vacuum-tube power-amplifier circuits, transistor power-amplifier circuits may be divided into two basic types: (1) single-ended, and (2) push-pull. In addition they may also be classified according to their mode of operation, such as Class A, Class B, Class AB, etc. Except for the fact that the output current in a transistor is the collector current and in a vacuum tube it is the plate current, these modes of operation have the same meaning for a transistor amplifier circuit as they have for a vacuum-tube amplifier circuit. Thus, a Class A transistor amplifier circuit indicates that the transistor is biased so that (1) collector current flows for the entire cycle, and (2) the signal current swings an equal amount in both positive and negative directions about the bias point.

Maximum Collector Power Dissipation. A transistor used as a power amplifier should be capable of safely handling the power dissipated as heat at the collector. Three methods used to aid in the removal of this heat are: (1) equipping the transistor with radiating fins, (2) enclosing the transistor in a metal case, and (3) attaching the transistor case directly to the chassis. The maximum power that a transistor can safely dissipate at its collector is listed in transistor manuals for definite ambient temperatures, and its symbol is P_c. A constant-collector power-dissipation curve can be drawn across a family of $V_c I_c$ characteristic curves to show the collector-voltage and collector-current intersections for its maximum rated power dissipation. The constant-collector power dissipation for the transistor having the $V_c I_c$ characteristics shown in Fig. 14-22 is indicated by the dashed-line curve which bends across the entire family of curves. The rated collector power dissipation for this transistor at an ambient temperature of 55°C is 35 mw. Hence, at any point of intersection of the constant-collector power-dissipation curve with the abscissa and ordinate, the product of V_c and I_c as indicated by this intersection equals 35 mw. The area of the graph below and to the left of this curve encloses all points of operation which are within the collector power-dissipation rating of the transistor.

Class A Amplifiers. A typical single-ended Class A power amplifier that may be used to drive a loudspeaker is shown in Fig. 14-23. In this circuit: (1) C_1 couples this circuit to the previous circuit, (2) R_1 biases the base-emitter circuit in a forward direction by making the emitter positive with respect to the base, (3) R_2 stabilizes the emitter current, (4) C_2 bypasses the a-c component in the emitter resistor, (5) C_3 reduces the a-c feedback through the common power supply, (6) C_4 limits the frequency bandwidth in order to minimize high frequency distortion, and (7) T_1 matches the output resistance of TR_1 to the resistance of the voice coil in the loudspeaker. The power output of this type of amplifier circuit is limited, as the maximum efficiency of a Class A transistor amplifier without a stabilizing resistor is 50 per cent. Adding a stabilizing resistor to the circuit causes a portion of the d-c collector power to be dissipated across this resistor, thus decreasing the over-all efficiency of the amplifier circuit. The decrease in the over-all efficiency depends on the ratio of the stabilizing resistance to the load resistance. For maximum protection against variations in the collector current R_2 is made equal to R_L, and the over-all efficiency is reduced to 25 per cent. However, for normal operation of transistors this high percentage of stability is usually not required, and a 5 to 10 per cent loss of d-c collector power across the stabilizing resistor is usually sufficient to produce satisfactory results. The value of the coupling capacitor is determined by the lowest frequency to be amplified. As with vacuum-tube circuits, a loss of 3 db is usually permitted; at this value the reactance of C_1 is equal to the input resistance of the amplifier stage.

Example 14-19. A PNP transistor is connected as shown in Fig. 14-23 and has the following circuit parameters; $V_{bb} = 9$ volts, $I_b = 150$ μa, $R_L = 560$ ohms. Find: (a) the value of the biasing resistor R_1, (b) the value of the stabilizing resistor R_2 if a 20 per cent power loss is permitted.

Given: $V_{bb} = 9$ volts   Find: (a) R_1
         $I_b = 150$ μa   (b) R_2
         $R_L = 560$ ohms
         $P_{R2} = 20$ per cent loss

Solution:

(a) $R_1 = \dfrac{V_{bb}}{I_b} = \dfrac{9}{150 \times 10^{-6}} = 60{,}000$ ohms

(b) $R_2 = \dfrac{\% \text{ loss} \times R_L}{100 - \% \text{ loss}} = \dfrac{20 \times 560}{100 - 20} = 140$ ohms

Example 14-20. The amplifier of Example 14-19 has an input resistance of 2,000 ohms. Find the value of coupling capacitance C_1 required when the lowest frequency to be passed is (a) 100 cycles, (b) 20 cycles.

Given: $R_i = 2000$ ohms   Find: (a) C_1
         (a) $f_l = 100$ cycles   (b) C_1
         (b) $f_l = 20$ cycles

Solution:

$$C_1 = \frac{10^6}{2\pi f X_{C1}} \quad \text{and} \quad X_{C1} = R_i \quad \text{Therefore, } C_1 = \frac{10^6}{2\pi f R_i}$$

(a) $C_1 = \dfrac{10^6}{6.28 \times 100 \times 2{,}000} \cong 0.8$ μf

(b) $C_1 = \dfrac{10^6}{6.28 \times 20 \times 2{,}000} \cong 4.0$ μf

Example 14-21. The bypass capacitor C_2 in the emitter circuit of Fig. 14-23 and Example 14-19 is 150 μf. What percentage of the a-c signal is bypassed by this capacitor at: (*a*) 100 cycles? (*b*) 20 cycles?

Given: Circuit Fig. 14-23   Find: (*a*) Per cent signal bypassed
           $C_2 = 150\ \mu$f   (*b*) Per cent signal bypassed
           $R_2 = 140$ ohms
        (*a*) $f = 100$ cycles
        (*b*) $f = 20$ cycles

Solution:

(*a*) $X_{C2} = \dfrac{10^6}{2\pi f C_2} = \dfrac{10^6}{6.28 \times 100 \times 150} = 10.6$ ohms

     Per cent signal bypassed $= \dfrac{R_2 \times 100}{\sqrt{R_2^2 + X_{C2}^2}} = \dfrac{140 \times 100}{\sqrt{140^2 + 10.6^2}} = 99.7$

(*b*) $X_{C2} = \dfrac{10^6}{6.28 \times 20 \times 150} = 53$ ohms

     Per cent signal bypassed $= \dfrac{140 \times 100}{\sqrt{140^2 + 53^2}} = 93.7$

The a-c power output of a transistor amplifier circuit can be calculated in a similar manner as the a-c power output of vacuum-tube circuits and is equal to one-eighth the product of the peak-to-peak collector voltage and collector current.

$$P_{ac} = \frac{V_{pp}I_{pp}}{8} \tag{14-30}$$

Example 14-22. The transistor amplifier circuit used in Examples 14-19, 14-20, and 14-21 has a d-c operating point of $V_c = -4$ volts and $I_c = -7$ ma. The input signal current causes the collector voltage to vary from -0.5 volt to -7.5 volts, and the collector current to vary from -1.0 ma to -13.0 ma. Find the efficiency of the circuit (*a*) disregarding the stabilizing resistor, (*b*) taking the stabilizing resistor into consideration.

Given: $V_c = -4$ volts   Find: (*a*) Efficiency
        $I_c = -7$ ma   (*b*) Efficiency
        $R_2 = 140$ ohms
        $V_{pp} = -0.5$ to -7.5 volts
        $I_{pp} = -1.0$ to -13.0 ma

Solution:

$$P_{d\text{-}c} = V_c I_c = 4 \times 7 \times 10^{-3} = 28 \times 10^{-3}\ \text{watt} = 28\ \text{mw}$$

$$P_{a\text{-}c} = \frac{V_{pp}I_{pp}}{8} = \frac{(7.5 - 0.5)(13.0 - 1.0)}{8} = 10.5\ \text{mw}$$

(*a*)   Efficiency $= \dfrac{P_{a\text{-}c} \times 100}{P_{d\text{-}c}} = \dfrac{10.5 \times 100}{28} = 37.5\%$

(*b*)   Efficiency $= \dfrac{P_{a\text{-}c} \times 100}{P_{d\text{-}c} + I_c^2 R_2} = \dfrac{10.5 \times 100}{28 + (7 \times 7 \times 10^{-3} \times 140)} \cong 30\%$

Load Lines. A load line can be constructed as shown in Fig. 14-22. A straight line is drawn from the collector-current ordinate when $V_c = 0$ to the collector voltage abscissa when $I_c = 0$. The values of V_c and I_c used in drawing the load line should not exceed the maximum ratings for the transistor used, and the load line should be to the left and under the constant-collector power-dissipation curve. It is thus possible to draw an infinite number of load lines that would conform to these conditions. However, the load line is also dependent on three circuit parameters, (1) the d-c operating point, (2) the supply voltage, and (3) the load resistance. These parameters are interrelated, and the selection of any two automatically determines the third, or

$$R_L = \frac{V_{cc} - V_c}{I_c} \tag{14-31}$$

Example 14-23. The transistor whose characteristics are shown in Fig. 14-22 has a d-c operating point of $V_c = -4$ volts and $I_c = -7$ ma. Find the value of load resistance when the supply voltage is -9 volts.

Given: $V_{cc} = -9$ volts   Find: R_L
        $V_c = -4$ volts
        $I_c = -7$ ma

Solution:

$$R_L = \frac{V_{cc} - V_c}{I_c} = \frac{9 - 4}{7 \times 10^{-3}} = 715 \text{ ohms}$$

The load line for a 715-ohm resistance (line A) would then be drawn by extending a straight line from the point where $V_c = -9$ volts and $I_c = 0$, through the point where $V_c = -4$ volts and $I_c = -7$ ma, and continuing to the $V_c = 0$ ordinate.

Maximum Power Output of a Class A Amplifier. The maximum power output is important in the operation of transistor amplifiers as their output capabilities are normally low compared with vacuum-tube amplifiers. In order to obtain the maximum power output, the load line should enclose as large an area as possible within the maximum current, voltage, and power-dissipation ratings of the transistor. Theoretically, maximum power output would be obtained for a load whose load line is identical to the constant-collector power-dissipation curve. Such a load line is curved and is therefore impractical to use, as the load line for an amplifier must be straight. A practical load line for maximum power output may be obtained by drawing a straight line tangent to the constant-collector power-dissipation curve. Since this curve is nonlinear, several load lines may be drawn that will produce a practical maximum power output. The signal requirement of the circuit is the determining factor as to which load line to use. Thus in Fig. 14-22 load line B represents the optimum load for a circuit with a small signal output, while load line C represents the optimum load for a circuit with a large signal output.

In order to produce maximum power output with maximum signal input for a Class A amplifier, it is necessary that the load line be drawn tangent to the constant-collector power-dissipation curve at the point

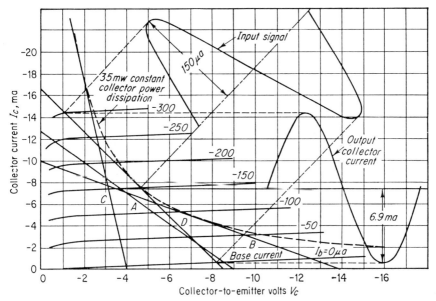

FIG. 14-22. Construction of load lines for a grounded-emitter transistor amplifier (using curves of an ideal transistor).

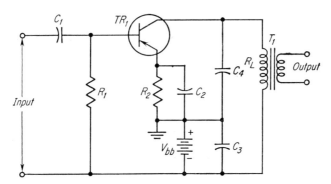

FIG. 14-23. Class A single-ended power amplifier.

where the mid-value of bias current crosses the curve, line D, Fig. 14-22. In this case, the d-c operating point is midway between the extreme limits of the base currents (0 and 300 μa) or at the point where $V_c = -4.6$ volts and $I_c = -7.5$ ma. With an input signal having a peak value of 150 μa, the output current varies uniformly from $I_c = -0.6$ ma to $I_c = -14.4$ ma. By using this graph and the equations previously discussed, it is possible to determine (1) the maximum input signal, (2) the value of the d-c biasing resistors, (3) the load resistance, (4) efficiency, etc.

Example 14-24. Using load line D in Fig. 14-22 and the circuit of Fig. 14-23, determine the required value of (*a*) load resistance, (*b*) biasing resistance, and (*c*) stabilizing resistance for a 15 per cent loss.

Given: Load line D (Fig. 14-22)   Find: (a) R_L
         Circuit diagram (Fig. 14-23)   (b) R_1
         $P_{R2} = 15$ per cent loss   (c) R_2

Solution:

From Fig. 14-22

$$V_{cc} = -8.5 \text{ volts} \qquad V_c = -4.6 \text{ volts} \qquad I_c = -7.5 \text{ ma}$$

(a) $R_L = \dfrac{V_{cc} - V_c}{I_c} = \dfrac{8.5 - 4.6}{7.5 \times 10^{-3}} = 520 \text{ ohms}$

From Fig. 14-22

$$V_{cc} = -8.5 \text{ volts} \qquad I_b = 150 \text{ }\mu\text{a}$$

(b) $R_1 = \dfrac{V_{cc}}{I_b} = \dfrac{8.5}{150 \times 10^{-6}} = 56{,}600 \text{ ohms}$

(c) $R_2 = \dfrac{\% \text{ loss} \times R_L}{100 - \% \text{ loss}} = \dfrac{15 \times 520}{100 - 15} \cong 92 \text{ ohms}$

Example 14-25.   Find the efficiency of the amplifier circuit of Example 14-24 with an input signal having a peak value of 150 μa; (a) disregarding the stabilizing resistor, (b) taking the stabilizing resistor into consideration.

Given: Circuit Example 14-24   Find: (a) Efficiency
                                                 (b) Efficiency

Solution:

From Fig. 14-22

$$V_{pp} = -1.0 \text{ to } -8.2 \text{ volts} \qquad I_{pp} = -0.6 \text{ to } -14.4 \text{ ma}$$
$$P_{\text{d-c}} = V_c \times I_c = 4.6 \times 7.5 = 34.5 \text{ mw}$$
$$P_{\text{a-c}} = \frac{V_{pp} \times I_{pp}}{8} = \frac{(8.2 - 1.0)(14.4 - 0.6)}{8} = 12.42 \text{ mw}$$

(a)   Efficiency $= \dfrac{P_{\text{a-c}} \times 100}{P_{\text{d-c}}} = \dfrac{12.42 \times 100}{34.5} \cong 36\%$

(b)   Efficiency $= \dfrac{P_{\text{a-c}} \times 100}{P_{\text{d-c}} + I_c{}^2 R_2} = \dfrac{12.42 \times 100}{34.5 + (7.5 \times 7.5 \times 10^{-3} \times 92)} \cong 31\%$

Distortion.   The family of transistor collector characteristic curves of Fig. 14-22 does not represent the characteristics of an actual transistor but is a modification of the collector characteristic curves for an actual transistor as shown in Fig. 14-24.   The spacings between the collector-current–collector-voltage curves of Fig. 14-22 were made the same for equal changes in base current for clarity in drawing and explanation of load lines.   In the characteristic curves of Fig. 14-24 the spacing between the collector-current–collector-voltage curves is not equal, as it becomes smaller for the higher values of collector current.   This is indicated on Fig. 14-24 by the unequal distances between AB and BC for equal changes in base current.   This crowding effect at the higher values of collector current, which is typical of most transistors, results in harmonic distortion.   The per cent of harmonic distortion can be calculated from the collector characteristic curves for a given operating condition by using the

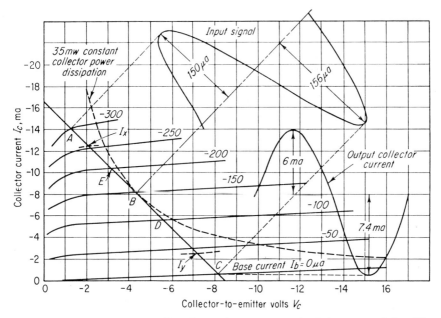

FIG. 14-24. Construction of load lines for maximum power output, for a grounded-emitter transistor amplifier (using curves of an ideal transistor).

same methods described for vacuum-tube circuits (Art. 7-10). Substituting equivalent transistor notations for the vacuum-tube notations, Eqs. (7-17), (7-18), and (7-19) may be written as

$$\% \text{ 2nd harmonic distortion} = \frac{i_{c.\text{max}} + i_{c.\text{min}} - 2I_c}{i_{c.\text{max}} - i_{c.\text{min}} + 1.41(I_x - I_y)} \times 100 \tag{14-32}$$

$$\% \text{ 3rd harmonic distortion} = \frac{i_{c.\text{max}} - i_{c.\text{min}} - 1.41(I_x - I_y)}{i_{c.\text{max}} - i_{c.\text{min}} + 1.41(I_x - I_y)} \times 100 \tag{14-33}$$

where I_x = collector current at $1.707I_b$

I_y = collector current at $0.293I_b$

% total (2nd + 3rd) harmonic distortion

$$= \sqrt{(\% \text{ 2nd harmonic distortion})^2 + (\% \text{ 3rd harmonic distortion})^2} \tag{14-34}$$

Example 14-26. The transistor whose collector characteristics are shown in Fig. 14-24 is to be operated along a load line that will produce maximum signal output with maximum signal input. Find the value of the required load resistance.

    Given: Collector characteristics (Fig. 14-24)   Find: R_L
Solution:

Draw a straight line from the collector-current ordinate when $V_c = 0$ to the collector-voltage abscissa when $I_c = 0$ that is tangent to the constant-collector power-dissipation curve at the point where this curve crosses the collector characteristic curve for $I_b = -150\ \mu a$. This load line indicates that $V_{cc} = -8.5$ volts, $V_c = -4.4$ volts, and $I_c = -8.0$ ma.

$$R_L = \frac{V_{cc} - V_c}{I_c} = \frac{8.5 - 4.4}{8 \times 10^{-3}} = 512 \text{ ohms}$$

Example 14-27. The transistor of Example 14-26 is to be operated with a signal input whose peak value is 150 μa. Find the per cent of (a) second harmonic distortion, (b) third harmonic distortion, (c) total second and third harmonic distortion.

Given: Collector characteristics 　　　Find: (a) % 2nd harmonic distortion
(Fig. 14-24) 　　　　　　　　　　　　　　(b) % 3rd harmonic distortion
$i_{b\cdot\max} = 150\ \mu a$ 　　　　　　　　　　　(c) % total 2nd and 3rd har-
　　　　　　　　　　　　　　　　　　　　　　monic distortion

Solution:

From Fig. 14-24

$$i_{c\cdot\max} = 14 \text{ ma} \qquad i_{c\cdot\min} = 0.6 \text{ ma}$$
$$I_c = 8.0 \text{ ma}$$
$$I_x = 12.4 \text{ ma} \qquad I_y = 2.6 \text{ ma}$$

(a) 2nd harmonic distortion $= \dfrac{i_{c\cdot\max} + i_{c\cdot\min} - 2I_c}{i_{c\cdot\max} - i_{c\cdot\min} + 1.41(I_x - I_y)} \times 100$

$$= \frac{14 + 0.6 - 2 \times 8}{14 - 0.6 + 1.41(12.4 - 2.6)} \times 100 = 5.15\%$$

(b) 3rd harmonic distortion $= \dfrac{i_{c\cdot\max} - i_{c\cdot\min} - 1.41(I_x - I_y)}{i_{c\cdot\max} - i_{c\cdot\min} + 1.41(I_x - I_y)} \times 100$

$$= \frac{14 - 0.6 - 1.41(12.4 - 2.6)}{14 - 0.6 + 1.41(12.4 - 2.6)} \times 100 = 1.47\%$$

(c) Total (2nd + 3rd) harmonic distortion
$$= \sqrt{(\% \text{ 2nd harmonic distortion})^2 + (\% \text{ 3rd harmonic distortion})^2}$$
$$= \sqrt{5.15^2 + 1.47^2} = 5.35\%$$

The equations used for calculating harmonic distortion assume that the source resistance is negligible. The resistance of the signal source in a transistor amplifier circuit is not negligible, and its value cannot be ignored. In vacuum-tube circuits the source resistance does not affect the calculations for harmonic distortion as the grid current is zero. The equivalent parameter in a transistor circuit is the base-emitter voltage which is not zero. The effect of this voltage and the source resistance is to increase the over-all harmonic distortion approximately 1 per cent more than the amount obtained by the use of the equations. Thus, the transistor of Example 14-27 would have a total harmonic distortion of approximately 6.35 per cent.

Another type of distortion is caused by the effect of the variations in the output circuit on the input circuit of an amplifier when the input resistance is high compared to its source impedance. In this type of amplifier circuit, the input resistance decreases in the high region of the collector a-c cycle, thus causing the amplitude of the input signal to increase. In the low region of the collector a-c cycle, the input resistance increases, thus causing the amplitude of the input signal to decrease. This type of

nonlinear distortion is shown in Fig. 14-25b.   Since the effect of this type
distortion is opposite to that caused by harmonic distortion (Fig. 14-25a),
it is possible to counteract one with the other by adjusting the source
impedance.

14-10. Push-pull Amplifiers. *General Considerations.* The collector
characteristic curves of Fig. 14-24 show that the output signal will be
linear if the input signal is operated between the base currents $I_b = 0$ and
$I_b = 200$ μa.   Any greater swing in the base current will cause distortion
of the output signal, as a result of the clipping effect on the input signal.
The maximum undistorted power output will therefore only be approximately 6.125 mw, which is considerably less than the 12.42 mw calculated

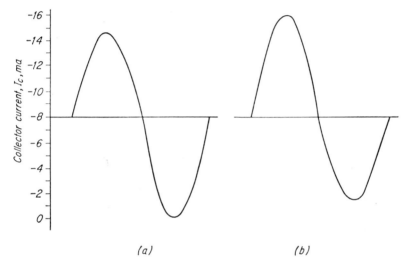

FIG. 14-25. Distortion of collector current: (a) Effect of variation of spacing between collector characteristic curves.   (b) Effect of variation in input circuit impedance.

in Example 14-25.   When a higher power output with a minimum amount
of distortion is required, a push-pull circuit arrangement is used.

Transistors are operated in a push-pull circuit in the same manner as
vacuum tubes; hence a push-pull transistor audio amplifier has the same
desirable characteristics as a push-pull vacuum-tube audio amplifier
(Arts. 7-16–7-21).   The elimination of the even order of harmonics in a
push-pull amplifier permits each transistor to be driven into the higher
collector current region.   For a given percentage of distortion it is possible for each transistor in a push-pull amplifier to deliver more power to
the load than the amount that each transistor could deliver to the same
load with the same percentage of distortion when connected as a single-
ended Class A amplifier.

Class A Push-pull Amplifier.   A Class A push-pull circuit consists of
two similar transistors having the same d-c operating point, load resist-

ance, and biasing resistors as were required for a single-ended Class A
amplifier.   The Class A push-pull audio-amplifier circuit of Fig. 14-26
shows that this circuit actually consists of two similar single-ended Class A
amplifiers, similar to Fig. 14-23, whose input signals are 180 degrees out
of phase with each other.   The d-c operating point, load resistance, and
biasing resistors are determined for each transistor circuit in the same
manner as a single-ended Class A amplifier stage.   When an input transformer is used to provide the 180 degree phase relation between the input
signals, one-half of its secondary impedance matches the input impedance
of TR_1 while the other half matches the input impedance of TR_2.   One half

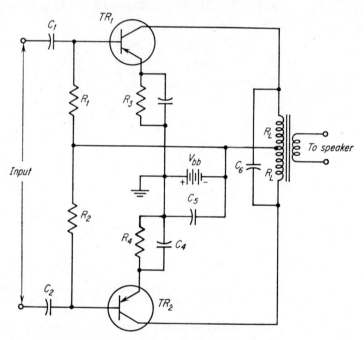

Fig. 14-26. Class A push-pull grounded-emitter power-amplifier circuit.

the impedance of the primary winding of the output transformer T_1 is
used to match the output impedance of TR_1, while the other half matches
the output impedance of TR_2.   Thus, both the input and output impedances are twice the value required for single-ended operation.

Phase Inverters.   The required balanced input for a push-pull transistor
amplifier may be obtained from a phase-inverter circuit in the same manner as with vacuum-tube circuits.   In the basic transistor phase-inverter
circuit of Fig. 14-27, two output signals of opposite phase and approximately the same amplitude are obtained from a single input signal by
using only one transistor stage.   The base-emitter circuit is biased in
a forward direction by the voltage-divider circuit R_1 and R_2.   One output

voltage V_3 is taken from across the collector load resistor R_3, while the other output voltage V_4 of opposite phase is taken from across the emitter resistor R_4.   Although the resistances of R_3 and R_4 are equal, the voltages V_3 and V_4 across the two resistors are not exactly the same.   The dif-

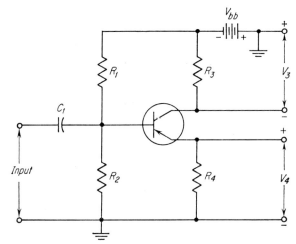

FIG. 14-27. Single-stage transistor phase-inverter circuit.

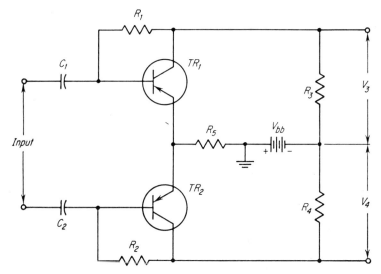

FIG. 14-28. Two-stage transistor phase-inverter circuit.

ference in voltage is small and is due to the difference between the collector and emitter currents.   A better balance may be obtained by altering the values of R_3 and R_4.

A more exact balance is obtained with two transistor stages as shown in Fig. 14-28.   In this circuit TR_1 provides an output signal of one polarity across R_3, and TR_2 takes a portion of this signal, amplifies it, inverts it, and provides a second signal of opposite polarity across R_4. TR_1 is connected as a grounded-emitter amplifier, with the emitter grounded through the parallel circuit formed by the emitter resistor R_5 and the path formed by the emitter-base resistance of TR_2, the negligible impedance of C_2, and the grounded side of the input signal source. As r_e is generally less than 50 ohms, practically all the a-c emitter current from TR_1 will flow through r_e. In order to maintain the equality of the alternating current in the emitter circuits of both transistors, very little of this current should be permitted to flow through the emitter resistor. This is accomplished by making the value of R_5 ten or more times greater than the emitter-base resistance of TR_2. The output resistance of each transistor is equal to the resistance of the parallel circuit formed by its collector resistance and the biasing resistor R_1 or R_2, which are equal. Since r_c is much greater than R_1, the output resistance is practically equal to R_1. For proper operation of this type of phase-inverter circuit, the load resistors R_3 and R_4 should be small compared with the output resistance; R_1 is generally made ten or more times greater than R_3. This circuit does not require the use of identical or matched transistors. Proper circuit operation may be obtained with transistors having current-transfer ratios in the range of 0.92 to 0.98.

Complementary Symmetrical Push-pull Circuit. The principles of complementary symmetry may be applied to push-pull amplifiers. Since the output of PNP and NPN transistors are 180 degrees out of phase with each other, these two types of transistors can be made to operate with their input circuits in parallel as a push-pull stage of amplification (Fig. 14-29). In achieving push-pull operation, the complementary action of the PNP and NPN transistors eliminates the need for an input transformer or a phase inverter. When an input signal is applied to this circuit, the collector current will increase in one transistor unit and decrease in the other, thus producing push-pull amplification. This circuit operates in the same manner as a Class A push-pull transistor amplifier and produces a high voltage gain with a high impedance load. When the transistors are exactly symmetrical, the direct collector currents cancel each other; thus no d-c component flows through the load impedance.

Power Output Class A Push-pull Amplifier. The power output, efficiency, and collector-to-collector load resistance of a Class A push-pull transistor amplifier can be calculated for a given operating condition from the collector characteristic curves. The maximum power output for two transistors in push-pull may be expressed as

$$P_{o \cdot \max} = \frac{V_{pp}I_{pp}}{2} \qquad (14\text{-}35)$$

and the collector-to-collector resistance as

$$R'_L = \frac{4V_{c\text{·max}}}{i_{c\text{·max}}} \tag{14-36}$$

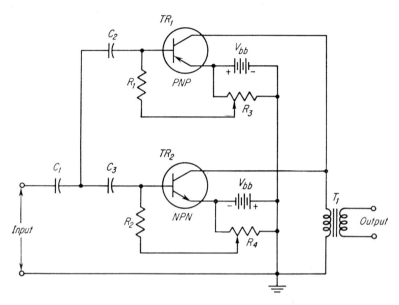

Fig. 14-29. Push-pull amplifier circuit using complementary symmetry.

Example 14-28. Two transistors having the collector characteristics shown in Fig. 14-24 are to be operated as a Class A push-pull amplifier. Each transistor is to be operated so as to produce maximum undistorted power output when connected as a single-ended Class A amplifier. Determine (*a*) the efficiency of the circuit, (*b*) the collector-to-collector load resistance.

Given: Transistor characteristics (Fig. 14-24)   Find: (*a*) Efficiency
                                                                   (*b*) R'_L

Solution:

From Fig. 14-24, the d-c operating point will be at D, and the 100 μa input signal will vary between points C and E.

$$V_{pp} = -3.15 \text{ to } -8.15 \text{ volts} \qquad I_{pp} = -0.6 \text{ to } -10.4 \text{ ma}$$
$$V_c = -5.6 \text{ volts} \qquad\qquad I_c = -5.6 \text{ ma}$$

$$P_{\text{d-c}} = 2V_cI_c = 2 \times 5.6 \times 5.6 = 62.7 \text{ mw}$$
$$P_{\text{a-c}} = \frac{V_{pp}I_{pp}}{2} = \frac{(8.15 - 3.15)(10.4 - 0.6)}{2} = 24.5 \text{ mw}$$

(*a*)   Efficiency $= \dfrac{P_{\text{a-c}} \times 100}{P_{\text{d-c}}} = \dfrac{24.5 \times 100}{62.7} \cong 39\%$

(*b*) $R'_L = \dfrac{4v_{c\text{·max}}}{i_{c\text{·max}}} = \dfrac{4 \times 8.15}{10.4 \times 10^{-3}} \cong 3{,}140 \text{ ohms}$

The power output of 24.5 mw obtained from the Class A push-pull arrangement of Example 14-28 is considerably higher than twice the value of 6.125 mw obtained for the single-ended Class A operation.   Because push-pull operation eliminates the even order of harmonics, the transistors can be operated in the higher collector-current region of unequal spacing between collector-current–collector-voltage curves to produce higher power outputs with higher efficiency and a minimum of distortion.   Thus if two transistors having the characteristics shown in Fig. 14-24 were operated with the input and output currents indicated on this diagram, a power output of approximately 45 mw can be obtained.

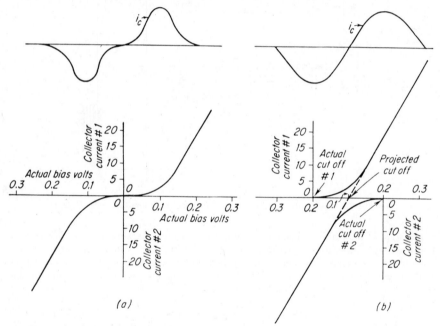

FIG. 14-30. Effect of projected cutoff on output current waveform: (a) when d-c bias is zero volt.   (b) When d-c bias is greater than zero volt.

14-11. Class B Push-pull Amplifiers.   *Advantages.*   Class B push-pull amplifiers are used when higher power output is required than can be obtained from the same transistors operated as a Class A push-pull amplifier.   A disadvantage of Class A amplifiers is that direct collector current flows at all times.   The collector dissipation is therefore high, even when no alternating signal current is flowing through the transistor. This dissipation can be greatly reduced by operating the two transistors in push-pull as Class B amplifiers.   Each transistor supplies one-half the cycle of the output waveform, and during its nonconducting half-cycle is in an OFF (or low dissipating) condition.   With zero input signal, neither transistor is in a conducting state and hence dissipates very little power.

Because of these favorable operating conditions, the d-c output efficiency is very high and may be of the order of 80 per cent.

Circuit Operation.   When a PNP transistor is operated as a Class B amplifier, the collector current flows only during negative signal inputs. When an NPN transistor is operated as a Class B amplifier the collector current flows only during positive signal inputs.   The distortion resulting from Class B operation can be reduced to a minimum by using two transistors as a push-pull amplifier.   The amplifier circuits of Figs. 14-26 and 14-29 may be operated as Class B push-pull amplifiers by using biasing resistors that cause the base-emitter circuit to be biased near cutoff.   Theoretically, the transistor should be operated at cutoff so that no collector current flows and hence there would be no collector power

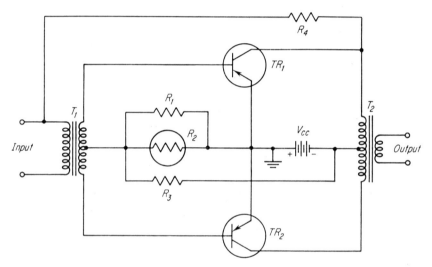

Fig. 14-31. Class B push-pull amplifier circuit using a thermistor in the biasing circuit.

dissipation with zero signal input.   However, if the transistors were operated at such a d-c operating point, the output signal would be distorted at low input signals (Fig. 14-30a).   This type of distortion is called *crossover distortion*, and most of it can be eliminated by biasing the transistors so that a small amount of collector current flows at all times. The manner in which the coincidence of the projected cutoff points minimizes crossover distortion is shown in Fig. 14-30b   Any crossover distortion that is not eliminated may be reduced by using negative feedback.

A push-pull Class B transistor amplifier circuit using an input transformer is shown in Fig. 14-31.   Proper biasing of TR_1 and TR_2 is obtained by means of the divider network consisting of R_1, R_2, and R_3.   A thermistor R_2 is used in this network to minimize the effects of the variations in the d-c operating point, collector current, and collector dissipation with changes in ambient temperature. A constant voltage is maintained across the biasing network by R_2 whose resistance decreases with increases in temperature, and increases with decreases in temperature. The thermistor is capable of stabilizing the d-c operating point over a wide range of ambient temperatures. Negative feedback is obtained by means of R_4.

A Class B push-pull amplifier circuit that eliminates the need of an input transformer is shown in Fig. 14-32. A reverse bias that is produced by the input signal would cause the transistor to operate at or beyond cutoff. This action is prevented by the two crystal diodes CD_1 and CD_2 which effectively short out any bias induced by the signal. This bypass

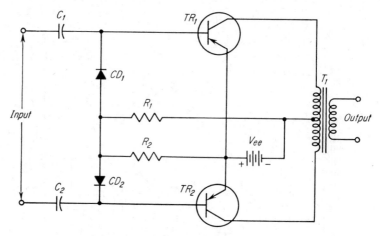

Fig. 14-32. Class B push-pull amplifier circuit using diodes in place of an input transformer.

action occurs at a point determined by the values of the biasing resistors R_1 and R_2.

Power Output Class B Push-pull Amplifiers. As with Class A power amplifiers the power output, efficiency, and collector-to-collector load resistance of a Class B push-pull transistor amplifier can be calculated for a given operating condition from the collector characteristic curves. Equation 14-35 can be used to obtain the maximum power output, and Eq. 14-36 to obtain the collector-to-collector load resistance. An exact equation for obtaining the d-c collector power input is rather complex as a high current flows through each transistor only during the half-cycle that it is conducting, while very little or no current flows during the other half-cycle. An approximate value of the d-c collector power can be obtained by using the following equation.

$$P_{\text{d-c}} \cong \frac{V_c I_c}{1.57} \qquad (14\text{-}37)$$

Where V_c = maximum collector voltage, corresponding to point C, Fig. 14-24

I_c = maximum collector current, corresponding to point A, Fig. 14-24

Example 14-29.   Two transistors having the collector characteristics shown in Fig. 14-24 are being operated as a Class B push-pull amplifier to produce maximum power output.   Determine (a) the efficiency of the circuit, (b) the collector-to-collector load resistance.

Given: Transistor characteristics, Fig. 14-24   Find: (a) Efficiency
                                                                          (b) R'_L

Solution:

From Fig. 14-24,

$$V_{pp} = -1.25 \text{ to } -8.15 \text{ volts} \qquad I_{pp} = -0.6 \text{ to } -14.0 \text{ ma}$$
$$V_c = -8.15 \text{ volts} \qquad I_c = -14.0 \text{ ma}$$

$$P_{\text{d-c}} \cong \frac{V_c I_c}{1.57} \cong \frac{8.15 \times 14.0}{1.57} \cong 72.5 \text{ mw}$$

$$P_{\text{a-c}} = \frac{V_{pp} I_{pp}}{2} = \frac{(8.15 - 1.25)(14.0 - 0.6)}{2} \cong 46 \text{ mw}$$

(a)   Efficiency $= \dfrac{P_{\text{a-c}} \times 100}{P_{\text{d-c}}} = \dfrac{46 \times 100}{72.5} \cong 63.5\%$

(b) $R'_L = \dfrac{4 V_{c \cdot \text{max}}}{i_{c \cdot \text{max}}} = \dfrac{4 \times 8.15}{14.0 \times 10^{-3}} \cong 2,320 \text{ ohms}$

QUESTIONS

1. Compare the basic operations of a transistor amplifier circuit with the basic operations of a vacuum-tube amplifier circuit.

2. Compare the following circuit parameters of a transistor amplifier circuit with their comparable vacuum-tube amplifier circuit parameters: (a) input impedance, (b) output impedance, (c) voltage gain, (d) current gain, (e) power gain.

3. Compare the basic operation of the grounded-emitter transistor amplifier circuit with the basic operation of its comparable vacuum-tube amplifier circuit.

4. Describe in detail the operation of a PNP grounded-emitter transistor amplifier circuit.

5. Describe in detail the operation of a NPN grounded-emitter transistor amplifier circuit.

6. Describe the following characteristics of the grounded-emitter transistor-amplifier circuit: (a) current gain, (b) voltage gain, (c) power gain, (d) stability.

7. Compare the basic operation of the grounded-base transistor-amplifier circuit with the basic operation of its comparable vacuum-tube amplifier circuit.

8. Describe in detail the operation of a PNP grounded-base transistor amplifier circuit.

9. Describe in detail the operation of a NPN grounded-base transistor amplifier circuit.

10. Describe the following characteristics of the grounded-base transistor amplifier circuit: (a) current gain, (b) voltage gain, (c) power gain, (d) stability.

11. Compare the basic operation of the grounded-collector transistor amplifier circuit with the basic operation of its comparable vacuum-tube amplifier circuit.

12. Describe in detail the operation of a PNP grounded-collector transistor amplifier circuit.

13. Describe in detail the operation of a NPN grounded-collector transistor amplifier circuit.

14. Describe the following characteristics of the grounded-collector transistor-amplifier circuit: (*a*) current gain, (*b*) voltage gain, (*c*) power gain, (*d*) stability.

15. Describe the applications of the grounded-emitter amplifier.

16. Describe the applications of the grounded-base amplifier.

17. Describe the applications of the grounded-collector amplifier.

18. Describe the following terms: (*a*) fixed bias, (*b*) self-bias, (*c*) stabilized bias.

19. Describe the following terms: (*a*) collector current, (*b*) collector cutoff current, (*c*) reverse leakage current.

20. Describe the need for stabilization with (*a*) a common-emitter amplifier, (*b*) a common-base amplifier.

21. How is fixed bias obtained in (*a*) a grounded-emitter amplifier? (*b*) a grounded-base amplifier?

22. What are the disadvantages of fixed bias?

23. Describe in detail the method of obtaining self-bias in a grounded-emitter amplifier (*a*) with degeneration, (*b*) without degeneration.

24. What are the advantages of using both fixed and self-bias?

25. Describe in detail how bias and stabilization are obtained in the transistor amplifier circuit of Fig. 14-8.

26. Describe in detail how bias and stabilization are obtained in the transistor-amplifier circuit of Fig. 14-9.

27. Describe in detail how bias and stabilization are obtained in the transistor-amplifier circuit of Fig. 14-10.

28. What circuit parameters of a transistor amplifier circuit should be known in order to solve for its current, voltage, or power amplification?

29. Name six variable factors that affect the current, voltage, and power amplification of a transistor amplifier.

30. What do the following types of equivalent transistor-amplifier circuit equations, as presented in this chapter, represent: (*a*) complex equation? (*b*) simplified equation?

31. (*a*) Describe the difference between α and α_L. (*b*) For what circuit conditions is α used? (*c*) For what circuit conditions is α_L used?

32. Draw a four-terminal equivalent circuit for a common-emitter PNP transistor amplifier circuit showing the direction of electron flow in the input and output circuits.

33. Draw a four-terminal equivalent circuit for a common-emitter NPN transistor amplifier circuit showing the direction of electron flow in the input and output circuits.

34. Draw a four-terminal equivalent circuit for a common-base PNP transistor amplifier circuit showing the direction of electron flow in the input and output circuits.

35. Draw a four-terminal equivalent circuit for a common-base NPN transistor amplifier circuit showing the direction of electron flow in the input and output circuits.

36. Draw a four-terminal equivalent circuit for a common-collector PNP transistor amplifier circuit showing the direction of electron flow in the input and output circuits.

37. Draw a four-terminal equivalent circuit for a common-collector NPN transistor amplifier circuit showing the direction of electron flow in the input and output circuits.

38. (*a*) What are the advantages of transformer coupling? (*b*) Describe some of the applications of transformer coupling.

39. (*a*) What are the advantages and disadvantages of *RC* coupling? (*b*) Describe some of the applications of *RC* coupling.

40. (*a*) What are the advantages of impedance coupling? (*b*) Describe some of the applications of impedance coupling.

41. (*a*) What are the advantages of direct coupling? (*b*) Describe some of the applications of direct coupling.

42. (*a*) What is meant by complementary symmetry? (*b*) Describe the operation of a complementary symmetrical coupled amplifier circuit.

43. What two circuit conditions determine the maximum power gain of a transistor amplifier circuit?

44. Describe the general rules for determining the type of common electrode amplifier circuit to use in the various stages of a multistage *RC*-coupled amplifier.

45. Describe the operation of a decoupling circuit as used in a multistage *RC*-coupled amplifier.

46. What factors affect the frequency response of an *RC*-coupled amplifier at (*a*) the low a-f range? (*b*) the medium a-f range? (*c*) the high a-f range?

47. (*a*) What methods are used for dissipating the heat at the collector of a transistor? (*b*) What is meant by a constant-collector power-dissipation curve?

48. Describe the operation of a single-ended Class A power-amplifier circuit.

49. Describe the characteristics of a single-ended Class A power-amplifier circuit.

50. What three circuit parameters determine the manner of drawing a load line?

51. What factors determine the maximum power output of a Class A single-ended amplifier?

52. Describe how distortion is produced in a Class A amplifier by (*a*) the variation in spacing between the collector-current–collector-voltage characteristic curves, (*b*) the resistance of the source generator, (*c*) the variation in impedance of the output circuit.

53. Describe the operation of a Class A push-pull amplifier.

54. Describe the operation of a single-stage transistor phase-inverter circuit.

55. Describe the operation of a two-stage transistor phase-inverter circuit.

56. What are the advantages of the two-stage transistor phase-inverter circuit?

57. Describe the principles of operation of a complementary symmetrical push-pull amplifier circuit.

58. What are the advantages of a complementary push-pull amplifier circuit?

59. What are the advantages and disadvantages of a Class A push-pull amplifier circuit?

60. Describe the operation of a Class B push-pull amplifier.

61. What are the advantages and disadvantages of a Class B push-pull amplifier circuit?

62. (*a*) What is meant by crossover distortion? (*b*) How may this type of distortion be eliminated?

63. Why is a thermistor used in the biasing network of a Class B push-pull amplifier?

PROBLEMS

1. A junction transistor is connected as shown in Fig. 14-4, with $V_{CC} = 3$ volts. What approximate value of base resistance is required to operate the transistor with a base current of (*a*) 25 μa? (*b*) 50 μa?

2. A junction transistor is connected as shown in Fig. 14-4, with $I_b = 0.9$ ma. What approximate value of base resistance is required to operate the transistor with V_{CC} equal to (*a*) 14.4 volts? (*b*) 9 volts?

3. A junction transistor is connected as shown in Fig. 14-5, with $I_b = 25$ μa. What approximate value of base resistance is required to operate the transistor with an emitter bias of (*a*) 1.0 volt? (*b*) 0.5 volt?

4. A junction transistor is connected as shown in Fig. 14-5, with $V_E = 0.3$ volt. What approximate value of base resistance is required to operate the transistor with a base current of (*a*) 500 μa? (*b*) 200 μa?

5. A transistor connected with a grounded base has the following circuit parameters: $\alpha = 0.97$, $I_e = 1$ ma, I_{co} at 25°C = 2 μa, I_{co} at 100°C = 10 μa. Find (*a*) I_c at 25°C, (*b*) I_c at 100°C.

6. A transistor connected with a common base has the following circuit parameters:

$\alpha = 0.98$, $I_e = 0.5$ ma, I_{co} at $25°C = 1.5$ μa, I_{co} at $100°C = 8.0$ μa. Find (a) I_c at $25°C$, (b) I_c at $100°C$.

7. A transistor connected with a grounded emitter has the following circuit parameters: $\beta = 50$, $I_b = 50$ μa, I_{co} at $25°C = 2$ μa, I_{co} at $100°C = 10$ μa. Find (a) I_c at $25°C$, (b) I_c at $100°C$.

8. A transistor connected with a common emitter has the following circuit parameters: $\beta = 30$, $I_b = 25$ μa, I_{co} at $25°C = 1.5$ μa, I_{co} at $100°C = 8$ μa. Find (a) I_c at $25°C$, (b) I_c at $100°C$.

9. A certain PNP junction transistor when connected as a grounded emitter has the following circuit parameters: $r_e = 34$ ohms, $r_b = 976$ ohms, $r_c = 3.45$ megohms, $R_L = 20,000$ ohms, $\alpha = 0.982$. Find the input resistance using (a) the complex equation, (b) the simplified equation.

10. A certain PNP junction transistor when connected as a common emitter has the following circuit parameters: $r_e = 21$ ohms, $r_b = 580$ ohms, $r_c = 1.82$ megohms, $R_L = 20,000$ ohms, $\alpha = 0.977$. Find the input resistance using (a) the complex equation, (b) the simplified equation.

11. The internal resistance of the source generator of Prob. 9 is 100 ohms. Find the output resistance using (a) the complex equation, (b) the simplified equation.

12. The internal resistance of the source generator of Prob. 10 is 100 ohms. Find the output resistance using (a) the complex equation, (b) the simplified equation.

13. Find the voltage gain of the transistor circuit of Prob. 9 using (a) the complex equation, (b) the simplified equation.

14. Find the voltage gain of the transistor circuit of Prob. 10 using (a) the complex equation, (b) the simplified equation.

15. Find the power gain (in decibels) of the transistor circuit of Prob. 9 using (a) the complex equation, (b) the simplified equation.

16. Find the power gain (in decibels) of the transistor circuit of Prob. 10 using (a) the complex equation, (b) the simplified equation.

17. The PNP junction transistor used in Prob. 9 is connected with a grounded base and has an output load resistance of 500,000 ohms. Find the input resistance using (a) the complex equation, (b) the simplified equation.

18. The PNP junction transistor used in Prob. 10 is connected with a common base, and an output load resistance of 500,000 ohms. Find the input resistance using (a) the complex equation, (b) the simplified equation.

19. The internal resistance of the source generator of Prob. 17 is 50 ohms. Find the output resistance using (a) the complex equation, (b) the simplified equation.

20. The internal resistance of the source generator of Prob. 18 is 100 ohms. Find the output resistance using (a) the complex equation, (b) the simplified equation.

21. Find the voltage gain of the junction transistor circuit of Prob. 17 using (a) the complex equation, (b) the simplified equation.

22. Find the voltage gain of the junction transistor circuit of Prob. 18 using (a) the complex equation, (b) the simplified equation.

23. Find the power gain (in decibels) of the junction transistor circuit of Prob. 17 using (a) the complex equation, (b) the simplified equation.

24. Find the power gain (in decibels) of the junction transistor circuit of Prob. 18 using (a) the complex equation, (b) the simplified equation.

25. The PNP junction transistor used in Prob. 9 is connected with a grounded collector, and an output load resistance of 10,000 ohms. Find the input resistance using (a) the complex equation, (b) the simplified equation.

26. The PNP junction transistor used in Prob. 10 is connected with a common collector, and an output load resistance of 18,000 ohms. Find the input resistance using (a) the complex equation, (b) the simplified equation.

27. The internal resistance of the source generator of Prob. 25 is 100 ohms. Find the output resistance using (a) the complex equation, (b) the simplified equation.

28. The internal resistance of the source generator of Prob. 26 is 100 ohms. Find the output resistance using (a) the complex equation, (b) the simplified equation.

29. Find the voltage gain of the junction transistor circuit of Prob. 25 using (a) the complex equation, (b) the simplified equation.

30. Find the voltage gain of the junction transistor circuit of Prob. 26 using (a) the complex equation, (b) the simplified equation.

31. Find the power gain (in decibels) of the junction transistor circuit of Prob. 25 using (a) the complex equation, (b) the simplified equation.

32. Find the power gain (in decibels) of the junction transistor circuit of Prob. 26 using (a) the complex equation, (b) the simplified equation.

33. Three transistors, each one similar to the transistors used in the odd-numbered problems from Prob. 9 to 31, are connected as shown in Fig. 14-19. Find the over-all power gain (in decibels) of this three-stage amplifier circuit.

34. Three transistors, each one similar to the transistors used in the even-numbered problems from Prob. 10 to 32, are connected as shown in Fig. 14-19. Find the over-all power gain (in decibels) of this three-stage amplifier circuit.

35. Two similar transistors each having a base bias current of 300 μa are connected as shown in Fig. 14-20. The maximum amount of voltage that can be lost across the decoupling resistor is 0.36 volt. Find the value of decoupling capacitance required to pass (a) 28 cycles, (b) 110 cycles.

36. Two similar transistors each having a base bias current of 160 μa are connected as shown in Fig. 14-20. The maximum amount of voltage that can be lost across the decoupling resistor is 0.24 volt. Find the value of decoupling capacitance required to pass (a) 22 cycles, (b) 74 cycles.

37. Find the over-all current gain at the medium frequencies for the first stage of amplification in Prob. 33. The collector load resistance is 5,000 ohms.

38. Find the over-all current gain at the medium frequencies for the first stage of amplification in Prob. 34. The collector load resistance is 3,900 ohms.

39. When the coupling capacitance is equal to 20 μf, find the over-all current gain of the transistor and circuit of Prob. 37 at (a) 100 cycles, (b) 20 cycles.

40. When the coupling capacitance is equal to 10 μf, find the over-all current gain of the transistor and circuit of Prob. 38 at (a) 120 cycles, (b) 30 cycles.

41. A PNP transistor is connected as shown in Fig. 14-23 and has the following circuit parameters: $V_{bb} = 6$ volts, $I_b = 240$ μa, $R_L = 820$ ohms. Find (a) the value of the biasing resistor R_1, (b) the value of the stabilizing resistor R_2, if a 12 per cent power loss is permitted.

42. A PNP transistor is connected as shown in Fig. 14-23 and has the following circuit parameters: $V_{bb} = 4.5$ volts, $I_b = 200$ μa, $R_L = 350$ ohms. Find (a) the value of the biasing resistor R_1, (b) the value of the stabilizing resistor R_2, if a 20 per cent power loss is permitted.

43. The amplifier of Prob. 41 has an input resistance of 4,250 ohms. Find the value of coupling capacitance C_1 required when the lowest frequency to be passed is (a) 75 cycles, (b) 25 cycles.

44. The amplifier of Prob. 42 has an input resistance of 1,330 ohms. Find the value of coupling capacitance C_1 required when the lowest frequency to be passed is (a) 120 cycles, (b) 30 cycles.

45. The bypass capacitor C_2 in the emitter circuit of Fig. 14-23 and Prob. 41 is 100 μf. What percentage of the a-c signal is bypassed by this capacitor at (a) 75 cycles, (b) 25 cycles.

46. The bypass capacitor C_2 in the emitter circuit of Fig. 14-23 and Prob. 42 is 50 μf. What percentage of the a-c signal is bypassed by this capacitor at (a) 120 cycles, (b) 30 cycles.

47. The transistor amplifier circuit used in Probs. 41, 43, and 45 has a d-c operating point of $V_c = -3.5$ volts and $I_c = -5.2$ ma. The input signal current causes the collector voltage to vary from -1.2 to -5.8 volts, and the collector current to vary from -0.4 to -10.0 ma. Find the efficiency of the circuit (*a*) disregarding the stabilizing resistor, (*b*) taking the stabilizing resistor into consideration.

48. The transistor amplifier circuit used in Probs. 42, 44, and 46 has a d-c operating point of $V_c = -2.5$ volts and $I_c = -2.2$ ma. The input signal current causes the collector voltage to vary from -1.0 to -4.0 volts, and the collector current to vary from -0.8 to -3.6 ma. Find the efficiency of the circuit: (*a*) disregarding the stabilizing resistor, (*b*) taking the stabilizing resistor into consideration.

49. The transistor whose characteristics are shown in Fig. 14-22 has a d-c operating point of $V_c = -3.5$ volts and $I_c = -5.2$ ma. Find the value of load resistance when the supply voltage is -6 volts.

50. The transistor whose characteristics are shown in Fig. 14-22 has a d-c operating point of $V_c = -2.5$ volts and $I_c = -2.2$ ma. Find the value of load resistance when the supply voltage is -4.5 volts.

51. Using load line A in Fig. 14-22 and the circuit of Fig. 14-23, determine the required value of (*a*) load resistance, (*b*) biasing resistance, and (*c*) stabilizing resistance for a 20 per cent loss.

52. Using load line C in Fig. 14-22 and the circuit of Fig. 14-23, determine the required value of (*a*) load resistance, (*b*) biasing resistance, and (*c*) stabilizing resistance for a 10 per cent loss.

53. Find the efficiency of the amplifier circuit of Prob. 51 with an input signal having a peak value of 125 μa, (*a*) disregarding the stabilizing resistor, (*b*) taking the stabilizing resistor into consideration.

54. Find the efficiency of the amplifier circuit of Prob. 52 with an input signal having a peak value of 150 μa: (*a*) disregarding the stabilizing resistor, (*b*) taking the stabilizing resistor into consideration.

55. The transistor of Prob. 51 is to be operated with a signal input whose peak value is 125 μa. Find the per cent of (*a*) second harmonic distortion, (*b*) third harmonic distortion, (*c*) total second and third harmonic distortion.

56. The transistor of Prob. 52 is to be operated with a signal input whose peak value is 150 μa. Find the per cent of (*a*) second harmonic distortion, (*b*) third harmonic distortion, (*c*) total second and third harmonic distortion.

57. Two transistors having the collector characteristics shown in Fig. 14-22 are to be operated as a Class A push-pull amplifier. Each transistor is to be operated so as to produce maximum undistorted power output when connected as a single-ended Class A amplifier operated along load line D. Determine (*a*) the efficiency of the circuit, (*b*) the collector-to-collector load resistance.

58. Two transistors having the collector characteristics shown in Fig. 14-22 are to be operated as a Class A push-pull amplifier. Each transistor is to be operated so as to produce maximum undistorted power output when connected as a single-ended Class A amplifier operated along load line A. Determine (*a*) the efficiency of the circuit, (*b*) the collector-to-collector load resistance.

59. Two transistors having the collector characteristics shown in Fig. 14-22 are to be operated as a Class B push-pull amplifier along load line D to produce maximum power output. Determine (*a*) the efficiency of the circuit, (*b*) the collector-to-collector load resistance.

60. Two transistors having the collector characteristics shown in Fig. 14-22 are to be operated as a Class B push-pull amplifier along load line A to produce maximum power output. Determine (*a*) the efficiency of the circuit, (*b*) the collector-to-collector load resistance.

CHAPTER 15

TRANSISTOR RECEIVER CIRCUITS

Transistors may be used in a manner similar to vacuum tubes to construct the various types of radio-receiver circuits. Transistors (1) are very small, (2) require very little power, and (3) are solid in construction. A transistor radio receiver circuit compared to a similar type of vacuum-tube receiver circuit will therefore: (1) be much smaller, (2) use less power, and (3) be more rugged. Because of these characteristics transistors are extensively used in small portable radio receivers.

15-1. Basic Radio Receiver Circuits. *Simple Radio Receiver.* The simple radio-receiver circuit shown in Fig. 15-1 consists of (1) a tuning

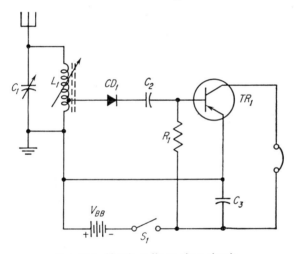

Fig. 15-1. Simple radio receiver circuit.

circuit, (2) a detector, and (3) a single stage of audio amplification. The variable capacitor C_1 and the tapped loopstick L_1 are used to select the desired signal. Loading of the tuned circuit is minimized by using the tap on L_1 to match the tuned circuit to the low input impedance of the detector-amplifier circuit. After detection by the crystal diode CD_1, the demodulated signal is coupled through the d-c blocking capacitor C_2 to the base of the transistor amplifier TR_1. Proper forward bias current for TR_1 is supplied through the biasing resistor R_1. Operating power is obtained from a single battery V_{BB}, that is controlled by S_1, and the a-c signal is bypassed around the power supply by C_3. The amplified audio signal appearing in the collector circuit of the transistor is used to drive the magnetic headphones.

Two-transistor Radio Receiver.   When greater sensitivity is required, the gain of a receiver may be increased by adding one or more stages of

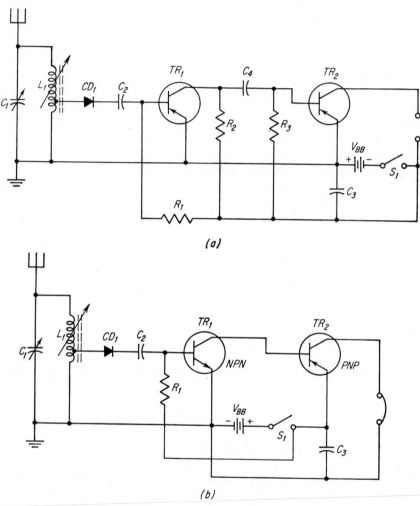

Fig. 15-2. Two-transistor radio-receiver circuits.   (*a*) *RC*-coupled circuit.   (*b*) Direct-coupled circuit.

radio or audio amplification.   The two-transistor radio receiver circuit shown in Fig. 15-2*a* is obtained by adding a stage of audio amplification to the simple circuit shown in Fig. 15-1.   The second a-f stage is *RC* coupled to the first a-f stage; the additional components used are (1) the collector load resistor R_2, (2) the interstage coupling capacitor C_4, (3) the biasing resistor R_3, and (4) the transistor TR_2. A single battery V_{BB} supplies the power to both stages.

The two-transistor radio receiver circuit shown in Fig. 15-2b is also obtained by adding a stage of audio amplification to the simple circuit of Fig. 15-1. In this circuit the two stages are directly coupled using the complementary symmetry of the NPN transistor TR_1 and the PNP transistor TR_2. The only additional component used is the NPN transistor TR_1. The circuits shown in Fig. 15-2 are relatively simple radio receiver circuits, and for clarity the following circuit elements have been omitted, (1) the resistors normally used to provide voltage division and stabilization, and (2) the capacitors normally used to prevent or minimize degeneration and signal attenuation.

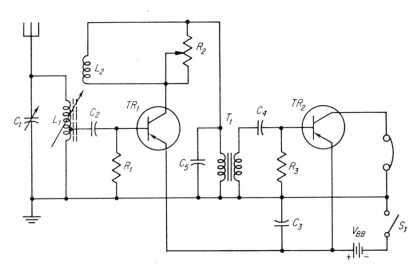

FIG. 15-3. Regenerative radio receiver circuit.

Regenerative Radio Receiver. In simple radio receiver circuits greater sensitivity may also be obtained by adding regeneration to a basic detector circuit. The regenerative radio receiver circuit shown in Fig. 15-3 consists of two stages: (1) a regenerative detector amplifier, and (2) a common-emitter audio amplifier. The desired signal is selected by the resonant circuit C_1L_1, and it is then coupled to the base of TR_1 through the d-c blocking capacitor C_2. Transistor TR_1 serves as both a detector and amplifier, and the small amount of base bias current required is supplied through R_1. The amplified signal in the collector circuit of TR_1 contains both r-f and a-f components. A portion of the r-f component is coupled back to the input circuit through the feedback coil L_2, thus increasing the strength of the original signal. The variable resistor R_2, connected across L_2, is called the *regeneration control* as it controls the portion of the r-f signal in the output circuit (collector-emitter) that is returned to the input circuit (base-emitter). The audio portion of the detected and amplified signal is coupled through an impedance-matching transformer T_1 to the input circuit of TR_2. Any r-f signals at the primary winding of T_1 are bypassed by capacitor C_5. The audio signal appearing across the secondary of T_1 is coupled through the d-c blocking capacitor C_4 to the base of TR_2. The forward base bias current for this transistor is supplied through R_3. Operating power is obtained from a single battery V_{BB}, that is controlled by S_1 and bypassed for a-f by C_3.

Tuned Radio Frequency Radio Receiver. A tuned radio-frequency (trf) radio receiver circuit can be obtained by adding an r-f amplifier stage to either of the two-transistor receiver circuits shown in Fig. 15-2. Both the sensitivity and selectivity of a circuit having only one stage of r-f amplification and two stages of a-f amplification are not too good, and in order to meet the demands of modern broadcast radio receivers, additional stages of both r-f and a-f amplification must be used. The number of r-f and a-f amplifier stages required to produce satisfactory reception is economically impractical and the trf circuit is seldom used with transistor radio receivers.

Superheterodyne Radio Receiver. In general, the superheterodyne radio receiver circuit offers better selectivity and greater sensitivity than any other radio receiver circuit, and is therefore the most popular type of circuit. The basic superheterodyne circuit is used in: (1) virtually all the receivers used for f-m reception, (2) the audio and video sections of television receivers, (3) most of the radio receivers used in the home and in automobiles, and (4) the various portable-type radio receivers. The fundamental principles of superheterodyne receivers have been taken up in Arts. 11-4 and 11-5. The basic principles and circuit applications using vacuum tubes, described in Chap. 11, for (1) the local oscillator, (2) the frequency converter, and (3) the i-f amplifier, may be applied to similar transistor circuit applications.

15-2. Oscillator Circuits. *Types of Oscillators.* An oscillator is basically an amplifier with an in-phase signal coupled from its output circuit back into its input circuit. As many of the properties of transistors and basic transistor amplifier circuits are practically equivalent to those of vacuum tubes and basic vacuum-tube amplifier circuits, any vacuum-tube oscillator circuit has an equivalent transistor circuit. Two basic types of feedback oscillator circuits will be described in this chapter, one using tuned reactive circuits and the other using RC networks.

Hartley Oscillator. One of the forms of the Hartley vacuum-tube oscillator circuit is shown in Fig. 15-4a. An equivalent transistor circuit using the grounded-emitter connection is shown in Fig. 15-4b. In the same manner as the vacuum-tube circuit (Art. 8-7), positive feedback is obtained by connecting the tuned L_1C_1 tank circuit so that it is common to both the input and output circuits of TR_1. The section AB of coil L_1 is in the base-emitter circuit, and the remainder AC is in the collector-emitter circuit. A portion of the amplified r-f energy in the collector circuit is returned to the base circuit by means of C_3 and the inductive coupling of L_1. The r-f current in the collector section of L_1 induces a voltage in the base section of L_1. The voltage developed across section

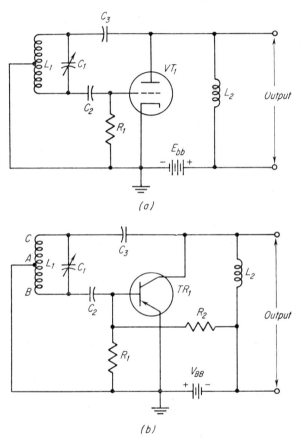

(a)

(b)

FIG. 15-4. Hartley oscillator circuits. (a) Vacuum-tube circuit with grounded cathode. (b) Transistor circuit with grounded emitter.

AB causes an r-f current to be fed to the base of TR_1. The collector and base voltages are 180 degrees out of phase with each other as they are taken from opposite ends of L_1 with respect to the common lead connected to the emitter. The current being returned to the base of the transistor will therefore be in phase with the input current to the base, thus maintaining oscillation. The amount of inductive feedback between the two sections of L_1 will depend on the number of turns in each section. Decreasing the number of turns in the collector section, while the number of turns in the base section is not changed, increases the amount of voltage induced in the base section, thus also increasing the amount of current feedback.   The other circuit parameters are as follows: (1) the base biasing resistors R_1 and R_2, (2) the d-c blocking capacitor C_2, (3) the d-c blocking and r-f coupling capacitor C_3 which prevents the direct current from entering the tank circuit L_1C_1, and (4) the r-f blocking and d-c coupling choke L_2 which prevents the r-f from entering the power supply.

A major difference between the two oscillator circuits shown in Fig. 15-4 is the lower operating Q of the transistor tank circuit.   This decrease in circuit Q is caused by the loading effect of the emitter-base resistance of TR_1 on coil L_1.   The emitter-base resistance is reflected into the tank circuit and as it is much smaller than the collector-emitter resistance, which also shunts a section of the tank circuit, it acts as the equivalent shunting resistance of the tank circuit.   The point on L_1 to be tapped to obtain optimum performance may be determined by

$$n = \frac{\alpha}{2} \tag{15-1}$$

Where n is the ratio of feedback turns to total turns, $\dfrac{N_{AB}}{N_{BC}}$

As oscillation will start when the equivalent shunting resistance of the tank circuit is counterbalanced by the reflected resistance of the emitter, the transistor will oscillate at its highest efficiency (assuming $\alpha = 1$) with a center-tapped coil.   For this condition the equivalent shunting resistance is equal to

$$R_{eq} = \frac{4r_e}{\alpha^2} \tag{15-2}$$

and
$$Q = \frac{2\pi f_r L \alpha^2}{4r_e} \tag{15-3}$$

where L = inductance of tank coil, mh
     f_r = resonant frequency of tank circuit, kc

The circuit shown in Fig. 15-5 is used as the local oscillator for an a-m broadcast radio receiver and is an adaptation of the basic Hartley circuit. The voltage developed across section AB of L_1 represents the energy returned to the base input circuit through capacitor C_2.   The collector terminal of TR_1 is connected to tap C on coil L_1: (1) to decrease the effect of the collector-emitter capacitance, (2) to provide a better impedance match between the transistor and the tank circuit (AD of L_1 and $C_1C_3C_4$), and (3) to improve the frequency stability and tracking.   Three-point tracking of the tuning capacitor C_1 is obtained by adjusting: (1) the slug S in the oscillator coil for the center portion of the band, (2) the padder capacitor C_4 for the low-frequency end, and (3) the trimmer capacitor C_3 for the high-frequency end.   Energy from the oscillator circuit is inductively coupled to the mixer circuit by means of the winding L_2 on the oscillator coil.   The other circuit parameters are: (1) the base biasing resistors R_1 and R_2, (2) the r-f blocking and d-c coupling choke L_3, (3) the d-c blocking and r-f coupling capacitor C_5, (4) the stabilizing resistor R_3, and (5) the emitter bypass capacitor C_6.

Colpitts Oscillator.   One of the basic forms of the Colpitts vacuum-tube oscillator circuit is shown in Fig. 15-6a.   An equivalent transistor circuit using the grounded-emitter connection is shown in Fig. 15-6b. Positive feedback is obtained by connecting the tuned $L_1C_1C_2$ tank circuit so that it is common to both the input and output circuits of TR_1. Capacitor C_1 is in the base-emitter circuit, and capacitor C_2 is in the collector-emitter circuit.   By varying the capacitance of C_1 and/or C_2,

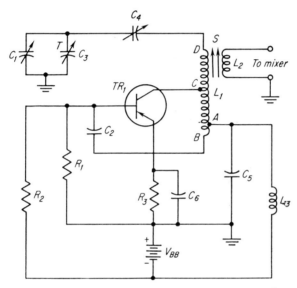

FIG. 15-5. Local oscillator circuit used in an a-m broadcast radio receiver.

the voltage across the tank circuit may be divided to produce the required voltage across C_1 to maintain oscillation.   The other circuit parameters are as follows: (1) the base biasing resistors R_1 and R_2, (2) the d-c blocking capacitor C_4, (3) the battery-supply decoupling r-f choke L_2, and (4) the battery-supply a-c feedback bypass capacitor C_3.

Clapp Oscillator.   The resonant load in the collector circuit of both the Hartley and Colpitts oscillators is a parallel arrangement; hence their circuits are voltage-controlled.   As a transistor is a current-operated device, greater frequency stability can be obtained by using a current-controlled oscillator circuit.   The Clapp oscillator circuit shown in Fig. 15-7 is current-controlled and is obtained by modifying the basic Colpitts circuit of Fig. 15-6b by connecting an additional capacitor C_5 in series with L_1.   The effective capacitance across L_1 is equal to C_1, C_2, and C_5 in

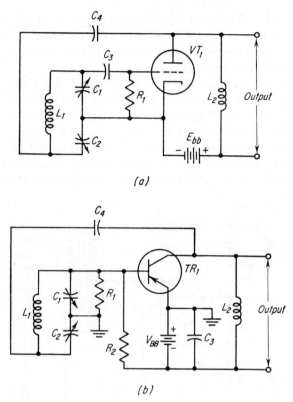

(a)

(b)

FIG. 15-6. Colpitts oscillator circuits.   (a) Vacuum-tube circuit with grounded cathode. (b) Transistor-circuit with grounded emitter.

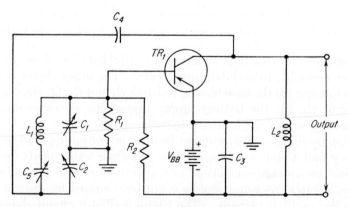

FIG. 15-7. Clapp oscillator circuit.

series. The capacitance of C_5 is made much smaller than C_1 or C_2, and therefore the resonant frequency of the tank circuit will be determined primarily by the values of L_1 and C_5. Since these two components are in series, the oscillator circuit will be current-controlled. The effect of changes in the interelectrode capacitances of the transistor on the oscillator frequency is minimized by making the value of C_1 much greater than the emitter-base capacitance, and the value of C_2 much greater than the emitter-collector capacitance.

RC Oscillators. Negative-resistance circuits using *RC* and/or *RL* networks and transistors may be used to produce relaxation-type oscillators in a similar manner to using these networks with vacuum tubes. A basic plate-to-grid–coupled vacuum-tube multivibrator circuit is shown in Fig. 15-8a. An equivalent transistor circuit using a collector-to-base–coupled common-emitter arrangement is shown in Fig. 15-8b. As there is a 180-degree phase shift in the grounded-emitter connection, the feedback signal from the collector output of TR_2 will be in phase with the input signal to the base of TR_1. The interstage coupling capacitor will be either C_1 or C_2, depending on the operating characteristics of the two transistors, and the other capacitor serves to provide feedback. The base biasing resistors are R_3 and R_4, and the collector load resistors R_1 and R_2. Power is supplied from a single battery V_{BB}. The approximate operating frequency is determined by the R_3C_1 and the R_4C_2 time constants. The transistor collector-current cutoff point (I_{co}) and the emitter-current cutoff point (I_{eo}) must also be considered in determining the exact operating frequency.

The normal operation of a PNP transistor is for the base to be negative with respect to its emitter. Cutoff can therefore be obtained by making the voltage at the base positive with respect to its emitter. At the instant when the operating voltages are applied, the collector current in one transistor will be slightly higher than the other. If the output current i_{c1} from TR_1 is slightly higher than the output current i_{c2} from TR_2, then i_{c1} will rise very rapidly to its maximum value, and i_{c2} will decrease very rapidly to zero. In other words TR_1 will be conducting and TR_2 will be at cutoff. The positive potential required by the base of TR_2 for effective cutoff is developed across C_2 and applied across R_4. This transistor will remain at cutoff until the positive charge on C_2 has discharged through R_4. TR_2 will then start conducting, and i_{c2} will rise very rapidly to its maximum value and i_{c1} will decrease very rapidly to cutoff. The controlling circuit for TR_1 is C_1 and R_3. The above actions will be repeated at a rate determined by the resistance and capacitance values of the circuit, and the current cutoff characteristics of the transistors. The output signal is basically a series of sharpsided pulses, or a *rectangular wave.* A saw-tooth wave can be obtained by connecting a capacitor across the output, or from the collector of TR_2 and ground.

15-3. Converter Circuits. *Mixer Circuit.* A transistor converter circuit consists of an oscillator and a mixer section. As with vacuum-tube circuits the functions of these two sections can be performed by using either a single transistor, or two transistors. When two transistors are

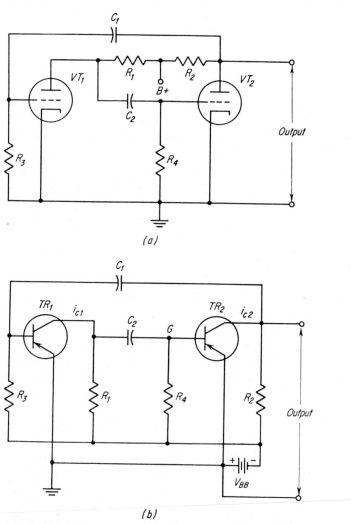

(a)

(b)

Fig. 15-8. *RC* multivibrator circuits. (a) Vacuum-tube circuit with plate to grid coupled. (b) Transistor circuit with collector to base coupled.

used, one is connected in an oscillator circuit similar to Fig. 15-5, and the other in a mixer circuit similar to Fig. 15-9. The tuned signal from the r-f stage is applied to the base of the common-emitter transistor circuit by means of C_1 and R_2. The required positive base voltage is obtained from the voltage divider R_1R_2. Stabilization of the collector current is

provided by R_3.   The oscillator signal is coupled to the emitter through
C_2.   This additional alternating voltage drives the emitter into conduction during the negative half-cycles, and into cutoff during the positive
half-cycles.   The rectified emitter voltage that is produced causes the
mixer transistor to operate nonlinearly near the Class B operating point.
The collector current will vary at both the r-f and oscillator signal rates
as both voltages modulate the base-emitter diode current.   The difference
frequency is selected by the i-f transformer T_1.

Autodyne Converter.   Many transistor radio receivers use a single
transistor to provide the functions of the local oscillator and mixer.   This
type of circuit is called an *autodyne converter*.   In a vacuum-tube pentagrid
converter the oscillator signal is coupled to the input circuit within the

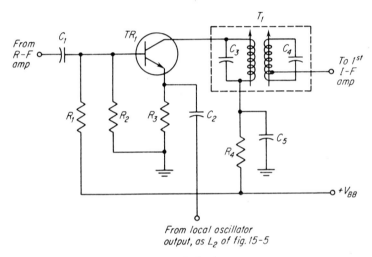

FIG. 15-9. Typical mixer stage.

tube.   In the autodyne converter circuit the oscillator signal must be
coupled to the input circuit outside the tube or transistor.   In the usual
transistor radio-broadcast frequency converter the r-f signal is fed to the
base, and the i-f signal is taken from the collector.   The oscillator feedback may be either to the base in series with the r-f signal, or to the
emitter.   A typical autodyne converter circuit using an NPN transistor
is shown in Fig. 15-10*a*.   This circuit may be redrawn to illustrate
separately the oscillator circuit Fig. 15-10*b*, and the mixer circuit Fig.
15-10*c*.

Operation of the Oscillator Circuit.   In the circuit of Fig. 15-10, a
slight variation in base-emitter current, produced for example by random
noises, is amplified to a larger variation of collector-emitter current.
The signal flowing through the primary winding of L_2 induces an alternating voltage in the secondary winding of L_2, which is tuned to the desired

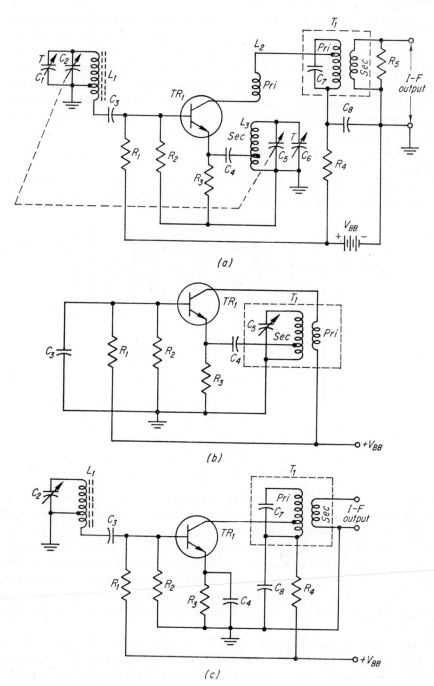

FIG. 15-10. Autodyne converter circuit using an NPN transistor. (*a*) The complete converter circuit. (*b*) Portion of the circuit performing the oscillator functions. (*c*) Portion of the circuit performing the mixer functions.

oscillator frequency by the relative values of L_2C_5. The resonant frequency signal is then coupled back into the emitter circuit through C_4. The feedback (or tickler) winding of L_2 is phased to produce a positive (regenerative) feedback signal of proper magnitude to sustain oscillations. To achieve a proper impedance match between the high-impedance tank circuit of $L_2C_5C_6$ and the relatively low impedance of the emitter circuit, the secondary winding of L_2 is tapped to operate as an autotransformer. As C_3 effectively bypasses any r-f signal around the biasing resistors R_1 and R_2 through a portion of L_1 to ground, the base of the transistor is a-c grounded, and the oscillator operates as a grounded-base configuration.

Operation of the Mixer Circuit. In the circuit of Fig. 15-10 the ferrite-core antenna L_1 is exposed to the radiation field of the entire frequency spectrum, and is tuned by C_2 to the desired frequency. The transistor is biased in a relatively low-current region so that it produces nonlinear characteristics. The incoming signal will mix with the oscillator signal, and the following four frequencies will be present in the transistor, (1) the local oscillator signal frequency, (2) the received incoming signal frequency, (3) the sum of these two frequencies, and (4) the difference between these two frequencies. The i-f load impedance T_1 is tuned to the difference between the oscillator and incoming signal frequencies. This frequency is maintained at a fixed value, as C_2 and C_5 are mechanically ganged together and designed to produce an oscillator frequency that is always greater than the incoming frequency by an amount equal to the intermediate frequency. Undesirable currents are kept out of the collector-emitter circuit by the filter network R_4C_8. The biasing and stabilizing resistor R_3 is essentially bypassed to ground for r-f signals through a few turns of the secondary winding of L_2 and the capacitor C_4. As the emitter is grounded and the incoming signal is injected into the base of the transistor, the mixer operates as a grounded-emitter configuration.

15-4. Intermediate-frequency Amplifier Circuits. *Frequency of Operation.* The frequency of operation of i-f amplifiers varies with the manufacturer and the type of receiver. In general the frequency is of the order of 455 kc, which is the same as the frequency used in vacuum-tube superheterodyne receivers. A much lower frequency of the order of 262 kc is sometimes used to provide greater gain and more stability. However, the low frequency of operation has the disadvantage of causing the receiver to be more susceptible to image-frequency pickup (see Art. 11-9).

Single-stage I-F Amplifier. A single stage of i-f amplification is shown in Fig. 15-11. Proper matching of the high output impedance of one stage to the low input impedance of the following stage is obtained by: (1) tuning the primary circuit of each transformer to the desired frequency by the variable inductance and the fixed capacitors C_1 and C_4,

(2) having the secondary circuits of each transformer untuned, and
(3) tapping the primary winding of each transformer. Peaking of
each i-f coil is achieved by varying the position of the iron-core slug.
Proper forward bias is obtained by the voltage-divider network R_1R_2,
and degeneration is minimized by C_2 and C_3. Connecting a transformer
to an i-f amplifier stage whose output resistance is less than 100,000 ohms
lowers its circuit Q, thus causing poor selectivity. The output resistance
of a 455-kc transformer-coupled transistor amplifier stage is approximately 25,000 ohms. Thus, if the collector circuit was shunted across the
entire primary winding of T_1 or T_2 the selectivity would be seriously
affected. This problem is remedied by connecting the collector to a tap
on the primary winding of the i-f transformer. The desired circuit Q,

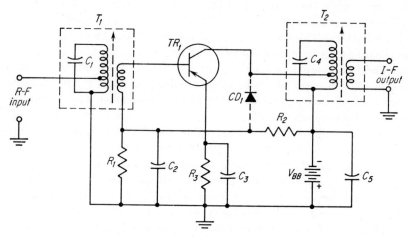

Fig. 15-11. Single-stage i-f amplifier circuit.

and proper impedance match, can be obtained by selecting the correct
tap for the collector connection.

A transistor i-f amplifier stage is more complex than a vacuum-tube i-f
amplifier stage because: (1) the low-input impedance of a transistor must
be matched to the relatively high output impedance of the preceding
circuit, and (2) the collector-base capacitance of the transistor may be
high enough to cause undesirable feedback, which makes it necessary to
use neutralizing circuits to achieve the desired stability. The i-f transformer is basically used to select the desired signal frequency having
sidebands capable of passing a specified range of audio frequencies, and
rejecting the adjacent signal frequencies. The i-f amplifier circuit
parameters must therefore be such that the Q of the transformers is
maintained. Usually it is impractical to use a single stage of i-f amplification (1) because of bandwidth considerations, and (2) because of the
low detection efficiency at low-signal levels. However, it is possible to use

a single stage of i-f amplification in circuits designed for maximum gain. In these circuits a crystal diode CD_1, called an *overload diode*, is connected from the collector to the biasing resistor R_2 as shown in Fig. 15-11. Because transistors have sharp-cutoff characteristics, this diode can provide automatic gain control. Conduction through the diode is delayed by the voltage across R_2 until the signal voltage is greater than the delay voltage. Conduction through the diode lowers the Q of transformer T_2, thus decreasing the power gain.

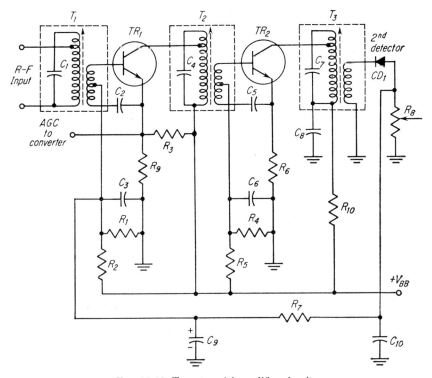

Fig. 15-12. Two-stage i-f amplifier circuit.

Multistage I-F Amplifier. The gain, bandwidth, and selectivity requirements of an i-f amplifier can usually be obtained with two stages of amplification. More than two stages are seldom used because of the relatively low gain obtained from an additional i-f amplifier stage compared to the gain that can be obtained from an additional stage of audio amplification. The low gain is partly due to the high stability factor required by the additional i-f amplifier stages. The bandwidth and selectivity requirements may necessitate the use of one or more double-tuned i-f transformers.

The i-f amplifier circuit shown in Fig. 15-12 uses two stages of i-f amplification to provide the required gain and selectivity with a high degree of circuit stability.　The two transistors are identical, and their load impedance is approximately 30,000 ohms.　The impedance of the resonant input circuits of T_1, T_2, and T_3 is of the order of 700,000 ohms. Maximum power transfer and high circuit Q are obtained by means of the tapped primary windings of these transformers.　The input impedance of the transistor is of the order of 40 ohms, and maximum power transfer is obtained by using a step-down transformer to match this low impedance.　The split input provided by the tapped secondary windings of T_1 and T_2 increases the circuit stability.　The two NPN transistors are biased in a forward direction by the two voltage-divider networks

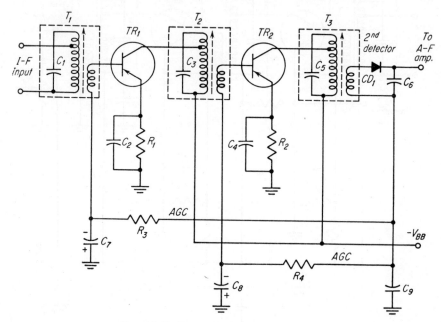

Fig. 15-13. Two-stage i-f amplifier circuit.

R_1R_2 and R_4R_5, and degeneration in the emitter circuits is minimized by C_3 and C_6.

A simpler and more economical two-stage PNP transistor i-f amplifier circuit is shown in Fig. 15-13.　A slight increase in the transistor load impedance plus higher values of resistance in the emitter circuits are two methods used to obtain circuit stability, rather than the tapped secondary i-f transformers in the circuit of Fig. 15-12.

Neutralizing Circuits.　Neutralization of i-f stages is usually necessary because high-frequency transistors are inherently regenerative in the grounded-emitter connection.　Whether or not an i-f amplifier requires neutralizing depends upon the collector-to-base capacitance of the transistor being used.　When this internal capacitance is low enough so that it does not cause a noticeable amount of feedback, neutralization is not required. However, when the capacitance is large enough to produce undesirable feedback, some form of neutralization must be used. In the circuit shown in Fig. 15-12 the secondary of the i-f transformer is tapped for neutralization purposes in a similar manner to the i-f transformers used with triode vacuum tubes. Capacitors C_2 and C_5 provide a feedback path from the emitter to base of their corresponding transistor. The value of this capacitance is not too critical as the amount of feedback is controlled by the transformer design. However, the capacitance must be high enough to provide a low-impedance path for the signal currents.

Neutralization of i-f amplifier stages may also be accomplished by connecting a feedback capacitor from the base of the transistor in one stage to the base of the transistor in the preceding stage. The value of this capacitance is dependent on the value of the collector-base capacitance of the transistor used, and the turns ratio between the tap on the primary and the turns on the secondary of the i-f transformer. The value of this capacitance may be approximated by

$$C_n = nC_c \tag{15-4}$$

where C_n = neutralizing capacitance, $\mu\mu f$

n = turns ratio

C_c = collector-base capacitance, $\mu\mu f$

Example 15-1. A transistor has a collector-base capacitance of 10 $\mu\mu f$. There are 54 turns to the tap on the primary and 18 turns on the secondary of the i-f transformer. What approximate value of neutralizing capacitance should be used?

Given: $C_c = 10\ \mu\mu f$ Find: C_n
$N_P = 54$
$N_S = 18$

Solution:

$$C_n = \frac{N_P}{N_S} C_c = \frac{54}{18} \times 10 = 30\ \mu\mu f$$

The collector-base capacitance of the transistors used in the i-f amplifier circuit of Fig. 15-13 is very low, and no form of neutralization is required.

Automatic Gain Control. A constant-strength r-f signal output is maintained at the detector of a radio receiver by a system called an *automatic-gain-control circuit*, abbreviated agc, that automatically varies the amplification of the i-f and/or r-f signal with changing input signal strength. This system is also referred to as an *automatic-volume-control circuit*, abbreviated avc. An agc system consists in taking a voltage proportional to the incoming signal from the second-detector output, and applying it to the base of one or more transistors in the i-f or converter stages to oppose the base current of the transistor.

In the circuit shown in Fig. 15-12, automatic gain control is applied to the first i-f stage and to the converter. Automatic gain control is obtained by indirectly controlling the emitter current of TR_1 by applying the control voltage to the base of this transistor. When the incoming signal is strong, the voltage returned through the agc line decreases the base voltage, thus reducing the emitter current and emitter voltage. As a decrease in the emitter current also decreases the collector current and the resultant signal developed across the tuned primary circuit of T_2, the gain of the stage is reduced. When the incoming signal is weak, the amount of control voltage returned is reduced, thus permitting an increase in the emitter current with the resultant increase in stage gain. The values of the agc line resistor R_7 and capacitor C_9 are opposite to those used in vacuum-tube circuits. The time constant of the RC circuit is the same for vacuum-tube and transistor circuits. In a vacuum-tube circuit the resistance is relatively high with a corresponding low value of capacitance. However, because of the relatively low values of resistance that can be used in transistor circuits, a high value of capacitance is required.

Example 15-2. What value of agc capacitance must be used with a resistance of 5,200 ohms to obtain a time constant of approximately 0.1 sec?

$$\text{Given:} \quad T = 0.1 \text{ sec} \qquad\qquad \text{Find:} \quad C$$
$$R = 5,200 \text{ ohms}$$

Solution:

$$C = \frac{T}{R} = \frac{0.1}{5,200} = 0.0000192 \text{ farad (use 20 } \mu\text{f)}$$

In the circuit shown in Fig. 15-13, agc is applied to the two i-f amplifier stages by means of R_3C_7 and R_4C_8, rather than to the converter and one i-f amplifier stage.

Reflex Amplifier Circuit. A reflex amplifier is one that amplifies two signals of different frequencies, usually the intermediate and audio frequencies. In this circuit the transistor is used concurrently as both an i-f and a-f amplifier. After detection, the audio portion of the signal is returned to the i-f amplifier where it is again amplified. Since the signals to be amplified are of widely different frequencies there is no regenerative effect. The input and output circuits of this stage can therefore have split i-f and a-f loads.

In the circuit shown in Fig. 15-14, the i-f signal is fed through the i-f transformer T_2 to the detector circuit. The audio signal developed across the volume control R_5 is returned through C_3 to the ground side of the secondary of T_1, and then on to the input circuit of TR_1. The secondary is essentially a short circuit at audio frequencies as the winding consists of only a few turns. The primary of T_2 is also essentially a short circuit at audio frequencies, and the audio signal developed across R_4 is fed to the audio-output stage. The i-f signals appearing across the parallel combination of R_1 and R_2 are bypassed by C_2. The i-f and a-f signals appearing across the emitter resistor R_3 are bypassed by C_4. The audio output resistor R_4 is bypassed for i-f by C_6.   This type of circuit can contribute in the order of 30 db i-f gain, and 35 db a-f gain.

An advantage of this type of circuit is that one amplifier stage can produce the gain usually obtained from two stages of amplification, with the resulting saving in cost, space, and battery drain.   A disadvantage is the distortion at minimum volume, caused by the balancing out of the fundamentals of the normal signal and the out-of-phase audio component.

15-5. Detector Circuits.   *Diode Detector.*   The simplest type of detector circuit is obtained by using a crystal diode as a half-wave rectifier (Art. 1-17).   The operation of a crystal diode as an a-m detector for transistor radio receivers is the same as for vacuum-tube radio receivers. The crystal diode consists of a germanium or silicon pellet called the

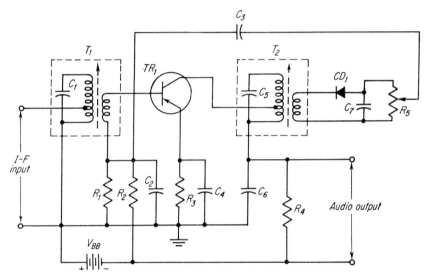

Fig. 15-14. Reflex amplifier circuit.

cathode, and a catwhisker called the *anode,* mounted in a tubular ceramic form with connecting leads at each end.   The cathode and anode of the crystal diode are connected in the same manner as the cathode and plate of a vacuum-tube diode (Fig. 3-2).   Crystal diodes are used extensively in transistor radio receivers because they offer these advantages: (1) small size, (2) low internal capacitance, (3) low dynamic resistance, (4) high efficiency, and (5) the fact that they require no operating power.

Class B Power Detector Circuit.   A transistor may be operated as a Class B power detector in a manner similar to plate detection with triode vacuum tubes (Art. 3-3).   The basic principle of this type of circuit is to use the current cutoff characteristics of transistors (I_b versus I_c, where V_c is constant) to accomplish the necessary rectification.   The base-collector transfer characteristic curves can be obtained by cross-plotting the values on the grounded-emitter collector characteristic curves of Fig. 13-18. There are several reasons why this type of circuit is used in some receivers rather than the simple crystal diode: (1) It may provide about 10 db of audio amplification in addition to performing the functions of detection, whereas a diode introduces a loss of gain.   (2) Essentially linear detection is obtained at lower power levels than with diodes.   (3) It has the ability to supply more agc power than a diode.   (4) Transistor characteristic curves are more linear than vacuum-tube characteristic curves and therefore introduce less distortion.

In the Class B detector circuit of Fig. 15-15 the i-f signal is applied to the base of the NPN transistor TR_1.   Since the circuit does not use any

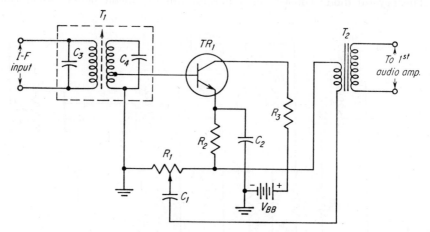

FIG. 15-15. Class B power-detector circuit.

forward biasing network, the transistor will remain at cutoff until a signal is applied.   During the positive half-cycles of the applied i-f signal the base of TR_1 is driven in a forward direction causing the transistor to conduct.   During the periods of conduction the collector and base currents will flow through R_1, causing a voltage to be developed across the volume control.   During the negative half-cycles of the applied i-f signal, the base of TR_1 is driven in a reverse direction, causing the transistor to operate at cutoff, and the voltage across the volume control will be zero.   Distortion of the rectified signal is minimized by bypassing the i-f signal around the detector load resistance (R_1 and R_2) through C_2 to ground.   The value of this capacitance depends on the value of the load resistance, and should be of a value that will filter the i-f signal to ground, and allow the a-f signal to remain.   A capacitance of the order of 0.05 μf is generally used.   The audio signal at the volume control is coupled through C_1 to the primary winding of interstage audio transformer T_2.

15-6. Radio Receiver Circuits. *Basic Design.* The superheterodyne circuit is the basic configuration used in commercial transistor radio receivers, and its basic functions are the same as those for a comparable vacuum-tube radio receiver (see Fig. 15-16). The typical basic circuit uses four transistors and a crystal diode to provide for: (1) a converter, (2) one stage of i-f amplification, (3) a detector, (4) an audio driver stage, and (5) a single stage of audio output. The more elaborate circuit may (1) substitute a transistor for the crystal diode, (2) use two or more stages of i-f amplification, (3) use two or more stages of a-f amplification, (4) use a diode or transistor in the i-f amplifier circuit as an overload limiter, or (5) use two transistors in a push-pull audio output stage. The crystal diode is generally used as the detector because of its simplicity and economy. Some receivers use a transistor operated as a Class B power detector because of the additional gain that can be obtained with this arrangement. The transistors are usually connected in the grounded-emitter configuration in which the signal is applied to the base-emitter,

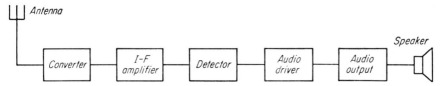

FIG. 15-16. Block diagram of a typical transistor superheterodyne radio receiver.

and the amplified output is taken from the collector-emitter. Both PNP and NPN transistors are used, and it is not uncommon to find both types being used in the same receiver circuit.

15-7. RCA Radio Receiver. *Circuit Operation.* The circuit shown in Fig. 15-17 is for an RCA broadcast-band radio receiver using eight transistors and a crystal diode in a superheterodyne circuit consisting of: (1) a 2N412 converter, (2) two 2N410 i-f amplifier stages, (3) a 3458 overload limiter, (4) a 1N60 second detector, (5) a 2N406 a-f amplifier stage, (6) a 2N406 audio driver stage, and (7) two 2N408's as a Class B push-pull audio output. The transistors are all of the PNP type, and are connected in the grounded-emitter configuration except (1) the oscillator section of TR_1 which operates as a grounded base, and (2) TR_4 which is connected as a grounded collector. The tuning range is from 540 to 1,600 kc, with an intermediate frequency of 455 kc. The receiver is capable of delivering a maximum output of 400 mw, and an undistorted output of 300 mw. A 3½-inch p-m speaker is provided for normal listening, and an audio jack J_1 for a high-impedance (2,000 ohms) earphone use. The receiver operates from a 6-volt power source having a no-signal current consumption of 9 ma. The power source may be either (1) four Penlite cells, which provide approximately 22 hours of

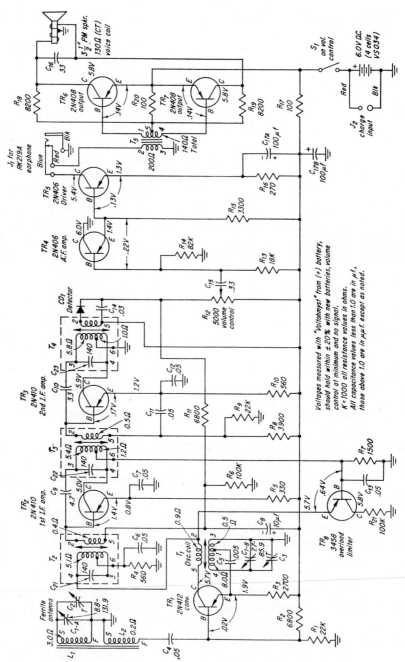

FIG. 15-17. RCA eight-transistor radio receiver circuit. (Courtesy of RCA.)

Voltages measured with "Voltohmyst" from (+) battery,
should hold within ± 20% with new batteries, volume
control at minimum and no signal.
K=1000 all resistance values in ohms.
All capacitance values less than 1.0 are in μf.,
those above 1.0 are in μμf. except as noted.

intermittent service, or (2) four rechargeable cells, which provide approximately 25 hours of intermittent service from one overnight charge. A socket is provided for connecting the charger unit to the battery. This is a personal type of radio, and the complete receiver including the batteries weighs approximately one and a half pounds. Light weight and compactness are obtained by using a printed circuit chassis (Fig. 15-18).

Converter Circuit. The incoming signal is received with a relatively high pickup, and good image rejection, by the ferrite-rod antenna L_1L_2. The desired signal is selected by adjusting the tuning capacitor C_{1A}. The selected r-f signal is fed to the base of the converter transistor TR_1 through C_4. The a-c signal flowing through the primary winding of the oscillator transformer T_1 causes an alternating current to flow through the secondary winding of this transformer, which is tuned to the desired oscillator frequency by the variable capacitor C_{1B} (mechanically ganged to C_{1A}). The resonant frequency signal is coupled back into the emitter through C_5 causing the transistor to operate nonlinearly near its Class B operating point. The collector current will vary at both the r-f and oscillator signal frequencies as both voltages modulate the base-emitter diode current. The difference frequency is selected by the i-f transformer T_2. The secondary winding of T_1 is tapped to operate as an autotransformer, thus providing a relatively good impedance match between the resonant circuit and the emitter circuit. The oscillator section of the converter operates as a grounded-base configuration as the base is a-c grounded by C_4, which bypasses the biasing resistors R_1 and R_2 to ground. Oscillator signal injection to the emitter of TR_1 is obtained through C_5. The mixer section of the converter operates as a grounded-emitter configuration, as the emitter is grounded and the incoming signal is injected into the base. Resistor R_3 provides the operating bias and stabilization for the converter TR_1.

I-F Amplifier Circuit. This receiver uses two stages of i-f amplification. Proper impedance matching between stages is obtained by using transformers with tuned primary circuits, untuned secondary circuits, and a tapped primary winding. Maximum stability and high gain are obtained from the permeability-tuned i-f transformers T_2, T_3, and T_4. The voltage-divider networks R_6R_{11} and R_8R_9 provide the required operating bias for TR_2 and TR_3 respectively. Degeneration is minimized by bypassing the biasing and stabilizing resistors R_5R_{10} to ground by C_7 and C_{12}. Neutralization is applied to the first and second i-f amplifier stages, and the feedback paths are provided by C_9 and C_{10}. Automatic gain control is only applied to the first i-f amplifier stage, and the control voltage is taken from the output of the crystal-diode detector CD_1 and is applied to the base of TR_2 through the $R_{11}C_8$ network. The sharp-cutoff characteristics of transistor TR_2 are counteracted by the overload limiter TR_8. The operating potential for the overload limiter is provided

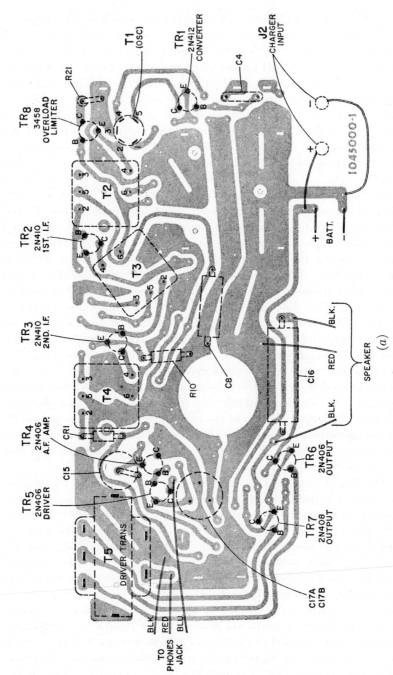

FIG. 15-18. Chassis wiring and components of receiver circuit shown in Fig. 15-17. (a) View from wiring side.

by R_4. When a strong signal is applied to the primary of the first i-f transformer T_2, it causes the voltage across R_4 to increase. This increase in voltage causes the overload limiter to conduct, thus decreasing the Q of T_2 and the resulting power gain of the stage.

Detector and Volume-control Circuits. A crystal diode detector CD_1 is used to drive the first stage of a-f amplification. The load resistance for the diode detector is the volume control R_{12}. The strength of the output signal is controlled by varying the amount of rectified signal voltage across R_{12} that is applied to the base-emitter circuit of TR_4. The resistance of the volume control is relatively low, compared with the

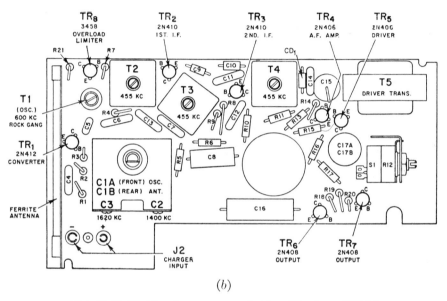

(b)

FIG. 15-18. (b) View from component side.   (*Courtesy of RCA.*)

resistance used in vacuum-tube circuits, because of the low-input impedance of TR_4. In order to maintain the time constant required to prevent the audio signal from blocking the transistor operation, a relatively high value of capacitance, compared with that used in vacuum-tube circuits, is required for the demodulating capacitor C_{14}, and the d-c blocking capacitor C_{15}.

Audio-amplifier Section. The a-f output from the detector is coupled by means of C_{15} to the base of TR_4 in the a-f amplifier stage. This stage operates as a grounded-collector configuration, or cathode-follower circuit, and serves as an excellent impedance match between the detector output and the base-emitter input of the audio-driver stage TR_5. A relatively high current output is taken from the emitter of TR_4 and is fed directly to the base of TR_5. The audio-driver stage is operated as a Class

A amplifier.   The output of TR_5 is coupled to the base-emitter circuits of TR_6 and TR_7 by the stepdown interstage transformer T_5.   Transistors TR_6 and TR_7 are operated as a Class B push-pull audio amplifier.   The correct forward bias, which is critical in this type of circuit, is provided for both transistors by R_{20} in conjunction with R_{18} and R_{19}.   The voice coil of the p-m speaker is center-tapped to ground, and has a resistance of 130 ohms.   Since this resistance compares favorably to the output impedance of TR_6 and TR_7, no impedance-matching transformer is required, and the voice coil is connected directly to the collectors of these two transistors.

15-8. Auto-radio Receivers.   *Hybrid Radio Receiver.*   A radio receiver that uses both vacuum tubes and transistors is called a *hybrid radio receiver.*   Hybrid auto-radio receivers using low-plate-voltage tubes in the front end and transistors in the audio output are very popular.   The use of low-plate-voltage tubes eliminates the need of a vibrator and its associated circuit elements as the 12-volt plate supply can be taken directly from the car battery.   The tube circuits and the transistor circuits each draw approximately 1 amp (or a total of 2 amp) from a 12-volt battery.   The efficiency of this type of receiver is relatively high as with less power consumption the hybrid radio receiver retains the 6-watt maximum power output, and the 4-watt undistorted (less than 10 per cent distortion) power output of high-plate-voltage all-tube receivers.

GMC Hybrid Radio Receiver.   The hybrid radio receiver circuit shown in Fig. 15-19 uses three low-plate voltage vacuum tubes and two transistors.   A superheterodyne circuit is used, and the operations of (1) the 12DZ6 r-f amplifier, (2) the 12AD6 oscillator modulator, and (3) the 12SD7 detector amplifier are similar to the operation of these circuits in an all-tube superheterodyne receiver, as explained in Chap. 11.   A sufficient amount of gain is obtained from a single stage of i-f amplification using a PNP transistor DS-22 in a grounded-emitter configuration.   The i-f signal from the converter stage is transformer-coupled to the input of the i-f stage in a conventional manner.   The two i-f transformers T_2 and T_3 are tuned to 262 kc, and both have tuned-primary and tuned-secondary circuits.   To obtain a good impedance match, the secondary winding of T_2 and the primary winding of T_3 are tapped.   The output of the i-f amplifier is applied to the diode section of VT_3.   After detection and a-f amplification, the output signal of VT_3 is coupled through the interstage transformer T_4 to the base-emitter input of TR_2.   This transistor is connected in the grounded-emitter configuration and operates as a Class A amplifier.   To compensate for variations in transistor characteristics, the bias potentiometer R_{16} is adjusted to obtain the correct collector voltage at TR_2.   The output of TR_2 is coupled through the output transformer T_5 to the 4-ohm voice coil of the p-m speaker.   The usual

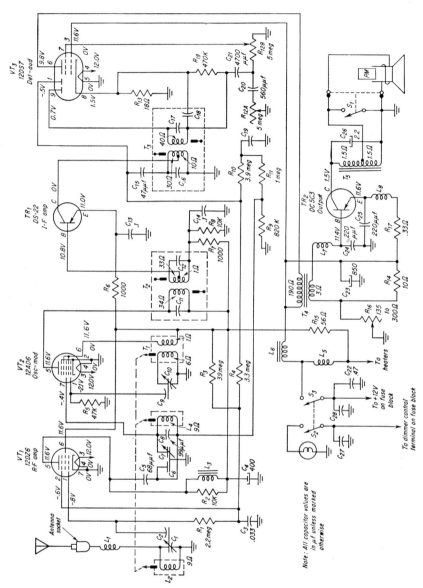

FIG. 15-19. GMCybrhid radio receiver circuit.   (*Courtesy of Delco Radio Division, General Motors Corporation.*)

interference obtained in auto-radio receivers is minimized by (1) the hash choke L_5, (2) the power-supply input choke L_6, (3) the transistor base choke L_7, and the transistor emitter choke L_8.   The receiver is a compact unit and employs printed circuits.   The power consumption is low, as the receiver only draws 1.4 amp at 12 volts.

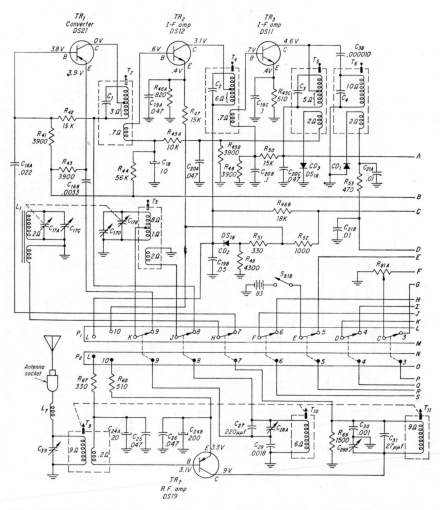

Fig. 15-20. Buick all-transistor radio receiver circuit.   (*Courtesy of Delco Radio Division,*

Buick All-transistor Portable Auto-radio Receiver.   This radio receiver may be used in the car or away from the car, as at the beach or in the home.   It consists of two sections, a portable unit and a car unit.   The portable unit is easily removed and provides a maximum power output of 100 mw with a 3-inch p-m speaker.   When the portable unit is connected with the car unit, an additional stage of r-f amplification and a power-output stage are added to boost the maximum power output to 6 watts with a 4- by 10-inch elliptical p-m speaker.   The car unit has its own permeability pushbutton tuner, and the portable unit a separate manual tuner.   The tuning range is from 540 to 1,600 kc, with an intermediate frequency of 465 kc.   The current drain as a car unit is 1.3 amp at

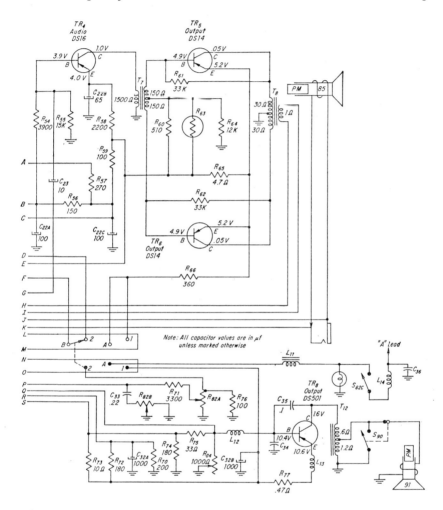

General Motors Corporation.)

12 volts, and as a portable unit 6 ma at 5.2 volts.   An audio jack J_1 is provided in the portable unit for earphone use.   The receiver is a compact unit and employs printed circuits in both the portable and car sections.

The circuit diagram for this receiver (Fig. 15-20) consists of two sections.   The portable unit uses six transistors and three crystal diodes in a superheterodyne circuit and is represented by the section of the circuit diagram above the terminal board P_1. The car unit uses two transistors and is represented by the section of the circuit diagram below the terminal board P_2. The signals to be received are picked up by a ferrite-core antenna L_1 when used as a portable receiver, and by a regular auto antenna when used as a car radio.

Using the portable unit, the received signal is selected by manually varying capacitor C_{17A} and is fed to the base of the converter transistor TR_1. The local oscillator signal, produced by the resonant tank circuit formed by the primary of $T_3C_{17B}C_{17D}$, is fed to the emitter of TR_1. The i-f signal output from the converter is fed to the primary winding of transformer T_2. Two stages of i-f amplification are used, employing i-f transformers T_2, T_4, and T_5T_6, and transistors TR_2 and TR_3. The first stage uses an NPN transistor TR_2, and the second stage an NPN transistor TR_3. Both transistors are connected in the grounded-emitter configuration. A crystal diode detector CD_1 is used to drive the audio amplifier TR_4. The load resistance for the diode detector consists of a portion of the secondary winding of the third i-f transformer T_6, R_{53}, and the volume control R_{81A}. Automatic gain control is obtained by use of the two overload crystal diodes, CD_2 in the first i-f amplifier stage, and CD_3 in the second i-f amplifier stage. The audio-amplifier stage uses a PNP transistor connected in a grounded-emitter configuration, and operated as a Class A amplifier. The output of this amplifier is coupled through the push-pull input transformer T_7 to the power-output stage, which operates as a Class B push-pull audio amplifier using matched PNP transistors TR_5 and TR_6. Accurate forward bias is provided for both transistors by R_{60}, R_{65}, and the thermistor R_{63}. The signal from the power-output stage is coupled through the push-pull output transformer T_8 to the voice coil of the p-m speaker.

When the receiver is to be used as a car radio, a multipoint switch connects and disconnects appropriate circuits. The antenna and oscillator circuits of the portable receiver are disconnected. In their place a permeability-tuned pushbutton tuning circuit consisting of the antenna coil T_9, the oscillator coil T_{10}, and the r-f coil T_{11} are connected into the circuit. The signal received by the car antenna is amplified by the PNP transistor TR_7, which is connected in the common-emitter configuration. The amplified r-f signal and the local oscillator signal are fed respectively to the base and emitter of TR_1. The output of the push-pull amplifiers TR_5 and TR_6 is transformer-coupled through T_8 to the base-emitter input of TR_8. This is a PNP transistor that is operated as a Class A audio power amplifier, and its output signal is coupled through the output transformer T_{12} to the voice coil of the p-m speaker in the car unit. The interference obtained in auto radios is minimized by (1) the power-supply input choke L_{11}, (2) the transistor base choke L_{12}, (3) the transistor emitter choke L_{13}, and (4) the hash choke L_{14}.

15-9. Auxiliary Circuits for Auto-radio Receivers. *Electronic Vibrator Power Supply.* A radio receiver may use standard vacuum tubes requiring high plate voltages in some of its stages. In such receivers transistors can be used to produce a square-wave output to replace the commonly used vibrator and its associated circuits. The relaxation, or blocking-oscillator, circuit shown in Fig. 15-21 uses two PNP transistors in a common-emitter connection. The operation of this circuit is similar to the *RC* oscillator explained in Art. 15-2. At the instant when the operating voltages are applied, one of the transistors will conduct through

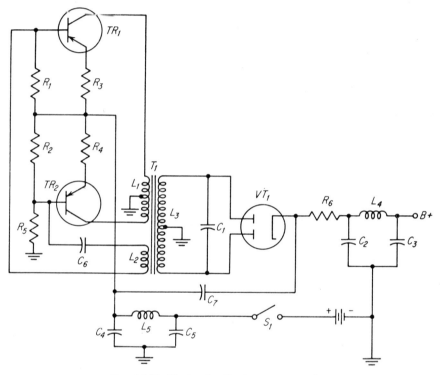

FIG. 15-21. Electronic-vibrator power supply.

L_1 and the other will remain at cutoff. After the collector current of the conducting transistor reaches saturation, the field in L_1 collapses and induces a voltage of opposite polarity in L_2. This induced voltage drives the conducting transistor to cutoff, and the other transistor into conduction. The current through L_1 will now flow in a direction opposite to the direction of flow when the first transistor was conducting, thus producing a square wave. These operations will be repeated at a frequency primarily determined by the inductance of T_1. A frequency of 20,000 cycles is commonly used as it is above the audio range, and the hum or buzz produced cannot be heard. The square-wave voltage on the primary of T_1 is increased to 200 to 250 volts by the step-up secondary winding L_3. The alternating voltage produced is rectified by the duo-diode vacuum tube VT_1, and filtered by the π network $L_4C_2C_3$.

Trigger Amplifier. When a radio receiver uses automatic tuning, a circuit is included to trigger the tuner relay. The source for the trigger voltage is taken from the output of the audio detector. In Fig. 15-22 this voltage is taken from the negative terminal of volume control R_1 and is applied directly to the base of the trigger amplifier TR_2. This NPN

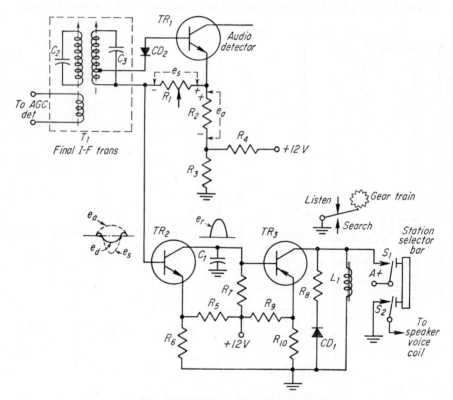

FIG. 15-22. Automatic tuning circuit.

transistor operates as a grounded emitter and is biased by the voltage-divider network R_3R_4. The purpose of this bias voltage is to maintain TR_2 at conduction until the negative trigger voltage is applied. The agc delay voltage e_a is in series with the detector output voltage e_s, and since e_a is positive, it adds to the d-c bias. However, the agc voltage opposes the signal voltage e_s to prevent the triggering action from occurring too quickly on strong signals. The manner in which e_a and e_s vary the base voltage is shown by the composite voltage curves drawn at the base of TR_2. Since both voltages increase in opposite directions in proportion to the signal strength, a constant difference voltage e_d is produced.   This difference voltage, which is always negative, drives TR_2 toward cutoff. The collector current through R_7 decreases causing the voltage at the collector to increase.   The output signal e_r is an amplified and inverted replica of the input signal e_d, and is used to trigger the relay-control stage.

Relay Control.   The positive signal from the trigger amplifier is filtered by C_1.   The direct voltage obtained is applied to the base of the relay-control PNP transistor TR_3, and the reverse bias causes the transistor to operate at cutoff.   Since no current flows in the collector circuit, the tuner relay L_1 deenergizes and blocks the motor, thus stopping the tuner. Once the tuner has been stopped, it is necessary to send current through the relay in order for it to resume searching.   This is accomplished by depressing the station selector bar, which closes S_1, causing a positive voltage to be applied to one end of the tuner relay.   Since the other end of the relay is grounded, current will flow through the winding and energize the tuner relay, thus starting the tuner to search.   No arcing noise is heard as the speaker is shorted to ground by S_2 at the same time that S_1 is closed.   Transistor TR_3 is protected by R_8 and CD_1 from excessive surge voltages that are developed as the relay is suddenly deenergized at station points.   The negative surge voltage causes the crystal diode CD_1 to conduct and leak the excess surge current through R_8 to ground.

QUESTIONS

1. Describe the operation of the one-transistor radio receiver circuit of Fig. 15-1.

2. Describe the operation of the two-transistor radio receiver circuit of Fig. 15-2a.

3. Describe the operation of the two-transistor radio receiver circuit of Fig. 15-2b.

4. Describe the operation of the two-transistor regenerative radio receiver circuit of Fig. 15-3.

5. Describe the operation and characteristics of a simple trf transistor radio receiver circuit.

6. Describe the operation of the basic Hartley oscillator circuit of Fig. 15-4b.

7. Describe the factors that determine the highest efficiency of oscillation of the tank circuit of Fig. 15-4b.

8. Describe the operation of the adaptation of the basic Hartley oscillator circuit shown in Fig. 15-5.

9. Describe the operation of the basic Colpitts oscillator circuit of Fig. 15-6b.

10. Describe the operation of the Clapp oscillator circuit of Fig. 15-7.

11. (a) With which type of oscillator circuit is it possible to obtain the greatest frequency stability?   (b) Explain your answer.

12. Describe the operation of the RC multivibrator circuit of Fig. 15-8b.

13. What four factors determine the exact operating frequency of the RC multivibrator circuit of Fig. 15-8b?

14. Describe the operation of the mixer circuit of Fig. 15-9.

15. Describe the operation of the oscillator section of the autodyne converter circuit of Fig. 15-10a.

16. Describe the operation of the mixer section of the autodyne converter circuit of Fig. 15-10*a*.

17. Describe three methods used for matching the high-output impedance of an i-f amplifier stage to the low-input impedance of the following stage.

18. Describe why the operation of a transistor i-f amplifier is more complex than the operation of a vacuum-tube i-f amplifier.

19. Why is it usually impractical to use only a single stage of i-f amplification?

20. Describe the operation of a single stage of i-f amplification using an overload diode.

21. Explain why only two stages of i-f amplification are generally used in radio receiver circuits.

22. Describe the operation of the two-stage i-f amplifier circuit of Fig. 15-12.

23. (*a*) Why is it necessary to neutralize a transistor i-f amplifier circuit?   (*b*) Describe the operation of the neutralizing circuit of Fig. 15-12.

24. Describe the operation of the agc circuit of Fig. 15-12.

25. (*a*) What is meant by a reflex amplifier?   (*b*) What are the advantages and disadvantages of the reflex amplifier?

26. Describe the operation of the reflex-amplifier circuit of Fig. 15-14.

27. Why are crystal diode detectors used extensively in transistor radio receivers?

28. (*a*) What is meant by a Class B power detector?   (*b*) What are the advantages of the Class B power detector?

29. Describe the operation of the Class B power detector circuit of Fig. 15-15.

30. (*a*) Name the basic circuits of a four-transistor radio receiver.   (*b*) Name the circuits that may be included in a more complex radio receiver.

31. (*a*) Describe the operation of the oscillator section of the converter circuit of Fig. 15-17.   (*b*) Describe the operation of the mixer section of the converter circuit of Fig. 15-17.

32. Describe how the following functions are obtained in the two-stage i-f amplifier circuit of Fig. 15-17: (*a*) impedance matching, (*b*) stability, (*c*) operating bias, (*d*) neutralization, (*e*) automatic gain control, (*f*) overload limiting.

33. In the radio receiver circuit of Fig. 15-17 describe the operation of (*a*) the detector circuit, (*b*) the volume-control circuit.

34. Describe the operation of the audio-amplifier section of the radio receiver circuit of Fig. 15-17.

35. Why is no output transformer required in the loudspeaker circuit of Fig. 15-17?

36. (*a*) What is meant by a hybrid radio receiver circuit?   (*b*) Where are hybrid radio receiver circuits generally used?   (*c*) What are the advantages of the hybrid radio receiver?

37. In the hybrid radio receiver circuit of Fig. 15-19 describe the operation of (*a*) the i-f amplifier circuit, (*b*) the power-amplifier circuit, (*c*) the noise-elimination circuits.

38. In the portable unit of the radio receiver circuit of Fig. 15-20 describe the operation of (*a*) the converter circuit, (*b*) the i-f amplifier circuit, (*c*) the detector circuit, (*d*) the agc circuit, (*e*) the audio amplifier stage.

39. In the car unit of the radio receiver circuit of Fig. 15-20 describe the operation of (*a*) the r-f tuning circuit, (*b*) the audio amplifier stage, (*c*) the noise-elimination circuits.

40. In the vibrator power supply of Fig. 15-21 describe the operation of (*a*) the RC oscillator circuit, (*b*) the rectifier circuit.

41. Describe the essential functions of automatic tuning.

42. Describe the operation of the trigger amplifier circuit of Fig. 15-22.

43. Describe the operation of the relay-control circuit of Fig. 15-22.

CHAPTER 16

TEST EQUIPMENT

Modern radio and electronic equipment consists of a large number and variety of circuit elements having a wide range of values. The circuit elements are connected to one another, either directly or indirectly, to form a complex circuit. The testing of such circuits can best be accomplished by the use of definite test procedures and adequate test equipment. As a detailed discussion of all types of test instruments and their various applications is beyond the scope of this text, only a brief description of the more common types of test instruments and an outline description of three basic systems of testing will be given in this chapter.

16-1. Systems of Testing. *Purpose of Testing.* There are two fundamental reasons for testing radio or electronic equipment: (1) to check the operation of a complete unit, or any of its sections, against a set of recommended operating values; (2) to locate a faulty section or part. Although there are numerous methods for testing radio and electronic equipment only three basic systems will be considered in this text, namely, (1) point-to-point resistance measurement, (2) measurement of voltage and current for each individual circuit, (3) signal tracing. The procedure to be followed in testing radio and electronic equipment may be based upon one or more of these three systems.

The ultimate aim of all systems of testing is to locate as quickly as possible the portion of a circuit that is not functioning properly. This is best accomplished by individually testing each section of the complete circuit. For a particular circuit, each system has certain characteristics and advantages that are not common to the other systems. Because of this, the best one to use will depend upon the circuit to be tested and the operating characteristics that are to be checked. A complete test procedure may be a combination of any two systems, or all three systems of testing.

Point-to-point Resistance Measurement. In this system, the resistance between the various elements of a circuit is measured with the power off. The values obtained are checked for any variation from the known correct values. The only instrument required for this system of testing is an ohmmeter. The point-to-point resistance measurement method of testing has a number of disadvantages, among which are: (1) a source of reference containing the known correct values of resistance between

627

certain points is required for every make and model of radio and electronic equipment that is to be tested; (2) care must be exercised when measuring the resistance between two points that all elements in the circuit are taken into consideration; (3) this is only a static test and cannot be used for determining the dynamic characteristics of the circuit; (4) a circuit element may appear satisfactory under static conditions but may be faulty under dynamic conditions. An advantage of this system is that, when the fault has been traced to a definite circuit, this system provides a convenient method for quickly locating a shorted or open-circuited element.

Voltage and Current Measurement. In this system, voltage and current measurements are taken at various points in the circuit. The values obtained are checked for any variation from the known correct values. The instruments used for this system may include one or more of the following: (1) a multirange d-c voltmeter, (2) a multirange a-c voltmeter, (3) a multirange d-c ammeter, (4) a vacuum-tube voltmeter, (5) a set analyzer. The voltage-current measurement system of testing also has a number of disadvantages, among which are: (1) a source of reference indicating the correct operating values of voltage and current is required for each make and model of radio and electronic equipment to be tested; (2) care must be exercised when taking voltage and current measurements so that the resistance of the instrument used does not affect the circuit. An advantage of this system of testing is that it provides a quick method of checking the operating voltages at the electrodes of a tube or transistor with the values recommended by the manufacturer for that type of tube or transistor. The only reference required is a tube and/or a transistor manual. Any major variation from the recommended values will ordinarily localize the faulty circuit. However, vacuum tubes and transistors are not always operated at the control voltages recommended by the manufacturer, and this fact must be taken into consideration before declaring a circuit to be faulty.

Signal Tracing. In this system the progress of a signal is traced through each stage of the equipment. With signal tracing it is possible to analyze the signal at each stage of a radio receiver for both its quality and strength. The circuit for each stage can then be adjusted so that the strength and quality of its output signal are at their optimum value. As the signal is the ultimate factor in a radio receiver, the signal-tracing system of testing is the most effective method of checking and aligning a receiver. There are various methods of signal tracing, and the equipment required will depend upon the method used. In general, the following equipment is required: (1) a source of signal input that is capable of producing an r-f, a-f, and modulated r-f signal output; (2) an output indicating device such as an output meter, electron-ray indicator tube, or cathode-ray oscilloscope.

One method of signal tracing is to test each stage individually by applying the proper signal to its input circuit and checking the strength and quality of its output signal.　Another method is to apply a modulated r-f signal to the input circuit of the first r-f stage and to check the output from each succeeding stage.　There are many variations of these two methods.

The signal-tracing method of radio circuit testing has many advantages, among which are: (1) a radio receiver may be tested under actual operating conditions; (2) the intermittent faulty operation of a circuit element can quite readily be located; (3) variation in the amount of gain, bypassing, and coupling (above or below the optimum value) can be detected.

16-2. Combination Meters.　*Requirements.*　In testing radio or electronic equipment a wide range of voltage, current, and resistance measurements may be required.　If individual meters having only one range were to be used for these tests, the number of instruments required would be too large and the cost too great to be practical.　One method of reducing the number of instruments needed is to use a multirange meter for each type of indication that is required.　A more practical method is to combine a multirange voltmeter, ammeter, and ohmmeter into one instrument having an adequate switching arrangement that permits only one range of one type of measurement to be used for each of its settings. This type of instrument is called a *combination meter* and is also known as a *multimeter, multitester, multipurpose tester,* and *volt-ohm-milliammeter.*

Since a combination meter consists of a voltmeter, ammeter, and ohmmeter, its principles of operation and use can best be understood by first studying the principles of operation and use of each individual type of meter.　Only a brief description of the individual meters and the combination meter will be given in this text.　For a detailed description of the principles of operation, construction, and use of these instruments the reader is referred to the chapter on meters in the authors' text *Essentials of Electricity for Radio and Television.*

Ammeters.　An ammeter is an instrument used for measuring electric currents.　It is always connected in series with that part of a circuit whose current is to be measured.　In Fig. 16-1, ammeter 1 is connected in series with the line to indicate the line current, and ammeters 2 and 3 are connected in series with the lamp and fixed resistor, respectively, to obtain their individual currents.　Being connected in series, an ammeter must carry the current passing through that part of the circuit in which it is connected.　In Fig. 16-1, if ammeter 1 has a resistance of 1 ohm and the line current is 10 amp, it follows that there would be a 10-volt drop across the ammeter.　This would cause the appliance and lamp to operate on a voltage of 10 volts less than intended, thus giving incorrect indications of the currents flowing in the circuit under normal conditions (that is, without the ammeters).   The voltage drop across the meter is decreased by making the resistance of the meter as low as possible.   The value of this drop varies with the type of meter and the manufacturer. Most ammeters are designed for 50-mv drop at full rated current.

Since the resistance of an ammeter is very low, the current that would flow through the ammeter, if through error it was connected directly across a source of power, would be excessively large, and the meter would be damaged.   *An ammeter should therefore never be connected across any source of power.*   The current due to an unforeseen short circuit or over-load may be large enough to injure an ammeter if it is left connected in a circuit.   To prevent damage due to factors that cannot be foreseen, *the ammeter should always be protected by connecting a short-circuiting switch across it*, as illustrated by S_1, S_2, and S_3 in Fig. 16-1.   The switch is always kept closed except when a reading is to be taken.   If, upon opening the switch, the needle swings backward or completely across the

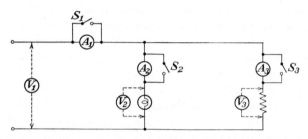

Fig. 16-1. Correct method of connecting ammeters and voltmeters.

meter scale, it should be closed instantly to prevent damaging the meter. The necessary changes to the circuit should then be made before the switch is again opened.

Voltmeters.   A voltmeter is an instrument used for measuring voltage. It is always connected across that part of the circuit whose voltage is to be measured.   Figure 16-1 shows the correct way to connect voltmeters in a parallel circuit.   Voltmeter 1 indicates the line voltage; voltmeter 2 shows the voltage across the lamp; and voltmeter 3 shows the voltage across the fixed resistance.

As a voltmeter is connected directly across the line, it is desirable that it take as little current as is practicable.   Because of its comparatively low resistance, approximately 20 ohms, the moving element of a volt-meter cannot be connected directly across the line.   It is therefore neces-sary to connect a high resistance in series with it.   The value of this resistance depends on the resistance of the coil, or moving element, and the full-scale voltage desired.

Voltmeters do not form a definite part of a circuit as do ammeters.

It is therefore not necessary to connect voltmeters permanently in the circuit. If a voltmeter is connected only when a reading is to be taken, a single meter can be used to take readings at various points in a circuit. A voltmeter may be damaged by excessive voltage, but as it is connected only when a reading is to be taken, it may therefore be disconnected instantly if any overvoltage condition is apparent.

Voltmeters are often rated in terms of their sensitivity in ohms per volt, which is obtained by dividing the resistance in ohms for any range by the full-scale voltage rating of that particular range. In testing radio and electronic equipment, it is generally desirable and in many cases necessary to use a voltmeter with a very high sensitivity. Although a voltmeter with a sensitivity of 100 ohms per volt may be excellent for industrial power circuits, it would prove very unsatisfactory for testing radio and electronic circuits. For testing these circuits it is recommended that the voltmeters have a sensitivity of 20,000 ohms per volt or higher.

A-C and D-C Instruments. Ammeters and voltmeters are usually designed to measure either direct or alternating currents or voltages and in order to obtain accurate readings an ammeter or voltmeter should only be used to measure the type of current or voltage for which it is designed. The ordinary a-c ammeter cannot be used to measure accurately currents whose frequencies are greatly in excess of 500 cycles. Thermal-type instruments measure the effective value of the current and therefore can be used on direct, a-f, and r-f currents. Instruments giving accurate readings at frequencies up to 100 mc can be obtained.

Rectifier-type meters are used for measuring alternating currents of small magnitude such as milliamperes and microamperes, and voltages ranging from values in millivolts up to 1,000 volts. A rectifier-type meter consists of a copper oxide rectifier and a permanent magnet moving-coil instrument. Rectifier-type meters can be used to measure alternating currents of frequencies up to 20,000 cps.

Shunts. The current that may safely be led into an ammeter movement is limited by the current-carrying capacity of the moving element, which must necessarily be small. To increase the range of such instruments, shunts and current transformers are used, the former with direct currents and the latter with alternating currents. It has previously been stated that the movement's current must be very small, generally about 1 ma. This is accomplished by connecting a low resistance in parallel with the meter. The ammeter is in reality now a voltmeter (see Fig. 16-2) indicating the voltage drop across a resistance. This resistance is called a *shunt* and forms a definite part of all ammeters. The resistance of the shunt may be calculated by use of the following equation.

$$R_{SH} = \frac{I_M \times R_M}{I_{SH}} \qquad (16\text{-}1)$$

where R_{SH} = shunt resistance, ohms
       R_M = meter resistance, ohms
       I_{SH} = shunt current, amp
       I_M = meter current, amp

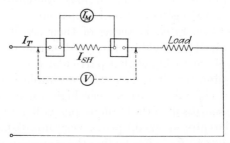

FIG. 16-2. Proper connections for the use of a shunt with a millivoltmeter.

When the shunt is connected permanently inside an ammeter, it is called an *internal shunt.* When it is desired to increase the range of an ammeter, an *external shunt* may be used. Multirange ammeters use a series of shunts and a rotary switch for connecting the proper shunt across the ammeter in order to obtain a desired range.

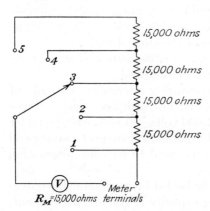

FIG. 16-3. Circuit showing the application of a rotary switch and resistors to increase the voltmeter range by multiples of 2, 3, 4, or 5.

Multipliers. The range of a voltmeter having its resistance incorporated within the instrument may be increased by the use of an external resistance connected in series with the instrument. External resistances used in this manner are called *multipliers.* Multirange voltmeters use a series of resistors and a rotary switch for connecting the proper value of resistance in series with the voltmeter in order to obtain a desired range (see Fig. 16-3). The multiplying power of a resistor may be calculated by use of the following equation.

$$M = \frac{R_{EX} + R_M}{R_M} \qquad (16\text{-}2)$$

where M = multiplying power
     R_{EX} = multiplier resistance, ohms
     R_M = meter resistance, ohms

Multipliers may be used for alternating voltages up to 1,000 volts as

well as for direct voltages.   For all alternating voltages higher than this,
it is advisable to use potential transformers.

Ohmmeters.   An ohmmeter is an instrument that indicates the resistance of a circuit directly in ohms without any need for calculations.
Figure 16-4 is a schematic diagram of the circuit of a simple ohmmeter.
In this instrument, unit cells are used as the power source.   The milliammeter scale is calibrated to indicate the resistance in ohms directly.
When terminals T_1 and T_2 are short-circuited and the battery cells are
new, so that $E_B = 4.5$ volts, then $R_1 + R_A$ must be equal to 4,500 ohms in
order to have the milliammeter indicate its full-scale deflection, which will
occur when the current in the circuit is 1 ma.   Inserting an unknown
resistance R_X between terminals T_1 and T_2 will decrease the indication on
the meter.   For example, if the value of R_X is 4,500 ohms, the reading on
the milliammeter will decrease to one-half of its full-scale deflection.   As
the meter scale is calibrated to indicate the value in ohms of the unknown

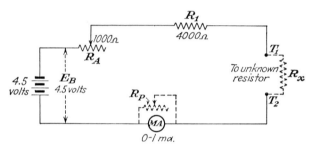

Fig. 16-4. Circuit of a simple ohmmeter.

resistor R_X, the point at the center of the meter scale is marked 4,500
ohms.   The remainder of the points on the meter scale may be calibrated
in the same manner.

As the voltage of the unit cells decreases with age, the resistor R_A is
made adjustable to compensate for variations in the voltage of the battery.   Resistance values obtained with an ohmmeter using this method of
compensation for decreases in battery voltage will not be accurate when
the voltage drops because of the aging of the cells.   A more accurate
method is to connect an adjustable resistor in parallel with the milliammeter as indicated by the dotted lines in Fig. 16-4.   The resistor R_P is
used in place of resistor R_A, which had been connected in series with the
meter and the fixed resistor R_1.

By using different values of resistance and battery voltage, an ohmmeter can be made to indicate any value of resistance.   In Fig. 16-5,
resistors R_1, R_2, R_3, and R_4 are whole-number multiples of one another,
therefore it is only necessary to have a single scale marked to indicate the
values of the lowest range and to use a multiplying factor for each of the

other ranges.   A rotary switch *SR* is used to connect or disconnect these resistors into the circuit.

Combination Meters.   The construction of a voltmeter, ammeter, or ohmmeter does not differ materially from one another as far as the movements and magnets are concerned.   With the proper switching and circuit

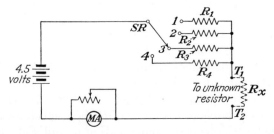

FIG. 16-5. Circuit showing the use of resistors and a switch to select the range of a multirange ohmmeter.

arrangements it is therefore possible to use a single d-c milliammeter as a multirange d-c voltmeter, d-c ammeter, or ohmmeter.   A meter used in this manner is called a *combination meter.*   The values of the multipliers, shunts, and ohmmeter resistors can be calculated by using the methods explained under multipliers, shunts, and ohmmeters. By using a copper-oxide rectifier, the d-c milliammeter can also be used to measure alternating currents and voltages. In radio and electronic test and service work, combination meters are very useful, since one instrument is made to take the place of several meters.

FIG. 16-6. A combination a-c/d-c voltmeter, d-c ammeter, and ohmmeter. (*Courtesy of Simpson Electric Co.*)

Figure 16-7 shows the circuit diagram of a combination meter. This instrument has six direct- and alternating-voltage ranges, 2.5, 10, 50, 250, 1,000, and 5,000 volts; five d-c ranges, 100 μa, 10 ma, 100 ma, 500 ma, and 10 amp; and three ranges of resistance, 2,000 ohms, 200,000 ohms, and 20 megohms. The direct voltage readings have a sensitivity of 20,000 ohms per volt, and the alternating voltage readings 1,000 ohms per volt. D-c readings are taken with a 250-mv drop across the instrument for fullscale deflection. This instrument may also be used to measure volume levels in volts and in decibels, and is calibrated for a reference level of 500 ohms and 6 mw.

In using combination meters, care should be taken to see that all switches and dial settings are in their correct positions before the instru-

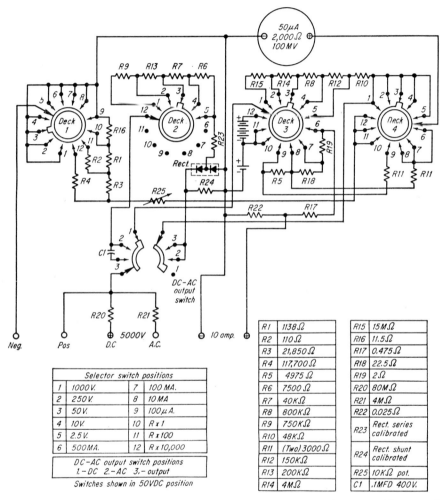

Selector switch positions		
1	1000 V.	7	100 MA.
2	250 V.	8	10 MA
3	50 V.	9	100 μ A.
4	10 V.	10	R x 1
5	2.5 V.	11	R x 100
6	500 MA.	12	R x 10,000

DC-AC output switch positions
1.-DC   2.-AC   3.- output

Switches shown in 50VDC position

R1	1138 Ω		R15	15 M Ω
R2	110 Ω		R16	11.5 Ω
R3	21,850 Ω		R17	0.475 Ω
R4	117,700 Ω		R18	22.5 Ω
R5	4975 Ω		R19	2 Ω
R6	7500 Ω		R20	80 M Ω
R7	40 K Ω		R21	4 M Ω
R8	800 K Ω		R22	0.025 Ω
R9	750 K Ω		R23	Rect. series calibrated
R10	48 K Ω			
R11	(Two) 3000 Ω		R24	Rect. shunt calibrated
R12	150 K Ω			
R13	200 K Ω		R25	10 K Ω pot.
R14	4 M Ω		C1	.1 MFD 400 V.

Fig. 16 7. Schematic diagram of the combination meter shown in Fig. 16-6. (*Courtesy of Simpson Electric Co.*)

ment is connected into the circuit. If this is not done, the instrument may be damaged, as it might be set for use on direct current and be used on alternating current, or set to be used as an ammeter and connected for use as a voltmeter, or set in some other incorrect fashion.

16-3. Special-purpose Voltmeters. *Output Meter.* The power level of radio or speech equipment is usually measured by means of an output meter.   This instrument consists of two parts: (1) a device for rectifying the a-f current, (2) a meter that is calibrated to indicate the strength of the rectified current in volts, decibels, or both.   Commercial instruments, such as the one shown in Fig. 16-8, generally use a copper oxide rectifier; however, a vacuum-tube rectifier can also be used.   An output meter is essentially the same as a rectifier-type a-c voltmeter.   Combination meters that employ a rectifier-type instrument generally make provision for using this instrument as an output meter.

The most convenient place to connect an output meter is across the voice-coil leads of the loudspeaker.   However, for very weak signals it is generally necessary to connect the output meter between the plate of the output tube and the chassis.   Because of the high direct voltage between these two points, a capacitor should always be connected in series with an output meter used in this manner in order to prevent damage to the instrument.   If the output stage consists of two tubes connected in push-pull, the output meter may be connected from the plate of either tube to the chassis.

Fig. 16-8. Output or power-level meter. (*Courtesy of Weston Electrical Instrument Corporation.*)

Electronic Voltmeter.   Because of its comparatively low internal resistance, the ordinary voltmeter cannot be used to measure the direct voltage at a number of circuits in radio or electronic equipment without affecting the operation of the unit.   Some of the circuits in which the use of a low-sensitivity voltmeter affects the circuit operation are: (1) vacuum-tube grid and plate, (2) transistor base, collector, and emitter, (3) oscillator, (4) f-m discriminator and ratio detector, (5) automatic gain control, (6) automatic frequency control, and (7) high-fidelity audio amplifier.   The electronic voltmeter has a relatively high input resistance and can be used to measure the control or operating voltage at any point in an electronic circuit where a signal is present without affecting the operation of the unit.

A popular type of electronic voltmeter is the *RCA Master Voltohmyst* illustrated in Fig. 16-9.   This instrument can be used to measure directly (1) direct voltage, (2) direct current, (3) peak-to-peak values of sine-wave and complex waveform voltages, (4) the rms values of sine-wave voltages, and (5) resistance.   To ensure high sensitivity and high stability, vacuum tubes are employed for all functions except current measurements.   This meter has seven ranges of direct and alternating voltages, namely 1.5, 5, 15, 50, 150, 500, and 1,500 volts. There is practically no loading of a circuit when taking direct voltage readings, as the instrument has a constant input resistance of 11 megohms for all direct-voltage readings. The sensitivity on the 1.5-volt range is 7.3 megohms per volt, and the over-all accuracy is ±3 per cent of the full-scale reading. A 1-megohm isolating resistor is connected in series with the probe for dynamic direct-voltage measurements. The instrument is frequency-compensated for the alternating-voltage ranges up to 500 volts and can be used at frequencies up to approximately 3 mc. On measurements of direct current, the meter will read from 10 μa to 15 amp with an accuracy of ±3 per cent

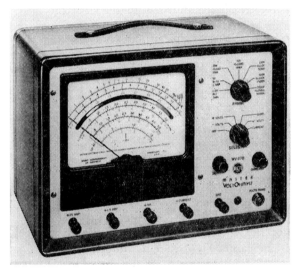

Fig. 16-9. A master Voltohmyst.   (*Courtesy of RCA.*)

of full-scale reading. Current is read on nine ranges of from 0 to 0.5, 1.5, 5, 15, 50, 150, and 500 ma, and from 0 to 1.5 and 15 amp. When used to measure resistance, the instrument will read from 0.2 ohm to 1,000 megohms in seven ranges. Mid-scale values are 10, 100, 1,000, 10,000, and 100,000 ohms and at 1 and 10 megohms. Provision is made for zero-centering of the meter pointer so that the instrument may be used for checking f-m discriminator alignment, and the polarity and condition of bias cells.

The Vacuum-tube Voltmeter. Another form of electronic voltmeter is the vacuum-tube voltmeter, commonly abbreviated *vtvm*. This instrument utilizes the characteristics of a vacuum tube for measuring voltages with minimum effect on the circuit to which it is connected. Some types of vacuum-tube voltmeters can be used to obtain the peak, trough, effective, or average values of voltage.   By using a known value of capacitance as a shunt, the vtvm can also be used to measure small values of current in high-frequency circuits.

There are various types of vtvm circuits, the choice of circuit used being determined by the type of measurement for which it is to be used. A simple form of vtvm circuit using plate rectification is shown in Fig. 16-10, and its principle of operation is as follows.   When an alternating voltage is applied to the grid circuit of the tube, the alternating voltage will be rectified by the tube.   The d-c component of the rectified output is indicated on the d-c meter, which is calibrated to indicate the proper voltage.   The value of the tube's operating voltages are determined by the value of the alternating voltage input and the range of voltage to be indicated.   The grid bias should be greater than the peak value of the input voltage in order that the grid be maintained at a negative value throughout the entire cycle.   The value of the plate voltage is determined

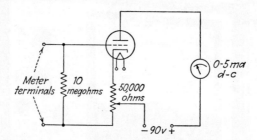

Fig. 16-10. A simple vacuum-tube voltmeter circuit.

by the range of the voltage that is to be indicated by the instrument. The plate voltage should be of such a value that it will produce a maximum deflection of the instrument for the prescribed input voltage.

A commercial vacuum-tube voltmeter that measures (1) direct voltage, (2) alternating voltage, (3) direct current, (4) resistance, (5) capacitance, and (6) inductance is illustrated in Fig. 16-11.   This instrument is capable of measuring voltages having frequencies up to 100 mc.   There are six alternating- and direct-voltage ranges of 3, 12, 30, 120, 300, and 1,200 volts, with peak-to-peak readings on all ranges up to 300 volts.   Direct current is indicated for six ranges of 3, 12, 30, 120, 300, and 1,200 ma. Resistance values of 1 ohm to 10,000 megohms may be read using eight ranges.   Capacitances of 0 to 10,000 $\mu\mu$f may be read using two ranges, and 0 to 1,000 μf using five ranges.   Inductance measurements of 50 mh to 100 henrys may be obtained by use of a conversion chart.   When using a vacuum-tube voltmeter for r-f measurements, care should be exercised to keep the input capacitance of the instrument and its leads at a minimum.   For extreme accuracy, the input capacitance should be approximately 1 $\mu\mu$f and under no circumstances should it exceed 6 $\mu\mu$f.   In order to obtain this low input capacitance, some vacuum-tube voltmeters make provisions for connecting the control grid of the vtvm tube directly to the point in the circuit being investigated. In order to prevent the circuit under test from being affected when the vacuum-tube voltmeter is connected to the circuit, the input impedance of the instrument should be very high. The meter illustrated in Fig. 16-11 has an input impedance of approximately 12 megohms shunted by $6\mu\mu f$. In order to obtain greater accuracy, some vacuum-tube voltmeters employ a voltage-regulator tube

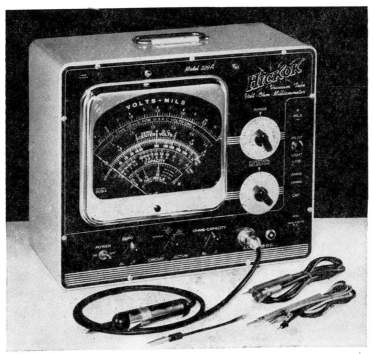

Fig. 16-11. A vacuum-tube voltmeter.   (*Courtesy of Hickok Electrical Instrument Co.*)

to isolate any fluctuations in the line voltage from the instrument circuit.

16-4. Signal Generators. *Audio Oscillator.* An a-f oscillator is an instrument employing vacuum-tube, or transistor, circuits that can be adjusted to generate a low-voltage signal having any desired value of audio frequency. This instrument is commonly called an *audio oscillator*. Commercial audio oscillators may be divided into two classes: (1) the beat-frequency type (see Art. 8-2), (2) the resistance-capacitance type. A simple a-f oscillator circuit of the *RC* type is shown in Fig. 16-12. The output of this oscillator circuit has a constant amplitude and constant frequency. However, by the proper connection of additional circuit elements to this simple circuit, a variable frequency and variable amplitude output can be obtained.   The *RC* type a-f oscillator has two advantages over the beat-frequency type: (1) it has higher frequency stability; (2) its circuit is much simpler.

Although an a-f oscillator may be obtained as a separate unit, it is generally combined with an r-f oscillator into a single unit.   Such an instrument is called a *signal generator,* and usually has two a-f oscillator circuits: one provides a fixed a-f output of 400 cycles and the other a variable a-f output of 50 to 20,000 cycles.   The r-f signal may be modulated by the output of either a-f oscillator.

The a-f oscillator has many uses in testing radio and electronic equipment and may be used to test (1) the amplifying characteristics at various frequencies of all forms of a-f coupling such as transformers, chokes, *RC* networks, etc., (2) the tone-reproducing qualities of all forms of loud-speakers, (3) the gain per stage and the overall gain of an a-f amplifier.   When an a-f oscillator is used for these tests the oscillator supplies the a-f signal and a vacuum-tube voltmeter or cathode-ray oscilloscope is used to determine the output characteristics of the unit being tested.   The a-f oscillator may also be used for the following purposes: (1) to supply a sound signal for a-c bridge measurements, (2) to trace open and grounded circuits with the aid of a pair of earphones, (3) for modulating an r-f oscillator, (4) for locating loose mechanical connections in electronic equipment.

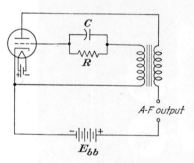

Fig. 16-12. A simple a-f oscillator circuit.

R-F Test Oscillator.   An r-f test oscillator is an instrument capable of generating the various r-f voltages required for alignment and servicing of radio and electronic equipment.   The circuit of this type instrument is based on the principles of oscillator circuits as they were explained in Chap. 8.   Since an r-f oscillator is generally designed for a particular type of service, its output requirements will determine the circuit arrangement and the values of the circuit elements.   The general requirements of an r-f test oscillator are: (1) good frequency and amplitude stability, (2) a wide range of frequency output, which can readily be adjusted and whose frequency value can easily be read, (3) an attenuator for adjusting the amplitude of the output signal from zero to maximum value without affecting the frequency, (4) provision for amplitude-modulation of the r-f signal, (5) provision for frequency-modulation of the r-f signal.

The r-f test oscillator can be obtained as either a separate unit or combined with an a-f oscillator.   However, as many tests require a modulated r-f signal, an audio oscillator is usually combined with an r-f oscillator to form a single test instrument, as shown in Fig. 16-13. This instrument has seven r-f ranges, which makes it possible to obtain a continuous variable radio frequency from 100 kc to 133 mc. The r-f output may be used as unmodulated, amplitude modulated, or frequency modulated. For amplitude modulation, the r-f output may be modulated with either a constant 60-cycle or 400-cycle signal, or with a variable 50- to 15,000-cycle signal. For frequency modulation, bandwidths of 30, 150, or 750 kc can be obtained. The r-f output is modulated with either the

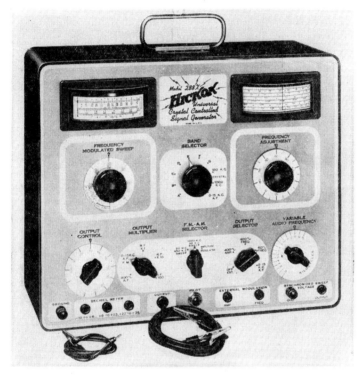

Fig. 16-13. A universal crystal-controlled signal generator. (*Courtesy of Hickok Electrical Instrument Co.*)

60-cycle, 400-cycle, or 50- to 15,000 cycle signal, depending upon which range and bandwidth is being used. In addition to the r-f output, this instrument can also provide the following: (1) a-f outputs of 60 cycles, 400 cycles, or 50 to 15,000 cycles, (2) a synchronized sweep voltage for oscilloscope use, (3) a power-level meter having three ranges that are capable of indicating outputs from −10 to +38 db. This instrument is very useful for testing and aligning all types of a-m radio equipment, f-m radio equipment, television equipment, and various types of electronic equipment.

Uses of the Signal Generator. A signal generator is required for accurate alignment and measurements of all tuned circuits in electronic equipment. In general, the output of the signal generator may be applied to the input circuit of the unit being tested or to any one of its individual stages. The signal generator should first be adjusted to produce a signal output of the desired frequency and amplitude. It is then possible to align each stage, measure the gain per stage, or the over-all gain, by connecting an output meter or electronic voltmeter in the proper circuit. The distortion characteristics of each stage can be observed by using a signal generator to supply an input signal and a cathode-ray oscilloscope for observing the waveform of the output from each stage. Four oscilloscope patterns illustrating the frequency-response characteristics of an i-f amplifier stage are shown in Fig. 16-14.

For proper alignment or testing of a radio receiver, the following procedure is suggested: (1) Ascertain the proper operation, applications, and limitations of the test equipment from the manufacturer's instructions. (2) Much time can be saved in the aligning of each stage or the

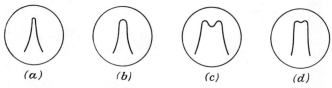

(a)   *(b)*   *(c)*   *(d)*

FIG. 16-14. Oscilloscope patterns illustrating the frequency characteristics of an i-f amplifier. *(a)* Single peak which is too narrow. *(b)* Single peak of the proper width. *(c)* Double peak which is too broad and whose depression is too deep. *(d)* Double peak of the proper width and depth.

localizing of a fault or faults by referring to the circuit diagram of the equipment being tested. (3) The output of the test oscillator should be maintained at a low level. (4) When aligning the i-f stage, short-circuit the local oscillator or in some other manner prevent any part of the oscillator output from entering the i-f stage. (5) The output of the signal generator should be fed through a dummy antenna as specified by the manufacturer of the receiver. (6) Follow the instructions for alignment suggested by the manufacturer of the receiver. (7) After the complete unit has been aligned, recheck the alignment of each stage and make any readjustment that may be required due to the reaction from the alignment of the other stages.

16-5. Cathode-ray Tubes. *General Principles.* The cathode-ray tube forms an essential part of the cathode-ray oscilloscope, and it is therefore best to understand the principles of operation of this tube before studying the operation of the oscilloscope. In a cathode-ray tube, the electrons emitted by the cathode are attracted by an anode carrying a high positive potential. The tube elements are arranged so that the electrons form a narrow beam, and their velocity is so great that they continue in a forward motion until they strike a fluorescent screen. This beam follows a straight line unless diverted by an electric or magnetic field. Characteristics of the electron beam are as follows: (1) it is attracted to positive-charged objects and repelled by negative-charged objects, (2) it is subject to forces from magnetic fields. Therefore, an electron beam may be focused and/or deflected by either electric or magnetic fields. Because an electron beam is not visible to the naked eye, the beam is directed at a treated screen which fluoresces when hit by an electron beam. The basic

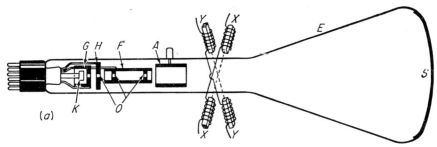

A = High-voltage electrode (anode no. 2)
B = Inner set of deflecting electrodes
C = Outer set of deflecting electrodes
E = Glass envelope
F = Focusing electrode (anode no. 1)
G = Control electrode (grid no. 1)
H = Accelerating electrode (grid no. 2)

K = Cathode
O = Apertures
S = Fluorescent screen
XX = Pair of coils for producing magnetic field
YY = Pair of coils for producing magnetic field at right angles to that produced by the pair of coils XX

Note: Electrodes K, G, H, F, and A constitute an "electron gun"

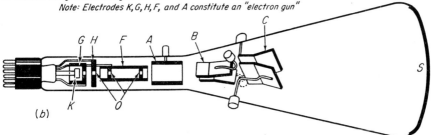

Fig. 16-15. Schematic arrangement of the electrodes in a cathode-ray tube. (a) Electromagnetic deflection. (b) Electrostatic deflection. (*Courtesy of RCA.*)

cathode-ray tube consists of a highly evacuated glass envelope containing an electron gun, a fluorescent screen, and some means of deflecting the electron beam.

The Electron Gun. In Fig. 16-15 the electrodes K, G, H, F, and A constitute the electron gun. The cathode K produces the electrons that are formed into an electron beam. The control electrode G, which performs the same function as the control grid in a vacuum tube, consists of a metal cylinder having a metal disk with a small aperture that is placed directly in front of the electron-emitting surface of the cathode. When the potential on the control electrode is made negative with respect to the cathode, the electrons are prevented from diverging from the cathode and are bent inward along the axis of the tube.

The accelerating electrode *H* is a cylinder containing one or more disks, each with an aperture, and is placed in front of the control electrode.   The apertures within cylinder *H* remove any strongly divergent electrons that would otherwise produce a fuzzy spot.   In order to accelerate the electron beam, this electrode has a positive charge of several hundred volts with respect to the cathode.

The focusing electrode *F* is of a larger diameter than the accelerating electrode and is kept at a higher potential than the accelerating electrode, often several thousand volts.   The focusing action occurs in the electrostatic field between the accelerating and focusing electrodes, and the beam is focused by varying the voltage on the focusing electrode.   The high-voltage or accelerating electrode *A* controls the brilliance and fineness of the trace.

The Fluorescent Screen.   The screen consists of a chemical deposited on the inside surface of the flat end of the tube and fluoresces when hit by the fast-moving electrons.   The emission of light from the fluorescent screen persists for only a small fraction of a second after the electron beam has been removed.

Electrostatic Deflection.   In this system, the electron beam is passed between two sets of parallel plates (see Fig. 16-15*b*).   When a voltage is applied to either set of plates, the electron beam will be deflected toward the positive plate.   The set of parallel plates mounted in a mechanically vertical position causes the beam to be deflected in a horizontal direction. The other set of plates, mounted mechanically horizontal, causes the beam to be deflected in a vertical direction.   The deflection of the electron beam is directly proportional to the voltages applied to the plates.

Electromagnetic Deflection.   In this system, two pairs of coils are mounted around the neck of the tube and are arranged with their respective axes perpendicular to each other and perpendicular to the axis of the tube (see Fig. 16-15*a*).   The coils *YY* have a common vertical axis, and any current applied to them will cause a horizontal deflection of the electron beam.   The coils *XX* have a common horizontal axis, and any current flowing through them will cause a vertical deflection of the beam.

16-6. Cathode-ray Oscilloscope.   *General Principle.*   The oscilloscope is an instrument for indicating the waveform of a variable voltage or current.   The operation of an oscilloscope is based on the principle that a moving fluorescent spot can be made to appear as a stationary figure by properly adjusting the frequencies of the voltages across the two sets of deflection plates.   A visible curve of one electrical quantity plotted as a function of another electrical quantity can therefore be produced on the screen of the oscilloscope.   If the quantities vary recurrently, persistence of vision will cause the electron beam to appear on the fluorescent screen as a continuous curve. The voltage whose waveform is to be investigated is applied to the vertical-deflection plates, and a voltage that acts as a timing wave is applied to the horizontal-deflection plates. Circuits incorporated within the oscilloscope generate a saw-tooth-shaped voltage that is used as a timing wave. This voltage is applied to the horizontal deflection plates so that the voltage being investigated will vary as a linear function of time.

Oscilloscopes are designed primarily for the analysis of electrical circuits by studying the waveform of voltage and current at various points in the network. However, such an instrument may be applied to the study of any variable quantity which may be translated into electrical voltages by such means as a vibrator pickup unit, pressure pickup, etc.

The cathode-ray tube may be considered as an indicating device with a pointer of negligible inertia, and for this reason its frequency range is practically unlimited. However, the frequency range of an oscilloscope

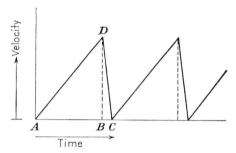

Fig. 16-16. A saw-tooth linear time base.

is limited by the frequency-response characteristics of the vacuum-tube amplifiers that are incorporated within the instrument.

Sweep. Applying a voltage that has a constant time rate of change to the horizontal deflection plates will cause the beam to move straight across the face of the oscilloscope. Applying a varying voltage to the vertical deflection plates will then cause the electron beam to trace the waveform of the applied voltage on the fluorescent screen. The circuit that supplies the timing voltage to the horizontal deflection plates is arranged so that as soon as the timing voltage moves the beam all the way across the screen, the beam is returned to the starting point and the process is repeated. This movement is called the *sweep* and the number of times it is repeated per second is referred to as the *sweep frequency.* Although the horizontal sweep is used practically universally for radio applications, a horizontal, vertical, or radial sweep may be used for electronic applications.

Sweep Oscillator. The sweep oscillator generates a saw-tooth wave, as shown in Fig. 16-16. The interval A to C constitutes *one period*, in which

AB is the *sweep* or *scanning time* and *BC* is the *flyback time.* The sweep portion *AD* should be linear, and the flyback time should be of short duration. To prevent the return trace from affecting the pattern on the screen, the electron beam is cut off during each flyback period.

Sweep Generator Circuits. The most common method of obtaining a saw-tooth wave is by charging a capacitor through a resistor from a high direct voltage source and using only the linear portion of the varying voltage. As a relatively small portion of the charging curve is linear, the capacitor should be allowed to charge for only this small interval of time. This can be accomplished by connecting the capacitor across the plate and cathode of a gas diode or triode. When the voltage charge on the capacitor reaches an amount equal to the breakdown voltage of the tube, the tube will start to conduct. Because of its low internal resistance while conducting, the tube will practically short-circuit the capacitor, thereby causing it to discharge almost instantaneously. The interval of time during which the capacitor is charging will be determined by the breakdown voltage of the gas tube and the values of the capacitor and resistor. A simple sweep generator circuit using these principles is shown in Fig. 16-17.

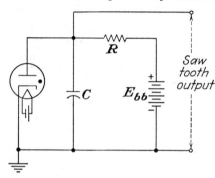

Fig. 16-17. A simple sweep-generator circuit.

The useful frequency range of a sweep generator circuit using a gas tube is limited by the deionization time of the tube. At frequencies above 50 kc, the deionization time of gas tubes becomes an appreciable portion of the total cycle, thus prohibiting their use at high frequencies. A number of vacuum-tube circuits have been developed that employ the trigger characteristic of a triode or pentode. The trigger action is produced by causing a slight change in a circuit constant to suddenly change the plate or screen-grid current. This sudden change in current is used to charge or discharge a capacitor. A simple multivibrator circuit using this principle is shown in Fig. 16-18. Linear time bases having a frequency range of from 2 to 1 million cps can be obtained with vacuum-tube sweep-generator circuits.

Synchronization. A stationary pattern is obtained on the fluorescent screen when the frequency of the sweep voltage is synchronized with the frequency of the voltage being investigated. Figure 16-19 illustrates the manner in which a single sine-wave pattern is produced on the screen when the frequency of the sine-wave input signal and the sweep frequency are equal. Figure 16-20 illustrates the manner in which a double sine-wave pattern is produced on the screen when the frequency of the sine-wave input signal is twice the value of the sweep frequency.

Operation of an Oscilloscope. The operation of an oscilloscope will vary with the manufacturer and model. However, the following general

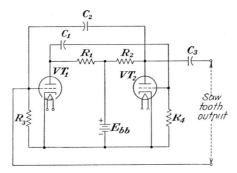

Fig. 16-18. A simple multivibrator circuit.

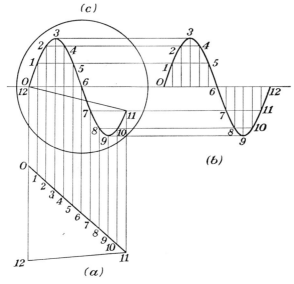

Fig. 16-19. Pattern produced on the oscilloscope screen when the frequencies of the input voltage and the time base are equal and in phase. (*a*) Linear time base. (*b*) Sine-wave input voltage. (*c*) Pattern on the oscilloscope screen.

instructions will apply to most oscilloscopes. The arrangement of the various controls of a commercial cathode-ray oscilloscope is shown in Fig. 16-21.

All normal operating controls and terminals for most oscilloscopes are located on the front panel. Direct connection to the deflecting plates can generally be made by proper connection to terminals located on the back or front panel of most oscilloscopes.

Related controls are grouped together and the groups are plainly marked. Controls for the vertical deflection are on one side of the instrument panel and the controls for the horizontal deflection are on the other side.

The beam controls comprise those which adjust the intensity, focus, and position of the fluorescent spot on the screen. The intensity control adjusts the bias on the control grid and thus varies the intensity of the beam current. The focus control adjusts the voltage on the focusing

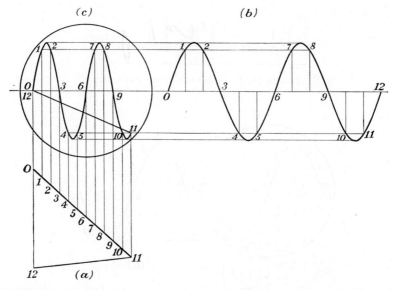

Fig. 16-20. Pattern produced on the oscilloscope screen when the frequency of the input voltage is twice the frequency of the time base and both frequencies are in phase. (a) Linear time base. (b) Sine-wave input voltage. (c) Pattern on the oscilloscope screen.

electrode. The vertical and horizontal centering controls adjust the location of the spot or trace in the vertical and horizontal directions respectively.

The linear-time-base or sweep-generator controls include the frequency-range and frequency-vernier controls, synchronizing signal selector, synchronizing signal amplitude control, and provision for applying an external synchronizing signal. The frequency-range selector determines the frequency range of the timing base. The frequency-vernier control adjusts the frequency to any value within the range of a particular frequency-range-selector setting. The position of the synchronizing selector switch determines whether the signal is to be synchronized internally, externally, or from the 60-cycle power line. A synchronizing adjustment for varying the amplitude of the synchronizing voltage is also usually provided.   The resultant signal, as seen on the screen, may be amplified either vertically or horizontally by adjusting their respective gain controls.

Uses of the Oscilloscope.   The uses of the cathode-ray oscilloscope are many and varied.   It is beyond the scope of this text to give a detailed discussion of each application.   A few of the many uses as applied to

FIG. 16-21. A cathode-ray oscilloscope.   (*Courtesy of Allen B. Dumont Laboratories, Inc*)

radio receivers are (1) to align the i-f stages of a radio receiver when used in conjunction with a signal generator, (2) to align the r f stages of a radio receiver when used in conjunction with a signal generator, (3) to determine the quality of the a-f signal by examination of its waveform at various points in the audio section of the receiver.

16-7. Tube Testers.   Since vacuum tubes form an essential part of radio and electronic equipment, it is necessary to have a test instrument that is capable of indicating the condition of a tube.   There are a number of types of commercial tube testers, and a good instrument should have provisions for checking (1) the emission of grid-type tubes, (2) the transconductance of grid-type tubes, (3) the plate current of rectifiers and diodes, (4) interelectrode short circuits in all types of tubes. For a quick appraisal of the condition of a tube, most tube testers provide an emission test with the results indicated on a scale marked *Bad-Fair-Good* or *Replace-Doubtful-Good*. Some instruments have provision for testing the power output of power-amplifier tubes. A tube tester may measure either the static or dynamic operating characteristics of a tube or both. An instrument that indicates the static characteristics uses one of two forms of emission testing. In one form, all the tube elements except the cathode are connected together. An alternating voltage is applied between the cathode and the other elements through a protective resistor and a d-c meter. The condition of the tube is indicated by the current through the d-c meter. The other form differs from the one just described in that the d-c meter is connected to read only the plate current. The other elements are connected elsewhere in the circuit. Since static tests are taken without a signal input they do not give a true indication of a tube's operating characteristics. With a dynamic tube tester, a signal, distinct from the bias, is applied to the input grid of the tube being tested. This grid signal produces a definite change in the plate current of the tube. The plate-current change, in microamperes, is divided by the grid signal, in volts, to produce a quotient which expresses in micromhos the transconductance of the tube. A dynamic tube tester therefore indicates the amount of plate modulation, or the dynamic transconductance of the tube, and hence is the preferred type of test.

The operation of a tube tester will vary with the manufacturer and model, and for best results should be operated in accordance with the instructions accompanying the instrument. A commercial tube tester illustrating the arrangement of the various controls and tube sockets is shown in Fig. 16-22. In addition to providing means for testing all types of vacuum tubes, this instrument also provides means of testing gas tubes, ballast resistors, and electron-ray indicator tubes. The dynamic mutual conductance is indicated in micromhos, and this test can be taken simultaneously with the emission test. The gas content of a tube can be measured and the condition of the tube indicated directly on the meter.

16-8. Transistor Testers. Transistors are usually tested for (1) their leakage-to-gain ratio, and (2) the amount of noise generated. A circuit diagram for a simple transistor checker that can be used for testing PNP or NPN transistors is shown in Fig. 16-23. The leakage-to-gain switches S_1 and S_2 are shown in their leakage position. Since the base lead is not connected for this part of the test, only a very small value of current should be indicated on the meter. The amount of current flowing in the collector circuit is a function of (1) temperature, (2) the resistivity of the

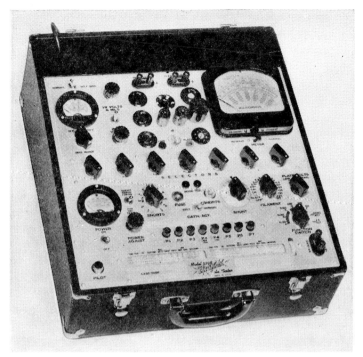

FIG. 16-22. A commercial tube tester.   (*Courtesy of Hickok Electrical Instrument Co.*)

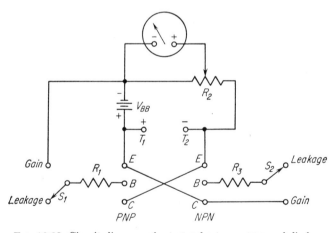

FIG. 16-23. Circuit diagram of a tester for transistors and diodes.

germanium, or silicon, and (3) the applied voltage. Any contamination on the surface of the germanium, or a short circuit, will be indicated by a high reading on the meter. Closing S_1 for testing PNP transistors, or S_2 for testing NPN transistors, applies a voltage to the base through a 220,000-ohm resistor. The collector current of a good transistor will show an increase, and the amount of increase will be determined by the quality as well as the beta gain factor.   To determine the leakage-to-gain ratio of a transistor the calibration control R_2 is adjusted for full-scale deflection with either S_1 or S_2 in the gain position.   To test for noise, a pair of sensitive headphones with a 0.05 μf capacitor in series with one of the leads is connected to terminals T_1 and T_2.   With switch S_1 or S_2 in the leakage position the amount of noise generated can be heard through the headphones.

This test circuit can also be used for checking crystal diodes and metallic rectifiers.   To test crystal diodes, follow these steps: (1) Connect the cathode of the diode (usually marked with a dot or k) to the negative terminal T_2.   (2) Connect the other lead to terminal T_1.   (3) Adjust the calibration control for full-scale deflection.   (4) Reverse the connections to the diode and note meter deflection.   (5) A diode that exhibits a high forward-to-reverse conduction ratio (50 to 1) is considered good.

To test metallic rectifiers, take the following measures: (1) The rectifier is biased in the forward direction by connecting its positive lead to the negative terminal T_2, and its negative lead to the positive terminal T_1.   (2) Adjust calibration control for full-scale meter deflection.   (3) The rectifier is biased in the reverse direction by reversing the connections to T_1 and T_2, and noting the meter deflection.   (4) Most good rectifiers will indicate a very small deflection, and a 20 to 1 ratio is considered good for a multiple plate rectifier.

To measure the forward-voltage drop, the following operations are performed: (1) The terminals T_1 and T_2 are shorted and the circuit adjusted for full-scale deflection.   (2) The rectifier is connected to T_1 and T_2 for forward bias, and the meter deflection noted.   A rectifier that is rated at 65 ma or less should indicate at least 90 per cent of full-scale deflection.   A rectifier rated at more than 65 ma should indicate at least 95 per cent of full-scale deflection.   A rectifier that does not meet these specifications has too high an internal resistance and is defective.

Although a transistor and/or a diode tester may be obtained as a separate instrument, it is included as part of some tube testers.

QUESTIONS

1. (*a*) State two reasons for testing radio and electronic equipment.   (*b*) Name three systems of testing radio and electronic equipment.

2. (*a*) Describe the point-to-point resistance measurement system of testing. (*b*) What are the advantages and disadvantages of this system?

3. (*a*) Describe the voltage and current measurement system of testing.   (*b*) What are the advantages and disadvantages of this system?

4. (*a*) Describe the signal-tracing system of testing.   (*b*) What are the advantages and disadvantages of this system?

5. (*a*) What is meant by a combination meter? (*b*) What are four other names for a combination meter? (*c*) What are the advantages of combination meters?

6. (*a*) What is an ammeter? (*b*) What is a milliammeter? (*c*) How is an ammeter connected in a circuit? (*d*) How should an ammeter be protected when readings are not being taken?

7. (*a*) What is a voltmeter? (*b*) How is a voltmeter connected in a circuit? (*c*) How should a voltmeter be protected when readings are not being taken?

8. (*a*) How is the sensitivity of a voltmeter usually expressed? (*b*) What sensitivity is considered desirable for a voltmeter used in testing electronic circuits?

9. (*a*) Name a few applications of thermal-type instruments. (*b*) Name a few applications of rectifier-type instruments.

10. (*a*) What is the purpose of a shunt? (*b*) What is the difference between an internal and external shunt?

11. (*a*) What is the purpose of a multiplier? (*b*) Can multipliers be used with alternating current and direct current instruments?

12. (*a*) What is an ohmmeter? (*b*) Explain its basic principle of operation.

13. (*a*) How does a decrease in battery voltage affect the reading of an ohmmeter? (*b*) What provison may be made to compensate for decreases in the battery voltage?

14. (*a*) Describe the basic construction of a combination meter designed to measure direct and alternating voltages, direct currents, and resistance. (*b*) What precautions are necessary in using combination meters?

15. (*a*) What is an output meter? (*b*) Describe the basic construction of an output meter.

16. (*a*) How is the output meter connected to a radio receiver? (*b*) What is the purpose of using a capacitor with an output meter?

17. (*a*) What are some applications of the electronic voltmeter? (*b*) What are the advantages of the electronic voltmeter?

18. Describe the principle of operation of a vacuum-tube voltmeter using plate rectification.

19. What are some applications of the vacuum-tube voltmeter?

20. What are the advantages of the vacuum-tube voltmeter?

21. What precautions are necessary in using a vacuum-tube voltmeter?

22. (*a*) What is an audio oscillator? (*b*) Name two types of circuits used in commercial audio oscillators. (*c*) What are the advantages of the *RC* type of audio oscillator?

23. Describe seven uses of the audio oscillator.

24. (*a*) What is an r-f test oscillator? (*b*) What are the requirements of a good r-f test oscillator?

25. (*a*) What is a signal generator? (*b*) What are two other names for a signal generator?

26. Describe some applications of the signal generator.

27. What procedure is suggested for aligning a radio receiver?

28. Describe the two characteristics of an electron beam that enable the beam to be deflected or focused by use of electric or magnetic fields.

29. Describe the general construction of a cathode-ray tube.

30. Explain the operation of an electron gun.

31. Explain the electrostatic field method of focusing the electron beam.

32. Explain the operation of the electrostatic deflection of an electron beam.

33. Explain the operation of the electromagnetic deflection of an electron beam.

34. (*a*) What is an oscilloscope? (*b*) What is the basic principle of the oscilloscope? (*c*) Describe the operation of an oscilloscope.

35. Define: (*a*) sweep, (*b*) sweep frequency, (*c*) sweep time, (*d*) flyback time.

36. (*a*) What is a sweep oscillator?   (*b*) What is the advantage of a saw-tooth wave?

37. Describe the operation of a simple sweep generator circuit that is capable of producing a saw-tooth wave.

38. (*a*) What is the disadvantage in using a gas tube in sweep-generator circuits? (*b*) What is meant by a trigger circuit?   (*c*) What is the advantage of trigger circuits?

39. How is a stationary pattern obtained on the fluorescent screen of an oscilloscope?

40. Describe the operation of a commercial cathode-ray oscilloscope.

41. Describe several applications of the oscilloscope.

42. (*a*) State four requirements of a good commercial tube tester.   (*b*) What is the advantage of a dynamic mutual conductance test?

43. Describe two forms of testing for the static characteristics of a vacuum tube.

44. Describe one form of testing for the dynamic characteristics of a vacuum tube.

45. Describe the operation of the test circuit of Fig. 16-23 for checking the leakage-gain ratio of a transistor.

46. Describe the operation of the test circuit of Fig. 16-23 for checking the amount of noise generated by a transistor.

47. Describe the operation of the test circuit of Fig. 16-23 for checking crystal diodes.

48. Describe the operation of the test circuit of Fig. 16-23 for checking metallic rectifiers.

DRAWING SYMBOLS USED IN RADIO AND ELECTRONICS

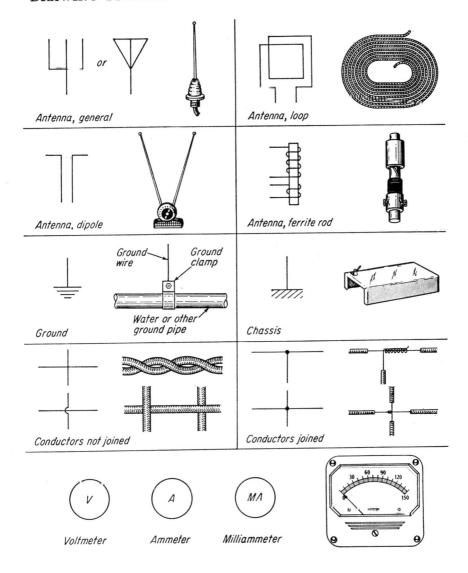

Antenna, general

Antenna, loop

Antenna, dipole

Antenna, ferrite rod

Ground
wire

Ground
clamp

Water or other
ground pipe

Ground

Chassis

Conductors not joined

Conductors joined

Voltmeter

Ammeter

Milliammeter

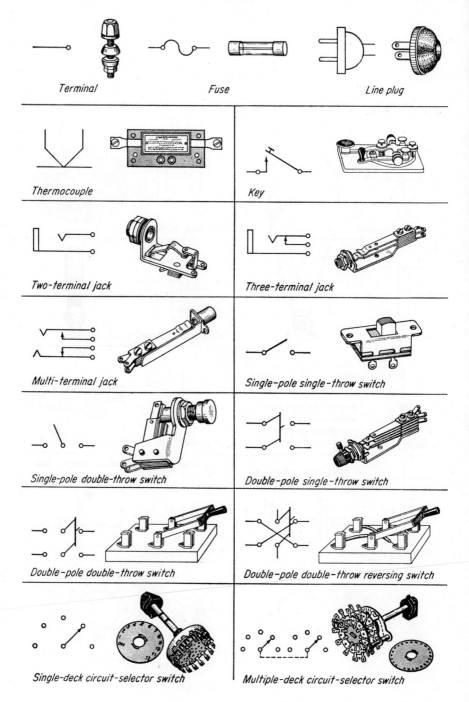

Terminal   Fuse   Line plug

Thermocouple   Key

Two-terminal jack   Three-terminal jack

Multi-terminal jack   Single-pole single-throw switch

Single-pole double-throw switch   Double-pole single-throw switch

Double-pole double-throw switch   Double-pole double-throw reversing switch

Single-deck circuit-selector switch   Multiple-deck circuit-selector switch

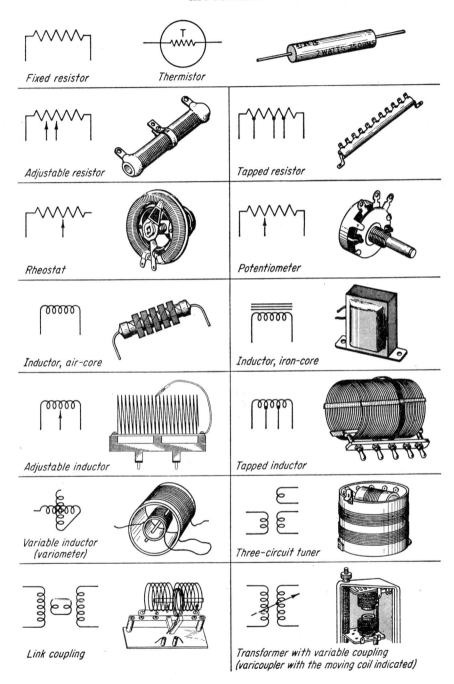

Fixed resistor   Thermistor

Adjustable resistor   Tapped resistor

Rheostat   Potentiometer

Inductor, air-core   Inductor, iron-core

Adjustable inductor   Tapped inductor

Variable inductor
(variometer)   Three-circuit tuner

Link coupling   Transformer with variable coupling
(varicoupler with the moving coil indicated)

*Transformer,
air-core*

*Transformer
bifilar
winding*

*Variable-core
transformer*

*Transformer
iron-core*

*Push-pull
transformer*

Power transformer

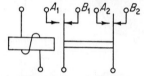

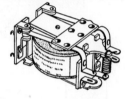

*Relay; circuit A
open when deenergized*

*Relay; circuits B_1 and B_2
closed when deenergized*

*Fixed capacitor
(Paper, mica,
or ceramic)*

(Padder)   *(Trimmer)*
Adjustable capacitors

Electrolytic capacitor

Variable capacitor

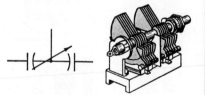

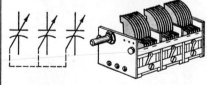

Split-stator variable capacitor

*Ganged variable capacitors,
mechanical linkage*

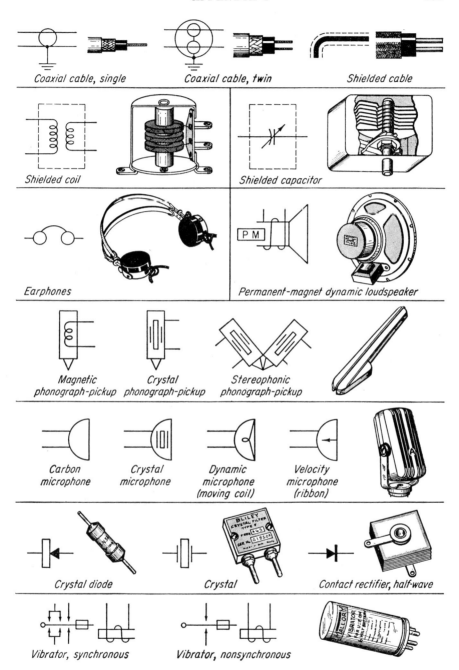

Coaxial cable, single   Coaxial cable, twin   Shielded cable

Shielded coil   Shielded capacitor

Earphones   Permanent-magnet dynamic loudspeaker

Magnetic phonograph-pickup   Crystal phonograph-pickup   Stereophonic phonograph-pickup

Carbon microphone   Crystal microphone   Dynamic microphone (moving coil)   Velocity microphone (ribbon)

Crystal diode   Crystal   Contact rectifier, half-wave

Vibrator, synchronous   Vibrator, nonsynchronous

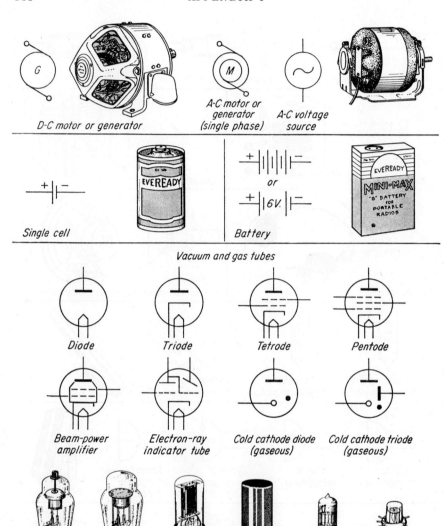

D-C motor or generator

A-C motor or generator (single phase)

A-C voltage source

Single cell

Battery

Vacuum and gas tubes

Diode

Triode

Tetrode

Pentode

Beam-power amplifier

Electron-ray indicator tube

Cold cathode diode (gaseous)

Cold cathode triode (gaseous)

See also art 2-17 and figure 2-29

Transistors

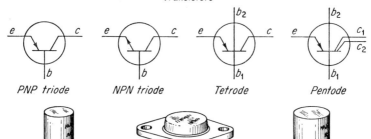

PNP triode   NPN triode   Tetrode   Pentode

Cathode-ray tubes

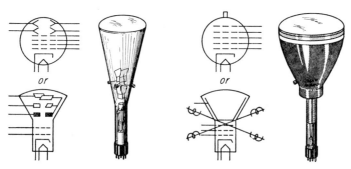

or   or

Electrostatic-deflection type   Electromagnetic-deflection type

---

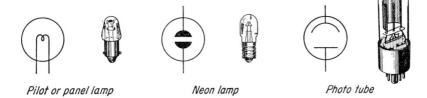

Pilot or panel lamp   Neon lamp   Photo tube

APPENDIX II

FORMULAS COMMONLY USED IN RADIO AND ELECTRONICS

NOTE: The numbers appearing opposite some of the equations correspond to the numbers of the same equations in the text or to the equations from which they were derived. These numbers are included to facilitate reference to figures, text, and nomenclature when such reference is desirable. Those equations having no number listing do not appear in this text, but were taken from the authors' text *Essentials of Electricity for Radio and Television*.

DIRECT CURRENT

Ohm's Law

$$\text{Voltage} = IR = \frac{P}{I} = \sqrt{RP}$$

$$\text{Current} = \frac{E}{R} = \frac{P}{E} = \sqrt{\frac{P}{R}}$$

$$\text{Resistance} = \frac{E}{I} = \frac{P}{I^2} = \frac{E^2}{P}$$

$$\text{Power} = EI = I^2R = \frac{E^2}{R}$$

Series Circuit

$$R = r_1 + r_2 + r_3 \cdots$$
$$E = e_1 + e_2 + e_3 \cdots$$
$$I = i_1 = i_2 = i_3 \cdots$$
$$P = p_1 + p_2 + p_3 \cdots$$

Parallel Circuits

Two resistors in parallel

$$R = \frac{r_1 r_2}{r_1 + r_2}$$

$$r_2 = \frac{R r_1}{r_1 - R}$$

Any number of resistors in parallel

$$R = \frac{1}{\dfrac{1}{r_1} + \dfrac{1}{r_2} + \dfrac{1}{r_3} \cdots}$$
$$E = e_1 = e_2 = e_3 \cdots$$
$$I = i_1 + i_2 + i_3 \cdots$$
$$P = p_1 + p_2 + p_3 \cdots$$

ALTERNATING CURRENT

Maximum, Effective, and Average Values of Sine Wave Currents and Voltages

$$\text{Maximum value} = \sqrt{2} \text{ effective value} = 1.414 \text{ effective value}$$
$$= \frac{\text{effective value}}{0.707} = 1.57 \text{ average value}$$
$$\text{Effective value} = \frac{\text{maximum value}}{\sqrt{2}} = \frac{\text{maximum value}}{1.414}$$

Effective value = 0.707 maximum value = 1.11 average value

Average value = 0.637 maximum value = 0.9009 effective value

Ohm's Law

$$\text{Voltage} = IZ = \frac{P}{I \times \text{PF}}$$

$$\text{Current} = \frac{E}{Z} = \frac{P}{E \times \text{PF}}$$

$$\text{Impedance} = \frac{E}{I} = \frac{R}{\text{PF}} = \sqrt{R^2 + (X_L - X_C)^2}$$

$$\text{Power} = I^2R = E \times I \times \text{PF}$$

$$\text{Power factor} = \frac{P}{EI} = \frac{R}{Z} = \cos\theta$$

Series Circuit

Resistance and inductance

$$Z = \sqrt{R^2 + X_L^2}$$
$$R = \sqrt{Z^2 - X_L^2}$$
$$X_L = \sqrt{Z^2 - R^2}$$

Resistance and capacitance

$$Z = \sqrt{R^2 + X_C^2}$$
$$R = \sqrt{Z^2 - X_C^2}$$
$$X_C = \sqrt{Z^2 - R^2}$$

Inductance and capacitance
  when $X_L > X_C$

$$Z = X_L - X_C$$
$$X_L = X_C + Z$$
$$X_C = X_L - Z$$

  when $X_C > X_L$

$$Z = X_C - X_L$$
$$X_L = X_C - Z$$
$$X_C = X_L + Z$$

Resistance, inductance, and capacitance
  when $X_L > X_C$

$$Z = \sqrt{R^2 + (X_L - X_C)^2}$$
$$R = \sqrt{Z^2 - (X_L - X_C)^2}$$
$$X_L = X_C + \sqrt{Z^2 - R^2}$$
$$X_C = X_L - \sqrt{Z^2 - R^2}$$

  when $X_C > X_L$

$$Z = \sqrt{R^2 + (X_C - X_L)^2}$$
$$R = \sqrt{Z^2 - (X_C - X_L)^2}$$
$$X_L = X_C - \sqrt{Z^2 - R^2}$$
$$X_C = X_L + \sqrt{Z^2 - R^2}$$

Any number of resistors, inductors, and capacitors in series

$$Z = \sqrt{(r_1 + r_2 + r_3 \cdots)^2 + (x_{L1} + x_{L2} + x_{L3} \cdots - x_{C1} - x_{C2} - x_{C3} \cdots)^2}$$

$$E = e_1 + e_2 + e_3 \cdots \qquad \text{(added vectorially)}$$

$$I = i_1 = i_2 = i_3 \cdots$$

$$P = p_1 + p_2 + p_3 \cdots$$

$$\text{Power factor (PF)} = \frac{P}{EI}$$

Parallel Circuits

Two circuits in parallel

$$Z = \frac{Z_1 Z_2}{Z_1 + Z_2}$$

Resistance and inductance

$$Z = \frac{R X_L}{\sqrt{R^2 + X_L^2}}$$

Resistance and capacitance

$$Z = \frac{R X_C}{\sqrt{R^2 + X_C^2}}$$

Inductance and capacitance

$$Z = \frac{X_L X_C}{X_L - X_C}$$

Resistance, inductance, and capacitance

$$Z = \frac{R X_L X_C}{\sqrt{X_L X_C)^2 + R^2 (X_L - X_C)^2}}$$

For parallel circuits listed above

$$E = e_1 = e_2 = e_3 \cdots$$

$$I = i_1 + i_2 + i_3 \cdots \qquad \text{(added vectorially)}$$

$$P = p_1 + p_2 + p_3 \cdots$$

$$\text{PF} = \frac{P}{EI}$$

Combination Parallel-series Circuit

Also see equations listed under Parallel Resonant Circuit. No single equation is available for this type of circuit. For solution see Art. 10-9 of the authors' *Essentials of Electricity for Radio and Television.*

INDUCTORS

Inductance of a Coil

Multilayer coil

$$L = \frac{0.8 a^2 N^2}{6a + 9b + 10c}$$

Flat or pancake coil

$$L = \frac{a^2 N^2}{8a + 11c}$$

Solenoid

$$L = \frac{a^2 N^2}{9a + 10b}$$

Inductive Reactance

$$X_L = 2\pi f L$$

Impedance of a Coil

$$Z = \sqrt{R_L{}^2 + X_L{}^2}$$

Power Factor

$$\cos \theta_L = \frac{R_L}{Z_L} = \frac{R_L}{\sqrt{R_L{}^2 + X_L{}^2}}$$

Coil Q

$$Q = \frac{X_L}{R} = \frac{2\pi f L}{R}$$

Inductors in Series

$$L_T = L_1 + L_2 + L_3 \cdots \qquad \text{(no flux linkage between coils)}$$
$$X_{LT} = x_{L1} + x_{L2} + x_{L3} \cdots \qquad \text{(no flux linkage between coils)}$$
$$L_T = L_1 + L_2 \pm 2K \sqrt{L_1 L_2} \qquad \text{(flux linking the coils)}$$
$$X_{LT} = 2\pi f L_T$$

Inductors in Parallel

$$L_T = \frac{1}{\dfrac{1}{L_1} + \dfrac{1}{L_2} + \dfrac{1}{L_3} \cdots} \qquad \text{(no flux linkage between coils)}$$

$$X_{LT} = \frac{1}{\dfrac{1}{x_{L1}} + \dfrac{1}{x_{L2}} + \dfrac{1}{x_{L3}} \cdots} \qquad \text{(no flux linkage between coils)}$$

Mutual Inductance

$$M = K \sqrt{L_1 L_2}$$
$$L_1 = \frac{M^2}{K^2 L_2}$$

Coefficient of Coupling

$$K = \frac{M}{\sqrt{L_1 L_2}}$$

Energy Stored

$$W = \frac{L I^2}{2}$$

CAPACITORS

Capacitance

$$C = \frac{22.45 K A (N - 1)}{10^8 t}$$

Capacitive Reactance

$$X_C = \frac{10^6}{2\pi f C} = \frac{159,000}{f C} \qquad (C \text{ in microfarads})$$

Impedance of a Capacitor

$$Z = \sqrt{R_C{}^2 + X_C{}^2}$$

Power Factor

$$\cos \theta_C = \frac{R_C}{Z_C} = \frac{R_C}{\sqrt{R_C{}^2 + X_C{}^2}}$$

Capacitors in Series

Two capacitors

$$C_T = \frac{C_1 C_2}{C_1 + C_2}$$

$$C_2 = \frac{C_T C_1}{C_1 - C_T}$$

Any number of capacitors

$$C_T = \frac{1}{\dfrac{1}{C_1} + \dfrac{1}{C_2} + \dfrac{1}{C_3} \cdots}$$

$$X_{CT} = x_{C1} + x_{C2} + x_{C3} \cdots = \frac{10^6}{2\pi f C_T}$$

Capacitors in Parallel

$$C_T = C_1 + C_2 + C_3 \cdots$$

$$X_{CT} = \frac{1}{\dfrac{1}{x_{C1}} + \dfrac{1}{x_{C2}} + \dfrac{1}{x_{C3}} \cdots} = \frac{10^6}{2\pi f C_T}$$

Energy Stored

$$W = \frac{CE^2}{2}$$

RESONANCE

Resonant Frequency, Inductance, and Capacitance

(Applies to both series and parallel resonant circuits)
Frequency of resonance

$$f_r = \frac{159}{\sqrt{LC}} \tag{1-9}$$

Inductance required for resonance

$$L = \frac{25{,}300}{f_r^2 C} \tag{1-11}$$

Capacitance required for resonance

$$C = \frac{25{,}300}{f_r^2 L} \tag{1-10}$$

Series Resonant Circuit

At any frequency

$$Z = \sqrt{R^2 + (X_L - X_C)^2}$$

At resonant frequency

$$Z = R \qquad \text{(minimum value possible)}$$

$$I = \frac{E}{R} \qquad \text{(maximum value possible)}$$

$$E_L = E_C = EQ$$

Parallel Resonant Circuit

At any frequency

$$Z_T = \frac{Z_1 Z_2}{Z_1 + Z_2}$$

$$Z_T = X_C \sqrt{\frac{R^2 + X_L^2}{R^2 + (X_L - X_C)^2}}$$

when R is much smaller than X_L,

$$Z_T \cong \frac{X_C X_L}{\sqrt{R^2 + (X_L - X_C)^2}}$$

At resonant frequency

$$Z_T = Q X_L = \frac{X_L{}^2}{R} \quad \text{(maximum possible value)}$$

$$I_L = I_C = IQ$$

$$I = \frac{E}{Q X_L} \quad \text{(minimum possible value)}$$

Width of Frequency Band for a Single Resonant Circuit at 0.707 of the Maximum Response

$$f_2 - f_1 = \frac{f_r}{Q} = \frac{R}{2\pi L}$$

COUPLED RESONANT CIRCUITS

Width of Bandpass

$$f_2 - f_1 = K f_r \tag{5-18}$$

Critical Coupling

$$K_c = \frac{1.5}{\sqrt{Q_P Q_S}}$$

$$Q_P Q_S = \frac{2.25}{K_c{}^2}$$

TANK CIRCUIT

Energy Stored

$$W_t = \frac{L i_t{}^2}{2} + \frac{C e_t{}^2}{2} \tag{8-11}$$

$$W_t = C E_t{}^2 \tag{8-18c}$$

$$W_t = L I_t{}^2 \tag{8-19}$$

TRANSFORMERS

Ratio

$$n = \frac{N_S}{N_P} = \sqrt{\frac{L_S}{L_P}} = \sqrt{\frac{R_S}{R_P}}$$

Secondary Voltage

$$E_S = n E_P$$

Reflected Impedance

$$Z_{P-S'} = \frac{Z_S}{n^2}$$

$$Z_{S-P'} = n^2 Z_P$$

Reflected Resistance

$$R_{P-S'} = \frac{R_S}{n^2}$$

$$R_{S-P'} = n^2 R_P$$

Reflected Reactance

$$X_{P-S'} = \frac{X_S}{n^2}$$

$$X_{S-P'} = n^2 X_P$$

Reflected Inductance

$$L_{P-S'} = \frac{L_S}{n^2}$$

$$L_{S-P'} = n^2 L_P$$

Reflected Capacitance

$$C_{P-S'} = n^2 C_S$$

$$C_{S-P'} = \frac{C_P}{n^2}$$

TIME CONSTANT

Resistance-inductance Circuit

$$t = \frac{L}{R}$$

Resistance-capacitance Circuit

$$t = CR$$

DECIBELS

In Terms of Power

$$\mathrm{db} = 10 \log \frac{P_1}{P_2} \tag{6-1}$$

In Terms of Voltage and Resistance

$$\mathrm{db} = 20 \log \frac{E_1 \sqrt{R_2}}{E_2 \sqrt{R_1}} \tag{6-2c}$$

In Terms of Voltage

$$\mathrm{db} = 20 \log \frac{E_1}{E_2} \tag{6-4}$$

In Terms of Current and Resistance

$$\mathrm{db} = 20 \log \frac{I_1 \sqrt{R_1}}{I_2 \sqrt{R_2}} \tag{6-3c}$$

In Terms of Current

$$\mathrm{db} = 20 \log \frac{I_1}{I_2} \tag{6-5}$$

In Terms of Pressure Levels

$$\mathrm{db} = 20 \log \frac{F}{F_R} \tag{10-1}$$

VACUUM-TUBE CONSTANTS

Amplification Factor

$$\mu = \frac{de_b}{de_c} \, (i_b - \text{constant}) \tag{2-1}$$

$$\mu = g_m r_p \tag{2-5}$$

Dynamic Plate Resistance

$$r_p = \frac{de_b}{di_b} \, (e_c - \text{constant}) \tag{2-2}$$

$$r_p = \frac{\mu}{g_m} \tag{2-5b}$$

Grid-plate Transconductance (Mutual Conductance)

$$g_m = \frac{di_b}{de_c} \; (e_b - \text{constant}) \qquad (2\text{-}3)$$

$$g_m = \frac{\mu}{r_p} \qquad (2\text{-}5a)$$

VOLTAGE AMPLIFICATION

Triode with Resistance Load

$$\text{VA} = \frac{\mu R_o}{R_o + r_p} \qquad (2\text{-}10)$$

Triode with Reactance Load

$$\text{VA} = \frac{\mu Z_o}{Z_o + r_p} \qquad (5\text{-}5)$$

Pentode with Reactance Load

$$\text{VA} = g_m \frac{r_p Z_o}{Z_o + r_p} \qquad (5\text{-}8)$$

when $r_p >> Z_0$

$$\text{VA} \cong g_m Z_o \qquad (5\text{-}11)$$

Multistage Amplifier Circuit

$$\text{VA}_T = \text{VA}_1 \times \text{VA}_2 \times \text{VA}_3, \; \cdots \qquad (6\text{-}40)$$
$$\text{db}_T = \text{db}_1 + \text{db}_2 + \text{db}_3, \; \cdots \qquad (6\text{-}41)$$

VOLTAGE AMPLIFICATION OF R-F AMPLIFIERS

Tuned Impedance

$$\text{VA} \cong g_m 2\pi f L Q \qquad (5\text{-}13)$$

$$\text{VA} \cong g_m \frac{(2\pi f L)^2}{R_L} \qquad (5\text{-}14)$$

Transformer with Untuned Primary and Tuned Secondary

$$\text{VA} \cong g_m 2\pi f M Q \qquad (5\text{-}17)$$

Transformer with Tuned Primary and Tuned Secondary (Bandpass Amplifier)

$$\text{VA} = g_m K \frac{2\pi f_r \sqrt{L_P L_S}}{K^2 + \dfrac{1}{Q_P Q_S}} \qquad (5\text{-}19)$$

Cascode Amplifier

$$\text{VA} = \frac{\mu (1 + \mu)}{1 + \dfrac{r_p}{R} (2 + \mu)} \qquad (12\text{-}3)$$

VOLTAGE AMPLIFICATION OF A-F AMPLIFIERS

Resistance-Capacitance Coupling

Medium frequencies
Triodes

$$\text{VA}_M = \frac{\mu}{\dfrac{r_p (R_c + R_g)}{R_c R_g} + 1} \qquad (6\text{-}8)$$

Pentodes

$$VA_M = \frac{g_m}{\dfrac{1}{R_c} + \dfrac{1}{R_g} + \dfrac{1}{r_p}} \tag{6-10}$$

Low frequencies

$$VA_L = K_L VA_M \tag{6-12}$$

$$K_L = \frac{1}{\sqrt{1 + \left(\dfrac{X_c}{R}\right)^2}} \tag{6-11}$$

High frequencies

$$VA_H = K_H VA_M \tag{6-15}$$

$$K_H = \frac{1}{\sqrt{1 + (2\pi f C_T R_{eq})^2}} \tag{6-14a}$$

Transformer Coupling

$$VA = \mu n \frac{Z_P}{Z_P + r_p} \tag{6-27}$$

Medium frequencies

$$VA_M = \mu n = \frac{X_{LP}}{\sqrt{R_1{}^2 + X_{LP}{}^2}} \tag{6-29}$$

Low frequencies

$$VA_L = K_L VA_M \tag{6-12}$$

$$K_L = \frac{X_{LP}}{\sqrt{R_1{}^2 + X_{LP}{}^2}} \tag{6-31}$$

High frequencies

$$VA_H = K_H VA_M \tag{6-15}$$

$$K_H = \frac{X_T}{\sqrt{R_{1-2}{}^2 + (X_P'' - X_T)^2}} \tag{6-32}$$

Cathode Follower

Triodes

$$VA = \frac{\mu R_k}{r_p + R_k(\mu + 1)} \tag{6-33}$$

Pentodes

$$VA = \frac{g_m R_k}{1 + g_m R_k} \tag{6-34}$$

Phase Splitter

$$VA = \frac{\mu R_2}{(\mu + 2)R_2 + r_p} \tag{6-39}$$

FEEDBACK AMPLIFIERS

Output Voltage

$$e_o' = e_s \frac{A}{1 - A\beta} \tag{6-47}$$

Distortion Voltage

$$D' = e_o' \frac{d}{1 - A\beta} \tag{6-48}$$

Negative Feedback

Voltage amplification

$$VA' = \frac{e_o'}{e_s} = \frac{A}{1 - A\beta} = -\frac{1}{\beta}\left(\frac{1}{1 - \dfrac{1}{A\beta}}\right) \tag{6-50}, \tag{6-50a}$$

when $A\beta > 1$

$$VA' \cong -\frac{1}{\beta} \qquad (6\text{-}51)$$

Feedback factor

$$A\beta = \frac{e_f}{e_s - e_f} \qquad (6\text{-}49)$$

Portion of output voltage being fed back

$$\beta \cong \frac{R_f}{R_f + R_2} \qquad (6\text{-}54a)$$

POWER AMPLIFIERS

Single-tube Triodes, Class A

Power output

$$P_o = R_o \frac{(\mu E_g)^2}{(R_o + r_p)^2} \qquad (7\text{-}9)$$

Maximum power output

$$P_{o\cdot\text{max}} = \frac{(\mu E_g)^2}{4r_p} = \frac{(\mu E_{g\cdot\text{max}})^2}{8r_p} \qquad (7\text{-}11), \ (7\text{-}11a)$$

Single-tube Pentode or Beam Power, Class A

Power output

$$P_o = R_o(g_m E_g)^2 \left(\frac{r_p}{R_o + r_p}\right)^2 \qquad (7\text{-}21)$$

Maximum power output

$$P_{o\cdot\text{max}} = \frac{R_o(g_m E_g)^2}{4} \qquad (7\text{-}23)$$

Push-pull Operation, Triodes, Class A

Power output

$$P_o = \frac{4(\mu E_g)^2 R_o'}{(2r_p + R_o')^2} \qquad (7\text{-}28)$$

Maximum power output

$$P_{o\cdot\text{max}} = \frac{(\mu E_{g\cdot\text{max}})^2}{4r_p} \qquad (7\text{-}29)$$

Decoupling Factor

Single stage

$$K_D \cong \frac{X_D}{R_D} \qquad (6\text{-}55a)$$

Multistage

$$K_{DT} \cong K_{D1} K_{D2} K_{D3} \cdots \qquad (6\text{-}56)$$

Per cent Harmonic Distortion

Triodes

$$\% \text{ 2nd harmonic distortion} = \frac{(i_{b\cdot\text{max}} + i_{b\cdot\text{min}}) - 2I_{b\cdot o}}{2(i_{b\cdot\text{max}} - i_{b\cdot\text{min}})} \times 100 \qquad (7\text{-}6)$$

Pentodes

$$\% \text{ 2nd harmonic distortion} = \frac{i_{b\cdot\text{max}} + i_{b\cdot\text{min}} - 2I_b}{i_{b\cdot\text{max}} - i_{b\cdot\text{min}} + 1.41(I_x - I_y)} \times 100 \qquad (7\text{-}17)$$

$$\% \text{ 3rd harmonic distortion} = \frac{i_{b\cdot\text{max}} - i_{b\cdot\text{min}} - 1.41(I_x - I_y)}{i_{b\cdot\text{max}} - i_{b\cdot\text{min}} + 1.41(I_x - I_y)} \times 100 \qquad (7\text{-}18)$$

$$\% \text{ 2nd harmonic plus 3rd harmonic distortion} =$$
$$\sqrt{(\% \text{ 2nd harmonic distortion})^2 + (\% \text{ 3rd harmonic distortion})^2} \qquad (7\text{-}19)$$

Characteristics

Triodes

$$\text{Plate efficiency} = \frac{P_o}{E_b \times I_b} \times 100 \qquad (7\text{-}13)$$

$$\text{Plate circuit efficiency} = \frac{P_o}{E_{bb} \times I_b} \times 100 \qquad (7\text{-}12)$$

$$\text{Power sensitivity} = \frac{P_o}{E_g{}^2} \times 10^6 \qquad (7\text{-}15)$$

Pentodes

$$\text{Plate efficiency} = \frac{P_o}{E_{b\cdot\text{av}}(I_{b\cdot\text{av}} + I_{g2\cdot\text{av}})} \times 100 \qquad (7\text{-}25)$$

$$\text{Plate circuit efficiency} = \frac{P_o}{E_{bb}(I_b + I_{g2})} \times 100 \qquad (7\text{-}24)$$

$$\text{Power sensitivity} = \frac{P_o}{E_g{}^2} \times 10^6 \qquad (7\text{-}15)$$

TRANSISTOR CONSTANTS

Current Gain

Common base

$$\alpha = h_{FB} = \frac{dI_C}{dI_E} \qquad (V_C \text{ is constant, } R_o = 0) \qquad (13\text{-}3)$$

$$\alpha_L = \frac{\beta}{1 + \beta} \qquad (14\text{-}9)$$

Common emitter

$$\beta = h_{FE} = \frac{dI_C}{dI_B} \qquad (V_C \text{ is constant, } R_o = 0) \qquad (13\text{-}4)$$

$$\beta = \frac{\alpha}{1 - \alpha} \qquad (13\text{-}5)$$

Mutual Characteristic

$$r_m = \alpha r_c \qquad (14\text{-}8)$$

TRANSISTOR AMPLIFIER CIRCUITS

Common Emitter

Input resistance

$$R_i = r_e + r_b + \frac{r_e(r_m - r_e)}{R_L + r_e + r_c - r_m} \qquad (14\text{-}10a)$$

Output resistance

$$R_o = r_e + r_c - r_m + \frac{r_e(r_m - r_e)}{R_G + r_b + r_e} \qquad (14\text{-}11a)$$

Current gain

$$I_g = \frac{r_e - r_m}{R_L + r_e + r_c - r_m} \qquad (14\text{-}12a)$$

Voltage gain

$$V_g = \frac{R_L(r_m - r_e)}{(r_b + r_e)(R_L + r_c) - r_b(r_m - r_e)} \qquad (14\text{-}13a)$$

Power gain

$$P_g = \frac{V_g{}^2 R_i}{R_L} \qquad (14\text{-}14a)$$

Common Base

Input resistance

$$R_i = r_b + r_e - \frac{r_b(r_b + r_m)}{R_L + r_b + r_c} \tag{14-15a}$$

Output resistance

$$R_o = r_b + r_c - \frac{r_b(r_b + r_m)}{R_G + r_b + r_e} \tag{14-16a}$$

Current gain

$$I_g = \frac{r_b + r_m}{r_b + r_c + R_L} \tag{14-17a}$$

Voltage gain

$$V_g = \frac{R_L(r_b + r_m)}{(R_G + r_b + r_e)(R_L + r_b + r_c) - r_b(r_b + r_m)} \tag{14-18a}$$

Power gain

$$P_g = \frac{V_g^2 R_i}{R_L} \tag{14-19a}$$

Common Collector

Input resistance

$$R_i = r_b + r_c - \frac{r_c(r_c - r_m)}{R_L + r_e + r_c - r_m} \tag{14-20a}$$

Output resistance

$$R_o = r_e + \frac{(r_b + R_G)(r_c - r_m)}{R_G + r_b + r_c} \tag{14-21a}$$

Current gain

$$I_g = \frac{r_c}{R_L + r_e + r_c - r_m} \tag{14-22a}$$

Voltage gain

$$V_g = \frac{r_c R_L}{(R_G + r_b + r_c)(R_L + r_e + r_c - r_m) - r_c(r_c - r_m)} \tag{14-23a}$$

Power gain

$$P_g = \frac{V_g^2 R_i}{R_L} \tag{14-24a}$$

Multistage Gain

Current gain

$$I_{gt} = I_{g1} \times I_{g2} \times I_{g3} \times \cdots \tag{14-25}$$

Power gain

$$P_{gt} = \frac{4 R_{G1} R_{Lf} I_{gt}^2}{(R_{G1} + R_{i1})^2} \tag{14-26}$$

Frequency Response Factor

Mid-frequency range (audio)

$$K_M = \frac{R_c}{R_c + R_i} \tag{14-28}$$

Low-frequency range (audio)

$$K_L = \frac{R_c}{\sqrt{(R_c + R_i)^2 + X_c^2}} \tag{14-29}$$

HIGH FIDELITY

Deemphasis Circuit

Response at high frequencies

$$R_{f \cdot h} = 20 \log \frac{1}{\sqrt{(2\pi f_h)^2 t^2 + 1}} \tag{10-3}$$

Response at low frequencies

$$R_{f \cdot l} = 20 \log \frac{1}{\sqrt{\dfrac{1}{(2\pi f_l)^2 t^2} + 1}} \tag{10-4}$$

Preemphasis Circuit

Response at high frequencies

$$R_{f \cdot h} = 20 \log \sqrt{(2\pi f_h)^2 t^2 + 1} \tag{10-5}$$

Response at low frequencies

$$R_{f \cdot l} = 20 \log \sqrt{\frac{1}{(2\pi f_l)^2 t^2} + 1} \tag{10-6}$$

Crossover Network

Parallel or series LC network, Fig. 10-30

$$L_1 = \frac{R_o}{2\pi f_c} \tag{10-12a}$$

$$L_2 = 1.6 L_1 \tag{10-12b}$$

$$L_3 = 0.625 L_1 \tag{10-12c}$$

$$C_1 = \frac{1}{2\pi f_c R_o} \tag{10-13a}$$

$$C_2 = 0.625 C_1 \tag{10-13b}$$

$$C_3 = 1.6 C_1 \tag{10-13c}$$

Noise

Equivalent noise resistance
Triodes, r-f tubes

$$R_n \cong \frac{2.5}{g_m} \tag{12-4}$$

Pentodes, r-f tubes

$$R_n \cong \frac{I_b}{I_b + I_{c \cdot 2}} \left(\frac{2.5}{g_m} + \frac{20 I_{c \cdot 2}}{g_m{}^2} \right) \tag{12-5}$$

Noise Figure

R-f vacuum-tube circuits

$$F_n = 20 \log \sqrt{\frac{2R_i + 4R_n}{R_i}} \tag{12-6}$$

Transistor circuits
One-cycle bandwidth at 1,000 cycles

$$F_n = 10 \log \frac{P_{no}}{P_{ni}} \tag{13-8}$$

One-cycle bandwidth at any frequency

$$F_{nf} = F_n + 10 \log \frac{1,000}{f} \tag{13-9}$$

POWER SUPPLY

Output Voltage, without Filter, Resistance Load

Full-wave

$$E_o = 0.9E_{\text{a-c}} \qquad (9\text{-}1)$$

Half-wave

$$E_o = 0.45E_{\text{a-c}} \qquad (9\text{-}2)$$

Per Cent of Ripple Voltage

General equation

$$\% \, E_r = \frac{E_r}{E_{\text{d-c}}} \times 100 \qquad (9\text{-}4)$$

Capacitor input filter circuit
  At the output of first capacitor

$$\% \, E_{r\cdot1} \cong \frac{2,245 \times 10^4}{f_r R_o C_1} \qquad (9\text{-}5)$$

  At the output of second capacitor
    Using a filter choke

$$\% \, E_{r\cdot2} \cong \frac{\% \, E_{r\cdot1}}{[10^{-6}(2\pi f_r)^2 L_1 C_2] - 1} \qquad (9\text{-}6)$$

  Using a filter resistor

$$\% \, E_{r\cdot2} \cong \frac{\% \, E_{r\cdot1} \times 10^6}{2\pi f_r C_2 R_1} \qquad (9\text{-}7)$$

Choke input filter circuit, $f_r = 120$ cycles
  At the output of first capacitor

$$\% \, E_{r\cdot1} \cong \frac{100}{L_1 C_1} \qquad (9\text{-}9a)$$

  At the output of second capacitor

$$\% \, E_{r\cdot2} \cong \frac{650}{L_1 L_2 (C_1 + C_2)^2} \qquad (9\text{-}10a)$$

Voltage Regulation

$$\mathrm{VR} = \frac{E_{NL} - E_L}{E_L} \times 100 \qquad (9\text{-}8)$$

Critical Inductance of Input Choke

$$L_c = \frac{R_o}{1,000} \qquad (9\text{-}11)$$

RELATION BETWEEN WAVELENGTH AND FREQUENCY

Wavelength

$$\lambda = \frac{300,000,000}{f(\text{cps})} = \frac{300,000}{f(\text{kc})} = \frac{300}{f(\text{mc})} \qquad (1\text{-}1)$$

Frequency

$$f = \frac{300,000,000}{\lambda} \, (\text{cps}) = \frac{300,000}{\lambda} \, (\text{kc}) = \frac{300}{\lambda} \, (\text{mc}) \qquad (1\text{-}2)$$

CRYSTALS

Resonant Frequency

$$f_r = \frac{k}{t} \qquad (8\text{-}22)$$

APPENDIX III

STANDARD COLOR CODING FOR RESISTORS

For the identification of resistance values of small carbon-type resistors, numbers are represented by the colors indicated in Table A-1. Three colors are used on each

TABLE A-1. COLOR CODE USED FOR THE IDENTIFICATION OF CARBON RESISTORS AND FLAT-PAPER, MICA, AND CERAMIC CAPACITORS

1—Brown	4—Yellow	7—Violet
2—Red	5—Green	8—Gray
3—Orange	6—Blue	9—White
	0—Black	

resistor to identify its value in ohms according to the methods indicated in Table A-2. Two systems are used for placing the color band or dot on a resistor, one for the axial-lead resistor Fig. A-1 and the other for the radial-lead resistor Fig. A-2. The tolerance is indicated by a fourth band or dot having the following color identification.

Gold, 5 per cent   Silver, 10 per cent   None, 20 per cent

TABLE A-2. METHODS USED FOR IDENTIFYING THE COLOR BANDS OR DOTS AND THEIR FUNCTIONS AS USED WITH CARBON RESISTORS AND FLAT-PAPER, MICA, AND CERAMIC CAPACITORS

Axial leads	Identification	Radial leads
Band A	First significant figure	Body A
Band B	Second significant figure	End B
Band C	The number of zeros following the first two figures	Band C or dot C
Band D	Per cent tolerance above and below the nominal resistance value	End D

For resistors having fractional values of more than 1 ohm but less than 10 ohms *band C is gold* indicating a multiplying factor of 0.1. For resistors having a value of less than 1 ohm *band C is silver* indicating a multiplying factor of 0.01.

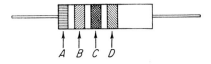

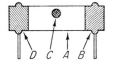

Axial-lead resistor

FIG. A-1

Radial-lead resistor

FIG. A-2

TABLE A-3. APPLICATION OF RESISTOR COLOR CODE

Resistance, ohms	A	B	C	Resistance, ohms	A	B	C
0.27	Red	Violet	Silver	10 k	Brown	Black	Orange
0.56	Green	Blue	Silver	11 k	Brown	Brown	Orange
1.0	Black	Brown	Black	12 k	Brown	Red	Orange
1.8	Brown	Gray	Gold	18 k	Brown	Gray	Orange
3.3	Orange	Orange	Gold	22 k	Red	Red	Orange
10	Brown	Black	Black	27 k	Red	Violet	Orange
12	Brown	Red	Black	33 k	Orange	Orange	Orange
22	Red	Red	Black	39 k	Orange	White	Orange
39	Orange	White	Black	47 k	Yellow	Violet	Orange
47	Yellow	Violet	Black	56 k	Green	Blue	Orange
68	Blue	Gray	Black	68 k	Blue	Gray	Orange
82	Gray	Red	Black	82 k	Gray	Red	Orange
100	Brown	Black	Brown	100 k	Brown	Black	Yellow
120	Brown	Red	Brown	120 k	Brown	Red	Yellow
180	Brown	Gray	Brown	150 k	Brown	Green	Yellow
220	Red	Red	Brown	180 k	Brown	Gray	Yellow
270	Red	Violet	Brown	200 k	Red	Black	Yellow
330	Orange	Orange	Brown	220 k	Red	Red	Yellow
470	Yellow	Violet	Brown	390 k	Orange	White	Yellow
560	Green	Blue	Brown	470 k	Yellow	Violet	Yellow
820	Gray	Red	Brown	510 k	Green	Brown	Yellow
1,000	Brown	Black	Red	680 k	Blue	Gray	Yellow
1,200	Brown	Red	Red	1.5 meg	Brown	Green	Green
1,500	Brown	Green	Red	2.2 meg	Red	Red	Green
1,800	Brown	Gray	Red	3.9 meg	Orange	White	Green
2,700	Red	Violet	Red	5.6 meg	Green	Blue	Green
3,900	Orange	White	Red	6.8 meg	Blue	Gray	Green
4,700	Yellow	Violet	Red	12 meg	Brown	Red	Blue
5,600	Green	Blue	Red	18 meg	Brown	Gray	Blue
6,800	Blue	Gray	Red	22 meg	Red	Red	Blue

APPENDIX IV

STANDARD COLOR CODING FOR CAPACITORS

The basic system used for color coding capacitors is the same as for resistors, and all capacitance values are in micromicrofarads. (1) The colors used for identifying the numbers are the same as those used for resistors (Table A-1). (2) The methods used for identifying the color bands or dots are also the same as those used for resistors (Table A-2). Capacitors may also use one or more of four additional identification bands or dots, namely,

    E—third significant figure, used in some of the old systems
    F—d-c working voltage
    G—Classification or operating characteristic
    H—Temperature coefficient

It is impossible to describe a standard method of identification that would be applicable to all of the many different types of capacitors manufactured. Typical examples of the identification systems for commonly used capacitors are illustrated in Fig. A-3, and the method of identifying the values of the multiplying factor, tolerance, and d-c working voltage in Tables A-4 and A-5.

TABLE A-4. IDENTIFICATION OF MOLDED PAPER AND MOLDED MICA CAPACITORS

Color of band or dot	*C* Multiplying factor	*D* Tolerance paper, %	*D* Tolerance mica, %	*F* Working voltage
Black	1	20	20	. . .
Brown	10	. . .	. . .	100
Red	10^2	. . .	2	200
Orange	10^3	30	3	300
Yellow	10^4	40	. . .	400
Green	10^5	5	5 (EIA)	500
Blue	10^6	. . .	. . .	600
Violet	. . .	. . .	. . .	700
Gray	. . .	. . .	. . .	800
White	. . .	10(EIA)	. . .	900
Gold	0.1	10(MIL)	5(MIL)	1,000
Silver	0.01	. . .	10	2,000

678

TABLE A-5. IDENTIFICATION OF CERAMIC CAPACITORS

Color of band or dot	C Multiplying factor	D Tolerance, capacitance less than $10\mu\mu f$, $\mu\mu f$	D Tolerance, capacitance more than $10\mu\mu f$, %	F Working voltage
Black	1	2	20	...
Brown	10	0.1	1	150
Red	10^2	...	2	...
Orange	10^3	...	2.5	350
Yellow	10^4	...	...	...
Green	...	0.5	5	500
Blue	...	...	...	...
Violet	...	...	...	...
Gray	0.01	0.25	...	...
White	0.1	1.0	10	...

TABLE A-6. APPLICATION OF CAPACITOR COLOR CODE

System Fig. A-3	Capacitance, $\mu\mu f$	Tolerance, %	Working voltage	A	B	C	D	E	F
(a)	54	1	...	Green	Yellow	Black	Brown		
(a)	270	2.5	...	Red	Violet	Brown	Orange		
(b)	8	0.5 $\mu\mu f$	...	Black	Gray	Black	Green		
(b)	820	5	...	Gray	Red	Brown	Green		
(c)	1,200	10	150	Brown	Red	Red	White	...	Brown
(c)	6,800	20	350	Blue	Gray	Red	Black	...	Orange
(d)	33	0.25 $\mu\mu f$	...	Orange	Orange	White	Gray		
(d)	0.68	0.1 $\mu\mu f$	...	Blue	Gray	Blue	Brown		
(d)	91	2	...	White	Brown	Black	Red		
(e)	7.5	...	...	Violet	Green	White			
(e)	4,700	...	...	Yellow	Green	Red			
(f)	500	10	...	Green	Black	Brown	White		
(f)	2,300	20	...	Red	Orange	Red	Black		
(g)	560	5	...	Green	Blue	Brown	Green		
(g)	10,000	20	...	Brown	Black	Orange	Black		
(h)	62	2	...	Blue	Red	Black	Red		
(h)	7,500	10 (EIA)	...	Violet	Green	Red	White		
(h)	910	3	...	White	Brown	Brown	Orange		
(i)	5	2	...	Black	Green	Black	Red		
(j)	33	3	300	Orange	Orange	Black	Orange	...	Orange
(k)	2,200	10 (MIL)	600	Red	Red	Red	White	...	Blue
(l)	360	2	1000	Orange	Blue	Brown	Red	...	Gold
(m)	8,200	20	400	Gray	Red	Red	Black	...	Yellow
(n)	430	5 (MIL)	2000	Yellow	Orange	Black	Gold	Black	Silver

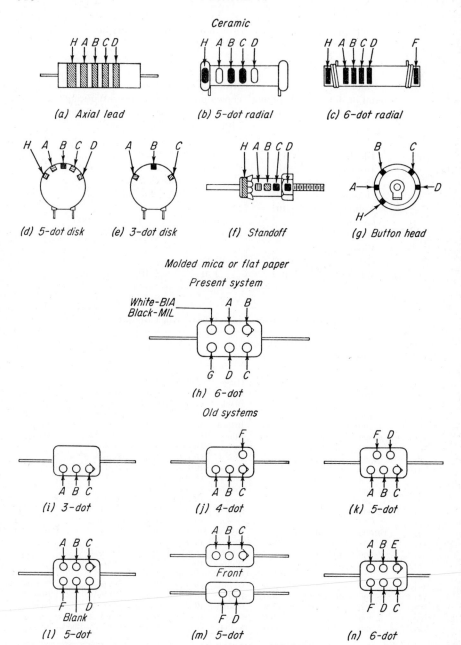

Fig. A-3. Identification systems used for ceramic, molded mica, and flat-paper capacitors.

STANDARD COLOR CODING FOR TRANSFORMER LEADS

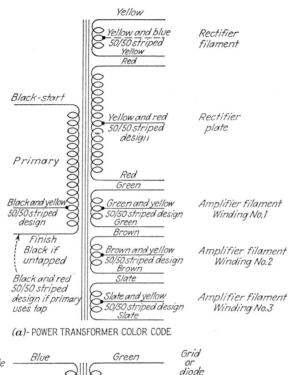

(a)- POWER TRANSFORMER COLOR CODE

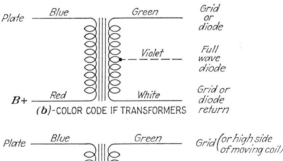

(b)-COLOR CODE IF TRANSFORMERS

The upper portion *(that code above the dotted line)*
for single primary and/or secondary transformers

(c)-COLOR CODE-AUDIO TRANSFORMERS

FIG. A-4. Color coding for transformer leads.

APPENDIX VI

TRIGONOMETRY

The solution of a-c problems frequently involves adding or subtracting quantities such as voltages, currents, and ohmages by means of vectors. The mathematical solution of these problems requires the use of trigonometry. The method of solution presented in the text makes it possible to solve all such problems by the use of right triangles. The following statements apply to any right triangle and are illustrated in Fig. A-5.

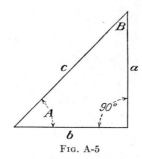

Fig. A-5

1. A right triangle is one in which one of the angles is a right angle (90 degrees).
2. The hypotenuse is the side opposite the right angle.
3. The legs of a right triangle are the two sides that form the right angle.
4. The sine of any angle θ is equal to the side opposite that angle divided by the hypotenuse.
5. The cosine of any angle θ is equal to the side adjacent to that angle divided by the hypotenuse.
6. The square of the hypotenuse is equal to the sum of the squares of the two legs of the triangle. (This is also commonly known as the *theorem of Pythagoras*.)

$$\sin A = \frac{a}{c} \qquad a = c \sin A \qquad c = \frac{a}{\sin A}$$

$$\cos A = \frac{b}{c} \qquad b = c \cos A \qquad c = \frac{b}{\cos A}$$

$$\sin B = \frac{b}{c} \qquad b = c \sin B \qquad c = \frac{b}{\sin B}$$

$$\cos B = \frac{a}{c} \qquad a = c \cos B \qquad c = \frac{a}{\cos B}$$

$$c^2 = a^2 + b^2 \qquad a^2 = c^2 - b^2 \qquad b^2 = c^2 - a^2$$

The tables of Appendix VII list the values of sine and cosine for angles between 0 and 90 degrees. In some instances, it is desired to obtain the sine of angles greater than 90 degrees, and they may be obtained in the following manner:

682

When θ is between 90 and 180 degrees

$$\sin \theta = \cos (\theta - 90)$$

Example: What is the sine of 137 degrees?

$$\sin 137° = \cos (137 - 90) = \cos 47° = 0.682$$

When θ is between 180 and 270 degrees

$$\sin \theta = - \sin (\theta - 180)$$

Example: What is the sine of 218 degrees?

$$\sin 218° = - \sin (218 - 180) = - \sin 38° = - 0.616$$

When θ is between 270 and 360 degrees

$$\sin \theta = - \cos (\theta - 270)$$

Example: What is the sine of 336 degrees?

$$\sin 336° = - \cos (336 - 270) = - \cos 66° = - 0.407$$

APPENDIX VII

SINE AND COSINE TABLES

Degrees	sin	cos	Degrees	sin	cos
0.0	0.000	1.000	21.5	0.366	0.930
0.5	0.009	1.000	22.0	0.374	0.927
1.0	0.017	0.999	22.5	0.383	0.924
1.5	0.026	0.999	23.0	0.391	0.920
2.0	0.035	0.999	23.5	0.399	0.917
2.5	0.043	0.999	24.0	0.407	0.913
3.0	0.052	0.998	24.5	0.415	0.910
3.5	0.061	0.998	25.0	0.422	0.906
4.0	0.070	0.997	25.5	0.430	0.902
4.5	0.078	0.997	26.0	0.438	0.899
5.0	0.087	0.996	26.5	0.446	0.895
5.5	0.096	0.995	27.0	0.454	0.891
6.0	0.104	0.994	27.5	0.462	0.887
6.5	0.113	0.993	28.0	0.469	0.883
7.0	0.122	0.992	28.5	0.477	0.879
7.5	0.130	0.991	29.0	0.485	0.875
8.0	0.139	0.990	29.5	0.492	0.870
8.5	0.148	0.989	30.0	0.500	0.866
9.0	0.156	0.988	30.5	0.507	0.862
9.5	0.165	0.986	31.0	0.515	0.857
10.0	0.173	0.985	31.5	0.522	0.853
10.5	0.182	0.983	32.0	0.530	0.848
11.0	0.191	0.981	32.5	0.537	0.843
11.5	0.199	0.980	33.0	0.544	0.839
12.0	0.208	0.978	33.5	0.552	0.834
12.5	0.216	0.976	34.0	0.559	0.829
13.0	0.225	0.974	34.5	0.566	0.824
13.5	0.233	0.972	35.0	0.574	0.819
14.0	0.242	0.970	35.5	0.581	0.814
14.5	0.250	0.968	36.0	0.588	0.809
15.0	0.259	0.966	36.5	0.595	0.804
15.5	0.267	0.963	37.0	0.602	0.798
16.0	0.275	0.961	37.5	0.609	0.793
16.5	0.284	0.959	38.0	0.616	0.788
17.0	0.292	0.956	38.5	0.622	0.783
17.5	0.301	0.954	39.0	0.629	0.777
18.0	0.309	0.951	39.5	0.636	0.772
18.5	0.317	0.948	40.0	0.643	0.766
19.0	0.325	0.945	40.5	0.649	0.760
19.5	0.334	0.942	41.0	0.656	0.755
20.0	0.342	0.940	41.5	0.663	0.749
20.5	0.350	0.937	42.0	0.669	0.743
21.0	0.358	0.933	42.5	0.675	0.737

Degrees	sin	cos	Degrees	sin	cos
43.0	0.682	0.731	67.0	0.920	0.391
43.5	0.688	0.725	67.5	0.924	0.383
44.0	0.695	0.719	68.0	0.927	0.375
44.5	0.701	0.713	68.5	0.930	0.366
45.0	0.707	0.707	69.0	0.934	0.358
45.5	0.713	0.701	69.5	0.937	0.350
46.0	0.719	0.695	70.0	0.940	0.342
46.5	0.725	0.688	70.5	0.943	0.334
47.0	0.731	0.682	71.0	0.945	0.326
47.5	0.737	0.675	71.5	0.948	0.317
48.0	0.743	0.669	72.0	0.951	0.309
48.5	0.749	0.663	72.5	0.954	0.301
49.0	0.755	0.656	73.0	0.956	0.292
49.5	0.760	0.649	73.5	0.959	0.284
50.0	0.766	0.643	74.0	0.961	0.276
50.5	0.772	0.636	74.5	0.964	0.267
51.0	0.777	0.629	75.0	0.966	0.259
51.5	0.783	0.622	75.5	0.968	0.250
52.0	0.788	0.616	76.0	0.970	0.242
52.5	0.793	0.609	76.5	0.972	0.233
53.0	0.798	0.602	77.0	0.974	0.225
53.5	0.804	0.595	77.5	0.976	0.216
54.0	0.809	0.588	78.0	0.978	0.208
54.5	0.814	0.581	78.5	0.980	0.199
55.0	0.819	0.574	79.0	0.982	0.191
55.5	0.824	0.566	79.5	0.983	0.182
56.0	0.829	0.559	80.0	0.985	0.174
56.5	0.834	0.552	80.5	0.986	0.165
57.0	0.839	0.544	81.0	0.988	0.156
57.5	0.843	0.537	81.5	0.989	0.148
58.0	0.848	0.530	82.0	0.990	0.139
58.5	0.853	0.522	82.5	0.991	0.130
59.0	0.857	0.515	83.0	0.992	0.122
59.5	0.862	0.507	83.5	0.994	0.113
60.0	0.866	0.500	84.0	0.994	0.104
60.5	0.870	0.492	84.5	0.995	0.096
61.0	0.875	0.485	85.0	0.996	0.087
61.5	0.879	0.477	85.5	0.997	0.078
62.0	0.883	0.469	86.0	0.997	0.070
62.5	0.887	0.462	86.5	0.998	0.061
63.0	0.891	0.454	87.0	0.998	0.052
63.5	0.895	0.446	87.5	0.999	0.043
64.0	0.899	0.438	88.0	0.999	0.035
64.5	0.903	0.430	88.5	0.999	0.026
65.0	0.906	0.423	89.0	0.999	0.017
65.5	0.910	0.415	89.5	1.000	0.009
66.0	0.913	0.407	90.0	1.000	0.000
66.5	0.917	0.399			

COMMON LOGARITHMS OF NUMBERS

Logarithms of Numbers

N	0	1	2	3	4	5	6	7	8	9
10	0000	0043	0086	0128	0170	0212	0253	0294	0334	0374
11	0414	0453	0492	0531	0569	0607	0645	0682	0719	0755
12	0792	0828	0864	0899	0934	0969	1004	1038	1072	1106
13	1139	1173	1206	1239	1271	1303	1335	1367	1399	1430
14	1461	1492	1523	1553	1584	1614	1644	1673	1703	1732
15	1761	1790	1818	1847	1875	1903	1931	1959	1987	2014
16	2041	2068	2095	2122	2148	2175	2201	2227	2253	2279
17	2304	2330	2355	2380	2405	2430	2455	2480	2504	2529
18	2553	2577	2601	2625	2648	2672	2695	2718	2742	2765
19	2788	2810	2833	2856	2878	2900	2923	2945	2967	2989
20	3010	3032	3054	3075	3096	3118	3139	3160	3181	3201
21	3222	3243	3263	3284	3304	3324	3345	3365	3385	3404
22	3424	3444	3464	3483	3502	3522	3541	3560	3579	3598
23	3617	3636	3655	3674	3692	3711	3729	3747	3766	3784
24	3802	3820	3838	3856	3874	3892	3909	3927	3945	3962
25	3979	3997	4014	4031	4048	4065	4082	4099	4116	4133
26	4150	4166	4183	4200	4216	4232	4249	4265	4281	4298
27	4314	4330	4346	4362	4378	4393	4409	4425	4440	4456
28	4472	4487	4502	4518	4533	4548	4564	4579	4594	4609
29	4624	4639	4654	4669	4683	4698	4713	4728	4742	4757
30	4771	4786	4800	4814	4829	4843	4857	4871	4886	4900
31	4914	4928	4942	4955	4969	4983	4997	5011	5024	5038
32	5051	5065	5079	5092	5105	5119	5132	5145	5159	5172
33	5185	5198	5211	5224	5237	5250	5263	5276	5289	5302
34	5315	5328	5340	5353	5366	5378	5391	5403	5416	5428
35	5441	5453	5465	5478	5490	5502	5514	5527	5539	5551
36	5563	5575	5587	5599	5611	5623	5635	5647	5658	5670
37	5682	5694	5705	5717	5729	5740	5752	5763	5775	5786
38	5798	5809	5821	5832	5843	5855	5866	5877	5888	5899
39	5911	5922	5933	5944	5955	5966	5977	5988	5999	6010
40	6021	6031	6042	6053	6064	6075	6085	6096	6107	6117
41	6128	6138	6149	6160	6170	6180	6191	6201	6212	6222
42	6232	6243	6253	6263	6274	6284	6294	6304	6314	6325
43	6335	6345	6355	6365	6375	6385	6395	6405	6415	6425
44	6435	6444	6454	6464	6474	6484	6493	6503	6513	6522
45	6532	6542	6551	6561	6571	6580	6590	6599	6609	6618
46	6628	6637	6646	6656	6665	6675	6684	6693	6702	6712
47	6721	6730	6739	6749	6758	6767	6776	6785	6794	6803
48	6812	6821	6830	6839	6848	6857	6866	6875	6884	6893
49	6902	6911	6920	6928	6937	6946	6955	6964	6972	6981
50	6990	6998	7007	7016	7024	7033	7042	7050	7059	7067
51	7076	7084	7093	7101	7110	7118	7126	7135	7143	7152
52	7160	7168	7177	7185	7193	7202	7210	7218	7226	7235
53	7243	7251	7259	7267	7275	7284	7292	7300	7308	7316
54	7324	7332	7340	7348	7356	7364	7372	7380	7388	7396

N	0	1	2	3	4	5	6	7	8	9
55	7404	7412	7419	7427	7435	7443	7451	7459	7466	7474
56	7482	7490	7497	7505	7513	7520	7528	7536	7543	7551
57	7559	7566	7574	7582	7589	7597	7604	7612	7619	7627
58	7634	7642	7649	7657	7664	7672	7679	7686	7694	7701
59	7709	7716	7723	7731	7738	7745	7752	7760	7767	7774
60	7782	7789	7796	7803	7810	7818	7825	7832	7839	7846
61	7853	7860	7868	7875	7882	7889	7896	7903	7910	7917
62	7924	7931	7938	7945	7952	7959	7966	7973	7980	7987
63	7993	8000	8007	8014	8021	8028	8035	8041	8048	8055
64	8062	8069	8075	8082	8089	8096	8102	8109	8116	8122
65	8129	8136	8142	8149	8156	8162	8169	8176	8182	8189
66	8195	8202	8209	8215	8222	8228	8235	8241	8248	8254
67	8261	8267	8274	8280	8287	8293	8299	8306	8312	8319
68	8325	8331	8338	8344	8351	8357	8363	8370	8376	8382
69	8388	8395	8401	8407	8414	8420	8426	8432	8439	8445
70	8451	8457	8463	8470	8476	8482	8488	8494	8500	8506
71	8513	8519	8525	8531	8537	8543	8549	8555	8561	8567
72	8573	8579	8585	8591	8597	8603	8609	8615	8621	8627
73	8633	8639	8645	8651	8657	8663	8669	8675	8681	8686
74	8692	8698	8704	8710	8716	8722	8727	8733	8739	8745
75	8751	8456	8762	8768	8774	8779	8785	8791	8797	8802
76	8808	8814	8820	8825	8831	8837	8842	8848	8854	8859
77	8865	8871	8876	8882	8887	8893	8899	8904	8910	8915
78	8921	8927	8932	8938	8943	8949	8954	8960	8965	8971
79	8976	8982	8987	8993	8998	9004	9009	9015	9020	9025
80	9031	9036	9042	9047	9053	9058	9063	9069	9074	9079
81	9085	9090	9096	9101	9106	9112	9117	9122	9128	9133
82	9138	9143	9149	9154	9159	9165	9170	9175	9180	9186
83	9191	9196	9201	9206	9212	9217	9222	9227	9232	9238
84	9243	9248	9253	9258	9263	9269	9274	9279	9284	9289
85	9294	9299	9304	9309	9315	9320	9325	9330	9335	9340
86	9345	9350	9355	9360	9365	9370	9375	9380	9385	9390
87	9395	9400	9405	9410	9415	9420	9425	9430	9435	9440
88	9445	9450	9455	9460	9465	9469	9474	9479	9484	9489
89	9494	9499	9504	9509	9513	9518	9523	9528	9533	9538
90	9542	9547	9552	9557	9562	9566	9571	9576	9581	9586
91	9590	9595	9600	9605	9609	9614	9619	9624	9628	9633
92	9638	9643	9647	9652	9657	9661	9666	9671	9675	9680
93	9685	9689	9694	9699	9703	9708	9713	9717	9722	9727
94	9731	9736	9741	9745	9750	9754	9759	9763	9768	9773
95	9777	9782	9786	9791	9795	9800	9805	9809	9814	9818
96	9823	9827	9832	9836	9841	9845	9850	9854	9859	9863
97	9868	9872	9877	9881	9886	9890	9894	9899	9903	9908
98	9912	9917	9921	9926	9930	9934	9939	9943	9948	9952
99	9956	9961	9965	9969	9974	9978	9983	9987	9991	9996

APPENDIX IX

UNIVERSAL TIME-CONSTANT CURVES AND TABLES

TABLE A-7. ASCENDING CURVE

k time constants	Per cent of maximum value	k time constants	Per cent of maximum value	k time constants	Per cent of maximum value
0.00	0.000	0.70	50.3	2.50	91.8
0.05	4.9	0.80	55.1	3.00	95.0
0.10	9.5	0.90	59.3	3.50	97.0
0.15	14.0	1.00	63.2	4.00	98.2
0.20	18.1	1.20	69.9	4.50	98.9
0.30	25.9	1.40	75.3	5.00	99.3
0.40	33.0	1.60	79.8	5.50	99.6
0.50	39.3	1.80	83.5	6.00	99.8
0.60	45.1	2.00	86.5	7.00	99.9

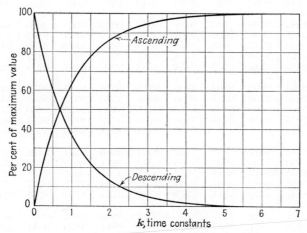

FIG. A-6. Universal time-constant curves.

TABLE A-8. DESCENDING CURVE

k time constants	Per cent of maximum value	k time constants	Per cent of maximum value	k time constants	Per cent of maximum value
0.00	100	0.70	49.7	2.50	8.2
0.05	95.1	0.80	44.9	3.00	5.0
0.10	90.5	0.90	40.7	3.50	3.0
0.15	86.0	1.00	36.8	4.00	1.8
0.20	81.9	1.20	30.1	4.50	1.1
0.30	74.1	1.40	24.7	5.00	0.7
0.40	67.0	1.60	20.2	5.50	0.4
0.50	60.7	1.80	16.5	6.00	0.2
0.60	54.9	2.00	13.5	7.00	0.1

APPENDIX X

ANSWERS TO PROBLEMS

NOTE 1. Answers are provided for approximately 50 per cent of the problems. Instructors using this text may obtain a complete answer book from the publisher.
NOTE 2. As far as is practicable, all answers are accurate to three significant figures.
NOTE 3. Answers to problems involving values obtained from curves are generally difficult to check accurately because of variations in reading the curves. In preparing the answer book, enlarged drawings of the curves were used to aid in obtaining greater accuracy. In most cases the values obtained from the curves for use in solving the problems have been included with the answers.

Chapter 1

1. (a) 4.414 ft
   (b) 1.345 meters
3. (a) 0.177 sec
   (b) 0.00268 sec
   (c) The radio listener
5. 5.65 to 0.141 ft
7. 0.941 ft
9. 570 kc
11. 4.89 to 4.56 meters
13. 7.81 mc
15. (a) 3.01 meters
   (b) 9.87 ft
17. 19,100 cycles
19. 11,000 cycles
21. 0.0134 sec
23. 15.1 ft
25. 80 per cent
27. (a) 80 kc
   (b) 120 kc
29. (a) 8 kc
   (b) 12 kc
31. 1,500 kc
33. 5 μh
35. 247 $\mu\mu$f
37. 9,938 kc
38. 546 kc
41. 10 μh
43. (a) 334 $\mu\mu$f
   (b) 550 to 813 kc
45. (a) 389 μh
   (b) 43.3 μh
   (c) 6.08 μh
47. 39.4 $\mu\mu$f

Chapter 2

1. (a) 181 ohms
   (b) 10 watts
3. (b) 2.66 ohms
   (c) $\frac{1}{2}$ watt
5. (b) 21 ohms
   (c) 5 watts
7. (b) $R_1 = 42$ ohms, $R_2 = 724.6$ ohms
   (c) $R_1 = 2$ watts, $R_2 = 30$ watts
11. 20 ($de_b = 40$ volts)
   ($de_c = 2$ volts)
13. 19 ($de_b = 38$ volts)
   ($de_c = 2$ volts)
16. 7,270 ohms
   ($de_b = 40$ volts)
   ($di_b = 5.5$ ma)
18. 12,500 ohms
   ($de_b = 40$ volts)
   ($di_b = 3.2$ ma)
21. (a) 2,600 μmhos
   ($di_b = 5.2$ ma)
   ($de_c = 2$ volts)
   (b) 2,750 μmhos
23. (a) 1,950 μmhos
   ($di_b = 3.9$ ma)
   ($de_c = 2$ volts)
   (b) 1,520 μmhos
25. 1,400 μmhos
27. 32,000 ohms
29. (a) 3.42
   (b) 10.1
   (c) 13.4
   (d) 5.28, 15.2, 20.1 volts

31. (a) 2.04
   (b) 7.12
   (c) 10.3
   (d) 3.06, 10.7, 15.4 volts
33. (a) 15,000 ohms
   (b) 45,000 ohms
35. 21,000 ohms

Chapter 3

1. (a) 50 μsec
   (b) 0.667 μsec
   (c) 75
2. (a) 1,060 ohms
   (b) R is 471 times greater than X_C
   (c) The capacitor path
   (d) 3,180,000 ohms
   (e) X_C is 6.36 times greater than R
   (f) The resistor path
5. (a) 3,419 ohms
   (b) Path through C_1
   (c) Yes. A small amount
   (d) 3,419 ohms
   (e) Additional i-f filtering
   (f) 80 per cent
7. (a) 1,060 ohms
   (b) 282,600 ohms
   (c) 3,180,000 ohms
   (d) 94.2 ohms
   (e) The capacitor path
   (f) The inductor path
9. (a) -4 to -2 volts
   (b) -6 to 0 volts
   (c) -7 to 1 volt
11. 17 volts
13. (a) 120,000 ohms
   (b) 48,000 ohms
   (c) 30,000 ohms
15. (a) 85,000 ohms
   (b) 0.0034 watt
   (c) ¼ watt
17. (a) 0.374 μf
   (b) 0.5 μf, 200-volt, paper capacitor
19. 317 volts
21. (a) 1,067 ohms ($I_b = 7.5$ ma)
   (b) 800 ohms ($I_b = 7.5$ ma)
   (c) 533 ohms ($I_b = 7.5$ ma)
   (d) 364 ohms ($I_b = 5.5$ ma)
23. (a) 0.06, 0.045, 0.03, 0.011 watt
   (b) ¼ watt (each)
25. (a) 172 ohms ($I_b = 5.3$ ma)
   (b) 476 ohms ($I_b = 3.7$ ma)
   (c) 757 ohms ($I_b = 2.8$ ma)
   (d) 1,200 ohms ($I_b = 2$ ma)

27. (b) 3 cycles
   (c) 4 to 12 volts
28. (a) 2,000.5 or 1,999.5 kc
   (b) 2,001 or 1,999 kc
   (c) 2,001.5 or 1,998.5 kc
29. (a) 0.1 sec
   (b) 9.5 per cent
   (c) 10 cycles
31. (a) 3 volts (D_2 is negative with respect to the cathode)
   (b) Zero
   (c) Zero, or ground potential
   (d) Above 3 volts
   (e) 2 volts, A will be negative
   (f) -2 volts
   (g) 0.1 sec
   (h) 125 μsec
   (i) 318 ohms
   (j) Keeps the grid bias from varying due to the a-f signal current variations in the cathode circuit

Chapter 4

1. (a) 278 μh
   (b) 2,383 kc
2. (a) 540 to 1,870 kc
   (b) 533 to 1,590 kc
5. (a) 258 $\mu\mu$f
   (b) 4,664 kc
7. 300 μh, 37.4 μh, 5.27 μh, 0.697 μh
8. 2,295 kc, 6.50 mc, 17.3 mc, 47.6 mc
9. 517 to 1,650 kc, 1.46 to 4.67 mc, 3.90 to 12.4 mc, 10.7 to 34.2 mc
13. 42.8 $\mu\mu$f
14. 1,500 to 2,100 kc
15. 517 to 1,300 kc, 1,270 to 1,570 kc
19. (a) 5.4 $\mu\mu$f
   (b) Reduced to 494 kc
21. (a) 710 $\mu\mu$f
   (b) 2,088 kc
22. (a) 945 $\mu\mu$f
   (b) 2,560 kc

Chapter 5

1. (a) 2 volts
   (b) 3 volts
   (c) 4 volts
   (d) 4 volts
3. (a) 1,400 ohms
     0.035 watt ($I_b = 5$ ma)
   (b) 800 ohms
     0.045 watt ($I_b = 7.5$ ma)

(c) 500 ohms
   0.05 watt ($I_b = 10$ ma)
(d) 308 ohms
   0.052 watt ($I_b = 13$ ma)
5. (a) 1 volt
   (b) 9.2 and 3.3 ma
   (c) 80 ohms
   (d) 0.0125 watt
7. (a) 35
   (b) 35 volts
   (c) 0.603 ma
9. 104,400 ohms
11. (a) 70.5
   (b) 56.4 volts
   (c) 1.41 ma
13. 390
15. (a) 114 ($Q = 145$)
   (b) 180 ($Q = 143$)
   (c) 185 ($Q = 98$)
   (d) 113 ($Q = 45$)
17. (a) 12.9
   (b) 35.1
   (c) 42.1
19. 181
21. (a) 226
   (b) 228
   (c) 226
23. (a) 72.9 per cent
   (b) 21.6 per cent
26. 135
27. (a) 0.0125
   (b) 150
28. (a) 170
   (b) 148
   (c) 25,200

Chapter 6

1. (a) 2.2553
   (b) 3.4393
   (c) 0.942
   (d) 1.0969
   (e) 0.699
   (f) 1.9934
   (g) 4.5441
   (h) $7.4871 - 10$
   (i) 1.2625
   (j) $6.9938 - 10$
3. (a) 300
   (b) 755,000
   (c) 6
   (d) 0.0107
   (e) 0.000218
5. 2.9 db

7. 7.56 db
9. (a) 39 db·m
   (b) 41.7 db·m
   (c) 44.7 db·m
   (d) 48.4 db·m
10. 1.99 watts
12. -35.2 db·m
13. (a) 0.07 μw
   (b) 5.9 mv
15. (a) 40 db
   (b) 20 db
   (c) 30 db
   (d) 22.2 db·6m
   (e) 30 db·m
17. (a) -0.9 db
   (b) -6 db
19. (a) 52.8
   (b) 49.6
   (c) 49.8
20. (a) -0.53 db at low frequency
   (b) -0.51 db at high frequency
23. (a) 284
   (b) 262
   (c) 236
24. -0.7 db at low frequency
   -1.6 db at high frequency
27. (a) 81.7
   (b) 74
   (c) 81
   (d) 81.7, 74, 81
29. (a) 6674
   (b) 5476
   (c) 6561
31. (a) 450 ohms
   (b) 0.5 megohm
   (c) 450 ohms
   (d) 0.5 megohm
   (e) 8.3 μf
   (f) 0.07 μf
   (g) 8.3 μf
   (h) 0.07 μf
33. 91 cps
35. (a) $C_b = 0.016$ μf
      $R_1 = 530$ ohms
      $C_1 = 10.9$ μf
      $R_2 = 370,000$ ohms
      $C_2 = 0.09$ μf
   (b) 110, 105.6, 109.4
   (c) 40.8 db, 40.4 db, 40.7 db
37. (a) 304 volts
   (b) 15.1, 19.7, 19.4
   (c) 2.3 db

39. (a) 50 volts
(b) 300 volts
(c) 188,400 ohms
1,884,000 ohms
9,420,000 ohms
(d) 189,500 ohms (inductive)
4,623,000 ohms (inductive)
682,000 ohms (capacitive)
41. (a) 0.0079 μf
(b) 79 μμf
(c) 3.16 μμf
43. (a) 250.64 volts
(b) 39, 50, 59.7
(c) 23,900 cps
(d) 148
(e) −2.1 db
(f) 1.5 db
45. 0.933
47. (a) 0.86
(b) 0.983
(c) 0.99
49. 308 ohms
51. (a) 3.74 μμf
(b) 40.8 μμf
53. (a) 2.5 megohms
(b) 5 megohms
(c) 10 megohms
55. 910,000 ohms
57. 0.896
59. (a) 24,000
(b) 26 db, 38.1 db, 23.5 db
(c) 87.6 db
61. (a) 22.5 volts
(b) 1.125 volts
(c) 45
(d) 0.489 volt
(e) 0.00053 volt
(f) 0.23 volt
(g) 0.0244 volt
63. (a) 40 volts
(b) −0.8
(c) 22.2 volts
(d) 4.5 volts
65. (a) 0.1
(b) 9.4
(c) 10
67. (a) 140,000 ohms
(b) R_f = 20,000 ohms
R_2 = 120,000 ohms
(c) 0.5 μf
69. (a) 0.424
(b) 0.18

Chapter 7

1. (a) 18.2
(b) 50 ma
3. (a) −20 to −67 volts
(b) 175 to 320 volts
(c) 90 to 30 ma
5. (a) 0 to −87 volts
(b) 103 to 378 volts
(c) 108 to 15 ma
7. (a) 430 volts
(b) 3.2 watts
(c) 1.61 per cent
9. (a) 60 ma
(b) 190 ma
(c) 0 ma
(d) 18.4 per cent
11. (a) 57.7 ma
(b) 136 volts
(c) 57.7 vs. 55 ma
136 vs. 130 volts
(d) 40.8 ma, 96 volts
(e) 3.92 watts
3.92 vs. 3.57 watts
13. 3.95 watts
3.95 vs. 3.92 vs. 3.57 watts
15. 5.23 watts
17. (a) 0.0043 watt
(b) 912
19. (a) 12.4 per cent
(b) 21.3 per cent
21. 11.8 watts
23. 3380 μmhos
25. (a) 3.3 watts
(b) 10.8 watts
(c) 25.8 watts
(d) 12.8 per cent
(e) 22 per cent
27. (a) 81 ma
(b) 5 ma
(c) 35.5 ma
(d) 68.5 ma
(e) 10.5 ma
(f) 315 volts
28. (a) 9.5 per cent
(b) 3.66 per cent
(c) 10.1 per cent
(d) 3.88 watts
(e) 28.1 per cent
31. (a) 38.8 ma
(b) 194 volts
33. (a) 3.76 watts
(b) 3.76 vs. 3.88 watts

35. 13,200
37. 7.32 watts, 2.62 watts
39. 28,600 μmhos
41. (a) 35.5 ma
  (b) 5 ma
  (c) 315 volts
  (d) 495 volts
  (e) 90 volts
42. (a) 4 watts
  (b) 6.4 watts
  (c) 2.62 watts
  (d) 17.6 watts
  (e) 19.8 per cent
  (f) 28.9 per cent
  (g) 8.2 watts
45. (a) 13.6 watts
  (b) 33.5 db
47. (a) 4.8 watts
  (b) 24 watts
  (c) 36 db
49. 15.9 watts
51. 54 watts
52. 10 watts
53. (a) 100,000 ohms
  (b) 41
  (c) 11,460 ohms
  (d) 458,540 ohms
54. 9.09 per cent
55. R_1 = 90,000 ohms
  R_2 = 10,000 ohms

Chapter 8

1. 877 to 2,833 kc
3. 997 to 2,400 kc
5. (a) 2.5 μsec
  (b) 0.227 μsec
  (c) Long
7. (a) X_L = 34,540 ohms
    X_C = 14.5 ohms
  (b) 2,380
  (c) X_L = 15,700 ohms
    X_C = 31.8 ohms
  (d) 494
9. 1,227 kc
11. 15
13. (a) 112 $\mu\mu$f
  (b) 15.9 ma
  (c) 7.96 mw
  (d) 0.0252 mw per cycle
  (e) 4.77
15. (a) 11.4 μh
  (b) 88.7 $\mu\mu$f
  (c) 4.16 amp

(d) 0.332 amp
(e) 199 μw per cycle
17. (a) 41.6 μh
  (b) 5.94 μh
  (c) 236 $\mu\mu$f
  (d) 4.62 μw per cycle
19. (a) 750 cycles (decrease)
  (b) 375 cycles (increase)
  (c) 750 cycles (increase)
21. 4,239
23. (a) 0.45 inch
  (b) 0.045 inch
  (c) 0.01126 inch
25. (a) 400 kc
  (b) 2,200 kc
  (c) 4,000 kc

Chapter 9

1. (a) 135 volts
  (c) 90 volts
  (d) 423 volts
3. (a) 405 volts
  (c) 330 volts
  (d) 634.5 volts
5. (a) 169 volts
  (b) 338 volts
  (c) 338 volts
  (d) 338 volts
6. (a) 169 volts at C_1
    338 volts at C_2
    507 volts at C_3
    676 volts at C_4
    845 volts at C_5
    1,014 volts at C_6
  (b) 338 volts at each tube
7. (b) 200 volts for C_1
    350 volts for C_2
    500 volts for C_3
  (c) 250 volts for C_1
    450 volts for C_2
    600 volts for C_3
  (d) 330 volts at each tube
9. 400 ma
11. (a) 117 volts
  (b) 111 per cent
  (c) 60 cps
13. 0.09 volt
15. 3.74 per cent
17. (a) 1500 ohms
  (b) 998 μf
  (c) 2.66 ohms
  (d) 62 amp

(e) Burn it out

(f) No

19. (a) 7.48 per cent

(b) 0.134 per cent

21. (a) 12.4 per cent

(b) 0.452 per cent

(c) 0.022 per cent

23. (a) 7.8 per cent

(b) 0.517 per cent

25. 40 per cent

28. 96 volts

29. 2 per cent

31. (a) 1 per cent

(b) 0.0162 per cent

33. 11.5 per cent

35. (a) 1.44 per cent

(b) 0.0337 per cent

36. 4.8 henrys

38. $R_1 = 3,367$ ohms

$R_2 = 1,900$ ohms

$R_3 = 5,618$ ohms

$R_4 = 24.7$ ohms

$R_5 = 6.2$ ohms

$R_6 = 191$ ohms

39. (a) 145.6 watts

(b) 6 watts, 2.6 watts

3.6 watts, 0.324 watt

0.081 watt, 2.5 watts

42. (a) Reduces the voltage for grid 2 of VT_1 and VT_3, and for grids 2 and 4 of VT_2

(b) 100 volts

(c) 1.88 watts

44. 420 volts

Chapter 10

1. (a) 378 μw

(b) 3.78 μw

(c) 3.78 μw

(d) 1.51 mw

3. 0.316 volt

5. (a) 0.0025 μw

(b) 0.062 μw

7. (a) 50 μv

(b) 2 mv

9. (a) 3.98 mv

(b) 39.8 mv

11. 24.5

13. (a) 44.7

(b) 22.3

(c) 14.1

15. (a) 50

(b) 100

17. (a) 0.00632 μf

(b) 100 ohms

(c) 5,385 ohms

(d) 3,434 ohms

19. 109.6 db

20. (a) 79 db below 6 mw

(b) 38 db below 6 mw

21. (a) 6 watts

(b) 3 watts

23. -3 db, -3 db

25. (a) 1,250 cps

(b) -3 db, falling

(c) -7 db

27. (a) -3.8 db

(b) 8.2 db

29. (a) 7,450 $\mu\mu$f

(b) 455 cps

(c) -3 db

(d) 47,000 ohms

31. (a) -3.6 db

(b) -13.3 db

33. (a) 1,250 cps

(b) 3 db, rising

(c) 7 db

35. (a) 3.8 db

(b) 8.2 db

37. (a) 7,450 $\mu\mu$f

(b) 455 cps

(c) 3 db

(d) 47,000 ohms

39. (a) 3.6 db

(b) 13.3 db

41. -5 db

43. -30.3 db

45. (a) 12.6 db

(b) -17.7 db

47. -0.6 db

49. -30 db

51. (a) -14.7 db

(b) -15.3 db

53. $L_1 = 1.16$ mh, $L_2 = 1.86$ mh

$C_1 = 18$ μf, $C_2 = 11.3$ μf

55. $L_1 = 1.16$ mh, $L_3 = 0.725$ mh

$C_1 = 18$ μf, $C_3 = 28.8$ μf

Chapter 12

1. 0.216 to 0.326 μh

3. (a) 16.3 and 10.8 $\mu\mu$f

(b) 4.5 to 10 $\mu\mu$f

5. (a) 1,200 ohms

(b) 1,810 ohms

(c) 35,000 ohms

(d) 14 megohms

7. (*a*) 333 ohms
   (*b*) 6
9. 8.35
11. (*a*) 500 ohms
   (*b*) 833 ohms
13. 2,230 ohms
15. 9.36 db
17. 4.9
19. (*a*) 15.5 μsec
   (*b*) 0.0934 μsec
   (*c*) Long

Chapter 13

1. (*a*) 1,900
   (*b*) 1,805
3. (*a*) 191
   (*b*) 517
5. 0.95 ($I_{C1} = 0.95$)
        ($I_{C2} = 1.9$)
7. 14 ($I_{C150} = 2.4$)
       ($I_{C100} = 1.7$)
9. 0.974
11. (*a*) 10 db
   (*b*) -4 db

Chapter 14

1. (*a*) 120,000 ohms
   (*b*) 60,000 ohms
3. (*a*) 40,000 ohms
   (*b*) 20,000 ohms
5. (*a*) 0.972 ma
   (*b*) 0.98 ma
7. (*a*) 2.6 ma
   (*b*) 3.01 ma
9. (*a*) 2,450 ohms
   (*b*) 2,456 ohms
11. (*a*) 164,000 ohms
   (*b*) 169,000 ohms
13. (*a*) 346
   (*b*) 347
15. (*a*) 41.67 db
   (*b*) 41.67 db
17. (*a*) 172 ohms
   (*b*) 173 ohms
19. (*a*) 330,000 ohms
   (*b*) 330,000 ohms
21. (*a*) 1,930
   (*b*) 1,925
23. (*a*) 31 db
   (*b*) 32 db
25. (*a*) 490,000 ohms
   (*b*) 550,000 ohms

27. (*a*) 52.7 ohms
   (*b*) 53.3 ohms
29. (*a*) 0.994
   (*b*) 1
31. (*a*) 16.9 db
   (*b*) 17 db
33. 81 db
35. (*a*) 59.5 μf
   (*b*) 15.1 μf
37. 44.5
39. (*a*) 44.5
   (*b*) 44.4
41. (*a*) 25,000 ohms
   (*b*) 112 ohms
43. (*a*) 0.5 μf
   (*b*) 1.5 μf
45. (*a*) 98.2 per cent
   (*b*) 86.9 per cent
47. (*a*) 30.3 per cent
   (*b*) 26 per cent
49. 480 ohms
51. (*a*) 693 ohms
   (*b*) 72,000 ohms
   (*c*) 173 ohms ($V_{cc} = -9$ volts)
             ($V_c = -4.5$ volts)
             ($I_c = -6.5$ ma)
             ($I_b = 125$ μa)
53. (*a*) 37.3 per cent
   (*b*) 29.8 per cent ($V_c = -4.5$ volts)
       ($V_{pp} = -0.65$ to -8.5 volts)
       ($I_c = -6.5$ ma)
       ($I_{pp} = -0.7$ to -11.8 ma)
55. (*a*) 2.2 per cent
   (*b*) 2.63 per cent
   (*c*) 3.42 per cent ($i_{c\cdot\mathrm{max}} = -11.8$ ma)
       ($i_{c\cdot\mathrm{min}} = -0.7$ ma)
       ($I_c = -6.5$ ma)
       ($I_x = -10.5$ ma)
       ($I_y = -2.2$ ma)
57. (*a*) 72 per cent
   (*b*) 2,245 ohms
       (max signal = 150 μa)
       ($V_c = -4.6$ volt)
       ($I_c = -7.6$ ma)
       ($V_{pp} = -1$ to -8.2 volts)
       ($I_{pp} = -0.6$ to -14.6 ma)
59. (*a*) 66.1 per cent
   (*b*) 2,246 ohms
       ($V_c = -8.2$ volts)
       ($I_c = -14.6$ ma)
       ($V_{pp} = -1$ to -8.2 volts)
       ($I_{pp} = -0.6$ to -14.6 ma)

INDEX